Analecta Gregoriana 312

G. AULETTA - M. LECLERC - R.A. MARTÍNEZ (eds.)

BIOLOGICAL EVOLUTION: FACTS AND THEORIES

A Critical Appraisal 150 Years After "The Origin of Species"

With an Address of Cardinal Levada

G&BP PRESS
GREGORIAN & BIBLICAL PRESS

"STOQ Research Project"

Supported by a grant from the John Templeton Foundation.
The opinions expressed in this book are those of the authors and do not necessarily reflect the views of the John Templeton Foundation or Institutions involved in STOQ Project.

Prooceedings of the International Conference
Held at the Pontifical Gregorian University
Rome, 3-7 March 2009

Cover Image: *Simulation of a colony of P. dendritiformis bacteria*,
courtesy of Eshel Ben-Jaob's Bacterial Cybernetics Group - School of Physics and Astronomy
Tel Aviv University, 69978 Tel Aviv, Israel

Cover: Serena Aureli

Layout: Lisanti srl - Roma

Piazza della Pilotta 35, 00187 - Roma
www.gbpress.net - books@biblicum.com

ISBN 978-88-7839-**180**-2

The organizers of the Conference would like
to thank very much the Pontifical Council for Culture
for its High Patronage and all the support provided.
In particular, the organizers express their gratitude
to the President of the Council, Cardinal Gianfranco Ravasi,
who also addressed a benevolent and insightful discourse
to the community of scholars involved in the event.

TABLE OF CONTENTS

Address by His Eminence Cardinal William Levada 7

Introduction 11

Contributors 19

I. Darwin and Evolution

1. Darwin's revolution — *Francisco J. Ayala* 25
2. History of evolution theories — *Jean Gayon* 49

II. Key concepts on evolution

Introduction — *Saverio Forestiero* 75

3. Bio-molecular evidences — *Werner Arber* 81
4. Taxonomic issues: What we learn from comparisons among species — *Douglas J. Futuyma* 95
5. Origin of evolutionary novelty by symbiogenesis — *Lynn Margulis* 107
6. The mystery of speciation — *Jeffrey L. Feder* 117
7. Evolution through developmental change: How alterations in development cause evolutionary changes in anatomy — *Scott F. Gilbert* 153
8. The evolution of evolutionary mechanisms: A new perspective — *Stuart A. Newman* 169

III. Man in evolution

Introduction — *Fiorenzo Facchini* 195

9. Human evolution: A brief history of the research since Darwin's time — *Giorgio Manzi* 199
10. The origin of man — *Yves Coppens* 217
11. The origin of man: Molecular approach — *Olga Rickards* and *Gianfranco Biondi* 225
12. The human lineage: Macro-evolutionary process acting during embryogeny with emergent macro-evolutionary implications — *Anne Dambricourt Malassé* 243
13. How humans came to be so different to other monkeys and apes — *Robin I. M. Dunbar* 275
14. Paleocultural approach in the hominization and different implications — *Fiorenzo Facchini* 291
15. The concept of evolution as applied to the development of human cultures — *Colin Renfrew* 307

IV. Epistemological aspects of evolution

Introduction — *Paolo D'Ambrosio* and *Massimo Stanzione* 317

16. Process and ontological priorities in evolution — *Robert E. Ulanowicz* 321
17. Can evolutionary theory be epistemologically closed? — *Stuart Kauffman* 337

18. Some epistemological questions concerning the use of mathematics in evolution theory — *Dominique Lambert* 347
19. Why methodological naturalism? — *Elliott Sober* 359
20. The recurrent debate: Darwinism and philosophy — *Massimo Stanzione* 379

V. Evolution, metaphysics and teleology
Introduction — *Gennaro Auletta* 403
21. Evolution and the conditio humana: A philosophical introduction — *Jürgen Mittelstrass* 407
22. The metaphysical significance of creation — *Georges Cottier* 419
23. Why teleological principles are inevitable for reason. Natural theology after Darwin — *Vittorio Hösle* 433
24. Accident, adaptation, and teleology in Aristotle, Empedocles, and Darwinism — *David J. Depew* 461
25. Emergence, directionality and finality in an evolutionary universe — *William R. Stoeger sj* 479
26. Teilhard de Chardin: Moving towards humankind? — *Lodovico Galleni* 493
27. Scientific creationism and intelligent design — *Ronald Numbers* 517

VI. The theological debate
Introduction — *Rafael A. Martínez* 539
28. The creation in the Old Testament: Some pointers — *André Wénin* 543
29. The various meanings of the word evolution in science, philosophy and theology — *Jean-Michel Maldamé* 557
30. Creationism, Intelligent design and evolution: A theological perspective — *Jacques Arnould* 573
31. The reception of evolutionary theories in the Church — *Rafael A. Martínez* 589
32. Evolution according to Teilhard de Chardin — *Georges Chantraine sj* 613
33. Theological debates around evolution — *Robert John Russell* 645

Appendix
34. From method to ideology: The resistible ascent of reductionism in biology — *Marcello Buiatti* 667
35. Evolutionary developmental biology and intelligent design — *Scott F. Gilbert* 691
36. Evolution and the emergence of the human person — *William B. Hurlbut* 701

Author Index 719

Subject Index 735

Address by His Eminence Cardinal William Levada

Prefect of the Congregation for the Doctrine of the Faith

(March the 3rd 2009)

* *Eminences,*
Excellences,
Most reverend Prelates,
Illustrious Professors,
Respectable Ladies and Gentlemen,
Dear Students,

I would like to cordially welcome the illustrious participants to this important international academic event on "Biological Evolution: Facts and Theories", wishing a profitable time of study and reflection. The event takes place in this ancient Pontifical University of Rome, which is the University of the Society of Jesus, with the collaboration of the University of Notre Dame and under the High Patronage of the Pontifical Council for Culture.

The field of investigation of the Conference arouse lively interest since it is related to the reflection on the ultimate origin of beings, on their cause and end, on the meaning of the history of mankind as well as of the entire universe. These reflections coincide with the human most profound questions, addressed by philosophy and especially metaphysics since centuries, and – in the light of Jesus Christ Revelation – engaging the intelligibility of faith and the related theological science.

The Congregation for the Doctrine of the Faith, over which I am honored to preside, does follow with interest this prestigious Meeting of Scientists representing different disciplines, which may provide a favorable occasion for a new and qualified exchange about the theory of biological evolution free from any reduction to an ideological system claiming to become the key of interpretation of the whole of reality.

In this spirit, the perspective of a peaceful interdisciplinary engagement concerning evolutionary theories, as envisaged by the Conference, seems very appropriate. I wish it may lead science, theology and philosophy to a mutual enrichment. [*Editorial translation from the original Italian text**]

Since there are many presentations in English, I would like to say a word of greeting also to the English-speaking participants:

---

* *Eminenze,*
Eccellenze,
Reverendissimi Presuli,
Illustri Professori e Professoresse,
Egregi Signori e Gentili Signore,
Cari Studenti,

desidero esprimere un cordiale saluto con l'augurio di un proficuo tempo di studio e di riflessione a tutti gli Illustri Convenuti a questo importante evento accademico internazionale sul tema "Biological Evolution: Facts and Theories", che si svolge in quest'antica Università Pontificia dell'Urbe, che è l'Università della Compagnia di Gesù, in collaborazione con l'University of Notre Dame e sotto l'Alto Patronato del Pontificio Consiglio della Cultura.

L'area d'indagine del Convegno solleva sempre vivo interesse in quanto legata alla riflessione sull'origine ultima degli esseri, sulla loro causa e fine, sul significato della storia dell'uomo e dell'intero universo. Sono riflessioni che coincidono con le domande più profonde dell'animo umano e che da secoli coinvolgono la filosofia e specialmente la metafisica, e – alla luce della Rivelazione di Gesù Cristo – impegnano l'intelligenza della fede e la relativa scienza teologica.

La Congregazione per la Dottrina della Fede, che ho l'onore di presiedere, segue con interesse questa prestigiosa Assise di Scienziati, rappresentanti di varie discipline, che potrà offrire l'occasione propizia per un nuovo e qualificato confronto sulla teoria dell'evoluzione biologica, libera da ogni riduzionismo ad un sistema ideologico con la pretesa di diventare chiave d'interpretazione dell'intera realtà.

In questo spirito, sembra assai opportuno il progetto di un sereno confronto interdisciplinare sulle teorie dell'evoluzione, che il Convegno si prefigge. Mi auguro che esso porti un arricchimento reciproco tra scienza, Teologia e filosofia.

In treating the question "Faith and science" the Catechism of the Catholic Church (n. 159) resorts to two conciliar texts: from the First Vatican Council's Constitution on the Catholic faith, *Dei Filius* (DS 3017), it gives the following quotation: "Though faith is above reason, there can never be any real discrepancy between faith and reason. Since the same God who reveals mysteries and infuses faith has bestowed the light of reason on the human mind, God cannot deny himself, nor can truth ever contradict truth." This conciliar teaching formed the intellectual ground for the studies of generations of Catholic students, including my own, where science and religion intersect.

And secondly, since the Catechism intended to provide a comprehensive overview of Catholic teaching that would also include the teachings of the Second Vatican Council, it cites as well this text from the "Pastoral Constitution on the Church in the Modern World" *Gaudium et Spes,* n. 36: "Consequently, methodical research in all branches of knowledge, provided it is carried out in a truly scientific manner and does not override moral laws, can never conflict with the faith, because the things of the world and the things of faith derive from the same God. The humble and persevering investigator of the secrets of nature is being led, as it were, by the hand of God in spite of himself, for it is God, the conserver of all things, who made them what they are."

While people of faith accept these theological certitudes, we also recognize that science rests on empirical foundations. The theme of the Conference which begins today is an important one both for faith and science, albeit in different ways. From the point of view of faith, the spectrum is quite broad: we are familiar with the views of fundamentalist Christians who insist that their children be taught "creationism" alongside evolution in public schools in the country I come from, and we hear from time to time the apologists for atheism, like Richard Dawkins, who draw philosophical conclusions from scientific evolution that not only leave no room for a Creator God, but definitively exclude Him.

Hence the theme which this Conference addresses is one of great value and importance for the contemporary mind. I am happy to join in a word of gratitude to my Alma Mater, the Pontifical Gregorian University, for hosting this Conference, and to all those who have organized this Conference, who participate in it, and who are its patrons and hosts. In particular I want to thank Archbishop

Gianfranco Ravasi, President of the Pontifical Council for Culture, Very Reverend Father Gianfranco Ghirlanda, S.J., Magnificent Rector of this University, and the distinguished presenters whose insights can contribute to the ongoing dialogue between faith and reason that has marked the Christian tradition from its beginnings.

May God bless the work begun here this morning.

CONVEGNO INTERNAZIONALE
"BIOLOGICAL EVOLUTION: FACTS AND THEORIES"
presso la Pontificia Università Gregoriana
sotto l'Alto Patronato del Pontificio Consiglio della Cultura
nell'ambito del Progetto "Scienza, Teologia
e la Questione Ontologica (STOQ)"
Roma, 3-7 marzo 2009

INTRODUCTION

The year 2009 marked the 150th anniversary of the publication of *The Origin of Species* by Charles Darwin. This celebration was the occasion for the organization by the Pontifical Gregorian University of the conference of which this volume contains the proceedings. The initiative, that also saw the collaboration of the University of Notre Dame, comes within the framework of the STOQ Project (Science, Theology, and the Ontological Quest), under the patronage of the Pontifical Council for Culture; the Gregorian has participated in this project from its inception. The occasion itself however provided only one of a whole complex of motivations that first led to the organization of the conference and then convinced us as of the possibility of publishing its proceedings. These motivations come from the natural sciences themselves, above all biology, from philosophical reflections on the theory of evolution and from the importance that evolution is progressively acquiring for theological discourse.

Although numerous debates continue to rage in the scientific arena in which supporters aim to validate it while skeptics raise questions about its validity, the Darwinian theory still constitutes the framework for studies of the living world. The basic conceptual structure of this theory has in fact shown itself to possess an exceptional solidity; it has been able to take on board the various discoveries that have gradually been accumulating with respect to past history and to the typical functioning of living beings, as the disciplines of relevance for the biological sphere have grown ever more specific and more numerous, and evolution itself has gone from the status of a promising hypothesis to that of fact verified by incontrovertible evidence. The adoption of new and more sophisticated instruments of research, made possible by the vertiginous technological and scientific advance characteristic of the last fifty years, has often ended up by confirming and reinforcing, rather than disproving, the validity of the arguments put forward by Darwin as to the general mechanisms responsible for the dynamics of evolution. Today, the broad area of research that we call "evolutionary biology" is coming to constitute one of the greatest and most solid fields of research by virtue of the contribution of a series of scientific disciplines and subdisciplines whose studies are coming to cover the various fields of the biological domain: from biomolecules to organisms (of the present and the past) as wholes, up to populations and ecosystems. All these approaches, down to the specifics of their respective methodologies, share a theoretical perspective of a clearly Darwinian mold that comes to define the kind of interpretative approach they have all adopted to the various phenomena under examination.

The Darwinian theory of evolution has had of course to undergo a series of updates and modifications in order to be in a position to deal on the explanatory level with a range of phenomena of such breadth. The conceptual nucleus outlined in *The Origin of Species* continues to be considered generally valid and the so-called "synthetic theory" — also known as "neo-Darwinism" — that subsumes Darwinian evolutionism and Mendelian genetics, still provides a privileged frame of reference in the field of evolutionary biology; but only in more recent times has it been possible to clarify the key role, on the evolutionary level, of bio-molecular processes (of a not necessarily genetic type), of the particular intrinsic complexity of biological systems and of their capacity to manifest self-organization, of symbiogenesis, of evolutionary convergences, of connections between the evolution of species and the processes of development of organisms (which has led to the so-called Evolutionary Developmental Biology), and of the construction of ecological niches. All these aspects belong in the general framework of the Darwinian theory, but they enrich it in directions not yet entirely made clear in the first phases of evolutionary biology and even in part unforeseeable. This can provide us with a clear example of how the Darwinian research programme is in a position to remain valid (and therefore in some way faithful to itself) in being able to integrate not only new discoveries, but straightforward intelligibility relative to entire areas of phenomena, proving to be in this way in its turn "in evolution", as suggestively indicated by some experts in the area. The desire to bring all these fascinating new things into a context that could contribute to their integration and that could therefore facilitate a better comprehension of what is today evolutionary biology was the first fundamental motivation for the conference, we could say its "scientific motivation". On the other hand, from the very heart of the natural sciences, the need for interdisciplinary research that can articulate perspectives and results in distinct but contiguous fields of scientific work becomes more and more pressing: and the Pontifical Gregorian University, being among other things the heir of the Collegium Romanum, cannot but look with favor on the emergence of cognitive perspectives that are maximally integrated and, in some cases, even less reductionist in their methodologies.

Living being among living beings, man, our species, is also the fruit of biological evolution, and the contribution of evolutionary biology leads us to consider the physical and biological dimension of the human being as essential for the elaboration of an anthropology consistent with what science is in a position to reveal to us. Man, nonetheless, is also the creator of culture and the subject of cultural evolution. Culture has its own peculiarities that are not reducible to the purely biological dimension. From this point of view, the need for interdisciplinarity above mentioned becomes even more compelling in the case of the human being. What we might call the "anthropological motivation" of the conference, then, resides precisely in the belief that it is necessary to link the biological dimension into the specifically cultural one.

Biological evolution, then, brings with it a series of philosophical questions of crucial importance. The first that comes to mind is the one connected to the issues of reductionism and mechanism, and of the extent to which such approaches are still adequate to the status of current knowledge of living beings and their evolution. The study of evolutionary biology surely raises questions about naturalism. From the methodological point of view, naturalism maintains in an epistemologically specific manner the need to circumscribe scientific explanations within the sphere of the natural, that is of phenomena, effects, and factors connected to experience as accessible directly or indirectly to us — thus excluding methodologically any recourse to the supernatural as a factor that can modify it. A more radical understanding of naturalism on the other hand reduces reality to the empirical sphere, therefore denying the existence of any supernatural dimension. A rejection of this radical understanding is ultimately quite compatible with the point of view of natural science: for if it is to be responsible, natural science must admit that it does not have the necessary instruments for coming to a conclusion as to the existence or otherwise of the supernatural. But naturalism according to the first of these two versions is by contrast assumed as a fundamental methodological imperative for scientific research. This methodological imperative is regularly violated by all those upholders of "intelligent design" who maintain the legitimacy of supernatural explanations in science. Basically, with regard to philosophical issues, the conference was born out of the need to go beyond the polarity of two opposed fundamentalisms, that of scientism and the one represented today by intelligent design theories; significantly, these opposing views agree in maintaining that scientific explanations and the recourse to the supernatural are in some way competing and therefore at odds. Perhaps current developments at the scientific level in particular represent a unique opportunity to go beyond these misunderstandings in the direction of a more free, more frank, and more open circulation of ideas.

To continue with this line of thought, philosophical analysis is also important to point up the fact that natural science cannot attribute to the evolution of life or indeed of the universe itself in its entirety a global finality, conceived as a project or plan imposed on the world from the outside. Philosophical reflection today can itself also be facilitated by the most recent acquisitions of the biological sciences: without denying the presence of genuinely contingent aspects (a denial that would make genuinely creative evolutionary novelties wholly incomprehensible), it can however point up the fact that we observe a certain directionality at the evolutionary level manifested by the development towards greater complexity of organisms and of the solutions that have gradually emerged in the course of evolution such as to be able even to canalize evolution towards the emergence of general types of possible solutions. However, these philosophical reflections must be articulated in full acceptance of the intrinsic dynamics of the evolutionary processes and of the constraints that are occasionally placed on possible evolutionary solutions. In this way epistemological, philosophical-natural, and therefore metaphysical reflection can facilitate the effort to grasp the constitution and the meaning of reality and of our relationship with it. The third motivation of the conference, the "philosophical motivation" as we might call it, lies then in the desire to create a space for dialogue between evolutionary biology (and the natural sciences in general) and philosophy. If responsible philosophical reflection can certainly help to clarify such conceptual questions, it can do so only by taking serious care to listen to the lessons that biological science has to offer us today.

There is also a "theological motivation" for the conference. What is the relationship between evolution and creation? How is the problem of the human soul (or of the ontological discontinuity between man and animal) to be approached in relation to evolutionary biology? These questions touch on central themes of Christian and Catholic theology, questions whose study today needs to be pursued in the light of a careful and considered comprehension of evolutionary biology and fortified by philosophical, methodological, and fundamental reflection on this latter. We have already touched on the problem of culture. One of man's fundamental characteristics lies in his capacity to interpret the universe that surrounds him and therefore to open up to the transcendent dimension (a datum of historical and phenomenological fact that ought to be recognized even by those who are not believers). Even if the solutions that different cultures and epochs find to such problems are fallible and therefore our comprehension of God, however much it may be assisted by Revelation and faith, is imperfect, it remains nonetheless true that these solutions and interpretations have a value that transcends the initial biological or anyway contingent context out of which they are born, and so they acquire a universal value potentially capable of being shared by other cultures and therefore by other intelligent beings. All this raises the theological meaning of such questions in the light of science and of contemporary culture. The conclusions of the modern natural sciences constitute at the same time a challenge and a possibility of development, not just for philosophy but also for theology, above all when the latter puts creation and the place occupied by man in it at the center of its research. From this point of view, therefore, the conference aimed at making constructive proposals as a contribution to the field of studies known as "Science and Religion" — a field of studies which is in a state of lively ferment today.

Inspired by such motivations, a group of individuals of vastly differing backgrounds, scientists, philosophers and theologians, lay persons and clerics, Christians and agnostics, gathered together to establish a structured program that would provide an itinerary for working through the many questions involved, while at the same time promoting a constructive dialogue that would be fully respectful of the specificity and autonomy of the various disciplines. The actual experience of the five days of the conference surpassed our expectations. The present volume is intended to provide a lasting testimony of an event that first of all represented a unique encounter, characterized by a profound mutual respect, between persons of extremely diverse backgrounds and types of formation. This diversity is surely reflected in the contributions in this volume, both in terms of their object and content and in terms of the form of their presentation itself and the wide diversity of opinions formulated. As usual in a scientific conference, each speaker expressed his own opinions and views and of course, each takes personal responsibility for those opinions and views, which do not of course necessarily reflect those of the organizers themselves (also very diverse).

The way the conference panned out has led us to organize the nine sessions in which it was structured into six clearly-defined sections. Each section has been entrusted to a specific coordinator and will be provided with a short descriptive introduction. We can present only the most general outlines of our itinerary here. The first short section includes two papers of a fairly historical character, they are more directly connected to the double celebration of 2009 that was the occasion for the conference: "Darwin and evolution". The second section brings together six texts that are purely scientific in the strict sense of the term, under the title "Key concepts on evolution". This constitutes a rigorous starting point for any serious reflection on biological evolution. In the third section, also a scientific one, are to be found seven papers on human evolution, which open up various perspectives that are sometimes complementary and sometimes divergent, on an argument of particular complexity: "Man in evolution". Here the views of palaeontology, molecular biology, and cultural anthropology intersect. The fourth section, "Epistemological aspects of evolution", is dedicated to the epistemological analysis of the theory of evolution; it contains five papers which introduce us to a more philosophical type of analysis of the theory of evolution. Philosophical rationality is brought to bear again in the following section, "Evolution, metaphysics and teleology", where seven papers introduce us to the extreme complexity of this field of study. The sixth and last section is devoted to theological debate on evolution: "The theological debate", where another six papers of an exegetical, historical, critical, or speculative nature allow us to deal smoothly with the highly debated question concerning the theological attitude towards the theory of evolution.

Finally, the volume contains a further three papers in an appendix: these were presented in a preliminary workshop on the mechanisms of evolution held in Rome in September 2008, within the preparatory phases of the conference.

It is clear that the conference tackled issues of extreme difficulty and complexity. The fact that such issues arise out of historical (scientific, philosophical, and theological) journeys and traditions means moreover that within them many erroneous or ideological prejudices have accumulated, as we have already mentioned; this makes dealing with them even more problematic, and it often makes coolheaded debate within them more difficult. Furthermore, the multiplicity of languages employed in each of the specific fields involved often exacerbates difficulties of mutual comprehension. Perhaps the most important result achieved through the conference then was the provision of a concrete opportunity for such issues to be aired in a single gathering.

Each speaker was able to profit from the points of view, the ideas, the results and the arguments of all the other colleagues — thirty-five in number and coming from fields of research and professional backgrounds that were extremely variegated. The conference certainly constituted an opportunity for each participant to be able to absorb points of view other than his own, an opportunity to try to attain to a deeper understanding of the motivations behind different positions, an opportunity for serious discussion of open questions and future perspectives and possibilities.

The participants in the conference, several hundred in number and very different, were able to appreciate the richness of so many points of view from which biological evolution was presented and discussed, which helped them to avoid remaining locked into a single point of view that would necessarily be myopic.

As we have indicated, the theory of evolution is often perceived as the bastion of materialism and hostility to religion. Many of the questions involved in the subject of the conference undoubtedly also touch on the sensitivities of the wider public: and especially so when there is a failure to tackle them in a sufficiently critical spirit and with openness to debate, with the result that there is a focus on an opposition between science and faith, between scientific culture and humanistic culture, between evolution and creation. The conference provided a practical demonstration that these very delicate subjects can be tackled without *a priori* polarizations and in a constructive manner, each one profiting from the point of view of the other while always remaining fully respectful of distinctions of levels and methodologies and of the autonomy of each discipline over against the others. The conference thus helped to show that the apparent grounds for the "conflict between science and faith" are more illusory and more dependent on ideological stances of various kinds than might be thought at first sight.

Our hope is that we have been able to open up a space for debate and for a genuinely interdisciplinary dialogue: interdisciplinary not only in the sense of involving diverse but contiguous scientific disciplines but in the fullest sense of a useful collaboration between such distinct rational undertakings as the scientific, the philosophical and the theological. The conference could be seen as an important step towards the establishment of a common language for those disciplines which, with time, are in a position not only to mitigate and go beyond present conflicts but also to incentivize research in each of the areas concerned.

We cannot here list by name all the individuals whose dedication and competence has enabled us to complete this work, the fruit of a long and extended collective effort: but we want to thank them all most warmly. There is one person to whom we do however want to express a particular gratitude, someone who dedicated herself assiduously to the work of preparation only to die less than four days before the opening of the conference: Doctor Marilena Amerise. A recognized historian of late Christian Antiquity, she had devoted herself body and soul to a conference that was apparently remote from her usual concerns, because she was convinced of the vital importance of cultural mediation between such diverse spheres as the natural sciences, philosophy, and theology, and especially in a difficult question like that of biological evolution. We would like to dedicate the fruits of our labors to this person who made such a great contribution to the interdisciplinary dialogue that inspired and inspires our project.

Finally, we would like to mention all of the associations that made this event possible thanks to their financial and technical support. A significant part of the funds was made available by the John Templeton Foundation. Other funds were made available by the Fondazione Cariplo and the Associazione Scienza e Fede. Microsoft Europe provided technical support. We are very much indebted to their kind assistance.

Gennaro Auletta, Marc Leclerc, and *Rafael A. Martínez*

I wish to thank my co-worker Paolo D'Ambrosio for his careful work on the indexes and proof correction. I wish also to thank my friend Sir Brian Heap for his kind support.

Gennaro Auletta

CONTRIBUTORS

Werner Arber
Nobel Prize in Medicine 1978
University of Basel, Switzerland

Jacques Arnould op
Centre National d'Études Spatiales (CNES), France

Francisco J. Ayala
University of California, Irvine, USA

Gianfranco Biondi
Università dell'Aquila, Italy

Marcello Buiatti
Università di Firenze, Italy

Georges Chantraine sj (†)
Theological Faculty of Lugano, Switzerland

Yves Coppens
Collège de France, France

Card. Georges Cottier op
Pontifical Academy of Sciences, Vatican City

Anne Dambricourt Malassé
Musée National d'Histoire Naturelle, France

David J. Depew
University of Iowa, USA

Robin I. M. Dunbar
University of Oxford, UK

Fiorenzo Facchini
Università di Bologna, Italy

Jeffrey L. Feder
University of Notre Dame, USA

Douglas J. Futuyma
State University of New York, Stony Brook, USA

Lodovico Galleni
Università di Pisa, Italy

Jean Gayon
Université Paris 1 – Panthéon Sorbonne, France

Scott F. Gilbert
Swarthmore College, USA

Vittorio Hösle
University of Notre Dame, USA

William B. Hurlbut
Stanford University Medical Center, USA

Stuart Kauffman
University of Calgary, Canada

Dominique Lambert
Facultés Universitaires N.-D. de la Paix, Namur, Belgium

Jean-Michel Maldamé op
Institut Catholique de Toulouse, France

Giorgio Manzi
Sapienza – Università di Roma, Italy

Lynn Margulis
University of Massachusetts-Amherst, USA

Rafael A. Martínez
Pontifical University of the Holy Cross, Rome, Italy

Jürgen Mittelstrass
University of Konstanz, Germany

Stuart A. Newman
New York Medical College, USA

Ronald Numbers
University of Wisconsin, USA

Colin Renfrew
University of Cambridge, UK

Olga Rickards
Università di Roma Tor Vergata, Italy

Robert John Russell
Center for Theology and Natural Sciences (CTNS), Berkeley, USA

Elliott Sober
University of Wisconsin, USA

Massimo Stanzione
Università di Cassino, Italy

William R. Stoeger sj
University of Arizona, Tucson, USA

Robert E. Ulanowicz
University of Maryland, USA

André Wénin
Université Catholique de Louvain, Belgium

Section I
Darwin and Evolution

Darwin's revolution

Francisco J. Ayala

Darwinism appeared, and under the guise of a foe, did the work of a friend. It has conferred upon philosophy and religion an inestimable benefit, by shewing us that we must choose between two alternatives. Either God is everywhere present in nature, or He is nowhere.

(Aubrey Moore 1891)

What I find even more surprising, and less understandable, is the way in which the "disguised friend" of Darwinism, and more generally of evolutionary ideas, has been admitted (if at all) only grudgingly ... into the parlors of Christian theology. I believe it is vital for this churlishness to be rectified ... if the Christian religion (indeed any religion) is to be believable and have intellectual integrity.

(Arthur Peacocke 1998)

New scientific knowledge has led us to realize that the theory of evolution is no longer a mere hypothesis. It is indeed remarkable that this theory has been progressively accepted by researchers, following a series of discoveries in various fields of knowledge. The convergence, neither sought nor fabricated, of the results of work that was conducted independently is in itself a significant argument in favor of this theory.

(John Paul II, 22 October 1996)

Nothing in biology makes sense except in the light of evolution.

(Theodosius Dobzhansky 1973)

Abstract

Darwin is deservedly given credit for the theory of evolution. He accumulated evidence demonstrating that organisms evolve over eons of time and diversify as they adapt to environments that are enormously diverse. Most important, however, is that he discovered natural selection, the process that accounts for the evolution of organisms and for their adaptive features; that is, their "design."

The argument from design asserts that organisms evince to have been designed by God. The argument from design was advanced, in a variety of forms, in Classical Greece and early Christianity. Its most extensive formulation is due to William Paley in his *Natural Theology* (1802). The eye — as well as all sorts of organs, organisms, and their interactions — manifest to be outcomes of design and not of chance, thus showing to have been created by God. In the 1990s, the design argument was revived in the United States by several authors.

Darwin's theory of evolution by natural selection disposed of Paley's arguments: the adaptations of organisms are not outcomes of chance, but of a process that, over time, causes the gradual accumulation of features beneficial to organisms. There is "design" in the living world: eyes are designed for seeing, wings for flying, and kidneys for regulating the composition of the blood. But the design of organisms is not intelligent, as it would be expected from an engineer, but imperfect and worse: defects, dysfunctions, oddities, waste, and cruelty pervade the living world.

Evolution can be seen as a two-step process. First, hereditary variation arises by mutation; second, selection occurs by which useful variations increase in frequency and those that are less useful or injurious are eliminated over the generations. Natural selection is much more than a "purifying" process, for it is able to generate novelty by increasing the probability of otherwise extremely improbable genetic combinations. Natural selection in combination with mutation becomes, in this respect, a creative process. Moreover, it is a process that has been occurring for many millions of years, in many different evolutionary lineages and in a multitude of species, each consisting of a large number of individuals. Evolution by mutation and natural selection has produced the enormous diversity of the living world with its wondrous adaptations.

Darwin and other nineteenth-century biologists found compelling evidence for biological evolution in the comparative study of living organisms, in their geographic distribution, and in the fossil remains of extinct organisms. Recent biological disciplines — genetics, biochemistry, ecology, animal behavior (ethology), neurobiology, and especially molecular biology — have supplied powerful additional evidence and detailed confirmation.

Molecular biology, a discipline that emerged in the second half of the twentieth century, provides the strongest evidence of the evolution of organisms. Molecular biology proves evolution by showing the unity of life in the nature of DNA and the workings of organisms at the level of enzymes and other protein molecules. Moreover, molecular evolution makes it possible to reconstruct evolutionary relationships that were previously unknown, and to confirm, refine, and time all evolutionary relationships from the universal common ancestor up to all living organisms. The precision with which these events can be reconstructed is one reason why the evidence from molecular biology is so useful to evolutionists and so compelling.

As the great evolutionist Theodosius Dobzhansky asserted in 1973, "Nothing in biology makes sense except in the light of evolution."

1. Darwin in the history of ideas

Darwin occupies an exalted place in the history of Western thought, deservedly receiving credit for the theory of evolution. In *The Origin of Species*, he laid out the evidence demonstrating the evolution of organisms. Darwin did not use the term "evolution," which did not have its current meaning, but referred to the evolution of organisms by the phrase "common descent with modification" and similar expressions. However, Darwin accomplished something much more important for intellectual history than demonstrating evolution. Indeed, accumulating evidence for common descent with diversification may very well have been a subsidiary objective of Darwin's masterpiece. Darwin's *Origin of Species* is, first and foremost, a sustained argument to solve the problem of how to account scientifically for the design of organisms. Darwin seeks to explain the adaptations of organisms, their complexity, diversity, and marvelous contrivances as the result of natural processes. Darwin brings about the evidence for evolution because evolution is a necessary consequence of his theory of design.

There is a version of the history of the ideas that sees a parallel between the Copernican and the Darwinian revolutions. In this view, the Copernican Revolution consisted in displacing the Earth from its previously accepted locus as the center of the universe, moving it to a subordinate place as just one more planet revolving around the sun. In congruous manner, the Darwinian Revolution is viewed as consisting of the displacement of humans from their exalted position as the center of life on earth, with all other species created for the service of humankind. According to this version of intellectual history, Copernicus had accomplished his revolution with the heliocentric theory of the solar system. Darwin's achievement emerged from his theory of organic evolution.[1]

What the standard version of the Copernican and Darwinian version of the two revolutions says is correct but inadequate. It misses what is most important about these two intellectual revolutions, namely that they ushered in the beginning of science in the modern sense of the word. These two revolutions may jointly be seen as the one Scientific Revolution, with two stages, the Copernican and the Darwinian.

The Copernican Revolution was launched with the publication in 1543, the year of Nicolaus Copernicus' death, of his *De revolutionibus orbium celestium* (On the Revolutions of the Celestial Spheres), and bloomed with the publication in 1687 of Isaac Newton's *Philosophiae naturalis principia mathematica* (The Mathematical Principles of Natural Philosophy). The discoveries by Copernicus, Kepler, Galileo, Newton, and others, in the sixteenth and seventeenth centuries, had gradually ushered in a conception of the universe as matter in motion governed by natural laws. It was shown that Earth is not the center of the universe, but a small planet rotating around an average star; that the universe is immense in space and in time; and that the motions of the planets around the sun can be explained by the same simple laws that account for the motion of physical objects on our planet. Laws such as $f = m \times a$ (force = mass × acceleration); or the inverse-square law of attraction, $f = g(m_1 m_2)/r^2$ (the force of attraction between two bodies is directly proportional to their masses, but inversely related to the square of the distance between them).

These and other discoveries greatly expanded human knowledge. The conceptual revolution they brought about was more fundamental yet: a commitment to

---

[1] Sigmund Freud refers to these two revolutions as "outrages" inflicted upon humankind's self-image and adds a third one, his own: "Humanity in the course of time had to endure from the hands of science two great outrages upon its naïve self-love. The first was when it realized that our earth was not the center of the universe, but only a tiny speck in a world-system of a magnitude hardly conceivable; this is associated in our minds with the name of Copernicus, although Alexandrian doctrines taught something very similar. The second was when biological research robbed man of his peculiar privilege of having been specially created, and relegated him to a descent from the animal world, implying an ineradicable animal nature in him: this transvaluation has been accomplished in our own time upon the instigation of Charles Darwin, Wallace, and their predecessors, and not without the most violent opposition from their contemporaries. The third and most bitter blow upon man's craving for grandiosity" was meted out in the twentieth century by psychoanalysis, revealing that man's ego "is not even master in his own house." (Freud 1920, 562)

the postulate that the universe obeys immanent laws that account for natural phenomena. The workings of the universe were brought into the realm of science: explanation through natural laws. All physical phenomena could be accounted for as long as the causes were adequately known.

The advances of physical science brought about by the Copernican Revolution had driven mankind's conception of the universe to a split-personality state of affairs, which persisted well into the mid-nineteenth century. Scientific explanations, derived from natural laws, dominated the world of nonliving matter, on the Earth as well as in the heavens. Supernatural explanations, which depended on the unfathomable deeds of the Creator, were accepted as explanations of the origin and configuration of living creatures. Authors, such as William Paley in his *Natural Theology* of 1802, had developed the "argument from design," the notion that the complex design of organisms could not have come about by chance, or by the mechanical laws of physics, chemistry, and astronomy, but was rather accomplished by an Omnipotent Deity, just as the complexity of a watch, designed to tell time, was accomplished by an intelligent watchmaker.

It was Darwin's genius to resolve this conceptual schizophrenia. Darwin completed the Copernican Revolution by drawing out for biology the notion of nature as a lawful system of matter in motion that human reason can explain without recourse to supernatural agencies. Darwin's greatest accomplishment was to show that the complex organization and functionality of living beings can be explained as the result of a natural process — natural selection — without any need to resort to a Creator or other external agent. The origin and adaptations of organisms in their profusion and wondrous variations were thus brought into the realm of science (Ayala 2007).

2. Argument from design

The English clergyman and author William Paley (1743–1805) was intensely committed to the abolition of the slave trade and had become by the 1780s a much sought public lecturer against slavery. He was also an influential writer of works on Christian philosophy, ethics, and theology. *The Principles of Moral and Political Philosophy* (1785) and *A View of the Evidence of Christianity* (1794) earned him prestige and well-endowed ecclesiastical benefices, which allowed him a comfortable life. Illness forced him in 1800 to give up his public speaking career, which provided ample time to study science, particularly biology, and write *Natural Theology; or, Evidences of the Existence and Attributes of the Deity* (1802), the book by which he would become best known to posterity and which would greatly influence Darwin. With *Natural Theology*, Paley sought to update John Ray's *Wisdom of God Manifested in the Works of the Creation* (1691), taking advantage of one century of additional scientific knowledge.

William Paley's *Natural Theology* is a sustained "argument from design" claiming that the living world provides compelling evidence of being designed by an omniscient and omnipotent Creator. The argument has two parts: first, that organisms give evidence of being designed; second, that only an omnipotent God could account for the perfection, multitude, and diversity of designs. Paley's keystone claim is that "There cannot be design without a designer; contrivance, without a contriver; order, without choice; … means suitable to an end, and executing their office in accomplishing that end, without the end ever having been contemplated." (Paley 1802, 15–16)

Paley elaborates the argument-from-design with greater cogency and more extensive knowledge of biological detail than it has ever been done by any other author, before or since. Paley brings in all sorts of biological knowledge, from the geographic distribution of species to the interactions between predators and their prey, the interactions between the sexes, the camel's stomach and the woodpecker's tong, the compound eyes of insects and the spider's web. He explores and rejects the possibility of a sort of "natural selection": organisms may have come about by chance in an endless multiplicity of forms; those now in existence are those that happened to be functionally organized because they are the only ones able to survive and reproduce. (Empedocles, the fifth-century BC philosopher of Classic Greece, had already speculated that living organisms arose as disjointed body parts — heads, trunks, and limbs — that combined at random, with only viable combinations surviving.) Paley's evidence for intelligent design and against chance derives from a notion that some contemporary authors have named "irreducible complexity," that he calls "relation": the presence of a great variety of parts interacting with each other to produce an effect, which cannot be accomplished if any of the parts is missing.

Paley's *Natural Theology* has chapters dedicated to the complex design of the human eye; to the human frame, which displays a precise mechanical arrangement of bones, cartilage, and joints; to the circulation of the blood and the disposition of blood vessels; to the comparative anatomy of humans and animals; to the digestive system, kidneys, urethras, and bladder; to the wings of birds and the fins of fish; and much more. For 352 pages, *Natural Theology* conveys Paley's expertise: extensive and accurate biological knowledge, as detailed and precise as it was available in the year 1800. After detailing the precise organization and exquisite functionality of each biological object or process, Paley draws again and again the same conclusion, that only an Omniscient and Omnipotent Deity could account for these marvels of mechanical perfection, purpose, and functionality, and for the enormous diversity of inventions that they entail.

Christian philosophers and theologians had argued, in the Middle Ages, that the functional organization of living beings evinces the existence of an Omnipotent and Omniscient Designer. In the thirteenth century St. Thomas Aquinas had used an argument-from-design as his "fifth way" to demonstrate the existence of God. Aquinas' argument was based on the "harmony" of the universe and all its parts.

A few years after the publication of *Natural Theology*, the eighth Earl of Bridgewater endowed the publication of treatises that would set forth "the Power, Wisdom and Goodness of God as manifested in the Creation." Eight treatises were published during 1833–1840, several of which artfully incorporate the best science of the time and had considerable influence on the public and among scientists. William Buckland, professor of geology at Oxford University, notes in *Geology and Mineralogy* (1836) the world distribution of coal and mineral ores and proceeds to point out that they had been deposited in a remote part, yet obviously with the forethought of serving the larger human populations that would come about much later. Another geologist, Hugh Miller in *The Testimony of the Rocks* (1858), would argue that it is not only the perfection of design but also the beauty of natural structures found in rock formations and in mountains and rivers that manifests the intervention of the Creator.

One of the Bridgewater Treatises, *The Hand, Its Mechanisms and Vital Endowments as Evincing Design*, was written by Sir Charles Bell, a distinguished anatomist and surgeon, famous for his neurological discoveries, who became professor of surgery in 1836 at the University of Edinburgh. Bell follows Paley's manner of argument, examining in considerable detail the wondrously useful design of the human hand, but also the perfection of design of the forelimb used for different purposes in different animals, serving in each case the particular needs and habits of its owner: the human's arm for handling objects, the dog's leg for running, and the bird's wing for flying.

In the 1990s, several authors, notably biochemist Michael Behe (1996), theorist William Dembski (1995), and law professor Phillip Johnson (2000), among others, revived the argument from design. Often, however, these authors sought to hide their real agenda, and, thus, typically avoid explicit reference to God, so that the "theory" of intelligent design (ID) could be taught in the public schools, as an alternative to the theory of evolution, without incurring conflict with the U.S.

Constitution, which forbids the endorsement of any religious beliefs in public institutions.

On December 20, 2005, after a five-week trial, John E. Jones III, federal judge for the Middle District of Pennsylvania, issued a 130-page-long decision (Kitzmiller vs. the Dover Area School District) declaring that "The overwhelming evidence at trial established that ID [intelligent design] is a religious view, a mere re-labeling of creationism, and not a scientific theory … ID is not supported by any peer-reviewed research, data, or publications."

3. Darwin's theory

The discovery of natural selection, Darwin's awareness that it was a greatly significant discovery because it was science's answer to Paley's "argument from design," and Darwin's designation of natural selection as "my theory" can be traced to Darwin's Transmutation Notebook D, written between July and October of 1838, not long after returning (on October 2, 1836) from his five-year voyage on the Beagle (Eldredge 2005). Darwin considered natural selection, rather than his demonstration of evolution, his most important discovery, and designated it as "my theory," a designation he never used when referring to the evolution of organisms.

The evolution of organisms was commonly accepted by naturalists in the middle decades of the nineteenth century. The distribution of exotic species in South America, in the Galápagos Islands, and elsewhere, and the discovery of fossil remains of long-extinguished animals, confirmed the reality of evolution in Darwin's mind. The intellectual challenge was to explain the origin of distinct species of organisms and how they adapted to their environments, that "mystery of mysteries," as it had been labeled by Darwin's older contemporary, the prominent scientist and philosopher Sir John Herschel (1792–1871).

Since the "Transmutation Notebooks," written from 1837 to 1839, until his death in 1882, Darwin's life would be dedicated to substantiating natural selection and its companion postulates, mainly the pervasiveness of hereditary variation and the enormous fertility of organisms, which much surpassed the capacity of available resources. Natural selection became for Darwin "a theory by which to work." He relentlessly pursued observations and performed experiments in order to test the theory and resolve presumptive objections.

In his Autobiography, Darwin wrote, "The old argument of design in nature, as given by Paley, which formerly seemed to me so conclusive, falls, now that the law of natural selection has been discovered. We can no longer argue that, for instance, the beautiful hinge of a bivalve shell must have been made by an intelligent being, like the hinge of a door by a man." (Barlow 1958)

Darwin's focus in *The Origin* was the explanation of design, with evolution playing the subsidiary role of supporting evidence. The Introduction and Chapters I through IX of *The Origin* explain how natural selection accounts for the adaptations and behaviors of organisms, their "design". The extended argument starts in Chapter I, where Darwin describes the successful selection of domestic plants and animals and, with considerable detail, the success of pigeon fanciers seeking exotic "sports." The success of plant and animal breeders manifests how much selection can accomplish by taking advantage of spontaneous variations that occur in organisms but happen to fit the breeders' objectives. A sport (mutation) that first appears in an individual can be multiplied by selective breeding, so that after a few generations that sport becomes fixed in a breed, or "race". The familiar breeds of dogs, cattle, chickens, and food plants have been obtained by this process of selection practiced by people with particular objectives.

The ensuing chapters (II–IX) of *Origin* extend the argument to variations propagated by natural selection for the benefit of the organisms themselves, rather than by artificial selection of traits desired by humans. As a consequence of natural selection, organisms exhibit design, that is, exhibit adaptive organs and functions. The design of organisms as they exist in nature, however, is not "intelligent design," imposed by God as a Supreme Engineer or by humans; rather, it is the result of a natural process of selection, promoting the adaptation of organisms to their environments. This is how natural selection works: Individuals that have beneficial variations, that is, variations that improve their probability of survival and reproduction, leave more descendants than individuals of the same species that have less beneficial variations. The beneficial variations will consequently increase in frequency over the generations; less beneficial or harmful variations will be eliminated from the species. Eventually, all individuals of the species will have the beneficial features; new features will arise over eons of time.

Organisms exhibit complex design, but it is not, in current language, "irreducible complexity," emerging all of a sudden in full bloom. Rather, according to Darwin's theory of natural selection, the design has arisen gradually and cumulatively, step by step, promoted by the reproductive success of individuals with incrementally more adaptive elaborations.

It follows from Darwin's explanation of adaptation that evolution must necessarily occur as a consequence of organisms becoming adapted to different environments in different localities, and to the ever-changing conditions of the environment over time, and as hereditary variations become available at a particular time that improve, in that place and at that time, the organisms' chances of survival and reproduction. *The Origin*'s evidence for biological evolution is central to Darwin's explanation of design, because this explanation implies that biological evolution occurs, which Darwin therefore seeks to demonstrate in most of the remainder of the book (Darwin 1859, chapters X–XIV).

In the concluding Chapter XV of *Origin*, Darwin returns to the dominant theme of adaptation and design. In an eloquent final paragraph, Darwin asserts the "grandeur" of his vision: "It is interesting to contemplate an entangled bank, clothed with many plants of many kinds, with birds singing on the bushes, with various insects flitting about, and with worms crawling through the damp earth, and to reflect that these elaborately constructed forms, so different from each other, and dependent on each other in so complex a manner, have all been produced by laws acting around us ... Thus, from the war of nature, from famine and death, the most exalted object which we are capable of conceiving, namely, the production of the higher animals, directly follows. There is grandeur in this view of life, with its several powers, having been originally breathed into a few forms or into one; and that, whilst this planet has gone cycling on according to the fixed law of gravity, from so simple a beginning endless forms most beautiful and most wonderful have been, and are being, evolved." (Darwin 1859, 489–490; emphasis added).

4. Natural selection

Darwin's Origin of Species addresses the same issue as Paley: how to account for the adaptive configuration of organisms and their parts, which are so obviously designed to fulfill certain functions. Darwin argues that hereditary adaptive variations ("variations useful in some way to each being") occasionally appear, and that these are likely to increase the reproductive chances of their carriers. The success of pigeon fanciers and animal breeders clearly shows the occasional occurrence of useful hereditary variations. In nature, over the generations, Darwin's argument continues, favorable variations will be preserved, multiplied, and conjoined; injurious ones will be eliminated.

This is how Darwin summarizes natural selection: "Can it, then, be thought improbable, seeing that variations useful to man have undoubtedly occurred, that other variations useful in some way to each being in the great and complex battle of life, should sometimes occur in the course of thousands of generations? If such do occur, can we doubt (remembering that many more individuals are born than can possibly survive) that individuals having any advantage, however slight, over others, would have the best chance of surviving and of procreating their kind? On the other hand, we may feel sure that any variation in the least degree injurious would be rigidly destroyed. This preservation of favorable variations and the rejection of injurious variations, I call Natural Selection. Variations neither useful nor injurious would not be affected by natural selection." (Darwin 1859, ch. IV)

Evolution affects all aspects of an organism's life — morphology (form and structure), physiology (function), behavior, and ecology (interaction with the environment). Underlying these changes are changes in the hereditary materials. Hence, in genetic terms, evolution consists of changes in the organisms' hereditary makeup.

Evolution can be seen as a two-step process. First, hereditary variation arises by mutation; second, selection occurs by which useful variations increase in frequency and those that are less useful or injurious are eliminated over the generations. "Useful" and "injurious" are terms used by Darwin in his definition of natural selection. The significant point is that individuals having useful variations "would have the best chance of surviving and procreating their kind." As a consequence, useful variations increase in frequency over the generations, at the expense of those that are less useful or injurious.

Natural selection is much more than a "purifying" process, for it is able to generate novelty by increasing the probability of otherwise extremely improbable genetic combinations. Natural selection in combination with mutation becomes, in this respect, a creative process. Moreover, it is a process that has been occurring for many millions of years, in many different evolutionary lineages and a multitude of species, each consisting of a large number of individuals. Evolution by mutation and natural selection has produced the enormous diversity of the living world with its wondrous adaptations.

Natural selection is an incremental process, operating over time and yielding organisms better able to survive and reproduce than others. Individuals of a given species differ from one another at any one time only in small ways; for example, the difference between bacteria that have or lack an enzyme able to synthesize the sugar lactose or between moths that have light or dark wings. These differences typically involve one or only a few genes, but they can make the difference between survival or death, as in beneficial resistance to DDT or to antibiotics. Adaptations that involve complex structures, functions, or behaviors involve numerous genes.

Several hundred million generations separate modern animals from the early animals of the Cambrian geological period (542 million years ago). The number of mutations that can be tested, and those eventually selected, in millions of individual animals over millions of generations is difficult for a human mind to fathom, but we can readily understand that the accumulation of millions of small, functionally advantageous changes could yield remarkably complex and adaptive organs, such as the eye.

Natural selection produces combinations of genes that would seem highly improbable because natural selection proceeds stepwise over long periods of time. Consider the evolution of the eye in humans and other vertebrates. Perception of light, and later vision, were important for the survival and reproductive success of their ancestors, because sunlight is a predominant feature of the environment. Accordingly, natural selection favored genes and gene combinations that increased the functional efficiency of the eye. Such mutations gradually accumulated, eventually leading to the highly complex and efficient vertebrate eye.

5. About artists and typing monkeys

Natural selection, as discovered by Darwin, does not operate in the manner of Paley's unacceptable hypothesis, acting on randomly formed organisms, allowing the functional ones to survive while the great majority die. Natural selection does not operate, either, as a sieve that retains the rarely arising useful genes and lets go the more frequently arising harmful mutants; at least, not only. Natural selection acts in the filtering way of a sieve, but it is much more than a purely negative process, for it is able to generate novelty by increasing the probability of otherwise extremely improbable genetic combinations. Natural selection is thus a creative process. It does not "create" the entities upon which it operates, but it produces adaptive (functional) genetic combinations that could not have existed otherwise.

The creative role of natural selection must not be understood in the sense of the "absolute" creation that traditional Christian theology predicates of the Divine act by which the universe was brought into being *ex nihilo*, or in the manner of creation in which Paley assumes God, the Supreme Engineer had created the adaptations of organisms. Natural selection may rather be compared to a painter who creates a picture by mixing and distributing pigments in various ways over the canvas. The canvas and the pigments are not created by the artist but the painting is. It is inconceivable that a random combination of the pigments might result in the orderly and complex whole of a work of art, such as Rembrandt's Night Watch. In the same way, the combination of genetic units that carries the hereditary information responsible for the formation of the vertebrate eye could have never been produced by a random process like mutation. Not even if we allow for the three-billion-years-plus during which life has existed on earth. The complicated anatomy of the eye like the exact functioning of the kidney are the result of a nonrandom process — natural selection.

Critics have sometimes alleged as evidence against Darwin's theory of evolution examples showing that random processes cannot yield meaningful, organized outcomes. It is thus pointed out that a series of monkeys randomly striking letters on a typewriter would never write *The Origin of Species*, even if we allow for millions of years and many generations of monkeys pounding at typewriters.

This criticism would be valid if evolution would depend only on random processes. But natural selection is a non-random process that promotes adaptation by selecting combinations that "make sense," i.e., that are useful to the organisms. The analogy of the monkeys would be more appropriate if a process existed by which, first, meaningful words would be chosen every time they appeared on the typewriter; and then there would also be typewriters with previously selected words rather than just letters in the keys, and again there would be a process to select meaningful sentences every time they appeared in this second typewriter. If every time words such as "the," "origin," "species," and so on, appeared in the first kind of typewriter, they each became a key in the second kind of typewriter, meaningful sentences would occasionally be produced in this second typewriter. If such sentences became incorporated into keys of a third type of typewriter, in which meaningful paragraphs were selected whenever they appeared, it is clear that pages and even chapters "making sense" would eventually be produced. The end product would be an "irreducibly complex" text.

We need not carry the analogy too far, since the analogy is not fully satisfactory, but the point is clear. Evolution is not the outcome of purely random processes, but rather there is a "selecting" process, which picks up adaptive combinations because these reproduce more effectively and thus become established in populations. These adaptive combinations constitute, in turn, new levels of organization upon which the mutation (random) plus selection (non-random or directional) process again operates. The complexity of organization of animals and plants is "irreducible" to simpler components in one or very few steps, but not thorough the millions and millions of generations and the multiplicity of steps and levels made possible by eons of time.

6. Chance and necessity

The critical point is that evolution by natural selection is an incremental process, operating over eons of time and yielding organisms better able to survive and reproduce than others, which typically differ from one another at any one time only in small ways; for example, the difference between producing more or fewer progeny or between having or lacking an enzyme able to catalyze the synthesis of one particular amino acid. Notice also that increased complexity is not a necessary outcome of natural selection, although complexity increases from time to time in some lineages of descent, so that, although rare, they are very conspicuous over time's eons. That is, increased complexity is not a necessary consequence of evolution by natural selection, but rather emerges occasionally as a matter of statistical bias. The longest living organisms on Earth are the microscopic bacteria, which have continuously existed on our planet for three and a half billion years and yet exhibit no greater complexity than their old time ancestors. More complex organisms came about much later, without the elimination of their simpler relatives. For example, the primates appeared on earth some fifty million years ago and our species, *Homo sapiens*, came about two hundred thousand years ago.

There is an important respect in which an artist makes a poor analogy of natural selection. A painter or a sculptor has a preconception of what he or she wants to paint and will consciously modify the painting or sculpture so that it represents what the artist wants. Similarly, an engineer has a preconception of what the design of a contrivance of structure is supposed to achieve, and will select suitable materials and arrange them in a preconceived manner so that it fulfills the intended function. On the contrary, natural selection has no foresight, nor does it operate according to some preconceived plan. Rather it is a purely natural process resulting from the interacting properties of physicochemical and biological entities. Natural selection is simply a consequence of the differential multiplication of living beings, as pointed out. It has some appearance of purposefulness because it is conditioned by the environment: which organisms reproduce more effectively depends on what variations they possess that are useful in the place and at the time where the organisms live. But natural selection does not anticipate the environments of the future; drastic environmental changes may be insuperable to organisms that were previously thriving. Species extinction is the common outcome of the evolutionary process. The species existing today represent the balance between the origin of new species and their eventual extinction. More than 99 percent of all species that ever lived on Earth have become extinct without issue. These may have been more than one billion species; the available inventory of living species has identified and described less than two million out of some ten million estimated to be now in existence.

The team of typing monkeys is also a bad analogy of evolution by natural selection, because it assumes that there is "somebody" who selects letter combinations and word combinations that make sense. In evolution there is no one selecting adaptive combinations. These select themselves because they multiply more effectively than less adaptive ones.

The process of natural selection can explain the adaptive organization of organisms, as well as their diversity and evolution, as a consequence of their adaptation to the multifarious and ever changing conditions of life. The fossil record shows that life has evolved in a haphazard fashion. The radiations, expansions, relays of one form by another, occasional but irregular trends, and the ever-present extinctions, are best explained by natural selection of organisms subject to the vagaries of genetic mutation and environmental challenge. The scientific account of these events does not necessitate recourse to a preordained plan, whether imprinted from without by an omniscient and all-powerful Designer, or resulting from some immanent force driving the process towards definite outcomes. Biological evolution differs from a painting or an artifact in that it is not the outcome of preconceived design.

Natural selection accounts for the "design" of organisms, because adaptive variations tend to increase the probability of survival and reproduction of their carriers at the expense of maladaptive, or less adaptive, variations. The arguments of Paley against the incredible improbability of chance accounts of the adaptations of organisms are well taken as far as they go. But neither Paley nor any other author before Darwin was able to discern that there is a natural process (namely, natural selection) that is not random, but rather is oriented and able to generate order or "create." The traits that organisms acquire in their evolutionary histories are not fortuitous but determined by their functional utility to the organisms, "designed" as it were to serve their life needs.

Chance is, nevertheless, an integral part of the evolutionary process. The mutations that yield the hereditary variations available to natural selection arise at random, independently of whether they are beneficial or harmful to their carriers. But this random process (as well as others that come to play in the great theater of life) is counteracted by natural selection, which preserves and multiplies what is useful and eliminates the harmful. Without hereditary mutation, evolution could not happen because there would be no variations that could be differentially conveyed from one to another generation. But without natural selection, the mutation process would yield disorganization and extinction because most mutations are disadvantageous. Mutation and selection have jointly driven the marvelous process that starting from microscopic organisms has yielded orchids, birds, and humans.

The theory of evolution conveys chance and necessity jointly intricated in the stuff of life; randomness and determinism interlocked in a natural process that has spurted the most complex, diverse, and beautiful entities in the universe: the organisms that populate the earth, including humans who think and love, endowed with free will and creative powers, and able to analyze the process of evolution itself that brought them into existence. This is Darwin's fundamental discovery, that there is a process that is creative though not conscious, a process that creates design without necessitating a Designer (Ayala 1999; 2007).

7. In praise of imperfection

When confronted with the cruelties, imperfections and dysfunctions commonly found in the living world, Paley advances a defense of the argument-from-design that is unsatisfactory. It is evasive to say of imperfections and deficiencies that "they are of little or no weight," or that "apparent blemishes ... ought to be referred to some cause, though we be ignorant of it" (Paley 1802, 46). Michael J. Behe, who has reformulated Paley's argument-from-design, makes a similarly evasive claim: "Clearly, designers who have the ability to make better designs do not necessarily do so ... The argument from imperfection overlooks the possibility that the designer might have multiple motives, with engineering excellence oftentimes relegated to a secondary role." (Behe 1996, 223).

But this will not do. Behe's account is not only evasive, but it totally destroys his claim that intelligent design is a scientific hypothesis, because it provides it with an empirically impenetrable shield. If we cannot reject intelligent design because the designer may have reasons that we could not possibly ascertain, there would seem to be no way to test intelligent design. Scientific hypotheses are tested by interrogating nature about predictions logically derived from the hypothesis that are expected to be observed in the world of experience. But it follows from Behe's evasion that we cannot draw out any predictions that could possibly falsify the hypothesis of intelligent design. Intelligent design as an explanation for the adaptations of organisms could be (natural) theology, as Paley would have it, but, whatever it is, it would not be a scientific hypothesis, because there is no way of ascertaining whether it is false by showing the falsity of its logically necessary consequences.

Moreover, the issue is not only that organisms and their parts are less than perfect, but also that deficiencies and dysfunctions are pervasive, evidencing defective "design." Consider the human jaw. We have too many teeth for the jaw's size, so that wisdom teeth need to be removed and orthodontists make a decent living straightening the others. Would we want to blame God for such defective design? A human engineer could have done better. Evolution gives a good account of this imperfection. Brain size increased over time in our ancestors, and the remodeling of the skull to fit the larger brain entailed a reduction of the jaw. Evolution responds to the organisms' needs through natural selection, not by optimal design but by "tinkering," as it were, by slowly modifying existing structures. Consider now the birth canal of women, much too narrow for easy passage of the infant's head, so that thousands upon thousands of babies die during delivery. Surely we do not want to blame an Intelligent Designer for this defective design or for the children's deaths. Science makes it understandable, a consequence of the evolutionary enlargement of our brain. Females of other primates do not experience this difficulty.

Examples of deficiencies and dysfunctions in all sorts of organisms can be endlessly multiplied, reflecting the opportunistic, tinkerer-like character of natural selection, rather than intelligent design. The world of organisms also abounds in characteristics that might be called "oddities," as well as those that have been characterized as "cruelties," an apposite qualifier if the cruel behaviors were designed outcomes of a being holding on to human or higher standards of morality. But the "cruelties" of biological nature are only metaphoric "cruelties" when applied to the outcomes of natural selection.

Examples of "cruelty" involve not only the familiar predators (say, a chimpanzee) tearing apart their prey (say a small monkey held alive by a chimpanzee biting large flesh morsels from the screaming monkey), or parasites destroying the functional organs of their hosts, but also, and very abundantly, between organisms of the same species, even between individuals of different sexes in association with their mating. A well-known example is the female praying mantis that devours the male after coitus is completed. Less familiar is that, if she gets the opportunity, the female will eat the head of the male before mating, which thrashes the headless male mantis into spasms of "sexual frenzy" that allow the female to connect his genitalia with hers (Lawrence 1992; Elgar 1992). In some midges (tiny flies), the female captures the male as if he were any other prey and with the tip of her proboscis she injects into his head her spittle that starts digesting the male's innards that are then sucked by the female; partly protected from digestion are the relatively intact male organs that break off inside the female and fertilize her (Downes 1978). Male cannibalism is known in dozens of species, particularly spiders and scorpions. Diverse sorts of oddities associated with mating behavior are described in the delightful, but accurate and documented, book by Olivia Judson, *Dr. Tatiana's Sex Advice to All Creation* (2002).

Some proponents of the argument-from-design, including Behe, claim, at times, that their hypothesis is scientific, because postulating an "Intelligent Designer" does not implicate any supernatural causes, such as God. But this is dissembling.

William Paley and Thomas Aquinas were adamant that the conclusion of the argument from design was the existence of an Omnipotent, Omniscient, and Benevolent God. But current biology has disclosed that the argument misfires. As the distinguished American philosopher, David Hull (1992), has pointed out: "What kind of God can one infer from the sort of phenomena epitomized by the species on Darwin's Galapagos Islands? The evolutionary process is rife with happenstance, contingency, incredible waste, death, pain and horror ... Whatever the God implied by evolutionary theory and the data of natural selection may be like, he is not the Protestant God of waste not, want not. He is also not the loving God who cares about his productions. He is not even the awful God pictured in the Book of Job. The God of the Galapagos is careless, wasteful, indifferent, almost diabolical. He is certainly not the sort of God to whom anyone would be inclined to pray."[2]

8. From Darwin to molecular biology

Darwin and other nineteenth-century biologists found compelling evidence for biological evolution in the comparative study of living organisms, in their geographic distribution, and in the fossil remains of extinct organisms. As pointed out above, in *The Origin of Species* Darwin dedicates five chapters to the evidence for evolution: two chapters to the geological record, or, as we are more likely to say nowadays, to paleontology; two chapters to biogeography, and one chapter to comparative anatomy and embryology. Since Darwin's time, the evidence from these sources has become stronger and more comprehensive, while biological disciplines that emerged more recently — genetics, biochemistry, ecology, animal behavior (ethology), neurobiology, and especially molecular biology — have supplied powerful additional evidence and detailed confirmation.

Darwin surely would have been pleased by the enormous accumulation of paleontological evidence, including the discovery of fossils of organisms intermediate between major groups, such as *Archaeopteryx*, intermediate between reptiles (dinosaurs) and birds, and *Tiktaalik*, intermediate between fish and tetrapods (Ahlberg and Clack 2006) and the numerous fossils and diverse species of ho-

[2] The great Argentine fabulist Jorge Luis Borges conveys an utterly pessimistic view of the world by quoting the damming words of Ugbar's sage: "Mirrors and fatherhood are abominable because they multiply and disseminate the universe." The second-century Gnostic Basilides would have approved of science's exclusion of God from misshapen creation, as he wrote: "What greater glory for a God than to be absolved of the world?" (quoted in Bloom 2002, 682; see also Ayala 2003).

minins, intermediate between apes and *Homo sapiens* (e.g., White *et al.* 2006; Dalton 2006; Cela-Conde and Ayala 2007). But there are good reasons to believe that Darwin would have been most pleased and most impressed with the overwhelming evidence for evolution and precise information about evolutionary history provided by molecular biology, a source of evidence and document of history that Darwin could not have even imagined.

Molecular biology, a discipline that emerged in the second half of the twentieth century, nearly 100 years after the publication of *The Origin of Species*, undoubtedly provides the strongest evidence yet of the evolution of organisms. Molecular biology proves evolution in two ways: first, by showing the unity of life in the nature of DNA and the workings of organisms at the level of enzymes and other protein molecules; second, and most important, by making it possible to reconstruct evolutionary relationships that were previously unknown, and to confirm, refine, and time all evolutionary relationships from the universal common ancestor up to all living organisms. The precision with which these events can be reconstructed is one reason why the evidence from molecular biology is so useful to evolutionists and so compelling.

DNA and proteins have been called "informational macromolecules" because they are long linear molecules made up of sequences of units — nucleotides or amino acids — that embody evolutionary information. Comparing the sequence of the components in two macromolecules establishes how many units are different. Because evolution usually occurs by changing one unit at a time, the number of differences is an indication of the recency of common ancestry. Thus, the inferences from paleontology, comparative anatomy, and other disciplines that study evolutionary history can be tested in molecular studies of DNA and proteins by examining the sequences of nucleotides and amino acids. The authority of this kind of test is overwhelming: each of the thousands of genes and thousands of proteins contained in an organism provides an independent test of that organism's evolutionary history.

Molecular evolutionary studies have three notable advantages over comparative anatomy and the other classical disciplines: *precision*, *universality*, and *multiplicity*. First, the information is readily quantifiable. The number of units that are different is easily established when the sequence of units is known for a given macromolecule in different organisms. It is simply a matter of aligning the units (nucleotides or amino acids) between two or more species and counting the differences. The second advantage, universality, is that comparisons can be made between very different sorts of organisms. There is very little that comparative anatomy can say when, for example, organisms as diverse as yeasts, pine trees, and human beings are compared, but there are numerous DNA and protein sequences that can be compared in all three. The third advantage is multiplicity. Each organism possesses thousands of genes and proteins, every one of which reflects the same evolutionary history. If the investigation of one particular gene or protein does not satisfactorily resolve the evolutionary relationship of a set of species, additional genes and proteins can be investigated until the matter has been settled.

The resourcefulness of molecular biology to study evolution can be noted in other ways as well. The widely different rates of evolution of different sets of genes opens up the opportunity for investigating different genes in order to achieve different degrees of resolution in the tree of evolution. Evolutionists rely on slowly evolving genes for reconstructing remote evolutionary events, but increasingly faster evolving genes for reconstructing the evolutionary history of more recently diverged organisms.

Genes that encode *ribosomal RNA* molecules are among the slowest evolving genes. They have been used to reconstruct the evolutionary relationships among groups of organisms that diverged very long ago: for example, among bacteria, archaea, and eukaryotes (the three major divisions of the living world), which diverged more than 2 billion years ago, or among the protozoa compared with plants and with animals, groups of organisms that diverged about 1 billion years ago. Cytochrome *c* evolves slowly, but not as slowly as the ribosomal RNA genes. Thus, it is used to decipher the relationships within large groups of organisms, such as among humans, fishes, and insects. Fast-evolving molecules, such as the fibrinopeptides involved in blood clotting, are appropriate for investigating the evolution of closely related animals — the primates, for example: macaques, chimps and humans.

It is now possible to make an assertion that would have delighted Darwin and would surely shock creationists and other anti-evolutionists, and perhaps startle many scientists and most of the general public: Gaps of knowledge in the evolutionary history of living organisms no longer need to exist. Molecular biology has made it possible to reconstruct the "universal tree of life," the continuity of succession from the original forms of life, ancestral to all living organisms, to every species now living on Earth. The main branches of the tree of life have been reconstructed on the whole and in great detail. More details about more and more branches of the universal tree of life are published in scores of scientific articles every month. The virtually unlimited evolutionary information encoded in the DNA sequence of living organisms allows evolutionists to reconstruct all evolutionary relationships leading to present-day organisms, with as much detail as wanted. Invest the necessary resources (time and laboratory expenses) and one can have the answer to any query, with as much precision as wanted.

9. Darwin and Wallace: A historical footnote

Alfred Russel Wallace (1823–1913) is famously given credit for discovering, independently of Darwin, natural selection as the process accounting for the evolution of species. On June 18, 1858, Darwin wrote to Charles Lyell that he had received by mail a short essay from Wallace such that "if Wallace had my [manuscript] sketch written in [1844] he could not have made a better abstract." Darwin and Wallace had started occasional correspondence in late 1855. At the time Wallace was in the Malay Archipelago collecting biological specimens. In his letters, Darwin would offer sympathy and encouragement to the occasionally dispirited Wallace for his "laborious undertaking". In 1858, Wallace came upon the idea of natural selection as the explanation for evolutionary change and he wanted to know Darwin's opinion about this hypothesis, since Wallace, as well as many others, knew that Darwin had been working on the subject for years, had shared his ideas with other scientists, and was considered by them as the eminent expert on issues concerning biological evolution.

Darwin was uncertain how to proceed about Wallace's letter. He wanted to credit Wallace's discovery of natural selection, but he did not want altogether to give up his own earlier independent discovery. Eventually, Sir Charles Lyell and Joseph Hooker proposed, with Darwin's consent, that Wallace's letter and two of Darwin's earlier writings would be presented at a meeting of the Linnean Society of London. On July 1, 1858, three papers were read by the society's undersecretary, George Busk, in the order of their date of composition: Darwin's abbreviated abstract of his 230-page essay from 1844; an "abstract of abstract" that Darwin had written to the American botanist Asa Gray on September 5, 1857; and Wallace's essay, "On the Tendency of Varieties to Depart Indefinitely from Original Type; Instability of Varieties Supposed to Prove the Permanent Distinctness of Species." (Wallace 1858)

The meeting was attended by some thirty people, who did not include Darwin or Wallace. The papers generated little response and virtually no discussion, their significance apparently lost to those in attendance. Nor was it noticed by the president of the Linnean Society, Thomas Bell, who, in his annual address the following May, blandly stated that the past year had not been enlivened by "any of those striking discoveries which at once revolutionize" a branch of science.

Wallace's independent discovery of natural selection is remarkable. But the lesser credit given to Wallace than to Darwin for this discovery may not be misplaced. Wallace was not interested in explaining design, but rather in accounting for the evolution of species, as indicated in his paper's title: "On the Tendency of Varieties to Depart Indefinitely from the Original Type." Wallace thought that evolution proceeds indefinitely and is progressive. Wallace (1858) writes: "We believe that there is a tendency in nature to the continued progression of certain classes of varieties further and further from the original type — a progression to which there appears no reason to assign any definite limits. This progression, by minute steps, in various directions..."

Darwin, on the contrary, did not accept that evolution would necessarily represent progress or advancement, nor did he believe that evolution would always result in morphological change over time; rather, he knew of the existence of "living fossils," organisms that had remained unchanged for millions of years. For example, "some of the most ancient Silurian animals, as the Nautilus, Lingula, etc., do not differ much from living species." (Darwin 1859, 306) In 1858, Darwin was at work on a multivolume treatise, intended to be titled "On Natural Selection." Wallace's paper stimulated Darwin to write *The Origin*, which would be published the following year. Darwin saw *The Origin* as an abbreviated version of the much longer book he had planned to write.

References

Ahlberg E., Clack J. A., 2006, "Palaeontology: a firm step from water to land", *Nature* 440: 747.

Ayala F. J., 1999, "Adaptation and novelty: teleological explanations in evolutionary biology," *History and Philosophy of the Life Sciences* 21: 3–33.

Ayala F. J., 2003, "Intelligent Design: the original version", *Theology and Science* 1: 9–32.

Ayala F. J., 2007, *Darwin's Gift to Science and Religion*, Washington, DC: Joseph Henry Press.

Barlow N. (ed.), 1958, *The Autobiography of Charles Darwin,* London: Collins.

Behe M. J., 1996, *Darwin's Black Box. The Biochemical Challenge to Evolution,* New York: Touchstone, Simon & Schuster.

Bloom H., 2002, *Genius: A Mosaic of One Hundred Exemplary Creative Minds,* New York: Warner Books.

Cela-Conde C. J., Ayala F. J., 2007, *Human Evolution. Trails from the Past,* Oxford: Oxford University Press.

Dalton R., 2006, "Feel it in your bones", *Nature* 440: 1100–1101.

Darwin C. R., 1859, 1967, *On the Origin of Species. A Facsimile of the First Edition, with an Introduction by Ernst Mayr,* New York: Athenaeum.

Dembski W., 1995, *The Design Inference. Eliminating Chance through Small Probabilities,* Cambridge: Cambridge University Press.

Downes J. A., 1978, "Feeding and mating in the insectivorous Ceratopogoninae (Diptera)", *Memoirs of the Entomological Society of Canada* 104: 1–62.

Eldredge N., 2005, *Darwin: Discovering the Tree of Life*, New York: Norton.
Elgar M. A, 1992, "Sexual cannibalism in spiders and other invertebrates", in Elgar, M. A., Crespi B. J. (eds.), *Cannibalism: Ecology and Evolution among Diverse Taxa*, Oxford: Oxford University Press.
Freud S. A., 1920, "a general introduction to psycho-analysis", in Adler M. J. (ed.), 1993, *Freud*, Great Books of the Western World, vol. 54, Chicago: Encyclopaedia Britannica, Inc.
Hull D. L., 1992, "God of the Galapagos", *Nature* 352: 485–86.
Johnson P., 2000, *The Wedge of Truth*, Downers Grove, Ill.: InterVarsity Press.
Judson O., 2002, *Dr. Tatiana's Sex Advice to All Creation,* New York: Holt.
Lawrence S. E., 1992, "Sexual cannibalism in the praying mantis, Mantis religiosa: a field study", *Animal Behaviour* 43: 569–83.
Paley W., 1802, *Natural Theology*, London: R. Fauldner.
Wallace A. R., 1858, "On the tendency of varieties to depart indefinitely from the original type", *Journal of the Proceedings of the Linnean Society of London (Zoology)* 3: 53–62.
White T. D., WoldeGabriel G., Asfaw B. *et al.*, 2006, "Asa Issie, Aramis and the origin of Australopithecus", *Nature* 440: 883–89.

History of evolution theories

Jean Gayon

1. Introduction

In this paper, I examine the question of the "continuity" of Darwinism as a theoretical tradition on a large historical scale. For nearly one-and-a-half centuries, evolutionary biologists have been haunted by the question of whether their conceptions are or are not "Darwinian". This persistent positioning of new scientific developments in reference to a single, pioneering figure is something quite exceptional in the history of modern science. Physicists working in the domain of relativity or quantum theory may refer to Einstein or Bohr (or other individuals). But their debates have not been massively structured through reference to these individual's names in the way that evolutionary theory has been and continues to be today (Gayon 1990; Lewens 2007). I know of nothing like "Einsteinianism" or "Bohrism" in physics, whereas "Darwinism" has been a keyword among evolutionary biologists since it was first coined. What can account for such a fact? One explanation might be that Darwinism, just as Marxism, is not only the name of an ordinary scientific tradition, but also of a massive historical phenomenon, ranging across many areas of modern culture and politics. In such cases it is common to identify a tradition of thinking with a person's name. But although this is perfectly true, we cannot be satisfied with that explanation. The fact is that biologists working on evolution have never stopped returning to Darwin and reading his masterpieces. It is still the case today. And, again, we do not observe anything like this in other areas of the natural sciences. My conviction is that evolutionary biologists' uninterrupted interest in Darwin since 1859 indicates a strong form of connection, such as that between a model (Darwin) and its successive copies (the successive versions of "Darwinism"). I maintain that there exists some kind of isomorphism between Darwin's Darwinism and historical Darwinism. By this I do not mean that everything present in Darwin's evolutionary views has been passed onto the "Darwinian" tradition. I want to say merely that something crucial in Darwin's own contribution has durably constrained the development of evolutionary theory. This something is not so much a list of particular hypotheses or doctrines (such as a particular interpretation of natural selection, or heredity, or variation, or even descent), but rather a general view related to the structure of evolutionary theory.[1] Or, to be more precise: a general view of the general framework within which evolutionary research makes sense. What matters is the overall structure, not the truth of Darwin's personal hypotheses about this or that element of this structure.

Let us first recall that Darwin referred to his theory in *The Origin of Species* as "Descent with modification through natural selection." (Darwin 1859, 459) This formula[2] suggests that Darwin's theory was made of two components or even subtheories: the first component is "descent with modification", that is a theory about what we would call today the general "pattern" of evolution; the second part ("through natural selection") refers to the conjecture that natural selection is the main explanation for the modification of species.

Darwin organized the whole book in such a way that the two parts are always tightly connected to each other. Nevertheless there are good reasons for taking this distinction seriously. First, Darwin was perfectly aware of its importance as he wrote by letter to Asa Gray, dated May 11, 1863: "Personally, of course, I care much about natural selection, but that seems to be utterly unimportant compared to the *Creation* **or** *Modification*." (Burkhardt 1985–2007, vol. XI, 403; italicized and bold characters in text) Another reason is that the scientific community did not react to the two parts of Darwin's theory in the same way. "Descent with modification" was quickly renamed "evolution" by Darwin's readers and by Darwin himself, who accepted this term and introduced it in the last edition of *The Origin* (1972), where it appears seven times, as an abbreviation for the "descent with modification" of the previous editions. This part of the theory was so successful and so rapidly adopted in the last third of the nineteenth century[3] that it became common to call it "the general fact of evolution", whereas the explanation of this "general fact" through natural selection has been subject to unending controversies

[1] I have been trying for some time to develop this structural view of the history of Darwinism (Gayon 1990, 1995, 1997, 1998, 1999, 2003, 2008, 2009).

[2] This formula remained the same till the sixth and last edition, where Darwin added one word: "Descent with modification through *variation* and natural selection".

[3] Peter Bowler says that this may well have been one of the most spectacular and rapid example of "paradigm shift" in the history of science (Bowler 2003).

since 1859. For these two reasons, the issue of the continuity vs. non-continuity of Darwinism over time deserves being examined from these two different perspectives, and not only from the viewpoint of the theory of natural selection.

2. Descent with modification or "the general fact of evolution"

2.1. Darwin

Darwin did not provide in the *Origin of species* an explicit definition of what he meant by "descent with modification", which he probably found self-evident: species "descend" from one another and they are "modified". Nevertheless he had a rather restrictive interpretation of this idea — we might well say a model of it. This model is clearly expressed in the unique illustration given in *The Origin*, the famous branching diagram given in chapter 4. This diagram (see *fig. 1*) was so important for Darwin that he commented on it twice in his book, in chapter 4 (devoted to natural selection) and in chapter 13 (devoted to classification and embryology) (Darwin 1859, 116–126 and 420–422). This diagram expresses a series of conjectures, none of which is trivial:

[1] Species are subject to modification.
[2] This modification is slow and gradual.
[3] Many species go extinct.
[4] The species that do not go extinct usually split into other species.
[5] Once two or more species have split, they diverge indefinitely.
[6] The diagram applies to all taxonomical levels from the lowest local varieties to the most encompassing groups (e.g. "A" may be a variety that leads to a species with eight varieties, or a species that leads to a genre with eight species, or a genre that leads to a family containing eight genres, etc.).
[7] Postulate [6] implies that classification is entirely determined by genealogy. Therefore there are as many "ranks" as there are branching events of ramification, and all taxonomic categories used in systematics (species, genres, families, etc.) are arbitrary.
[8] Postulate [6] also implied for Darwin that the entire process of diversification reduces to diversification at the lowest level (varieties and species); supra-specific taxa have no particular role in evolution, they are just an outcome.

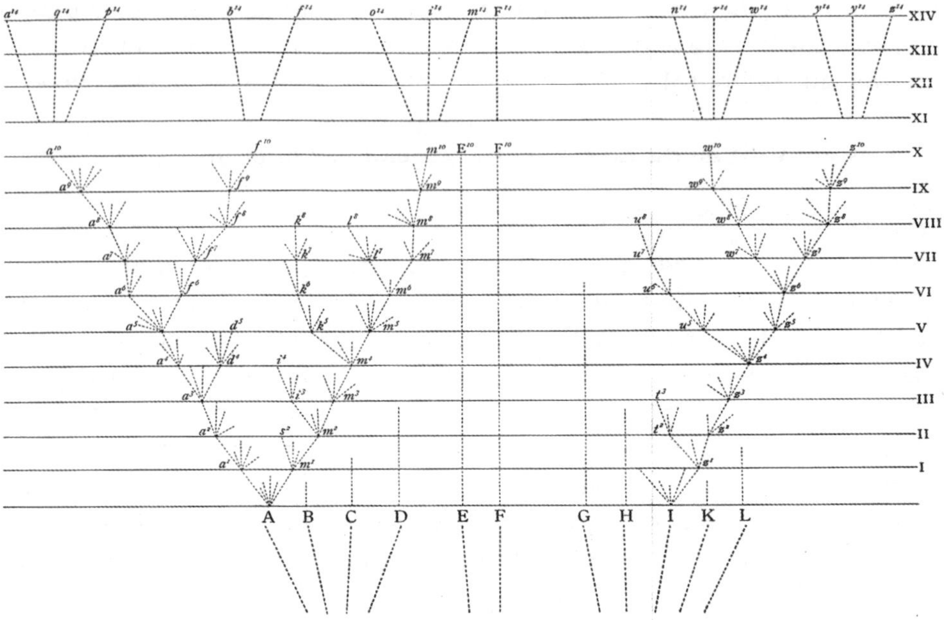

Figure 1. Darwin 1859: contra p. 117. Darwin comments on this figure twice in his book. In chap. IV, the letters "A, B, C … L" represent "the species of a genus large in its own country". The roman numbers represent intervals between an arbitrary high number of generations. They produce local varieties which modify and diverge during many generations. At the end of the process, species 1 (for example) leads to eight species. In chapter XIII, the letters "A, B, C … L" represent "allied genera" which will lead at level XIV to a greater number of genera (for instance A has differentiated into eight genera forming a family). The same reasoning can be applied to any desired level of taxonomic categories. Darwin explains that what is natural is the genealogical "arrangement"; "ranking" (variety, species, genus, sub-family, or any taxonomic category) is arbitrary.

Darwin's tree-like diagram had an almost immediate effect upon the entirety of the biological community. Within a short period of time, it became the paradigmatic representation of organic evolution as an established fact, the "general fact of evolution". The sudden and dramatic effect of this diagram constitutes one of the most spectacular examples of a paradigm-shift in the history of science, to use Thomas Kuhn's famous terminology. One could not say the same for the theory of natural selection, which opened a long series of controversies.

From a methodological point of view, the characterization of Darwin's tree-of-life as a "general fact" of nature is misleading. This diagram, with the non-trivial postulates that it illustrates, is a theoretical construction. The kind of theory involved is not a causal theory, although it may be used for explanatory purposes in certain contexts. It is a theory that postulates the general existence in nature of several classes of phenomena. It is a heuristic device, a plausible bet about the general form and pace of the phenomena that constitute evolution as a result, and that call for explanation. Darwin's theory of "descent with modification", as summarized in his diagram, is a low level theory, but nevertheless a theory. It is a conjecture about the general pattern of evolution.

2.2. After Darwin

Now the question is: has the scientific consensus about the so-called "fact of evolution", as represented by Darwin's diagram, been significantly threatened since Darwin's time? There have been at least three serious challenges to Darwin's tree-of-life picture. Each of these challenges, or classes of criticisms, relates to the shape of the tree. The first two have been recurrent since the time of Darwin. The third is recent.

The first kind of criticism can be found among those biologists or paleontologists who have advocated a non-gradual or saltationist representation of the origin of species. Since Thomas Henry Huxley, Francis Galton and the early Mendelians, saltationist theories of the emergence of species have been numerous and varied.[4] Scientists belonging to very different biological sub-disciplines have defended them: morphologists, paleontologists, biometricians, geneticists, and developmental biologists. Today, the most famous and most convincing example is represented by Eldredge's and Gould's (1972) theory of "punctuated equilibria". This theory, first formulated in 1972, postulates that evolutionary change occurs primarily during events of speciation or cladogenesis (i.e. the splitting of a given species into two or more species). Although it may take many generations, this change is rapid at a geological scale.

Whatever the particular form they take, saltationist theories of evolution entail a significant alteration of Darwin's tree. Instead of branches that progressively diverge, the tree will have the shape of successive candelabras. The most famous example is Eldredge's and Gould's "tree" in their 1972 paper (*fig. 2*). This kind of criticism, which focuses on the level of species (or at least on low taxonomic levels) challenges the second, the fifth and the sixth Darwinian postulates listed above: gradualness of change, divergence of species as such, and indefinite divergence after splitting.

[4] For a systematic examination of the early forms of saltationist theories, see Bowler (1983).

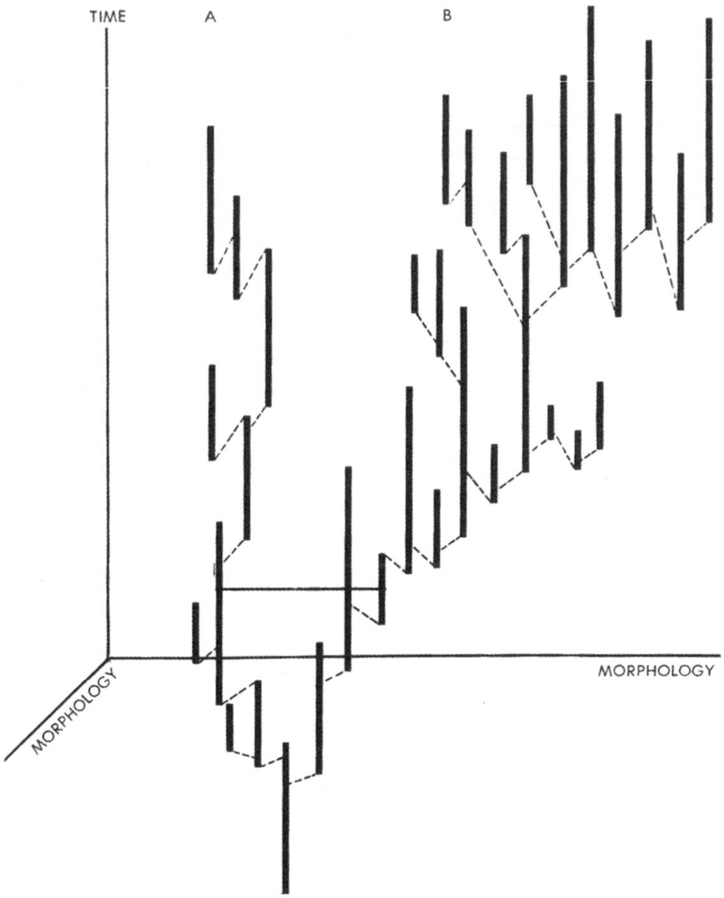

Figure 2. Eldredge and Gould 1972, 113. Original legend: "Three-dimensional sketch contrasting a pattern of relative stability (A) with a trend (B), where speciation (dashed lines) is occurring in both major lineages. Morphological change is depicted here along the horizontal axes, while the vertical axis is time. Though a retrospective pattern of directional selection might be fitted as a straight line in (B), the actual pattern is stasis within the species, and differential success of species exhibiting morphological change in a particular direction."

A second class of criticism emphasizes the view that evolution is something that does not happen only at the level of varieties and species, but also takes place at higher taxonomic levels. This kind of criticism has also been extremely popular since Darwin, especially among morphologists and paleontologists. In its commonest expression, it states that higher taxa tend to appear and diversify rather suddenly, giving rise to morphological types that remain stable ever after. In such a view, modification and multiplication of species are superficial phenomena, not the core of the evolutionary process. Biologists and paleontologists who adopt this view do not accept Darwin's "tree"; they refer to the image of a "bush" with parallel twigs arising from a given level. In 1929, the French morphologist Louis Vialleton gave a vivid summary of the shift from traditional Darwinian trees to "bushes" in systematics at the end of the nineteenth century and at the beginning of the twentieth century:

> Genealogical trees, by illustrating in a simple way a number of infinitely complex relations in the real world, were very useful to the transformist theory; they provided the concrete examples of transformation that Lyell demanded from Darwin. ... The first diagrams were quite satisfying. They strikingly represented the evolutionary unfolding of beings and seemed to be able to provide their natural relationships that had been sought for such a long time. But disillusion came fast... Taking into account the time of appearance of various forms [and not only this or that morphological criterion], it soon appeared that genealogical trees looked quite different from how they looked before. Instead of showing a more or less elongated trunk during the period when only the supposed initial form existed, they had the shape of bushes whose twigs stemmed from approximately the same height above the ground, and moved away in a parallel way from each other, each of them being more or less ramified (Vialleton 1929).

To illustrate these doubts about Darwinian "trees", Vialleton gave two different representations of the genealogical tree of Mammals: one was borrowed from the American paleontologist Henry Fairfield Osborn (*fig. 3*), the other was Vialleton's representation of exactly the same data and interpretation (*fig. 4*). Vialleton's representation emphasizes the idea that something decisive occurs at high taxonomic ranks. To be fair, Vialleton was as anti-Darwinian as one can imagine: he thought that "transformism" (his name for Darwinian evolution — i.e., gradual modification and splitting of species) was only a superficial phenomenon, whereas genuine "evolution" was the generation or "creation" of major morphological novelties in groups that suddenly appeared and remained approximately stable thereafter.

More recently, Eldredge's and Gould's punctuated equilibria have provided a totally new version of the old idea that evolution is not confined to the level of species. The famous diagram reproduced in *fig. 2* illustrates the hypothesis that a given evolutionary trend in a taxonomic group (either stasis or directional evolution) might well result from changes above the level of species. For instance, in lineage "B", the evolutionary trend results from differential success of (static) species exhibiting morphological change in a particular direction. Although this way of thinking is far from Vialleton's typological approach (and, in that sense, much closer to Darwin), it exemplifies another manner of advocating evolution above

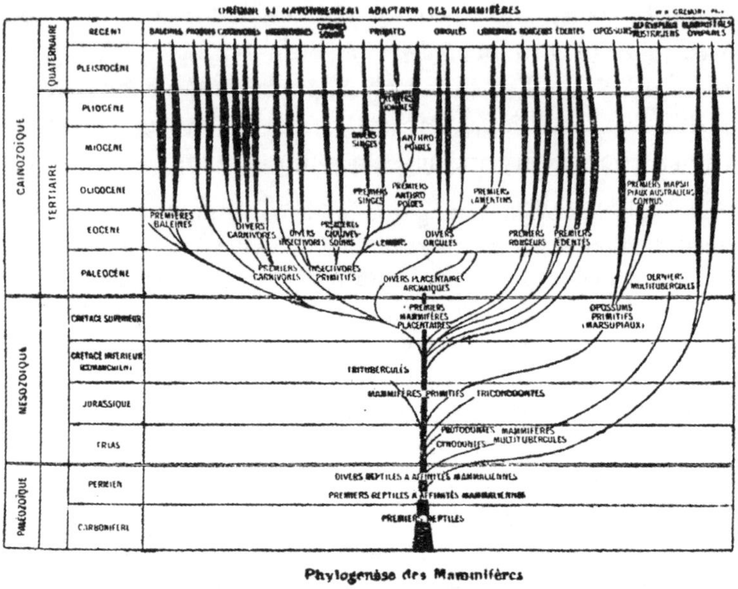

Figure 3. Vialleton 1929, 183. Original legend "Arbre généalogique des mammifères (d'après Osborn)" [Genealogical tree of Mammals (from Osborn)]. Vialleton refers to Osborn 1917.

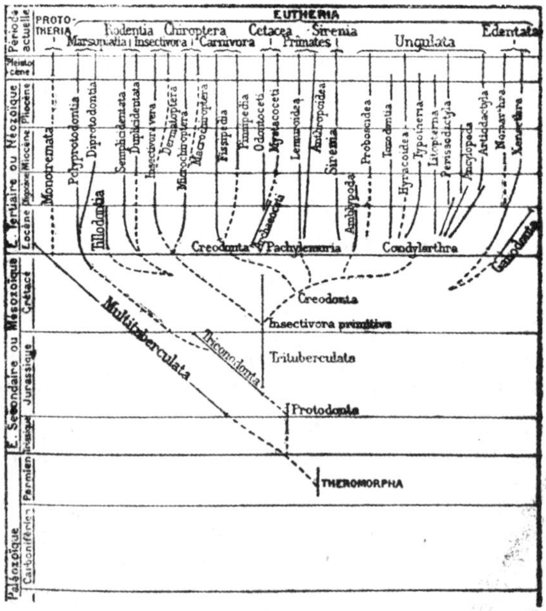

Figure 4. Vialleton 1929, 181. Original legend "Arbre généalogique des mammifères" [Genealogical tree of Mammals].

the species level, and consequently of challenging the Darwinian postulates [6] and [8] listed above (and, also, [5]: the entities that "diverge" are not necessarily species). Most of the literature in paleobiology since the mid 1970's illustrates a similar search for specifically macroevolutionary patterns (For a comprehensive review, see Stanley 1998). The third class of criticism of Darwin's diagram is more recent and, in a sense, more radical. The suspicion is that this diagram represents only a fraction of the real pattern of the history of life. Darwin's tree involves the conviction that a tree of irreversible divergence *of species* adequately and sufficiently represents the history of biological diversity. Although Darwin did not explicitly say this, his diagram has been understood as meaning that there actually exists one single "tree of life", the branches of which (namely, species) become, at some time or another, totally isolated from any other in terms of reproduction and hereditary transmission. An increasing number of phenomena challenge this view of species as (material) monads "without any doors or windows", to freely pastiche Leibniz's famous formula. These two phenomena are lateral gene transfer and symbiosis.

Since the end of the 1990s, lateral gene transfer has been known to be a major phenomenon among prokaryotes (Eubacteria and Archaea). Prokaryotes are single-celled organisms without a membrane-bound nucleus. Since Carl Woese's pioneering work in the 1970s and 1980s, they are divided into two "domains": Bacteria and Archaea. Eukaryotes constitute the third "domain": they include all single-cell and multicellular organisms with a membrane-bound nucleus. At the beginning, the application of molecular techniques to prokaryotes generated the hope of reconstructing the unique and universal tree of life (Woese 1987). But an increasing number of anomalies led to the discovery that gene transmission among prokaryotes is not only vertical, it is also horizontal. The magnitude of this phenomenon has been fiercely debated for almost twenty years now. It is now clear that it is important enough to cast serious doubts on the very existence of a unique tree representing the natural relationships among all cellular organisms, or, at least, on the possibility of reconstituting this tree if it exists (for a review, see McInerney, Cotton, and Pisani 2008). This is due to the fact that almost all (if not all) prokaryotic genes have been subject to lateral gene transfer. Whether this turnover of genes in Eukaryotes is significant remains a debated issue. Nevertheless, we now observe an intense controversy between two ways of representing the genealogical pattern of the history of life: tree and network.

It is presently uncertain whether the "tree of life hypothesis" will be able to successfully combat the "network of life hypothesis". In reality, it is likely that both the "tree" and the "network" will have to coexist in a pluralistic view of evolutionary patterns, where *several* histories are told by several methodologies. In fact, evolutionary biologists seem now to admit that Darwin's diagram, which focuses exclusively at the level of species, encapsulates only one fraction of the history of life. In the case of prokaryotes at least, lateral gene transfer has been important enough to obliterate, and, perhaps, overcome "vertical transmission" at a large historical scale.

Another process that jeopardizes Darwin's representation of genealogy is symbiosis. Symbiosis has probably been a major evolutionary process, especially at the level of cell evolution (Margulis and Sagan 2002). Symbiosis is much more rare than lateral gene transfer; but when it occurs, it may have dramatic effects because it implies the coexistence of two full genomes. Symbiosis seems to have been a key process in the emergence of a number of major groups of unicellular eukaryotic organisms, and of eukaryotes themselves. Symbiosis does not challenge the notion of irreversible divergence of *species*, because separate organisms that fuse initiate new species that diverge from others. But fusion events between organisms introduce complications that cannot be assimilated by an exclusively tree-like pattern of phylogeny. *Fig. 5* illustrates the kind of anastomosis that result from such conjectures.

I now conclude on the first aspect of Darwin's theory (Descent with). Darwin's diagram has proved immensely fertile. It remains the basic intuition behind most of the work on phylogenies. But it is not true that the diagram has remained unchanged. Several of the postulates embodied in Darwin's representation appear not to be false, but rather a simplified view. At least four of the seven postulates that I have listed have been seriously challenged: [2] (gradualness of modification), [5] (irreversible divergence of species), [6] (uniform validity of the tree-like diagram at all levels), and [8] (evolution above the species level). Modern evolutionary biologists are not in agreement regarding the assertion that modification of species is *always* slow and gradual (against [2]). Indefinite divergence after splitting is not strictly false, but it is only one part of the story of life: important horizontal relations must be taken into account (against [5] and [6]). Finally, many contemporary biologists (especially paleobiologists) deny that evolution is nothing more than modification of species (against [8]).

To summarize in a single sentence, then: Darwin was successful in imposing a view of the history of life in terms of genealogy and modification, but the particular model that he offered has been repeatedly discussed.

3. Natural selection as an explanatory principle

The second half of "Darwin's theory" was the theory of modification through natural selection. Whereas "descent with modification" was a first-level theory about the general pace of evolution, natural selection was for Darwin an explanatory hypothesis that aimed at unifying the entirety of evolutionary phenomena under a common set of law-like statements. Just as in the case of "descent with modification", I will first remind the reader of Darwin's views about the theoretical status of natural selection. Then I will compare the post-Darwinian history of evolutionary biology with Darwin's conception of his explanatory theory.

3.1. Darwin

In *The Origin of Species*, Darwin used a double strategy for justifying the principle of natural selection. The first strategy consisted in providing inductive arguments in favor of the existence of natural selection as a process. The second strategy was hypothetico-deductive: it consisted in showing that natural selection explains and unifies various classes of independent facts. Strangely, Darwin did not make this double strategy explicit in the text of the *Origin,* although the overall structure of the book corresponds more or less to it: on the whole, chapters 1 through 5 of the first edition establish the plausibility of the existence of variation and natural selection on the basis of a mixture of empirical generalizations and deductions; chapters 7 through 13 are devoted to establishing its explanatory power (and also its limits: chap. 8 on hybridism).

In fact, it is only in the introduction to *The Variation of Animals and Plants Under Domestication*, published nine years after the first edition of *The Origin of Species*, that Darwin made perfectly explicit his argumentative strategy:

> In scientific investigations it is permitted to invent any hypothesis, and if it explains various large and independent classes of facts it rises to the rank of a well-grounded theory. The undulations of the ether and even its existence are hypothetical, yet every one now admits the undulatory theory of light. The principle of natural selection may be looked at as a mere hypothesis, but rendered in some degree probable by what we positively know of the variability of organic beings in a state of nature, — by what we positively know of the struggle for existence, and the consequent almost inevitable preservation of favourable variations, — and from the analogical formation of domestic races. Now this hypothesis may be tested, — and this seems to me the only fair and legitimate manner of considering the whole question, — by trying whether it explains several large and independent classes of facts; such as the geological

> succession of organic beings, their distribution in past and present times, and their mutual affinities and homologies. If the principle of natural selection does explain these and other large bodies of facts, it ought to be received (Darwin 1868, 8–9).

In an extremely dense formulation, Darwin distinguishes two levels of justification of natural selection.

At the first level, natural selection is a "mere hypothesis" that is "rendered probable" by empirical arguments. Two different sorts of arguments are invoked. One is analogical: artificial selection, which shares some preconditions with the hypothetical process of natural selection (variation, heredity, biases in survival or reproductive rates) does modify species. This argument was crucial to Darwin, since he had no direct evidence for the existence of natural selection in nature (chap. 1 and 2 of *The Origin*). The second kind of argument invoked at this level is based upon empirical generalizations about the two specific preconditions of natural selection, relative to the rate of reproduction of organisms and the limitation of resources (chapter 3 of *The Origin*), and the notion of more or less heritable and "favourable" (or unfavourable) variations (chapter 4 of *The Origin*). Taken together, these arguments establish the plausibility of natural selection as a very general process existing in nature, in all species, for many characters and at all times.

At the second level, Darwin says, "this hypothesis may be tested" through its capacity to "explain" a large array of "independent classes of facts". This refers to chapters 7 through 13 of *The Origin*, and Darwin mentions some of the facts explained in the passage quoted. One might disagree about exactly what the list should contain; but, on the whole, Darwin claims that natural selection explains at least: adaptations (morphological and instinctual), the geographic distribution of species, the stratigraphic distribution of fossils, extinction, divergence, homologies, the relation between development and evolution, and taxonomic patterns. This second level of justification of natural selection was crucial to Darwin, since it is at this level that the "mere hypothesis" becomes "a well-grounded theory", a theory that deserves being "received".

The Newtonian style of the entire argument has been underlined by a number of commentators (Kavaloski 1974; Ruse 1975; Hodge 1987). What I want to stress here is that this argumentative structure provides a powerful tool for the understanding of the structure of evolutionary research since Darwin. The first level led to the development of population biology, with a careful exploration of the bases and modes of natural selection as a widespread principle in nature. The second level provided a theoretical principal for the reframing of the entire field of natural history. At that second level, natural selection is not only a common and fascinating process in nature, but also a principle that unifies many classes of facts, and, ultimately, biology as a whole. *Fig. 5* gives a schematic representation of this argumentative structure.

3.2. After Darwin

3.2.1 Brief historical survey (1859–1950)

Let us now assess the fate of this grand scheme. The first hundred years of evolutionary theory after Darwin may be summarized in a rather simple way. For a period of approximately seventy years after the publication of *The Origin* (1859), most of discussions and criticisms were targeted at the very existence of natural selection as a process, and of its capacity to account for adaptation. The reason for this was the absence of a convincing theory of variation and heredity. Of course, if this level was problematic, the second was even more so. In the 1920s and 1930s, it became clear that Mendelian genetics was able to reformulate Darwin's hypothesis (or a particular model of it, I would say) in a way that made it testable and predictive (on this complicated story, see Gayon 1998).

Then came the "modern synthesis", built in the period 1930–1950. As Julian Huxley said when he coined this expression, this period consisted in a "rebirth of Darwinism" (Huxley [1942] 1978, 26). Since population genetics was able to prove the existence and efficacy of natural selection, the time became ripe for developing the second aspect of Darwin's theory. The modern synthesis consisted in an attempt to show, in its modernized genetic version, that natural selection was indeed a major principle for the explanation of such phenomena as geographical variation, long-term modification of species (anagenesis), speciation (cladogenesis), extinction, the tempo of evolution and phylogenetic trends. This enterprise was genuinely collective. A number of different disciplines and scientists all over the world contributed to this "synthesis": theoretical population geneticists (Fisher, Haldane, Wright), field and experimental population geneticists (Dobzhansky, Teissier, L'Héritier, Timofeeff-Ressovsky), cytogeneticists (Darlington, White), animal and plant systematists (Mayr, Stebbins), paleontologists (Simpson, Newell), embryologists (Schmalhausen, Waddington), ecologists (Lack) and specialists in human evolution (Rensch), to mention just a few (Mayr and Provine 1980; Smocovitis 1996). In the 1950s the triumph of natural selection as an explanatory principle seemed to have no limit (see for instance Mayr 1959). These years corresponded to the time when evolutionary biology was institutionalized as a discipline. In subsequent years, however, evolutionary biologists became more circumspect. Two kinds of criticisms developed, at the levels of microevolution and macroevolution.[5]

3.2.2. Criticisms of natural selection at the level of microevolution

At the level of microevolution, what natural selection explains is first and above all adaptations. One should observe that since the modern synthesis, this ambition has never been seriously challenged. For almost all contemporary biologists, natural selection is the sole explanation for adaptations. This was true sixty years ago, and it is even truer today. Many modern textbooks on evolutionary biology can testify to this.[6] However, three main criticisms have been addressed to natural selection at the microevolutionary level over the past fifty years.

First, at the molecular level, the discovery of an unexpected amount of polymorphism in proteins and DNA has led to the recognition that most of this polymorphism is approximately neutral with respect to natural selection. This has been the central tenet of the "neutral mutation — random drift hypothesis of molecular evolution" (Kimura 1968; 1973). This theory has generated one of the most intense controversies in population genetics in the second half of the twentieth century, precisely because it challenged the idea that natural selection controls the evolution of biological phenomena at all levels.[7] The theory is now widely admitted. It states that natural selection does not control everything at the molecular level, especially DNA; it controls only the limited fraction of mutations that affect the phenotype at a given time in evolution.

Secondly, there has been a huge debate about group selection. Darwin and most population geneticists in the middle of the twentieth century denied the existence of group selection, or at least its ability to generate adaptations. Here also, things have changed. Group selection, which was first perceived as a threat to Darwinism (because it went against the particular model of natural selection that was favored by Darwin and by population biologists in the period 1930–1980) is now treated as a legitimate model of natural selection, which appears to be important for the

---

[5] For a more detailed exposition, see Gayon 1990, 1997, 2003, 2008, 2009.

[6] See for instance Ridley 1996, chap. 13, pp. 338–368. The first three sections of this chapter are entitled: "13.1. Natural selection is the only known explanation for adaptations"; "13.2. Pluralism is appropriate for the study of evolution, not of adaptation"; "13.3. Natural selection can, in principle, explain all known adaptation".

[7] The best image of this controversy can be found in Lewontin 1974.

explanation of the origin of major biological phenomena such as sex and a number of phenomena of cooperation.[8]

Thirdly, theoretical biologists in the name of complexity have formulated strong attacks against Darwinism at the microevolutionary level. The claim, here, is that complexity imposes strong limits to the modifying power of natural selection. Because of the large amount of connectivity existing in genetic systems, natural selection may fail to increase the mean fitness of the population, whatever the intensity of selection (Kauffman 1993).

These challenges to Darwinism at the level of microevolution are important. All of them have led to the development of new and fertile domains of research in evolutionary biology. The first and the third (neutral theory of molecular evolution and complexity theory) share a similar character: both of them relativize the idea of natural selection as an all-sufficient principle in evolution. Natural selection does not operate on a passive field of variation. Its action is limited by formal constraints such as stochasticity and generic properties of self-organized systems. The second challenge (group selection) is different: by admitting multilevel models of natural selection, it increases the theoretical content and explanatory power of natural selection rather than restricting it. It also makes natural selection much more complicated, much less intuitive, and probably difficult to test. Ultimately, none of these criticisms has challenged natural selection as the only acceptable explanation for adaptations. Therefore, on the whole, all attempts to challenge the explanatory power of natural selection at the microevolutionary level have resulted in enrichment rather than a decline of Darwinism.

3.2.3. Macroevolutionary criticisms of natural selection

Over the past forty years it is mainly at the level of macroevolution that major cracks in the Darwinian edifice have occurred. Remember here that for Darwin, the crucial test for the "acceptability" of natural selection was the ability of this hypothesis to explain "independent classes of facts": not only adaptations, but also extinction, divergence and other macroevolutionary phenomena (the boxes at the bottom of *fig. 5*). What has become problematic over the past forty years is precisely the claim that natural selection suffices for the explanation of *all* macroevolutionary phenomena. Here, I will just give two examples.

[8] For a synthesis, see Sober and Wilson 1998.

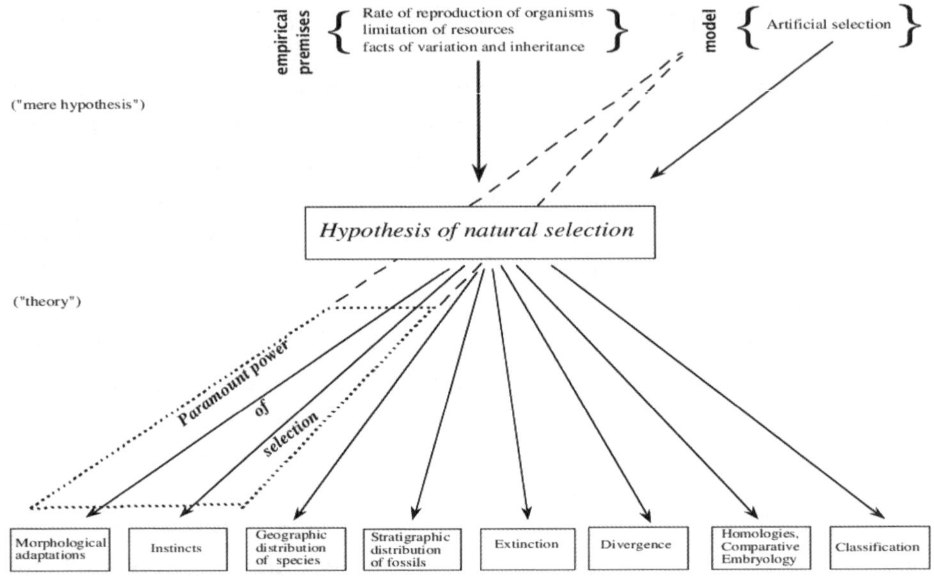

Figure 5. The two levels of the diagram (above and below "hypothesis of natural selection") represent the two levels of justification of Darwin's hypothesis of natural selection (from Gayon 1997, 270).

The first example is extinction. In the traditional Darwinian view (i.e. that of Darwin himself and of classical theories of extinction in the mid- twentieth century), extinction is just a large-scale effect of natural selection within species: as natural selection transforms biological species, some happen to be fitter in the competition with other species. The less favored species (those which evolve more slowly) decrease in numbers and go extinct. But paleobiologists have shown that such an explanation does not work in the case of mass extinction. In episodes of mass extinction (a phenomenon that has occurred many times in the history of life, at different scales), species go extinct not because of their relative success when competing with other species for places in the ecological theatre, but because they are ill equipped to face a brutal perturbation of their physical environment. For instance, if diatoms survived better than other planktonic forms during the great Cretaceous extinction, it is because they were endowed with a capacity of becoming encysted. This feature had evolved for reasons that had nothing to do with the environmental conditions that generated the mass extinction of the Cretaceous. The diatom's dormancy is an adaptation to life in cold water that had been acquired long before. As David Raup said in his magisterial book on extinction, when species go extinct in mass extinction events, this is not due do "bad genes" but to "bad luck" (Raup 1991).

Another example is divergence. Following Eldredge and Gould's seminal paper on punctuated equilibria, a number of paleobiologists have denied that morphological divergence is necessarily the result of continued selection within species. They have supported alternative explanations, according to which divergence can also be a consequence of repeated events of speciation and extinction of species, either because of species selection, or because of "species drift" (resulting itself from stochastic considerations or from some kind of developmental constraint). This kind of theory is now currently being debated among paleobiologists (Stanley 1975; 1998).

Another kind of criticism that deserves being mentioned has come from the "evo-devo" evolution-development) school. It emphasizes the importance of major morphological and embryological constraints that affect the evolutionary process. Evo-devo insists that the level of the elementary hereditary material (the genetical level) is insufficient for explaining the mechanisms responsible for evolutionary change: epigenetic heritable traits (such as developmental pathways) should also be taken into account. It is unclear whether advocates of evo-devo claim that such higher level traits canalize evolution or constitute another level of selection (Amundson 2005 gives a remarkable analysis of this school of thinking).

Other examples could be used, showing that the predominance of natural selection in the explanation of macroevolutionary phenomena is seriously contested today. This is not to say that contemporary evolutionary biologists deny that natural selection is a major force that should always be taken into account when studying the history of life. But the majority of them think that things are more complicated: other principles are needed, and perhaps there is no hope of deducing everything in the history of life from a short list of theoretical principles.

3. Conclusion: death or persistence of Darwinism?

Let us now come back to the question posed at the beginning of this paper. What do the modern evolutionary biologists' repeated and contradictory references to Darwin mean? I would like here to take the evolution of Stephen Jay Gould's attitude on this issue as a key example.

In 1980, Gould published a paper provocatively titled "Is a new and general theory of evolution emerging?" This paper was probably the most radical attack against Darwinism that Gould ever published. It provides an exceptional landmark for anyone who wants to understand what was at stake when a number of evolutionary biologists began to say, around 1980, that the classical "Darwinian" view of evolution was in deep crisis.

Let us first point out the precise target of Gould's attack. Although he often writes of "Darwinism", it is only a particular version of Darwinism that he has in mind — the modern synthesis. In the first section of the paper, entitled "The modern synthesis" (Gould 1980, 119–121), Gould quotes the following passage from Mayr:

> The proponents of the synthetic theory maintain that all evolution is due to the accumulation of small genetic changes, guided by natural selection, and that transpecific evolution is nothing but an extrapolation and magnification of the events that take place within populations and species (Gould 1980, 120).

And Gould adds:

> … if Mayr's characterization of the synthetic theory is accurate, then that theory, as a general proposition, is effectively dead, despite its persistence as text-book orthodoxy (Gould 1980, 121).

What were Gould's arguments supporting the "death" of the synthetic theory? The 1980 paper advocates a "hierarchical approach" to evolution, with three main levels: variation within populations, speciation, and patterns of macroevolution. Gould defends the claim that at each of these three levels, the conception of evolution as an accumulation of small genetic changes, guided by natural selection, has been refuted.

At the level of population genetics, Gould evokes the neutrality of most genetic change.

At the level of speciation, he refers to various models of speciation (especially White's parapatric model) showing that reproductive isolation is not a by-product of adaptation, but rather a rapid and stochastic event. He also emphasizes heavily the revival of Goldschmidt's "hopeful monster" (Gould 1980, 122–125).

Finally, at the level of macroevolution (Gould 1980, 125–128), Gould primarily discusses punctuated equilibria, and proposes that evolutionary trends are not a mere extrapolation of intra-specific evolution, but result from differential rates of speciation and extinction among species.[9]

---

[9] "… if species originate in geological instants and then do not alter in major ways, then evolutionary trends cannot represent a simple extrapolation of allelic substitution within a population. Trends must be the product of differential success among species (Eldredge and Gould 1972; Stanley 1975). In other words, species themselves may be inputs, and trends the result of their differential origin and survival. Speciation interposes as an irreducible level between change in local populations and trends in geological time. Macroevolution is, as Stanley argues (1975, 648), decoupled from microevolution." (Gould 1980, 125–126) The full reference of Stanley 1975 is given in the bibliography of the present paper.

Thus Gould's 1980 paper leaves the reader with the conviction that the traditional "Darwinian" or synthetic theory of evolution has broken down, and has been, or will soon be, replaced with another theory, where natural selection of small individual-level inherited variation is just one process among others, and definitely not a unifying principle for the evolutionary process as a whole.

If we now look at Gould's testament, the luxuriant book published in 2002, the very year of his own death, under the title *The Structure of Evolutionary Theory*, we observe a rather different attitude regarding Darwinism. The first chapter is wholly devoted to summarizing the entire argument of the book. Gould, again, explains the structure of the new theoretical framework through a comparison with the former classical, or purely Darwinian, framework. But this time he refers to Darwin himself rather than to the synthetic theory. He maintains that the new theoretical framework of evolutionary biology is neither an "extension" of the older Darwinian theory, nor a "destruction" or "replacement" of it. This vocabulary, which the author uses very carefully, is both puzzling and stimulating. In Gould's terms, "extension" would mean that the same principles have been applied to a wider spectrum of phenomena. "Replacement" would mean that other principles are now at the heart of evolutionary theory. Instead of these two terms, Gould prefers "expansion". "Expansion" means that the same principles remain central to the theory, but they have been "reformulated" in a way that gives a truly different aspect to the entire edifice (just as an architect makes a new edifice from an older one, by adding new parts to the existing unchanged ones, which themselves constrain the whole building).[10]

This notion of expansion is rather unorthodox in terms of current philosophy of science. "Extension" would make philosophers think of a wider explanatory scope. "Replacement" irresistibly evokes Thomas Kuhn's "paradigm shift". But, obviously, Gould wants to avoid these two classical ways of interpreting scientific change. What, then, does he mean precisely by "expansion"? He defines the "Darwinian framework" (which he also calls "the Darwinian logic") as a conjunction of three principles:

---

[10] "I do believe that the Darwinian framework, and not just the foundation, persists in the emerging structure of a more adequate theory. But I also hold ... that substantial changes, introduced during the last half of the twentieth century, have built a structure so expanded beyond the Darwinian core, and so enlarged by new principles of macroevolutionary explanation, that the full exposition, while remaining within the domain of Darwinian logic, must be construed as basically different from the canonical theory of natural selection, rather than simply extended." (Gould 2002, 3)

1. "Agency": natural selection works almost entirely at the level of organisms.
2. "Efficacy": natural selection is the mechanism responsible for the emergence of evolutionary novelty, the exclusive factor that shapes evolutionary change.
3. "Scope": the extrapolation of minute, incremental changes, can explain the entirety of the history of life (Gould 2002, 14–15).

Gould, of course, refuses these three principles in their crude formulation. Natural selection *does not* work only at the level of organisms: species selection or even clade selection is essential to him (this is indeed the major commitment of his 2002 book). Selection *is not* the only mechanism of evolutionary novelty: morphological constraints, developmental constraints and historical constraints canalize evolutionary pathways. Finally, Gould does not admit that the entirety of life is explained by extrapolating from small intra-specific changes to macroevolution.

Why, then, does Gould say that Darwin's principles have been "expanded" rather than "replaced"? "Expansion" seems to be a convenient word for the expression of two ideas:

1. Generalization of Darwinian (or more precisely synthetic) principles: species selection, species drift, clade selection and clade drift are obvious examples of this strategy. Gould's final judgment is that he and other paleobiologists have applied Darwinian schemes to entities that were not considered by Darwin (Gould 2002, 6).
2. Addition of new principles, such as morphological, developmental and historical constraints.

To make these ideas more intuitive, Gould makes use of an image. He compares the relationship between Darwinism and the new theoretical framework with a coral. At its basis, this coral has a trunk with three successive branches. The central trunk represents the theory of natural selection and cannot be severed. The branches represent the three "Darwinian principles" (agency, efficacy, scope). These principles remain at the basis of the new theoretical framework, but each of them has been "truncated" at a rather low level, so that they all "expand" and "develop" into an "organism" substantially different from the original one.

Thus, in Gould's terms, a new theory has not replaced the classical theory of evolution; rather, it has *developed* into a new one. What Gould means by this metaphor is the relationship between Darwin and the historical Darwinism is not only a question of genealogical influences. The claim is that there is a genuine conceptual similarity or "isomorphism" (not identity) between the two theories. In other words, the relation between Darwin, Darwinism and the modern evolutionary theory is not only historical; it is also a formal relation. This brings me back to the previous part of this paper, where I exposed my own view of the relationship between the original Darwinism (Darwin's own view of evolution) and Darwinism as historical tradition. Just like Gould, I am convinced that the repeated references made to Darwin by so many evolutionary biologists since 1859 are not only a matter of "genealogy". In a very strong sense, Darwin settled a framework that constrained the very structure of evolutionary theory for a long time, though certainly not the entirety of its empirical and theoretical content.

REFERENCES

Amundson R., 2005, *The Changing Role of the Embryo in the Evolutionary Thought. Roots of Evo-Devo*, Cambridge: Cambridge University Press.

Bowler P., 1983, *The Eclipse of Darwinism. Anti-Darwinian Evolution Theories in the Decades around 1900*, Baltimore & London: Johns Hopkins University Press.

Bowler P., 1984, *Evolution. The History of an Idea,* Berkeley: The University of California Press.

Burkhardt F. *et al.*, 1985–2007, *The Correspondence of Charles Darwin*, 15 vols., Cambridge: Cambridge University Press.

Darwin C. R., 1859, *On the Origin of Species by Means of Natural Selection, or the Preservation of Favoured Races in the Struggle for Life*, London: John Murray.

Darwin C. R., 1868, *The Variation of Animals and Plants Under Domestication,* London: Murray.

Doolittle W. F., Bapteste E., 2007, "Pattern Pluralism and the tree of life hypothesis", *Proceedings of the National Academy of Sciences USA* 104: 2043–49.

Eldredge N., Gould S. J., 1972, "Punctuated equilibria. An alternative to phyletic gradualism", in Schopf T. J. M. (ed.), *Models in Paleobiology*, San Francisco: Freeman, 82–115.

Gayon J., 1990, "Critics and criticisms of the Modern Synthesis: the viewpoint of a philosopher", *Evolutionary Biology* 24: 1–49.

Gayon J., 1995, "Neo-Darwinism", in Wolters G., Lennox J. G., McLaughlin P. (eds.), *Concepts, Theories, and Rationality in the Biological Sciences. The Second Pittsburgh-Konstanz Colloquium in the Philosophy of Science*, Konstanz and Pittsburgh: Universitätsverlag Konstanz & University of Pittsburgh Press, 1–25.

Gayon J., 1998, *Darwinism's Struggle for Survival: Heredity and the Hypothesis of Natural Selection*, Cambridge: Cambridge University Press [1st French edn.: 1992, *Darwin et l'après-Darwin*, Paris: Kimé].

Gayon J., 1999, "Sélection", in Lecourt D. (ed.), *Dictionnaire d'histoire et de philosophie des sciences*, Paris: Presses Universitaires de France, 854–61.

Gayon J., 1997, "The paramount power of selection: from Darwin to Kauffman", in Dalla Chiara M. L., Doets K. D., Mundici D., van Benthem J. (eds.), *Structures and Norms in Science, Volume Two of the Tenth International Congress of Logic, Methodology and Philosophy of Science*, Dordrecht: Kluwer, 265–82.

Gayon J., 2003, "From Darwin to today in evolutionary biology", in Hodge J., Radick G. (eds.), *The Cambridge Companion to Darwin*, Cambridge: Cambridge University Press, 240–64.

Gayon J., 2000, "History of the concept of allometry", *American Zoologist* 40: 748–58.

Gayon J., 2003, "Gould biométricien", *Comptes Rendus Palevol* 2: 353–60.

Gayon J., 2008, "'Is a new and general theory of evolution emerging?' A Philosophical Appraisal of Stephen Jay Gould's evaluation of contemporary evolutionary theory", in Gonzalez W. J. (ed.), *Evolutionism: Present Approaches*, La Coruña: Netbiblo, 77–105.

Gayon J., 2009, "Mort ou persistance du darwinisme ? Regard d'un épistémologue", *Comptes Rendus Palevol* 8: 321–340

Gould S. J., 1966, "Allometry and size in ontogeny and phylogeny", *Biological Reviews of the Cambridge Philosophical Society*, 41: 587–640.

Gould S. J., 1969, "An evolutionary microcosm: Pleistocene and recent history of the land snail *P. (Poecilozonites)* in Bermuda", *Bulletin of the Museum of Comparative Zoology* 138: 407–532.

Gould S. J., 1977, *Ontogeny and Phylogeny*, Cambridge, Mass.: Harvard University Press.

Gould S. J., 1980, "Is a new and general theory of evolution emerging?", *Paleobiology* 6: 119–30.

Gould S. J., 2002, *The Structure of Evolutionary Theory*, Cambridge, Mass.: The Belknap Press of Harvard University Press.

Hodge M. J. S., 1987, "Natural selection as a causal, empirical, and probabilistic theory", in Krüger L., Gigerenzer G., Morgan M. S. (eds.), *The Probabilistic Revolution*, Cambridge, Mass.: The MIT Press, vol. 2, 233–70.

Huxley J., 1978, *Evolution: The Modern Synthesis*, London, Allen & Unwin.

Kavaloski V. C., 1974, *The* vera causa *principle: a historico-philosophical study of a metatheoretical concept from Newton through Darwin,* PhD diss., University of Chicago.

Kimura M., 1968, "Evolutionary rate at the molecular level", *Nature* 217: 624–26.

Kimura M., 1973, "Mutation and evolution at the molecular level", *Genetics* (Supplement) 73: 19–35.

Lewens T., 2007, *Darwin,* London: Routledge.

Lewontin R. C., 1974, *The Genetic Basis of Evolutionary Change*, New York: Columbia University Press.

Margulis L., Sagan D., 2002, *Acquiring Genomes. A theory of the Origins of Species*, New York: Basic Books.

Mayr E., 1963, *Animal Species and Evolution*, Cambridge, Mass.: The Belknap Press of Harvard University Press.
Mayr E., 1959, "Where are we?", *Cold Spring Harbor Symposia on Quantitative Biology* 24: 409–40.
Mayr E., Provine W. B., 1980, *The Evolutionary Synthesis*, Cambridge, Mass.: Harvard University Press.
McInerney J. O., Cotton J. A., Pisani D., 2008, "The prokaryotic tree of life: past, present… and future", *Trends in Ecology and Evolution* 3: 276–81.
Osborn, H. F., 1917, *The Origin and Evolution of Life, on the Theory of Action, Reaction and Interaction of Energy*, New York: C. Scribner's Sons.
Ridley M., 2004, *Evolution*, 3rd edn., Malden, Mass.: Blackwell Scientific Publishing.
Ruse M., 1975, "Darwin's debt to philosophy: an examination of the influence of the philosophical ideas of John F.W. Herschel and William Whewell on the development of Charles Darwin's theory of evolution", *Studies in History and Philosophy of Science* 6: 159–81.
Smocovitis B. V., 1996, *Unifying Biology. The Evolutionary Synthesis and Evolutionary Biology*, Princeton: Princeton University Press.
Sober E., Wilson, D. S., 1998, *Unto Others: The Evolution and Psychology of Unselfish Behavior*, Cambridge, Mass.: Harvard University Press.
Stanley S. M., 1975, "A theory of evolution above the species level", *Proceedings of the National Academy of Sciences USA* 72: 646–50.
Stanley S. M., 1998, *Macroevolution. Pattern and Process*, Baltimore and London: The Johns Hopkins University Press.
Vialleton L., 1929, *L'Origine des êtres vivants. L'illusion transformiste*, Paris: Librairie Plon.
Woese C. R., 1987, "Bacterial evolution", *Microbiology Reviews* 51: 221–71.

Section II
Key concepts on evolution

INTRODUCTION

The six contributions in this section look at the main mechanisms with which biological evolution occurs, the causes of the outcomes, biodiversity and adaptation, phenomena sometimes associated with complexification of structures and processes.

Werner Arber shows how immediately after the discovery of the DNA structure molecular biology began to provide a series of proofs of biological evolution studied at the level of behavior of the informational molecules of prokaryotes. Molecular proofs of evolution concern specifically the individuation of mechanisms that produce inheritable genetic variation; mechanisms that can be recognized in three diverse modalities: local changes of the nucleotidic sequences of DNA, intragenomic re-arrangements of DNA, incorporation in the genetic sequence of a segment of extraneous DNA. As well as discussing these factors and their weight in evolution, Arber indicates how current molecular theory postulates the existence of true and proper "genes for evolution", with products that are able to cooperate with extragenetic factors to provide the necessary new genetic variation for Darwinian evolution. Balancing the rate of very high levels of mutation, favorable to the production of an indiscriminately high diversity of sequences and rates of mutation that are very low, conferring stability onto individual organisms and populations, is the correct "evolvability" of a population; that is the capacity of generating a certain degree of random genetic diversity and of evolving through natural selection in changing environments.

The entire awesome variety of products and typical processes of the world of life has its *raison d'être* in the form of evolution known as descent with modification. In light of the evolutionary thinking, **Douglas Futuyma** examines the intertwinement between evolutionary convergences and homologies, which marks the great processes of evolution, showing how the parental relations between species can be inferred, and a series of taxonomically calibrated molecular clocks can be set out in order to measure with a fair amount of accuracy the speed of evolution and the degree of affinity between different organisms. Phylogenetic analysis does not concern only the morphology of the species, as happened in the past, but it also reaches into the structure of the genome and allows us to elaborate phylogenetic trees that can be treated as hypotheses on the evolutionary parental relations of organisms. Futuyma emphasizes how the molecular genetics approach highlights the fact that proofs of parental relations between species can be polluted by phenomena such as reticular evolution, which provokes not divergence but fusion between branches of the tree of life. On the basis of a vast array of examples, the author also discusses phenomena such as the so-called *mosaic evolution*, the creative reusing of pre-existent material for new functions, adaptative convergence of different organisms for characters undergoing selection, the existence of evolutionary trends, and the use of certain genetic polymorphisms as reliable signals of relationship between living beings.

Lynn Margulis begins with an in-depth discussion and many examples of the unsuitability of the concept of species and its inapplicability to an enormous portion of the world of life. Even aside from prokaryotes (to which the notion of species is not applied as the individuals are characterized by very intense unidirectional exchange of genes, and so are marked by a sort of perennial evolutionary instability), Margulis observes that also the unicellular eukaryotes, which are important for biomass and variety of forms, remain excluded. Contrary to most shared opinions, Margulis holds that speciation did not start after the appearance of meiotic sexuality, but that, on the contrary, in many lineages the latter preceded the former, as seems to be shown by the enormous diffusion of new types of organism obtained via symbiogenesis. Margulis finds in Russian scientific literature from the twenties the first ideas of symbiogenesis, and then looks more generally at the history of defeats and victories of the symbiogenetic theories in biology. She observes that the foundations and lexicon of the evolutionary theory, as deriving from zoological and botanic research, come with some of the limits imposed by the object of study and by the related systems of belief, in spite of the fact that animals and plants are only a minority portion of the living organisms. Underlining repeatedly that the biological definition of species is inadequate to describe the life forms of the biosphere, Margulis complains that notwithstanding the diffusion and weight of symbiogenesis are well documented beyond any reasonable doubt, it is not yet acknowledged, as she thinks it should be, as a co-cause of speciation, and concludes affirming that there is a total lack today of detailed scenarios able to explain the origin of any species of eukaryote.

As is well known, *The Origin of Species,* despite its title, does not specifically address the formation of new species, but rather many other aspects of the evolution of organisms: from the existence of variation within the species and among them, to geographical distribution, biological adaptation, and natural selection. Speciation, the "mystery of mysteries", to which Darwin refers, remains one of the most studied and complex evolutionary phenomena. It is addressed by **Jeffrey Feder** who makes a rapid overview of the concept of species and uses the classification of different types of genetic flux barrier as a frame by which to individuate and separate the different mechanisms and circumstances which lead to speciation. Using comparative applied analysis on given examples, the author examines the different modalities with which the reproductive barriers to speciation can in turn evolve, referring as well to the role of hybridization and to the action of intrinsic (genetic) factors and extrinsic (ecological) factors in the speciation processes. After a synthetic treatment on how the conditions of the geographical distribution of populations influences the initial evolution of the barriers to genetic flux, Feder discusses in detail the role of the environment and of natural selection in sympatric speciation. This study concludes with a discussion of the possibility of individuating the genes responsible for the production of biological barriers to genetic flux that separate populations of a new species that is being produced from the species of origin. Having individuated the diverse methodological options and discussed the procedure to identify any genes of speciation, the author reports some cases in which modern genome analysis has led to the discovery of the first speciation genes.

The chapter written by **Scott Gilbert** concerns the role that changes occurring during developmental processes can play in evolution of organisms. That modifications of genes that are active during development could contribute to evolution of organisms was suspected since the time of the synthetic theory of evolution, but proofs were lacking. Indeed embryological knowledge was given little weight in the elaboration of the synthetic theory of evolution, whose theoretical nucleus was concentrated on the evidence of population genetics, a discipline focussed on the processes at the microevolutionary scale. Only with the introduction of the techniques of molecular analysis that allowed parallel studies of gene sequence and expression (even for genes involved in development), basic molecular mechanisms have been identified that are taxonomically transversal and shared by an enormous quantity and variety of animals. Gilbert shows how these deep homologies, whose discovery was allowed by developmental genetics and comparative genome studies, witness to the Darwinian idea of a relationship between living beings and evolution as descent with modification. Moreover, the entire new field of studies that derives from this, evolutionary developmental biology (evo-devo), confirms with new data and language the strategic role of genetic regulation in biological evolution. In particular, Gilbert shows through examples how the construction of animal organisms, obtained by reusing the same genetic materials, is nearer to the creative activity of a *bricoleur* than to the execution of a project by an engineer. In conclusion, the author shows how evo-devo research generally does not contradict the mainstays of the synthetic theory, but, rather individuating other mechanisms for the production of selectable variation, modifies the synthetic theory extending it to phenomena of morphological evolution.

Stuart Newman begins with the consideration that the properties of modern organisms are not necessarily all and the same as those of other organisms of the most remote past. This implies, Newman observes, that the rules for the production of the form of living beings would not be today the same as at the beginning of the history of the Metazoa. The basic idea is that the mechanisms with which evolution proceeds have their own history, that is that they are themselves evolved. The fact that there have been verifiable changes in the mechanisms and in the processes that have led to modern organisms allows to speak of an evolution of evolution. Naturally, the consequences of the existence of important differences between ancient and modern (in the sense of diversification of entire phyla in one case and of species in the other one) are, as is shown, of great importance for the general theory of evolution. Newman opens his contribution observing how the Darwinian explanation and the so-called neodarwinianism one, both populational in character and centred on the mutation-selection approach, are today challenged by a series of new acquisitions both in the paleontological field and in neontology, relative as much to unicellular as to multicellular organisms. He proceeds by giving details of a series of recent discoveries that show the existence of cellular mechanisms that may have allowed phenomena explainable by saltationism, orthogenesis and inheritability of acquired characters. All these explicative theories were rightly judged unacceptable when the synthetic theory was elaborated, but today they can no longer be excluded from the modern theory of evolution. Newman observes that Darwin and his successors' ignorance about mechanisms underlying the production of genetic variation, morphological innovation and what evo-devo is today discovering about the properties of developmental systems (phenotypical plasticity, one-to-many relation between genotypes and phenotypes, the weight of epigenetic factors, etc.) had to influence the theory of evolution by focusing it on the role of microevolutionary factors. As Newman and other academics maintain, the form of the ancient organisms that populated the environment of the Precambrian and the early Cambrian were probably directly shaped by factors of the habitat and can therefore be interpreted not in the classical adaptative sense, but as a consequence of the viscose elastic properties of cellular material. This fact is a second important consequence implicit in the author's position. This means that the mechanisms of microevolution, which develops largely in Darwinian mode via selection in a continuist and gradualist way, could not automatically be used to explain the diversification of the phyla. The author maintains, instead, that the morphogenetic explosion at the macroevolution level precedes the production of biodiversity at a lower level. To sum up, in Newman's view, evolution is a process in which there coexist more ancient phenomena typified by significant discontinuities, by phase transitions, by plasticity of living systems, as well as more recent phenomena characterized by continuity and relative graduality in their dynamics.

Saverio Forestiero

Bio-molecular evidences

Werner Arber

Abstract

Shortly after the so-called "modern evolutionary synthesis" leading to Neo-Darwinism around 1940, microbial geneticists discovered that DNA molecules are the carriers of genetic information. Together with the thereafter discovered double-helical structure of DNA molecules, this led to molecular genetics and recently also to molecular evolution. Experimental investigations on spontaneous mutagenesis mainly with bacteria have revealed a number of different, specific molecular mechanisms that contribute to the generation of genetic variants, the drivers of biological evolution. These mechanisms can be classified into three natural strategies to generate genetic variants: (1) local changes in the sequence of nucleotides of DNA, (2) intra-genomic rearrangements of DNA segments, and (3) acquisition of a segment of foreign DNA. A few examples for these mechanisms and strategies will be discussed, as well as their qualitative differences with regard to their contributions to evolution. The general relevance of these insights into biological evolution for all living beings will be discussed. On the basis of current knowledge, specific "evolution gene" products are postulated to provide novel genetic variants in cooperation with several non-genetic elements such as structural flexibilities and chemical instabilities of biomolecules, the impact of chemical and physical mutagens, and random encounter. In conclusion: Natural reality takes actively care of biological evolution of populations of organisms and it keeps the rates of mutagenesis low, ensuring a comfortable genetic stability to all organisms.

1. Roots and development of molecular Darwinism

Evolutionary biology has its roots 150 years ago when Charles Darwin published his ideas on the principles of biological evolution according to which all living organisms are steadily submitted to the pressure of natural selection. On the one hand, Darwin noted that different individual members of any given species often displayed different phenotypic properties. These forms of life were called variants. On the other hand, those variants that can deal well with their encountered environments were postulated to have a selective advantage allowing them to overgrow in the longer term less favored variants.

Only a few years later classical genetics took its origin with the publication of the first laws of inheritance by Gregor Mendel. However, it took about 70 years of independent scientific development of evolutionary biology and of classical genetics until it came to the so-called modern evolutionary synthesis (Mayr 1982). According to the resulting neo-Darwinian theory of biological evolution, the phenotypic variants are due to an alteration in the genetic information. At that time, the concept of the gene still did not have a known material basis; it was an abstract concept.

This situation changed shortly thereafter, thanks to experimental contributions from microbial genetics, when Avery, MacLeod and McCarty (1944) could show that genetic information is carried in desoxyribonucleic acid (DNA) molecules. The fundamental principles of the genetic language became obvious nine years later, when Watson and Crick (1953) explained the structure of DNA as long filamentous molecules whose building blocks are four different nucleotides. It is then the linear sequences of nucleotides that contain the genetic information, similar to the linear sequences of letters in our written language.

On the basis of this knowledge molecular genetics developed and eventually led to genomics. Time has now come for another synthesis: the one between molecular genetics and neo-Darwinism. The thereby resulting Molecular Darwinism will be explained in the following sections.

2. Definition of the term mutation

According to the definition used in classical genetics a mutant displays an altered phenotype (as compared with the one of its parents) that becomes transmitted to its progeny.

In contrast, it has become a habit in molecular genetics and more generally in genomics, to define a mutation as any kind of alteration in the nucleotide sequences.

It is important to be aware of this different use of language. In this paper we use the terms mutation and genetic variation as synonyms and we generally follow the definition used in molecular genetics.

We now know that altered phenotypes that become transmitted to the progeny are, in general, due to an alteration in the nucleotide sequence. But we also know that, by far, not all alterations in a nucleotide sequence result in an altered phenotype. These are then silent or neutral mutations, some of which may become evolutionarily relevant at some later time in conjunction with other, additional nucleotide sequence alterations.

3. The effectiveness of spontaneous mutations

It is a widely accepted general observation that only a minority of novel spontaneous mutations is favorable and provides an advantage to the mutant organism in its habitual living conditions. More often, a mutation is unfavorable and provides a selective disadvantage. As we have already mentioned, many mutations may have no immediate influence on life processes. Some of such mutations may occur within DNA sequences that do not carry essential genetic information. Other mutations, such as some nucleotide substitutions, may have no influence on the amino acid sequence, the structure and the functions of the protein gene products in question.

A spontaneous mutation occurring within an open reading frame may improve or impair the protein gene product in question. A mutation that occurs within an expression control signal may result in an increased or in a decreased amount of the gene product.

Since there are generally more unfavorable than favorable mutations, the natural reality must keep the rates of spontaneous mutagenesis at low levels in order to ensure a certain degree of genetic stability as required for the continuity of life. Another general conclusion from this situation is that there is no evidence available for a directedness of spontaneous mutations. In other words, living organisms have no reliable sensors that would enable them to identify environmental needs to modify relevant genes in specific directions.

4. The three pillars of biological evolution: genetic variation, natural selection, isolation

If genomes were absolutely stable, undergoing no genetic variation, there would be no biological evolution. Genetic variation is thus the driving force of biological evolution.

Natural selection, together with the forms of life available at any time (parental and variant forms), directs biological evolution by determining the directions in which the branches of the tree of evolution grow. The constraints that are exerted on living organisms by natural selection depend principally on the physico-chemical environment and on activities of any kind of organisms present in an ecosystem. As a general rule, it is not the DNA but rather the activities of the gene products that are the substrates for natural selection.

Both reproductive and geographic isolations modulate the process of biological evolution.

With the nowadays available experimental methodology it is possible to identify molecular mechanisms that contribute to genetic variation. This will be discussed in the following sections.

5. Experimental approaches to investigate molecular mechanisms of genetic variation

With today's facilities to determine DNA sequences and to analyze DNA sequences with bioinformatic tools, it is possible to compare DNA sequences of different natural sources. This can be applied to a particular gene for a specific function, or even to a part of a gene representing a functional domain. Such comparisons can reveal local sequence alterations affecting one or a few adjacent nucleotides. Sequence comparisons can also be applied to particular groups of genes, as well as to entire genomes of more or less closely related organisms. Such research can reveal past events of DNA rearrangements that are likely to have caused the observed alterations in the structural organization of the genomes. These DNA sequence comparisons can also reveal possible past events of horizontal gene transfer.

In order to study in a more specific way the molecular processes of genetic variation, it is best to work with relatively short genomes of organisms that propagate rapidly. Microorganisms such as bacteria and bacterial viruses respond to these requests. Microbial genetics can indeed provide deep insights into individual processes of genetic variations by comparing DNA sequences and expressed functions before and shortly after an event of mutagenesis.

Bacteria, such as *Escherichia coli* can be propagated in the laboratory as monocultures in appropriately chosen growth media with a generation time of about 30 minutes. This means that the cells divide every 30 minutes, they grow exponentially as long as there is enough growth medium available. Under these conditions, the progeny of each original cell will attain 1000 cells after only five hours of growth. These bacteria are haploid, they have all their genes in a genome that is just one long filamentous DNA molecule. Because of haploidy, novel, spontaneously occurring genetic variants express their altered functions relatively rapidly; effects of a mutation become quickly phenotypically manifested.

In these bacteria one can observe that in each generation one individual cell per about 200 to 300 cells in the population suffers a spontaneous, novel genetic variation. Novel genetic variants will subsequently give rise upon further growth to subclones of these variants. But some drastic mutations may be lethal and cause the death of the affected cell. Note that at any time, parental forms and all genetic variants in the growing population are exposed to and form the substrate for natural selection.

6. Overview of results from studies on genetic variation of microorganisms

An important conclusion from this kind of experimental investigations is that a number of different specific molecular mechanisms contribute to the overall genetic variation. Several of these mechanisms revealed by microbial genetics are now better understood. A certain degree of randomness in space and in time is characteristic for all of them. These molecular mechanisms of generating genetic variations can be assigned to three qualitatively different natural strategies of genetic variation. These are local sequence changes, intragenomic DNA rearrangements and DNA acquisition by horizontal gene transfer (Arber 2003; 2007).

7. The natural evolutionary strategy of local DNA sequence change

Molecular mechanisms that contribute to this strategy for genetic variation affect often only one nucleotide (by either substitution, deletion or additional insertion of one nucleotide), but sometimes also a few adjacent nucleotides (by short duplication, short deletion or scrambling of parental nucleotides). Many of these events occur during DNA replication. In the literature, such mutations are often seen as replication errors and accidents. In contrast to this view, we propose to take a more positive attitude. Since these replication infidelities largely depend on intrinsic natural properties of matter and more specifically of bioorganic molecules, we rather consider that natural reality makes actively use of properties, such as assuming short-living isomeric forms, in order to occasionally, and more or less randomly, procure a local sequence change. A thereby resulting nucleotide substitution can explain the cause of single nucleotide polymorphisms (SNP). By the way, the advent of these mutations generally serves in research as a molecular clock to measure evolutionary distances between partially homologous DNA sequences.

It has become obvious that DNA mispairing due to tautomeric forms of nucleotides could be expected to occur more frequently than observed. A limitation of this kind of mutagenesis to quite low rates finds its explanation in the activities of genetically encoded repair systems (Radman and Wagner 1986). Repair enzymes can identify nucleotide mispairings shortly after their appearance, and this leads rapidly to the reconstitution of the parental sequence. From an evolutionary point of view, it is fortunate that this repair is generally not 100% efficient. This allows reality to take occasionally advantage of the occurrence of a nucleotide substitution. This kind of mutagenesis, as well as other kinds of local sequence changes, serve in biological evolution for an improvement of available biological functions. We have to keep in mind that local sequence changes (like other kinds of genetic variation) are not specifically directed. It is natural selection that will influence their maintenance or eventual elimination from propagating populations. As we have already discussed, these evolutionary switches act at the level of the expressed genetic information, i.e., of the phenotype.

There are of course other causes than tautomery for local DNA sequence changes. These causes include a certain low degree of chemical instability of nucleotides, but also the impact of internal and environmental, mainly chemical, mutagens.

8. The natural evolutionary strategy of intragenomic DNA rearrangements

In this strategy of genetic variation, DNA segments of variable lengths can become deleted, duplicated and sometimes further amplified, inverted at their parental location, or translocated to another site in the genome. In general, specific, genetically encoded enzymes are involved in these activities that contribute to genetic variation. We will comment here on three types of relatively well studied microbial recombination systems.

General or homologous recombination has since long been known to promote occasional recombination between paternal and maternal chromosomes during meiosis. This contributes to the unicity of the individual genomes of eukaryotic sexual organisms. General recombination is also found in bacteria. One of their identified functions is to repair chromosomal breaks that can, e.g., be due to high energy irradiation. In the undisturbed growth of bacteria, enzymes for general recombination are present, at most, at very low concentrations. But their more intensive production becomes induced for so-called SOS repair in the presence of chromosomal damage such as breaks in DNA molecules. Remind that DNA replication produces from the parental haploid genome two daughter copies that will be distributed to the two daughter cells upon cell division. SOS repair attempts in cases of DNA breaks to reconstitute an intact genome by general recombination between fragments of the two already replicated sister chromosomes.

General recombination becomes effective between DNA segments of a high degree of sequence homology of a length of about 50 or more nucleotides. In the presence of enzymes for general recombination it can thus happen that repetitive, homologous DNA sequences that are carried at different chromosomal locations become substrates for recombination. This can lead to various types of DNA rearrangements, such as the deletion or the inversion of a stretch of DNA. These reactions may bring about the fusion of two different, previously independent functional domains or the juxtaposition of an open reading frame with an alternative expression control signal. The functional consequences of such occasional rearrangements are obvious. These will exert their influence on natural selection.

Mobile genetic elements have been well studied in microorganisms and to a lesser degree also in several higher organisms (Shapiro 1983). In the literature they are described as transposable elements. Such elements encode, in general, an enzymatic activity known as transposase. These enzymes are expressed quite rarely, so that they act at only low rates. Their occasional action can promote the translocation of the element (that carries at both ends identifying DNA sequences) to another location. In bacteria, many such mobile elements are known as IS elements (for inserted sequences). In bacterial genomes one often finds several different IS elements with one to several copies of each specific IS element. As we have already discussed, they occasionally transpose to another location. By inserting into a stretch of essential genetic information IS translocation can cause lethality for the concerned organism. In a relevant experimental study with a bacterial virus we could show that about 95% of lethal mutations affecting the production of infective viral particles were caused by the transposition of IS elements from the host genome into the viral genome. Only about 5% of independent lethal mutations were due to local DNA sequence changes (Sengstag and Arber 1983).

As it was seen in this, as well as in many other experiments, each independent IS element displays its own characteristic insertion strategy. While some IS elements highly prefer to insert into relatively short, specific nucleotide sequences, other IS elements, such as IS2, insert preferentially into relatively extended, so-called hot regions for insertion. But within such regions many independent insertion sites are used and these show no obvious sequence homology (Sengstag and Arber 1987).

As we will see in section 9, mobile genetic elements make their evolutionary contribution not exclusively by intragenomic DNA rearrangements, they are in fact also involved in mobilizing genetic information by transposition into natural gene vectors. More passively, if several copies of a given IS element are carried at different locations in a genome, their DNA sequences can occasionally become a substrate for general recombination as already described above.

Some bacteria and some bacterial viruses are equipped with site-specific recombination systems. Some of these promote preferentially the periodic inversion of a DNA segment that is flanked by so-called consensus crossover sites (Glasgow *et al.* 1989). In a well-studied system these largely homologous sites have a length of 26 base pairs and they are carried in inverted orientation. Active crossover sites are reconstituted upon each inversion event. This leads to a flip-flop system with two possible and reproducibly reappearing genetic information contents. In a bacterial virus this can affect the host range of the virus. However, since only two specific DNA sequences are possible, the flip-flop activities are not of immediate evolutionary relevance. But a high degree of evolutionary relevance of such enzymatic inversion systems became obvious, when it was found that the inversion reaction can also occur occasionally between a consensus crossover site and one of many possible secondary crossover sites that differ considerably from the consensus and from each other (Iida and Hiestand-Nauer 1987). Such inversions occur with much lower rates than the flip-flop reaction and the recombination stabilizes the reactivity, since the recombined hybrid crossover sites do not act as efficient substrates for inversion. The evolutionary interest of the rare use of secondary crossover sites lies in the possibility for novel fusions between previously independent functional domains of gene products and in the occasional provision of an open reading frame with an alternative expression control signal (Arber 1991).

In conclusion, living beings are well equipped with a number of enzyme systems capable of producing more or less by chance specific novel DNA sequences from preexisting elements. A small minority of such novel sequences may provide a selective advantage to the organism suffering the genetic variation. This can contribute significantly to the evolutionary progress. This kind of sequence reassembly can offer possible explanations for the sudden emergence of novel phenotypic properties by an improved use of available capacities. Other sources for such emergence are in horizontal gene transfers to be discussed in the following section.

9. The natural evolutionary strategy of DNA acquisition by horizontal gene transfer

Bacterial and bacteriophage genetics took its successful start around 1940 and this was followed by a very efficient and fruitful development. Bacteria are haploid and their propagation does not depend on sexual intercourse, they just propagate by cell division. However, parasexual gene exchange does sometimes occur. The relevant processes are: (1) transformation, where DNA molecules that had been set free by a donor cell are taken up either actively or passively by a recipient cell; (2)

conjugation where donor and recipient cells closely meet so that a conjugative plasmid can promote the transfer of genetic information from the donor to the recipient cell, and (3) virus-mediated transduction. In this latter case, a bacteriophage serves as a natural vector for the transfer of bacterial DNA from its former host to a newly infected recipient host (Arber 1994). Depending on the particular bacterial virus involved, a viral particle can be filled with a fragment of the donor genome (generalized transduction) or by an autonomously replicating hybrid between bacteriophage DNA and bacterial donor DNA (specialized transduction). Such hybrid DNA molecules can originate by picking up genetic information from the bacterial host through transposition and through other recombinational events.

In general, these different gene transfer processes are relatively efficient if donor and recipient bacteria belong to the same bacterial strain and may just differ by a few mutations. But the same kind of gene transfer can also occur between two different, although often related strains of bacteria. In these cases, several barriers that limit gene acquisition by the recipient cells to low rates have to be overcome. First, such barriers can limit the uptake of the foreign DNA by the recipient cell, its penetration through the cell wall into the cytoplasm. Second, one or even several different restriction-modification systems may identify the invading DNA as of foreign origin and subsequently cut it into DNA fragments that become later further degraded. Third, in order to be of evolutionary relevance, the invading DNA must become part of the recipient genome, normally by recombination, or it must become propagated by autonomous replication. Fourth, the acquired and eventually expressed biological functions must be compatible with the functional harmony of the invaded recipient cell. All of these natural barriers to be taken favor the success if acquisition occurs in small steps, i.e., involving DNA segments ranging from a functional domain to a few genes from the donor cell.

Nevertheless, DNA acquisition is a remarkably successful natural evolutionary strategy. By the impact of this strategy, a recipient organism may gain in just one evolutionary step a useful function that it had not yet possessed before the invasion by the foreign DNA. If all the described barriers are overcome, this process can allow the recipient strain to profit from a long term evolutionary progress made by another kind of organism. DNA acquisition is a sharing in successful developments made by others.

10. The evolutionary tree as a network

In the paradigmatic drawing of evolutionary trees, the occasionally forking branches symbolize the vertical flux of entire genomes suffering from time to time a genetic variation. This accommodates the mutational events due to local sequence changes and to intragenomic DNA rearrangements. In order to take also into account the natural evolutionary strategy of DNA acquisition by horizontal gene transfer, we draw into the classical trees of evolution randomly placed horizontal connectors between two branches that may be more or less close to each other. These connectors symbolize the occasional flux of a relatively small DNA segment of the donor branch into the recipient branch (Arber 1991). The resulting network can improve our comprehension of the evolutionary events and progress in the reality of the living world.

Darwin's insights into biological evolution propagated the idea of a common descent of different species of living organisms. By extending our knowledge to also include the natural strategy of DNA acquisition by horizontal gene transfer, we can extend Darwin's dictum to also include a common future. Long-term future evolutionary developments of any given kind of organisms may, at one time or another, profit from the acquisition of genetic capacities that have their roots in another branch of the evolutionary tree.

11. Genetic variations are generated by coordinated actions of products of evolution genes and of non-genetic elements

In the studies of molecular mechanisms of genetic variation it soon became obvious that products of certain genes are specifically involved in the generation of genetic variants (Arber 1991; 1999). Good examples are the various recombination enzymes described in Section 8 and restriction-modification systems encountered in Section 9. As far as we know, these enzymes are not essential for bacterial growth, for the propagation from one generation to the next. Their role is principally evolutionary. We therefore call their genetic information evolution genes (Arber 2003; 2005a; 2005b). Some evolution genes serve as variation generators. Recombination genes, such as those for the transposition of mobile genetic elements, are good examples. Other evolution genes keep the rates of specific types of genetic variation low. In Section 7 we encountered enzymatic repair activities and in Section 9 restriction enzymes. Repair systems modulate the rates of local sequence changes to low levels, while restriction enzymes keep the rate of DNA acquisition at low levels, and their DNA fragmentation activity favors the strategy of DNA acquisition in small steps.

Genetic variation also depends on a number of non-genetic elements (Arber 2003; 2007). These include structural flexibilities, such as tautomery of biological molecules, a certain degree of chemical instability of nucleotides, as well as environmental mutagens and random encounter, such as the encounter of a natural gene vector with a potential recipient cell.

One can postulate that the presently observed activities of evolution genes had in their own past evolution become fine-tuned to exert their functions in a fruitful cooperation with the available intrinsic non-genetic elements described. This postulated evolution of the evolution genes must have been influenced by second-order selection (Weber 1996). This means that maintenance of those genotypes had been favored which were able to provide occasional genetic variants but which could also keep the rates of genetic variation at low levels, affecting relatively rare individuals so that a good genetic stability was ensured at the population level. We can conclude that natural reality actively takes care of biological evolution by involving products of evolution genes and intrinsic non-genetic elements.

12. Duality of the genome and evolutionary fitness

Many of the genes carried in the genomes serve for the fulfillment of the life of each individual. These are the housekeeping genes, the accessory genes for use under particular life conditions and in higher organisms the developmental genes. In contrast, the evolution genes work for the benefit of the evolutionary development of populations. They serve for the expansion of life, for a rich biodiversity.

In reality, the here described duality of the genome does not allow us to strictly assign each particular gene to one of the described functional purposes; some gene products serve for both. Nevertheless, it is philosophically relevant to become aware that the genome takes care both, of the life of the concerned individual, and of the evolutionary progress of the population of organisms.

The critical reader may object against a general validity of the theory of molecular Darwinism as described in this paper, in which a majority of arguments and evidences were taken from microbial genetics. In this context, one can also raise the question on the historical path of evolution from prokaryotic microorganisms to higher eukaryotic organisms. In the absence of clear answers at this time, we will have to wait for results from future research. Let us thus just make a few speculations.

In microorganisms, evolutionary fitness seems to be reached if an organism is equipped with evolution genes for each of the three natural strategies for genetic variation that are described in Sections 7, 8 and 9. The thereby involved specific molecular mechanisms may vary, as long as each strategy can be followed successfully. Most likely, this rule will also apply to higher organisms. In multicellular eukaryotes, some of the here described evolution gene products, such as DNA repair functions, will of course also be relevant at the somatic level. A strict classification is thus less obvious in these cases.

Another relevant aspect is the evolutionary role of symbiotic cohabitation including endosymbiosis. As a matter of fact, this may favor occasional gene transfer over wide evolutionary distances, e.g. between a bacterium and a higher eukaryotic organism. While it is quite clear that the evolutionary strategy of local sequence changes and of intragenomic DNA rearrangements are also used in higher organisms, the relevance of DNA acquisition through horizontal gene transfer is at present not generally accepted by all biologists. More research is required and it would be of interest to know, for example, the evolutionary distances that horizontal gene transfer can span. The quasi-universality of the genetic code may facilitate the DNA acquisition. One may even wonder if the evolutionary importance of occasional horizontal gene transfer may have selected and stabilized a universal code for protein synthesis (Arber 2006).

13. Brief general conclusions

The here presented molecular Darwinism is based on the knowledge of a number of different molecular processes that contribute to the overall spontaneous genetic variation. Genetic variation drives biological evolution. It is thus interesting that genetic variation is jointly due to the activities of evolution genes and to non-genetic natural properties. From the point of view of molecular genetics, genetic variation, and thus biological evolution, is a self-organizing process. Spontaneous mutations should no longer be seen as errors and other accidents occurring to DNA upon its replication. Rather unexpectedly, we have realized that products of some evolution genes generate more or less randomly novel mutants. Hence, a certain degree of randomness remains a central attribute of biological evolution. It insures rich possibilities for adaptation to changing environmental living conditions. Natural selection continues to play its directive role, also in molecular Darwinism. Therefore, molecular evolution confirms the neo-Darwinian concept of central evolutionary roles of genetic variation and natural selection.

REFERENCES

Arber W., 1991, "Elements in microbial evolution", *Journal of Molecular Evolution* 33: 4–12.

Arber W., 1994, "Bacteriophage transduction", in Webster R. G., Granoff A. (eds.), *Encyclopedia of Virology*, London: Academic Press, 107–113.

Arber W., 1999, "Involvement of gene products in bacterial evolution", in Caporale L. H. (ed.), *Molecular Strategies in Biological Evolution, Annals of the New York Academy of Sciences* 870: 36–44.

Arber W., 2003, "Elements for a theory of molecular evolution", *Gene* 317: 3–11.

Arber W., 2005a, "Gene products with evolutionary functions", *Proteomics* 5: 2280–84.

Arber W., 2005b, "Duel nature of the genome: genes for the individual life and genes for the evolutionary progress of the population", *IUBMB Life* 57: 263–66.

Arber W., 2006, "The evolutionary strategy of DNA acquisition as a possible reason for a universal genetic code." *History & Philosophy of the Life Sciences* 28: 525–32.

Arber W., 2007, "Genetic variation and molecular evolution", in Meyers R. A. (ed.), *Genomics and Genetics*, Weinheim: Wiley-VCH, 385–406.

Avery O. T., MacLeod C., McCarty M., 1944, "Studies on the chemical nature of the substance inducing transformation of pneumococcal types. Induction of transformation by a desoxyribonucleic acid fraction isolated from pneumococcus type III", *Journal of Experimetnal Medicine* 79: 137–58.

Glasgow A. C., Hughes K. T., Simon M. I., 1989, "Bacterial DNA inversion systems", in Berg D. E., Howe M. M. (eds.), *Mobile DNA*, Washington: American Society for Microbiology, 637–59.

Iida S., Hiestand-Nauer R., 1987, "Role of the central dinucleotide at the crossover sites for selection of quasi-sites in DNA inversion mediated by site-specific Cin recombinase of phage P1", *Molecular and General Genetics* 208: 464–68.

Mayr E., 1982, *The Growth of Biological Thought: Diversity, Evolution and Inheritance*, Cambridge: Harvard University Press.

Radman M., Wagner R., 1986, "Mismatch repair in *Escherichia coli*", *Annual Review of Genetics* 20: 523–38.

Sengstag C., Arber W., 1983, "IS2 insertion as a major cause of spontaneous mutagenesis of the bacteriophage P1: non-random distribution of target sites", *The EMBO Journal* 2: 67–71.

Sengstag C., Arber W., 1987, "A cloned DNA fragment from bacteriophage P1 enhances IS2 insertion", *Molecular and General Genetics* 206: 344–51.

Shapiro J. A. (ed.), 1983, *Mobile Genetic Elements*, New York: Academic Press.

Watson J. D., Crick F. H. C., 1953, "Molecular structure of nucleic acids. A structure for deoxyribose nucleic acid", *Nature* 171: 737–38.

Weber M., 1996, "Evolutionary plasticity in prokaryotes: a panglossian view", *Biology and Philosophy* 11: 67–88.

Taxonomic issues: What we learn from comparisons among species

Douglas J. Futuyma

Summary

Common ancestry of diverse organisms is supported by numerous independent lines of research, and the common ancestry of all known forms of life is a scientific fact.

The principle of common ancestry, as embodied in phylogenetic relationships among specific organisms, makes predictions that are borne out by subsequent investigation. It provides a rationale for advancing medical research through research on other organisms.

Descent with modification from common ancestors makes explicable diverse facts of biology; in fact, it unifies observations across the biological sciences, from molecular biology to embryology to ecology, under a single explanatory framework. Evolution is the central unifying principle of the biological sciences.

I have been asked to discuss "taxonomic issues," which I take to mean the inferences about evolution that have been made from comparisons among and within species. Such comparisons — for example, of structure — have been the basis of efforts to classify organisms into different species and higher categories, such as families and phyla. These studies have also provided crucially important evidence for the fact of evolution and have continued to reveal the kinds of events that have transpired during evolutionary history (Futuyma 2005).

Traditionally, comparative and taxonomic biology was based on anatomical and other phenotypic characteristics. Today, DNA sequences and even entire genomes are the major focus. I will refer to a few such examples, but I will mostly describe phenotypic evolution.

The structural similarities among different organisms were called *homology* by anatomists before Darwin. Darwin hypothesized that the cause of homology is inheritance of the feature from the common ancestor of the several organisms, often with some modification since the common ancestor. At first, Darwin saw common ancestry as the cause of similarity among very similar species of birds, but he soon realized that by extrapolation of this argument, shared features indicate that all birds are derived from a common ancestor. Darwin went further: he concluded that all animals must be derived from a common ancestor, and finally, as he wrote in *The Origin of Species* (Darwin 1859), "probably all the organic beings which have ever lived on this earth have descended from some one primordial form."

Thus, Darwin hypothesized a branching process, represented by a phylogenetic tree, in which species that are formed by splitting of an ancestral species are at first very similar, but acquire more and greater differences as time passes. They may in turn split into similar species that again diverge in turn. Thus, branch points closer to the root of the tree represent more ancient splitting into different lineages, and the contemporary species on two deep branches are generally more different than species on two shallower branches, because they have had a longer history of independent, incremental evolutionary changes from each other.

Such common descent is the scientific rationale — the only scientifically explicable rationale — for studying cellular and other biological processes in model species such as fruit flies and mice, in order to understand function and malfunction of the human body.

How are we to infer which species have the closest genealogical relationships, and which are more remotely related? That is, how can we infer the structure of the phylogenetic tree? In general, more closely related species should be expected to have more similar characteristics. But there are many exceptions presented by individual characteristics. For example, a rook and a bear are each more closely related to other birds or bears, even if they are both black. The black color has evolved independently, or *convergently*, in these species. In this case it is obvious that coloration is not a good indicator of relationship, because so many other features indicate the actual relationships of the rook and the black bear. However, convergent evolution often causes considerably more uncertainty, especially in the among species that are all closely related to each other. For this and several other reasons, students of this subject have developed fairly complex methods of analysis that attempt to take convergent evolution and other misleading processes into account (Felsenstein 2004). To greatly oversimplify, the best estimate of phylogeny (the history of branching) is the one that requires us to postulate the fewest evolutionary changes, summed over all examined characteristics, compared to other possible phylogenies. This estimate constitutes the simplest hypothesis of evolutionary history that is consonant with the data.

A phylogenetic tree is a hypothesis about evolutionary relationships. Throughout science, we obtain confidence in a hypothesis if it is supported by multiple, independent sources of evidence. This is very often the case with phylogenetic estimates. For example, the relationships among the major groups of vertebrates that were inferred from anatomical features are almost all supported by evidence from DNA sequences as well (Meyer and Zardoya 2003). Similarly, the phylogeny of major groups of primates based on DNA is the same as had been postulated from anatomy (Page and Goodman 2001).

This consonance, moreover, confirms that evolution of these organisms has been one of lineages splitting, and then becoming different. We know what data look like when evolution occurs by other modes, e.g. by reticulation (merging of lineages) rather than by branching. For example, the DNA sequence of a certain gene makes it appear that the small species of cats are related to monkeys and unrelated to the leopard and other large cats. That is because the gene is a virus gene that was transferred from monkeys to the ancestor of these cats, probably when a cat ate an infected monkey (Li and Graur 1991). The consequence is a strongly different pattern of differences among species than is seen in other parts of the genome. If you sequence random pieces of DNA from an aphid, you may think it is a bacterium rather than an insect — because the aphid includes not only its own genome, but that of symbiotic bacteria in some of its cells. As Lynn Margulis describes in her contribution to these proceedings, similar events have resulted in merging of genomes and lateral gene transfer among lineages, rather than simple branching of distinct lineages.

I have noted that we might expect more distantly related species to be more different from each other. This correlation is often stronger for DNA sequences than for morphology. The differences in DNA sequence between pairs of species may be plotted against the time that has elapsed since those species had a common ancestor, determined in the fossil record. For example, the most recent common ancestor of a monkey and a mouse is set at the first definitive fossil primate or rodent. A plot of this kind often shows that DNA sequence difference is strongly correlated with time, so that it may be used as a "molecular clock" (e.g., Langley and Fitch 1974). This enables us to calibrate the rate of evolution of DNA sequences, and thus to determine the time of origin of related species that do not have a fossil record, by measuring the difference between their DNA sequence. It must be understood that molecular clocks are at best very approximate, usually estimating times of divergence on the scale of millions rather than thousands of years. Some DNA sequences evolve faster than others, so the genome provides a variety of different 'clocks," an instance of the phenomenon called mosaic evolution that is described below.

Approximate constancy of the rate of evolution of a DNA sequence follows from the mathematical theory of population genetics, as developed by Motoo Kimura (1983) in his "neutral theory of molecular evolution." Mutations such as single-base-pair substitutions occur at an approximately constant rate, but evolution occurs only if a mutation increases in frequency and replaces the previously prevalent form throughout an entire population or species. This replacement can take place either by random genetic drift or natural selection. Genetic drift is the stochastic fluctuation of the proportions (frequencies) of two or more forms of a gene in a population; natural selection is a consistent, or "biased," difference in the rate of increase in frequency of a genetic type, due to a consistent difference in survival or reproduction (fitness). A mutation may be advantageous (increases fitness), deleterious (decreases fitness) or neutral (having neither effect). Many mutations are deleterious, and are eliminated from populations by "purifying selection." Of the remainder, a small proportion are advantageous, and may be "positively selected" and so become a feature of the entire population or species. Other mutations are neutral, and change in frequency by genetic drift. Mutations in genes that encode proteins are especially likely to be neutral if they are "synonymous" mutations that alter the DNA sequence but not the amino acid sequence of the protein, whereas synonymous mutations that alter proteins are more likely to affect fitness.

It is the neutral fraction of mutations that, according to mathematical theory, results in an approximately constant rate of evolution of DNA sequences, because neutral mutations all have much the same probability of increasing in frequency by genetic drift, and are expected to arise at a fairly constant rate. The higher the fraction of neutral mutations, the higher the rate at which a DNA sequence will evolve. Positive natural selection of advantageous mutations will cause an elevation of the evolutionary rate in some gene regions or in some species, and may also be reflected in a higher ratio of nonsynonymous to synonymous nucleotide substitutions (since synonymous changes are those that can potentially enhance the organism's fitness by altering proteins). These deviations from the neutral evolutionary model provide evidence of positive natural selection at the DNA level.

Comparisons of DNA sequences among species have provided many examples of all these phenomena. A relatively constant rate of evolution is often shown by comparing DNA sequences between three species, where species A and B are more closely related to each other (stem from a more recent common ancestor) than to species C. If the evolutionary rate has been constant, the difference in DNA sequence between A and C should equal that between B and C, since the elapsed time since common ancestry has been the same in both cases. Many examples fit this pattern, showing that the hypothesis of a molecular clock need not rely on the fossil record. Natural selection on DNA sequences has also been inferred in many cases. For example, synonymous substitutions in the FOXP2 gene have occurred at a significantly higher rate than in other primate or mammal lineages (Enard *et al.* 2002). These advantageous mutations might have contributed to the evolution of the capacity for language, since deleterious mutations in this gene are known to diminish linguistic abilities.

An extremely important use of phylogenetic trees is to infer the most likely history of changes in specific characteristics (Futuyma 2004). For instance, the first toe of monkeys and apes (including the chimpanzees, the closest relatives of *Homo sapiens*) is opposable (like the thumb), but in humans it is not opposable. If we postulate the simplest hypothesis, namely that the common ancestor of chimpanzee and human had an opposable toe that became non-opposable during human evolution, only a single evolutionary change need be envisioned. The alternative, that the ancestor had a non-opposable toe that then became opposable in the chimpanzee, converging on the condition in other apes, is a needlessly complex hypothesis.

Like any good, strongly supported scientific hypothesis, the hypothesis of descent from common ancestors makes predictions.

For example, in general we expect that characteristics of closely related species will be similar. So when investigators examined the arrangement of different genes along the chromosomes, using modern techniques, they expected that human chromosomes would be most similar to those of chimpanzees, and this is exactly what they found (e.g., Goidts *et al.* 2005).

Paleontologists seek intermediate fossils that exemplify the transition from ancestral forms to highly different descendants. Because of their phylogenetic hypothesis, they *know what they are looking for*: they predict many of the features such an intermediate must have. In 1967, Wilson, Carpenter, and Brown, authorities on ants, reported the discovery of the earliest (Cretaceous) fossil ant. Almost all its features matched the features that these authors had proposed, some years before, to be the critical features that an intermediate between ants and their wasplike ancestors should have. Neil Shubin went to northern Canada to look for fossils that would show the transition from lobe-finned fishes to the first terrestrial four-legged vertebrates, and after a year or more of search, he and his coworkers found the new fossil, named *Tiktaalik*, that fills in the gap between previously known forms (Daeschler *et al.* 2006).

A marvelous prediction of this kind was made by Darwin in *The Descent of Man*: "In each great region of the world the living mammals are closely related to the extinct species of the same region. It is therefore probable that Africa was formerly inhabited by extinct apes closely allied to the gorilla and chimpanzee, and as these two species are now man's nearest allies, it is somewhat more probable that our early progenitors lived on the African continent than elsewhere" (Darwin 1982 [1871]). Of course, after about 50 years, hominid fossils were found in Africa, and we now know that the entire early history of hominids occurred there.

We have learned a great deal about patterns of evolution from comparisons.

For instance, organisms have many nonfunctional characteristics — vestigial features — that were functional in their ancestors. Examples are the eye rudiments in blind cave-dwelling fishes, and the flower structure of the dandelion *Taraxacum officinale*, which is very similar to that of related plants, in which the yellow structures attract pollinators. But this species produces no viable pollen, and reproduces asexually, without cross-pollination. It has recently evolved the asexual life history from sexual ancestors. The genome has great numbers of nonfunctional DNA sequences (pseudogenes) that are not transcribed but are the decaying remains of once functional genes. Such features can be understood only in terms of evolutionary history.

Some features appear to be arbitrary, but once they evolved, deviation from the norm would be deleterious in descendant organisms (like driving on the dextral side of the road in England). The genetic code is presumably rather arbitrary, but is nearly universal. All proteins are composed of L amino acids, even though R isomers would have the same properties — but the two cannot be mixed together. Amphibians and reptiles have a pair of aortic arches that carry blood from the heart. Evidently it is advantageous for warm-blooded vertebrates to have only one such blood vessel, but it happens to be the right arch in birds and the left arch in mammals, in which homeothermy evolved independently. Such arbitrary differences are understandable as the consequence of history — of initial starting conditions.

To these signatures of history we can add many others, including many that are evident in embryological development. For example, adult birds and anteaters lack teeth, but tooth anlagen appear in the embryos, and then are absorbed. Horses, with one toe, evolved from five-toed ancestors. Side toes begin to develop in horse embryos, and then become fused with the central toe. As Darwin noted, many plants have female-functioning flowers in which vestiges of male sexual organs develop. In the ancestral condition, flowers possess the organs of both sexes. We also see many examples of features that exhibit apparent imperfection, or suboptimal "design," such as the blind spot in vertebrate eyes, caused by the "reverse wiring" of the retinal cells, the axons of which run over the retinal surface and converge into the optic nerve that passes through the retina to the brain.

Comparisons among species show us that different features evolve at different rates, more or less independently — a phenomenon called *mosaic evolution*. For example, two closely related species of sea urchins (*Heliocidaris tuberculata* and *H. erythrogramma*) are nearly identical in the adult stage, but their larvae are very different, so larval features have evolved faster (Wray and McClay 1989). The principle of mosaic evolution is extremely important. It enables us to study evolution by focusing on individual traits or genes, rather than having to understand coordinated change in great numbers of characteristics. And of course, if many or all of an organism's features had to evolve together, in perfect coordination, evolution would be extremely slow and difficult.

Comparisons among living species often show that characteristics can evolve gradually — incrementally by many small steps — because we see such steps represented as the typical condition among closely related species — as in the length and shape of the bill in sandpipers (Futuyma 2005). Because of such gradations, the higher taxa in classifications, such as genera and families, are often arbitrary divisions of a continuum, and taxonomists have frequently differed on the number of such taxa into which they would place the various species.

An extremely important principle is that most new features are modifications of ancestral features, in which form has evolved to serve a different function. To cite one example out of thousands, primitive insects, such as grasshoppers, have an ovipositor, a tube through which the female inserts eggs into soil or plants. Most hymenopterous insects are parasites that use the ovipositor to insert eggs into the body of a insect or spider. In one group of these insects, the wasps, bees, and ants (*Aculeata*), the ovipositor has been modified into a sting, with a defensive function. This explains, incidentally, why only female wasps and bees sting.

The same principle is seen at the molecular level. Proteins that serve novel functions must come from somewhere, so they must originate from proteins that had a different earlier function. The crystallin proteins that compose the lens of the eye are very similar to, and often exactly the same as, proteins that serve an entirely different function elsewhere in the body (Piatigorsky 2007). For example, the human eye lens is made of the enzyme aldehyde dehydrogenase, in crystalline form. Among various other animals, other proteins serve this function.

I have already described convergent evolution, in which a similar feature evolves independently in different evolutionary lineages. It often does so because natural selection favors a similar adaptation. Convergence, then is often an indicator of adaptation. For example, in at least five different bird families, a long, slender, curved bill has evolved, in every case for probing into flowers to obtain nectar. Lineages of plants in many families have, reciprocally, evolved special adaptations for pollination by birds, including long, tubular, red or orange flowers with copious amounts of sugar-rich nectar.

Convergent evolution was already familiar to Darwin, but now we know it occurs also at the molecular level. For example, ruminant hoofed mammals, langur monkeys, and a South American bird called the hoatzin all feed on leaves, and all have a stomach compartment with bacteria that digest the cellulose and other leaf material. These animals all obtain nourishment by digesting their bacteria, using the enzyme lysozyme to digest the bacterial cell wall. Certain amino acid changes in this enzyme have evolved independently in all three lineages (Kornegay *et al.* 1994).

Convergent evolution is very common, but it is far from universal, and there are countless examples of what surely count as singular, unparalleled evolutionary "innovations." Of course, almost every individual in a sexually reproducing species is genetically unique, and every species has genetic features that are unique, so it is only because we classify instances into categories (e.g, "*Homo sapiens*," or "leg") that we have the possibility of recognizing repeated instances in biology. So it is that we can characterize the slender bills of nectar-feeding hummingbirds and sunbirds as convergent. But nonconvergent, singular instances of evolution are equally plentiful: the elephant's nose, used as an arm; the single genus of fish-eating bats, restricted to the American tropics; the folding fangs of vipers; the nervous system of animals, the sole example of a multicellular nervous system in known evolutionary history. Gould (1989) may have exaggerated the role of contingency in evolution when he (following other evolutionary biologists) suggested that the tape of life, if replayed, would be a very different movie, but there surely has been far more contingency than some would suggest. Evolutionary biology does not support a deterministic view of history.

These considerations also bear on the issue of trends in evolution. It is often supposed that there has been a trend toward greater complexity, and the further inference has often been made that this trend manifests a determined drive toward greater complexity. This view presumes a teleological process, a final cause of evolution. Evolutionary biology provides no support for this view. It is useful to distinguish between "active" and "passive" trends (McShea 1994). Suppose, for example, that the common ancestor of a set of organisms is very small, near the lower bound that is compatible with life processes. If most or all of the subsequently arising lineages evolve toward greater size, we suspect some common, biasing process or force. (It might be natural selection for greater size, which of course is a nonteleological process.) Suppose, however, that there simply exists a process that generates variation, unbiased in either direction. Average size will necessarily increase, simply because the lower bound to size prevents the realization of a symmetrical distribution of the variation. And variation subsequently generated around the higher mean will produce some organisms of still greater size (as well as some smaller ones). The result is a passive trend. Now since the earliest known organisms were very small bacteria and archaea, and since we know of redwood trees, dinosaurs, and whales, we might say that life has displayed a trend toward greater body size; yet very small organisms, including bacteria and one-celled eukaryotes, constitute the vast majority of species and individuals on Earth. The variance has increased immensely, but there has not been a directed orientation toward greater size. (Of course, each evolutionary change in size, whether toward smaller or larger, may have resulted from natural selection in the particular species involved; but the distribution of selective advantages may have been more or less symmetrical.)

So it is with complexity. Since the earliest organisms were necessarily very simple, complexity could only increase. But not all lineages have done so, and many lineages have become less complex (by any measure of complexity) than their ancestors. Comparisons among species do not indicate an active trend toward complexity, and certainly provide no evidence of a necessary or finally determined "movement toward" any grand evolutionary outcome.

I have emphasized comparisons among diverse, often distantly related organisms, but even more has been learned about evolution by comparing closely related species and different geographic populations of the same species. Such comparisons have provided evidence of rapid adaptive evolution and of the causes of evolution. For example, many great studies have described genetic differences among geographically distinct populations of a species that are adaptations to differences in environmental factors. But for brevity, I will describe only two other lessons.

First, characteristics typically evolve gradually, by small steps. This is true of reproductive isolation, which many evolutionary biologists consider to be the defining feature of species. If two populations simply will not mate with each other, that is one form of reproductive isolation. Different populations of what is called a single species of salamander, *Desmognathus ochrophaeus*, show degrees of reproductive isolation when females are presented with males from various other populations. In general, a pair is less likely to mate, the farther apart their home localities, and the more genetically different they are, as measured by molecular genetic data (Tilley *et al.* 1990).

The same result has been shown by plotting the degree of reproductive isolation between many different pairs of populations and incipient species of fruit flies against the "genetic distance" between them, the level of molecular genetic difference that is an index of time since the populations began to diverge. (This is an application of the molecular clock.) Reproductive isolation, whether manifested as failure to mate or as sterility or inviability of hybrid offspring, varies gradually; it is not an all-or-none matter (Coyne and Orr 1989). Therefore, different species evolve by the gradual genetic differentiation of initially similar, reproductively and genetically compatible, populations.

My final point is that sometimes two species inherit not just a particular gene from their common ancestor — they may inherit two or more forms (alleles) of the same gene. This is called a genetic polymorphism — two or more alleles, such as A, B, O blood types, within a population. For example, two related species of geese, *Chen caerulescens* and *Chen rossii*, both have both a dark form and a white form, which are based on a single gene difference. So it appears that two forms of the gene arose in their common ancestor, and both forms were carried over into both descendant species. Another example is presented by two of the gene loci in the major histocompatibility complex (MHC), which encode proteins that present foreign molecules to the immune system. At each locus, human populations have several alleles, and the same holds in the chimpanzee (*Pan troglodytes*) (Nei and Hughes 1991). Several of the human alleles are genealogically more closely related (and are more similar in DNA sequence) to certain chimpanzee alleles than they are to other human alleles. The most likely explanation is that several forms of the gene have been maintained by natural selection because they are each capable of providing defense against different pathogens. This polymorphism, maintained by natural selection in the common ancestor of humans and chimpanzees, was carried over into both descendant species. It is hard to imagine more striking evidence that humans and chimpanzees have descended from a relatively recent common ancestor.

REFERENCES

Coyne J. A., Orr H. A., 1989, "Patterns of speciation in *Drosophila*", *Evolution* 43: 362–81.

Darwin C., 1859, *The Origin of Species by Means of Natural Selection, or The Preservation of Favoured Races in the Struggle for Life,* London: J. Murray.

Darwin C., 1982 (1871), *The Descent of Man, and Selection in Relation to* Sex, Princeton N.J.: Princeton University Press.

Daeschler E. B., Shubin N. H., Jenkins F. A. Jr., 2006, "A Devonian tetrapod-like fish and the evolution of the tetrapod body plan", *Nature* 440: 757-63.

Enard W., Przeworski M., Fisher S. E., Lai C. S., Wiebe V., Kitano T., Monaco A. P., Pääbo S., 2002, "Molecular evolution of *FOXP2,* a gene involved in speech and language", *Nature* 418: 869–72.

Felsenstein J., 2004, *Inferring Phylogenies,* Sunderland, Mass.: Sinauer.

Futuyma D. J., 2004, "The fruit of the tree of life: insights into evolution and ecology", in Cracraft J., Donoghue M. J. (eds.), 2004, *Assembling the Tree of Life*, Oxford and New York: Oxford University Press, 25–39.

Futuyma D. J., 2005, *Evolution*, Sunderland, Mass: Sinauer.

Goidts V., Szamalek J. M., de Jong P. J., Cooper D. N., Chuzhanova N., Hameister H., Kehrer-Sawatzki H., 2005, "Independent intrachromosomal recombination events underlie the pericentric inversions of chimpanzee and gorilla chromosomes homologous to human chromosome 16", *Genome Research* 15: 1232-42.

Gould S. J., 1989, *Wonderful Life: The Burgess Shale and the Nature of History*, New York: W. W. Norton.

Kimura M., 1983, *The Neutral Theory of Molecular Evolution,* Cambridge: Cambridge University Press.

Kornegay J. R., Schilling J. W., Wilson A. C., 1994, "Molecular adaptation of a leaf-eating bird: stomach lysozyme of the hoatzin", *Molecular Biology and Evolution* 11: 921–28.

Langley C. H., Fitch W. M., 1974, "An examination of the constancy of the rate of molecular evolution", *Journal of Molecular Evolution* 3: 161–77.

Li W.-H., Graur D., 1991, *Fundamentals of Molecular Evolution*, Sunderland, Mass.: Sinauer.

McShea D. W., 1994, "Mechanisms of large-scale evolutionary trends", *Evolution* 48: 1747–63.

Meyer A., Zardoya R., 2003, "Recent advances in the (molecular) phylogeny of vertebrates", *Annual Review of Ecology, Evolution and Systematics* 34: 311–38.

Nei M., Hughes A. L., 1991, "Polymorphism and evolution of the major histocompatibility loci in mammals", in Selander R. K., Clark A. G., Whittam T. S. (eds.), *Evolution at the Molecular Level,* Sunderland, Mass.: Sinauer, 222–47.

Page S. L., Goodman M., 2001, "Catarrhine phylogeny: noncoding DNA evidence for a diphyletic origin of the mangabeys and for a human-chimpanzee clade", *Molecular Phylogenetics and Evolution* 18: 14–25.
Piatigorsky J., 2007, *Gene Sharing and Evolution: The Diversity of Protein Functions*, Cambridge, Mass.: Harvard University Press.
Tilley S. G., Verrell P. A., Arnold S. J., 1990, "Correspondence between sexual isolation and allozyme differentiation: A test in the salamander *Desmognathus ochrophaeus*", *Proceedings of the National Academy of Sciences U.S.A.* 87: 2715–19.
Wilson E. O., Carpenter F. M., Brown W. L. Jr., 1967, "The first Mesozoic ants", *Science* 157: 1038–40.
Wray G. A., McClay D. R., 1989, "Molecular heterochronies and heterotopies in early echinoid development", *Evolution* 43: 803–813.

Origin of evolutionary novelty by symbiogenesis

Lynn Margulis

Speciation, the transformation from one species of organisms to a differently named species or the transformation of one species into two related descendant taxa (different species in the same genus) by "accumulation of random DNA mutations" has not been detailed in the professional literature. Accumulation of favorable DNA mutations sufficient in number and in temporal sequence should lead from one to a second or branch to a second and third species according to neo-Darwinian theory. Yet in no well-characterized species of nucleated organism (whether protoctist, fungus, animal or plant) (Margulis and Chapman 2009) has the "passage" from one to a second or third species been documented in the field, the fossil record or the laboratory. Such examples of "Darwinian gradual origin of a new species" by accumulation of gene mutations e.g., heritable random changes in DNA sequence should abound. Even speciation by accumulation of gene duplication ought to be defensible. Yet we still await detailed documentation of a case of speciation that satisfies naturalists, systematists, taxonomic specialists or paleontologists, scientists who formally name species.

I exclude prokaryotes in this discussion (bacteria, both eubacteria and archaebacteria) because the term "species" does not apply to this kingdom (Margulis and Chapman 2009). Bacteria donate and receive genes from each other (by unidirectional conjugation, transfer of plasmids or viruses, uptake of transforming principle DNA and other means of genetic exchange) across their so-called species and more inclusive taxa lines (genera, families and even domains). One way gene transfer in prokaryotes began very early in the history of life on Earth (Becerra *et al.* 2007). We infer from the modern world that the common ancestor of all extant life engaged in one-way genetic transfer. Gene donation and receipt from eu- to archaebacteria and vice versa, occurs in immediate response to environmental selective pressures on very short time scales from the geological perspective in prokaryotes and forms the basis of genetic engineering (Becerra *et al.* 2007). Since, in wanton abandon at high frequency, prokaryotes engage in such unidirectional sexual exchange, members of the entire kingdom may be considered to be one single global species. Or, an equivalently bacterial groups do not speciate as do eukaryotes: they do not form species that are stable over periods of time from 0.001 to 10 million years (Sonea and Matthieu 2000). An entire tube of 107/ml nitrogen-fixing *Rhizobium* bacteria can all convert to the genus *Azotobacter* in the refrigerator by the addition of ethidium bromide (the lose the nif-plasmid, or, even more impressive, *Bacillus cereus* can evolve to become *Bacillus anthracis* (the anthrax bacterium) overnight (by the incorporation of the correct two plasmids). The new genera and species propagate their kind in future generations or their heriditary characteristics change with conditions on rapid time scales in a process dismissed as Lamarckian by evolutionary biologists unfamiliar with 3/5ths of life's most inclusive taxa (i.e., microbes: bacteria, protoctists, fungi).

A plethora of high-quality scientific studies of nucleated organisms, by contrast, has unequivocally shown symbiogenesis to underlie the origins and stability of species and more inclusive taxa in members of the four Eukarya kingdoms. These are the two most inclusive embryo-forming groups (Animalia and Plantae), Fungi (haploids and dikarya that develop from conidiospores, zygospores, ascospores, basidiospores, soredia or other propagules) and the Protoctista, their ancestors. Protoctista are eukaryotes unified by their exclusion from the other three kingdoms.

Species, some 250,000 estimated by Corliss (1990), are identifiably protoctists. Names, independent of their sexuality, have been given to thousands of protoctists, whether unicellular amitochondriates, e.g., Archamoebae or Parabasalia (Dolan in press), ciliates, euglenas, slime nets, water molds, green algae or giant phaeophytes (photosynthetic brown marine algae such as kelp). The Protoctist kingdom comprises some 36 phyla (Margulis and Chapman 2009). In many species no vestige or hint of meiotic sex exists. In others, far more than the two standard genders (male/female) are known. "Protoctista" is the taxon from which descended the three lineages of larger eukaryotes (animals, fungi and plants). This is not disputed: plants from chlorophytes, animals from choanomastigote protoctists and fungi from chytrids. All members of the Superkingdom Eukarya have recognizably speciated; this fact is attested to by the universality of taxonomic nomenclature, whether the species name itself be Latin, Greek, English-based or other.

Meiotic-fertilization sexual cycles are absent, that is, no sexuality associated with reproduction has ever been found. Such microbes and their descendants reproduce uniparentally (from a single parent). The absence of meiotic sex holds for protoctistan taxa more inclusive than species and genera (e.g., most Archaeprotista such as archamoebae, metamonads, calonymphids; rhizopods, the heterolobosean and testate amoebae; kinetoplastids/euglenids and hyphochytrids as well as for fungi imperfecti. Members of these groups reproduce in the absence of gendered meiotic-fertilization sexuality. Therefore the requirement that members of the same species be sexually compatible is inappropriate, untested and untestable in thousands of named species. The uniparental mode of reproduction (no genders, no meiotic-fertilization sex) is the rule in nearly all species of Protoctista, in many fungi and even in some embryo-formers. The widely held assumption that fertile offspring must be produced as prerequisite to assign individuals to the same species is a legacy of zoocentrism. Furthermore this criterion defies testing if the putative members of the same species are widely separated by either time or geographical location.

Our alternative symbiogenetic view is that the accumulation of random mutations alone is not adequate to mandate assignment of living organisms to a newly named species. We claim that the process of speciation itself evolved with eukaryosis and preceded the origin of meiotic sex and fertilization in several lineages. The evolutionary appearance of the nucleated cell by symbiogenesis correlates with the onset of speciation. Members of the same species share the same genetically integrated bionts, we assert. New individuals (holobionts) evolved by integration such that their dissimilar component bionts, are identifiable in principle by genomic and proteomic techniques. We detailed our symbiogenetic concept with many examples (Margulis and Sagan 2002) and employ it as the basis for holarchical classification of extant phyla (Margulis and Chapman 2009).

The claim that species originate by symbiogenesis was announced in an abstract in 1921 and elaborated in his book by a young Boris Mikhailovich Kozo-Polyansky (1924) (Khakhina 1992). Wallin, an anatomist in New York City, ignorant of the Russian "symbiogenesis" literature, published (1927) a virtually equivalent concept of species origin for which he invented the word "symbionticism". Both seminal books were ignored or dismissed in English-language publications until Armin Geus (Marburg an der Lahn) and Ekkehard Höxtermann (Berlin) with Irmgard Müller (Bochum) all of whom are Professor-Doctors produced their Marburg art exhibit and catalogue (1998).

The critical scientific literature that demonstrates the appearance of new species of more complex larger organisms correlated with fusion of members of earlier, smaller species of beings has been ignored, denied or suppressed by neoDarwinists. The German Anton de Bary (1831-1888) invented the terms "*symbiosis*" and "*symbionts*"; the Swiss botanist Simon Schwendener (1829-1910) established lichenology yet until the publication of the >700-page work by Geus and Höxtermann (2007), the importance of symbiogenesis in species origin was recognized only outside the mainstream. The solid scientific literature on symbiosis (an ecological phenomenon) and symbiogenesis (its evolutionary consequence that can be documented in the fossil record) has been brilliantly reviewed by Sapp (1994; 2003). Much of the primary science he reviews, of course, was published in British and North American journals. Yet the importance of symbiogenesis in evolutionary novelty especially in the appearance of new species is primarily written in languages other than English, e.g., French, German and Russian. Although symbiogenesis is documented as a major source of evolutionary novelty (Margulis and Fester 1991) (Margulis and Sagan 2002), it still is not taught as a coherent explanation of speciation in English-language institutions or in evolutionary biology texts.

Zoologists, and to a much lesser extent, botanists have dominated the funding and therefore the literature on evolution. Unfortunately both animal and plant scientists play with less than one-third of the biological deck of cards. Their neoDarwinian terminology reflects belief-systems that are permeated with anthropocentric, predominantly iatrocentric, language (e.g., daughter cells, female choice, inclusive fitness, reciprocal altruism, parental investment, sperm competition). Not only today's research physicians but authors who were trained as medical practitioners, in Samuel Butler's (1840-1914) terms are the "high-priests of our modern age" (Margulis and Sagan 1998). The definition of these purse-string holders, bolstered by the professional academics whom they support, has been that "organisms may be assigned to the same species if they can mate and produce fertile offspring". Expressed succinctly "species" according to such a definition, are groups of reproductively isolated individuals. But uniparental reproduction is the rule in protoctist species, in all but one small group of fungi (Laboulbenomycetes), and an estimated 50% of the species of extant plants. In fact biparental sexuality may be conditionally bypassed in the reproduction of certain species in more than 90% of the 37 animal phyla. The standard "biological species definition", blatantly zoocentric, is grossly inadequate to describe life forms in general.

Reproductive isolation was first induced in the laboratory by intense selection in population cages of both extremes: cold and hot tolerant flies (*Drosophila paulistorium*) over a period of about two years in the middle of the last century in classical studies by Theodosius Dobzhansky. Recent work, in the field and laboratory, by Wolfgang Miller at the University of Vienna has confirmed Dobzhansky. Although called "a mycoplasma" the microbe lost in the heat tolerant flies and retained in the cold tolerant ones, a strain of *Wolbachia*, asymmetrically distributed in the mating pair, induces sexual incompatibility. If one of the mates carries the particular *Wolbachia* and the other lacks it reproductive isolation ensues. Many fine examples of the influence of intracellular *Wolbachia* and other bacterial strains on gender determination, cytoplasmic incompatibility, reproductive isolation, nutritionally-related species definitions and other factors important to arthropod speciation are detailed by Bourtzis & Miller (2007).

Contemporary green, chlorophyll-bearing photosynthetic animals (e.g., the shell-less gastropod mollusk (*Elysia viridis* or *E. chlorotica*), and the platyhelminth worm *Convoluta roscoffensis*) inform our search for the history of oxgenic phototrophy in nucleated organisms including the chlorophyte ancestors of extant algae and plants. In these cases hungry translucent animals ingest but fail to digest photosyntheizers. Green photosynthetic animal species result.

Nitrogen-fixing fungi develop today from fusions that indicate their evolutionary history. One case of parasexual merger can be mentioned here between members of *different kingdoms*! In central European riparian habitats notably in the Hartz mountains near Neustadt, Germany such wonders live among hornworts and mosses. *Nostoc* cyanobacteria that act as "males" "fertilize" by endosymbiotic merger an *Endogone*-like vesicular-arbuscular fungus. The fungus acts as a "female" in that it receives penetration by the "male" cyanobacterial hormogonium. The fusion leads to the development of *Geosiphon pyriforme*, recognized as a distinctive new macroscopic photosynthetic, nitrogen-fixing taxon worthy of its own name. *Geosiphon pyriforme,* that resembles in its size and ecological niche a bryophyte, is vastly larger than its heterogenous microscopic "parents". It is not a plant in any way. The filament, a nostocalean blue-green gliding cyanobacterium, after it is taken in, is literally ingested but not at all digested by the absorptive thread of the heterotrophic fungus. The *Nostoc* filament resides inside the hypha as the fungus dramatically swells to become a monstrous "plant analogue". The once-microscopic thread of the white, fuzzy, translucent fungus metabolically "converses" with its bacterium, now an internalized nitrogen-provider with supernumerary heterocysts. The *Nostoc* supplies organic compounds to the plant-like holobiont (Mollenhauer and Mollenhauer 1997). Here morphological development emerges from behavioral, metabolic and genetic interactions between the dissimilar bionts that cyclically fuse. The fusion of nitrogen-fixing oxygenic phototrophic bacterium with rapidly growing desiccation resistant translucent phosphate accumulating fungus forms a new, differently named holobiontic species. That the new species derives directly by symbiogenesis is patently obvious to anyone willing to study *Geosiphon pyriforme* and a related burgeoning literature.

Examples of novel speciation underlain by symbiogenesis include: grass feeding cellulose-digesting mammals such as members of the inclusive taxon Bovidae (e.g., ruminating cows) or wood-feeding insects including cockroaches and termites (such as *Mastotermes darwiniensis*) that harbor cellulolytic communities in hypertrophied intestines called " the paunch"). *Gunnera manicata*, named by Ecuadorians and Brazilians "parasol de los pobres" ("umbrella of the poor people") because a single one of its meter-sized leaves can protect a person from torrential downpour. This equatorial jungle herbaceous flowering plant reaches the size of a shrub and evolved from much smaller ground cover members of the monogeneric family Gunneraceae. Conspicuous leaf-derived symbio-organs are produced by *G. manicata*. These symbio-organs permit migration of more than a meter upwards of the "male" filamentous cyanobacteria. The *Nostoc* cells glide against gravity to "fertilize" the pared glands on sprouted seeds that germinate and cling to the huge aerial inflorescences of the mother plant. These symbio-organs are long slimy red "bracts", also called stipules. They distinguish the tall *G. manicata* from its several ground-dwelling conspecifics including *Gunnera monoeica* (Benson and Margulis 2002).

Such examples, and myriad more, of symbiogenetic novelty make us virtually certain that Kozo-Polyansky's (1890-1957) analysis (1924) was and still is correct (Khakhina 1992). Symbiogenesis accounts for the origin of hereditary variation of importance to natural selection in eukaryotic life's major lineages. A far-flung multinational literature attests to the correlation of symbiogenesis with speciation not only at the cell level (Margulis 1993) but also at the level of both kingdoms of embryo formers: animals and plants (Margulis and Sagan 2002). A given species' biotic potential is infinite. The number of organisms produced per unit generation (convertible, of course, to number of organisms per unit time) that can be produced (born, hatched, from gemmae, germlings sprouted from spore or seed, propagated by root, shoot or scion, budded or released as chromidia), is always greater than the number that can be maintained by any environment in a specific place and time. The prodigious biotic potential, in other words, can not be reached for long in this "real world". The intrinsic limits for any population to reach its potential size in nature is the measurable phenomenon that Charles Darwin recognized with the phrases "natural selection" or "artificial selection". Behaviorally, metabolically and genetically integrated symbiotic complexes are maintained and perpetuated by Darwin's "natural selection" limitations. The failure of a live population of any species within its extant community in nature to reach its species-specific omnipresent biotic potential is another way of describing "*natural selection as a process of elimination*". The expansive habitual behaviors (of fellow community members including their predatory predilections and weedy growth patterns) exclude other given populations from the incessant growth and reproduction of which they are inherently capable. Thus, Darwin's "natural selection", in itself, is neither the source of heritable novelty nor is it the entire evolutionary process. When "natural selection" eliminates unassociated or poorly integrated symbiotic partners it hones and redirects the propagating partnership. Such selection pressures on co-evolving at first separate bionts lead to further partnership integration at behavioral, metabolic and ultimately genetic levels of organization in merged holobionts. Symbionts may become so entirely fused, so integrated that the path of their history is obscured. "Individuals" at larger levels of complexity and size result underlying the emergence of new taxa. Species thus originate and evolve as partners irreversibly and permanently fused and together leave descendant. Nature does make leaps. "Macromutations" are realities.

The process of speciation by integration did not end with evolution of protoctista and fungi. Rather in Animalia a variation on symbiogenesis ("called hybridogenesis") led to the origin of myriad larvae-producing holometabolous mollusks, echinoderms, crutaceans, insects and other more obscure inclusive taxa. These are animals that dramatically metamorphose from one to another larval or adult form in their predictable maturation process as described for echinoderms by Williamson and Vickers 2007.

The particular example of the origin by symbiogenesis of the earliest nucleated cells ("eukaryosis") equivalent to the evolution of the first protoctist or *LECA*, the *L*ast *E*ukaryotic *C*ommon *A*ncestor is outlined. On the vast scale of geological time, the eukaryotic, earliest protoctist ancestor and its animal, fungal and plant descendants still can be recognized and illustrated in videos by their live co-descendants. Few missing links occur in this robust symbiogenetic scenario that correlates with the Proterozoic eon (2500-542 mya) fossil record. Members of at least two prokaryotic domains (a sulfidogenic archaebacterium, and a sulfide-oxidizing motile eubacterium) merged in the origin of the first nucleated organisms (first anaerobic amitochondriate protoctist) to form the earliest eukaryote, by this reckoning. The seme, the defining trait of evolutionary importance, of eukaryosis is the membrane-bounded nucleus. We posit that this organelle began as a component of the karyomastigont organellar system. The first nucleus was attached to the kinetosome-centrioles by a proteinaceous "nuclear connector" called "the rhizoplast" in some protozoological literature (Margulis et al. 2006). Whether or not the karyomastigont of the nucleo-cytoskeletal system evolved symbiogenetically is discussed by Chapman et al. (in prep.). *A Thermoplasma*-like archaebacterium and *Perfilievia-Spirochaeta*-like eubacterium formed the earliest eukaryote in our model. The co-descendant *Thermoplasma* archaebacterium and *Perfilievia* spirochete, independent organisms, still thrive in organic-rich anoxic, microo-oxic and/or sulfidic habitats where they are amenable to study. We have videographed some of them (MacAllister and Margulis, 2004). No "missing links" interrupt our scenario. The symbiogenetic evolution of the nucleated ancestral cell from its free-living prokaryotic ancestors took so long and generated so many of Darwin's "oddities and imperfections" that we still can reconstruct the history of the cell from clues we take to be representational of past events.

Our testable scenario of eukaryosis outlines the Proterozoic eon history of the karyomastigont and later symbiogenetic organelles in relation to the origin of the highest (most inclusive) taxa. From the symbiogenetic origin of the four kingdoms of eukaryotic life from the bacterial world plotted vertically against geological time, modified from an illustration by Kathryn Delisle, one sees the implications for the most inclusive extant taxa. The animated, abbreviated sonorous version of Eukaryosis hypothesis was presented in our video at the Pontifical University of Rome (MacAllister and Margulis 2004).

Although I realize of course that symbiogenesis must have been complemented and refined by random DNA mutation and maintained by natural selection in these scenarios for the origin of eukaryotic species and higher taxa I suggest that my verbose, numerologist "evolutionary biologist" detractors, like the great C.R. Darwin himself in spite of the title of his great book, lack detailed scenarios for the origin of any species of eukaryotes. Neither in the field, the laboratory or in nature has one named species transmuted to another been tracked in detail through the "accumulation of random mutations". Indeed the so-called mechanism of species change: "accumulation of random mutations" in the absence of symbiogenesis is inadequate. The death-knell tolls for neoDarwinism in this year of the celebration of the great man's contribution. Although the last part of the title of his good book (*On the Origin of Species* "*by Means of Natural Selection, or the Preservation of Favoured Races in the Struggle for Life*") was aptly demonstrated in Darwin's work, without the symbiogenesis the promise promulgated by his title, until now, has been unfulfilled.

REFERENCES

Bourtzis K., Miller T. A. (eds.), 2007, *Insect Symbiosis*, vol. 2, Boca Raton: CRC Press/Taylor & Francis Group.

Becerra A., Delaye L., Islas S., Lazcano A., 2007, "The very early stages of biological evolution and the nature of the last common ancestor of the three major cell domains", *Annual Review of Ecology, Evolution and Systematics* 38: 359–77

Benson J., Margulis L., 2002, "The *Gunnera manicata-Nostoc* symbiosis: is the red stipulate tissue symbiogenetic?", *Biology and Environment: Proceedings of the Royal Irish Academy* 102B: 45–48.

Chapman M. J., Alliegro M. A., Dolan M. F., in prep., "The karyomastigont seme in the origin of the nucleus".

Corliss J., 1990, "Phylum Zoomastigina Class Pseudociliata", in Margulis L., Corliss J. O., Melkonian M., Chapman D. J. (eds.), *Handbook of Protoctista: The Structure, Cultivation, Habitats and Life Histories of the Eukaryotic Microorganisms and Their Descendants Exclusive of Animals, Plants and Fungi*, Boston: Jones and Bartlett Publishing Company, 211–14.

Dolan M. F., in press "Parabasalids", in Margulis L., Chapman M. J. (eds.), *Handbook of Protoctista: The Structure, Cultivation, Habitats and Life Histories of the Eukaryotic Microorganisms and Their Descendants Exclusive of Animals, Plants and Fungi*, 2nd edn., Boston: Jones and Bartlett Publishing Company.

Geus A., Höxtermann E., 1998, *Bacterienlicht & Wurzelpilz: Endosymbiosen in Forschung und Geschichte*, Rangsdorf: Basilisken-Presse.

Geus A., Höxtermann E., 2007, *Evolution durch Kooperation und Symbiosis*, Rangsdorf: Basilisken-Presse.

Khakhina L. N., 1992, *Symbiogenesis: A Historical-Critical Study of the Russian Botanists*, New Haven: Yale University Press.

Kozo-Polyansky B. M., 1924, *Symbiogenesis: A New Principle of Evolution* (in Russian), Fet V. (English tran. and anno.), Margulis L., Fet V. (eds.), in press, Cambridge: Harvard University Press.

MacAllister J., Margulis L., 2004, *Eukaryosis: Origin of eukaryotic cells*, color digital video, 14min., Amherst: University of Massachusetts.

Margulis L., 1993, *Symbiosis in Cell Evolution*, 2nd edn., New York: W. H. Freeman company.

Margulis L., Fester R. (eds.), 1991, *Symbiosis as a Source of Evolutionary Innovation: Speciation and Morphogenesis*, Cambridge, Mass.: MIT Press.

Margulis L., Sagan D., 1998, *What is Life?* Berkeley: University of California Press.

Margulis L., Sagan D., 2002, *Acquiring Genomes: A Theory of the Origins of Species*, New York: Basic Books.

Margulis L., Chapman M., Guerrero R., Hall J., 2006, "The *L*ast *E*ukaryotic Common Ancestor (LECA): Acquisition of cytoskeletal motility from aerotolerant spirochetes in the Proterozoic eon", *Proceedings of the National Academy of Sciences USA* 103: 13080–85.

Margulis L., Chapman M. J., 2009, *Kingdoms & Domains: An Illustrated Guide to the Phyla of Life on Earth*, 4th edn., San Diego, London: Elsevier.

Mollenhauer D., Mollenhauer R., 1997, *Endosymbiosis between Nostoc and Geosiphon pyriforme*. Scientific film, Göttingen: Institut für den Wissenschaftlichen Film, Film No. C1955.

Sapp J., 1994, *Evolution by Association: History of Symbiosis Research*, New York: Oxford University Press.

Sapp J., 2003, *Genesis: The Evolution of Biology*, New York: Oxford University Press.

Sonea S., Mathieu L. G., 2000, "*Prokaryotology: A Coherent View*", Montréal: Les Presses de L'Université de Montréal.

Wallin I. E., 1927, *Symbioticism and the Origin of Species*, New York: Williams and Wilkins.

Williamson D. I., Vickers S. E., 2007, "Origins of larvae", *American Scientist* 93: 509–17.

The mystery of speciation

Jeffrey L. Feder

Abstract

Charles Darwin once described speciation as that "mystery of mysteries". However, this is one mystery that is now arguably solved. In this essay, I will discuss advances in our understanding of how new species form since Darwin first posed the problem. I will outline current views regarding what constitutes a species, how to study population divergence, and the geographic context and genetic bases for speciation. I will also examine how and why barriers to gene flow evolve — the crux of the speciation problem — highlighting a few case studies demonstrating incipient speciation in action. I will conclude by discussing current research directions in the field. Although we may understand the general mechanisms generating new species, much remains to be learned. In particular, we are entering an exciting new period of synthesis in which the ecological, physiological, developmental, and genetic bases for population divergence can now be fully integrated for model and non-model organisms alike. From this body of work, case studies are accumulating to soon allow broad patterns to be identified concerning the relative importance of different mechanisms for the genesis of biodiversity. Although the mystery surrounding speciation may be gone, the thrill is not. Speciation is as important and as fascinating a question now, as ever, for understanding life.

What causes biodiversity and generates Darwin's (en)tangled bank of life? A principle cause is speciation, the process by which new species are formed. There are currently about 1.8 million species that have been described by scientists and reasonable estimates put the total number at perhaps 10 million, with an upper limit of 100 million (May and Harvey 2009). The world would be a rather dull place without speciation. It is the greatest show on Earth.

In this essay, I will explore our current understanding of speciation. Although the title of the essay is "The Mystery of Speciation", there is no mystery or doubt as to whether speciation occurs. Rather, I use the word mystery in the original sense that Darwin intended to convey his wonderment at the mystery of how new life forms were created by worldly processes. Aside from controversies and misunderstandings in the public and certain religious spheres, speciation is as accepted a scientific fact as there is. There is no scientific debate about the reality and materialistic basis for speciation. The study of speciation is as hypothesis driven and as testable a question as there is in science. There is little intrigue about how to go about sleuthing to unravel Darwin's "mystery of mysteries". Indeed, in broad outline, we shall see that the mystery is largely solved. But this does not mean that the study of speciation is now just anticlimactic filler. We are entering a period of extreme excitement in the field where new technical advances in genomic techniques can be applied to non-model genetic, but speciation model systems. The prospects for gaining an integrative understanding of how natural history and genetic changes in populations are transformed through development and physiology into new species have never been greater. Will general rules emerge or is each speciation event unique? The goal of this essay is to highlight how we got to this point and where we are going. In essence, what have we learned about speciation since Darwin and what would he be doing if he were a young student today. As long as life persists, speciation is the wonder that will never cease to amaze.

1. The tree of life

To understand the origins of species, we must first place speciation within the proper evolutionary framework. A good way to do this is Darwin's analogy of the tree of life. One of Darwin's most profound insights was that the discontinuities we currently observe among organisms in the ways they look and behave do not represent fixed typological differences, but rather are the products of a dynamic evolutionary process. Extant species are the growing twigs at the tips of the tree of life. Moving backward through time, we can descend downward from the twigs to the branches of the tree. As we do so, the seemingly unbridgeable gaps separating taxa disappear and species merge. As more species merge, they form the great systematic boughs of the tree, starting with genera and families and continuing to phyla, kingdoms, and eventually domains. When we reach the base of our tree, we have converged on the first life form. Thus, all organisms share a common ancestor as we trace their histories back into the past.

Darwin's next brilliant insight was to climb back up the tree. He had postulated that natural selection was a fundamental process causing populations to evolve through time. Ascending up the tree of life, he reasoned that an inevitable consequence of the changes occurring within populations would be the evolution of differences between populations. As time passed, varietal differences would appear within species that could become accentuated into more pronounced racial differences and eventually be manifested as species differences. Understanding the origins of species in Darwin's mind therefore equated with discerning how variation within populations became transformed into differences between species. An important take home message is therefore that if we want to understand speciation and the genesis of large-scale "macroevolutionary" patterns, we must go to those growing buds in the tree of life at the "microevolutionary" population level where new species are actively forming and branching off.

2. What exactly are species?

The next critical question about speciation is to ask what must be resolved at these branching points to discern the origins of new species. Most primers on speciation fall flat out of the tree at this point by trying to define exactly what a species is. After all, if we cannot precisely define what the essence of a species is then how can we ever know when and how speciation is occurring. Unfortunately, perhaps never in the field of human conflict, scientific or otherwise, has so much confusion been owed by so many to so few. This need not be the case and we must be careful in not sidetracking ourselves from productive science into philosophical sideshow. The key point is that there is no invariant or fixed "essence" of a species; this is exactly what the Darwin revolution overturned. Imposing a rigid species criterion is divorced from the reality of evolution. As we saw from Darwin's tree of life, all life forms are interconnected by past common ancestors and are, to varying degrees, in the process of actively diverging from each other. Nevertheless, we can still identify and study the important population genetic process underlying the origins of species: the evolution of inherent barriers to gene flow separating taxa.

Although it has been written to the contrary, Darwin was clear in his thinking about species. His pragmatic answer to the question was that species represent groups of individuals that maintain observable discontinuities when and where they co-occur in sympatry. Following the Modern Synthesis in the first half of the twentieth century in which Mendelian genetics was wedded with Darwinian evolutionary principles, it has become increasingly accepted in practice, if not theory, to think of species differences as being maintained by inherent barriers to gene flow (a.k.a., reproductive isolation; Dobzhansky 1935; 1937; Mayr 1942).

The root cause for much of the confusion concerning species stems from the need for different groups of biologists to develop their own species concept based on their own particular requirements and uses for the word (Hey 2001). Thus, conservationists interested in preserving biodiversity describe species based on any observable population difference, whether or not the difference(s) have any bearing on gene flow (i.e., whether or not they would or could be maintained in sympatry). Systematists want to define species as distinct taxonomic entities (units) based on fixed and non-arbitrary differences to assist and standardize classification. But nature always has a way of defying human attempts to straight jacket preconceived pattern on process. Because evolution (speciation) is a continuous process, there is usually no specific point at which we can say that a new species has come into being. As a result, every definition of a species, whether it is based on the presence of diagnostic differences or reproductive isolation is, in practice, subjective. Exactly how many or how large the differences or barriers to gene flow must be for populations to be considered species, as opposed to races or varieties, is arbitrary. We cannot just snap off a growing twig on the tree of life and think that we have somehow accurately defined a species. For this reason, Darwin considered species to be non-natural, human constructs, no different from any other taxonomic category (e.g., why should a certain bough on the tree of life be consider a family vs. a genus?). This does not mean that species are not real. Mice and elephants are clearly different species that maintain their differences in sympatry and are separated by substantial (insurmountable) gene flow barriers. It is just that at the active branching points of speciation that we are most interested in studying, there is no objective breakpoint for decision making. One man's species is another man's race.

3. Once more unto the speciation breach

The uncertainties in defining an event horizon separating races and species does not deter students of speciation from understanding how new species are formed. There may not usually be a single point in time when a new species is born but we can still recognize the stuff that sexual species are made of: inherent barriers to gene flow. This realization is critical because it establishes a hypothesis driven research program for studying speciation; by understanding how genetically based traits evolve that restrict exchange between populations we can understand the origins of new species. The wall need not close up with species concepts dead.

Much of the remainder of the essay will therefore be organized around the theme of gene flow barriers. I will first classify the different types of gene flow barriers that can evolve between populations. I will then discuss the various geographic contexts (modes) under which these barriers can evolve and speciation can occur. A major question about speciation modes has been whether divergent natural selection is ever strong enough to allow new species to form in the absence of complete geographic isolation and in the face of gene flow. A related and broader issue stemming from this debate concerns the importance of adaptation to speciation: is speciation primarily the by-product of natural selection or do stochastic processes (i.e., genetic drift) often play an important role, as well? An increasing number of biologist, including myself, are ascribing to a Darwinian view that selection underlies much of speciation. To us, the question then becomes whether this selection is based primarily on the external ecological environment, on antagonistic sexual interactions, or on internal genomic conflicts (e.g., meiotic drive and the suppression of selfish genes). As we explore the issue of gene flow barriers, I will highlight a few relevant case studies illustrating key points. As this essay is designed to be more of an introduction to the field (Speciation 101) rather than an exhaustive survey, I will not cover several important topics such as the role of hybridization and chromosomal rearrangements in the genesis of new species (see Arnold 1992, 1997; Rieseberg 1997, 2001; Rieseberg *et al.* 2003; Noor *et al.* 2001; Seehausen 2004; Schwartz 2005; Kirkpatrick and Barton 2006). I will also not delve deeply into the significance of mathematical models for understanding speciation. The synergism of theory and empirical work is one of the strengths of speciation research, providing a foundation for making testable predictions and interpreting empirical results (see Turelli *et al.* 2001; Kirkpatrick and Ravigne 2002; Gavrilets 2004). A rich literature on laboratory based speciation experiments also exists that I will only briefly touch upon in the essay (see Rice and Hostert, 1993; Coyne and Orr 2004). I will conclude by discussing the long-view of where speciation research is headed. I believe, and I feel many of my colleagues agree, that speciation research has matured to the stage where we now recognize and accept that organisms can travel many different paths on the road to speciation. Whatever possible ways gene flow barriers can arise they likely will. We are now at a stage in which we can more critically evaluate the relative importance of different routes for different groups of organisms. Concomitant with these advances in the comparative "meta-analysis" of speciation, mass sequencing and genotyping technologies are making the genomes of an increasing number of organisms more assessable to discerning the genetic, developmental, and physiological underpinnings of adaptation and its relationship to population divergence.

4. Classifying barriers to gene flow

Theodosious Dobzhansky, building on the insights from E. B. Poulton (1904) and with a nudge from Alfred Sturtevant (see Lewontin *et al.* 1981), was the first to crystallize thought on how to study speciation in the wild. In his 1937 book *Genetics and the Origin of Species*, Dobzhansky emphasized that the path to speciation enlightenment lay in understanding the evolution of genetically based barriers to gene flow between populations. Note the use of the word genetically-based. Physical barriers such as mountain ranges, rivers, or oceans may separate populations and impede gene flow. Although such spatial barriers may often be an important initial step for allowing populations to independently evolve into different species, they are not the inherent reproductive barriers that Dobzhansky talked about. For if the geographic barriers were to disappear and the populations were to come back into contact, there would be nothing to stop them from freely interbreeding and fusing. Physical barriers are therefore not the stuff of speciation. It is the genetically-based barriers that evolve that keep populations separate when in contact that are the key.

Dobzhansky's categorization of inherent gene flow barriers is summarized with some modification in *Table 1*. A major division can be seen between barriers that restrict male and female gametes from coming together and fusing to form a fertilized zygote (termed prezygotic) vs. barriers that manifest themselves after the zygote forms (termed postzygotic). Prezygotic isolation refers to behavioral, ecological, mechanical, and mating system differences that impede courtship, copulation, and gamete fusion between individuals from different populations (see *Table 1*). In comparison, postzygotic barriers are manifested after fertilization and include phenomena such as hybrid offspring inviability and sterility. A useful subdivision of gene flow barriers not emphasized by Dobzhansky is whether the barrier is extrinsic or intrinsic to the organism. Extrinsic barriers occur when the fitness of a migrant or hybrid is reduced due to negative interactions of its phenotype (as determined by the genes it possesses) with external environmental conditions, like how hot or wet it is or how many competitors or predators are present. In contrast, intrinsic barriers reflect reduced hybrid fitness due to internal genomic incompatibilities that cause developmental irregularities independent of the environment (Coyne and Orr 2004).

5. Measuring gene flow barriers in nature

Now that we appreciate the different ways and types of traits that can generate gene flow barriers between populations, we can go into the field and study the nature of barriers separating species. As an example, consider two species of monkey flowers, *Mimulus lewisii* and *M. cardinalis* from western North America (Ramsey *et al.* 2003). These two plants are each other's closest relatives (i.e., they are "sister species") and they tend to grow at different elevations in the Sierra Nevada Mountains. The two species are therefore separated by both physical distance and ecological differences related to differential adaptation to climate and soil, the latter representing extrinsic barriers to gene flow. However, the two plants do co-occur in certain areas. Here, they are further isolated by having different and nearly non-overlapping sets of pollinators (prezygotic isolation), strong con-specific pollen preference, and partial inviability and sterility of hybrids when they form (the latter representing intrinsic, postzygotic isolation).

The example of the monkey flower illustrates that as populations diverge many different gene flow barriers can evolve to separate them. Are certain types of barriers more important and tend to arise earlier than others for initiating speciation? Are certain forms of reproductive isolation more effective than others in restricting gene flow? These can be difficult questions to answer, as differences arising after two taxa are completely reproductively isolated might not be germane to speciation but could nevertheless substantially restrict gene flow at the present time. The resolution of these questions lies at the heart of understanding speciation.

In the grand scheme of things, prezygotic barriers are generally more effective at reducing genetic exchange than postzygotic barriers because they act earlier in the reproductive life history of organisms (Coyne and Orr 2004). Without mating, hybrid inviability and sterility are of little significance. However, this does not mean that ecological, behavioral, pollinator, or mechanical forms of premating isolation are always the first or most prominent barriers separating species.

One avenue to investigate the question of chronological order is to study populations that are in the incipient stages of divergence and are only partially reproductively isolated and test whether any particular pre- or postzygotic barrier consistently evolves first. My own particular bias is that behavioral differences tied to ecological adaptation will often prove to be "the pacemakers of evolution" (Mayr 1963), but the jury is still out. However, just because populations differ in certain characteristics partially restricting gene flow does not mean that they will eventually evolve into distinct species. Thus, early changes can potentially be misleading and not always represent the critical precursors leading to speciation.

6. Comparative approaches to speciation

The difficulty in drawing definitive conclusions from single case studies highlights the importance of comparative methods for analyzing speciation. In comparative approaches, data from a large number of case studies are compiled to examine whether consistent patterns emerge for taxa at various different stages of divergence. The strategy of comparing different stages of divergence to draw inferences about speciation can, like many things in the field, be traced back to Charles Darwin. Darwin, when faced with the problem that speciation can be a slow and gradual process and that he might never see a new species form in his lifetime, was drawn to his knowledge of geology and the birth and death process of coral reefs. On his voyage on the Beagle in the South Pacific, Darwin noted the various stages in the life of coral reefs, beginning as fringing reefs close to the shores of newly created volcanic islands, moving to mid-life as barrier reefs defining the outreaches of lagoons, and ending as atolls ringing submerged islands. Darwin reasoned that although he could not observe the complete ontogeny of a reef from start to finish in his lifetime, he could nevertheless logically order the stages of their development through time in an interconnected sequence. Darwin extended this logic to species; although he may not see a specific speciation event from start to finish, he could observe all the relevant stages of divergence in various phases of completion and connect the dots in chronological order.

Darwin's reasoning concerning stages of divergence forms the intellectual cornerstone for comparative studies of speciation and can be used to assess the relative contributions of different types of gene flow barriers to the process, as well as for inferring the likelihood for different geographic modes of speciation (see below). A particular influential study in this vein was conducted by Jerry Coyne and Allen Orr (1989a, 1997). These two scientists compiled data on the degree of reproductive isolation (behavioral premating and postzygotic hybrid sterility and inviability) displayed between 171 pairs of *Drosophila* fruit flies. They also estimated the extent of genetic divergence between these flies as a metric for time of separation. The authors then plotted the degree of reproductive isolation between fly taxa against genetic divergence (time) to determine whether different forms of isolation tended to evolve more quickly than others.

Several general patterns emerged from Coyne and Orr's study which have since been observed, in general, in subsequent comparisons involving butterflies and moths (Presgraves 2002), frogs (Sasa 1998), birds (Price and Bouvier 2002; Lijtmaer *et al.* 2003), fish (Russell 2003), and plants (Moyle *et al.* 2004). First, intrinsic hybrid sterility and inviability evolve gradually and rise slowly through time. Recently diverged taxa rarely show complete postzygotic isolation. Second, intrinsic hybrid sterility generally appears to evolve faster than inviability. Third, J.B.S. Haldane's rule (1922) generally holds: when intrinsic hybrid sterility or inviability is found to affect only one sex, it is usually the heterogametic sex that is affected (For flies this is males, while for butterflies and birds, it is females). Fourth, postzygotic isolation evolves at different rates in different groups. For example, hybrid inviability appears to evolve fastest in frogs, more slowly in butterflies and moths, even slower in birds, and slowest in plants. Fifth, in *Drosophila*, when Coyne and Orr separated fly taxa into pairs showing no geographic overlap (allopatric) and pairs displaying some range overlap (sympatric), they found that sympatric taxa developed behavioral isolation significantly earlier than allopatric taxa. These results imply the evolution of reinforcement in sympatric, a subject that I will elaborate on below. The details of these comparative studies may seem overwhelming and cause eyes to glaze over. But this is exactly the point. From such compilations of comparative data it is possible to uncover many large scale patterns concerning the evolution of gene flow barriers and speciation and see how general they are across different groups of organisms.

7. How do gene flow barriers evolve?

The question of how gene flow barriers evolve may now seem rather straightforward. Evolution occurs and as a consequence different traits arise in populations with some of these traits generating gene flow barriers between taxa. For traits involved in ecological adaptation it could well be this simple. For a population residing in a particular habitat, a new mutation arises that confers a fitness advantage to its bearers. The favored gene (mutant allele) increases in frequency with time, eventually fixing with every individual in the population coming to possess it. In a different population in a different habitat natural selection acts to fix a different adaptive mutation, either at the same or a different gene. If individuals from the two populations happened to mate, they would produce offspring that will generally fair poorly in both parental habitats because they do not have the full complement of genes favored in the alternate environments. The result is an extrinsic ecological barrier to gene flow, especially if the trait conferring higher fitness in one habitat directly results in lower fitness in the other habitat (i.e., a fitness "tradeoff").

The same logic does not apply to intrinsic postzygotic isolation, however, for what good is a new mutation causing inviability or sterility? How could such an unfavorable mutation when it arises spread and become differentially fixed between populations? This dilemma caused both Charles Darwin and Alfred Russel Wallace to shy away from defining species strictly on the basis of reproductive isolation (intrinsic incompatibilities). How could such reproductive isolation be the product of natural selection they reasoned? One answer to Darwin and Wallace's dilemma is what is termed the Bateson-Dobzhansky-Muller model for hybrid incompatibilities. Bateson (1909), Dobzhansky (1934), and Muller (1939; 1940; 1942) proposed that the solution to the problem lay in understanding that at no time in their evolution did the specific genes causing hybrid incompatibilities reside together in the same individual/population (see Orr 1996 for an overview). In one isolate, a new mutation(s) fixes in the genetic background present in the population. In a second isolate, a different mutation at a separate gene arises and fixes in the genomic background of the second population. When hybrids are formed, these two new mutations are brought together for the first time. One potential outcome is that the two mutations negatively interact with each other and cause inappropriate gene expression disrupting development, derailing gametogenesis, or altering proper functioning of biochemical pathways. The end result is intrinsic hybrid inviability or sterility. (Note that this mechanism can also work if the two derived mutation occur in the same population. In this case the second mutation to arise interacts negatively in hybrids with the ancestral allele present at the other locus.) An important message to highlight here is how intrinsic postzygotic isolation is not directly selected for but is the inadvertent by-product of independent evolution in different demes.

The difference between extrinsic (ecological) vs. intrinsic (negative epistatic) forms of reproductive isolation has historically tended to separate speciation researchers into different schools of study. One school is more field oriented and is largely focused on measuring the strength and establishing the chronology of gene flow barriers in nature. Since this task is best accomplished studying taxa that geographically overlap so that genetic exchange can be directly assessed, this ecological school of speciation tends to concentrate on divergence-with-gene-flow systems. In contrast, the second school is more interested in mapping speciation genes that cause gene flow barriers. For practical reasons, this usually necessitates using model laboratory genetic systems and focusing on intrinsic postzygotic reproductive isolation, especially with respect to *Drosophila* fruit flies. It could be argued that this division to some degree reflects the success and failure of Dobzhansky's ambitious Genetics of Natural Populations series (see Lewontin *et al.* 1981). Dobzhansky initially set out to define how the demographics, structure, and ecology of populations influence their potential for genetic drift, adaptation, and speciation. But an inability to resolve how *Drosophila pseudoobscura* make a living in nature turned the speciation aspects of his work into mapping the genetics of postzygotic isolation for *Drosophila*, a phenotype that he could characterize.

In the discussions to follow of the geographic modes of speciation, the ecology of population divergence, and the mapping of speciation genes, the reader will observe how this dichotomy and tension between field and laboratory approaches has played out and still exists in speciation research today. In my opinion, one of the most significant long-term effects of the application of genomic techniques to the study of speciation will come from fostering a better integration of complementary fields and gene mapping approaches to understanding the origins of species.

8. Geographic modes of speciation

Modes of speciation have traditionally been defined by the geographic circumstances under which the initial barriers to gene flow evolve between populations (Poulton 1904). Three major categories are recognized: allopatric (other fatherland), sympatric (same fatherland), and parapatric (alongside fatherland; *Figure 1*). Other forms of allopatric speciation such as peripatry (around fatherland) have been postulated in which founder effects, drift, and natural selection interact in peripheral populations to trigger speciation (*Figure 1*). Although feasible (see Templeton 1980), many feel that peripatric modes of speciation may be the least common mode of divergence for organisms (Barton and Charlesworth 1984; Barton 1989; Gavrilets 2004), and I will not discuss founder effect models further.

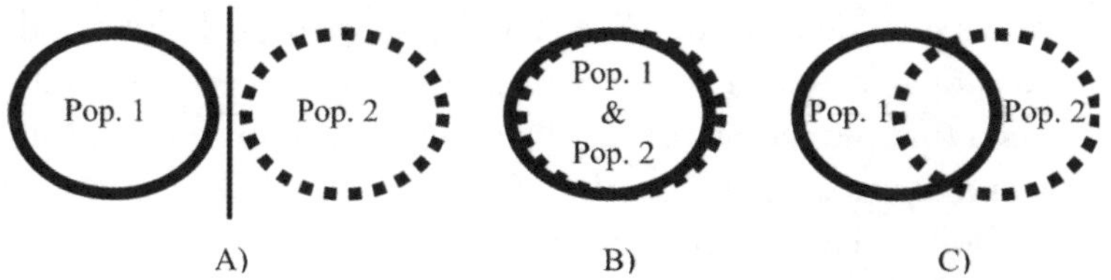

Figure 1. The major modes of geographic speciation. A) Allopatric, B) Sympatric, C) Parapatric. See text for detailed discussion of these modes.

The first stage of allopatric speciation involves the geographic isolation of populations. The physical separation of populations restricts migration and allows for their independent evolution. As populations diverge, some of the genetic differences they accumulate can inadvertently cause reproductive isolation if and when the populations were ever to come back into "secondary" contact. Note that the changes that occur in allopatry could be due to the action of natural or sexual selection or even due to stochastic processes (genetic drift). In addition, the changes may represent responses to differences in environmental conditions in the two areas or represent different evolved responses to similar environmental conditions. The restricted gene flow between populations allows for any and all of these evolutionary routes to diversification; all types and combinations of pre- and postzygotic gene flow barriers can potentially evolve in any particular order and lead to speciation.

If allopatric populations happen to reestablish contact, then it is possible for them to diverge further through a process Dobzhansky termed reinforcement. Given that postzygotic isolation has evolved between populations, individuals from different populations that tended to refrain from mating with each other could be favored because they would avoid wasting their reproductive effort producing hybrid offspring of reduced fitness. The idea of reinforcement has until relatively recently, been met with skepticism. But several examples, including the case of *Drosophila pseudoobscura* and *Drosophila persimilis* (Noor 1995) — the very organisms that Dobzhansky studied to pioneer the genetic analysis of natural populations — have provided evidence for the process (see reviews by Howard 1993 and Noor 1999).

The process of reinforcement is important for two reasons. First, it is an example showing how in certain circumstances, prezygotic isolation can be directly selected for. Second, reinforcement demonstrates that the geographic context of speciation can be mixed. During speciation, diverging populations can, through time, display varying degrees of geographic isolation and overlap with one another. Important gene flow barriers can evolve during all these times. Since speciation is usually not instantaneous, it is erroneous to think that a given speciation event necessarily occurs exclusively within only one geographic context. Mixed speciation modes are also possible.

In contrast to allopatry, sympatric speciation is initiated when populations physically overlap such that individuals of the diverging taxa are in the "cruising range" of one another and can potentially interbreed (Mayr 1963). The central premise of sympatric speciation is that selection on the biology of the organisms themselves, rather than geographic isolation, generates the initial gene flow barrier triggering speciation. This selection can be manifested in the form of differential adaptation to alternate ecological habitats present in an area, as intraspecific competition causing individuals to specialize on different aspects of a limiting resource, or simply as variation in the preferences of individuals to mate with others sharing similar features. The latter works best when the trait(s) experiencing disruptive sexual selection are also associated with differential adaptation to the environment, termed magic traits by Gavrilets (2004). For example, big and small fish often tend to assortatively mate by size. When big and small fish are also adept at utilizing different resources found in different zones of a lake, this could provide the impetus for speciation.

Sympatric speciation has been a controversial topic over the years because it implies that divergence can be initiated in the face of gene flow between populations. Migration is generally thought to be a very effective force in homogenizing gene frequencies between populations and preventing divergence (Mayr 1963). The key to sympatric speciation is that disruptive selection and assortative mating can sometimes be strong enough to counter the homogenizing effects of migration and favor further differentiation. In this sense, sympatric speciation most closely encapsulates

Darwin's original thinking about the prominent role natural selection and ecological adaptation play in speciation. Note that once a degree of differentiation is established and hybrids between individuals from different populations have reduced fitness, the completion of sympatric speciation and reinforcement following secondary contact can be similar processes. Selection can act to directly favor enhanced prezygotic isolation to avoid wasting reproductive effort which, by reducing gene flow, feeds back to allow for further differential adaptation to the environment and increased postzygotic isolation. The question concerning sympatric speciation is therefore whether selection can be strong enough to kick start the process and establish populations sufficiently diverged in the absence of geographic isolation to favor their continued differentiation into species. If the requisite prestanding genetic variation was present or fortuitously happened to arise by mutation, then sympatric speciation could potentially occur quite rapidly. But if this was not the case or if the habitats were not very different and/or were changing slowly, then sympatric speciation could take an extended period of time to complete.

The sympatric debate reached a peak in the late 1980's when evidence from host plant specialist insects such as *Rhagoletis pomonella* (Feder *et al.* 1988; McPheron *et al.* 1988), *Eurosta soligenesis* (Craig *et al.* 1993; Abrahamson and Weis 1997) and *Enchenopa binotata* (Wood 1993) began to change conventional wisdom about the possibility of sympatric speciation in animals. Other evidence soon followed from examples such as cichlid fish in volcanic crater lakes in Cameroon, Africa (Schliewen *et al.* 1994) and in Nicaragua, Central America (Barluenga *et al.* 2006), and for palm trees on Lord Howe Island in the South Pacific (Savolainen *et al.* 2006) that strengthened that case for sympatric divergence.

The apple maggot fly, *Rhagoletis pomonella*, is a model for sympatric speciation in action and serves to illuminate several key points concerning divergence-with-gene-flow speciation (Bush 1966, 1969). The recent shift of the fly from its native host plant hawthorn to introduced, domesticated apple ~150 years ago in the United States provides a real time example of how adaptation to alternate habitats can rapidly cause incipient speciation for phytophagous specialist insects. Apples were introduced to North America by European settles ~400 years ago providing an historical timeframe for the *R. pomonella* host shift. Two key traits are responsible for ecologically isolating the hawthorn and apple "host races" of *R. pomonella* flies, the first incipient forms on the road to sympatric speciation. The first trait is host specific mating. *R. pomonella* flies mate exclusively on or near the fruit of their respective host plants (Prokopy *et al.* 1971; 1972; Feder *et al.* 1994). Host choice therefore translates directly into mate choice for these flies and prezygotic reproductive isolation. Volatile chemical compounds emitted from the surface of ripening fruit are key olfactory cues that apple and hawthorn flies use to find and discriminate between their respective host plants (Linn *et al.* 2003). Hawthorn flies prefer the odor of hawthorn fruit and tend to avoid apple fruit volatiles, while apple flies show the opposite behaviors (Linn *et al.* 2003; Forbes *et al.* 2005; Forbes and Feder 2006). Most interestingly, hybrids between apple and hawthorn flies fail to respond to either apple or hawthorn fruit volatiles in flight tunnel tests (Linn *et al.* 2004; Dambroski *et al.* 2005). This result shows the potential for host specific mating to generate both pre- and postzygotic ecological isolation — the impaired chemosensory abilities of hybrid flies may make it difficult for them to find host plants in nature for mating and oviposition, resulting in them suffering a form of intrinsic behavioral sterility (Feder and Forbes 2007, 2008).

Diapause life history differences are the second major trait ecologically isolating apple and hawthorn flies (Feder *et al.* 1993, 1994). Apples fruit an average of ~3 weeks earlier in the season than hawthorns. *R. pomonella* flies are univoltine (they have 1 generation per year) and live for only a short time as adults in the field (~1 month maximum). Apple and hawthorn flies must therefore differentially time when they break their overwintering diapauses to emerge as adults at times matching when their respective host plants have maximal amounts of ripe fruit (Feder *et al.* 1993; Filchak *et al.* 2000). This results in a degree of temporal offset (allochronic isolation) between the host races. Hybrids between apple and hawthorn flies also have life history profiles that reduce their ability to utilize the parental host plants, generating an additional postzygotic ecological barrier to gene flow (Filchak *et al.* 2000).

Apple and hawthorn flies are part of a larger group of related sibling species including the blueberry fly, *R. mendax*, and the snowberry fly, *R. zephyria* (Bush 1966; Berlocher 2000). The outstanding feature of the larger *R. pomonella* group of flies is the contrast between their close morphological similarity and overlapping geographic ranges vs. distinct host requirements (each species attacks a different non-overlapping set of host plants). These considerations led Guy Bush (1966; 1969) to propose that the entire *R. pomonella* group radiated through sympatric host plant shifts. The *R. pomonella* sibling species display the same type of host-related ecological adaptation in habitat choice and diapause as the apple and hawthorn host races, only writ large.

The *R. pomonella* group provides another nice example of how to apply Darwin's stages of speciation concept to studying divergence; we can logically extrapolate that the partial ecological isolation observed for the host races can become further enhanced through time to generate distinct species. A similar line of reasoning can also be used to distinguish between sympatric and allopatric speciation, in general, as the two modes of divergence make different predictions about the how population ranges should change through time. For allopatrically diverging groups, taxa in the earliest stages of speciation should be the most geographically separated. As speciation advances, the potential for spatial overlap increases as range expansion occurs and diverged taxa have greater capacity for ecologically co-existing together without fusing or being competitively excluded. In contrast, the reverse is true for sympatric speciation, with younger taxa showing the greatest degree of spatial overlap. Applications of such comparative tests have to date yielded interesting but mixed results (Lynch 1989; Barraclough and Vogler 2000; Berlocher and Feder 2002; Fitzpatrick and Turelli 2006).

The *R. pomonella* group is a case study of how disruptive ecological selection can generate reproductive isolation and initiate population divergence in real time in our backyards. As host specific insects (phytophagous and parasitic) are the most diverse group of organisms on the planet (close to 1 million described species; Gullan *et al.* 2004) there is a world of opportunity for host shifting to generate new life forms. Indeed, it has recently been found that a parasitoid wasp that attacks *R. pomonella* has also recently formed a new incipient species on the apple fly race, showing how ecological effects can cascade through the ecosystem and sequentially induce divergence across trophic levels (Forbes *et al.* 2009). Biodiversity can begat more biodiversity.

There are, however, theoretical reasons to suspect that sympatric speciation is, for many groups of organisms, less likely than divergence involving at least some degree of geographic isolation. For *R. pomonella*, the initial gene flow barriers are established by a combination of host choice and diapause life history adaptation. Host volatiles and fruiting time therefore set apple and hawthorn trees apart as different chemical and seasonal resources islands for flies, in much the same manner as geography separates allopatric populations on different spatial islands. Biology therefore serves the same role for *Rhagoletis* flies that physical separation does in allopatric speciation. *Rhagoletis* flies mate on preferred hosts while allopatric organisms are confined by geographic barriers to mate with other individuals from the same population in a particular area. But for organisms not possessing the biology or genetic variation for habitat specific mating that *Rhagoletis* has, the assortative mating built into allopatry (Kirkpatrick and Ravigne 2002) makes geographic speciation a more tractable mode.

Parapatric speciation contains elements of both allopatric and sympatric speciation and occurs when taxa overlap in part, but not all, of their respective ranges. Again, as in sympatry, parapatric speciation occurs in the face of gene flow. But in this case, there is a degree of geographic isolation adding to the ecological differences that exist and assisting divergence. Because clines in the frequency of phenotypes and genes can be created by differential selection across the transition zones between taxa, parapatric divergence is often also referred to as clinal speciation (Fisher 1930; Endler 1977; Slatkin 1982). Although parapatric speciation may in principle be fairly common, it is the most difficult mode to verify (Coyne and Orr 2004). The recent demonstration of color morph differentiation with water depth (ecological zones) for cichlids in Lake Victoria, Africa (Seehausen *et al.* 2008) associated with a cline in opsin allele frequencies may represent an example of incipient parapatric speciation in action due to sensory drive. In more cases than not, however, we have little or no direct historical information about past species ranges. As a result, current parapatric distributions for many taxa may represent secondary contact and differential introgression following allopatry rather than the de-novo generation of a primary cline. On the flip side, however, geographic barriers completely separating taxa are unlikely to form overnight. Many species assumed to have undergone allopatric speciation may have actually evolved a significant number, if not all, of their key gene flow barriers during a period of parapatry preceding allopatry. Such a mixed para-allopatric mode of divergence could conceivably be the rule rather than the exception.

9. The importance of natural selection and ecology to speciation

The debate about sympatric (divergence-with-gene-flow) speciation helped spark broader consideration of the general importance natural selection and ecology play in speciation, irrespective of geography. Although ecological adaptation was always thought to be involved to some degree in speciation, Ernst Mayr, the champion of allopatric speciation, generally considered it to be of secondary importance to the process. Mayr (1947, 1963) viewed ecological divergence to be most important in allowing allopatrically formed species to coexist in a region if and when range expansion occurred. In Mayr's view, ecological adaptation was more often like speciation finishing school than primary origins school.

A reawaking in thinking about the central role of ecology in speciation began in the 1980's with the accumulating evidence for sympatric speciation (for reviews see Via 2001; Berlocher and Feder 2002; Dres and Mallet 2002), seminal work in model systems of adaptive radiation including Darwin's finches (Grant 1999; Grant and Grant 2008) and three-spine stickleback fish (Schluter 1994; 1995), and sophisticated laboratory selection experiments involving varying environmental conditions (Rice and Salt 1988; 1990; Rice and Hostert 1993). Suddenly, the mantra again became ecology, ecology, ecology.

Through the ensuing years, substantial additional evidence has accumulated pointing to an important role of disruptive selection in the early formative stages of speciation for many organisms (Schluter 2000; 2009; Berlocher and Feder 2002;

Coyne and Orr 2004; Rundle and Nosil 2005). We now have many studies from a diverse array of organisms in which phenotypic traits under natural selection have been identified and the effects that these traits have on ecologically isolating taxa measured, including plants (e.g., *Artemisia, Chamaecrista, Costus, Gelsemium, Gilia, Helianthus, Ipomopsis, Iris, Penstemon, Pedicularis* and *Phlox* [summarized in Lowry *et al.* 2008]), and animals (amphibians [*Bombina*]), fish (sticklebacks and cichlids), spiders (*Agelenopsis*), and insects (*Rhagoletis, Timema, Acyrthosiphon, Eurosta, Galerucella*, and *Heliconius* [summarized in Nosil *et al.* 2005]). Several examples implying that new ecological forms have evolved multiple times in parallel in groups (e.g., stickleback fish [Rundle, *et al.* 2000; Taylor and McPhail 2000; Colosimo, *et al.* 2005], *Timema* walking sticks [Nosil *et al.* 2002, 2008], *Littorina* marine snail [Butlin *et al.* 2008], *Anolis* lizards [Losos *et al.* 1998], and mosquito fish [Langerhans, *et al.* 2007]) also attest to the significance of ecology for population divergence. Other convincing evidence for ecological isolation has come from transplant experiments in which parental taxa and hybrids are reciprocally moved between habitats to test for environmentally related reduction in fitness (reviewed in Nosil *et al.* 2005 and Lowry *et al.* 2008). In a clever enumeration of this approach, Rundle and Whitlock (2001) noted that extrinsic ecological isolation could be potentially distinguished form intrinsic isolation by constructing F1 and F2 backcross hybrids and examining the resulting pattern of hybrid fitness. If F1 fitness was intermediate to parental types across all habitats and F2 offspring resulting from backcrosses to the taxa from the natal habitat have higher fitness than the alternate cross, this would indicate extrinsic ecological isolation. If F1 and F2 hybrids had uniform low fitness in all crosses, this would signify intrinsic isolation.

Behavioral studies also support a role for ecology in speciation. In particular, a number of studies have documented assortative mating and habitat choice for a variety of organisms (reviewed in Schluter 2001; Via 2001; Berlocher and Feder 2002; Lowry *et al.* 2008). In this regard, it is possible that many cases of sexual selection may also ultimately be tied to (channeled by) ecological differences (see Kirkpatrick 1982; Endler 1983; Luyten and Liley 1991; Endler and Théry 1996; Endler and Basolo 1998). Researchers have also proposed modifying the comparative approach of Coyne and Orr (1989a) to test for a role of ecology in speciation (Funk *et al.* 2002). In a demonstration of the power of this strategy, Funk *et al.* (2006) found significant correlations between the degree of ecological differentiation and the strength of reproductive isolation among plants, invertebrate, and vertebrate taxa.

The overarching message to be gleaned from the recent work is that ecology may often be king when it comes to initiating speciation, whether it be due to differential adaptation to alternate or similar environments (the former in sympatry, parapatry or allopatry and the latter in allopatry, where independent evolution in geographic isolates can produce different solutions to the same environmental conditions that could lead to intrinsic genomic incompatibilities in hybrids). In this respect, I differ from the suggestion of Schluter (2009) that ecological speciation only includes cases involving disruptive selection. I consider all instances in which adaptation to the external environment initiates divergence to represent ecological speciation.

10. Speciation genes

A penultimate goal of speciation research is the mapping and identification of "speciation genes". By "speciation genes" I am referring to genes that are responsible for the physiological/phenotypic barriers to gene flow separating populations. The ultimate goal is to discern the action of these genes and integrate this information with the cause for their differentiation and the natural history of the study system (i.e., merging ecological and speciation gene approaches).

There are two general means to identify speciation genes. One strategy involves genetic surveys of populations and the second conventional mapping techniques based on test crosses. In the first approach, a genome scan is performed in which a large number of representative marker loci assumed to be neutral are screen for differentiation between geographically overlapping populations likely to be experiencing at least some gene flow. The rationale is that gene flow will homogenize regions of the genome between the taxa not associated with reproductive isolation. But genes coding for or linked to loci contributing to reproductive isolation should display differences. Once candidate regions are identified, the arduous task of discerning the actual gene(s) responsible for reproductive isolation commences through a combination of expanded population surveys, finer scale mapping experiments, and the use of manipulative techniques (when developed for a system) such as gene knockout, RNAi interference, and genetic transformation (reviewed in Noor and Feder 2006). This secondary search and verification phase is greatly facilitated when the genome of the target organisms has been sequenced.

In contrast to the gene survey approach that seeks to link genetic divergence with phenotype, the second approach starts with the phenotype associated with reproductive isolation in hand and attempts to dissect its genetic basis using conventional mapping methods. Test crosses are performed between lines of the populations (often isofemale) or specially constructed chromosomal lab lines that display partial hybrid sterility or inviability or that differ in a trait (physiological or morphological) associated with restricting gene flow. Marker loci or chromosomal regions are then scored in F1, F2, or backcross progeny to demarcate the positions of the genome contributing to the phenotype. When several different loci contribute to the trait (the phenotype is polygenic) the resulting genomic regions containing putative candidate genes affecting the trait are referred to as quantitative trait loci (QTL). At this point, population survey and gene mapping approaches converge, and the non-trivial task of indentifying the specific gene(s) in a region causing reproductive isolation begins. In some sense, trying to map intrinsic reproductive isolation is like trying to map the unmappable, as the phenotype itself is the lack of viable and fertile offspring. Nevertheless, analyses can be conducted where reproductive isolation is not complete. Also, in taxa following Haldane's rule, hybrids of the heterogametic sex can suffer reduced fitness, while hybrids of the other sex can remain relatively viable and fertile and be used to construct backcrosses for genetic study. Finally, hybrid rescue genes such as *Lethal hybrid rescue* (*Lhr*; Watanabe 1979), *Hybrid male rescue* (*Hmr*; Hutter *et al.* 1990; Orr and Irving 2000; Barbash and Ashburner 2003), *maternal hybrid rescue* (*mhr*; Sawamura *et al.* 1993a, b), *Zygotic hybrid rescue* (*Zhr*; Sawamura *et al.* 1993 b, c) and *167.7* (Davis *et al.* 1996) have been found in *Drosophila* that restore viability and/or fertility in certain crosses and allow for genetic mapping of postzygotic isolation (Barbash, *et al.* 2000, 2003; Orr and Irving 2000). Indeed, wildtype forms of these rescue mutations may often be speciation genes themselves, contributing to hybrid incompatibilities (Barbash *et al.* 2000, 2003; Orr and Irving 2000).

To date, 16 or so genes from a variety of organisms have been identified that could qualify as "speciation genes" in that they cause reproductive isolation in one guise or another (see *Table 2*), and the list is increasing daily. Some of the loci have been associated with ecological/pollinator adaptation and some with sperm/egg recognition (*Table 2*). But by far, most speciation genes have been identified in *Drosophila* and are associated with intrinsic postzygotic isolation (Noor and Feder 2006). Although we still know relatively little about the genetics of ecological speciation, several general patterns appear to be emerging concerning intrinsic postzygotic isolation. As outlined by Tang and Presgraves (2009):

1. Much of the genetic basis for hybrid inviability and sterility appears to involve seeming ordinary genes that function normally in parental taxa.
2. Hybrid incompatibilities tend to accumulate gradually as species diverge. In cases where pronounced hybrid inviability and sterility are observed, many loci can cause reproductive isolation. For example, Presgraves (2003) has estimated that over 190 genes cause incompatibilities between *Drosophila melanogaster* and *D. simulans*. Indeed, the number of genes contributing to postzygotic isolation can "snowball" and increase non-linearly with the square of the time of separation (Orr and Turelli 2001).

3. Hybrid inviability and sterility follow Haldane's rule with the heterogametic (XY or ZW) sex differentially affected.

4. Many hybrid incompatibilities behave as partial recessives in hybrids. This finding is consistent with a general model for Haldane's rule based on the exposure of the recessive negative epistatic effects of these alleles in hybrids of the heterogametic sex vs. their masking in the homogametic sex (Muller 1942; Orr 1993; Turelli and Orr 1995, 2000). However, other factors can also affect Haldane's rule, such as a general tendency for hybrids of the heterogametic sex to be afflicted by incompatibilities more often that the heterogametic sex. For example, hybrid male sterility may be more common because spermatogenesis could be a more inherently sensitive process that is easily perturbed or sexual selection might cause the faster evolution of male expressed genes (Wu and Davis 1993). This has been termed the faster-male theory (Wu *et al.* 1996) and it can work to either strengthen (when males are the heterogametic sex) or weaken (when females are the heterogametic sex) Haldane's rule. Evidence from *Drosophila* (True *et al.* 1996; Hollocher and Wu 1996; Tao and Hartl 2003), *Anopheles* and *Aedes* mosquitoes (Presgraves and Orr 1998), and *Xenopus* frogs (Kobel 1996; Kobel *et al.* 1996) support the faster male hypothesis.

5. Hybrid incompatibilities often map to the sex determining chromosome in species where sex is chromosomally determined such as *Drosophila*, mice, butterflies, and birds (referred to as the large X-effect; Coyne and Orr 1989b; Tucker *et al.* 1992; Dod *et al.* 1993; Hagen and Scriber 1989; Pashley-Prowell 1998; Saetre *et al.* 2003; Payseur and Nachman 2005).

6. The majority of incompatibility genes identified to date show molecular evolution signatures of recurrent adaptive evolution, consistent with hybrid sterility and inviability evolving as incidental byproducts of natural selection.

The finding that many genes often cause complete or near complete hybrid incompatibility between species (e.g., over 190 have been estimated between *Drosophila melanogaster* and *D. simulans*, two flies that have been estimated to have been separated for ~3 million years; Presgraves 2003) highlights the difficulties of directly equating genes causing postzygotic isolation with speciation genes. Given that just a single pair could effectively cease gene flow between the diverging taxa, the majority of the differences must have accumulated after the flies were completely reproductively isolated species. Consequently, most of these genes did not initiate or contribute to the speciation process itself and should not be considered speciation genes in the strict sense. How then does the geneticist know that a particular incompatibility being characterized was important to speciation at all? The answer is that she/he does not. Nevertheless, a reasonable case could be made that for taxa that display only partial postzygotic isolation any gene contributing to isolation has a bearing on speciation and should be considered a speciation gene. Moreover, even if a particular genetic difference arose after speciation was complete, there is no reason to suspect that the types of changes or factors involved should be different from those that preceded it, at least for allopatrically separated taxa. Thus, valid inferences about the genetics of speciation can still be drawn from the analysis of intrinsic hybrid sterility and inviability.

In a similar vein, the evidence for positive selection acting on most speciation genes, while suggestive, does not prove that natural selection was responsible for the specific change(s) in the gene(s) that generated postzygotic isolation during speciation. The substitution(s) indicative of selection may have accumulated after speciation was complete or the specific substitution causing isolation could have been fixed due to genetic drift. The molecular data only indicate that the preponderance of changes to a particular gene deviate from the pattern predicted by neutral evolution. Moreover, these tests do not reveal the source of the selection. Therefore, even if selection was responsible for the genetic differences inadvertently generating postzygotic isolation, we would not know if the selection was due to extrinsic ecological factors or intrinsic causes like intersexual conflict or meiotic drive (Willis 2009). Unfortunately, very little is known about the ecology of most *Drosophila* species to know what to test for with regards to environmental effects, with some notable exceptions like *D. sechellia* (a specialist on the rotting fruit of the Morinda tree on the Seychelles Islands in which host adaptation appears to have contributed to differentiation; Jones 1998) and the *D. mojavensis* group (a radiation associated with host shifting onto a variety of different cactus species in the Southwestern United States and Mexico; Fellows and Heed 1972; Etges *et al.* 1999; Makow and O'Grady 2007). But recent work on hybrid male sterility between *D. pseudoobscura* subspecies from Bogata, Colombia and the United States, as well as evidence from *D. subobscura* (Hauschteck-Jugen 1990) and *D. mauritiana* x *D. simulans* crosses (Tao *et al.* 2001), imply that inproper suppression of meiotic drive can cause reproductive isolation. It is not always clear, however, whether meiotic drive was the evolutionary cause for the substitutions that ultimately resulted in postzygotic isolation or represents a pathology expressed in hybrids (Coyne and Orr 2004).

Another area in which we are just beginning to acquire insight concerns the number, roles, and nature of the squabbling loci underlying genomic incompatibility. Recall that in the Bateson-Muller-Dobzhansky model it takes at least a pair of negatively interacting genes to generate postzygotic sterility and inviability. Yet we still have few, if any, examples in which all of the co-conspirators have been identified. Does the genetic architecture of postzygotic isolation typically involve many genes (complex interactions) or just a few (simple interactions of major effect loci)? Recent evidence indicates that the Nup96 and Nup160 autosomal loci negatively interact with each other in and possibly another Nup component on the X chromosome (Nup153) to cause hybrid lethality between *D. melanogaster* and *D. simulans* (Tang and Presgraves 2009). In this case, perhaps only a few players are involved and the change is structural in nature to a rather typical gene, at least on the surface. Will this prove to be a common pattern or will we find that regulatory changes and/or mutations in regulatory genes will be the most common types of changes leading to postzygotic isolation? Will the identity of these genes matter, with some genes tending to cause incompatibilities and others not (Wu *et al.* 1996) or will the factors involved be interchangeable (i.e., will the identities of the genes be of little significance and their numbers matter most; Naveira and Maside 1998)? In addition, more information is needed concerning the number and identity of genes causing F1 hybrid sterility/inviability vs. backcross or F2 isolation, as well as for sterility vs. inviability, in general. With respect to the latter issue, some general patterns are emerging in that hybrid inviability is caused by genes that are expressed in both sexes and often involve maternally expressed genes. In contrast, genes causing hybrid sterility usually affect one sex only and do not often show maternal effects (Coyne and Orr 2004). Thus, while it may seem that finding a new speciation gene for *Drosophila* is the same old same old, there really are a number of interesting and important issues still to be resolved concerning postzygotic isolation and its role in speciation, that, for the time being, can be most readily approached in the model genetic systems.

11. Sailing into the future

What does the future of speciation research hold? Genes, genes, and more genes is the easy answer. If Darwin where a graduate student today, then he might well be tempted to quit his day job as a naturalist on the Beagle and rush into the lab and become a genomicist. There is little doubt that a large void to fill is the genetics of speciation, especially with respect to ecological differentiation. We still have only a relatively few speciation genes identified, know little about how they act and interact to cause reproductive isolation, do not know whether they are primarily regulatory or structural in their nature, and although many speciation genes appear to show the signature of selection, we cannot say what the source of this selection is. I feel that the external ecology will be a large part of the answer, perhaps combined with a share of intersexual conflict and a dose of meiotic drive, when more extensive information becomes available from other organisms in addition to *Drosophila*.

Nevertheless, Darwin should not jump ship. If anything, I would hope I have convinced the reader that the ultimate answers to the mystery of speciation rests with combining what can be learned behind the laboratory doors with the natural history of organisms in the world beyond. Genomics provides one means, and a powerful one, to help understand the natural history of the how and why of speciation. Genomics will quicken our pace of discovery, will lead to a wealth of new information on the molecular and genetic bases of gene flow barriers, and will hasten the compilation of comparative data sets for meta-analyses testing for broad patterns and possible rules of speciation. However, this information loses significance divorced from the proper ecological, historical, and phylogenetic context of speciation. We must climb back into the tree of life to fully understand the endless forms speciation generates and its consequences. Thus, students of speciation today must be both ardent field naturalists (or at least not adverse to tree climbing) and proficient genome and gene mapping jocks (or have good friends that are). The future forward still lies on the Beagle, refitted with modern genomic rigging, to realize the full speciation research program initiated by Darwin and advanced by Dobzhansky. There is still mystery and the spirit of great adventure and discovery to come for those who make the voyage.

Acknowledgements

I would like to thank the Wissenschaftskolleg zu Berlin and my colleagues there Axel Meyer, James Mallet, and Patrik Nosil for insightful discussions about speciation during the 2008-2009 academic year. Also to my *Rhagoletis* mentors and friends Guy Bush, the late Ron Prokopy, and Stewart Berlocher, to whom I owe any success in studying speciation.

REFERENCES

Abrahamson W. G., Weis A. E., 1997, *Evolutionary Ecology Across Three Trophic Levels: Goldenrods, Gallmakers, and Natural Enemies*, Princeton, N.J.: Princeton University Press.

Alt S., Ringo J., Talyn B., Bray W., Dowse H, 1998, "The *period* gene controls cycles in courtship song of *Drosophila melanogaster*", *Animal Behavior* 56: 87–97.

Arnold M. L., 1992, "Natural hybridization as an evolutionary process", *Annual Review of Ecology and Systematics* 23: 237–61.

Arnold M. L., 1997, *Natural Hybridization and Evolution*, New York: Oxford University Press.

Barbash D. A., Ashburner M., 2003, "A novel system of fertility rescue in *Drosophila* hybrids reveals a link between hybrid lethality and female sterility", *Genetics* 163: 217–26.

Barbash D. A., Roote J., Ashburner M., 2000, "The *Drosophila melanogaster Hybrid male rescue* gene causes inviability in male and female species hybrids", *Genetics* 154: 1747–71.
Barbash D. A., Siino D. F., Tarone A. M., Roote J. A., 2003, "A rapidly evolving MYB-related protein causes species isolation in *Drosophila*", *Proceedings of the National Academy of Sciences USA* 100: 5302–5307.
Barluenga M., Stölting K. N., Salzburger W., Muschick M., Meyer A., 2006, "Sympatric speciation in Nicaraguan crater lake cichlid fish", *Nature* 439: 719–23.
Barraclough T. G., Vogler A. P., 2000, "Detecting the geographical pattern of speciatin from species-level phylogenies", *American Naturalist* 155: 419–34.
Barton N. H., 1989, "Founder effect speciation", in Otte D., Endler J. A. (eds.) *Speciation and its Consequences,* Sunderland, Mass.: Sinauer Assoc. Press, 229–56.
Barton N. H., Charlesworth B., 1984, "Genetic revolutions, founder effects, and speciation", *Annual Review of Ecology and Systematics* 15:133–64.
Bateson W., 1909, "Heredity and variation in modern lights", in Seward A. C. (ed.) *Darwin and Modern Science*, Cambridge: Cambridge University Press, 85–101.
Berlocher S. H., 2000, "Radiation and divergence in the *Rhagoletis pomonella* species group: Inference form allozymes", *Evolution* 54: 543–57.
Berlocher S. H., Feder J. L., 2002, "Sympatric speciation in phytophagous insects: Moving beyond controversy?", *Annual Review of Entomology* 47: 773–815.
Bradshaw H. D., Schemske D. W., 2003, "Allele substitution of a flower color locus produces a pollinator shift in monkeyflowers", *Nature* 426: 176–78.
Brideau N. J., Flores H. A., Wang J., Maheshwari S., Wang X., Barbash D. A., 2006, "Two Dobzhansky-Muller genes interact to cause hybrid lethality in *Drosophila*", *Science* 314: 1292–95.
Bush G. L., 1966, "The taxonomy, cytology, and evolution of the genus *Rhagoletis* in North America (Diptera, Tephriditae)", *Bulletin of the Museum of Comparative Zoology* 134: 431–562.
Bush G. L., 1969, "Sympatric host race formation and speciation in frugivorous flies of the genus *Rhagoletis* (Diptera, Tephritidae)", *Evolution* 23: 237–51.
Butlin R. K., Galindo J., Grahame J. W., 2008, "Sympatric, parapatric or allopatric: the most important way to classify speciation?", *Philosophical Transactions of the Royal Society of London, Series B: Biological Sciences* 363: 2997–3007.
Colosimo P. F., Hosemann K. E., Balabhadra S., Villarrea G. Jr., Dickson M., Grimwood J., Schmutz J., Myers, R. M., Schluter D., Kingsley D. M., 2005, "Widespread parallel evolution in sticklebacks by repeated fixation of ectodysplasin alleles", *Science* 5717: 1928–33.
Coyne J. A., Orr H. A., 1989a, "Patterns of speciation in *Drosophila*", *Evolution* 43: 362–81.
Coyne J. A., Orr H. A., 1989b, "Two rules of speciation", in Otte D., Endler J. A., (eds.) *Speciation and its Consequences,* Sunderland, Mass.: Sinauer Assoc. Press, 180–207.
Coyne J. A., Orr H. A., 1997, ""Patterns of speciation in *Drosophila*" revisited", *Evolution* 51: 295–303.
Coyne J. A., Orr H. A., 2004, *Speciation*, Sunderland, Mass.: Sinauer Assoc. Inc.

Coyne J. A., Elwyn S., 2006, "Does the *desaturase-2* locus in *Drosophila melanogaster* cause adaptation and sexual isolation?", *Evolution* 60: 279–91.

Craig T. P., Itami J. K., Abrahamson W. G., Horner J. D., 1993, "Behavioral evidence for host-race formation in *Eurosta solidaginis*", *Evolution* 47:1696–1710.

Dambroski H. R., Linn C. Jr, Berlocher S. H., Forbes A. A., Roelofs W., Feder J. L., 2005, "The Genetic Basis for fruit odor discrimination in *Rhagoletis* flies and its significance for sympatric host shifts", *Evolution* 59: 1953–64.

Dallerac R., Labeur, C., Jallon J. M., Knipple D. C., Roelofs W. L., Wicker-Thomas C., 2000, "A $\Delta 9$ desaturase gene with a different substrate specificity is responsible for the cuticular diene hydrocarbon polymorphism in *Drosophila melanogaster*", *Proceedings of the National Academy of Sciences USA* 97: 9449–54.

Davis A. W., Wu C. I., 1996, "The broom of the sorcerer's apprentice: the fine structure of a chromosomal region causing reproductive isolation between two sibling species of *Drosophila*", *Genetics* 143: 1287–98.

Dobzhansky T., 1934, "Studies on hybrid sterility. I. Spermatogenesis in pure and hybrid *Drosophila pseudoobscura*", *Zeitschrift für Zellforschung und mikroskopische Anatomie* 21: 169–221.

Dobzhansky T., 1935, "A critique of the species concept in biology", *Philosophy of Science* 2: 344–55.

Dobzhansky T., 1937, *Genetics and the Origin of Species,* New York: Columbia University Press.

Dod B., Jermiin L. S., Boursot P., Chapman V. H., Nielsen J. T., Bonhomme F., 1993, "Counterselection on sex chormosomes in the *Mus musculus* European hybrid zone", *Journal of Evolutionary Biology* 6: 529–46.

Drès M., Mallet J., 2002, "Host races in plant-feeding insects and their importance in sympatric speciation", *Philosophical Transactions of the Royal Society of London, Series B: Biological Sciences* 357: 471–92.

Endler J. A., 1977, *Geographic variation, speciation and clines,* Princeton, N.J.: Princeton University Press.

Endler J. A., 1983, "Natural and sexual selection on color patterns in poeciliid fishes", *Environmental Biology of Fishes* 9: 173–90

Endler J. A., Théry M., 1996, "Interacting effects of lek placement, display behavior, ambient light, and color patterns in three Neotropical forest-dwelling birds", *American Naturalist* 148: 421–52.

Endler J. A., Basolo A. L., 1998, "Sensory ecology, receiver biases and sexual selection", *Trends in Ecology & Evolution* 13: 415–20.

Etges W. J., Johnson W. R., Duncan G. A., Huckins G., Heed W. B., 1999, "Ecological genetics of cactophilic *Drosophila*", in Robichaux R. (ed.), *Ecology of Sonoran Desert Plants and Plant Communities,* Tucson: University of Arizona Press, 164–214.

Fang S., Takahashi A., Wu C. I., 2002, "A mutation in the promoter of desaturase 2 is correlated with sexual isolation between *Drosophila* behavioral races", *Genetics* 162: 781–84.

Feder J. L., Chilcote C. A., Bush G. L., 1988, "Genetic differentiation between sympatric host races of *Rhagoletis pomonella*", *Nature* 336: 61–64.

Feder J. L., Hunt T. A., Bush G. L., 1993, "The effects of climate, host plant phenology and host fidelity on the genetics of apple and hawthorn infesting races of *Rhagoletis pomonella*", *Entomologia Experimentalis et Applicata* 69: 117–35.

Feder J. L., Opp S., Wlazlo B., Reynolds K., Go W., Spisak S., 1994, "Host fidelity is an effective pre-mating barrier between sympatric races of the Apple Maggot Fly", *Proceedings of the National Academy of Sciences USA* 91(17): 7990–94.

Feder J. L., Forbes A. A., 2007, "Habitat avoidance and sympatric speciation", *Functional Ecology* 21: 589–97.

Feder J. L., Forbes A. A., 2008, "Host fruit-odor discrimination and sympatric host-race formation", in Tilmon K.J. (ed.), *Specialization, Speciation, and Radiation: The Evolutionary Biology of Herbivorous Insects*, Berkeley: University of California Press, 101–116.

Fellows D. P., Heed W. B., 1972, "Factors affecting host plant selection in desert-adapted cactophilic *Drosophila*", *Ecology* 53: 850–58.

Filchak K. E., Roethele J. B., Feder J. L., 2000, "Natural selection and sympatric divergence in the apple maggot, *Rhagoletis pomonella*", *Nature* 407: 739–42.

Fitzpatrick B. M., M., Turelli, 2006, "The geography of mammalian speciation: mixed signals from phylogenies and range maps", *Evolution* 60: 601–615.

Fisher R. A., 1930, *The Genetical Theory of Natural Selection*, Oxford: Clarendon Press.

Forbes A. A., Fisher J., Feder, J. L., 2005, "Habitat avoidance: overlooking an important aspect of host specific mating and sympatric speciation?", *Evolution* 59: 1552–59

Forbes A. A., Feder J. L., 2006, "Divergent preferences of *Rhagoletis pomonella* host races for olfactory and visual fruit cues", *Entomologia Experimentalis et Applicata* 119: 121–27.

Forbes A. A., Powell T. H. Q., Stelinski L. L., Smith J. J., Feder J. L., 2009, "Sequential sympatric speciation across trophic levels", *Science* 323: 776–79.

Funk D. J., Filchak K. E., Feder J. L., 2002, "Herbivorous insects: model systems for the comparative study of speciation ecology", *Genetica* 116: 251–67.

Funk D. J., Nosil P., Etges W. J., 2006, "Ecological divergence exhibits consistently positive associations with reproductive isolation across disparate taxa", *Proceedings of the National Academy of Sciences USA* 103: 3209–13.

Gavrilets S., 2004, *Fitness Landscapes and the Origin of Species,* Princeton, N.J.: Princeton University Press.

Grant P. R., 1999, *Ecology and Evolution of Darwin's Finches,* Princeton, N.J.: Princeton University Press.

Grant P. R., Grant B. R., 2008, "Pedigrees, assortative mating and speciation in Darwin's finches", *Proceedings of the Royal Society of London. Series B: Biological Sciences* 275: 661–68.

Gullan P. J., Cranston P. S., Hansen-McInnes K., 2005, *The Insects: An Outline of Entomology,* 3rd edn., Oxford: Blackwell.

Hagen R. H., Scriber J. M., 1989, "Sex-linked diapause, color, and allozyme loci in *Papilio glaucus*: Linkage analysis and significance in a hybrid zone", *Journal of Heredity* 80: 179–85.

Haldane J. B. S., 1922, "Sex ratio and the unisexual sterility in animal hybrids", *Journal of Genetics* 12: 101–109.

Hauschteck-Jungen E., 1990, "Post-mating reproductive isolation and modification of the sex-ratio trait in *Drosophila pseudoobscura* induced by the sex-chromosome gene arrangement A2+3+5+7", *Genetica* 83: 31–44.

Hey J., 2001, *Genes, Categories, and Species*, Oxford: Oxford University Press.

Hollocher H., Wu. C. I., 1996, "The genetics of reproductive isolation in the *Drosophila simulans* clade: X versus autosomal effects and male vs. female effects", *Genetics* 143: 1243–55.

Howard D. J., 1993, "Reinforcement: origins, dynamics, and fate of an evolutionary hypothesis", in Harrison R.G. (ed.) *Hybrid Zones and the Evolutionary Process*, New York: Oxford University Press., 46–69.

Hutter P., Roote J., Ashburner M., 1990, "A genetic basis for the inviability of hybrids between sibling species of *Drosophila*", *Genetics* 124: 909–20.

Jones C. D., 1998, "The genetic basis of *Drosophila sechellia*'s resistance to a host plant toxin", *Genetics* 149: 1899–1908.

Kirkpatrick M., 1982, "Sexual selection and the evolution of female choice", *Evolution* 36: 1–12.

Kirkpatrick M., Ravigne V., 2002, "Speciation by natural and sexual selection: models and experiments", *American Naturalist* 159: S22–S35.

Kirkpatrick M., Barton N., 2006, "Chromosome inversions, local adaptation and speciation", *Genetics* 173: 419–434.

Kobel H. R., 1996, "Reproductive capacity of experimental *Xenopus gilli* x X. *l. laevis* hybrids", in Tinsley R. C., Kobel H. R. (eds.), *The Biology of Xenopus*, Oxford: Clarendon Press, 73–80.

Kobel H. R., Loumont C., Tinsely R. C., 1996, "The extant species", in Tinsley R. C., Kobel H. R. (eds.), *The Biology of Xenopus*, Oxford: Clarendon Press, 9–33.

Langerhans R. B., Gifford M. E., Joseph E. O., 2007, "Ecological speciation in *Gambusia* fishes", *Evolution* 61: 2056–74.

Lee H. Y., Chou J. Y., Cheong L., Chang N. H., Yang S. Y., Leu J. Y., 2008, "Incompatibility of nuclear and mitochondrial genomes causes hybrid sterility between two yeast species", *Cell* 135: 1065–73.

Lewontin R. C., Moore J. A., Provine W. B., Wallace B., 1981, *Dobzhansky's "Genetics of natural populations" I–XLIII,* New York: Columbia University Press.

Lijtmaer D. A., Mahler B., Tubaro P. L., 2003, "Hybridization and postzygotic isolation patterns in pigeons and doves", *Evolution* 57: 1411–18.

Linn C. Jr., Feder J. L., Nojima S., Dambroski H. R., Berlocher S. H., Roelofs W., 2003, "Fruit odor discrimination and sympatric host race formation in *Rhagoletis*", *Proceedings of the National Academy of Sciences USA* 100: 11490–93.

Linn C. Jr, Dambroski H. R., Feder J. L., Berlocher S. H., Nojima S., Roelofs W. L., 2004, "Postzygotic isolating factor in sympatric speciation in *Rhagoletis* flies: reduced response of hybrids to parental host-fruit odors", *Proceedings of the National Academy of Sciences USA* 101: 17753–58.

Losos J. B., Jackman T. R., Larson A., de Queiroz K., Rodriguez-Schettino L., 1998, "Contingency and determinism in replicated adaptive radiations of island lizards", *Science* 279: 2115–18.
Lowry D. B., Modliszewski J. L., Wright K. M., Wu C. A., Willis J. H., 2008, "The strength and genetic basis of reproductive isolating barriers in flowering plants", *Philosophical Transactions of the Royal Society of London. Series B: Biological Sciences* 363: 3009–21.
Luyten P. H., Liley N. R., 1991, "Sexual selection and competitive mating success of males guppies (*Poecilia reticulata*) from four Trinidad populations", *Fishes* 9: 173–90.
Lynch J. D., 1989, "The gauge of speciation: on the frequencies of modes of speciation", in Otte D., Endler J. A. (eds.) *Speciation and its Consequences,* Sunderland, Mass.: Sinauer Assoc. Press, 527–56.
Markow T. A., O'Grady P. M., 2007, "Drosophila biology in the Genomic Age", *Genetics* 177: 1269–76.
May R. M., Harvey P. H., 2009, "Species uncertainties", *Science* 323: 687.
Mayr E., 1942, *Systematics and Origin of Species*, New York: Columbia University Press.
Mayr E., 1947, "Ecological factors in speciation", *Evolution* 1: 263–88.
Mayr E., 1963, *Animal Species and Evolution*, Cambridge, Mass.: Harvard University Press.
McCartney M. A., Lessios H. A., 2004, "Adaptive evolution of sperm bindin tracks egg incompatibility in neotropical sea urchins of the genus *Echinometra*", *Molecular Biology and Evolution* 21: 732–45.
McPheron B. A., Smith D. C., Berlocher S. H., 1988, "Genetic-differences between host races of *Rhagoletis pomonella*", *Nature* 336: 64–66.
Mihola O., Trachtulec Z., Vlcek C., Schimenti J. C., Forejt J., 2009, "A mouse speciation gene encodes a meiotic histone H3 methyltransferase", *Science* 323: 373–75.
Moyle L. C., Olson M. S., Tiffin P., 2004, "Patterns of reproductive isolation in four angiosperm genera", *Evolution* 58: 1195–1208.
Muller H. J., 1939, "Reversibility in evolution considered from the standpoint of genetics", *Biological Reviews of the Cambridge Philosophical Society* 14: 261–80.
Muller H. J., 1940, "Bearing of the *Drosophila* work on systematics", in Huxley J. (ed.), *The New Systematics*, London: Oxford University Press, 185–268.
Muller H. J., 1942, "Isolating mechanisms, evolution and temperature", *Biology Symposium* 6: 71–125.
Naveira H. F., Maside X. R., 1998, "The genetics of hybrid male sterility in *Drosophila*", in Howard D. J., Berlocher S. H. (eds.), *Endless Forms: Species and Speciation*, Oxford: Oxford University Press, 330–38.
Noor M. A. F., 1995, "Speciation driven by natural selection in *Drosophila*", *Nature* 375: 674–75.
Noor M. A. F., 1999, "Reinforcement and other consequences of sympatry", *Heredity* 83: 503–508.
Noor M. A. F., Grams K. L., Bertucci L. A., Reiland J., 2001, "Chromosomal inversions and the reproductive isolation of species", *Proceedings of the National Academy of Sciences USA* 98: 12084–88.

Noor M. A. F., Feder J. L., 2006, "Genetics of speciation", *Nature Genetics* 7: 851–61.

Nosil P., Vines T. H., Funk D. J., 2005, "Perspective: reproductive isolation caused by natural selection against immigrants from divergent habitats", *Evolution* 59, 705–19.

Nosil P., Crespi B. J., Sandoval C., 2002, "Host-plant adaptation drives the parallel evolution of reproductive isolation", *Nature* 417: 441–43.

Nosil P., Egan S. P., Funk D. J., 2008, "Divergent selection plays multiple roles in generating heterogeneous genomic differentiation between walking-stick ecotypes", *Evolution* 62: 316–36.

Orr H. A., 1993, "A mathematical model of Haldane's rule", *Evolution* 47: 1606–11.

Orr H. A., 1996, "Dobzhansky, Bateson, and the genetics of speciation", *Genetics* 144: 1331–35

Orr H. A., Irving S., 2000, "Genetic analysis of the hybrid male rescue locus of *Drosophila*", *Genetics* 155: 225–31.

Orr H. A., Turelli M., 2001, "The evolution of postzygotic isolation: accumulating Dobzhansky-Muller incompatibilities", *Evolution* 55: 1085–94.

Palumbi S. R., 1999, "All males are not created equal fertility differences depend on gamete recognition polymorphisms in sea urchins", *Proceedings of the National Academy of Sciences USA* 96: 12632–37.

Pashley-Prowell D., 1998, "Sex linkage and speciation in Lepidoptera", in Howard D. J., Berlocher S. H. (eds.), *Endless Forms: Species and Speciation*, Oxford: Oxford University Press, 309–19.

Payseur B. A., Nachman M. W., 2005, "The genomics of speciation: investigating the molecular correlates of X chromosome introgression across the hybrid zone between *Mus domesticus* and *Mus musculus*", *Biological Journal of the Linnean Society* 84: 523–34.

Phadnis N., Orr H. A., 2009, "A single gene causes both male sterility and segregation distortion in *Drosophila* hybrids", *Science* 323: 376–79.

Poulton E. B., 1904, "What is a species?", *Proceedings of the Entomological Society of London* (1903), lxxvii–cxvi.

Presgraves D. C., 2002, "Patterns of postzygotic isolation in Lepidoptera", *Evolution* 56: 1168–83.

Presgraves D. C., 2003, "A fine-scale genetic analysis of hybrid incompatibilities in *Drosophila*", *Genetics* 163: 955–72.

Presgraves D. C., Orr H. A., 1998, "Haldane's rule in taxa lacking a hemizygous X", *Science* 282: 952–54.

Presgraves D. C., Balagopalan L., Abmayr S. M., Orr H. A., 2003, "Adaptive evolution drives divergence of a hybrid inviability gene between two species of *Drosophila*", *Nature* 423: 715–19

Price T. D., Bouvier M. M., 2002, "The evolution of F1 postzygotic incompatibilities in birds", *Evolution* 56: 2083–2089.

Prokopy R. J., Bennett E. W., Bush G. L., 1971, "Mating behavior in *Rhagoletis pomonella* (Diptera: Tephritidae), I. Site of assembly", *Canadian Entomologist* 103: 1405–1409.

Prokopy R. J., Bennett E. W., Bush G. L., 1972, "Mating behavior in *Rhagoletis pomonella* (Diptera: Tephritidae), II. Temporal organization", *Canadian Entomologist* 104: 97–104.
Ramsey J., Bradshaw H. D. Jr., Schemske D. W., 2003, "Components of reproductive isolation between the monkeyflowers *Mimulus lewissii* and *M. cardinalis* (Phrymaceae)", *Evolution* 57: 1520–34.
Rawson P. D., Burton R. S., 2002, "Functional coadaptation between cytochrome *c* and cytochrome *c* oxidase within allopatric populations of a marine copepod", *Proceedings of the National Academy of Sciences USA* 99:12955–58.
Rice W. R., Salt G., 1988, "Speciation via disruptive selection on habitat preference: experimental evidence", *American Naturalist* 131: 911–917.
Rice W. R., Salt G., 1990, "The evolution of reproductive isolation as a correlated character under sympatric conditions: experimental evidence", *Evolution* 44: 1140–52.
Rice W. R., Hostert E .E., 1993, "Laboratory experiments on speciation: what have we learned in 40 years?", *Evolution* 47: 1637–53.
Ritchie M. G., Halsey E. J., Gleason J. M., 1999, "*Drosophila* song as a species-specific mating signal and the behavioural importance of cycles in *D. melanogaster* song", *Animal Behavior* 58: 649–57.
Ritchie M. G., Kyriacou C. P., 1994, "Reproductive isolation and the *period* gene of *Drosophila*", *Molecular Ecology* 3: 595–99.
Rieseberg L. H., 1997, "Hybrid origins of plant species", *Annual Review of Ecology and Systematics* 28: 359–89.
Rieseberg L. H., 2001, "Chromosomal rearrangements and speciation", *Trends in Ecology & Evolution* 16: 351–58.
Rieseberg L. H., Raymond D. M., Lai Z., Livingstone K., Nakazato T., Durphy J. L., Schwarzbach A. E., Donovan L. A., Lever C., 2003, "Major ecological transitions in wild sunflowers facilitated by hybridization", *Science* 301: 1211–16.
Rundle H. D., Nagel L., Boughman J. W., Schluter D., 2000, "Natural selection and parallel speciation in sympatric sticklebacks", *Science* 287: 306–308.
Rundle H. D., Whitlock M. C., 2001, "A genetic interpretation of ecologically dependent isolation", *Evolution* 55: 198–201.
Rundle H. D., Nosil P., 2005, "Ecological speciation", *Ecology Letters* 8: 336–52.
Russell S. T., 2003, "Evolution of intrinsic post-zygotic reproductive isolation in fish", *Annales Zoology Fennici* 40: 321–329.
Saetre G. P., Moum T., Bures S., Kral M., Adamjan M., Moreno J., 1997, "A sexually selected character displacement in fly catchers reinforces premating isolation", *Nature* 387: 589–592.
Sasa M. M., Chippindale P. T., Johnson N. A., 1998, "Patterns of postzygotic isolation in frogs", *Evolution* 52: 1811–1820 .
Savolainen V., Anstett M. C., Lexer C., Hutton I., Clarkson J. J., Norup M. V., Powell, D., Springate M. P., Salamin N., Baker W. J., 2006, "Sympatric speciation in palms on an oceanic island", *Nature* 439: 719–723.

Sawamura K., Taira T., Watanabe T. K., 1993, "Hybrid lethal systems in the *Drosophila melanogaster* species complex, I. The *maternal hybrid rescue* (*mhr*) gene of *Drosophila simulans*", *Genetics* 133: 299–305.

Sawamura K., Watanabe T. K., Yamamoto M. T., 1993b, "Hybrid lethal systems in the *Drosophila melanogaster* species complex", *Genetica* 88: 175–85.

Sawamura K., Yamamoto M. T., Watanabe T. K., 1993c, "Hybrid lethal systems in the *Drosophila melanogaster* species complex, II. The *Zygotic hybrid rescue* (*Zhr*) gene of *D. melanogaster*", *Genetics* 133: 307–313.

Schliewen U., Tautz D., Pääbo S., 1994, "Sympatric speciation suggested by monophyly of crater lake cichlids", *Nature* 368: 629–32.

Schluter D., 1994, "Experimental evidence that competition promotes divergence in adaptive radiation", *Science* 266: 798–801.

Schluter D., 1995, "Adaptive radiation in sticklebacks: trade-offs in feeding performance and growth", *Ecology* 76: 82–90.

Schluter D., 2000, *The Ecology of Adaptive Radiation*, Oxford: Oxford University Press.

Schluter D., 2001, "Ecology and the origin of species", *Trends in Ecology & Evolution* 16: 372–80.

Schluter D., 2009, "Evidence for ecological speciation and its alternative", *Science* 323: 737–41.

Schwartz D., Matta B. M., Shakir-Botteri N. L., McPheron B.A., 2005, "Host shift to an invasive plant triggers rapid animal hybrid speciation", *Nature* 436: 546–49.

Seehausen O., 2004, "Hybridization and adaptive radiation", *Trends in Ecology & Evolution* 19: 198–207.

Seehausen O., Terai Y., Magalhaes I. S., Carleton K. L., Mrosso H. D.J., Miyagi R., van der Sluijs I., Schneider M. V., Maan M. E., Tachida H., Imai H., Okada N., 2008, "Speciation through sensory drive in cichlid fish", *Nature* 455: 620–26.

Shaw A., McRee D. E., Vacquier V. D., Stout C. D., 1993, "The crystal structure of lysin, a fertilization protein", *Science* 262: 1864–67.

Slatkin M., 1982, "Pleiotropy and parapatric speciation", *Evolution* 36: 263–70.

Sun S., Ting C. T., Wu C. I., 2004, "The normal function of a speciation gene, *Odysseus*, and its hybrid sterility effect", *Science* 305: 81–83.

Swanson W. J., Vacquier V. D., 1997, "The abalone egg vitelline envelope receptor for sperm lysin is a giant multivalent molecule", *Proceedings of the National Academy of Sciences USA* 94: 6724–29.

Tang S., Presgraves D. C., 2009, "Evolution of the *Drosophila* nuclear pore complex results in multiple hybrid incompatibilities", *Science* 323: 779–82.

Tao Y., Hartl D. L., Laurie C. C., 2001, "Sex ratio distortion associated with reproductive isolation in *Drosophila*", *Proceedings of the National Academy of Sciences USA* 98: 13183–88.

Tao Y., Hartl D. L., 2003, "Genetic dissection of hybrid incompatibilities between *Drosophila simulans* and *Drosophila mauritiana*, III. Heterogeneous accumulation of hybrid incompatibilities, degree of dominance, and implications for Haldane's rule", *Evolution* 57: 2580–98.

Taylor E. B., McPhail J. D., 2000, "Historical contingency and ecological determinism interact to prime speciation in sticklebacks, *Gasterosteus*", *Proceedings of the Royal Society of London, Series B: Biological Sciences* 267: 2375–84.
Templeton A. R., 1980, "Theory of speciation via the founder principle", *Genetics* 94: 1011–38.
Ting C. T., Tsaur S. C., Wu M. L., Wu C. I., 1998, "A rapidly evolving homeobox at the site of a hybrid sterility gene", *Science* 282: 1501–1504.
Tucker P. K., R. D., Sage R. D., Warner J., Wilson A. C., Eicher E. M., 1992, "Abrupt cline for sex chromosomes in a hybrid zone between two species mice", *Evolution* 46: 1146–63.
True J. R., Weir B. S., Laurie C. C., 1996, "A genome-wide survey of hybrid incompatibility factors by introgression of marked segments of *Drosophila mauritiana* chromosomes into *Drosophila simulans*", *Genetics* 142: 819–37.
Turelli M., Orr H. A., 1995, "The dominance theory of Haldane's rule", *Genetics* 140: 389–402.
Turelli M., Orr H. A., 2000, "Dominance, epistasis, and the genetics of postzygotic isolation", *Genetics* 154: 1663–79.
Turelli M., Barton N. H., Coyne J. A., 2001, "Theory and speciation", *Trends in Ecology & Evolution* 16: 330–43.
Via S., 2001, "Sympatric speciation in animals: the ugly duckling grows up", *Trends in Ecology & Evolution* 16: 381–90.
Walter R. B., Kazianis S., 2001, "*Xiphophorus* interspecies hybrids as genetic models of induced neoplasia", *Institute for Laboratory Animal Research Journal* 42: 299–321.
Watanabe T. K., 1979, "A gene that rescues the lethal hybrids between *Drosophila melanogaster* and *D. simulans*", *Japanese Journal of Genetics* 54: 325–31.
Wheeler D. A., Kyriacou C. P., Greenacre M. L., Yu Q., Rutila J. E., Rosbash M., Hall J. C., 1991, "Molecular transfer of a species specific behavior from *Drosophila simulans* to *Drosophila melanogaster*", *Science* 251: 1082–85.
Willett C. S., 2006, "Deleterious epistatic interactions between electron transport system protein-coding loci in the copepod *Tigriopus californicus*", *Genetics* 173: 1465–77.
Willett C. S., Burton R. S., 2001, "Viability of cytochrome *c* genotypes depends on cytoplasmic backgrounds in *Tigriopus californicus*", *Evolution* 55: 1592–99.
Willett C. S., Burton R. S., 2003, "Environmental influences on epistatic interactions: viabilities of cytochrome *c* genotypes in interpopulation crosses", *Evolution* 57: 2286–92.
Willis J. H., 2009, "Origin of species in overdrive", *Science* 323: 350–51.
Wood T. K., 1993, "Speciation of the *Enchenopa binotata* complex (Insecta: Homoptera: Membracidae)", in Lees D. R. and Edwards D. (eds.), *Evolutionary Patterns and Processes*, London: Academic Press, 299–317.
Wu C. I., Davis A. W., 1993, "Evolution of post-mating reproductive isolation - the composite nature of Haldane's rule and its genetic basis", *American Naturalist* 142: 187–212.
Wu C. I., Johnson N. A., Palopoli M., 1996, "Haldane's rule and its legacy: why are there so many sterile males?" *Trends in Ecology & Evolution* 11: 281–84.

I. Prezygotic barriers to gene flow (barriers arising before the zygote is formed)

a) Extrinsic ecological prezygotic isolation.

1. Habitat isolation - taxa use different habitats or occupy different niches in same region.
2. Seasonal or temporal isolation - taxa mate or flower at different times.
3. Pollinator isolation - plants pollinated by different means restricting cross fertilization.

b) Behavioral/Ethological isolation - different taxa are not sexually attracted to one another.

c) Mechanical isolation - non-correspondence of genitalia impedes copulation.

d) Gamete isolation - insemination occurs but gametes do not come into contact (sperm may be inviable in sexual duct of other taxon) or recognize each other for fertilization to occur.

e) Mating system - evolution of partial or complete self fertilization or asexual reproduction

II. Postzygotic barriers to gene flow (barriers arising after the zygote is formed)

a) Extrinsic postzygotic isolation - dependent on the external biotic or abiotic environment.

1. Ecological inviability - hybrids develop normally but suffer lower fitness because they do not reside in an appropriate habitat for their intermediate phenotypes.
2. Behavioral sterility - hybrids are less fertile than parental species because they cannot obtain mates due to their intermediate phenotypes or courtship behaviors.

b) Intrinsic postzygotic isolation - hybrids develop abnormally independent of environment.

1. Hybrid inviability - developmental difficulties cause full or partial hybrid lethality.
2. Hybrid sterility - problems in reproductive system/gamete development or neurological/ physiological abnormalities rendering hybrids incapable of successful courtship.

Table 1. Inherent gene flow barriers generating reproductive isolation between taxa (modified from Dobzhansky 1937 and Coyne and Orr 2004).

Ods Odysseus (*Ods*) is a homeobox-containing gene that is associated with hybrid male sterility between the geographically isolated species *D. simulans* and *D. mauritiana* (Ting *et al.* 1998). Its effect has been shown to require several gene interactions (Sun *et al.* 2004).
Overdrive (Ovd) Crosses between female *Drosophila pseudoobscura* flies from Bogotá, Columbia and males from the United States result in fertile female hybrids and sterile males. However, as hybrid males age they can become weakly fertile. These males produce mostly female daughters, indicative of true drive of the X-chromosome during meiosis or inactivation (killing) of Y-chromosome carrying gametes. A gene named Overdrive (*Ovd*) mapping to the right arm of the X chromosome has been shown to contribute to both hybrid sterility and meiotic drive (Phadnis and Orr 2009). DNA sequence comparisons of the two subspecies in the vicinity of Overdrive indicate rapid evolution in the Bogotá lineage, consistent with an evolutionary history of sex-ratio distortion. How *Ovd* interacts with other genes to cause meiotic drive and hybrid male sterility is not yet known.
Prdm9 The *Prdm9* gene encodes a methyltransferase that epigenetically modifies a chromatin structural protein, histone H3. This activates the promoters of genes involved in meiosis and its disruption in hybrids of different subspecies of the mouse *Mus musculus* results in male sterility (Mihola *et al.* 2009).
per The period (*per*) gene contributes to differences in courtship song between *D. melanogaster* and *D. simulans* that are associated with behavioral discrimination (Alt *et al.* 1998; Ritchie *et al.* 1999). It also functions as a regulator of circadian rhythms. When transformed between species, the gene changes the important courtship song difference (Wheeler *et al.* 1991) but, curiously, no behavioral discrimination is apparent (Ritchie and Kyriacou 1994).
Xmrk2 The *Xmrk2* gene is overexpressed in some hybrid classes that are produced by *Xiphophorus maculatus* platyfish and *Xiphophorus helleri* swordtails, and causes invasive melanomas that can be fatal (Walter and Kazianis 2001). Functional tests have confirmed the role of this gene in tumor formation, but its effects are inconsistent when different strains of platyfish are used.
YUP The YELLOW UPPER (YUP) region controls the presence or absence of yellow pigments in the petals of pink-flowered *Mimulus lewisii*, which is pollinated by bumblebees, and its red-flowered sister species *Mimulus cardinalis*, which is pollinated by hummingbirds (Bradshaw and Schemske 2003). This study identifies a single major-effect locus that controls a pigmentation difference causing differential pollination by bees versus hummingbirds. This region could have a role in mediating a species barrier through differential pollinator attraction.

Table 2. A short list of putative speciation genes that have been isolated and characterized to date (modified from Noor and Feder 2006).

AEP2 and OLI1
In one hybrid line resulting from a cross between the yeast *Saccharomyces cerevisiae* and *S. bayanus*, the protein product of the AEP2 gene on chromosome 13 from *S. bayanus* has been shown to be incapable of regulating the translation of OLI1 messenger RNA encoded by the mitochondrial genome of *S. cerevisiae* (Lee *et al.* 2008), contributing to reduced hybrid fitness and sterility. The difference may represent adaptation to using non-fermentable carbon sources.

Bindin
Bindins are acrosomal proteins on the outside of sperm that promote adhesion to eggs, and their role in species-specific binding has been studied extensively in Echinometra sea urchins (Palumbi 1999; McCartney and Lessios 2004).

cytc
Co-adaptation between cytochrome c, cytochrome c oxidase, and other parts of the mitochondria electron transport system appear to mediate differences in hybrid viability and development time among *Tigriopus californicus* copepod populations (Rawson and Burton 2002; Willett and Burton 2001, 2003; Willett 2006).

desat2
The desaturase2 (desat2) gene is associated with female cuticular hydrocarbon differences between some African and most non-African *Drosophila melanogaster* (Dallerac *et al.* 2000). The African allele is also associated with behavioral discrimination by African strains (Fang *et al.* 2002; Coyne and Elwyn 2006), although Caribbean strains possess the African allele while being behaviorally indiscriminate.

Hmr and Lhr
Lethal hybrid rescue (*Lhr*) has functionally diverged in *Drosophila simulans* and interacts with Hybrid male rescue (*Hmr*), which has functionally diverged in *D. melanogaster*, to cause lethality in F1 hybrid males (Brideau *et al.* 2006). *Hmr* encodes a rapidly evolving protein with sequence similarity to the myb/SANT-like domain in Adf-1 (MADF) class of DNA binding proteins. The function of LHR is not known, but it appears to interact with Heterochromatin Protein 1 (HP1) a chromodomain-containing protein that localizes to heterochromatic regions of chromosomes and is required to maintain heterochromatic states. LHR has diverged extensively in sequence between *D. simulans* and *D. melanogaster* in a manner consistent with positive selection.

Lysin
Within acrosomes of abalone (*Haliotis*) sperm lysin interacts with VERL (vitelline envelope receptor for lysine) on eggs to mediate species-specific binding of sperm to eggs (Shaw *et al.* 1993; Swanson and Vacquier 1997). Lysins could also mediate fertilization specificity in other gastropods.

Nup96 and Nup 160
Hemizygous (and presumably homozygous) copies of the *D. simulans Nup96* and *Nup160* genes bear sequence variants that cause inviability when associated with the *D. melanogaster* X chromosome, possibly involving the *Nup153* allele in *D. melanogaster* (Presgraves *et al.* 2003; Tang and Presgraves 2009). However, the effects of these genes are only noticeable in deficiency lines, as F1 hybrids of these species are typically fully sterile.

Evolution through developmental change: How alterations in development cause evolutionary changes in anatomy

Scott F. Gilbert

Nature interests me because it's beautiful, complex, and robust. Evolutionary theory interests me because it explains why nature is beautiful, complex and robust.

(David Quammen, 2007)

...a study of the effects of genes during development is as essential for an understanding of evolution as are the study of mutation and that of selection.

(Julian Huxley 1942, 8)

1. Introduction: The modern synthesis and its critiques

For the past half-century, the mechanisms of evolution have been explained by the fusion of genetics and evolutionary biology called "the Modern Synthesis." The tenets of the Modern Synthesis have been generally formulated as such:

1. There is genetic variation within the population.
2. There is competition for survival, with most of the population not reproducing.
3. This leads to the differential survival and reproduction of those organisms with genetic variants that make them more likely to succeed in the particular environment. (The "survival of the fittest.")
4. The offspring have a high likelihood of inheriting those genetic variants that enabled the differential survival of their parents.

This synthesis explains natural selection within a species, and it explains it remarkably well. Moreover, the Modern Synthesis explains evolutionary change in populations throughout the natural kingdoms. For example, drug resistance in bacteria, coat colors in rodents, mimicry in moths and butterflies, carbon metabolism in plants, and malarial resistance in humans has each been shown to be caused by the selection of randomly produced genetic variation by factors within the environment. These genetic variants can be assigned to gene mutations, and the mathematical modeling of the selection on the variations has shown the power of this model (see, for example, Rice 2004; Futuyma 2009; Gilbert and Epel 2009).

Just a few examples will be mentioned here: The ability of mice to blend into the background and avoid predators is provided by a mutation in the *Agouti* gene, which is involved in hair pigmentation (Linnen *et al.* 2009). The ability of human populations to avoid succumbing to malaria can be acquired by a change in the hemoglobin gene. This genetic change causes the replacement of the amino acid glutamate by valine at the sixth position, and the malarial parasite cannot multiply in the red blood cells having this variant of hemoglobin (see Gilbert and Epel 2009). Humans have acted as selective agents on mosquitoes, such that the *Anopheles* mosquito (that transmits the malarial parasite) is becoming less sensitive to DDT. By widely spraying DDT, we have selected for mosquitoes that have evolved enzymes that are resistant to the pesticide (in that their amino acids do not bind the pesticide) or that actually destroy the pesticides before it can kill the mosquito (Raymond *et al.* 1998; Donnelly *et al.* 2009). Similarly, genetic changes in the opsin protein can cause it to be activated by a different wavelength of light, thereby allowing some species to see in the ultraviolet rather than violet light. This mutation enables kestrels to fly high and catch voles by seeing their urine trails, which are visible in the ultraviolet (Viitala *et al.* 1995).

But while such examples show that evolution can occur within a species, the species itself did not transform into anything else. Malaria- resistant humans and DDT-resistant mosquitoes were still humans and mosquitoes, respectively. The mice with the *Agouti* gene mutation persisted as better camouflaged mice, but they did not evolve into anything un-mousy. Until recently, one could study how the selection of certain genetic variants allowed a particular phenotype to persist and characterize a species; but the differences *between* species (that did not interbreed) could not be studied in this manner.

While there has been excellent evidence for evolution provided by comparative anatomy, biogeography, paleontology, and several other sciences, this inability to study the genetic mechanisms of evolution above the species level provided a space through which critics could claim that evolution had not been proven. Fr. Stanley Jaki (1992), for instance, wrote that "As to the claim … that the Darwinian evolutionary mechanism (the interplay of chance mutations with environmental pressure) has solved all basic problems, I hold it to be absurd and bordering at times on the unconscionable. While the mechanism in question provoked much interesting scientific research, it left unanswered the question of transition among genera, families, orders, classes, and phyla where the absence of transitional forms is as near-complete as ever."

2. Tools for answering the critiques: DNA sequencing and developmental genetics

This situation changed dramatically, starting in the mid-1970s and continuing into the present day. Here, two major advances contributed to a more complete evolutionary theory that could explain both evolution within a species (microevolution) and evolution between species (macroevolution). The advances were (1) DNA sequencing and (2) developmental genetics. DNA sequencing allows scientists to actually compare the gene sequences between species, and developmental genetics gives scientists a theory of body construction that enables us to see how the changes in the DNA can cause (selectable) changes in anatomy.

Genomics: Comparing genomes between species

DNA sequencing and the computer-enhanced science of genomics provided "the ultimate forensic record of evolution" (Carroll 2006). First, they have allowed us to see which animals are grouped together and which are not. DNA contains the documentary record of evolution. Just as a linguist can compare words to show that French, Romanian, Italian, and Spanish descended from a common ancestor (and that Spanish and Portuguese diverged rather recently), so can a biologist compare particular sequences of DNA to show how animals are related to one another (Gilbert 2003; Rice 2004). Descent with modification was shown not only in our bones but also in our genes.

Developmental genetics: How cell communication controls cell determination

The revolution in DNA technology enabled developmental biology to formulate a theory of body construction based on cell-cell communications. For decades, it had been known that interactions between vertebrate embryonic cells are told what type of cell to become; but the actual proteins involved were not able to be isolated until DNA sequencing enabled scientists to isolate the genes rather than the proteins. The common plan for communicating was that the inducing cell secreted a protein called a *paracrine factor.* These factors do not go through the blood (like endocrine factors) but, rather, they diffuse over a few cell diameters, working on neighboring cells. The responding cell has a receptor for the paracrine factor, and this receptor protein spans the cell membrane. Its outer domain can bind the particular paracrine factor, and when it does so, this changes the structure of the domain of the receptor residing inside the cell. Usually, the binding of a paracrine factor with its receptor gives the inner domain an enzymatic activity that it did not formerly possess. This enzymatic activity can (either directly or indirectly) modify a protein called a *transcription factor*. In many cases, the transcription factor protein is phosphorylated (has a phosphate group added to it) by the new enzymatic activity. The phosphate group activates the transcription factor enabling it to enter the nucleus, bind to a specific sequence of DNA, and activate the gene adjacent to that DNA sequence.

In this manner, one cell type can change gene expression in a neighboring cell, and thereby instruct the cell to become a pancreatic cell, a liver cell, or a neuron. By sequential interactions with other cells, this specification gets finer, such that the cell will be told to be an insulin-secreting beta cell of the pancreas, or an oval cell of the liver, or a dopaminergic neuron of the ventral brain. Cell adhesion and cell migration can also be controlled in this manner.[1] Moreover, changes in gene expression in the embryo can cause different types of anatomical structures to form.

One critical consequence of the combination of DNA sequencing and developmental genetics has been the discovery of "deep homologies" (Shubin *et al.* 1997; 2009). Here, the same gene is responsible for the generation of a particular organ, even if the organ differs greatly between animal groups. The eyes of insects, squids, and vertebrates are very different, and are even formed from different parts of the organism. Evolutionary developmental geneticists such as Walter Gehring (1998; 2005) demonstrated that there is a developmental genetic pathway specifying photoreceptor development and that this is common to all animals with eyes. The basic genetic pattern for eye development is specified by homologues of the *Pax6* gene. Pax6 is one of those transcription factor proteins mentioned earlier. The pathway initiated by Pax6 and its homologues probably evolved only once, and the eyes of all the vertebrates and arthropods, flatworms, and snails come from a modification of this central scheme. Indeed, if one activates a mouse *Pax6* gene

[1] This induction is conveniently ignored by Intelligent Design proponents. The eye, for instance, develops in such a way that the lens cells produce paracrine factors that induce the retina and the retina makes paracrine factors that induce the lens. In this way, the parts of the eye construct one another. There is no incomprehensible complexity in getting the lens to form in the same organ as the retina. The developing lens will transform adjacent cells into retina and vice versa.

in those cells of a fly larva destined to become its jaw, that jaw will develop into a fly eye. If one expresses the fly *Pax6* gene in frog skin, the fly *Pax6* initiates eye development in the frog skin and causes it to become an amphibian retina (Halder *et al.* 1995; Chow *et al.* 1999).

Indeed, several transcription factors have deep homology throughout the animal kingdom. A gene called *tinman* specifies cells to become the heart in both insects and mammals, and another set of genes, called *Hox* genes, specifies the anterior-posterior axis of all animals. Indeed, eyes, hearts, and limbs "arose by the modification of pre-existing genetic regulatory circuits established in early metazoans" (Shubin *et al.* 2009).

Deep homology has very important ramifications. First, it is remarkable evidence for the relatedness of all animals. We all came from the same ancestor. Second, it makes obsolete the notion that similarities in animals arose from convergence to a particular state. One can only have convergence when the two lineages were originally divergent. However, we now know that the various lineages have common genes that specify these "convergent traits." Although the notion of convergence has been brought up again by Simon Conway Morris (2005 and this conferencee), this idea is based on the disproved notion that homologous genes could not exist between greatly divergent phyla (see Amundson 2005, 217). Rather, DNA sequencing and developmental genetics has shown that there is descent with modification within the genome.

3. Evolutionary developmental biology

Evolutionary developmental biology holds that evolution occurs by changing the trajectory of development. In other words, the major processes in evolution are those involving the changes in the genetic regulatory genes that construct the organism. Rather than studying gene *frequency within* a population, evolutionary developmental biology focuses on the changes in gene *expression between* or within populations. To the evolutionary developmental biologist, evolutionary biology remains incomplete without a theory of body construction and how the construction of bodies can change (Amundson 2005). For instance, when confronted with the question of how arthropod body plans arose, Hughes and Kaufman (2002) state:

> To answer this question by invoking natural selection is correct — but insufficient. The fangs of a centipede … and the claws of a lobster accord these organisms a fitness advantage. However, the crux of the mystery is this: From what developmental genetic changes did these novelties arise in the first place?

The realization that a theory of body construction was needed to supplement the evolutionary theory of change was proclaimed in a well-publicized 1975 research paper by Mary-Claire King and Alan Wilson. Entitled "Evolution at Two Levels in Humans and Chimpanzees," this study showed that despite the large anatomical differences between chimps and humans, their DNA was almost identical. King and Wilson's solution to this conundrum that bodily change was outpacing gene changes harkened back to the ideas of Richard Goldschmidt, Conrad Waddington, and other developmental biologists who felt that evolutionary change was predicated upon specific changes in gene *regulation*. Specifically, the King and Wilson paper states:

> The organismal differences between chimpanzees and humans would … result chiefly from genetic changes in a few regulatory systems, while amino acid substitutions in general would rarely be a key factor in major adaptive shifts.

That is to say, the nucleotide substitutions in the DNA sequences that encode protein sequences — which seem to be pretty much the same for chimps and humans — are not what is important. The important differences are where, when, and how much the genes are activated. François Jacob's (1977) paper "Evolution by tinkering," suggested a new model, generalizing the work of King and Wilson, and focusing attention on how the same genes can create new types of body plans when their expression pattern is altered during development.

Over the next thirty years, it became possible to study gene expression and gene sequence together. Recombinant DNA technology, polymerase chain reaction, in situ hybridization, and high-throughput RNA analysis now enable us to look at gene sequences and to compare their expression between species. This ability has revolutionized the way we look at evolutionary processes. The new science of *evolutionary developmental biology* ("*evo-devo*") is looking at the mechanisms by which changes in gene regulation cause changes in anatomy. It looks at the mechanisms of tinkering.

Wallace Arthur (2004) has catalogued four ways in which Jacob's "tinkering" can take place at the level of gene expression:

- Heterotopy (change in location of gene expression)
- Heterochrony (change in time or duration of gene expression)
- Heterometry (change in amount of gene expression)
- Heterotypy (change in kind of gene being expressed)

These mechanisms are predicated on the modularity of enhancers (see Kirschner and Gerhart 2005; Gilbert and Epel 2009). Enhancers are the genetic sequences that tell a gene where and when (and how much) they are to be transcribed. Enhancers bind the above-mentioned transcription factors, thereby allowing the presence or absence of these factors to regulate gene expression. One of the most important discoveries of evolutionary developmental biology is that different enhancers can direct the transcription of a gene in different organs. Thus, by mutating enhancers, one can have organisms that express the same gene, but in different cells or for different durations.

Heterotopy

Changing the location of a gene's expression (i.e., altering which cells it is expressed in) can result in enormous amounts of selectable variation. For instance, one of the biggest differences between the duck and the chicken is the webbed feet of the duck. This is an obviously important adaptation to swimming (as are the duck's broad beak and the oil glands of its skin). Vertebrate embryonic limbs develop with their fingers and toes surrounded by a web of connective tissue and skin. In chicks (and humans), the cells that make up the webbing between the digits undergo apoptosis (programmed cell death) initiated by the paracrine factor BMP4. This cell death destroys the webbing and frees the digits. It turns out that ducks also express BMP4 in the webbing of their hindlimbs, but this protein is prevented from signaling cell death by the presence of another protein, the BMP inhibitor Gremlin (Laufer *et al.* 1997; Merino *et al.* 1999). Chick and mouse limbs also express Gremlin around the cartilaginous skeletal elements of the digits, but not in the webbing. Thus, the webbing in embryonic duck feet remains intact thanks to the heterotopy — the change in place of gene expression — of the Gremlin gene. Indeed, if one adds beads containing the Gremlin protein to the webbing of an embryonic chick limb, the limb will keep its webbing, just as a duck does.

Heterotopy is also seen to be responsible for the formation of bat forelimbs that extend and retain their webbing (Weatherbee *et al.* 2005), the formation of the carapace of the turtle shell (Cebra-Thomas *et al.* 2005), the evolution of the bird feather from the reptilian scale (Harris *et al.* 2002), appears to be the proximate mechanism for the Lord's curse on snake limbs (Cohn and Tickle 1999).

Heterochrony

Heterochrony is a shift in the relative timing of two developmental processes from one generation to the next. The elongated fingers in the dolphin flipper appear to be caused by heterochrony, wherein the region responsible for producing the paracrine factors for the developing limb (especially Fgf8) is present longer than in other mammals (Richardson and Oelschläger 2002). The enormous number of ribs formed in embryonic snakes (>500 in some species) is likewise due to heterochrony. In this case, the segmentation reactions cycle nearly four times faster relative to tissue growth in snake embryos than they do in related vertebrate embryos (Gomez *et al.* 2008).

Heterometry

Heterometry is the change in the *amount* of a gene product. One of the best examples of heterometry involves Darwin's celebrated finches, a set of 14 closely related birds collected by Charles Darwin and his shipmates during his visit to the Galápagos and Cocos Islands in 1835. These birds helped him frame his evolutionary theory of descent with modification, and they still serve as one of the best examples of adaptive radiation and natural selection (see Weiner 1994; Grant 1999; Grant and Grant 2007). Systematists have shown that these finch species evolved in a particular manner, with a major speciation event being the split between the cactus finches and the ground finches. The ground finches evolved deep, broad beaks that enable them to crack seeds open, whereas the cactus finches evolved narrow, pointed beaks that allow them to probe cactus flowers and fruits for insects.

Earlier research (Schneider and Helms 2003) had shown that species differences in the beak pattern were caused by changes in the growth of the neural crest-derived mesenchyme of the frontonasal process (i.e., those cells that form the facial bones). Moreover, they and others (Wu *et al.* 2004; 2006) showed that birds use BMP4 as a growth factor promoting cell division in their beaks, and that species-specific differences in beaks (such as those between the chick and the duck) were due to the placement of BMP4 expression.[2]

Abzhanov and his colleagues (2004) found a remarkable correlation between the beak shape of the finches and the amount of *Bmp4* expression. No other paracrine factor showed such differences. The expression of *Bmp4* in the mesenchyme of embryonic ground finch beaks starts earlier and is much greater than

[2] Yes, the same paracrine factor that induces cell death in the webbing of the birds' limb also induces cell division in the birds' beaks (see Gilbert and Epel 2009.

Bmp4 expression in cactus finch beaks. In all cases, the *Bmp4* expression pattern correlated with the breadth and depth of the beak.

The importance of these expression differences was confirmed experimentally by changing the *Bmp4* expression pattern in chick embryos (Abzhanov *et al.* 2004; Wu *et al.* 2004, 2006). When *Bmp4* expression was enhanced in the frontonasal process mesenchyme, the chick developed a broad beak reminiscent of the beaks of the ground finches. Conversely, when BMP signaling was inhibited in this region (by adding a BMP inhibitor to the developing beak primordium), the beak became narrow and pointed, like those of cactus finches. Thus, enhancers controlling the amount of beak-specific BMP4 synthesis may be critically important in the evolution of Darwin's finches.

Heterotypy

In heterochrony, heterotopy, and heterometry, the mutations affect the regulatory regions of the gene. The gene's product — the protein — remains the same, although it may be synthesized in a new place, at a different time, or in different amounts. In heterotypy, the changes affect the protein that binds to these regulatory regions. The changes of heterotypy affect the actual coding region of the gene, and thus can change the functional properties of the protein being synthesized.

For instance, insects have six legs, whereas most other arthropod groups (spiders, millipedes, centipedes, and crustaceans) have many more. How is it that the insects came to form legs only in their three thoracic segments and have no appendages in their abdominal regions? The answer seems to reside in the relationship between Ultrabithorax (Ubx) protein and the *Distal-less* gene.

The *Ubx* gene encodes a transcription factor that is used throughout the arthropods to activate those genes expressed in the lower thorax and abdomen. In most arthropod groups, Ubx does not inhibit the *Distal-less* gene. However, in the insect lineage, a mutation occurred in the *Ubx* gene wherein the original 3' end of the protein-coding region was replaced by a group of nucleotides encoding a stretch of about 10 alanine residues (Galant and Carroll 2002; Ronshaugen *et al.* 2002). This polyalanine region represses *Distal-less* transcription. When a shrimp *Ubx* gene is experimentally modified to encode this polyalanine region, it, too, represses the *Distal-less* gene. The ability of insect Ubx to inhibit *Distal-less* thus appears to be the result of a gain-of-function mutation that characterizes the insect lineage. One cannot explain insect evolution without a developmental explanation as to why limb development of this lineage has been so constrained.

Maize, commonly known in the U.S. as corn, is an organism whose domestication and subsequent propagation throughout the world may have been effected through a mutation of a transcription factor (Wang *et al.* 2005). There are many steps postulated between the ancestral teosinte plant and the modern cultivated maize plant *Zea mays*, and one of the most economically important steps in this was the liberation of the kernel from its hard protective envelope (the glume). This mutation exposed the kernel on the cob so that the kernels can be readily harvested for human consumption. This critical event in maize evolution and domestication is controlled by a single gene, *teosinte glume architecture* (*TGA1*). This gene encodes a transcription factor that is active in the development of the influorescence (the "ear") of the plant. When maize *TGA1* is placed into teosinte, the teosinte kernels become corn-like; and when teosinte *TGA1* becomes expressed in maize, the kernels become teosinte-like. These genes are expressed in the respective ears of the plant, and they appear to be expressed in the same amounts.

Within the *TGA1* gene, there is a mutational change resulting in a single amino acid difference between the maize and teosinte proteins the two alleles encode: a substitution of lysine in teosinte to asparagine in maize at amino acid position 6. (The lysine at this position is conserved in teosinte, wheat, and rice, suggesting that it is important for protein function or stability.) The mutation to asparagine appears to cause the TGA1 protein to be degraded faster, so perhaps there is a failure to make the glume surface that would normally cover the kernel.

Population genetic analysis shows that there is very little variation in the regions around the *TGA1* gene, suggesting that there was a "sweep" of this allele through the teosinte population. These results demonstrate that some of the complex differences between the teosinte and corn cobs are regulated by a single gene, and the differences are probably the result of a single amino acid substitution.

Heterocyberny

In addition to these four genetic ways of producing variation, a fifth type of developmental change involves the induction of genetic change by the environment (West-Eberhard 2005). Variously called "heterocyberny" (Gilbert and Epel 2009) or the "adaptability driver" (Bateson 2005), these processes denote changes occurring when a physiological response to the environment becomes incorporated into the genome. In its simplest form, this is a transfer of the trigger for a pre-existing phenotype that was originally triggered by the environment to a new initiator within the embryo. Both Waddington (1942, 1952) and Schmalhausen (1949) independently proposed this "genetic assimilation" to explain how some species have evolved rapidly in particular directions. For instance, both scientists were impressed by the calluses of the ostrich. Most mammalian skin has the ability to form calluses on areas that are abraded by the ground or some other surface. The skin cells respond to friction by proliferating. While such examples of environmentally induced callus formation are widespread, the ostrich is born with calluses where it will touch the ground. Waddington and Schmalhausen hypothesized that since the skin cells are already competent to be induced by friction, they could be induced by other signals as well. As ostriches evolved, a mutation (or a particular combination of alleles) appeared that enabled the skin cells to respond to some substance *within* the embryo. Waddington (1942) wrote:

> Presumably its skin, like that of other animals, would react directly to external pressure and rubbing by becoming thicker. … This capacity to react must itself be dependent upon genes. … It may then not be too difficult for a gene mutation to occur which will modify some other area in the embryo in such a way that it takes over the function of external pressure, interacting with the skin so as to "pull the trigger" and set off the development of callosities.

Genetic assimilation has been shown numerous times in the laboratory, and two of the most well-documented cases include selection for environmentally induced ectopic wings and environmentally induced coloration (see Gibson and Hogness 1996; Suzuki and Nijhout 2006; Gilbert and Epel 2009).

The advantages of heterocyberny are:

- The phenotype is not random. The phenotype was originally an adaptive physiological response to an environmental agent. Therefore, the phenotype has already been tested by natural selection. This would eliminate a long period of testing phenotypes derived by random mutations.
- This phenotype would have already existed in a major portion of the population. This model would solve the problem of how a new phenotype gets established in a population. In the Modern Synthesis, one of the problems of having a new phenotype is that the bearers of such a phenotype are "monsters" compared to the wild-type. How would such mutations, perhaps present in one individual or one family, get established and eventually take over a population? With genetic assimilation, this phenotype would have already been seen in a large percentage of the population, whenever the environmental conditions induced it. The phenotype just needs to be genetically stabilized by the selection of modifier genes (that already exist in the population.)

Indeed, West-Eberhard (2005) has proposed that studying gene expression and plasticity will help us understand the genetic divergence that gives rise to speciation. "Contrary to common belief, environmentally initiated novelties may have greater evolutionary potential than mutationally induced ones. Therefore, the genetics of speciation can profit from studies of changes in gene expression as well as changes in gene frequency and genetic isolation." This is not "Lamarckian" in that the inheritance of the variation is not based on physiological use-or-disuse, nor is there any "willing" of the phenotype. Yet, environmentally induced modifications can be selected (at least in the laboratory) to be incorporated into the genetic repertoire of the organism.

Evolutionary developmental biology has a program for explaining biodiversity through changes in gene expression during embryonic and juvenile phases. New forms appear to evolve largely by altering the timing, location, and amount of gene expression. The heterotopy, heterochrony, and heterometry is accomplished by changing the enhancer elements regulating gene expression. Heterotypy can alter the transcription factor that binds to these enhancer sequences, giving these proteins different properties. Heterocyberny can take a physiological response to the environment and make it part of normal development. Nothing about this goes against the Modern Synthesis. Rather, this is an expansion of the mechanisms of variation, and it is an expansion that allows small genetic changes to make large anatomical alterations. As Carroll (2008a) has noted, "This constitutes a developmental genetics theory of morphological evolution which can supplement and extend the Modern Synthesis of evolutionary biology and population genetics."

References

Abzhanov A., Protas M., Grant B. R., Grant P. R., Tabin C. J., 2004, "*Bmp4* and morphological variation of beaks in Darwin's finches", *Science* 305: 1462–65.

Amundson R., 2005, *The Changing Role of the Embryo in Evolutionary Thought: Structure and Synthesis*, Cambridge: Cambridge University Press.

Arthur W., 2004, *Biased Embryos and Evolution*, Cambridge: Cambridge University Press.

Bateson P., 2005, "The return of the organism", *Journal of Biosciences* 30: 31–39.

Carroll, S. B., 2008a, "Evo-Devo and an expanding evolutionary synthesis: a genetic theory of morphological evolution", *Cell* 134: 25–36.

Carroll, S. B., 2008b, *The Making of the Fittest: DNA and the Ultimate Forensic Record of Evolution*, Waltham, Mass.: Quercus Press.

Cebra-Thomas J., Tan F., Sistla S., Estes E., Bender G., Kim C., Gilbert, S. F., 2005, "How the turtle forms its shell: a paracrine hypothesis of carapace formation", *Journal of Experimental Zoology* 304B: 158–69.
Chow R. L., Altmann C. R., Lang R. A., Hemmati-Brivanlou A., 1999, "*Pax6* induces ectopic eyes in a vertebrate", *Development* 126: 4213-22
Cohn, M. J., Tickle C., 1999, "Developmental basis of limblessness and axial patterning in snakes," *Nature* 399: 474–79.
Conway Morris, S., 2005, *Life's solution: Inevitable Humans in a Lonely Universe,* Cambridge, UK: Cambridge University Press.
Donnelly M. J., Corbel V., Weetman D., Wilding C. S., Williamson M. S., Black W. C. 4th, 2009, "Does kdr genotype predict insecticide-resistance phenotype in mosquitoes?" *Trends In Parasitology* 5: 213–19.
Futuyma D., 2009, *Evolution*, 2nd edn., Sunderland, MA: Sinauer Associates.
Galant R., Carroll S. B., 2002, "Evolution of a transcriptional repression domain in an insect Hox protein", *Nature* 415: 910–13.
Gibson G., Hogness D. S., 1996, "Effect of polymorphism in the *Drosophila* regulatory gene *Ultrabithorax* on homeotic stability", *Science* 271: 200–203.
Gilbert S. F., 2003, "Opening Darwin's Black Box: teaching evolution through developmental genetics", *Nature Reviews: Genetics* 4: 735–41,
Gilbert S. F., 2006, *Developmental Biology*, 8th edn., Sunderland, Mass.: Sinauer Associates.
Gilbert S. F., Epel D., 2009. *Ecological Developmental Biology: Integrating Epigenetics, Medicine, and Evolution*, Sunderland, Mass.: Sinauer Associates.
Gomez C., Özbudak E. M., Wunderlich J., Baumann D., Pourquié O., "Control of segment number in vertebrate embryos", *Nature* 454: 335–38.
Grant P., 1999, *The Ecology and Evolution of Darwin's Finches*, Princeton, NJ: Princeton University Press.
Grant P. R., Grant B. R., 2007, *How and Why Species Multiply: The Radiation of Darwin's Finches*, Princeton, NJ: Princeton University Press.
Halder G. P., Callaerts P., Gehring W. J., 1995, "Induction of ectopic eyes by targeted expression of the *eyeless* gene in *Drosophila*", *Science* 267:1788-92
Harris M. P., Fallon J. F., Prum R. O., 2002, "Shh-Bmp2 signaling module and the evolutionary origin and diversification of feathers", *Journal of Experimental Zoology* 294: 160–76.
Hughes C. G., Kaufman T. C., 2002, "Hox genes and the evolution of the arthropod body plan," *Evolution & Development* 4: 459–99.
Huxley J., 1942, *Evolution: The Modern Synthesis*, London: Allen and Unwin.
Jacob F., 1977, "Evolution and tinkering", *Science* 196: 1161–66.
Jaki S., 1992, quoted in W. Dembski 1996, "What every theologian should know about creation and evolution, and design", http://www.origins.org.
King M. C., Wilson A. C., 1975, "Evolution at two levels in humans and chimpanzees", *Science* 188: 107–116.

Kirschner M. W., Gerhart J. C., 2005, *The Plausibility of Life: Resolving Darwin's Dilemma*, New York: Norton.

Laufer E., Pizette S., Zou H., Orozco O. E., Niswander L., 1997, "BMP expression in duck interdigital webbing: a reanalysis", *Science* 278: 305.

Linnen C. R., Kingsley E. P., Jensen J. D., Hoekstra H. E., 2009, "On the origin and spread of an adaptive allele in deer mice", *Science* 325: 1095–98.

Lynch M., 2005, "Simple evolutionary pathways to complex proteins", *Protein Science* 14: 2217–25.

Merino R., Rodríguez-León J., Macias D., Ganan Y., Economides A. N., Hurle J. M., 1999, "The BMP antagonist Gremlin regulates outgrowth, chondrogenesis and programmed cell death in the developing limb", *Development* 126: 5515–22.

Müller F., 1869, *Facts and Arguments for Darwin*, London: John Murray.

Newman, J. H., 1870, Letter to Brownlow, quoted in Ward W. P., *The Life of John Henry Cardinal Newman*, New York: Longmans, Green, and Co., vol 2, 269.

Quammen D., 2007, quoted in "Wordsmiths of Southwestern Montana", *Outside Bozeman* (Winter 2006/7), 21.

Raymond M., Chevillon C., Guillemaud T., Lenormand T., Pasteur N., 1998, "An overview of the evolution of overproduced esterases in the mosquito *Culex pipiens*", *Philosophical Transactions of the Royal Society of London, Series B: Biological Sciences* 353: 1707–11.

Richardson, M. D., Oelschläger H. H., 2002, "Time, pattern and heterochrony: a study of hyperphalangy in the dolphin embryo flipper", *Evolution & Developmen* 4: 435–44.

Rice S., 2004, *Evolutionary Theory*, Sunderland, Mass.: Sinauer Associates.

Ronshaugen M., McGinnis N., McGinnis W., 2002, "Hox protein mutation and macroevolution of the insect body plan", *Nature* 415: 914–17.

Schmalhausen I. I., 1949, *Factors of Evolution: The Theory of Stabilizing Selection*, Philadelphia: Blakiston.

Schneider R. A., Helms J. A., 2003, "The cellular and molecular origins of beak morphology", *Science* 299: 565–68.

Shubin N., Tabin C., Carroll S., 1997, "Fossils, genes and the evolution of animal limbs", *Nature* 388: 639–48.

Shubin N., Tabin C., Carroll S., 2009, "Deep homology and the origins of evolutionary novelty", *Nature* 457: 818–23.

Sturtevant A., 1923, "Inheritance of direction of coiling in *Limnaea*", *Science* 58: 269–70.

Suzuki Y., Nijhout H. F., 2006, "Evolution of a polyphenism by genetic assimilation", *Science* 311: 650–52.

Viitala J., Korpimäki E., Palokangas P., Koivula M., 1995, "Attraction of kestrels to vole scent marks visible in ultraviolet light", *Nature* 373: 425–27.

Waddington C. H., 1942, "The canalization of development and the inheritance of acquired characters", *Nature* 150: 563.

Waddington C. H., 1952, "Selection of the genetic basis for an acquired character", *Nature* 169: 278.

Wang H. *et al.*, 2005, "The origin of the naked grains of maize", *Nature* 436: 714–19.

Weatherbee S. D., Behringer R. R., Rasweiler J. J. 4th, Niswander L. A., 2006, "Interdigital webbing retention in bat wings illustrates genetic changes underlying amniote limb diversification", *Proceedings of the National Academy of Sciences USA* 103: 15103–107.

Weiner J., 1994., *The Beak of the Finch: A Story of Evolution in Our Time,* New York: Random House.

West-Eberhard M. J., 2005, "Phenotypic accommodation: adaptive innovation due to developmental plasticity", *Journal of Experimental Zoology, Part B: Molecular and Developmental Evolution* 304B: 610–18.

Wu, P., Jiang T. X., Suksaweang S., Widelitz R. B., Chuong C. M., 2004, "Molecular shaping of the beak", *Science* 305: 1465–66.

Wu, P. Jiang T. X., Shen J. Y., Widelitz R. B., Chuong C. M., 2006, "Morphoregulation of avian beaks: comparative mapping of growth zone activities and morphological evolution", *Developmental Dynamics* 235: 1400–12.

THE EVOLUTION OF EVOLUTIONARY MECHANISMS: A NEW PERSPECTIVE

Stuart A. Newman

1. Introduction

The Modern Evolutionary Synthesis, based on Charles Darwin's concept of natural selection in conjunction with a genetic theory of inheritance in a population-based framework, has been, for more than six decades, the dominant scientific perspective for explaining the diversity of living organisms. In recent years, however, with the growth in knowledge of the fossil record, the genetic affinities among different life forms and the roles played by non-genetic determinants of organismal shape and form, there have been challenges to the synthesis in the realms of both cellular (Margulis and Sagan 2002; Woese and Goldenfeld 2009) and multicellular (Alberch 1989; Newman 1994; Jablonka and Lamb 1995, 2005; Newman and Müller 2000; West-Eberhard 2003) evolution.

According to the eminent evolutionary biologist Ernst Mayr, "Nothing strengthened the theory of natural selection as much as the refutation, one by one, of all the competing theories, such as saltationism, orthogenesis, inheritance of acquired characters, and so forth." (Mayr 1982, 840) As I will argue below, a coherent account of the origination of the morphological motifs of multicellular organisms in fact requires bringing all three of these ideas back into evolutionary theory, though in a fashion that acknowledges their declining efficacy as evolution progresses. Put in other terms: evolutionary mechanisms themselves have evolved.

It is easy to see why the first two of Mayr's "competing theories" are antithetical to the standard picture. *Saltationism* is the idea that organismal phenotype can change from one generation to the next in a manner that is very large compared to the organism's usual range of phenotypic variation. Saltation is often associated with the concept of the "hopeful monster," a term invented by the geneticist Richard Goldschmidt to refer to novel phenotypes that might arise in a single generation by the mutation of a single broadly acting gene (a "macromutation"). Arguments developed by population geneticists such as Ronald Fisher (Fisher

1930) in the twentieth century contended that such jumps would be exceptionally rare and typically lethal occurrences not contributing significantly to the origin of species or higher taxa. But even before the Synthesis was formulated, Darwin expressed the incompatibility of saltationism with his own ideas in a statement in the first edition of *The Origin of the Species*, "[i]f it could be demonstrated that any complex organ existed, which could not possibly have been formed by numerous, successive, slight modifications, my theory would absolutely break down" (Darwin 1859, 158).

Orthogenesis is the doctrine that organisms change in preferred directions over the course of evolution. Although this is consonant with the fact that all material systems have inherent patterns of organization that may manifest themselves rapidly (as in the transition from waves to vortices in water) or over time (as in the generation of the chemical elements), and despite the purported examples of such inherencies in organismal development and evolution provided by biologists such as William Bateson (Bateson 1909) and W. D'Arcy Thompson (1942), orthogenesis was forcefully rejected by leading architects of the Synthesis, including George Gaylord Simpson (1944) and Mayr himself (Mayr 1974).

The most disparaged of all the competing theories listed by Mayr is the third one, the inheritance of acquired characteristics, or Lamarckism. This is ironic, considering that Darwin himself became increasingly receptive to this notion in successive volumes of *The Origin*. Indeed, a role for inheritance of characters acquired during an individual's lifetime was written out of the standard theory not because it conflicted with natural selection, but because it violated the tenet of chromosomal genes as the exclusive medium of inheritance. An added barrier to the operation of Lamarckian mechanisms for animal species first emphasized by the post-Darwinian biologist August Weismann is the sequestering of the germ line from the rest of the body during development.

It is important to recognize, however, that Darwin and most of his successors concerned themselves with multicellular forms, both plants and animals, but conspicuously avoided providing any account of the origination and innovation of the morphological motifs, e.g., body plans and organ forms, that are raw material for natural selection. The focus on this missing element by the emerging field of evolutionary developmental biology (Robert 2004; Müller and Newman 2005; Müller 2007; Callebaut *et al.* 2007; Moczek 2008; Fusco and Minelli 2008) has highlighted a number of properties of developmental systems that were previously marginal to evolutionary theory. These include developmental and phenotypic plasticity and genotype–phenotype discordances (Newman 1994; Trut *et al.* 2009;

Pigliucci 2001; West-Eberhard 2003; Badyaev 2005; Badyaev *et al.* 2005; Salazar-Ciudad 2006; Goodman 2008; Vedel *et al.* 2008), determination of form by physical and epigenetic factors (Müller and Streicher 1989; Newman and Comper 1990; Newman and Müller 2000), and inheritance systems that extend beyond the gene (Jablonka and Lamb 1995; 2005).

In the remainder of this chapter I will describe how our current inferences about the genetic endowment of the single-celled ancestors of the Metazoa (multicellular animals) combined with knowledge of the physical properties of viscoelastic materials (e.g., cell clusters and tissue primordia) have led to new understanding of the origination and early evolution of animal form that is simultaneously saltational, orthogenic and Lamarckian. The three erstwhile prohibited concepts converge in this revised explanatory framework precisely because the earliest multicellular animals were subject to external physical forces and effects that would have shaped and reshaped them in often nonlinear and abrupt fashions, and because the range of morphological outcomes (if not the particular outcome) emerging under these conditions were to a surprising extent physically inevitable.

We have proposed that this early period of physics-dependent morphogenesis is when much large-scale macroevolution took place (Newman and Müller 2000; Newman 2005; Newman *et al.* 2006). In particular, it was the era in which the major phyla were established. Only after extensive stabilizing and canalizing selection would the descendents of these early-diversifying organisms have settled into the mode described by the Modern Synthesis. In this current mode, gene mutation generally leads to deleterious or incremental alteration of the phenotype because generation of form is now guided by robust hierarchical programs of gene regulation. Any effects of the external environment on developmental outcomes, moreover, are either minimal or specifically incorporated into the generative program of the species (e.g., temperature-dependent sex determination in reptiles; Bowden *et al.* 2000).

2. Dynamical patterning modules vs. cell state switching mechanisms

Our concept is based on the presence of a set of molecules in certain modern unicellular organisms, the Choanozoa, and the inferred presence of the single-celled ancestors of the animals. The molecules in question (products of a subset of the genes of the "developmental-genetic toolkit"; Wilkins 2002; Carroll *et al.* 2005) were predisposed to assume novel functions in the multicellular state by mobilizing physical effects that were irrelevant to patterning on the scale of the individual cell. We refer to the joint effect of these ancient molecules and the physical processes they mobilize as "dynamical patterning modules" (DPMs). Most fundamentally, the multicellular state itself likely came into being by one class of ancestral molecules, the cadherins, forming a DPM by harnessing the physical effect of homophilic adhesion. Because of the limited set of relevant physical effects that apply to matter on the scale of cell aggregates, a set of inevitable morphological motifs arose in animal systems, constituting a deeply embedded orthogenetic principle in this class of organisms. These include the recurrent appearance of interior body cavities, multiple tissue layers, elongated bodies, segments, and appendages. The emergence of these structures thus required neither adaptationist nor incrementalist mechanisms.

Because physics can act with immediacy to shape and modify form, particularly in the developmental stages of organisms for which canalizing mechanism were not yet in place, early morphological diversification also need not have taken long periods of time. Genetic consolidation therefore likely followed, rather than accompanied, the rapid radiation of body plans referred to as the Cambrian explosion. In addition, while the limited number of physical effects that enter into DPMs are dictated by the laws of nature, the set of molecules involved in these modules is also limited, but this is just a fortuitous matter of their having been the only suitable ones present in the ancestral cells. The suggestion that the animal phyla emerged early and rapidly by means of the DPMs carries the implication that the associated molecules should be central to the developmental pathways of all extant animals, and this is borne out by the evidence.

The biosynthetic states of all cells are determined by the dynamics of transcription factor-mediated gene regulatory networks (GRNs) (Davidson 2006). Such networks, containing feedback and feed-forward loops by which the transcription factors promote and suppress their own and each other's synthesis, exhibit multistability (Forgacs and Newman 2005). The systems can thus switch among discrete states, the number of states always being much smaller than the total number of genes in the organism's genome. Since the genes that specify nontranscription factor proteins and regulatory RNAs are themselves subject to transcriptional control, the alternative stable states of the GRNs specify cell types distinguished by extensive biosynthetic differences. All unicellular organisms, be they bacteria, fungi, protists, or algae, exhibit alternative states of differentiation, both reversible and irreversible, under different conditions (e.g., Blankenship and Mitchell 2006; Vlamakis *et al.* 2008). This must also have been the case for the single-celled ancestors of the Metazoa, that is, the ancient and modern animals.

The transcription factors contained in the developmental-genetic toolkit form

GRNs that help generate distinct cell types during development. But an equally important aspect of development is the arrangement of cells into appropriately coherent spatiotemporal patterns (Salazar-Ciudad *et al.* 2003; Gilbert 2006). Unlike cell switching mechanisms, however, specific mechanisms of developmental pattern formation and tissue morphogenesis cannot have existed before multicellularity. In fact, the processes that generate spatial organization at the multicellular level are of an entirely different character from those which operate in individual cells (Newman and Bhat 2009; Newman *et al.* 2009).

3. The emergence of the Metazoa

The evolutionary history of the metazoans was initiated with remarkable rapidity during the late Precambrian and early Cambrian periods and is relatively well described (Rokas *et al.* 2005; Larroux *et al.* 2008). Development in all the metazoan phyla has been mediated by the same conserved developmental–genetic toolkit regulatory molecules for more than half a billion years (Carroll *et al.* 2005). The extant metazoans have classically been divided into the Eumetazoa, organisms which exhibit true tissues, epithelia with polarized cells, cell–cell junctions, a well-defined basement membrane, and neurons and muscle cells; and the sponges (Porifera) and Placozoa, which lack these features. *Trichoplax adhaerens*, the single known type of placozoan, contains several cell types and layers, but unlike the sponges (which exhibit gastrulation-like movements during development and complex labyrinthine morphologies) (Larroux *et al.* 2006), it has a simple, flat body without internal cavities. Surprisingly, on the basis of purely genetic criteria Placozoa may have greater genetic affinity to the Eumetazoa than the earlier-diverging sponges (Srivastava *et al.* 2008).

The eumetazoans, in turn, are divided into the diploblasts, consisting of the Cnidaria (e.g., hydroids and corals) and, traditionally, the Ctenophora (e.g., comb jellies), have two epithelial body layers and true lumens. The triploblasts (chordates, echinoderms, arthropods, mollusks, etc.), in contrast, have a third, mesenchymal, body layer. Genomic analysis has suggested that rather than being in the main line of the animals, the Ctenophora are actually a sister clade of the Metazoa (Dunn *et al.* 2008).

Sheetlike and hollow spherical forms (Yin *et al.* 2007), and budding and segmented tubes (Droser and Gehling 2008), possibly the most ancient metazoans, are seen beginning about 630 million years ago in fossil beds of the Precambrian Ediacaran period. Essentially all the triploblastic metazoan body plans then emerged within the space of no more than 20 million years, beginning about 535 million years ago (Conway Morris 2006), during the well-known Cambrian explosion. It has been suggested that the first Cnidaria (corals, hydroids) may have been holdovers from the Precambrian (Erwin 2008). Modern animals, and perhaps some of the Ediacaran forms, have a common ancestry in the Precambrian along with the Choanozoa, some of whose extant members are transiently colonial (Wainright *et al.* 1993; Lang *et al.* 2002; King *et al.* 2003; Philippe *et al.* 2004; Shalchian-Tabrizi *et al.* 2008).

Many of the toolkit genes, including some which have key roles in morphogenesis and pattern formation, are found in the genome of *Monosiga brevicollis*, an exclusively unicellular choanozoan (King *et al.* 2008). A few additional genes were added to the toolkit concomitant with the emergence of the sponges, and a few more arose with the simplest eumetazoans, the cnidarians. The Cambrian explosion followed with no more significant additions to the toolkit.

Metazoan complexity was thus achieved in a rapid fashion with an essentially unchanging set of ingredients. An essential evolutionary step for multicellularity was the acquisition by single-celled antecedents of the capacity to remain attached to one another after dividing. No new genes or gene products were required to mediate this function. The genome of *M. brevicollis* contains 23 putative cadherin genes, as well as 12 genes for C-type lectins (Abedin and King 2008; King *et al.* 2008), the protein products of which mediate cell attachment and aggregation in metazoan organisms in the presence of sufficient levels of extracellular calcium ion. Rising oceanic Ca2+ levels during the period in which multicellularity was established may have recruited these ancient proteins to new roles (Kazmierczak and Kempe 2004).

4. The developmental genetic toolkit: new contexts, new roles

Metazoan embryos employ a variety of patterning and shaping processes (Salazar-Ciudad *et al.* 2003), some of which are used in all the animal phyla and others of which are used in some of them. As noted above, the first DPM, designated ADH (*Table 1*), results in the formation of a multicellular cluster. Within such a cluster, any or all of the following can occur: the local coexistence of cells of more than one type, the formation of distinct cell layers, or an internal space or lumen, elongation of the cell cluster, the formation of repeated metameres or segments, the change in state or type of cells in one region of the cluster due to local or long-range signals from another region, the change in stiffness or elasticity of a cell layer, and the dispersal of cells while they continue to remain part of an integral tissue (reviewed in Forgacs and Newman 2005).

As the Metazoa emerged, DPMs mediated all of the above transformations by mobilizing physical forces and processes characteristic of viscoelastic, chemically active materials on the spatial scale of cell aggregates and tissue primordia (100 μm – 1 mm). Such materials are referred to by physicists as "soft matter" (De Gennes 1992) which is simultaneously an "excitable medium" (Mikhailov 1990).

DPM	Characteristic molecules	Physical principle	Morphogenetic Role
ADH	cadherins	adhesion	multicellularity
DAD	cadherins	differential adhesion	tissue multilayering
LAT	Notch	lateral inhibition	coexistence of alternative cell types
POLa	Wnt	cell surface anisotropy	lumen formation
POLp	Wnt	cell shape anisotropy	tissue elongation
OSC	Wnt + Notch	synchronized biochemical oscillation	morphogenetic fields; segmentation
MOR	TGF-β/BMP; Hh	diffusion	pattern formation
ASM	FGFs	diffusion	induction
TUR	MOR + Wnt + Notch	chemical waves	periodic patterning
ECM	collagen; chitin; fibronectin	stiffness; dispersal + cohesion	epithelial elasticity; skeletogenesis; epithelial-mesenchymal transformation

Table 1: Names, components and roles of major Dynamical Patterning Modules (DPMs)

Detailed descriptions of the properties of the major basic and combined DPMs and their developmental and proposed evolutionary roles can be found in earlier publications (Newman and Bhat 2008; 2009; Newman *et al.* 2009). Here these features will be characterized briefly. Each DPM is given a three-letter designation.

Adhesion and differential adhesion — As mentioned above, the emergence of Metazoan multicellularity depended on cadherins and C-type lectins of single-cell ancestors taking on the new function of cell–cell adhesion (ADH). If, in addition, subsets of cells within an aggregate contain sufficiently different levels of cell adhesion molecules on their surfaces, there will be a sorting into islands of more adhesive cells within lakes of less adhesive ones (Steinberg and Takeichi 1994). This constitutes a second DPM, differential adhesion (DAD). Since cells undergo random motion, small islands of like cell types will coalesce and an interface will be established, across which cells will not intermix (Steinberg 2003). This effect, which has the same physical basis as phase separation of two immiscible liquids such as oil and water (reviewed in Forgacs and Newman 2005), leads to the formation of nonmixing cell layers, an early stage of most animal embryogenesis.

Lateral inhibition and choice between alternative cell fates – Morphologically complex organisms always employs lateral inhibition (LAT) during embryogenesis, whereby early differentiating cells signal to cells adjacent to them to take on a different fate (Rose 1958; Meinhardt and Gierer 2000). Lateral inhibition in metazoans is mediated by the Notch signal transduction pathway, specifically, interaction of the cell surface receptor Notch with members of a class of other integral membrane proteins (Delta, Serrate/Jagged, and Lag2: the DSL proteins) that act as ligands for the receptor and mediators of Notch activity (Ehebauer *et al.* 2006). This mechanism does not determine the particular fate of any cell, but only enforces the coexistence of alternative fates in adjacent cells in the same cluster or aggregate.

The most basal metazoans to contain the Notch receptor are sponges (Nichols *et al.* 2006) though related protein modules probably existed in a choanaozoan ancestor (King *et al.* 2008). Lateral inhibition would have enabled basic cell pattern formation in these organisms and more complex animals.

Apical-basal and planar cell polarity – As noted above, cell aggregates behave like viscoelastic liquid droplets (reviewed in Forgacs and Newman 2005). This means that their default morphology is topologically solid (i.e., having no lumen), and spherical. Animal embryos defeat these morphological defaults by employing cell polarization. Cells can be polarized in one of two ways. When they become anisotropic along their surfaces (referred to as apical–basal (A/B) polarization; Karner *et al.* 2006b), interior spaces or lumens can arise within aggregates. Specifically, when A/B polarization leads cells to have lowered adhesiveness on one portion of their surface, they will preferentially attach to their neighbors on their more adhesive (lateral) portions, leaving the less adhesive (basal) portions adjoining an interior space (Newman 1998). Apical–basal polarity is also important in fostering layered tissue arrangements.

Tissue elongation may occur when cells individually polarize in shape (rather than surface properties), a phenomenon called planar cell polarity (PCP; Karner *et al.* 2006a). Planar-polarized cells can intercalate along their long axes, causing the tissue mass to narrow in the direction parallel to the cell's long axis, and consequently elongate in the orthogonal direction. This tissue reshaping is known as convergent extension (Keller *et al.* 2000; Keller 2002).

Both A/B polarity and PCP are mediated by secreted factors of the Wnt family (Karner *et al.* a,b). Which type of polarization occurs depends on the presence of different accessory proteins. The A/B- and PCP-inducing Wnt pathways are referred to, respectively, as the canonical and noncanonical Wnt pathways. In each case, the structural alterations of individual cells have novel consequences in a multicellular context, permitting multicellular aggregates to overcome the morphological defaults of solidity and sphericality. We designate the DPMs involving the Wnt pathway operating in a multicellular context as POLa and POLp (*Table 1*).

Some key intracellular components of the Wnt pathway have counterparts in fungi, where they also mediate cell polarity (Mendoza *et al.*, 2005). Their role in the shaping of metazoan embryos could only have emerged with the multicellular state, but the appearance of this new function would have been all but automatic. Sponges, which are characterized by many interior spaces, have genes for Wnt proteins and their ligands (Nichols *et al.* 2006). Such genes are also present in the placozoan *Trichoplax* (Srivastava *et al.* 2008), which despite containing only four cell types, has them arranged in three distinct layers, which is possible only if the cells are polarized.

Small, hollow, cell clusters identified in the Precambrian Doushantuo Formation in China and referred to as "embryos" (Chen *et al.* 2004; Hagadorn *et al.* 2006; Yin 2007) may actually have been the definitive forms of the earliest metazoans and metazoan-like organisms (Newman *et al.* 2006). The origination of these hollow forms at the transition between the Ediacaran biota and those of the Cambrian explosion was plausibly based on the presence of POLa. Specifically, the presence of the canonical Wnt pathway in a multicellular context could have readily led to the aggregates developing interior spaces.

Genes specifying components of the noncanonical Wnt pathway do not appear to be present in sponges and placozoans, which correspondingly show no sign of body elongation. They are, however, present in the morphologically more complex cnidarians (Guder *et al.* 2006), which display elongated body stalks and appendages.

Oscillations in cell state – As noted earlier, cell differentiated states are determined by intracellular transcription factor-based gene regulatory networks (GRNs). Such GRNs are typically multistable, but certain arrangements of positive and negative feedbacks will cause such systems to exhibit temporal oscillations in concentration of gene products (Goldbeter 1996; Reinke and Gatfield 2006).

In a single-celled organism, a periodic recurrence of cell state has no developmental consequences. If the oscillation involves downstream effectors of the Notch pathway (which normally enforces alternative cell fate decisions; see above), the result will be that cells will remain labile and uncommitted (Kageyama *et al.* 2008). Furthermore, if the oscillations become synchronized, the labile cell state will be coordinated across broad tissue domains, permitting concerted responses to a variety of developmental signals. Such synchronized Notch-associated oscillations have indeed been observed during early vertebrate development (Özbudak and Lewis 2008). Because of the near-ubiquity of oscillatory gene expression (Reinke and Gatfield 2006) and the inevitability of synchronization when oscillators are weakly interacting (as is typical for cells in a common tissue) (Garcia-Ojalvo 2004), we have proposed (Newman and Bhat 2009) that the oscillation DPM (OSC; *Table 1*) is at the basis of the ubiquitous but mechanistically elusive phenomenon of the "morphogenetic field" (Gilbert 2006).

Morphogen gradients and activator–inhibitor systems – While single-celled organisms can change their physiological state in response to molecules secreted into the microenvironment by other such cells (Luporini *et al.* 2006), this effect has novel developmental consequences when it occurs in a multicellular context and *gradients* can be formed. Secreted molecules that act as patterning signals in metazoan embryos by mediating concentration-dependent responses are termed morphogens (MOR; *Table 1*). Examples of morphogens are Wnt, discussed above, Hedgehog, BMP/TGF-β, and FGF (Zhu and Scott, 2004). The genome of marine sponges contains genes specifying examples of the first two of these categories of morphogens and their receptors (Nichols *et al.* 2006), whereas Placozoa contains components of the first three (Srivastava *et al.* 2008) and Cnidaria all four (Holstein *et al.* 2003; Rentzsch *et al.*, 2008).

The ability of one or a small group of cells to influence other cells via morphogens, either within a common tissue primordium, or, in a special case referred to as embryonic induction, asymmetrically across tissue boundaries (the ASM DPM; *Table 1*), enables the generation of nonuniform cellular patterns. Assuming that the function of morphogens is tied to the physical principle of molecular diffusion, Crick calculated that they would generate patterns over tens of hours on a spatial scale of 100 μm–1 mm, similarly to what is observed in embryos (Crick 1970). Building on this basic mechanism, evolution has often produced transport processes that are formally equivalent to diffusion, but which, by using additional cell-dependent modalities, are faster or slower than the simple physical process (Lander 2007).

When morphogens are positively autoregulatory, that is, directly or indirectly stimulatory of their own synthesis in target cells, they tend not to be maintained as gradients, since all cells eventually become morphogen sources. This tendency can be held in check, however, if the positively autoregulatory morphogen elicits a mechanism of lateral inhibition (such as the LAT DPM associated with Notch signaling). In this case, a zone will be induced around any peak of morphogen activity within which activation will not spread (Gierer and Meinhardt 1972; Meinhardt and Gierer 2000). Peaks of activation in such systems will form only at distances sufficiently far from one another so that the effects of the inhibitor are attenuated. This arrangement, termed local autoactivation–lateral inhibition (LALI) (Meinhardt and Gierer 2000; Nijhout 2003; Newman and Bhat 2007), can produce regularly spaced spots or stripes of morphogen concentration (TUR; *Table 1*). In contemporary metazoans the TUR DPM (named after the mathematician Alan Turing who first investigated such pattern-forming systems; Turing 1952), has been proposed to underlie pattern formation of the vertebrate limb skeleton (Newman and Frisch 1979; Hentschel *et al.* 2004), the dentition (Salazar-Ciudad and Jernvall 2002), feather germs (Jiang *et al.* 2004), and hair follicles (Sick *et al.* 2006).

The MOR DPM is used in conjunction with the OSC DPM in vertebrate *somitogenesis*, the process by which blocks of tissue, the primordia of vertebrae and associated muscles, form in a progressive spatiotemporal order along the central axis of vertebrate embryos. In the presomitic plate of vertebrate embryos, the expression of certain genes (including mediators of the Notch pathway, as discussed above) undergoes temporal oscillation with a period similar to the formation of the somites (Dequéant *et al.* 2008). These oscillations then become synchronized across the plate (Giudicelli *et al.* 2007; Kageyama *et al.* 2007; Riedel-Kruse *et al.* 2007). In conjunction with an FGF morphogen gradient with its source at one end of the extended embryo, a subset of the periodically expressed molecules provides the basis for the generation of somites in vertebrate embryos (Dequéant *et al.* 2008). The OSC DPM may have an analogous role in the segmentation of some arthropods (Salazar-Ciudad *et al.* 2001; Damen *et al.* 2005; Pueyo et al 2008).

Extracellular matrices – The DPMs ADH and DAD mediate the formation of "epithelioid" tissues and tissue layers, which are composed of cells that are directly attached to each other. In these tissues physical properties such as viscosity, elasticity, and cohesiveness are determined by the strength of cell-cell attachment and the rheology of the cytoplasm.

The other major cell aggregate or tissue type, "mesenchyme," is composed of cells that are embedded in a secreted macromolecular microenvironment, the extracellular matrix (ECM; Comper 1996; *Table 1*). In mesenchymal tissues physical properties are largely determined by the ECM, making them subject to a range of physical processes not seen in epithelioid tissues. The ECM molecules and the physics they mobilize thus constitute a novel DPM.

Metazoan ECMs consist largely of glycosaminoglycans (a category of polysaccharide) which are typically attached to proteins in the form of proteoglycans, and various fibrillar collagens, which occupy the interstitium between mesenchymal cells and the cells of more mature connective tissues. Metazoans also produce a network-type collagen and laminin, which are components of the basement membrane that attaches epithelial sheets to mesenchymal and connective tissues.

Genes specifying a variety of interstitial and basement membrane ECM proteins and cell surface receptors for ECMs are found in the *M. brevicollis* genome (King *et al.* 2008). It is unclear what function these molecules perform in the single-celled organism, or would have performed in its common ancestor with the Metazoa, but they have clearly been recruited to new roles in the multicellular context.

Sponges contain both epithelial-like and mesenchymal cells, which reside upon and within an ECM called the "mesohyl" (Wimmer *et al.* 1999). These organisms actively remodel the branched skeletal structures defined by their ECM by the continuous movement of their cells (Bond 1992), thus exhibiting environment-dependent morphological plasticity (Uriz *et al.* 2003), but only a limited array of morphological themes. It is only the triploblasts (arthropods, annelids, echinoderms, mollusks, chordates), which contain true epithelial and mesenchymal tissue types, that collectively exhibit the entire spectrum of DPM-generated motifs (Newman and Bhat 2009) (*Figure 1*).

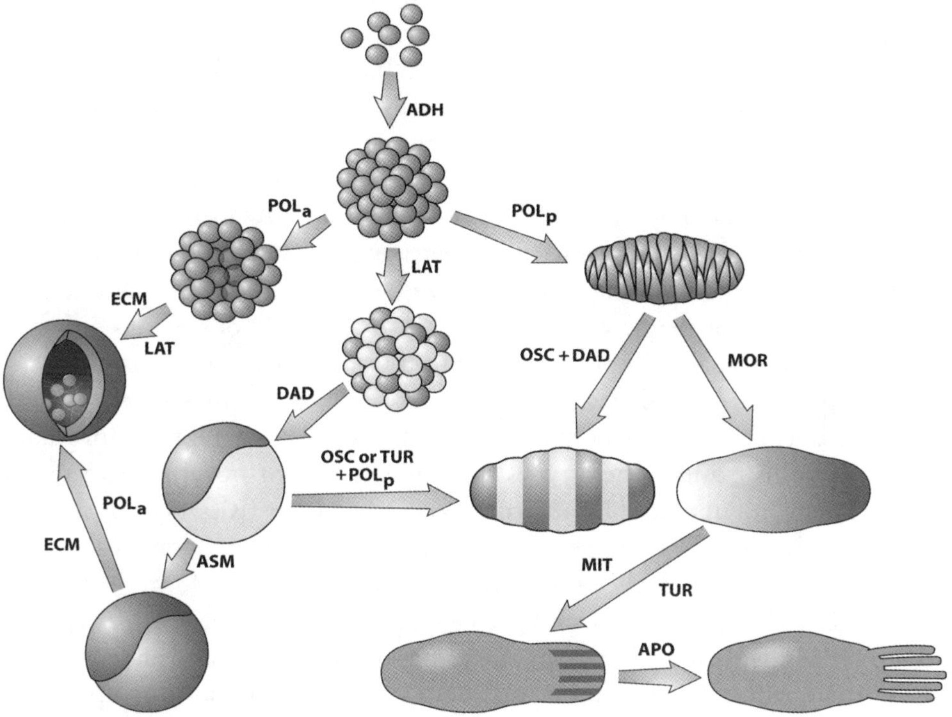

Figure 1. Schematic representation of single and combinatorial action of dynamical patterning modules (DPMs) in the generation of potential metazoan forms. Cells are represented individually in the upper tiers of the diagram, while the middle and lower tiers are shown at the scale of tissues. Beginning at the top, single cells form cell aggregates by the action of the ADH (adhesion, e.g., cadherins, lectins) module. The POL (polarity: Wnt pathway) module has two versions, apical-basal (POLa) and planar (POLp) polarity. POLa causes cells to have different surface properties at their opposite ends, leading to structurally polarized epithelial sheets and lumens within cell aggregates. POLp, in contrast, causes cells to elongate and intercalate in the plane, which leads to convergent extension and elongation of the cell mass. The LAT (lateral inhibition: Notch pathway) module transforms an aggregate of homotypic cells into one in which two or more cell types coexist in the same aggregate, while the expression of ADH molecules in different amounts lead to sorting out by the action of the differential adhesion (DAD) module. Production of diffusible molecules by cells capable of responding to these same molecules leads to morphogen (MOR, e.g., TGF-β/BMP, hedgehog) gradients, whereas morphogens can also act inductively and asymmetrically (ASM, e.g., FGFs) by being produced by one type of tissue and affecting a different type. Synchronous biochemical oscillation (OSC) of key components of the Notch and Wnt pathways, in conjunction with the DAD module, can generate segments. Appropriate feedback relationships among activating and inhibitory morphogens can lead to patterns with repetitive elements by Turing-type reaction-diffusion processes (TUR). The action of the MAPK signaling pathway in the context of multicellular aggregate containing morphogen gradients leads to nonuniform growth by the mitogenesis (MIT) module, whereas the apoptosis (APO) module leads to differential cell loss. The secretion of extracellular matrix (ECM, e.g., collagen, fibronectin) between cells or into tissue spaces creates novel mechanical properties in cell sheets or masses, or new microenvironments for cell translocation. (Adapted from Newman and Bhat, 2009, which can be referred to for additional details, including DPMs not discussed in the text.)

5. Niche construction and genetic accommodation

Models based on the ideas of the Modern Synthesis would predict that morphologically aberrant subpopulations brought into being by DPMs would be poorly adapted to the ecological niches inhabited by the originating species. But niches are not preexisting slots in the natural environment passively occupied by organisms that have the right set of characters. They are explored, selected by, and in many cases constructed by their inhabitants (Levins and Lewontin 1985; Odling-Smee *et al.* 2003). When novelties arise, particularly (as could be the case with DPM-based innovation mechanisms) in multiple members of a population, there is no requirement for the new forms to remain at their sites of origin.

If saltation inescapably implied maladaption we would expect invasive species to be less prevalent than they actually are (Carroll 2007; Stohlgren *et al.* 2008), and would not anticipate phenomena such as "transgressive segregation" in plants, whereby hybrids exhibit phenotypes that are extreme or novel relative to the parental lines wind up founding and colonizing new niches (Rieseberg *et al.* 1999; 2003). At early stages of a lineage's history, before a high degree of organism-niche coadaptation had evolved, the possibility of organisms "bolting" from their niches and setting up elsewhere would have been even greater than in present-day species.

Since DPMs by definition incorporate physical mechanisms and effects, the results of their action depend on externalities — ionic composition, temperature, pressure and so on. As long as DPMs were the major determinants of form, consistency of development outcome would thus require stability of the relevant environmental inputs. The niche to which a novel form would initially be best maintained would thus be the one that provided the conditions for its existence. The most effective way a phenotypically plastic organism can assume a morphotype independent of environmental lability, however, is via "consolidating" genetic or epigenetic change. Selection for persistence of an environmentally induced phenotype, variously termed stabilizing selection (Schmalhausen 1949), genetic assimilation (Waddington 1961), or genetic accommodation (West-Eberhard 2003), can convert body plans and morphological characters that started out as dependent on intrinsic physical properties of tissues and external conditions into products of lineage-specific developmental programs (Newman 1994; Newman and Müller 2000).

6. Conclusion: Darwinism as a limiting case

I have argued here that evolutionary mechanisms have themselves evolved. This analysis has involved positing roles for features that have been relegated to the margins, or worse, in the standard accounts of evolution associated with the Modern Synthesis: saltation, orthogenesis and inheritance of acquired characters. These features are all natural consequences of considering biological systems to be material, i.e., physical, systems.

To take them in order: the possibility of saltational change in an organism's phenotype arises from the propensity of virtually all complex physical systems to exhibit nonlinear behaviors and changes in state. Phase transitions, such as melting or vaporization, or bulk transformations, such as the change from undulatory to vortical motion, are examples that present themselves if we just confine our attention to water. Living tissues of course have many more latent possibilities.

Orthogenesis is just a reflection of the fact that any physical system will assume a limited set of characteristic forms based on its inherent dynamics or modes of behavior. Liquid water can be still, or form waves or whirlpools; it can break up into drops or form rivulets on a surface. It cannot form elongated or branched structures in three-dimensions, or enclose hollow spaces.

Inheritance of acquired characters is a bit more complicated, since the typical physical system does not reproduce and therefore does not exhibit inheritance in the biological sense. This property is based on *plasticity*, the propensity of any physical system to assume alternative forms based on its inherent properties and its external conditions. No organism, no matter how under the influence of its genes, is immune from some degree of phenotypic plasticity. That is to say, organisms are not unwavering implementations of rigid genetic programs. And if an organism undergoes development as all multicellular forms do, plasticity will be manifested at nearly all stages of its ontogeny.

Developmental processes, being functions of dynamical patterning modules, will necessarily exhibit combinations of the associated morphological motifs — multiple tissue layers, lumens, segments, appendages, rod- and nodule-like cell condensations and cells dispersions. These are also the motifs that appeared in the course of metazoan evolution, hence the orthogenic character of the early stages of this process. The physical nature of the DPMs ensured that development of the earliest multicellular animals was plastic and saltational. Their molecular-genetic nature guaranteed that developmental pathways were inheritable and could evolve. If the quantity and quality of the gene products change (by mutation, for example), this will not alter the basic material nature of cell aggregates and tissues but can influence which of their inherent structural modes they will assume.

Evolution of the DPM-associated and other genes can restrict, channel, and render increasingly stereotyped, the morphological outcomes of developmental processes. This would suppress the tendency of organisms to deviate dramatically in form from their progenitors, render variation incremental, with no obvious preferred directions, around canalized phenotypic norms that will exhibit little developmental plasticity. In other words, organisms would eventually evolve into entities whose further evolution will be largely non-saltational, non-orthogenic and non-Lamarckian.

Uniformitarianism is the supposition Darwin adopted from the geologists James Hutton and Charles Lyell, that similar forces drive morphological change at all stages of transformation (Gould 1987). It is clear from the previous discussion that the framework presented here is not uniformitarian. From our viewpoint, indeed, the Darwinian mode of microevolution is a late product of a more comprehensive process that at its earlier, more "physical" stages was generative of macroevolutionary, i.e., phylum-scale, transitions.

It should not be surprising that Darwin's mechanism would come to be seen as a special case of a broader theory of organismal change. Darwin (and Alfred Russel Wallace, the co-originator of natural selection), of necessity, based their evolutionary hypothesis on the properties of present-day organisms — Galapagos finches, domesticated animals, orchids — that are hundreds of millions of years removed from their phylogenetic origins. Not only, as invariably noted by historians of biology, were the mechanisms of inheritance unknown to Darwin and Wallace, but so were the mechanisms of development. Embryogenesis in animals and most plants lead to defined outcomes. The only readily observable inherited variations within populations are small ones. Selection based on adaptive advantage exerted over vast periods of time was thus an inspired guess for how large-scale morphological differences could be generated. Darwin's and Wallace's radically materialist theory of evolution was thus formulated as a uniformitarian and incrementalist one.

Our current understanding of the physical-molecular modules that constitute developmental mechanisms, including recognition of their pre-phylum origination and prolific dynamics, permits us to take a longer view than was available to Darwin and Wallace, while holding fast to their materialist philosophical perspective. Although present-day organisms are indeed subject to microevolution by natural selection, our new outlook suggests that their prodigious variety can only have been produced in a world that is largely lost to the distant past.

Acknowledgments

I thank Ramray Bhat for very helpful comments on an earlier version of this paper and the National Science Foundation for support.

REFERENCES

Abedin M., King N., 2008, "The premetazoan ancestry of cadherins", *Science* 319: 946–8.

Alberch P., 1989, "The logic of monsters: evidence for internal constraint in development and evolution", *Geobios* 19: 21–57.

Badyaev A. V., 2005, "Stress-induced variation in evolution: from behavioural plasticity to genetic assimilation", *Proceedings of the Royal Society of London. Series B: Biological Sciences* 272: 877–86.

Badyaev A. V., Foresman K. R., Young, R. L., 2005, "Evolution of morphological integration: developmental accommodation of stress-induced variation", *American Naturalist* 166: 382–95.

Bateson W., 1909, "Heredity and variation in modern lights", in Seward A. C. (ed.), *Darwin and modern science*, Cambridge: Cambridge University Press.

Blankenship J. R., Mitchell A. P., 2006, "How to build a biofilm: a fungal perspective", *Current Opinion in Microbiology* 9: 588–94.

Bond C., 1992, "Continuous cell movements rearrange anatomical structures in intact sponges", *Journal of Experimental Zoology* 263: 284–302.

Bowden R. M., Ewert M. A., Nelson C. E., 2000, "Environmental sex determination in a reptile varies seasonally and with yolk hormones", *Proceedings of the Royal Society of London. Series B: Biological Sciences* 267: 1745–9.

Callebaut W., Müller G. B., Newman S. A., 2007, "The organismic systems approach. Streamlining the naturalistic agenda", in Brandon S. R., Brandon R. N. (eds.), *Integrating Evolution and Development. From Theory to Practice*, Cambridge, Mass.: MIT Press.

Carroll S. B., Grenier J. K., Weatherbee S. D., 2005, *From DNA to Diversity: Molecular Genetics and the Evolution of Animal Design,* 2nd edn., Malden, Mass.: Blackwell Pub.

Carroll S. P., 2007, "Brave New World: the epistatic foundations of natives adapting to invaders", *Genetica* 129: 193–204.

Comper W. D., 1996, *Extracellular Matrix,* Amsterdam: Harwood Academic Publishers.

Conway Morris S., 2006, "Darwin's dilemma: the realities of the Cambrian 'explosion'", *Philosophical Transactions of the Royal Society of London, Series B: Biological Sciences* 361: 1069–83.

Crick F. H. C., 1970, "Diffusion in embryogenesis", *Nature* 225: 420–422.

Damen W. G., Janssen R., Prpic N. -M., 2005, "Pair rule gene orthologs in spider segmentation", *Evolution & Development* 7: 618–28.

Darwin C., 1859, *On the Origin of Species by Means of Natural Selection, or, The Preservation of Favoured Races in the Struggle for Life,* London: J. Murray.

Davidson E. H., 2006, *The Regulatory Genome: Gene Regulatory Networks in Development and Evolution,* Amsterdam, London: Elsevier/Academic Press.

de Gennes P. G., 1992, "Soft matter", *Science* 256: 495–497.

Dequéant M. L., Pourquié O., 2008, "Segmental patterning of the vertebrate embryonic axis", *Nature Reviews Genetics* 9: 370–82.
Droser M. L., Gehling J. G., 2008, "Synchronous aggregate growth in an abundant new Ediacaran tubular organism", *Science* 319: 1660–2.
Dunn C. W., Hejnol A., Matus D. Q., Pang K., Browne W. E., Smith S. A., Seaver E., Rouse G. W., *et al.,* 2008, "Broad phylogenomic sampling improves resolution of the animal tree of life", *Nature* 452: 745–9.
Ehebauer M., Hayward P., Arias A. M., 2006, "Notch, a universal arbiter of cell fate decisions", *Science* 314: 1414–5.
Erwin D. H., 2008, "Wonderful Ediacarans, wonderful cnidarians?" *Evolution & Development* 10: 263–4.
Fisher R. A., 1930, *The Genetical Theory of Natural Selection,* Oxford: Clarendon Press.
Forgacs G., Newman S. A., 2005, *Biological Physics of the Developing Embryo,* Cambridge: Cambridge University Press.
Fusco G., Minelli A., 2008, *Evolving Pathways: Key Themes in Evolutionary Developmental Biology,* Cambridge; New York: Cambridge University Press.
Garcia-Ojalvo J., Elowitz M. B., Strogatz S. H., 2004, "Modeling a synthetic multicellular clock: repressilators coupled by quorum sensing", *Proceedings of the National Academy of Sciences USA* 101: 10955–60.
Gierer A., Meinhardt H., 1972, "A theory of biological pattern formation", *Kybernetik* 12, 30–39.
Gilbert S. F., 2006, *Developmental Biology,* 8th edn., Sunderland, Mass.: Sinauer Associates.
Giudicelli F., Özbudak E. M., Wright G. J., Lewis J., 2007, "Setting the tempo in development: an investigation of the zebrafish somite clock mechanism", *PLoS Biology* 5: 1309-1323.
Goldbeter A., 1996, *Biochemical Oscillations and Cellular Rhythms: The Molecular Bases of Periodic and Chaotic Behaviour,* Cambridge: Cambridge University Press.
Goldschmidt R. B., 1940, *The Material Basis of Evolution,* New Haven: Yale University Press.
Goodman R. M., 2008, "Latent effects of egg incubation temperature on growth in the lizard *Anolis carolinensis*", *Journal of Experimental Zoology. Part A: Ecological Genetics and Physiology* 309: 525–33.
Gould S. J., 1987, *Time's Arrow, Time's Cycle: Myth and Metaphor in the Discovery of Geological time, Jerusalem-Harvard Lectures,* Cambridge, Mass.: Harvard University Press.
Guder C., Philipp I., Lengfeld T., Watanabe H., Hobmayer B., Holstein T. W., 2006, "The Wnt code: cnidarians signal the way", *Oncogene* 25: 7450–60.
Hagadorn J. W., Xiao S., Donoghue P. C., Bengtson S., Gostling N. J., Pawlowska M., Raff E. C., Raff R. A. *et al.,* 2006, "Cellular and subcellular structure of neoproterozoic animal embryos", *Science* 314: 291–4.
Hentschel H. G., Glimm T., Glazier J. A., Newman S. A., 2004, "Dynamical mechanisms for skeletal pattern formation in the vertebrate limb", *Proceedings of the Royal Society of London, Series B: Biological Sciences* 271: 1713–22.

Holstein T. W., Hobmayer E., Technau U., 2003, "Cnidarians: an evolutionarily conserved model system for regeneration?" *Developmental Dynamics* 226: 257–67.

Jablonka E., Lamb M. J., 1995, *Epigenetic Inheritance and Evolution,* Oxford, U.K.: Oxford University Press.

Jablonka E., Lamb M. J., 2005, *Evolution in Four Dimensions: Genetic, Epigenetic, Behavioral, and Symbolic Variation in the History of Life, Life and Mind,* Cambridge, Mass.: MIT Press.

Kageyama R., Masamizu Y., Niwa Y., 2007, "Oscillator mechanism of Notch pathway in the segmentation clock", *Developmental Dynamics* 236: 1403–9.

Kageyama R., Ohtsuka T., Shimojo H., Imayoshi I., 2008, "Dynamic Notch signaling in neural progenitor cells and a revised view of lateral inhibition", *Nature Neuroscience* 11: 1247–51.

Karner C., Wharton K. A. Jr., Carroll T. J., 2006a, "Planar cell polarity and vertebrate organogenesis", *Seminars in Cell & Developmental Biology* 17: 194–203.

Karner C., Wharton K. A. Jr., Carroll T. J., 2006b, "Apical-basal polarity, Wnt signaling and vertebrate organogenesis", *Seminars in Cell & Developmental Biology* 17: 214–22.

Kazmierczak J., Kempe S., 2004, "Calcium build-up in the Precambrian sea: A major promoter in the evolution of eukaryotic life", in Seckbach J. (ed.), *Origins*, Dordrecht: Kluwer.

Keller R., 2002, "Shaping the vertebrate body plan by polarized embryonic cell movements", *Science* 298: 1950–4.

Keller R., Davidson L., Edlund A., Elul T., Ezin M., Shook D., Skoglund P., 2000, "Mechanisms of convergence and extension by cell intercalation", *Philosophical Transactions of the Royal Society of London, Series B: Biological Sciences* 355: 897–922.

King N., Hittinger C. T., Carroll S. B., 2003, "Evolution of key cell signaling and adhesion protein families predates animal origins", *Science* 301: 361–3.

King N., Westbrook M. J., Young S. L., Kuo A., Abedin M., Chapman J., Fairclough S., Hellsten, U. *et al.,* 2008, "The genome of the choanoflagellate *Monosiga brevicollis* and the origin of metazoans", *Nature* 451: 783–8.

Lander A. D. 2007, "Morpheus unbound: reimagining the morphogen gradient", *Cell* 128: 245–56.

Lang B. F., O'Kelly C., Nerad T., Gray M. W., Burger G. 2002, "The closest unicellular relatives of animals", *Current Biology* 12: 1773–8.

Larroux C., Fahey B., Liubicich D., Hinman V. F., Gauthier M., Gongora M., Green K., Worheide G., Leys S. P., Degnan B. M., 2006, "Developmental expression of transcription factor genes in a demosponge: insights into the origin of metazoan multicellularity", *Evolution & Development* 8: 150–73.

Larroux C., Luke G. N., Koopman P., Rokhsar D. S., Shimeld S. M., Degnan B. M., 2008, "Genesis and expansion of metazoan transcription factor gene classes", *Molecular Biology and Evolution* 25: 980–96.

Levins R., Lewontin R. C., 1985, *The Dialectical Biologist,* Cambridge, Mass.: Harvard University Press.

Luporini P., Vallesi A., Alimenti C., Ortenzi C., 2006, "The cell type-specific signal proteins (pheromones) of protozoan ciliates", *Current Pharmaceutical Design* 12: 3015–24.

Margulis L., Sagan D., 2002, *Acquiring Genomes: A Theory of the Origins of Species,* New York: Basic Books.

Mayr E., 1974, "Teleological and teleonomic, a new analysis", in Cohen R. S., Wartowsky M. W. (eds.), *Methodological and Historical Essays in the Natural and Social Sciences,* Boston Studies in the Philosophy of Science 14, Dordrecht: Reidel, 91–117.

Mayr E., 1982, *The Growth of Biological Thought: Diversity, Evolution, and Inheritance,* Cambridge, Mass.: Belknap Press.

Meinhardt H., Gierer A., 2000, "Pattern formation by local self-activation and lateral inhibition", *Bioessays* 22: 753–60.

Mendoza M., Redemann S., Brunner D., 2005, "The fission yeast MO25 protein functions in polar growth and cell separation", *European Journal of Cell Biology* 84: 915–26.

Mikhailov A. S., 1990, *Foundations of synergetics I,* Berlin: Springer-Verlag.

Moczek A. P., 2008, "On the origins of novelty in development and evolution", *Bioessays* 30: 432–47.

Müller G. B., 2007, "Evo-devo: extending the evolutionary synthesis", *Nature Reviews Genetics* 8: 943–9.

Müller G. B., Newman S. A., 2005, "The innovation triad: an EvoDevo agenda", *Journal of Experimental Zoology. Part B: Molecular and Developmental Evolution* 304: 487–503.

Müller G. B., Streicher J., 1989, "Ontogeny of the syndesmosis tibiofibularis and the evolution of the bird hindlimb: a caenogenetic feature triggers phenotypic novelty", *Anatomy and Embryology* 179: 327–39.

Newman S. A., 1994, "Generic physical mechanisms of tissue morphogenesis: a common basis for development and evolution", *Journal of Evolutionary Biology* 7: 467–488.

Newman S. A., 1998, "Epithelial morphogenesis: a physico-evolutionary interpretation", in Chuong C.-M. (ed.), *Molecular Basis of Epithelial Appendage Morphogenesis,* Austin, TX: R. G. Landes.

Newman S. A., 2005, "The pre-Mendelian, pre-Darwinian world: shifting relations between genetic and epigenetic mechanisms in early multicellular evolution", *Journal of Biosciences* 30: 75–85.

Newman S. A., 2006, "The developmental-genetic toolkit and the molecular homology-analogy paradox", *Biological Theory* 1: 12–16.

Newman S. A., Bhat R., 2007, "Activator-inhibitor dynamics of vertebrate limb pattern formation", *Birth Defects Research, Part C: Embryo Today* 81: 305–319.

Newman S. A., Bhat R., 2008, "Dynamical patterning modules: physico-genetic determinants of morphological development and evolution", *Physical Biology* 5: 15008.

Newman S. A., Bhat R., 2009, "Dynamical patterning modules: a "pattern language" for development and evolution of multicellular form", *International Journal of Developmental Biology* 53: 693-705.

Newman S. A., Comper W. D., 1990, " 'Generic' physical mechanisms of morphogenesis and pattern formation", *Development* 110: 1–18.

Newman S. A., Forgacs G., Müller G. B., 2006, "Before programs: the physical origination of multicellular forms", *International Journal of Developmental Biology* 50: 289–99.

Newman S. A., Frisch H. L., 1979, "Dynamics of skeletal pattern formation in developing chick limb", *Science* 205: 662–668.

Newman S. A., Bhat R., Mezentseva N. V., 2009, "Cell state switching networks and dynamical patterning modules: complementary mediators of plasticity in development and evolution", *Journal of Biosciences* 34: 553-572.

Newman S. A., Müller G. B., 2000, "Epigenetic mechanisms of character origination", *Journal of Experimental Zoology (Molecular and Developmental Evolution)* 288: 304–17.

Nichols S. A., Dirks W., Pearse J. S., King N., 2006, "Early evolution of animal cell signaling and adhesion genes", *Proceedings of the National Academy of Sciences USA* 103: 12451–6.

Nijhout H. F., 2003, "Gradients, diffusion and genes in pattern formation", in Müller G. B., Newman S. A. (eds.), *Origination of Organismal Form: Beyond the Gene in Developmental and Evolutionary Biology*, Cambridge, Mass.: MIT Press.

Odling-Smee F. J., Laland K. N., Feldman M. W., 2003, *Niche Construction*, Princeton and Oxford: Princeton University Press.

Özbudak E. M., Lewis J., 2008, "Notch signalling synchronizes the zebrafish segmentation clock but is not needed to create somite boundaries", *PLoS Genetics* 4: e15.

Philippe H., Snell E. A., Bapteste E., Lopez P., Holland P. W., Casane D., 2004, "Phylogenomics of eukaryotes: impact of missing data on large alignments", *Molecular Biology and Evolution* 21: 1740–52.

Pigliucci M., 2001, *Phenotypic Plasticity: Beyond Nature and Nurture,* Baltimore: Johns Hopkins University Press.

Pueyo J. I., Lanfear R., Couso J. P., 2008, "Ancestral Notch-mediated segmentation revealed in the cockroach *Periplaneta americana*", *Proceedings of the National Academy of Sciences USA* 105: 16614–9.

Reinke H., Gatfield D., 2006, "Genome-wide oscillation of transcription in yeast", *Trends in Biochemical Sciences* 31: 189–91.

Rentzsch F., Fritzenwanker J. H., Scholz C. B., Technau U., 2008, "FGF signalling controls formation of the apical sensory organ in the cnidarian *Nematostella vectensis*", *Development* 135: 1761–9.

Riedel-Kruse I. H., Muller C., Oates A. C., 2007, "Synchrony dynamics during initiation, failure, and rescue of the segmentation clock", *Science* 317: 1911–5.

Rieseberg L. H., Archer M. A., Wayne R. K., 1999, "Transgressive segregation, adaptation and speciation", *Heredity* 83 (Pt 4): 363–72.

Rieseberg L. H., Widmer A., Arntz A. M., Burke J. M., 2003, "The genetic architecture necessary for transgressive segregation is common in both natural and domesticated populations", *Philosophical Transactions of the Royal Society of London, Series B: Biological Sciences* 358: 1141–7.
Robert J. S., 2004, *Embryology, Epigenesis, and Evolution: Taking Development Seriously*, Cambridge studies in philosophy and biology, Cambridge; New York: Cambridge University Press.
Rokas A., Kruger D., Carroll S. B., 2005, "Animal evolution and the molecular signature of radiations compressed in time", *Science* 310: 1933–8.
Rose S. M., 1958, "Feedback in the differentiation of cells", *Scientific American* 199: 36–41.
Salazar-Ciudad I., 2006, "On the origins of morphological disparity and its diverse developmental bases", *BioEssays* 28: 1112–22.
Salazar-Ciudad I., Jernvall J., 2002, "A gene network model accounting for development and evolution of mammalian teeth", *Proceedings of the National Academy of Sciences USA* 99: 8116–20.
Salazar-Ciudad I., Jernvall J., Newman S. A., 2003, "Mechanisms of pattern formation in development and evolution", *Development* 130: 2027–37.
Salazar-Ciudad I., Solé R., Newman S. A., 2001, "Phenotypic and dynamical transitions in model genetic networks, II. Application to the evolution of segmentation mechanisms", *Evolution & Development* 3: 95–103.
Schmalhausen I. I., 1949, *Factors of Evolution,* Philadelphia: Blakiston.
Shalchian-Tabrizi K., Minge M. A., Espelund M., Orr R., Ruden T., Jakobsen K. S., Cavalier-Smith T., 2008, "Multigene phylogeny of choanozoa and the origin of animals", *PLoS ONE* 3: e2098.
Sick S., Reinker S., Timmer J., Schlake T., 2006, "WNT and DKK determine hair follicle spacing through a reaction-diffusion mechanism", *Science* 314: 1447–50.
Simpson G. G., 1944, *Tempo and Mode in Evolution*, Columbia biological series, no. 15, New York: Columbia University Press.
Srivastava M., Begovic E., Chapman J., Putnam N. H., Hellsten U., Kawashima T., Kuo A., Mitros T. *et al.*, 2008, "The *Trichoplax* genome and the nature of placozoans", *Nature* 454: 955–60.
Steinberg M. S., 2003, "Cell adhesive interactions and tissue self-organization", in Müller G. B., Newman S. A. (eds.), *Origination of Organismal Form: Beyond the Gene in Developmental and Evolutionary Biology*, Cambridge, Mass.: MIT Press.
Steinberg M. S., Takeichi M., 1994, "Experimental specification of cell sorting, tissue spreading, and specific spatial patterning by quantitative differences in cadherin expression", *Proceedings of the National Academy of Sciences USA* 91: 206–9.
Stohlgren T. J., Barnett D. T., Jarnevich C. S., Flather C., Kartesz J., 2008, "The myth of plant species saturation", *Ecology Letters* 11: 313–222.
Thompson D. A. W., 1942, *On Growth and Form,* 2nd edn. Cambridge: Cambridge University Press.

Trut L., Oskina I., Kharlamova A., 2009, "Animal evolution during domestication: the domesticated fox as a model", *BioEssays* 31: 349–60.

Turing A. M., 1952, "The chemical basis of morphogenesis", *Philosophical Transactions of the Royal Society of London, Series B: Biological Sciences* 237: 37–72.

Uriz M. J., Turon X., Becerro M. A., Agell G., 2003, "Siliceous spicules and skeleton frameworks in sponges: origin, diversity, ultrastructural patterns, and biological functions", *Microscopy Research and Technique* 62: 279–99.

Vedel V., Chipman A. D., Akam M., Arthur W., 2008, "Temperature-dependent plasticity of segment number in an arthropod species: the centipede Strigamia maritima", *Evolution & Development* 10: 487–92.

Vlamakis H., Aguilar C., Losick R., Kolter R., 2008, "Control of cell fate by the formation of an architecturally complex bacterial community", *Genes and Development* 22: 945–53.

Waddington C. H., 1942, "Canalization of development and the inheritance of acquired characters", *Nature* 150: 563–565.

Waddington C. H., 1961, "Genetic assimilation", *Advances in Genetics* 10: 257–93.

Wainright P. O., Hinkle G., Sogin M. L., Stickel S. K., 1993, "Monophyletic origins of the metazoa: an evolutionary link with fungi", *Science* 260: 340–342.

Weismann A., 1892, *Das Keimplasma; eine Theorie der Vererbung,,* Jena: Fischer.

West-Eberhard M. J., 2003. *Developmental Plasticity and Evolution*, Oxford, New York: Oxford University Press.

Wilkins A. S., 2002, *The Evolution of Developmental Pathways,* Sunderland, Mass.: Sinauer Associates.

Wimmer W., Perovic S., Kruse M., Schröder H. C., Krasko A., Batel R., Müller W. E., 1999, "Origin of the integrin-mediated signal transduction. Functional studies with cell cultures from the sponge *Suberites domuncula*", *European Journal of Biochemistry* 260: 156–65.

Woese C. R., Goldenfeld N., 2009, "How the microbial world saved evolution from the scylla of molecular biology and the charybdis of the modern synthesis", *Microbiology and Molecular Biology Reviews* 73: 14–21.

Yin L., Zhu M., Knoll A. H., Yuan X., Zhang J., Hu J., 2007, "Doushantuo embryos preserved inside diapause egg cysts", *Nature* 446: 661–3.

Zhu A. J., Scott M. P., 2004, "Incredible journey: how do developmental signals travel through tissue?", *Genes and Development* 18: 2985–97.

Section III
Man in evolution

Introduction

150 years of fossil discoveries and genetic researches have thrown light on the origins of man, a theme (subject) only mentioned by Darwin in *The Origin of Species* and in the following *The Descent of Man* (1871) faced only on the basis of similarities between man and animals.

Currently we have a rich fossil documentation, even if ever scarce in relation to the long period of the evolutionary process. We can add the information we can obtain by molecular biology. The data at our disposal show a sufficient picture of information that brings to interpretations, theories and hypothesis about the course of the human evolution and the different factors of this process.

The Darwinian explanation is the most followed, but we should keep in mind the recent views of the evolutionary developmental biology (*evo-devo*), that look at other evolutionary models focused in genetic and epigenetic factors. They could be added to the environmental factors of natural selection. To these factors is to be joined human cultural behavior in competition with the environment and as a factor of co-evolution.

The development of paleoanthropology with a brief overview of the most important discoveries until about 30 years ago is presented by **Giorgio Manzi**. The synopsis is followed by an introduction to current methods of research, such as geometric morphometrics for the construction of phylogenetic trees. The author points out the process of encephalization which started with the emergence of the genus *Homo* and followed an exponential increase during the last 2 million of years (ontogenetic revolution). The origin of *Homo sapiens* is considered a convincing example of evolution by means of natural selection combined with the model of punctuated equilibria.

A general view of the evolutionary history of Hominids in the last 10 millions of years is briefly presented by **Yves Coppens** in four parts: the last common ancestors of Hominids and chimpanzees; the prehumans; the emergence of Man; the evolution and expansion of Man. The interpretative key of the appearance of man is the adaptation to more open and drier environment. These conditions, in the interpretation of the author, favored a larger size of the brain, the emergence of consciousness, and the building of a new environment, the cultural one. *Homo* is a product of nature, a necessary adaptation to a climatic change, and marks a discontinuity in a continuum.

Olga Rickards and **Gianfranco Biondi** present the main results of molecular biology in the studies of human evolution of the last ten years. They regard not only the DNA of the genome of apes and *Homo sapiens*, but also ancient DNAmt and DNA nuclear in fossils remains since Neanderthal remains. We have new information about the split between early humans and African apes and about the emergence of *Homo sapiens*, in particular about the possible relationships between Neanderthals and *Homo sapiens* during the period of their coexistence. Moreover a possible new taxonomy of apes and humans could be derived from the genetic similarity of chimpanzees and humans according to some authors.

The modalities of the acquisition of bipedality in human evolution are the object of the paper of **Anne Dambricourt Malassé**. According to her, losing arboricolism is not necessary to become verticalized because the process of verticalization is a dynamic embryonic acquisition. In particular, she considers the mandible in its basicranial context in the great Ape clade, in Australopithecine clade and in the oldest species of *Homo* clade. The bipedality is considered the result of a macroevolution explanable with new ontogenetic embryonic pattern, attested by basi-cranial contraction, and should be referred to homeotic genes and to the model of punctuated equilibria better than to the Darwinian gradualism in relation with environmental changes.

Paleocultural approach in human evolution is addressed by **Fiorenzo Facchini.** According to him culture is to be considered as planning capacity and symbolization, that are human characteristics and may be recognized in the products of technology, in linguistic communication, and in spiritual manifestations. Culture shows a development since *Homo habilis* until *Homo sapiens*, and represents a discontinuity with respect to the behavior of Australopithecus and apes. Different implications of culture are considered (evolutionary, ecological, phenomenological, cosmological, philosophical). Moreover, he discusses the ethical behavior which is not linked to natural selection, but to the values and freedom, that is, to culture in the specific sense.

After drawing the quantitative relationship between social group size and relative neocortex in primates, **R. I. M. Dunbar** points out different layers of social networks in human groups including approximately 50, 150, 500 and 1500 members. The dimension of 150 members would be the best for personal relations. Three specific behavioral mechanisms are suggested: laughter, music and religion, that are particularly good in humans at triggering the release of endorphins which are useful for the communal activity. The author links the social networks to the dimensional growth of the human brain. The social networks are favored by language, by intentionality, by the world of imagination.

The participation of human activity by culture to the process of human evolution, often indicated as co-evolution, is illustrated by **Colin Renfrew**, according to whom cultural evolution occurred from the human dispersal of 60.000 years ago and was essentially an increase of the different cultures or cultural systems. In this period, considered as tectonic phase, the genotype is broadly fixed, while essentially cultural innovations and cultural evolution took place. Most of these innovations, some of them very important and differentiated in various areas (as agriculture, metallurgy or urbanization), occurred independently along several trajectories of development in different continents.

In the light of the contributions offered in this section we see the complexity of the human evolution, which is not explainable by means of a unique model of interpretation. The Darwinian framework, based on the gradual mutations and natural selection, seems to be integrated with other models, such as evo-devo biology and the theory of punctuated equilibria, particularly for the emergence of bipedality and for the process of cerebralization. The growth of the brain and the development of culture are to be considered the main factors of the success of human species.

Fiorenzo Facchini

Human evolution: A brief history of the research since Darwin's time

Giorgio Manzi

Abstract

"*Light will be thrown on the origin of man and his history*": this was the single line that Charles Darwin devoted to human evolution in the *Origin of Species*, 150 years ago. Despite short, apparently incidental and innocuous, this sentence actually represented a "risky prediction". In other words, Darwin's claim was: whether natural selection is the basic mechanism of the origin of species, in general, there is no reason to exclude *Homo sapiens* from analogous circumstances. As a matter of fact, the debate that followed was focused on the phylogenetic relationship of our species with monkeys and apes. The circa twenty species of extinct hominids that we know at present demonstrate that Darwin's prediction was correct: light has been thrown. At present, the science of human origin (or paleoanthropology) appears to be able to shed light not only on our evolution, but also on mechanisms and patterns of the evolution in general. This paper contains a brief overview of the most important discoveries that characterised the history of paleoanthropology until about 30 years ago; this synopsis will be followed by an introduction to current methods of research, to briefly discuss the origin of our own species.

1. Introduction

This is the story of a young interdisciplinary field of scientific investigation commonly referred to as paleoanthropology. It appeared on the scene when Darwin was developing what he called, soon after his voyage on the Beagle, "*my theory*" (Darwin 1837, 214). Since then, it is widely recognized that the phenomenon of evolution is ingrained on that simple, although counterintuitive and perhaps cognitively uncomfortable, idea of natural selection shared by Charles R. Darwin and Alfred R. Wallace in 1858 (Moody 1971), then examined in great detail one year later in *The Origin of Species* (Darwin 1859). Today, whether the facts supporting the evolutionary approach to biology are many, whether many are the involved mechanisms, and many are the pertinent perspectives and approaches, the scientific theory explaining the phenomenon of biological evolution is just one, Darwin's; it represents the focus of the entire field of biology, as in the words of Theodosius G. Dobzhansky (1973, 125): "*nothing in biology makes sense except in the light of evolution*." Thus, it is clear that speaking about theories (using the plural) is — at least in my view — scientifically incorrect, epistemologically inappropriate, and may be philosophically misleading.

This paper starts with some of Darwin's expectations on human evolution: what he did, what he did not, and what he could not foresee with respect to our present knowledge on the issue. Then we will follow a historical journey to review the most important discoveries (*Figure 1, a-f*) and some debates that have characterised the history of paleoanthropology until recent times. This overview will be closed by a brief outline of current methods of research, focused on discussing the appearance of modern representatives of the genus *Homo* — i.e., the origin of our own species: *Homo sapiens* — including aspects of investigation that Darwin could have never imagined at his time nor we could only 20 or 15 years ago.

2. Darwin's predictions

It is well known that in *The Origin of Species* — a volume of about 500 pages — only a line toward the end of the book was devoted to the origin of man and his history (Darwin 1859, 488): "*light will be thrown on the origin of man and his history*." However, when all the 1250 copies went out of print and an intense debate started on both evolution as a phenomenon and the mechanism of natural selection that explains the phenomenon, the debate was not on orchids, chaffinches or turtles, but was focused on the relationships of monkeys and apes with *Homo sapiens*. In a sense, he was prudent because he was aware that this — the nature and origin of ourselves as a species — would have been the real target of the controversies he expected to emerge after a theoretical framework was given to the phenomenon of evolution. However, we will see how this concise and "risky" prediction was appropriate.

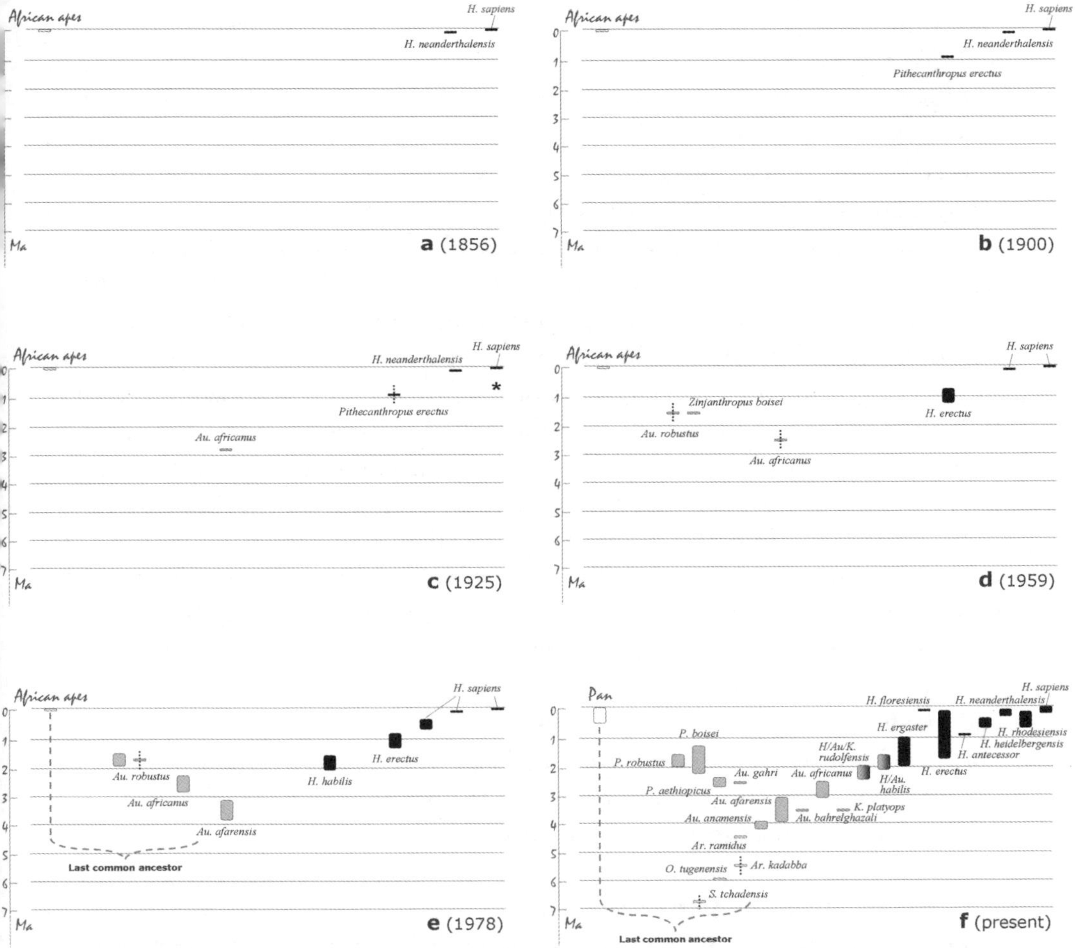

Figure 1. Our understanding of human evolution throughout the 150 years passed since Darwin's time is organized in six momentous steps. The fossil evidence known in 1856 (a) dealt only with the skeleton found in the Neanderthal, while in 1900 (b) we had many other Neanderthals, fossil representatives of our species (the "Cro-Magnons"), and the bones found in Java and ascribed to *Pithecanthropus erectus* (later included within *Homo erectus*); then, in 1925 (c), the description of the first australopithecine opened the gates of Africa to paleoanthropology; in the last one of these charts (c), the position of the species inferred after the Piltdown forgery is indicated by a star. Subsequently, part of the sub-Saharan Africa became the main theatre of field endeavours, leading to special improvements in 1959 (d), with the first discovery at Olduvai, and in 1978 (e) with the description of *Australopithecus afarensis*, which followed after about 15 years that of *Homo habilis* (1964); in the same period, the contribution of the so-called "molecular clock" added an indirect evaluation of the coalescence time (still subjected to adjustments due to new fossil and genetic data) between our evolutionary lineage and those of the extant African apes. At present, the knowledge on human evolution has enhanced enormously, with the identification of about 20 species, attributed to a number of genera that include: *Sahelanthropus* (S.), *Orrorin* (O.), *Ardipithecus* (Ar.), *Australopithecus* (Au.), *Kenyanthropus* (K.), *Paranthropus* (P.), and *Homo* (H.)

It is also well known that, on the spur of the moment and with great efficacy, the controversial target of the nature and origin of our species was faced by Thomas Henry Huxley, through a volume in three parts with the eloquent title *Evidence as to Man's Place in Nature* (Huxley 1863). It is worth mentioning that this book influenced the conference read by Filippo De Filippi in Turin, which represented the official entrance of the Darwinism in Italy (De Filippi 1864).

Only after a decade will Darwin devote an entire treaty in two volumes on both the topic of human evolution and the idea of sexual selection (Darwin 1871). Other risky predictions were put forward here. I would like to draw the attention of the reader to a couple of them. It was risky, for instance, to state that Africa was our more probable homeland, in view of the great affinities of our species with gorillas and chimpanzees that, as a matter of fact, live in Africa at present (page 199). It was risky because at Darwin's time the only discoveries of human fossils were from Europe; in addition, it was also the time when other scientists were suggesting to look toward the Far East, influenced by our affinities with the orangutangs from Borneo and Sumatra. Another risky prediction was to state that the fossil record and the science of paleontology as a whole could not shed light on the process of evolution, in general, and on our origins, in particular. Too dispersed in space and time is the hard evidence represented by the fossil record, he reasoned, to be a concrete support for our comprehension of the natural history of extinct and living creatures, and the evolutionary mechanism implied in it (page 201). Fortunately, he was wrong on this point. The discoveries and achievements of paleoanthropology — just a newborn science at Darwin's time — demonstrate that he was rather pessimistic about our heuristic approach.

3. Neanderthals and Cro-Magnons

A first spot of light on human evolution, furnished by a fossil specimen, was actually there three years before the publication of *The Origin of Species*. An incomplete skeleton had been discovered by workers in a cave in the Neander valley, or Neanderthal, not far from Dusseldorf in Germany. We will not go into details here, but we may recall that the morphology of that skeleton appeared diverse in some respects from the range of variation displayed by extant human populations; thus, the extinct species *Homo neanderthalensis* was introduced into the literature (King 1864). Other specimens pertaining to the same human type had been found in the previous decades at a site in Belgium and from a quarry in Gibraltar and many more Neanderthals will come later from other European localities in Belgium, Croatia, France, and Germany. Thus, at the beginning of the twentieth century, all these discoveries had largely demonstrated that the skeleton from the Feldhofer Grotto in the Neanderthal was not an isolate enigmatic finding, but the genuine representative of an extinct human species.

Nevertheless, according to some eminent scientists (e.g. Huxley 1863; 1864), such an extinct humanity failed to represent the expected evolutionary link between the apes and ourselves, particularly in view of the large cranial volumes and big brains showed by its representatives. Conversely, this feature suggested that the Neanderthals were a collateral and partially diverging side-branch, rather than an ancestor of *Homo sapiens.* It is extraordinary to see how this view is close to what we think about Neanderthals today (e.g., Stringer and Gable 1993; Trinkaus and Shipman 1993; Harvati and Harrison 2006), given that our present conclusion is based on an extraordinary variety of data, increasingly deepened and detailed, including those on Neanderthals' DNA (see Biondi and Rickards, this volume).

Neither was it possible to consider as plausible ancestral creatures in relationships with non-human primates the so-called Cro-Magnons, the well-known fossil representatives of *Homo sapiens* that were discovered in those same decades (e.g., Broca 1868). They displayed a morphology in many respects identical to extant human populations and were the probable authors of the magnificent manifestations of prehistoric rock art. Thus, these modern-looking specimens appeared at that time (as well as today) simply as fossil representatives of our own species.

Between 1908 and 1912 there were also the discoveries at Piltdown, in England (Dawson and Woodward 1913), which were due to a well known as well as deplorable episode of forgery that I will only mention here. Nevertheless, I think it is important to recall that it represented a useful test — unintentional of course — to evaluate the potential of the emerging science of paleoanthropology to reject a false evidence, as it successfully happened in the following decades until the definitive exposure of the fraud (Weiner, Oakley and Le Gros Clark 1953).

4. The missing link

In the second half of the nineteeth century, naturalists and palaeontologists were waiting for the discovery of a definitive link between apes and ourselves, usually referred to as the "missing link". Thomas Huxley had written: "*...the structural differences which separate Man from the Gorilla and the Chimpanzee are not so great as those which separate the Gorilla from the lower apes.*" (Huxley 1863, 123) What it missed was just a ring in the chain, capable of definitively connecting the greater apes with humans. This could have fastened *Homo sapiens* to other living creatures: in a word, to nature. The German morphologist and embryologist Ernst Haeckel came also to hypothesise the existence of such a missing link (Haeckel 1874) and to give it a name: *Pithecanthropus alalus* (that is, from Greek, "ape-man incapable to speak").

Influenced by these ideas, the Dutch anatomist Eugène Dubois went to the Island of Java, in Indonesia, toward the end of the century and really found... the missing link. In 1891 a molar tooth came from the excavations in the bank of the Solo River at Trinil, in central Java, followed by a calotte, which appeared extremely platycephalic, with heavy brow-ridges and a receding forehead, clearly more archaic than a Neanderthal. In 1892, then, his workers found a modern-looking femur and eventually, in 1894, he gave to these discoveries the name of *Pithecanthropus erectus* ("ape-man", because of the calotte; "upright walker", because of the morphology of the femur) (Dubois 1894; see also Campbell 1965, 15-16). It was only half a century later, that the great biologist Ernst Mayr proposed to include these fossils, together with other similar specimens found during the first half of the twentieth century in Java and China, within the genus *Homo*, representing a single species referred to as *Homo erectus* (Mayr 1950).

Now, we might consider that — while some light began to be thrown on the origin of man and his history — one of Darwin's predictions was still waiting to be verified, given that with the Neanderthals, the Cro-Magnons, and Dubois' *Pithecanthropus* we are still remained in Eurasia. In other words: what about Africa...?

5. Eventually in Africa

In 1924, a young anatomist teaching at the University of Witwatersrand in Johannesburg, Raymond Dart, disclosed the gates of Africa to the study of human evolution. The occasional discovery of an infantile cranium coming from the limestone quarries of Taung gave rise to the identification of a kind of extinct "ape-man" that he named *Australopithecus africanus* (Dart 1925), advocating the formidable intuition that the species represented by such a diminutive specimen was a small-brained, but bipedal ancestor of *Homo sapiens*.

The Taung discovery is among the "*most significant finds in the history of paleoanthropology (...) because of its status as the first recognised member of a totally new, previously unknown, major group of fossil hominids.*" (Meikle and Taylor Parker 1994, 52) However, Dart's claim did not persuade the academic community and he was severely criticised. Many were happy to admit that the Taung child was the representative of a new species of fossil primate, but they did not accept that it might be ancestral to ourselves. However, thanks to the efforts of the Scottish palaeontologist Robert Broom, between the '30s and the '40s, from the contiguous caves of Sterkfontein, Kromdraai, and Swartkrans (again near Johannesburg, in South Africa), a number of fossil hominids came to light (Broom 1936; 1938; 1949). It was therefore possible to know the aspect of the adult individuals of *Australopithecus africanus*, giving definitive support to Dart. A different type of hominid was also discovered in some of those caves: it appeared unquestionably more heavy and robust *than Australopithecus africanus* in many dental traits and skeletal features of the masticatory apparatus. This new type of South-African ape-man received various binomial appellatives — including that of *Paranthropus robustus* (Broom 1938, 378) that is still in use today — and suggested the existence of different contemporaneous lineages of australopithecines or, alternatively, a single evolutionary trajectory from a generalised type (*Australopithecus*) towards a more derived one (*Paranthropus*), in combination with an increasing masticatory, hence alimentary, specialization.

It was not possible at that time to have a clear picture of these South-African findings in terms of chronology, because of the complex process of deposition within the karstic cave systems were the hominids and other fossils had been found. More light — also in terms of absolute chronology — had to come from another extraordinary area for human paleontology: East Africa.

This region started to become interesting for many researchers and open to a number of international expeditions only after the end of the '50s. It was in 1959, in fact, that a new type of hominid came to bless the efforts of Louis and Mary Leakey, who were looking for human ancestors and Palaeolithic stone tools since 1931 in one of the most celebrated among the East-African hominid localities. This is the site known as Olduvai gorge, a very special area along the Great Rift Valley, in Tanzania. A robust type of early hominid, similar but even more specialized than the *Paranthropus robustus* found in South Africa, was discovered there in apparent association with some kind of early stone tools. At first it was named "*Zinjanthropus*" (Leakey 1959), but now it is commonly included within the genus *Paranthropus*, as *Paranthropus boisei*.

Was this australopithecine the first tool-maker? Was this massively-toothed and bipedal ape the author of the lithic artefacts that the Leakey found in the same stratigraphic levels at Olduvai? The answer is most probably negative and this became clearer in the following years: particularly when Louis Leakey, Phillip Tobias, and John Napier (1964) described a small sample of other fragmentary fossil pieces found at Olduvai, and gave them the name of *Homo habilis*. The relatively larger volume of the braincase of these fossils, among other features, was the evidence expected to pertain to some kind of an earliest human, a creature still close to the australopithecines, but which could be included within our own genus: the genus *Homo*. The debate about the taxonomic and phylogenetic status of *Homo habilis* is still open and controversial, but those and other fossils — found subsequently at Olduvai and elsewhere (see Coppens, this volume), particularly in localities around the Lake Turkana, in Kenya (e.g., Wood 1991) — are indeed there. These fossils demonstrate that around 2 million years ago (probably earlier) a new kind of hominids, ecologically interpreted as "scavengers", were capable to produce elementary stone tools and were at an earlier stage of what we may consider the fundamental process for our evolution: that is the process of progressive increase in endocranial volume — a process usually referred to as "encephalization" (Bruner, Manzi and Arsuaga 2003).

At the beginning of the '70s, our knowledge about the early African hominids was broadened again, in terms of geography, chronology, and taxonomy. Certainly, the pinnacle of this stage in the history of the research was the discovery by Don Johanson of a very popular fossil hominid: the skeleton nicknamed and known as "Lucy" (formally AL 128-1), which was found at Hadar, Ethiopia, in 1974 (Johanson and Taieb 1976). During the same decade and subsequently, other important discoveries were made in Ethiopia, in Kenya (around the Lake Turkana), and in Tanzania, including the footprint fossil track that came to light at Laetoli, not far from Olduvai. A new species was described merging the discoveries from Hadar and Laetoli, thanks to the collaboration between Don Johanson, Tim White, and Yves Coppens (1978): the name given to the new species is *Australopithecus afarensis*, probably the best known "pre-human" taxon that we know at present (e.g., Kimbel, Rak and Johanson 2004). It extended our knowledge of the australopithecines to the range between 4 and 3 million years before present. It also made clearer that these ancestral hominids were bipedal creatures, like us, but that they were more similar to apes as far as their endocranial volumes and body proportions are concerned. Like the other australopithecines, they already had a peculiar combination of dental traits, including the enlargement of molars and premolars and the reduction of canines and incisors, with a significant increase in enamel thickness, furnishing the best evidence of features and adaptations clearly distinct from either those of the extinct and extant apes or of the genus *Homo*.

6. Phylogenetic trees

It is worth noting that a fundamental phylogenetic tree was introduced in the literature toward the end of the '70s, in combination with the description of *Australopithecus afarensis*, viewed as the most ancient and the most archaic fossil hominid ever discovered (as far as I know, the most refined version of this phylogeny is that reported in Johanson and Edey's book of 1981). This scenario (see *Figure 1e*) put the new species at the origin of our evolutionary lineage, in the position of last common ancestor of both the other australopithecine and the genus *Homo*. It is also noteworthy that its geometry is very similar to what we think now — although the number of extinct species considered at that time (N=6) is almost nothing with respect to the 20 hominid taxa, or more, we recognize — showing a major divergence, at about 2.5 million years before present [compare *Figure 1f*], between the phylogenetic and adaptive trajectories of *Australopithecus* and *Paranthropus*, on one side, and that of the genus *Homo*, on the other.

With the exception of this major divergence, we may notice however that a sequential and linear pattern of evolution is the rule in such a scenario. As a model, in fact, the tree of the late '70s tends to foresee continuous and progressive anagenetic lineages, still following the gradualistic approach influenced by the "modern synthesis" of the '40s (e.g., Huxley 1942). By contrast, it is well known that during the same decade, two palaeontologists were suggesting a model (*punctuated equilibria*) that envisages a higher contribution of cladogenetic events in the course of the evolutionary process (Eldredge and Gould 1972). When applied to human evolution, this new model may imply a *bushy hominid tree* (Gould 1977a), richer in species and cladogenetic events, than those expected with the gradualistic model, which vice versa is strictly consistent with the micro-evolutionary paradigm of the synthesis.

As a consequence, according to a convinced *lumper* among my colleagues (White 2003; 1994), Gould's prediction could have pushed contemporary paleoanthropologists to "*herald each new fossil as evidence of ancient hominid biodiversity*", advocating these claimed new species and/or new genera in support of a "*many-branched bush of diversity*" (using the words of two recognized *splitters* among my colleagues: Tattersall and Schwartz 2000, 33).

Although there can be elements of truth in White's criticism and even if the currently recognized number of species may appear as overestimated, in the last thirty years an extraordinary number of paleoanthropological discoveries has greatly expanded our understanding of the past complexity [*Figure 1f*], and has substantially confirmed Gould's prediction. Viewed from another perspective, according to a seminal paper by Ian Tattersall (1986, 168):

> ...it might well be argued that it would be better for the comprehensiveness of our understanding of the human fossil record that, if err we must, we err (within reason!) on the side of recognizing too many rather than (...) too few species units. After all (...) even a subspecies has a history worthy of investigation.

Nevertheless, there are some crucial evolutionary passages until the appearance of *Homo sapiens* — including the emergence of our species, as I will briefly discuss later — that are apparently consistent with the pattern postulated by the model of punctuated equilibria.

7. Hominidae

Figure 1f describes the distribution in time of the approximately 20 species, or more, that are recognized at present by a great number of paleoanthropologists. While time is represented vertically, from 7 million years to the present, and the various bars display the chronological span of each species, these same species are disposed horizontally in a tentative order that would be indicative of both the morphological affinities between them and their respective, although hypothetical, phylogenetic relationships. Different grey tones represent different genera (or groups of affine genera): for instance we have the earliest hominids in white, australopithecines in mid-grey, and the genus *Homo* in black.

There is a side debate, at present, as far as the taxonomic status of the ensemble of genera and species represented here is concerned. It is debated in fact whether we and our extinct close relatives and ancestors should be considered a family (*Hominidae*), a subfamily (*Homininae*), a tribe (*Hominini*), or even a single genus; whether such was the case, all the species represented in *Figure 1f* should be included within the genus *Homo*, together with the two living species of *Pan* and, perhaps, also with the present genus *Gorilla* (e.g., Goodman *et al.* 1998). I do not wish to enter this debate now, but I should point out that the extended knowledge we have at present about human evolution is uncomfortably described by a restricted taxonomic approach.

Anyway, as a whole, a picture emerges that recalls the basic geometry of the phylogenetic tree put forward in the late '70s, with a major divergence at about 2.5 million years before the present, although this new representation exhibits a huge expansion in time as well as in diversity.

Obviously, crucial passages include the detachment of our lineage from the common ancestors with African apes, possibly around 6-7 million years before present or earlier, the origin of the genus *Homo* (despite a certain taxonomic incertitude, but together with the first appearance of Palaeolithic implements), slightly earlier than 2.5 million years ago, and eventually the emergence of modern humans, well established by a variety of data at about 200 thousand years before present. It is worth noting that all these crucial passages are African events.

From this general picture, it is also clear that various species were contemporaneous among them. For instance the species of *Paranthropus* were synchronic with the earliest *Homo* species and, at least in part, lived in sympatry with them. Moreover, there were species of later *Homo* that, given their extended geographical diffusion in Africa and Eurasia, overlapped in time for long periods, being either sympatric (living in the same geographic area) or, more frequently, allopatric. Some of them greatly diverged between each other over periods of allopatry, as for instance happened for either *Homo ergaster* and *Homo erectus*, respectively in Africa and in Eastern Asia, or the sequence of chronospecies *Homo heidelbergensis* – *Homo neanderthalensis*, in Europe, with respect to their counterparts (*Homo rhodesiensis*), in Africa, which in turn led to the appearance of *Homo sapiens*.

It seems clear to me that, after the divergence from the African apes, what we know at present about the natural history of the hominids depicts a very complex history. This story strongly recalls the evolutionary pattern implied by the model of punctuated equilibria — hence, a bushy tree — rather than a simple trajectory from non-human primates to "Man": what we might call a *ladder*, according to the Gould's (1977a) metaphor.

8. New methods, new paradigms

In the last couple of decades, the study of the fossil record has been supported by the development of geometric morphometrics (e.g., Slice 2005). This term includes a series of approaches based on the multivariate statistical analysis of Cartesian coordinate data, usually (but not always) limited to landmark point locations. These include the evaluation of a mean shape and the description of variations from this, which allows for the visualization of individual differences, as well as within or between group diversity estimates, in view of the morphology of the original specimens. Geometric morphometrics has been considered, not without emphasis, a *new synthesis* in the study of biological forms (Rohlf and Marcus 1993), since it is able to combine the description of different shapes (the field of morphology) with a rigorous numerical and multivariate approach (which is typical of morphometrics).

In addition, the utilization of electronic equipments imported from the highly sophisticated fields of both medical imagery (namely, computed tomography) and industrial imaging techniques (like microtomography or laser scanning), currently allows researchers to virtually penetrate into the intimate structure of each fossil specimen and compare it with other such specimens from an innovative perspective (e.g., Zollikofer *et al.* 1998; 2005; Spoor *et al.* 2003; Bruner and Manzi 2006; Macchiarelli *et al.* 2006). These new tools have been revolutionizing paleoanthropology in the last 20 years or so, by the development of what is commonly called "virtual paleoanthropology" (Weber 2001). As a matter of fact, virtual representations like these, combined with geometric morphometrics when appropriate, have opened a number of new possibilities for the analysis of the fossil record, including: a) the virtual extraction and reconstruction of anatomical elements, with the possibility to correct plastic distortions; b) the analysis of inaccessible internal and small-sized structures; c) the modelling of biomechanical properties, of the ontogenetic process, of evolutionary changes; etc. It is usually said that, in the near future (in some cases already at present), the anatomy of the fossil hominids, that usually represent rare and precious museum items, will no longer be studied on the specimens themselves but on their virtual representations.

Assisted by these new techniques and methodological improvements, and after the publication of a seminal book by Stephen Jay Gould (1977b), during the last decades we have assisted a number of successful attempts to bring human evolutionary studies towards the growing field known as *evo-devo*, which combines evolution (phylogeny) with developmental biology (ontogeny). Virtually all the morphological change we observe along evolutionary lineages reflects some modification of developmental processes; thus, we may ask when and how the modern human pattern of growth and development appeared (Moggi-Cecchi 2001). A number of studies have demonstrated that the australopithecines, and possibly the earliest species of the genus *Homo* too, had a growth and developmental pattern similar to those of extant great apes (Bromage and Dean 1985; Zeresenay *et al.* 2006; Lacruz *et al.* 2008). Moreover, although debated (see, e.g., Macchiarelli *et al.* 2006), there are researches strongly suggesting that even late representatives of the genus *Homo*, like the Neanderthals, did not have an ontogenetic pathway identical to ours (e.g., Ponce de León and Zollikofer 2001; Manzi 2003; Ramirez Rozzi and Bermudez de Castro 2004).

9. The "ontogenetic revolution"

Now, there is an anatomical district, where our pattern of growth and development deeply interacts with the process of encephalization that has characterised the evolution of the genus *Homo*. This anatomical district is the pelvis and, particularly, I am referring to the pelvic inlet, the bony channel through which the large cranium of the future newborn must necessarily pass in order to come to light. A strong selective pressure should have driven the co-evolution between our pelvis morphology — which is functionally consistent with the bipedalism acquired at the beginning of the evolutionary trajectory of the hominids, let's say between 7 and 4 million years before present — and the process of encephalization, which started with the emergence of the genus *Homo* and followed an exponential increase during the last 2 million of years.

Given a number of constraints imposed by both the morphology of the pelvis of our remote ancestors, as bipedal creatures, and the selective pressure acting in favor of progressively enlarged brain and cranial dimensions, the adaptive solution probably followed an *evo-devo* template, implying a rapid change in the regulation of growth and developmental pattern. This should have led — among other non irrelevant effects — to a more premature and altricial newborn in the new species than in other "archaic" *Homo* species. It is well known, in fact, that our babies are premature, with respect to the correlation between brain dimensions and gestation lengths observed among non-human primates, and follow foetal rates of growth for another 9 months after birth or so. Viewed in this perspective, the origin of *Homo sapiens* corresponds to what I call our "ontogenetic revolution". Such an expression echoes the observation that the encephalization process followed different trajectories along the various evolutionary lineages of the genus *Homo* (Bruner *et al.* 2003). As a matter of fact, despite the expansion of brain and cranial dimensions, a process shared by *Homo erectus* in Eastern Asia, *Homo neanderthalensis* in Europe, and the ancestors of *Homo sapiens* in Africa, this process neither followed the same evolutionary modalities nor led to the same evolutionary destinies (e.g., Manzi 1991; 2003; Manzi *et al.* 2000). Particularly, the Asian and European lineages had a tendency to maintain an archaic architecture of the braincase, similar in many respects to the putative early *Homo* ancestors from Africa, and eventually went extinct. The same model was preserved in Africa too until about 200 thousand years before present, when this architecture sharply changed with the appearance of modern humans (e.g., Lieberman *et al.* 2002; White et al. 2003).

The origin of *Homo sapiens* — which is obviously the pivotal event in human evolution — appears in a stimulating light when viewed from this perspective. Actually, according to an *evo-devo* approach, the emergence of our species appears as a convincing example of evolution by means of natural selection, combined with a model of punctuated equilibria, which in turn postulates both the fundamental role of exaptations (in this case a big brain) and the intervention of an innovative change (in this case a new regulation of our growth and developmental pattern), occurred within a restricted and isolated population, eventually leading to a punctuated event of speciation: our speciation. This was the ontogenetic revolution of *Homo sapiens.*

Given the above scenario, from different sources of data (e.g., Stringer 2002; Mellars *et al.* 2007) we know at present that modern humans appeared in Africa

— more probably in eastern Africa — at about 200 thousand years ago, when in Europe the Neanderthals were still at an early stage of their evolution. We also know that *Homo sapiens* started soon a worldwide diffusion, driven by a cultural potential that previous hominids had never experienced, including the symbolic and conceptual thought implied by Palaeolithic art expressions. A side-effect of this adaptive success was represented by the extinction of other representatives of the same genus *Homo*, as the fate of the Neanderthals demonstrates, according to the evolutionary principle of "exclusive competition" between closely related and sympatric species.

Then, the relatively rapid distribution of *Homo sapiens* across the entire planet followed. This was the prelude of the present gargantuan demographic expansion of a single species of a bipedal and highly encephalised primate that appears to me out of any natural control. We may just hope in our potential to evolve from the "cultural" viewpoint: in this perspective, science and philosophy should play their respective and relatively independent roles.

References

Broca P., 1868, "Sur les cranes et ossements des Eyzies", *Bulletins de la Société d'anthropologie de Paris* 3: 350–92, 416–46, 554–74.

Bromage T. G., Dean M. C., 1985, "Re-evaluation of the age at death of Plio-Pleistocene fossil hominids", *Nature* 317: 525–28.

Broom R., 1936, "A new fossil anthropoid skull from Sterkfontein, near Krugersdorp, South Africa", *Nature* 138: 486–88.

Broom R., 1938, "The Pleistocene anthropoid apes of South Africa", *Nature* 142: 377–379.

Broom R., 1949, "Another new type of fossil ape-man", *Nature* 163: 57.

Bruner E., Manzi G., 2006, "Digital tools for the preservation of the human fossil heritage: Ceprano, Saccopastore, and other case studies", *Human Evolution* 21: 33–44.

Bruner E., Manzi G., Arsuaga J. L., 2003, "Encephalisation and allometric trajectories in the genus *Homo*: evidence from the Neanderthal and modern lineages", *Proceedings of the National Academy of Sciences USA* 100: 15335–40.

Campbell B. G., 1965, "The nomenclature of the Hominidae", *Royal Anthropological Institute of Great Britain and Ireland, Occasional Paper* no.22: 1–33 [Reprinted 1994 in Meikle W. E., Taylor Parker S. (eds.), *Naming our Ancestors: An Anthology of Hominid Taxonomy*, Prospect Heights , Ill.: Waveland].

Dart R. A., 1925, "*Australopithecus africanus*: the Man-Ape of South Africa", *Nature* 115: 195–99.

Darwin C. R., 1837, *Notebook B [Transmutation of Species]*, in "Darwin Online", CUL-DAR121(http://darwin-online.org.uk/).

Darwin C. R., 1859, *On the Origin of Species by Means of Natural Selection, or the Preservation of Favoured Races in the Struggle for Life*, London: John Murray.

Darwin C. R., 1871, *The Descent of Man, and Selection in Relation to Sex*, London: John Murray.

Dawson C., Woodward A. S., 1913, "On the discovery of a Palaeolithic human skull and mandible in flint-bearing gravel overlying the Wealden (Hastings Beds) at Piltdown, Fletching (Sussex)", *Quarterly Journal of the Geological Society* 69: 117–23.

De Filippi F., 1864, *L'Uomo e le Scimie (lezione pubblica detta a Torino la sera dell'11 gennaio 1864)*, Milano: G. Daeli.

Dobzhansky T., 1973, "Nothing in biology makes sense except in the light of evolution", *The American Biology Teacher* 35: 125–29.

Dubois E., 1894, "*Pithecanthropus erectus*. Eine Menschenähnliche Ubergangsform aus Java", Batavia (Jakarta) [Reprinted 1895 in *Jaarboek van het Mijnwezen* 24: 5–77].

Eldredge N., Gould S. J., 1972, "Punctuated equilibria: an alternative to phyletic gradualism", in Schopf T. J. M. (ed.), *Models in Paleobiology*, New York: Freeman, 82–115.

Goodman M., Porter C. A., Czelusniak J., Page S. L., Schneider H., Shoshani J., Gunnell G., and Groves C. P., 1998, "Toward a phylogenetic classification of Primates based on DNA evidence complemented by fossil evidence", *Molecular Phylogenetics and Evolution* 9: 585–98. .

Gould S. J., 1977a, *Ever since Darwin: Reflections in Natural History,* New York: Norton.

Gould S. J., 1977b, *Ontogeny and Phylogeny*, Cambridge, Mass.: Belknap.

Haeckel E., 1874, *Anthropogenie oder Entwickelungsgeschichte des Menschen*, Leipzig: W. Engelmann.

Harvati K., Harrison T. (eds.), 2006, *Neanderthals Revisited*, Dordrecht: Springer.

Huxley J. S., 1942, *Evolution: The Modern Synthesis*, London: Allen and Unwin.

Huxley T. H., 1863, *Evidence as to Man's Place in Nature*, London: Williams & Norwood.

Huxley T. H., 1864, "Further remarks on the human remains from the Neanderthal", *Natural History Review (London)* 4: 429–46.

Johanson D. C., Edey M., 1981, *Lucy: The Beginning of Humankind*, New York: Warner Books.

Johanson D. C., Taieb M., 1976, "Plio-Pleistocene hominid discoveries in Hadar, Ethiopia", *Nature* 260: 293–97.

Johanson D. C., White T. D., Coppens Y., 1978, "A new species of the genus *Australopithecus* (Primates: Hominidae) from the Pliocene of Eastern Africa", *Kirtlandia* 28: 1–14.

Kimbel W. H., Rak Y., Johanson D. C., 2004, *The Skull of Australopithecus afarensis*, New York: Oxford University Press.

King W., 1864, "The reputed fossil man of the Neanderthal", *Quarterly Journal of Science* 1: 88–97.

Lacruz R. S., Dean M. C., Ramirez-Rozzi F., Bromage T., 2008, "Megadontia, striae periodicity and patterns of enamel secretion in Plio-Pleistocene fossil hominins", *Journal of Anatomy* 213: 148–58.

Leakey L. S. B., 1959, "A new fossil skull from Olduvai", *Nature* 184: 491–93.

Leakey L. S. B., Tobias P. V., Napier J. R., 1964, "A new species of the genus *Homo* from Olduvai Gorge", *Nature* 202: 7–9.

Lieberman D. E., McBratney B. M., Krovitz G., 2002, "The evolution and development of cranial form in *Homo sapiens*", *Proceedings of the National Academy of Sciences USA* 99: 1134–39.

Macchiarelli R., Bondioli L., Debenath A., Mazurier A., Tournepiche J.-F., Birch W., Dean C., 2006, "How Neanderthal molar teeth grew", *Nature* 444: 748–51.

Manzi G., 1991, "Braincase architectural morphology of the Neanderthal cranium from Grotta Guattari and the nature of the archaic/modern transition", *Quaternaria Nova* 1: 113–24.

Manzi G., 2003, "'Epigenetic' cranial traits, Neanderthals and the origin of *Homo sapiens*", *Rivista di Antropologia (Journal of Anthropological Sciences)* 81: 57–67.

Manzi G., Saracino B., Bruner E., Passarello P., 2000, "Geometric morphometric analysis of mid-sagittal cranial profiles in Neanderthals, modern humans, and their ancestors", *Rivista di Antropologia (Journal of Anthropological Sciences)* 78: 193–204.

Mayr E., 1950, "Taxonomic categories in fossil hominids", *Cold Spring Harbor Symposia on Quantitative Biology* 15: 109–18.

Meikle W. E., Taylor Parker S., 1994, *Naming our Ancestors: An Anthology of Hominid Taxonomy*, Prospect Heights, Ill.: Waveland

Mellars P., Boyle K., Bar-Yosef O., Stringer C. (eds.), 2007, *Rethinking the Human Revolution: New Behavioural and Biological Perspectives on the Origin and Dispersal of Modern Humans*, Cambridge: McDonald Institute for Archaeological Research.

Moggi-Cecchi J., 2001, "Human evolution: questions of growth", *Nature* 414: 595–97.

Moody J. W. T., 1971, "The reading of the Darwin and Wallace papers: an historical 'non-event'", *Journal of the Society for the Bibliography of Natural History* 5: 474–76.

Ponce de León M. S., Zollikofer C. P. E., 2001, "Neanderthal cranial ontogeny and its implications for late hominid diversity", *Nature* 412: 534–38.

Ramirez Rozzi F. V., Bermudez de Castro J. M., 2004, "Surprisingly rapid growth in Neanderthals", *Nature* 428: 936–39.

Rohlf F. J., Marcus L. F., 1993, "A Revolution in Morphometrics", *Trends in Ecology and Evolution* 8: 129–32

Slice D. E. (ed.), 2005, *Modern Morphometrics in Physical Anthropology*, New York: Kluwer Academic / Plenum Publishers.

Spoor F., Hublin J.-J., Braun M., Zonneveld F., 2003, "The bony labyrinth of Neanderthals", *Journal of Human Evolution* 44: 141–65.

Stringer C. B., 2002, "Modern human origins: progress and prospects", *Philosophical Transactions of the Royal Society of London, Series B: Biological Sciences* 357: 563–79.

Stringer C. B., Gamble C., 1993, *In Search of the Neanderthals: Solving the Puzzle of Human Origins*, New York: Thames and Hudson.
Tattersall I., 1986, "Species recognition in human paleontology", *Journal of Human Evolution* 15: 165–75.
Tattersall I., Schwartz J. H., 2000, *Extinct Humans*, Boulder, Col.: Westview.
Trinkaus E., Shipman P., 1993, *The Neandertals: Changing the Image of Mankind*, New York: A. A. Knopf.
Weber G. W., 2001, "Virtual anthropology (VA): a call for Glasnost in paleoanthropology" *The Anatomical Record (Part B)* 265: 193–201.
Weiner J. S., Oakley K. P., Le Gros Clark W. E., 1953, "The solution of the Piltdown problem", *Bulletin of the British Museum (Natural History) – Geology* 2: 139–46.
White T. D., 2003, "Early hominids: diversity or distortion?", *Science* 299: 1994–97.
White T. D., Asfaw B., DeGusta D., Gilbert H., Richards G. D., Suwa G., Howel F. C., 2003, "Pleistocene *Homo sapiens* from Middle Awash, Ethiopia", *Nature* 423: 742–47.
Wood B., 1991, *Koobi Fora Research Project – Volume 4: Hominid Cranial Remains*, New York: Oxford University Press.
Zeresenay A., Spoor F., Kimbel W. H., Bobe R., Geraads D., Reed D., Wynn J. G., 2006, "A juvenile early hominin skeleton from Dikika, Ethiopia", *Nature* 443: 296–301.
Zollikofer C. P. E., Ponce de León M. S., Martin R. D., 1998, "Computer-assisted paleoanthropology", *Evolutionary Anthropology* 6: 41–54.
Zollikofer C. P. E., Ponce de León M. S., Lieberman D. E., Guy F., Pilbeam D., Likius A., Mackaye H. T., Vignaud, P., Brunet M., 2005, "Virtual cranial reconstruction of *Sahelanthropus tchadensis*", *Nature* 434: 755–59.

THE ORIGIN OF MAN

Yves Coppens

Through the fossil record and its data I will try to say what I think could have been the History of Man, which means the History of the last ten million years.

It is well known today that bonobos, *Pan paniscus*, and chimpanzees, *Pan troglodytes*, are the creatures closest to us in nature.

In an evolutionary way of thinking, it means that they and we have common ancestors.

Because all Primates are tropical, and because bonobos and chimpanzees are African, there is some probability that those ancestors would have been tropical and African.

Moreover, the morphological, anatomical, physiological, genetic, molecular and even ethological distance between these cousins and ourselves allows to situate our last common ancestor somewhere in the upper Miocene, about ten million years ago.

So we have the birth place, *Africa*, and the geological age: *ten million years* of our last common ancestor.

We will travel through these ten million years, in a chronological order, in four parts :

1st part : Ten million years ago, the time of the last common ancestor we share with the chimpanzees

2nd part : before Man, the Prehumans

3rd part : the emergence of Man

4th part : the evolution and expansion of Man.

First part

What do we know about the Apes in Africa at these late Miocene times? And who do we know could pretend to be the last common ancestor to Man and chimpanzee.

We have three main candidates to answer these questions.

Chrororapithecus abyssinicus from Ethiopia, 10.7 to 10.1 million years old (Suwa *et al.* 2007), *Nakalipithecus nakayamai* from Kenya 9.88 to 9.80 million years old (Kunimatsu *et al.* 2007), and *Samburupithecus kiptalami* from Kenya, 9.6 million years old (Ishida *et al.* 1998).

This very modest fossil record, indeed, *does not* allow, of course, to tell who among these candidates is the last common ancestor or the closest to the last common ancestor of chimpanzee and Human.

Let us say that we have just started to find some remains of African Great Apes of the right geological age, which are giving us an idea of what these famous grand parents could have looked like.

We do not know either, as a matter of fact, where we stand, in the phylogeny, C*hororapithecus, Nakalipithecus* and *Samburupithecus* respectively; we do not know even if they precede the divergence *Homo-Pan*; or if they are already engaged on one of the two lines; of if they are engaged in another independent branch, having nothing to do with the *Pan* or with the *Homo* branches.

Second part

Let us forget this common ancestor, let us forget as well the slice of ten million years of prechimpanzees and chimpanzees, to focus our attention on our side, the Prehuman and Human side of the divergence.

We will divide the ten million years of our filiation into two major episodes: the Prehuman one, from ten million years ago to one million years ago and the Human one, from three million years ago to the present, which immediately shows that the last Prehumans have coexisted with the first Humans.

The Prehumans are magnificently documented by 7 genera and 14 species.

I call this diversity, which is reflecting a diversity of ecological niches, a bunch (a bunch is looking to me clearer than a bush).

These seven genera and fourteen species come from central, eastern and southern Africa, Chad, Ethiopia, Kenya, Tanzania, Malawi and South Africa.

Chronologically, they can be organized in three steps or grades :

The earliest, from 7 million to 4 million years ago, with *Sahelanthropus* (Brunet *et al.* 2002), *Orrorin* (Senut *et al.* 2001) and *Ardipithecus* (White *et al.* 1994; 1995; Haile Selassie 2001; Haile Selassie *et al.* 2004)).

A second step, from 4 million to 3 million years ago, with *Australopithecus afarensis* (Johanson *et al.* 1978), *Australopithecus anamensis* (Leakey *et al.* 1995), *Kenyanthropus playtops* (Leakey *et al.* 2001) and *Australopithecus bahrelghazali*

(Brunet *et al.* 1996), emerging after an opening of the landscape (Coppens 2006).

And a third step, between 3 million and 1 million years ago with *Australopithecus garhi* (Asfaw *et al.* 1999), *Australopithecus africanus* (Dart 1925; Clarke 1995), the famous robust Prehumans, *Zinjanthropus* (Leakey 1959; Arambourg *et al.* 1967; Tobias 1967) and *Paranthropus* (Broom 1938) and the genus *Homo* (Leakey *et al.* 1964; Tobias 1991), emerging as well after a drought and because of this drought, more dramatic than the 4 million years old "opening" (Coppens 1975).

In order to try to appreciate the bunch of the 14 Prehumans as a group, we will examine the traits that they shared and the few which divide them.

1. All these Prehumans were tropical, without any exception
2. All these Prehumans were African, without any exception
3. All of them, as far as we know, were permanently upright
4. All the earliest ones, as far as we know, were both bipedal and arboreal. This is the case of *Orrorin tugenensis, Ardipithecus ramidus* and the less ancient *Australopithecus afarensis.*
5. All of them, but at different speeds, seem to become exclusively biped. *Australopithecus anamensis*, 4 million years old, seems to have been the first not to climb anymore.
6. All of them have a (slow) increasing endocranial capacity as well as a (slow) more complex organization of the brain — more complex convolutions and more complex irrigation.
7. All of them show a tendency of the face to reduce its prognathism to reach a sort of orthognathism. *Kenyanthropus platyops* and *Australopithecus bahrelghazali*, 3,5 million years old, seemed to have been the first and the best in that reduction.
8. But as far as the teeth are concerned, we can follow two trends: a tendency to a reduction in the size of the cheek teeth and a tendency to an increased size of them, which means two different adaptations, two different diets and two different habitats.

In summary, the Prehumans are tropical, African and upright; they seem to have been arboreal and bipedal before adopting, for ecological reasons, an exclusively bipedal locomotion. They all show a brain of increasing size and complexity. They all show a trend to a reduced prognathism, at different speeds, according to their phylum. Some of them, at last, show a trend to a reduced size of cheek teeth, while others show the reverse tendency.

Third part

An important drought happened in tropical Africa between 3 million and 2 million years ago corresponding to a global cooling of the Earth. This climate change appears for instance in the oxygen isotopes ratios O^{18}/O^{16} of the tests of microorganisms collected in deep sea cores.

It is also particularly well visualized in the sediments of the lower Omo river basin in Ethiopia, because these sediments are the only ones in tropical Africa to offer a clear, continuous, deposit of these times.

The Omo sedimentary sequence is a superb stratigraphical column, more than 1 kilometer thick, particularly well exposed due to tectonic reasons, particularly rich in fossils, including Hominids, and particularly well calibrated by biostratigraphic, paleomagnetic and radio-isotopic cross checking dates.

And all the fossils collected (I collected 50 tons of them in 9 field seasons between 1967 and 1976) show this cooling.

All the animals, elephants, rhinos, pigs, horses, bovids, primates, rodents, show during that time adaptations to more open and dryer environment, grass eating adaptations and all the plants are showing the same tendency towards a less and less humid and dryer climate.

Having collected pollens in all these levels, we did a ratio with the number of pollens of trees on the number of pollens of grasses; the result for 3 million years was 0.4 and 2 million years, 0.01.

The answers of the Prehumans have been done in the same direction and they have been good enough to have given two answers :

1st answer: a Prehuman had chosen a larger size of the body and larger cheek teeth: it is the so-called robust Australopithecine.

2nd answer: another Prehuman had chosen larger size of the brain and teeth to eat a wider diet, meat included; it is the genus *Homo* (Coppens 1975).

The consequence of a larger brain is, of course, the emergence of consciousness.

And consciousness is starting to build a new environment, the cultural environment for the first time after 4 billion years of life and its natural environment.

For Prehistorians, Culture is everything which is not Nature.

Cultural environment can be technical but also intellectual, spiritual, ethic, aesthetic and so on.

I call this drought, this event, the *(H)Omo event* to remind the pioneer role of the Omo area in this demonstration (Coppens 1975; 1999; 2006).

In summary, around 3 million years ago, an important cooling appeared in the whole world, becoming an important drought in tropical areas, especially in Africa.

And the whole ecosystem tried to adapt itself to this climatic change to this new environment. The Prehumans were then a part of this ecosystem. And one of these successful adaptations is called Man.

Homo is looking like a product of nature, a necessary adaptation to a climatic change. But Human being with, for the first time, a capacity of "knowing that he knows", is human since this beginning.

It is a discontinuity in a continuum! "more is sometimes different".

Fourth part

For biological and cultural reasons, the very first species of the genus *Homo* has been more mobile than his ancestor because of his diet (he became a carnivore and had to run behind the game), more curious also because of his better brain and the beginning of his consciousness, more equipped, because of his new ability to make tools, more numerous because of his successful adaptation to the climatic change we talked about.

More mobile, more curious, more equipped, more numerous, it is this very first species (and not the second) which moved out of its tropical birth place and out of its ecological niche. Some environmental reasons can be also added to have pushed *Homo* out of Africa around 2 to 3 million years ago,

Theoretically, I would think that this first *Homo* would have moved as soon as 2.5–2 million years ago. Which means that I would not be surprised to meet this first species of our genus anywhere and everywhere in the Old World, in Africa, in Europe and in Asia at that geological time.

We know for instance stone tools in Israel, 2.2 to 2.3 million years old, stone tools in China, a little more than 2 million years old, in Algeria, 1.8 million years old, in Georgia, 1.8 million years old, in Java, around 1.8 million years as well, in Italy 1.6 million years old.

A problem is that there are two first species of the genus *Homo*, *Homo habilis* and *Homo rudolfensis* and I do not think that both moved.

Because of a certain number of reasons, I currently think that it is *Homo rudolfensis* who moved.

So 2 million years ago, Man was everywhere in Africa, and almost everywhere in Eurasia, except may be in the extreme north.

As he was not demographically numerous enough to exchange genes everywhere and all the time, he first became *Homo erectus* where and everywhere he was.

But this *Homo erectus* became *Homo sapiens*, in Africa and in Asia, or in Africa only, *Homo neandertalensis* in Europe, *Homo soloensis*, or Java Man, in Java, *Homo floresiensis* or Flores Man, in Flores and probably several other *Homo* in several other places, isolated by sea or by ice (Coppens 1999; 2006; 2008).

References

Arambourg C., Coppens Y., 1967, "Sur la découverte dans le Pléistocène inférieur de la vallée de l'Omo (Ethiopie) d'une mandibule d'Australopithécien", *Comptes rendus de l'Académie des Sciences (Paris)*, Series D, 265: 589–90.

Asfaw B., White T., Lovejoy O., Latimer B., Simpson S., Suwa G., 1999, "*Australopithecus garhi*: a new species of early hominid from Ethiopia", *Science* 284: 629–35

Broom R., 1938, "The pleistocene Anthropoid apes of South Africa", *Nature* 142: 377–79.

Brunet M., Beauvilain A., Coppens Y., Heintz E., Montaye A. H. E., Pilbeam D., 1996, "*Australopithecus bahrelghazali*, a new species of early hominid from Koro Toro region (Chad)", *Comptes rendus de l'Académie des Sciences (Paris)*, Series IIa, 322: 907–13.

Brunet M., Guy F., Pilbeam D., Mackaye H. T., Likius A., Ahounta D., Beauvilain A., Blondel C., Bocherens H., Boisserie J.-R, Bonis L. de, Coppens Y., Dejax J., Denys C., Duringer P., Eisenmann V., Fanone G., Fronty P., Geraads D., Lehmann T., Lihoreau F., Louchart A., Mahamat A., Merceron G., Mouchelin G., Otero O., Pelaez-Campomanes P., Ponce de León M., Rage J.-C., Sapanet M., Schuster M., Sudre J., Tassy P., Valentin X., Vignaud P., Viriot L., Zazzo A., Zollikofer C., 2002, "A new hominid from the Upper Miocene of Chad, Central Africa", *Nature* 418: 145–51.

Clarke R. J., Tobias P. V., 1995, "Sterkfontein Member 2 foot bones of the oldest South African hominid", *Science* 269: 521–24.

Coppens Y., 1975, "Evolution des Hominidés et de leur environnement au cours du Plio-Pléistocène dans la basse vallée de l'Omo en Ethiopie", *Comptes rendus de l'Académie des Sciences (Paris)* 281: 1693–96.

Coppens Y., 1999, *Le genou de Lucy*, Paris: Odile Jacob.

Coppens Y., 2006, *Histoire de l'Homme et changements climatiques*, Paris: Fayard.

Coppens Y., 2008. *L'Histoire de l'Homme, 22 ans d'amphi au Collège de France (1983-2005)*, Paris: Odile Jacob.

Dart R., 1925, "*Australopithecus africanus*, the man-ape of South Africa", *Nature* 115: 195–99.

Haile Selassie Y., 2001, "Late Miocene hominids from the Middle Awash, Ethiopia", *Nature* 412: 178–81.

Haile Selassie Y., Suwa G., White T. D., 2004, "Late Miocene teeth from Middle Awash, Ethiopia and early hominid dental evolution", *Science* 303: 1503–1505.

Ishida H., Pickford M., 1998, "A new late Miocene hominoid from Kenya; Samburupithecus kiptalami gen. et sp. Nov.", *Comptes rendus de l'Académie des Sciences (Paris)*, Series IIa, 325: 823–29.

Johanson D. C., White T. D., Coppens Y., 1978, "A new species of the genus *Australopithecus* (Primates: Hominidae) from the Pliocene eastern Africa", *Kirtlandia* 28: 1–14.

Kunimatsu Y., Nakatsukasa M., Sawada Y. Sakai, T., Hyodo M., Hyodo H., Itaya Y., Nakaya H., Saegusa H., Mazurier A., Saneyoshi M., Tsujikawa H., Yamanoto A., Mbua E., 2007, "A new late Miocene great ape from Kenya and its implications for the origins of African great apes and human", *Proceedings of the National Academy of Sciences USA* 104: 19220–25.

Leakey L. S. B., 1959, "A new fossil skull from Olduvai", *Nature* 184: 491–93.

Leakey L. S. B., Tobias P.V., Napier, J. R., 1964, "A new species of the genus Homo from Olduvai Gorge", *Nature* 202: 7–9.

Leakey M. G., Feibel C. S., McDougall I., Walker A., 1995, "New four million-year-old hominid species from Kanapoi and Allia Bay, Kenya", *Nature* 375: 565–71.

Leakey M. G., Spoor F., Brown F. H., Gothogo P. N., Kiairie Ch., Leakey L. N. McDougall I., 2001, "New hominin genus from eastern Africa shows diverse middle Pliocene lineages", *Nature* 410: 433–39.

Senut B., Pickford M., Gommery D., Mein P., Cheboi K., Coppens Y., 2001, "First hominid from the Miocene (Lukeino formation, Kenya)", *Comptes rendus de l'Académie des Sciences (Paris),* Series IIa, 332: 137–44.

Suwa G., Kono R. T., Katoh S., Asfaw B., Beyene Y., 2007, "A new species of great ape from the late Miocene epoch in Ethiopia", *Nature* 448 : 921–24.

Tobias P. V., 1967, *Olduvai Gorge.* vol. 2. *The cranium and maxillary dentition of Australopithecus (Zinjanthropus) boisei.* Cambridge: Cambridge University Press.

Tobias P. V., 1991, *Olduvai Gorge.* vol. 4 . *The Skulls, Endocasts and Teeth of Homo habilis.* Cambridge: Cambridge University Press.

White T. D., Suwa G., Asfaw B., 1994, "*Australopithecus ramidus,* a new species of early hominid from Aramis, Ethiopia", *Nature* 371: 306–12.

White T. D., Suwa G., Asfaw B., 1995, "*Australopithecus ramidus,* a new species of early hominid from Aramis, Ethiopia. Corrigendum", *Nature* 375: 88.

THE ORIGIN OF MAN: MOLECULAR APPROACH

Olga Rickards and *Gianfranco Biondi*

1. Introduction

Two events about 100 years apart from one another laid the foundation for the scientific study of human beings. In the mid seventeenth century Carl Linnaeus freed mankind from the bonds imposed by a supernatural or metaphysical interpretation of nature, gave humans a name, and assigned them a proper place in the classification of nature. That humans had prehistoric ancestors was an unthinkable notion at that time. Child of the Enlightenment and creationist, Linnaeus based his concept of man on direct observation and knowledge of historical and religious sources. In the first edition of *Systema naturae* (1735) he devised the name *Homo sapiens*, to which he associated the motto, *nosce te ipsum*, and placed humans, along with monkeys and apes, in the same zoological order: Anthropomorpha, which in the 10th edition (1758–1759) was renamed Primates. Classifying man as a primate meant operating a bold shift in the scientific and cultural paradigm of an era that set humans apart from other living beings and far greater than the distance perceived among other animals. This paradigm shift was best described a century later by Thomas Henry Huxley in his *Evidence as to Man's Place in Nature* (1863):

> Thus, whatever system of organs be studied, the comparison of their modifications in the ape series leads to one and the same result — that the structural differences which separate Man from the Gorilla and the Chimpanzee are not so great as those which separate the Gorilla from the lower apes.
>
> But in enunciating this important truth I must guard myself against a form of misunderstanding, which is very prevalent. I find, in fact, that those who endeavour to teach what nature so clearly shows us in this matter, are liable to have their opinions misrepresented and their phraseology garbled, until they seem to say that the structural differences between man and even the highest apes are small and insignificant. Let me take this opportunity then of distinctly asserting, on the contrary, that they are great and significant; that every bone of a Gorilla bears marks by which it might be distinguished from the corre-

> sponding bone of a Man; and that, in the present creation, at any rate, no intermediate link bridges over the gap between *Homo* and *Troglodytes*.
>
> It would be no less wrong than absurd to deny the existence of this chasm; but it is at least equally wrong and absurd to exaggerate its magnitude, and, resting on the admitted fact of its existence, to refuse to inquire whether it is wide or narrow. Remember, if you will, that there is no existing link between Man and the Gorilla, but do not forget that there is a no less sharp line of demarcation, a no less complete absence of any transitional form, between the Gorilla and the Orang, or the Orang and the Gibbon. I say, not less sharp, though it is somewhat narrower. The structural differences between Man and the Man-like apes certainly justify our regarding him as constituting a family apart from them; though, inasmuch as he differs less from them than they do from other families of the same order, there can be no justification for placing him in a distinct order (Huxley 1863, 123–124).

Recent advances in molecular studies have borne out the choice Linnaeus formulated. Humans and gorillas share 97.5% genome identity; and the similarity increases when humans and chimpanzees, and the pigmy chimpanzee or the bonobo in particular, are compared. In view of such close genetic identity, some scientists have proposed enlarging the genus *Homo* to make room beside *Homo sapiens* for the common chimpanzee (*Homo troglodytes*) and for the pigmy chimpanzee (*Homo paniscus*). Others have gone even further to suggest adding one more place for the gorilla (*Homo gorilla*). But looking back, these ideas may not be so farfetched as they appear: even Linnaeus classified under the genus *Homo*, particularly in *Homo nocturnus,* an ape. That said, these taxonomic adjustments should not be misunderstood. Scientists are not disclaiming the enormous difference between apes and us. The point they are making is that in the rest of the animal world that high degree of genetic identity is employed to bring species under the same genus.

Countless years passed while scientific research, working under the constraints of anti-evolutionary prejudice, focused on the features we share with apes and prepared the ground for the now almost indisputable claim that man is just another primate. Today, however, interest has turned to the perplexing question as to how so few genetic differences can determine such wide variation in anatomy, morphology, walking, susceptibility to disease, and in intellectual and behavioral capabilities. This new direction has been boosted by programs that sequence the human genome and that of other primate species, including the chimpanzee.

Earlier, we stated that classifying man among the primates is incontestable, yet many religious circles in Western society, not to mention others, continue to privilege a metaphysical interpretation of nature over scientific evidence to the contrary.

One example of this entrenched opposition to accepting knowledge based on reason and experiment appeared in an editorial published in *La Civiltà Cattolica* in 2003:

> Unquestionably, in the Christian view, there exists a radical difference between humans and animals that is not only a difference by degree (humans are *more* intelligent, *more* able than animals) but also by nature: man is not a superior animal, *more* perfect than other animals; instead, man *is not an animal* at all. Because, although sharing with animals certain anatomic features and brain functions, man alone possesses an immortal soul. Therefore, when one speaks of man, one cannot say, as the animalists do, "man and the *other* animals", as if man were another animal though superior in kind; instead, one should say "man and animals". Man is a "separate being", because created "in God's likeness" (Gen. 1: 26) he is the center and the purpose of creation (Editoriale 2003, 6).

The other founding event dates back to the mid nineteenth century when Charles Robert Darwin challenged creationism with the publication of *On the Origin of Species* (1859) and *The Descent of Man* (1871). Working in the natural sciences, Darwin demonstrated by empirical methods the inconsistencies in the theological reconstruction of life (creationism) with respect to a scientific explanation (evolutionism). He could not dispute the question of the creation insofar as "the creation" refers to a psychological need most of humanity nurtures toward an external and metaphysical power entrusted with our concerns about the present, the future, and inevitable decline and end. Creation may be conceived as a projection of our mind — the same mind that evolution granted us in all its complexity. And, as such, creation's value lies outside the interest of experimental science, as does, indeed, the value of art, for example. In brief, demonstrating the non existence of the creation or a creator is not the task of science. What science does strive to do is prove the extraneousness of any action or principle beyond nature related to the construction, function and explanation of the world, which is self-sufficient and self-responsible.

Darwinian evolutionism could not have been envisaged if our concept of time had not been radically changed from the idea of historical time, as derived from biblical genealogies, to the idea of deep time. In the 150 years since Darwin stretched life's timeline and opened new vistas, evolutionism itself has evolved, going from strength to strength, yet never losing sight of its proponent. The new vistas have been the discovery of genetic laws and the application of mathematics to these laws (the modern synthesis theory), the discovery of a second evolutionary mechanism (molecular neutrality), the abandoning of a linear sequence of evolution (punctuated equilibrium) and the co-existence of several lines of evolution in a group of living organisms (the bush theory).

In 2005 Giuseppe De Rosa corrected the editorial published in *La Civiltà Cattolica* in 2003, placing the problem of man's origin within the context John Paul II had drawn in his *Message to the Pontifical Academy of Sciences* in which evolutionism was attributed the rank of scientific theory, upgrading it from being a serious hypothesis, as Pius XII had admitted in the *Humani generis* encyclical (John Paul II 1996). It is worth noting, however, that at point 6 of his *Message*, John Paul II made a statement incompatible with Darwinian evolutionism, affirming that modern man represented "an ontological leap" (Biondi and Rickards 2008). In his article, De Rosa picked up on John Paul II's line of reasoning, stating that:

> The hominization is to be set within a wider frame of evolution of life on earth, after the creation by God (De Rosa 2005, 3).

and that only when modern man had appeared was the human "threshold" reached:

> But what does reaching the human "threshold" mean for *Homo sapiens sapiens*? It implies both continuity and discontinuity. Continuity in the sense that *Homo sapiens sapiens* is in evolutive continuity with the preceding human forms somatically and culturally; and discontinuity at the spiritual level. (...) This discontinuity at the spiritual level (...) means that in *Homo sapiens sapiens* there is a spiritual principle that can explain the inner presence of activities that transcend a purely material order. This spiritual principle, which unites all human activities, physical, sensory and intellectual, is the spiritual soul. (...) This means that the spiritual soul, as a principle of the spiritual activities of *Homo sapiens sapiens*, represents an ontological "leap", and is therefore qualitative, in the evolutionary process that science can recognize and observe but not explain according to the laws regulating the evolution of living beings (De Rosa 2005, 9–10).[1]

And again, in 2005 Benedict XVI underscored an unbridgeable distance from Darwinian evolutionism, stating that:

> We are not a random, aimless product of evolution. Each of us is the fruit of God's thought (Benedict XVI 2005, 5).

When Darwin postulated his theory of descent by natural selection and the transmission of advantageous variations, which has proved to be a fact, hereditary material was still unknown and traces of our prehistoric past were to be found

[1] The name *Homo sapiens sapiens* for modern humans is an old nomenclature. Molecular anthropology has proved that we do not belong to the subspecies *Homo sapiens sapiens,* we belong to the species *Homo sapiens.*

only in fossil remains. Today, we know that our past is contained in our DNA and that by analyzing that molecule we can read it.

Molecular analysis touches every corner of biology and medicine. It has also had an impact on research further afield in anthropology and paleoanthropology. And its impact has been practical as well as theoretical. For example, in resolving quandaries that beset classic comparative anatomic classification of fossil remains and living organisms, molecular anthropology has given authoritative answers to four anthropological questions:

1. when did the split occur between early human and African ape lineages?[2]
2. how did our species *Homo sapiens* originate?
3. why is the concept of "biological race" inapplicable to humans?
4. how should primates be classified?

2. The split between early humans and large African apes

In 1967 researchers at the University of California, Berkeley, Vincent Sarich and Allan Wilson, questioned the then almost universally accepted idea that *Ramapithecus*, a primate fossil dated between 15 and 25 million years ago, was the earliest member of the human line (Biondi and Rickards 2004). Morphologic analysis of its dentition and skeletal characteristics seemingly indicated that *Ramapithecus* had appeared on the evolutionary scene after the human clade had diverged from that of the large African apes and that the split must have occurred at some point before 25 million years ago. However, the idea that the two lines had branched away so early met with scepticism in some circles, because results of contemporary genetic studies showed remarkable similarities between man, gorilla and chimpanzee that did not fit the notion that an evolutionary split could have occurred about 30 million years ago (Biondi and Rickards 2004).

In their analysis, Sarich and Wilson calculated the amount of time that had passed since early humans and African apes had diverged by measuring the differences in the amino acid sequences in some of their blood proteins. Since these differences result from copying errors that occur during DNA replication before it is transmitted to the next generation, the errors (mutations) that accumulate from one generation to the next can be used to put a timescale on the divergence between apes and humans. Underlying this phenomenon is the concept of the molecular clock. The same idea lies behind the dating of Medieval manuscripts. With

[2] Large African apes include two species of chimpanzees: *Pan troglodytes*, or the common chimpanzee, and *Pan paniscus*, or the bonobo or the pygmy chimpanzee, and the gorilla, *Gorilla gorilla*.

each copy, scribes risked miscopying a word: the more the errors, the more times the manuscript was copied, and the more the copies made, the more time had elapsed between older and newer manuscript copies. By counting the number of differences in copies of the same text, the copies can be dated on a timeline.

In the same way, the number of changes between molecules can be taken to count time with the molecular clock and converted into evolutionary time. In brief, the more similar the proteins of different species, the more recent is their divergence from each other. Having determined for specific proteins that the mutations, which are essentially neutral, tend to accumulate at a regular, almost clocklike rate in various different ancestral lineages through time, the number of differences in mutations accumulated between the species will indicate the length of time since they split from a common ancestor.

By applying the molecular clock, Sarich and Wilson estimated that early humans and African apes had probably split approximately 5 million years ago, a finding which ran counter to then prevailing paleoanthropological views. Following the later discovery of more complete *Ramapithecus* remains, the general consensus now rates it as *Sivapithecus*[3] and therefore it represents the ancestor of the orangutan.[4] It wasn't until the 1980s, however, that the scientific community finally accepted the idea of a recent divergence between human and non-human lines. But Sarich and Wilson had left open the question about whether man, gorilla, and chimpanzee had diverged simultaneously or whether the split had occurred at different epochs.

In the early 1980s, two Yale University researchers, Charles Sibley and John Ahlquist, tackled this problem with another technique: DNA hybridization. The two strands of the DNA molecule are held together by bonds that form its complementary base pairs (adenine-thymine and cytosine-guanine). To separate the two strands and obtain single strands, the DNA molecule needs to be heated or treated with chemical agents. After treatment, the DNA molecule returns to its double helix structure. Applying this process to DNA mixtures taken from various different species yields hybrid double helixes with bonds that are stronger the more similar the nucleotide sequences between species, so that more heat is needed to separate the bonds. By measuring the amount of heat needed for separation, the degree of similarity between species can be determined and the evolutionary distance between them (Biondi and Rickards 2004).

---

[3] Primate fossils recovered from sites in Turkey and the Indo-Pakistan region, whose most recent remains have been dated to 8 million years ago.

[4] *Pongo pygmaeus*, the large Asiatic ape.

Using this technique Sibley and Ahlquist made the surprising discovery that in contrast with evidence from comparative anatomy, chimpanzees and humans are more genetically similar to each other than either is to the gorilla and that the orangutan is even more distant in evolutionary terms. During their evolutionary course, the first of the four to diverge was the orangutan, then the gorilla, while man and chimpanzee continued to share the same path until they split somewhat later.

The enormous advances in DNA technologies over the past two decades have enabled scientists to date fairly accurately when these lines branched off: the orangutan split away about 14 million years ago; the gorilla lineage diverged about 7 million years ago; the chimpanzee and human lineages were the most recent to diverge from the common ancestor, their split having occurred about 6 million years ago; the bonobo (or pygmy chimpanzee) diverged from the common chimpanzee only 3 million years ago (Biondi and Rickards 2006).

3. The origin of *Homo sapiens*

Thanks to DNA analysis, molecular anthropology has been able to shed light on the question of how *Homo sapiens* originated. But here, too, as in previous controversies about when African apes and hominins went on their separate evolutionary ways,[5] heated scientific debate arose between paleoanthropologists and molecular anthropologists. The controversy was even more complex because the paleoanthropological camp was divided over which of two different scenarios could best explain the emergence of *Homo sapiens* (*Figure 1*). The first, known as the Multiregional or Regional Continuity model, held that modern man had arisen simultaneously from archaic populations inhabiting different regions of the Old World which they had occupied after leaving the African cradle.[6] From its start about 2 million years ago, this gradual evolutionary process was accompanied by a high uninterrupted gene flow between local populations, which would have inhibited the emergence of different species in various regions.

---

[5] The zoological subfamily *Homininae* includes the human clade, i.e. humans and their direct and indirect ancestors that appeared on the evolutionary scene after they split from the chimpanzees.

[6] *Homo ergaster* was the first hominin to leave the African continent and then spread to Georgia where he gave rise to *Homo georgicus*, to Asia, where he gave rise to *Homo erectus*, and then to Europe where his descendants were *Homo antecessor*, *Homo cepranensis*, *Homo heidelbergensis* and *Homo neanderthalensis*.

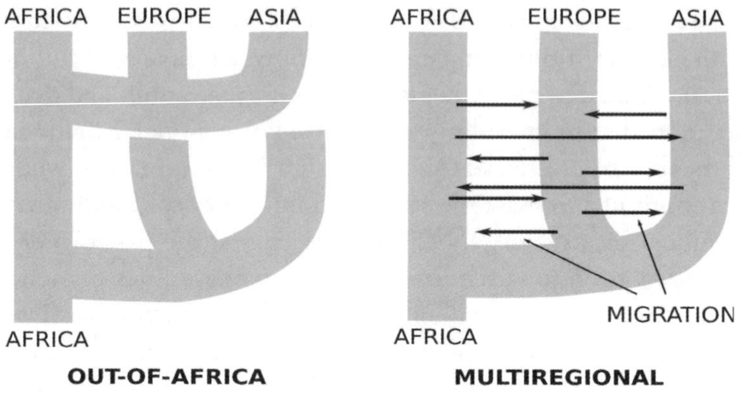

Figure 1. The two evolutionary models proposed on the basis of fossil remains to explain the emergence of Homo sapiens.

The second scenario which other paleoanthropologists embraced was the Out of Africa theory which holds that all mankind has a recent single African origin. In detail, the first modern humans appeared in Africa about 200,000 years ago following a sudden punctuated speciation event, after which they spread throughout the rest of the world, displacing the pre-existing archaic populations and without interbreeding with them. This scenario implies a genetic discontinuity between *Homo sapiens*, the new species, and the pre-existing hominins that belonged to different species. By contrast, the Multiregional model opts for genetic continuity starting about 2 million years ago and allows for a certain degree of hybridization between the various populations that had come into contact (Thorne and Wolpoff 1981; Stringer and Andrews 1988; Biondi *et al.* 2006).

In the attempt to determine which of the two models was more suitable to explain our origins, researchers looked to DNA analysis. In the late 1980s, three researchers at the University of California, Berkeley, Rebecca Cann, Mark Stoneking, and Allan Wilson published their study in *Nature* on the first molecular evidence of a single recent African origin in support of the Out of Africa theory (Cann *et al.* 1987). Studying mitochondrial DNA (mtDNA), a particular type of DNA,[7] they

[7] mtDNA is contained in the cytoplasmatic organelles called mitochondria where it provides the cell with energy. Its genome is tiny when compared with the nuclear genome (approximately 16,500 base pairs versus 3 billion base pairs). It is a circular, double-stranded DNA molecule which is abundantly found in all cells of an organism (on the order of thousands in somatic cells and hundreds of thousands in germinal cells). These characteristics, together with its location in the cytoplasm, makes mtDNA easy to analyze and to study the DNA in ancient remains. It has a high evolutionary rate, between 10 and 20 times that of portions of the nuclear genome with comparable function.

wanted to see whether in the genetic pool of current human populations there could be ancient genes that had been transmitted by the archaic *Homo* populations, as the Multiregional model assumed, or if ours was a completely modern genome, as predicted by the Out of Africa model.

Mitochondrial DNA is an ideal tool for answering this question. Since it is transmitted exclusively by the maternal line, mothers and daughters share the same mtDNA, without the mutations that may occur in the germinal line of each generation. By contrast, the genes in the nucleus, inherited from both parents, are susceptible to such phenomena as segregation and recombination that reshuffle the genes with each generation, making it more difficult to retrace the gene path back in time.

If, as proposed by the Out of Africa model, mankind had a recent origin, modern human populations would have mtDNA with very few mutations: the mutations would be fairly homogeneous because they would not have had enough time to accumulate. Furthermore, if modern humans had evolved from a single ancestral population living in a geographically restricted environment, the current population inhabiting the area, because older, would show greater genetic variation. If, on the other hand, mankind had evolved from remote and multiregional origins, the various populations would have many mutations and all in equal measure.

When the three Berkeley researchers arranged their results into a phylogenetic tree,[8] it neatly illustrated an African and recent mtDNA origin (the famous Mitochondrial Eve hypothesis). The tree showed two distinct main blocks: one grouped the more diverse types, as shown by the length of the branches proportional to the number of mutations, which resulted solely African, and the other showed all the sequences identified in the rest of the world, plus some African lineages. Since we know that the higher the number of mutations, the greater the length of time to accumulate them, the mtDNA of the African population turned out to be older, and thus represents the most recent common ancestor. The common ancestor's descendents (the African lineages in the non-African clusters) gradually spread across the world, giving rise to various different local populations through time. When the molecular clock was applied to the mutations in the two main blocks, it showed that the common ancestral mtDNA was recent and could be dated to about 200,000 years ago.[9]

---

[8] A phylogenetic tree is a graphic representation that permits the comparison between the genomes of various different individuals against the ancestor-descendant scheme.

[9] This clock beats at 2-4 mutations every 100 bases per million years. Another rate of ticking at 12-17 mutations per 100 bases per million years was estimated starting from the D-loop control region of the mtDNA. This region is highly variable and the date set for the origin of our species is between 250,000 and 170,000 years ago.

Additional analysis of the mitochondrial genome gave the same result (Ingram *et al.* 2000). Then in 2002 Alan Templeton, from work on diverse proteins and enzymes, pushed back the origin of *Homo sapiens* to about 600,000 years ago, opening the hypothesis for two migratory waves out of Africa to the rest of the Old World: the first dating soon after the origin and the second at around 100,000 years ago. The conclusions to be drawn from these studies agree with the hypothesis, based on paleoanthropological findings, of a recent single origin of our species.

Of note is that modern humans did not appear 600,000–200,000 years ago;[10] instead, this period marks a genetic divergence which does not necessarily coincide with a population divergence. Because the principles of molecular evolution predict that genetic divergence precedes population divergence, the mtDNA data collected so far are incompatible with the Multiregional model, which holds that the differences that led to the origin of our species began to accumulate about 2 million years ago and that genetic divergence would have been even earlier.

mtDNA holds enormous interest for research into human evolution, yet the molecule accounts for a minute portion of the entire genome (0.0006%). Owing to random or selective phenomena, reconstructing human evolutionary history solely from mtDNA is hazardous, since its pattern of variation may differ substantially from that of other portions of the genome, potentially falsifying the real picture. Accurate reconstruction will take into account an analysis of nuclear gene loci, such as those of Y-specific DNA, which permit retracing the male line of evolution and autosomal inheritance. The wealth of data from studies during the 1990s agree substantially with a recent African origin of modern man (Thomson *et al.* 2000; Underhill *et al.* 2000; Ke *et al.* 2001; Biondi and Rickards 2004, 2006).

Having established that we are a relatively young species and that our ancestors came out of Africa to occupy dry land, the dilemma of our phylogenesis appeared to have been settled. However, some suggested that the experiments on extant DNAs were not robust enough to rule out that *Homo sapiens*, on reaching Europe, had not interbred with the Neanderthals, and that there were two subspecies of the same species: *Homo sapiens neanderthalensis* and *Homo sapiens sapiens*, as proposed by the multiregionalists.

Again, advances in biotechnology helped molecular anthropologists to elucidate the question: the study of ancient DNA (aDNA). The temporal limit of DNA survival, estimated on theoretical calculations to be between 1 and 2 million years,[11]

---

[10] The oldest *Homo sapiens* fossils can be dated to about 200,000 years ago.

[11] In the early 1980s, initial studies on aDNA seemingly suggested that DNA extracted from dinosaur bones or insects preserved in amber, tens if not hundreds of millions of years ago,

was perfectly suited for definitively settling the Neanderthal controversy, given that Neanderthal man ranged across Europe, the Near East and western Asia between 300,000 and 27,000 years ago (Jobling *et al.* 2004).

In 1997, just 10 years after the publication of the Mitochondrial Eve hypothesis, a joint team of German and American researchers headed by Svante Pääbo studied the mtDNA of the most famous representative of the Neanderthals, the remains recovered from the Feldhofer Cave in the Neander valley in Germany, which gave the species its name. The team found that Neanderthal mtDNA differs significantly from ours. The comparison between any two sequences of modern mtDNA shows a mean of 8 nucleotide differences, whereas the number of differences between modern and Neanderthal mtDNA sequences is 3.5 times as high. Moreover, to render this value more eloquent, the data were extracted and the Neanderthal sequences were compared against those of current populations from various different continents. Had the Neanderthals contributed to the formation of our species, they would show a closer relationship with individuals currently inhabiting Europe than with those living in other geographic areas of the world. But no trace of a privileged kinship in the mtDNA could be found. The Neanderthal sequences showed 28 differences with respect to those of Europeans and Australians and 27 differences with respect to Asian, African and native American populations. So although the Neanderthals inhabited the same regions as the current European populations, the genetic distances do not support the idea that they are any more closely related to Europeans than to other modern populations. This would appear to further dismantle the theory of regional continuity: that Neanderthal man was a direct ancestor of modern Europeans. But Pääbo went a step further and constructed a phylogenetic tree with the Neanderthal sequences and those of modern man. The answer was even clearer. Neanderthal man located on a side branch of our evolutionary tree; in other words, he may be considered as a cousin of our species. Pääbo's team then applied the molecular clock to put the most recent common ancestor of modern humans and the Neanderthals on a timescale. The estimated period extending from between 690,000 and 555,000 years ago perfectly fit palaeontological and archaeological findings that the two lines had split around 300,000 years ago — set within the perspective that genetic divergence precedes population divergence — thus strengthening the African Eve hypothesis (Krings *et al.* 1997).

---

could be analyzed. Today, however, we know that those data were simply laboratory artefacts that resulted from contamination with modern DNA.

But with this came the usual problem of DNA studies: reproducibility demonstrated on other Neanderthal remains. That the Neanderthals were a different species from ours could not be proven with a single sequence. Generally, there is a wide range of genetic variation in primates, and if this were to be the same for the Neanderthals, the sequence Pääbo obtained would have been one of the most divergent. Besides studying diverse Neanderthal fossil remains spatially and temporally distant from the Feldhofer Cave site fossils, whose DNA would necessarily lie outside our interval of variation, scientists subsequently needed to turn their attention to the ancient remains of our species, i.e., those of Cro-Magnon man, whose DNA would be expected to be more similar to ours.

This line of research began in the early years of this century. Comparison of the genetic distance between the 12 Neanderthals studied so far, the 21 samples of ancient *Homo sapiens* recovered from known Palaeolithic sites in Italy, nearly all of which have been analyzed in the laboratories at the University of Rome, Tor Vergata, and samples from thousands of living individuals from different continents showed that the Neanderthals locate completely outside the known variation of our species, whereas no difference was found between ancient and modern *Homo sapiens* (Caramelli *et al.* 2003; Tarsi *et al.* 2006). A recent study estimated that the molecular data currently available for the two species overlap up to a maximum of 120 cross links between us and our Neanderthal cousins in over 12,000 years of co-existence (Currat and Excoffier 2004).

On 12 February 2009 it was announced that a first draft version of the Neanderthal genome had been completed by the Max Planck Institute for Evolutionary Anthropology, in Leipzig, Germany, and the 454 Life Sciences Corporation, in Branford, Connecticut. Obtaining nuclear sequences of the Neanderthals marked a turning point in the reconstruction of the evolutionary timeline of man and his prehistory. Knowing the complete sequence of the Neanderthal genome, it will be possible to compare human and recently obtained chimpanzee genomes and identify the characteristics that can be used to distinguish us from them. For example, we could determine definitively whether *Homo sapiens* and Neanderthals interbred during the period of co-existence, which mtDNA has categorically ruled out, and whether the skin and hair colors of our Neanderthal cousins were dark or fair, as initial studies on the *MC1R* gene regulating melanin has suggested for some (Lalueza-Fox et al., 2007). And we could even resolve the question whether our relatives used speech for communication. Pääbo is currently conducting research in this direction and has seen that the *FOXP2* gene is involved in various aspects of language development.[12] For instance, just a few point mutations is all it takes to seriously impair oral articulation, with subsequent effects on orofacial movement and efficient development of spoken language, including linguistic and grammatical reasoning. The normal form of our gene differs from that of the chimpanzee in two positions on the DNA molecule, which means the substitution of two amino acids in the related proteins. The results of the analysis of the *FOXP2* gene in two Neandertal specimens showed that they displayed a sequence identical to that of our species suggesting that our cousin could have had an advanced type of language like ours (Krause *et al.* 2007).

Therefore, we can state that modern man descended neither from the Neanderthals nor interbred with them: we are two different species, *Homo neanderthalensis* and *Homo sapiens* and we can also state that ancient DNA disproved the Regional Continuity model.

4. Human races: a false concept

In the 1950s, a review of "racial" classification by morphologic features using genetic polymorphisms revealed completely different results (*Figure 2*). Only two decades later did empirical evidence emerge against classifying human biological variation according to a concept of "biological race". Over 85% of genetic variation is encountered among individuals of the same population and only the remaining 15% characterizes different populations. This explains why among populations the distinct morphological differences arising from environmental factors are fleetingly few at the genetic level.

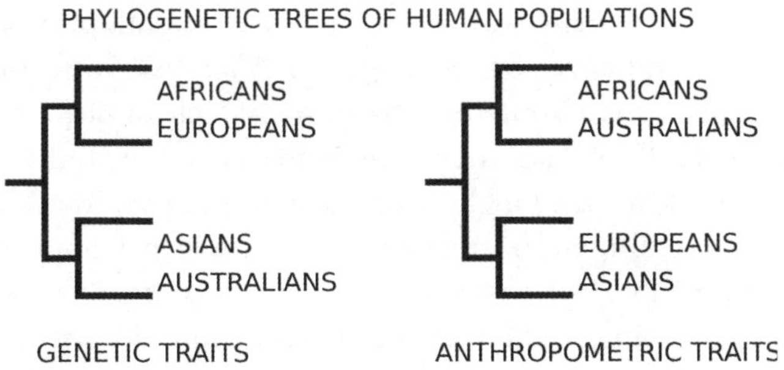

Figure 2. Phylogenetic trees of human populations based on anthropometric and genetic traits.

---

[12] The *FOXP2* gene is located on chromosome 7, near the areas associated with autism and other mental disorders.

Moreover, molecular anthropology has shown that we are a young species and therefore cannot be subdivided into "races". Reconstructing the phylogenetic tree of populations based on DNA sequences, specifically mtDNA sequences, resists "racial classification" and shows that no clear distinction can be drawn between "races" in classical anthropology. For example, the populations of the different continents, what were once called "races", are not characterized by a single mitochondrial lineage, so disproving any scientific evidence that races descend from different common ancestors. If, on the other hand, "races" were real taxonomic categories, the populations making them up would share the same line of ascendant (Barbujani 2006; Biondi and Rickards 2002, 2007).

Furthermore, if "race" were a real taxonomic category, it would permit the reconstruction of the phylogenesis of human populations according to an ancestor-descendant scheme. But the concept of "race" fails to do so; it allows only grouping human variation in relation to the environments where our species has lived and still lives. Used in this way, it may be helpful for tracing ecologic history, certainly an important part of biology, but it cannot be scientifically applied to reconstructing phylogenetic history.

5. Classification of the Primates

The fourth and most recent question molecular anthropology has sought to address regards the classification of the Primates. The genetic similarity (from 98.4, when synonymous DNA sites are included, to 99.4%, when non-synonymous sites are taken into account) shared between humans and chimpanzees estimated by Morris Goodman's group in 2003 (Wildman *et al.* 2003) reveals a surprisingly close relationship. This estimate was later confirmed by data published in September 2005 (The Chimpanzee Sequencing and Analysis Consortium) on the complete sequencing of the chimpanzee genome, which indicate that humans and chimpanzees share 97-99% of their genetic material. Given this high degree of affinity, Goodman decided to include in the genus *Homo*, in addition to our species, the common chimpanzee and the bonobo, which would be given the name of *Homo* (subgenus *Pan*) *troglodytes* and *Homo* (subgenus *Pan*) *paniscus*, respectively. This reordering of the Primate classification did not catch the academic community by surprise, since in 1998 Elizabeth Watson (Watson *et al.* 2001), and before her, Jared Diamond in his book, *The Third Chimpanzee: the Evolution and Future of the Human Animal*, had suggested a similar idea. Watson went one step further and included in our genus the gorilla as well: *Homo* (subgenus *Gorilla*) *gorilla*.

What may have appeared a bold new idea was actually anticipated by Carl Linnaeus. In 1747 in a letter to his botanist colleague Johann Gmelin, who had accused Linnaeus of daring to place man with the Primates, he wrote:

> If I had called man a monkey, or vice versa, I would have been scorned by churchmen. Perhaps as a naturalist I could not have done otherwise (Greene 1959).

The implication being that had he not been influenced by authorities, Linnaeus would have gone farther in his classification, given that his observations of nature had suggested a close tie between man and the large apes.

And once the chimpanzees and the gorilla were allowed into our genus, it became easier to give a new identity label to our ancestors who had appeared on the evolutionary scene after the divergence from the large apes. In Goodman's new classification all the familiar genera will be replaced by the single genus *Homo*. The switch from the old to the new taxa will be, respectively: *Sahelanthropus tchadensis* to *Homo* (subgenus *Homo*) *tchadensis*;[13] *Orrorin tugenensis* to *Homo* (*Homo*) *tugenensis*; *Ardipithecus kadabba* and *Ardipithecus ramidus* to *Homo* (*Homo*) *kadabba* and *Homo* (*Homo*) *ramidus; Australopithecus anamensis-afarensis-africanus-bahrelghazali-garhi* to *Homo* (*Homo*) *anamensis-afarensis-africanus-bahrelghazali-garhi*; *Kenyanthropus platyops* to *Homo* (*Homo*) *platyops; Paranthropus aethiopicus-boisei-robustus* to *Homo* (*Homo*) *aethiopicus-boisei-robustus*. Less dramatic will be the task of reclassifying our ancestors already included in the genus *Homo*, to which will be added the subgenus *Homo* (*Homo*) *rudolfensis-habilis-ergaster-georgicus-erectus-floresiensis-antecessor-cepranensis-heidelbergensis-neanderthalensis-sapiens*.

In our opinion, in the near future Goodman's classification although quite revolutionary will force us to rewrite palaeoanthropological textbooks.

REFERENCES

Barbujani G., 2006, *L' Invenzione delle razze*, Milano: Bompiani.

Benedict XVI, 2005, "Omelia alla Concelebrazione Eucaristica per l'assunzione del ministero Petrino", *L'Osservatore Romano*, 25 aprile: 4–5.

Biondi G., Rickards O., 2002, "The scientific fallacy of the human biological concept of race", *Mankind Quarterly* 42: 355–88.

[13] Not all anthropologists are inclined to agree that *Sahelanthropus* is a hominin species since they consider it as an ancestor of the gorilla.

Biondi G., Rickards O., 2004, *Uomini per caso*, Roma: Editori Riuniti.
Biondi G., Rickards O., 2006, *Il codice darwin*, Torino: Codice edizioni.
Biondi G., Rickards O., 2007, "Race: the extinction of a paradigm", *Annals of Human Biology* 34: 588–92.
Biondi G., Rickards O., 2008, "Darwin senza papa", *MicroMega-Almanacco di Scienze* 3: 129–46.
Biondi G., Martini F., Rickards O., Rotilio G., 2006, *In carne e ossa*, Roma-Bari: Editori Laterza.
Cann R. L., Stoneking M., Wilson A., 1987, "Mitochondrial DNA and human evolution", *Nature* 325: 31–36.
Caramelli D., Lalueza-Fox C., Vernesi C., Lari M., Casoli A., Mallegni F., Chiarelli B., Dupanloup I., Bertranpetit J., Barbujani G., Bertorelle G., 2003, "Evidence for a genetic discontinuity between Neanderthals and 24,000-year-old anatomically modern Europeans", *Proceedings of the National Academy of Sciences USA* 100: 6593–97.
Currat M., Excoffier L., 2004, "Modern humans did not admix with Neanderthals during their range expansion into Europe", *PLoS Biology* 2: 2264–74.
De Rosa R., 2005, "L'origine dell'uomo", *La Civiltà Cattolica* 3715: 3–14.
Editoriale, 2003, "I delitti contro gli animali", *La Civiltà Cattolica* 3679: 3–9.
Greene J. C., 1959, *The Death of Adam*, New York: Mentor.
Huxley T. H., 1863, *Evidence as to man place in nature*, New York: D. Appleton and Company.
Ingram M., Kaessmann H., Pääbo S., Gyllensten U., 2000, "Mitochondrial genome variation and the origin of modern humans", *Nature* 408: 708–13.
Jobling M. A., Hurles M., Tyler-Smith C., 2004, *Human Evolutionary Genetics*, New York: Garland Science Publishing.
John Paul II, 1996, "Message to Pontifical Academy of Sciences", *L'Osservatore Romano, Weekly Edition in English*, 30 October: 3, 7.
Ke Y., Su B., Song X., Lu D., Chen L., Li H., Qi C., Marzuki S., Deka R., Underhill P., Xiao C., Shriver M., Lell J., Wallace D., Wells R. S., Seielstad M., Oefner P., Zhu D., Jin J., Huang W., Chakraborty R., Chen Z., Jin L., 2001, "African origin of modern humans in East Asia: a tale of 12,000 Y chromosomes", *Science* 292: 1151–53.
Krause J., Lalueza-Fox C., Orlando L., Enard W., Green R. E., Burbano H. A., Hublin J. J., Hänni C., Fortea J., de la Rasilla M., Bertranpetit J., Rosas A., Pääbo S., 2007, "The derived *FOXP2* variant of modern humans was shared with Neanderthals", *Current Biology* 17: 1–5.
Krings M., Stone A., Schmitz R. W., Krainitzki H., Stoneking M., Pääbo S., 1997, "Neanderthal DNA sequences and the origin of modern humans", *Cell* 90: 19–30.
Lalueza-Fox C., Römpler H., Caramelli D., Stäubert C., Catalano G., Hughes D., Rohland N., Pilli E., Longo L., Condemi S., de la Rasilla M., Fortea J., Rosas A., Stoneking M.,

Schöneberg T., Bertranpetit J., Hofreiter M., 2007, "A melanocortin 1 receptor allele suggests varying pigmentation among Neanderthals", *Science* 318: 1453–55.
Stringer C. B., Andrews P., 1988, "Genetics and fossils evidence for the origin of the modern humans", *Science* 239: 1263–68.
Tarsi T., Noto F., Martinez-Labarga C., Giampaolo R., Babalini C., Scano G., Contini I., Lorente J. A, Lorente M., Pacciani E., Silvestrini M., Del Lucchese A., Maggi R., Lattanzi E., Formicola V., Mallegni F., Martini F., Rickards O., 2006, "Ricostruzione della storia genetica per via materna delle comunità paleolitiche delle grotte dei Balzi Rossi, della caverna delle Arene Candide e di grotta del Romito e di quelle neolitiche ed eneolitiche di Samari e di Fontenoce di Recanati", in Martini A. F. (a cura di), *La cultura del morire nelle società preistoriche e protostoriche italiane, Origines*, Firenze: Istituto Italiano di Preistoria e Protostoria, Progetti 3, 315–46.
Templeton A. R., 2002, "Out of Africa again and again", *Nature* 416: 45–51.
Thomson R., Pritchard J. K., Shen P., Oefner P. J., Feldman M. W., 2000, "Recent common ancestry of human Y chromosomes: evidence from DNA sequence data", *Proceedings of the National Academy of Sciences USA* 97: 7360–65.
Thorne A. G., Wolpoff M. H., 1981, "Regional continuity in Australian Pleistocene hominid evolution", *American Journal of Physical Anthropology* 55: 337–49.
Underhill P. A., Shen P., Lin A. A., Jin L., Passarino G., Yang W. H., Kauffman E., Bonné-Tamir B., Bertranpetit J., Francalacci P., Ibrahim M., Jenkins T., Kidd J. R., Mehdi S. Q., Seielstad M. T., Wells R. S., Piazza A., Davis R. W., Feldman M. W., Cavalli-Sforza L. L., Oefner P. J., 2000, "Y chromosome sequence variation and the history of human populations", *Nature Genetics* 26: 358–61.
Watson E. E., Esteal S., Penny D., 2001, "*Homo* genus: a review of the classification of humans and the great apes", in Tobias P. V., Raath M. A., Moggi-Cecchi J., Doyle G. A. (eds.), *Humanity from African Naissance to Coming Millennia*, Firenze-Johannesburg: Firenze University Press-Witwatersrand University Press, 307–18.
Wildman D. E., Uddin M., Liu G., Grossman L. I., Goodman M., 2003, "Implications of natural selection in shaping 99.4% nonsynonymous DNA identity between humans and chimpanzees: enlarging genus *Homo*", *Proceedings of the National Academy of Sciences USA* 100: 7181–88.

The human lineage: Macro-evolutionary process acting during embryogeny with emergent macro-evolutionary implications

Anne Dambricourt Malassé

I. Introduction: gradualism and punctuated equilibria

Palaeoprimatology, which includes human palaeontology, studies the fossilized species in the large majority of the cases in terms of taxonomic identification on one hand, and adaptation to their environment, on the other. Study of dental remains, by far the most numerous fossils, and of postcranial skeletons, reflect different adaptations by the process of speciation, the extra-uterine environment being the main factor taken into account to explain the emergence of a new phenotype. Study of the skull, on the other hand, raises more questions than it solves, especially retaining the shape of the vault and the endocranial endocast significant of the cephalisation. However, skull base is still little explored, although it is the crossroads between face, central nervous system and spine.

This part of the skeleton is fragile and it is rare to find skulls with their base, the only well preserved great anthropoid is *Sahelanthropus tchadensis*. However there is still no biometric study of its endocranial surface that allows observation of the brain endocast, the position of the cereballar fossa and comparison of the angle of the base with the different growth stages of Gorilla and Pan (a thesis is in progress since 2007). Indeed, this angle acquired during ontogeny determines the position of the cerebellar fossa and the neck (*Figure 1*). The paradigm to explain the origin of a vertical axial skeleton, i.e., from the neurocranium with the flexed basicranium to the sacrum, specific to *Homo* and Australopithecines species, is still totally hypothetico-deductive. This paradigm postulates a slow, gradual and mosaic acquisition of *adult* anatomical changes such as a small canine tooth, isolated from the ontogeny of the body, reaching the permanent erect standing posture, due to undetermined genetic mutations. However, since the sixties with the work of the Delattre and Fenart (1960), it is well known that the skull base of the great ape's species changes its orientation from the infantile stage, losing bipedal ability. Neither the famous studies on mammal, primate and hominid ontogeny well known in Clark Howell's American school (Rak and Clark Howell 1978), nor the works in palaeoembryology, take into account the works on *Sahelanthropus tchadensis* fossil bones. The discovery of fossils is not the discovery of evolutionary process, even their environment and adult specialisations are reconstructed (Jaeger 1997); world renowned palaeontologists such as Stephan Jay Gould and Jean Chaline demonstrated many years ago that gradualism and environment are insufficient to explain all the types of biological evolution: embryogeny, heterochronies, punctuated equilibria are major biological realities of the primate phylogenesis which leads to *Homo sapiens* (Dambricourt Malassé 1988; Chaline *et al.* 1996; 1998; 2000; Dambricourt Malassé *et al.* 1999).

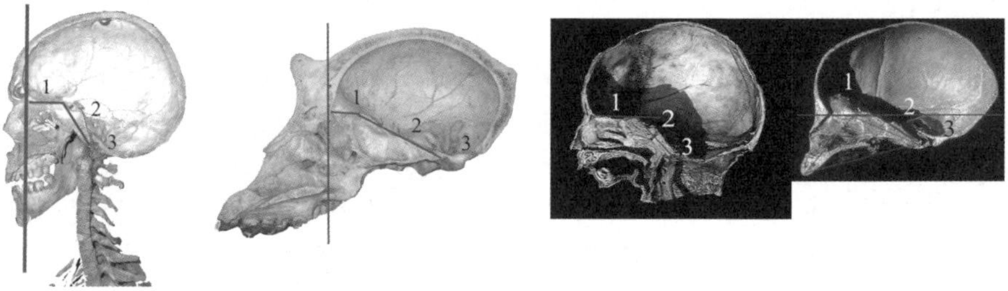

Figure 1. Skull base flexure: adulte *Homo sapiens* with hemi-mandible and Gorilla (left), *Homo sapiens* and Gorilla at late fetal stage.

Gradualism predefines the evolutionary identity of *Homo* genus as distinguished by the common features of the oldest fossil specie (*H. ergaster*, *H. habilis*, *H. erectus*, *H. rudolfensis, H. georgicus*…), while it is demonstrated that modes and tempo 1) concern the embryonic morphogenesis of the body plan, 2) are saltationist and macroevolutionary (Dambricourt Malassé 1988; 1993; 1996a; 2006; Caro 2004).

In such a paradigm *Homo* genus, or the erected basi-cranio-caudal posture and its corresponding psychomotor control, would be the result of cumulative post-natal locomotor adaptations, justified by a long climatic reheating and the accumulation of undetermined but advantaged mutations selected during 10 millions years (between 15 and 6–5 millions years). The body would gradually stand up and abandon arboreal locomotion, a bush of anthropoid species would emerge without possibility to distinguish filiations between exhumed fossils (Harcourt-Smith and Aiello 2004).

Concerning the environmental conditions of emergence, many palaeontologists among which the world-renowned Phillip V. Tobias abandoned the global warming hypothesis, because Australopithecines were arboreal primates reproducing in environments as diverse as Miocene great ape's biotope (mixed environment forest, gallery forest and arboreal savannah). The paradigm is abandoned in South Africa with the theory of chaos (the self-regulated dynamical systems) with McKee (1999). It is reconsidered in East Africa (Kenya) after the discovery in a wooded environment of a 6 mya femoral diaphysis corresponding to a post-cranial skeleton with occasional erected posture, *Orrorin tugenensis* (Senut and Pickford 2005). No anatomical data allows evidence of any evolutionary link with *Homo* (Richmond and Jungers 2008), so the femur does not imply its locomotor equilibrium was an ancestor of human permanent bipedality. But this proves 1) that wooded environments were always compatible with occasional bipedal equilibrium, as for extant Chimpanzees and 2) that there were no environmental necessities to develop a bipedal imbalance, as bipedality *sui generis* was adapted to the wooded East African biotopes.

Concerning anatomical emergence, one of the best proofs of the saltationist process is the semi-circular canal (inner ear) conformation of *Homo* genus, quite different from that common to australopithecines and fossil/living great ape species (Spoor and Zonneveld 1994). It implies a complete change of the central nervous system controlling the posture from the intra-uterine life. This anatomical detail emerges with a verticalization of the axial skeleton, a chain of bones from the skull to the sacrum.

Frequent speculative scenarios explain that vertebral column determines the erected posture of hominines, but this is not the case. The chain of bones starts with the posterior part of the skull base, the post-sphenoid and the basi-occipital, which are in continuity with the axial segmented part of the embryo, the future vertebral column. The verticalization occurred inside the skull base at the end of the embryonic period, between the post-sphenoid and the pre-sphenoid, i.e., the segmented and the non segmented parts of the embryo (see below).

The most ancient fossil bones of a verticalized axial skeleton are isolated fragments of the base: one basi-occipital and two petrous bones with forwarded foramen magnum, discovered in Ethiopia, in a wooded environment dating no more than 4, 4 mya (WoldeGabriel *et al.* 2004). Teeth similar to the chimpanzee anatomy, especially the huge canine tooth, were collected in other sedimentary deposits one kilometer away (White *et al.* 1994). The great ape's canine type of occlusion is typically that of a skull base in extension never observed in verticalized axial skeletons, typical of hominids, i.e., restricted to Australopithecines and *Homo* species (*Figure 1* and *8*). The skull bases of young gorilla and chimpanzee are less flexed than an adult Hominine, but more than their late ontogenetic stages, consecutively they frequently adopt an erected posture. But their occluso-postural imbalances, especially the canine tooth occlusion, are similar to the adult one and completely different from Hominines. Nevertheless authors associated the base and the teeth to create a new taxon, *Ardipithecus*. This artificial association does not allow concluding that australopithecine morphotype results from gradual and mosaic evolutionary modalities, since it was created on this premise. In fact, this fossil is the oldest hominine skull base ever excavated which proves once again with the isolated dentition, that not only great apes and Hominines shared the same wooded ecosystem, but that the acquisition of a vertical axis did not replace Mio-Pliocene great ape's species. This major axiom of neo-Darwinism is refuted by the paleontological data, *no ecological pressure are evident*.

The evolutionary identity of both *Homo* and Australopithecine is so strongly influenced by the paradigm of gradual ecological pressure, that the fossil bones are not understood for themselves, i.e. for the morphogenesis underlying their shape. The *Ardipithecus* genus can be attributed to this great ape's dental type of occlusion, especially since similar teeth and mandibles coming from North-East Ethiopia have been discovered in the oldest layers (4,5–5,8 mya) revealing a mixed environment of forest and grassland. But as we will see, a very rigorous attitude had do be adopted without any paradigmatic influence because 1) the flexure of the base and the occlusion between the mandible and the maxillar are very closed, as known in posturology, and 2) gradualism is genetically not possible for this type of homeotic mutations (genes which preside the elaboration of the body plan).

The femur of *Orrorin tugenensis* was discovered in association with the great ape's canine type of occlusion which corresponds to an occluso-postural imbalance typical of any great ape's species, that is to say, with a postnatal growth in extension (post-natal decreasing angulation of the base between pre and postsphenoid). In the opposite, "permanent bipedality" is the specific imbalance of an axial skeleton not only verticalized from the sphenoid to the sacrum, but also with a postnatal angle which keeps the flexion till the adult stage (20 years in *Homo sapiens*). No paleontological data allow us to conclude to a gradual acquisition of permanent bipedality that is totally in accordance with the embryonic origins of the verticalized axial skeleton. The evolutionary process is that of punctuated equilibria, but as embryogeny and ontogeny of the posture remain the big scientific gap of paleoprimatology, human paleontology develops in a paradigm more and more incompatible with both paleontological and embryonic data. Because bipedality is not a delimitating bound between Hominines and great anthropoids, it is very important to distinguish occasional erected postures of the inferior appendicular skeleton (the femoral diaphysis of *Orrorin* or the young panines) and the permanent erected posture of a verticalized axial skeleton, from the basicranium to the sacrum, from which the fossil bones are defined as Australopithecines and *Homo* genus.

The ontogenetic discontinuity is not reduced to the post-natal locomotion, in fact the unsolved problem for the gradualists vs punctualists, is that Australopithecines were tree-dwelling species, while their axial skeleton was already verticalized at birth, after a complete embryonic change of the Miocene great ape's pattern (Dambricourt Malassé 1988; 2006). The application of embryogeny in human palaeontology gives more precise information; the ontogenetic discontinuity concerns the entire skeleton and the neuromotor control of its imbalance. The evolutionary process is a macro-evolutionary event, a change of the embryonic pattern common to living and fossil great ape's species after 15 millons years of genetic stability. I have called this common embryonic pattern "Fundamental Ontogeny" (embryogenetic funds common to living and fossil species) (Dambricourt Malassé *et al.* 1999). In other words, permanent bipedality is not the anatomical evolution of arboricalism after losing this adaptation, *losing arboricolism is not necessary to become verticalized because the process of vertizalisation has nothing to do with the locomotion; this is a dynamical embryonic acquisition.* The cranial part which revealed this discovery is *the most frequent bone* of the fossil records, the mandible.

II. The mandible: a morphogenetic memory from embryogeny to post-natal occluso-postural imbalance. Towards a new paradigm

The twenty-five year morphogenetic study of the mandible involved provides the best illustration of macro and micro-evolutionary modalities of primate fossil species organogenesis, as it is the most documented part of the embryonic cranium. Actually the inferior jaw, formed by two symmetric bones, is not separable from the basicranium after embryogenesis, because of its posterior connexion close to the location of the basicranial flexure, one of the major characteristics of the primate macro-evolution from oldest primate species to modern *Homo sapiens*. The more the skull base is flexed, the more the mandible is short and enlarged, an ontogenetic process I called "cranio-facial contraction" (J. Chaline in Encyclopedia Universalis, 1996).

This research started in 1983 and revealed the embryonic origin of the dynamic flexure to be all the more significant as the cartilaginous skull is the first skeletal element to be formed. The consideration of embryonic chronology showed itself of major interest, considering the discovery of the homeogenes and the homeodomains. We know that the transcription factors are the mechanisms of gene regulation, which act in cascade and follow a rostro-caudal gradient.

A mutagenesis of homeotic genes controlling the rostro-caudal organization will have immediate implications on the cephalic morphogenesis including the teeth, and all along the axial and appendicular skeleton. This discovery changes the evolutionary identity of human lineage at a paradigmatic level: it was supported at the National Museum of Natural History and validated by the Academy of Paris Universities (Ph.D.), then presented to the French Academy of Sciences in April 1988 by the palaeontologist Jean Piveteau and published in the Comptes Rendus of the Academy (Dambricourt Malassé 1988). In 1989, Philipp Tobias described the phylogenetic process of "cranio-facial contraction" during the final address of the Head-Neck Sensori-Motor System International Conference organized by Alain Berthoz (CNRS- Collège de France) published by the Oxford University Press (Dambricourt Malassé and Deshayes 1992). In 1990 the European Science Foundation (ESF) created a network on "Quaternary Mammalian Faunas", in 1992 Jean Chaline organized the second workshop on "Modes and Tempos of Evolution in Quaternary" where the discovery was discussed then published in Quaternary International (Dambricourt Malassé 1993). In 1996, Yves Coppens proposed to compare the scientific synthesis of Pierre Teilhard de Chardin with the discovery and concluded that internal evolutionary process of memorization are frequent (Coppens 1996). In 2004 the Institut de France and Henry de Lumley, former Director of the National Museum of Natural History, organized an international conference with both the Academy of Sciences and the Académie des Inscriptions et Belles Lettres, with a session dedicated to the debate between gradualism and punctuated equilibria; Brigitte Senut (*Orrorin tugenensis*), Michel Brunet (*Sahelanthropus tchadensis*), Timothy White (*Ardipithecus ramidus*), Yves Coppens (Omo event) on one hand, and embryonic origin of permanent bipedality, punctated equilibria, on the other. After the debate, the discovery was again published in the Comptes Rendus of the Academy (Dambricourt Malassé 2006).

II.1 The mandible in its basi-cranial context

In profile section, the basicranium and the face of *Homo sapiens* are clearly distinguished from both paninae (Gorilla, Pan) and ponginae (Pongo) by the closed sphenoidal angle, the cerebellum fossa developed to the bottom, the verticalization of the three cerebral fossa according to the vertical axis, the consecutive shortening of the base, the upper and lower faces under the frontal floor and a radically different occlusal imbalance. Those differences are less marked at the fetal stage but the architectonic pattern is clearly established (*Figure 1*).

The occlusion is defined in *Homo sapiens* by the position of each mandibular tooth relative to the maxillar dentition, the whole forming a specific gearing, as for the deciduous dental arcade than the permanent one. Although having the same dental formula, the great ape develop an occlusion with intercuspidation quite distinct from all Plio-Pleistocene species (5 mya to 10,000 years) having acquired an intra-cranial verticalization with permanent cranio-sacral dynamics in flexion (whatever position, standing, sitting, lying, hanging, running, swimming...). These fossils are Australopithecines (*Australopithecus*, *Paranthropus*) and *Homo* species (*Figure 2* D, E).

Modern *Homo sapiens* is unique with its retromandibula, the mandibular dental row develops behind the upper teeth, while fossil and living great apes develop promandibula, locked by two large canine teeth, even at the deciduous stage with little crown (*Figure 2* A, B, C); the temporo-mandibular joints migrate in a postero-superior direction, but retromandibula is mechanically impossible. On the upper maxilla, a pre-canine diastem gives place to the inferior canine tooth, as for the little crown of the deciduous tooth and then for the large adult one (*Figure 2* B, C).

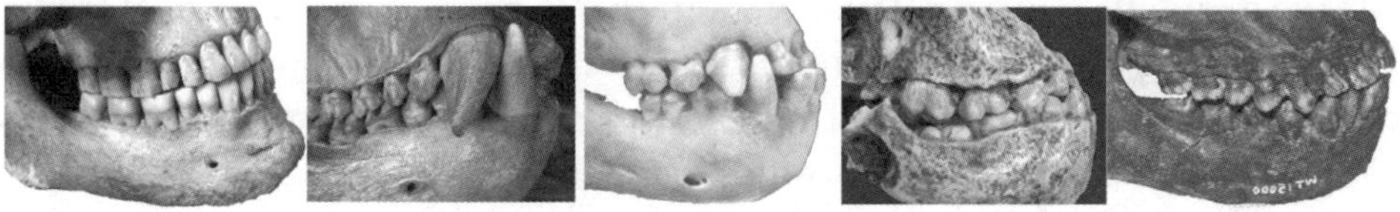

Figure 2. Canine tooth occlusion: A: adulte *Homo sapiens*, B: adult Gorilla, C: deciduous Gorilla, D: Australopithecus mixed dentition (juvenile), E: sub-adult *Homo rudolfensis.*

The retromandibula is a recent occlusal imbalance appeared with modern *Homo sapiens* during Holocene (10,000 years-present). Previously, the incisors teeth of robust *sapiens* (Qafzeh 90 000 years ago) and oldest *Homo* fossil species are in contact, their apical surface are eroded, but the inferior canine teeth are behind the superior. In contrast Australopithecines were still in promandibula, but without an occlusion locked by canine teeth nor temporo-mandibular joints growing towards the posterior pole of skull: stabilizing the face as with the great ape was not necessary because the axial skeleton was in flexion. This is the best anatomical observation to understand that, even the canine crown of *Sahelanthropus* is "little", the size of the crown has no taxonomical value outside the ontogenetical and dynamical context given by the intercuspidation.

The retrognathic motion of the mandibular alveolar arch is clearly correlated to the flexure of the basi-cranium where the mandible is attached by both its temporo-mandibular joints. The differences are found again in the transversal endocranial and exocranial organization of the skull base. The mandibular alveolar arch reflects the shape of the upper maxillar arch; it is long and upsiloide (U-shaped) in ponginae and paninae, whereas it is shorter and parabolic in *Homo sapiens*, with a frontalisation of incisor-canine rows (alignment in the frontal plane) (*Figure 3*).

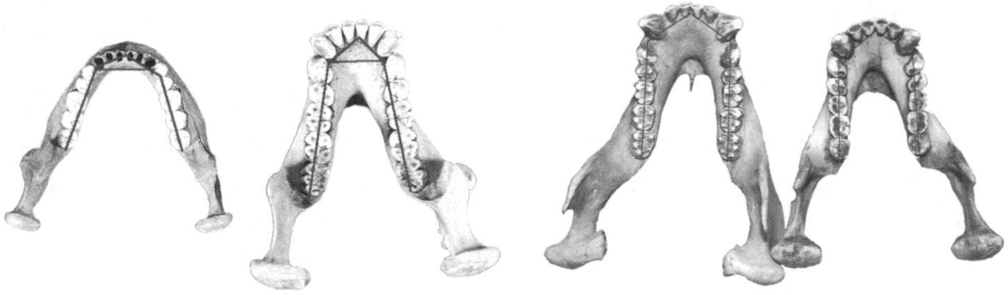

Figure 3. Adulte mandibles in superior view, comparison of alveolo-dental arcades: left *Homo sapiens* and Gorilla, right Gorilla and Pongo.

As the mandible is in articulation with the base, it reflects the kinetics of the basicranial growth in its three dimensions. These correlations are observed with clinical data known by some schools of dento-maxillo-facial infantile orthopaedics. Symphysis (chin area) is tilted forward, fused all along its axis in great ape, while it is straightened in *Homo sapiens* with a triangular disjunction in the lower part visible in early fetal ossified mandible. This structure forms the chin at birth with cartilaginous nodules. These differences are already visible at the late embryonic stages (symphyseal notch, orientation of mandibular osseous trabeculars and position of the foramen magnum), further intensified during the fetal period (transverse growth of neo-cortex in humans, resulting in enlargement between the two hemi-mandibles and frontalisation of the canino-incisor alveolar row). These differences are measured using protocols applied to the different stages of ontogeny and can be followed in all three endo and exocranial planes. Indeed, contrary to an old idea still published in some books on cranio-facial growth, when the occipito-sphenoidal chondrocranium forms in mammals, around the rostral segment of the notochord, it is not flexed. The human embryonic chondrocranium morphogenesis was described with very fine details by Levi in 1900 and it is flat, that is the original conformation of all mammals, the ancestral conformation (*Figure 4*). When the human embryo begins to develop, its axial skeleton is like in any other mammal, it is not verticalized. It was the same condition for Miocene great apes, as for the living ones; each species begins in a horizontal organizational plane. The axial skeleton begins to verticalize in the ape, in the great ape, and in humans, but the amplitude of the process increases. And there are fossils which reveal that the evolutionary process occurred with successive macro-evolutionary events. Big confusions occurred between micro-evolutions, i.e. the radiation of many species until living populations, and this evolution. *Homo* genus is not a species of great ape pattern, whereas after 20 millions years of genetic mutations, *Pongo*, *Gorilla* and *Pan* share the same pattern. This pattern is the ancestor of the *Homo* pattern.

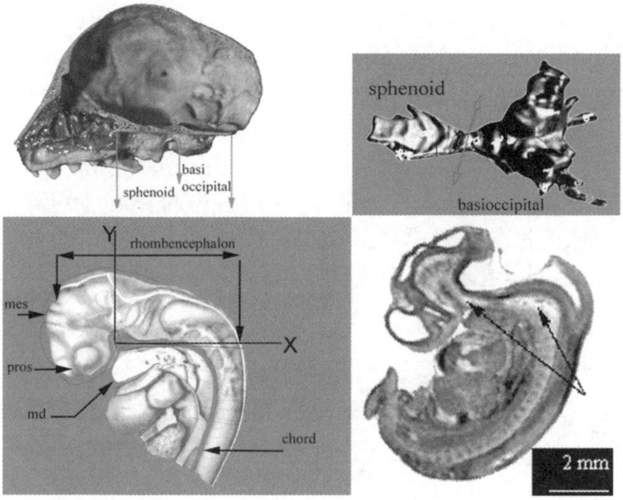

Figure 4. Flat chondrocranium: Bradypus (Sloth) at fetal stage and human embryo (Levi 1900, stage of Carnegie 16-17, ca 39 day); cephalic terminaison of human neural tube in a notochordal referential (stage of Carnegie 12-13, 4-6mm, ca. 32 day) and section of a later stage Carnegie 17-18 ("The Multi-Dimensional Human Embryo" NICHD on line).

It is therefore possible to quantify and compare growth trajectories by angles from a common reference which is the embryonic conformation. Understanding the morphological and angular differences therefore starts with the study of the bending origin.

II.2 Two discoveries with paradigmatic consequences

The skull base is probably the most complex part of the skeleton and of embryogeny, as it is, *at its centre*, the boundary between the segmentation of the body designated by the dorsal chord and the prechordal area (*Figure 5*). It represents an assemblage of cartilage articulated in the median plane by cartilaginous sutures (the synchondroses), which differentiate progressively during embryogenesis, in close relation with kinetics and rotations of the central nervous system laying above it (neural groove then neural tube). The chronology of neural kinetics is important to track the origin of the bending. The rostro-caudal axis of the embryo is organized into segments induced by homeodomain genes, active since the notochord (dorsal chord), inducing a regionalization of the central nervous system lying above it, including the hindbrain (the rhombomeres, the future rhombencephalon): "Analysis of mouse Hox mutants have clearly demonstrated that the genes have important roles in normal regionalization of the hindbrain and branchial arches, and this has lead to interest in how their early patterns are established in the nervous system. The Hox genes and their relation to hindbrain segmentation therefore provide a means of examining the cascade of events which regulates pattern formation in early neural development." (Krumlauf *et al.* 1993) "Neuronatin participates in the maintenance of segment identity in the hindbrain and pituitary development and maturation or maintenance of the overall structure of the nervous system." (Winjhols *et al.* 1995)

Thus, segmentation is extended into the skeletal tissues underlying the hindbrain. They form the cerebellar fossa with 1) the cartilaginous occipital scale, 2) the exo-

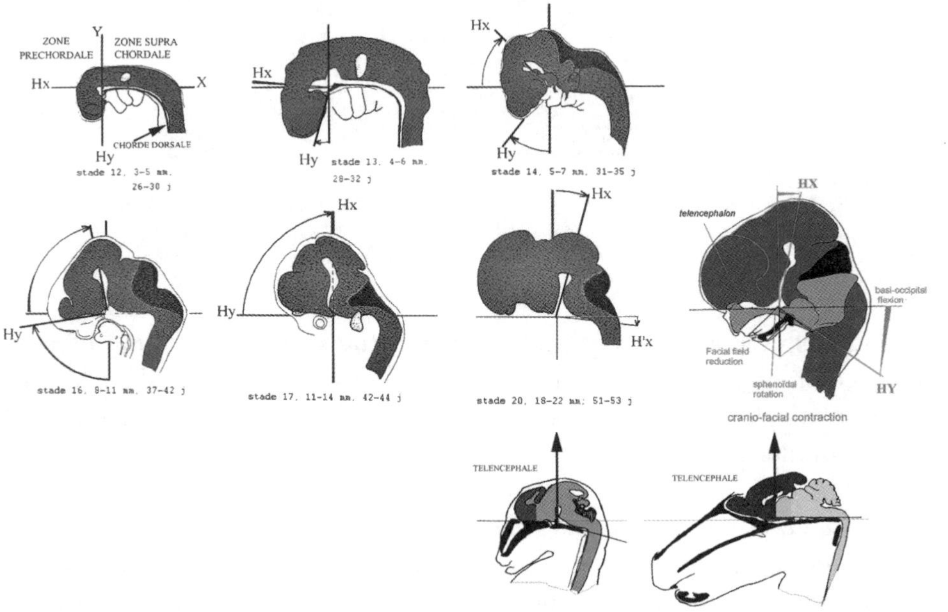

Figure 5. Opercularization of neural groove in notochordal referential and origin of the sphenoidal flexure in *Homo sapiens*, comparison with fetus and adult Cheirogaleus (bath).

occipital of either side of the cerebral trunk, 3) in-between, the basioccipital and 4) basisphenoid (or postsphenoid). Presphenoid territories are under the induction of prechordal area and derived tissues from the neural crests, some of them forming the future teeth buds. The dorsal chord stops between the basi and the presphenoid. We have shown that the basicranial flexure activates at the apex of the chord in relation to the migration of neural crest cells, antagonist of the neural groove opercularization (see below). As we see, the forward extremity of the segmented axial skeleton is incorporated into the skull, and not articulated under the skull base; the posterior part of the skull base is the extension of the segmented axial skeleton. Thus, the head posture is not reduced to a question of exoskeleton imbalance maintained by muscles, tendons and fascia; at first it is an endoskeletal organization between pre- and basiphenoid in relation to the dynamics of the neural groove.

1) The mandibular morphogenesis, a mirror-shape dynamics of the basi-cranial bending

It would take too long to get into dynamics of cephalic tissue generated by the kinetics of the sphenoid, the mandible is the guiding thread. Made of two pairs of bone joined at the symphysis, it is the bridge between the chordal territories (articulation with the basicranium bending) and pre-chordal areas (the upper level of the face). Its morphogenesis, both internal (orientation of collagen fibers and osseous trabecules) and topographical (architecture) reflects the high degree of integration of the two prechordal and chordal areas, not only genetic or epigenetic, but also dynamic.

The primitive lower jaw contains a cartilaginous axis which gives the general direction of the fibers orientation of the future mandible. Its posterior extremity is integrated into the skull base and forms the bones of the middle ear (transmission of sound vibrations), near the bending zone. The lower jaw is also in contact with the upper maxilla. However, according to some schools of orthodontics and dentofacial orthopaedics, maxillar kinetics are under the influence of presphenoid kinetics (upper/middle floor of the midsagittal endocranial cavity), in an opposite rotation to the flexure (inferior cerebral fossa).

All these dynamics can be modelled by a pantograph (*Figure 6*). We described the kinetic process said "cranio-facial contraction". For its relationship with the base, the mandibular morphogenesis develops in a complex context of kinetics, as a mirror-effect of the dual influence of the chordo-parachordal kinetics (the cerebellar fossa motion down and forwards), and inverse kinetics (premaxilla-maxilla influence).

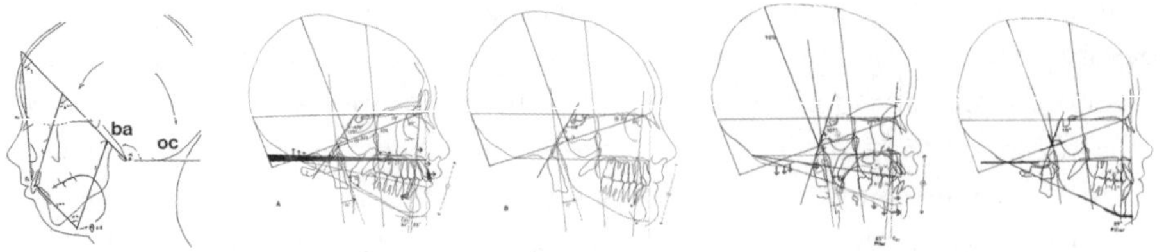

Figure 6. Gudin's pantograph and two orthodontics cases: class II or lower flexion and class III greater flexion. Both orthopaedic treatments lead to class I occluso-postural imbalance (Deshayes and Dambricout Malassé 1991).

In the early fetal stages (i.e., after 8 weeks), the osseous trabecules show a very consistent organization. They are straightened in the alveolar part of the symphysis with a triangular disjunction of the both basal parts, never observed in other living primate species in which the flexure is much less developed than in *Homo sapiens*. During the post-natal growth, a new articulation adds between the two floors of the face, that of the occlusion. Each mandibular tooth is characterized by a position with upper teeth, known as "Class I". This articulation is inseparable from the level of basal flexure. If the mirror effect between the two antagonist kinetics (basispheno-occipital and pre-sphenoid) is disturbed during morphogenesis, disorders between head-neck posture and occlusion may increase during the primary, mixed and permanent phase of dentition. The imbalance may be even more disturbed: when the embryonic dental germs were not differentiated (agenesis), the dental formula is no longer the same between the two arcades; occlusal contacts are modified and the constraints of balancing disorganized. The posturologists show the importance of malocclusion related to disorders of postural changes resulting from cellular kinetics of cartilaginous synchondrosis (*Figure 6*). Conferences and symposia have succeeded in understanding what therapeutic principles adapt to orthodontic and orthopaedic disorders (Lalauze *et al.* 2007). Some schools advocate the application of early extra-oral forces acting on the spheno-occipital synchodrosis, others preferring the osteopathic gesture, insisting on the interference with the metabolism of the pituitary gland located near the bending zone.

The 1987 discovery therefore concerns the relationship between the mandible and the basicranial flexure after embryogeny and during primate macro-evolution. The results were presented to the first annual meeting of the Société d'Anthropologie de Paris and the cranio-facial principles confirmed by the French Dento-Maxillo-Facial Orthopaedic School of Nantes University (Deshayes 1991; 1996; Deshayes and Dambricourt Malassé 1990). In 1992, the discovery again is confirmed by a report of Dr. Eugene Fisher from the New York College of Dentistry for the French National Centre of Scientific Research (C.N.R.S.). In 1999 Philippe Andrieux, Director of the Archaeological Laboratory of the Val de Marne's State, Djillali Hadjouis, palaeontologist and palaeoanthropologist, proposed a symposium under the patronage of the National Museum of Natural History (Pr. Henry de Lumley), the College de France (Pr. Yves Coppens) and the C.N.R.S. to inform palaeoanthropologists, palaeontologists and sectors of Public Health. We called it "Human identity in question". The Proceedings were awarded by the French National Academy of Dental Surgery in 2002.

2) The dynamic embryonic origin of the basi-cranio-facial contraction

The loss of the basi-sphenoid and basi-occipital alignment is described using *Figure 5*. Two axes were drawn from the dorsal chord to define the bi-dimensional orthonormal reference (*Figure 4*). The origin is the chordal apex; each stage described referred to this standard. A neural kinetics emerges: as the neural groove is opercularized to form the neural tube, it grows in front of the chord. Just above the chordal apex, neural cells follow migration from the upper part of the groove to its floor, an antagonist motion of the opercularization, forming the ventricle 5 (black in *Figure 5*).

A general kinetics emerges in the entire cephalic pre-chordal pole in a forward-backward rotation leading to the upright of the head. Among non-primate mammals, the neural rotation stops at the limit above the chordal apex, while it exceeds this limit in monkeys and more again increases in *Homo sapiens*. The rolling up is a spiral, followed during the seventh week by a rolling movement of the pre- and postsphenoid around their synchondroses. The basioccipital rotates later, during the eighth week, at the junction with the sphenoid (the spheno-occipital synchondrosis). The parachordal tissues such as temporo-mandibular joint follow the movement which tilts down and forward. At the same time, intra-sphenoidal antagonist constraints emerge between pre- and postsphenoid (or pre- and chordal partition of the body axis) leading to mirror kinetic effects: the backward-forward verticalisation of the basisphenoccipital provokes a downward-upward mandibular verticalisation in the symphyseal area.

In *Homo sapiens*, the rolling up kinetics continues after birth with the growth trajectory of the vault and the occiput in the same orientation. The bending growth area is the spheno-occipital synchondrosis, which ossifies around 20 years. So the primordial chondrocranium is not still bending when chondrogenesis starts, second, then this is not the pressure of the both telencephalic vesicles (the future hemispheres), which maintain the flexure in human, whereas it decreases in other primate species. To conclude, the second 1987 discovery is the embryonic origin of the sphenoidal rotation and its axial corollary, the caudal verticalisation of the skull base. The third 1987 discovery derives from its application to the living and fossil primate species, without modifying either their phylogenetic classification, nor their taxonomy. The locomotor postures said "permanent bipedality" correspond to post-natal prolonging of embryonic neural dynamics vectors, after an intra-cranial verticalization of its chordal segment. *Homo* and Australopithecines are the only genera concerned by these morphogeneses.

III. The recovery of morphogenetic basi-cranio-facial body plans of primate fossil species

III.1 The Great Ape clade

The ponginae (*Pongo*, Indonesia) and paninae (*Gorilla* and *Pan*, Africa), the last descendant of Miocene Dryopithecidae (*Proconsul africanus*, 20 million years) develop the same basi-cranio-mandibular trajectories, with the same spheno-occipital amplitude of flexion (Deshayes and Dambricourt Malassé 1990). They are distinguished from *Homo sapiens* organogenesis as shown by the comparison of different embryonic stages of man. The difference between both rotations is evident as shown by the lower position of the cerebellar fossa in an early human fetus of 43 mm compared to a chimpanzee fetus of 70 mm. The imbalance between neonate maxilla and mandible is not yet that of occlusion because of the lack of teeth, but it is identifiable in dento-maxillo-facial orthopaedics (Deshayes 1991). The growth trajectories of the occipital scale are compared with the inion position, an anatomical point located above the Frankfurt plane at birth. As the growth vectors reverse in apes, the inion is in its highest position at the adult stage relative to the base and a huge canine-tooth occlusion stabilizes the face. In the opposite, in Australopithecines and *Homo* genus adult stage, the inion is lowest, keeping the embryonic trajectory and the cranio-palatine imbalance (Dambricourt Malassé 2006b).

Divergences in the pre-sphenoidal kinetics are observed between Pongo and Gorilla. The intra-sphenoidal synchondroses ossified in Pongo during fetal growth, causing an entire shift up of the face. This is a microevolutionary divergence, late in the chronology of development, while spheno-occipital amplitude is so similar to Gorilla that it is not possible to distinguish this ontogenetic dichotomy on the mandible (*Figure 3*). At birth Pongo, Pan and Gorilla mandibular angle are similar, with the U alveolar shape quite divergent from *Homo sapiens*. This ontogenetic pattern was transmitted to living great apes from the oldest Dryopithecines during Miocene. Many species were living 15 millions years ago in Europe (from Spain to Turkey), in Himalayan foothills (Siwaliks) and beyond in Asia (China, Thailand, Vietnam). The mandibles attributed to Dryopithecines, Proconsulines, Ramapithecines, Sivapithecines, Lufengpithecine, Gigantopithecines show inter-specific variability in the mandibular arcade shape, as well as in the morphology of the occlusal surface: micro-evolutions occurred in relation with the dietary regime, the inhabitants and the type of society with an important role for sexual selection. No confusion is possible between Pongo and an Asian Ramapithecine, but each one developed with the same pattern. Starting from 5 millions years, a mass extinction occurred from Europe to the Himalayan foothills. These paleozones formed an entity apart from Asia where some species evolved to the ancestors of orang-utan (South-West China, Thailand, Vietnam and Indonesia). The East African species (*Proconsul nyanze, Proconsul major, Kenyapithecus)* extended to West Africa with species such as *Sahelanthropus tchadensis* (10–7 millions years). An interesting point is that, although the orang-utan (*Pongo*) share the same morphogenetic pattern with sub-Himalayan fossil species (18–7 myr), they differ in the dental arcade growth tempo and are very similar to *Gorilla*. The great proximity of size and conformation, well distinct from chimpanzee, raises questions about the date of the dichotomy between the two lineages. The geological period marking the dichotomy may be more recent than expected (18 mya), as East Africa and South Asia have several periods of communications through the Arabian Peninsula after Miocene (de Bonis 2008).

But much more critical is the classification in one family said Hominidae, with *Gorilla*, *Pan*, Australopithecinae and *Homo*, excluding *Pongo*. *Gorilla*, *Pan* and *Pongo* keep the same embryonic pattern, common with the oldest great ape species, without any change of the chordo-prechordal embryonic flexure. No embryological evolution occurred during long geological periods before the emergence of both new embryonic patterns. *Gorilla* and *Pan* are no more hominidae than *Pongo*, the oldest classification Panidae (*Gorilla*, *Pan*) and Hominidae (Australopithecines, *Homo*) is incomparably more relevant because it corresponds to an authentic dichotomy between the body plans. Unfortunately this biological level of the species does not exist in the hierarchic classification; the palaeontologist is reduced to the usual name “great ape”. Such a vision creates confusion between, on one hand, the genetic continuity from great ape’s fossil species to the first hominids which is not recognizable (Hartcourt-Smith and Aiello 2004) and, on the other, the biological level of evolution (common embryogeny), which is recognizable and belongs to the macro-evolutionary process (punctuated equilibria). Nevertheless, other anatomists who compare the embryonic conformations of fossil anthropoids reach the same conclusion such as Spoor (1997) "That all great apes, as well as the 10-Myr-old hominoid Dryopithecus, share the same basic morphological pattern suggests that this represents the ancestral human condition."

Australopithecines emerged from an African great ape species, in East Africa (Ethiopia, 4.4 mya) before reaching Central and South Africa (3.5 mya). On the other hand, numerous palaeontologists abandoned the australopithecine origins of human lineage (Deloison 1999) or never supported the hypothesis (Zuckerman, Oxnard); the first *Homo* species emerged separately. The challenge to find where and when the emergence occurred depends on the paradigm which gives the great orientation: gradual acquisition of bipedality from African great ape populations with ecological pressures because of global warming. Or integration of scientific knowledge developed over more than 25 years with embryogeny, punctuated equilibria, homeogenes, ontogeny, occluso-postural imbalance and dynamical systems without neglecting the environment (Chaline 1983; 2006; Chaline *et al.* 1986; 1990; 1993; 1996; 1998; 2000; David *et al.* 1989; Devillers, Chaline and Laurin 1990; Dambricourt Malassé *et al.* 1992a; 1999; 2000; Dubois, Chaline and Brunet-Lecomte 1992). In the second approach, water and protection against predators are the pre-required conditions chosen by the females; this is not environment which selects the species, but intelligent behavior emerging from the internal evolution of the central nervous system which select the biotopes, as estuaries for water, for corpses and hard stones to defend itself against predators and butchery. In the first paradigm, scientists are looking for adult populations exposed to open savannah, whereas I am looking for pregnant females with their amniotic liquid, and the best conditions for delivery and education.

Morphogenetic identification of Sahelanthropus tchadensis (10-7 mya)

The virtual reconstitution of the exocranial skull (Zollikofer *et al.* 2005) allows us to identify the major stages of its morphogenesis (Dambricourt Malassé 2006b). All the great ape characteristics are present: 1) the right canine crown is not so little, it is broken. Nevertheless its cusp goes far beyond the occlusal surface of the molar and premolar range, and the root visible with the dental canal, corresponds to a huge tooth, 2) contrary to what is published, it is not evident to observe the lack of a pre-canine diastem on the maxillar, because of alveolar alteration. Besides, even large adult gorilla can lack such a diastem, whereas in opposite, rare australopithecines developed this biomechanical space, as in modern

Homo sapiens (abnormal promandibula), 3) the occlusion can be deduced from the inferior P3 (first mandibular premolar) of an isolated mandible with its surface eroded by the contact with the upper canine tooth. So, the occlusion is typically of a basi-cranio-facial skeleton in extension. This is confirmed by the ontogenetic diagnosis: 4) the upsiloïd form of the maxilla, 5) the orientation of the foramen magnum quite similar to the decidual stage of pongines-panines and 6) the inion above the Frankfurt plane in the decidual position of pongines-panines (see below). *Sahelanthropus tchadensis* reflects the great ape pattern. Nothing indicates any hominine trend.

This is found again with the analysis of the base. The orientation of the foramen magnum (limit between the skull and the vertebral column) published in Nature does not correspond to the foramen magnum, this is not the protocol used by specialists (*Figure 7*). On the chimpanzee this is not the basion (midsagittal anterior point) which is detected on the radiograph, but the lateral mastoid where the nuchal muscles are attached. So the position of the neck seems more like in a quadrupedal mammal. In contrary, on the virtual fossil skull, this is not the opisthion (midsagittal posterior point) which is identified, because it is not detectable in exocranial view. In general this sagittal point is lightly above the lateral occipital scale. So the position of the foramen magnum is not so horizontal, and it looks like the deciduous stage of the chimpanzee (*Figure 7*).

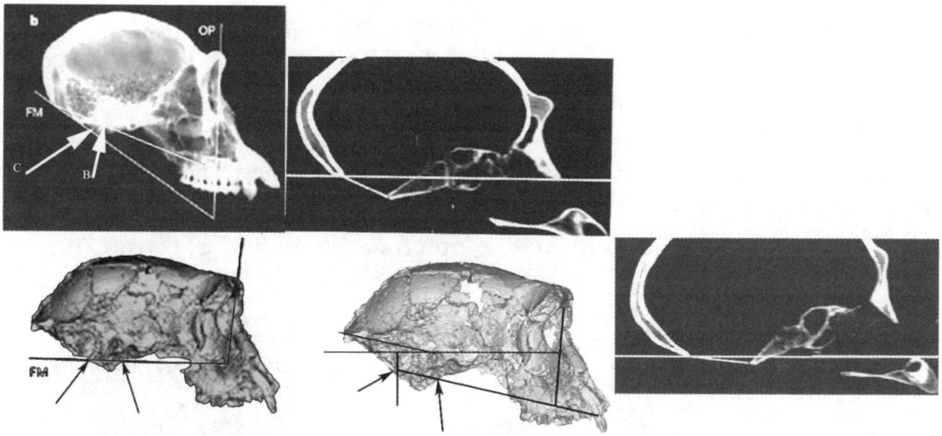

Figure 7. Corrected orientation of foramen magnum in Pan and *Sahelanthropus tchadensis.* Left: published analysis (Zollikofer *et al.* 2005) with corrected position of the basion (B); middle: CT scan of adult Pan in Frankfurt Plan and Sahelanthropus with corrected orientation of foramen magnum and inion position. Right: scan of deciduous Pan with orientation of the foramen magnum (line is parallel to the Frankfurt plane).

Once the skull oriented in the Frankfurt plane and centred on the porion (exocranial point near the axis of rotation), the exocranial curve can be compared with the ontogenetic curve of panines. It appears clearly that the occipital scale superposes with their ontogenetic trajectory. Both inion position and foramen magnum orientation correspond to the infantile ontogenetic stage of *Pan* and *Gorilla. In conclusion, the head posture and the occluso-postural imbalance of Sahelanthropus was no more bipedal than young Gorilla or Chimpanzee,* the anatomical features of the fossil results of the conservation of infantile occluso-postural imbalance. In other words, anatomical data described as hominines have no ontogenetic homologies with any Hominines; they are not derived. In conclusion this morphogenesis can be an ancestor of gorilla or chimpanzee by heterochronies (hypermorphosis) (Dambricourt Malassé 2006b) even of extinct species. This selection of anatomical features gives no proof of gradualism and their interpretation as hominines characters is refuted by the specialists of great apes and hominids fossils (Wood, Pickford, Senut, Wolpoff, Hawks), *Sahelanthropus* belongs to the great ape clade.

Even this fossil could be an ancestor of australopithecines lineage, and why not, authors give no explanation of the biological processes underlying not only the morphogenesis of Sahelanthropus, but also the emergence of Hominines, that is the finality of paleontology. *If Sahelanthropus is an ancestor, what I conclude is that, this is its embryonic development which evolved by punctuated equilibria.*

III.2 The Australopithecine clade and the oldest species of Homo *clade*

The application of the protocol to Australopithecines and *Homo* fossil species shows a net and radical dichotomy with all the species that preceded them. Whatever the variability between species of the two genera, all have a more important sphenoid rotation than can never be confused with the great ape's pattern, the mandibular symphysis is raised, the three cranial fossa are staged in the direction of the verticality and the occipital scale pursues the rolling-up up to the adult stage (*Figure 8*).

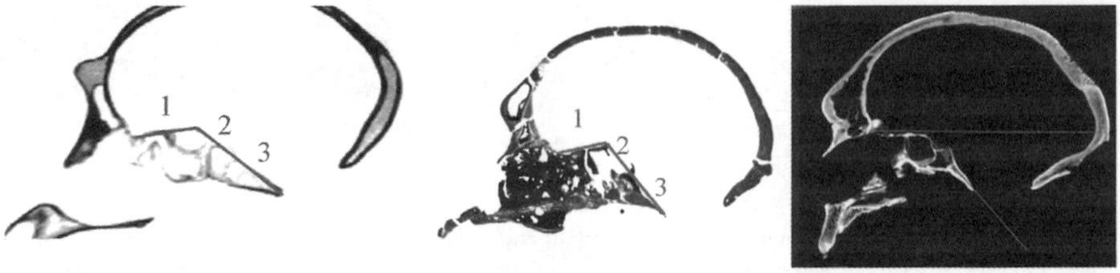

Figure 8. Ct scan of adult Pan, *Australopithecus africanus* StS5 (CT scan of the Witwatersrand University, Johannesburg, picture A. Dambricourt Malassé) and *Homo rhodesiensis* (British Museum).

In Australopithecines, the occlusion still in promandibula is reorganized because the mandibular dental arcade is shortened and widened, that explains the less prognathism, whereas the sphenoid keeps the dynamics of flexion. As in *Homo*, both deciduous molars developed with five cusps instead of two in great ape, the crown of the canine exceeds hardly the occlusal surface the other teeth, the pre-canine diastem is very rare. The cerebellum develops forwards and downwards, the central nervous system develops longer as shows the delay of the face growth, with the retardation of the supra-orbital torus, whereas the occlusion is already more advanced with the first permanent molar (Taung, Dikita). The australopithecine deciduous and adult occlusion is still in promandibula while *Homo* species are in retromandibula, even incisors are "end to end". CT Scan of the inner ear, have shown a very important detail between *Homo* on one hand, great ape and Australopithecine on the other, that is the acquisition of an angular distance between the posterior semi-circular canal and the lateral (horizontal) (Spoor and Zonneweld 1994). So, *Homo* is apart from all species which preceded its embryogeny.

It is admitted that a lower position of the posterior semicircular canal is a consequence of the rotation of the skull base, thus the origins of this angular opening is considered a dynamic effect of the flexure and important research program are seeking to understand the ontogenetic process. With Dr. Fabienne Lallouet, we carried out the study of an anencephaly, i.e. a human fetus without opercularization of the neural groove and without flexion (Dambricourt and Lallouet 2009). We have shown that this angle is independent of the sphenoid rotation; indeed, *semi-circular canals have the human conformation while the rotation of the sphenoid is absent.* Now it is shown that homeogenes of the Hox family govern the formation of semi-circular canals. It is thus clear that their morphogenesis is inscribed in a cascade of transcriptions from the rostral until the caudal pole all along the axis of the embryo.

It is well known that the post-cranial skeleton of Australopithecine and the oldest species of *Homo* genus are quite different, on one hand, and totally separated from great ape, on the other. Our conclusion is that all fossils of Australopithecines and *Homo* species indicate a mutagenesis of the homeogenes which control the great ape embryonic development, from the tooth buds to the differentiation of the appendicular skeleton. They are the only primates for which the expression "permanent bipedality" can be applied in the sense of morphogenetic and dynamic patterns completely news, emerged by saltationist processes. The discontinuity is even more radical with *Homo*, which have a unique conformation of semi-circular canals whereas that of the Australopithecines remains identical to the great ape.

Today the age of *Homo* genus emergence is at least 3 millions years ago, the oldest species have the same age of 1.8 millions years and are already well diversified with a huge dispersal in Europe (*Homo georgicus*), Indonesia (*Homo erectus*-Modjokerto), East Africa (*Homo ergaster, habilis, rudolfensis, erectus*), South Africa (*Homo habilis*) and artefacts in sub-himalayan foothills (Riwat in Pakistan). Moreover the artefacts are oldest, 2.55 mya in Ethiopia (Gona), 2.2–2.6 mya in China (Renzidong), 2.4–2.5 mya in Israel (Yiron). The paradigm is the African story, looking for hominids in Libya and oldest artefacts in North Africa. Nevertheless Chinese and Indian palaeontologists do not exclude an Asian origin, while Lufengpithecus survived until 2 mya (South West China, Yunnan Province) later after the Human dichotomy.

I have proposed an alternative, considering a "fertile crescent" between North-East Africa and South-West Indian Sub-Continent, through a corridor crossing the South of Arabic Peninsula, (even though no great ape's fossil species has yet been discovered) for the following reasons: 1) the ancestor is the genetic pool governing the ancestral embryonic body plan common to *Pongo*, *Pan* and *Gorilla*, 2) global warming is also observed in Asia ca 5 millions years ago, 3) whatever Africa or Asia, species of both body plans survived in tropical and sub-tropical biotopes, 4) gradualism is not the evolutionary process, 5) there is no reason to reduce the type of mutagenesis to the African species (Dambricourt Malassé 2008). Anyway, what is important first to understand is the level of information that evolved, because with this changing paradigm, the more the environment is rich in water, the more the longer pregnancy and period of feeding have chance to be preserved.

Discussion: theoretically, punctuated equilibria and the entire change of the rostro-caudal skeleton can be explained by natural evolution of homeogenes. Certainly congenital malformations occurred in large number as in the current human population; alterations are inherent to the great fragility of the inner dynamic complexity of combination, and entropy (second principle of thermodynamics) is an important parameter to integrate in the model, an internal constraint which cannot be ignored. But how can we explain the two evolutionary processes, *Homo* and Australopithecines emergence, in terms of *gradual* natural selection? Natural selection is seen as the motor of a natural process of adaptation acting gradually during numerous generations. With the two embryonic patterns and their occluso-postural equilibrium, there is no geological time for natural selection, the process is internal and corresponds to the recombination of the cascade transcriptions. With *Homo* genus, the evolutionary problematic is much clearer: even one question is the time necessary to obtain the same homozygoty between female and male, the recombination stays the most difficult evolutionary process to understand. *Homo* is the unique vertebrate in all the evolutionary story of the chordates which is verticalized from the face to the sacrum, with unique podal skeleton, gravity centres of the head and the trunk very close to the vertical axis of the body. If the psychomotor control is impossible with the canal's conformation common to the great ape and the Australopithecine, if the body cannot move without the angle acquired between both lateral and posterior canals, numerous questions emerge that natural selection can not solve: the probability that the angular formation appears between the two canals before the future axial skeleton adopt a straightened orientation at the end of the embryonic period.

It is noticeable that the organ which allows the brain to be informed about the angular movements of the head is formed by three morphologically identical structures, occupying each one of the three Euclidian plans. Thirteen years after *On the Origin of Species by Means of Natural Selection, or the Preservation of Favoured Races in the Struggle for Life*, Charles Darwin concluded in his correspondence with the French palaeontologist Albert Gaudry, that the origin of embryogenesis will remain for ever inaccessible to the human genius. One hundred fifty years later, the problematic of *Homo* lineage's origin is given by the misunderstanding of macro-evolution process. But rather than to conclude an "Intelligent Design" which is not a scientific discovery, it seems necessary to evolve in an enlarged scientific transdisciplinarity including all the sciences studying the natural origins of "life", at least the self replicated properties of RNA, DNA molecules, with the integration of quantum physics (Schrödinger 1944).

We do not know the natural laws and properties acting at the interface between quantum physics and biochemical properties during the duplication of RNA and DNA, we do not know what "information" means at this level, but we can admit that the evolution of such complex "information" is constrained by its own dynamical coherence and history engrammed in its own organization (and not programmed). Its history, or memory, is itself the organization of this "information".

III.3 Other primates species

The phylogenetic classification is the encyclopaedic reference "The Cambridge Encyclopedia of Human Evolution" edited by S. Jones, R. D. Martin and D. R. Pilbeam with a foreword by Richard Dawkins. The encyclopaedic illustration represents the phylogenesis of fossils and living primates by a selection of adult skulls with their mandible. However, this representation does not give information according to the X axis. On the other hand, if we clarify the appearance of clades such as positioned according to the Y axis, two major facts are observed: 1) a geometrical order of the skull base (the saltationist increasing chordo-prechordal flexion), 2) the equivalence of Australopithecines and *Homo* with the clades who preceded them. This report can surprise, however fossils and 634 living species built it (*Figure 9*).

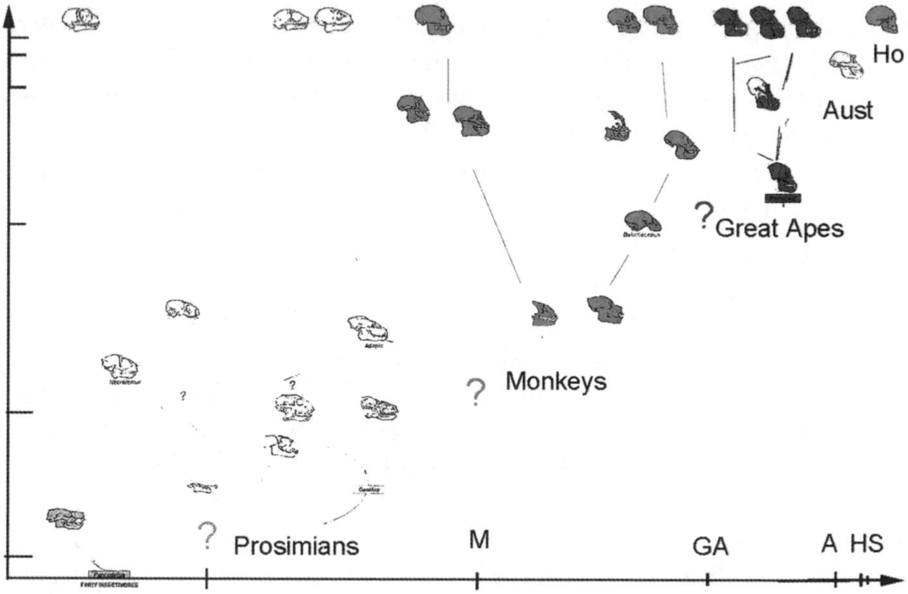

Figure 9. Primate phylogenesis according to "The Cambridge Encyclopedia of Human Evolution".

The distribution is not an increasing arborescence, but successive rostro-caudal morphogenetic body plans identified by the cephalic prechordal- chordal relations, printed in skull by the flexure. The increasing flexure reflects the evolution of these relations marked by an increasing complexity of the central nervous system, not only the hemispheres, but all the neural system including the cerebellum. The more embryonic organization is ancient, the more its arborescence widened with specializations, but no Strepsirhinii species (Prosimians), presents rotation of the sphenoid bone, the chondrocranium is long, narrow and flat. The fossil species appeared ca 50 millions years ago with Adapiformes, then Omomyidae that certain paleontologists place as ancestors of the clade Haplorrhini (extant Tarsiers [Tarsiiforms] and Simians [Simiiforms]). All species grouped in oldest Simiiforms, i.e. the monkeys, developed a shorter base with a new mandibular shape and a cerebellar fossa in a lower position (Parapithecidea, Propliopithecidae such as *Aegyptopithecus*, 35 millions years discovered in Egypt).

The comparison with living apes shows an increasing flexion between pre- and postsphenoid. There are no counter-examples: the base of adult *Alouatta* (Howler monkey, Central America) is opened to 180°, but its mandible is characteristic of a simian morphogenesis and the amplitude of flexion is visible at birth. The living *Tarsius* which is not a Simiiform (ape), reproduces in South East Asia, is characterized by an advanced position of the foramen magnum, but the mandible is comparable to the Loris, a living Strepsirhinii (Prosimian). The sphenoid flexure is not that of a Simian. The dynamics does not result from the neural groove rolling-up beyond the Prosimian amplitude; probably the flexure comes from the exceptional development of the embryonic optic capsules (future eyes). In 1972 Starck observed numerous embryonic characters common with Strepsirhinii, others with the Simiiform's embryogeny, nevertheless in no case *Tarsius* is considered as a Simian. The gradual and linear paradigm generates the obligation to create one clade (Suborder Haplorrhini) with two infra-orders, Tarsiiform and Simiiform, whereas two clades can be suggested evolving separately from the same ancestral embryonic pattern. *In both cases, the evolutionary mechanisms concern the strepsirhinii embryogeny considered as the primitive pattern* (plesiomorphon). With Simiiform, the mechanisms belong clearly to the register of the macro-evolutionary processes restructuring the internal organization. All living species preceding the great ape's clade including gibbons, and derived from the first Simiiforms do not exceed the sphenoid rotation of the Propliopithecidae (*Figure 10*).

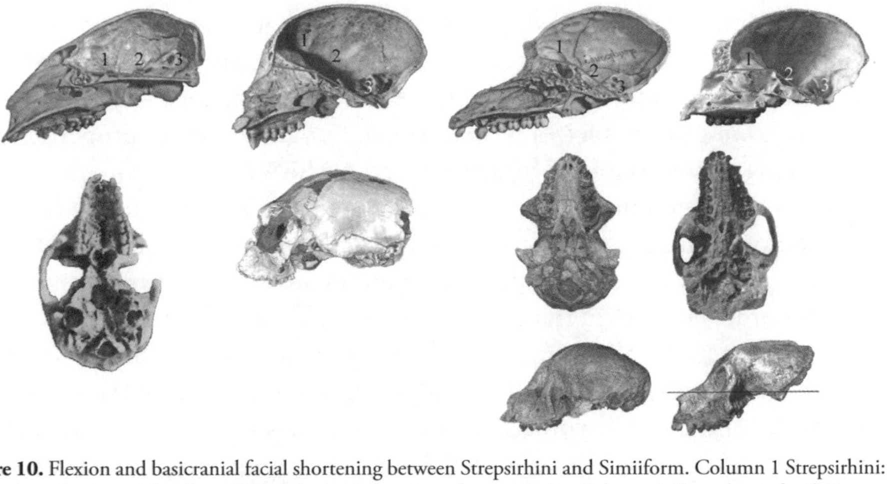

Figure 10. Flexion and basicranial facial shortening between Strepsirhini and Simiiform. Column 1 Strepsirhini: inferior view of extinct Adapiform (45 mya), sagittal section of extant Indri; Column 2 Platyrrhinii: fossil *Tremacebus* (15 mya) and living *Cebus*, Column 3-4 Catarrhinii: sagittal and inferior view of female and male *Propliopithecus* (35 mya), living *Papio* and *Hylobates*.

So, naturally, microevolutionary arborescences (or radiation) are clearly visible among each clade, with speciations and cephalization, but not with cerebellization (development of the cerebellum after its embryonic verticalization). The great rupture with great ape clade is for human and australopithecine lineages that are not arborescent microevolutions of Miocene Dryopithecines. Both are a new macro-evolution of the pre-chordal–chordal axis relations. So, from great ape clade (or embryonic pattern), two macro-evolutionary events can be identified, the extinct Australopithecine lineage, and the *Homo* lineage characterized by an acceleration of this evolution. *Homo* fossil species developed a rostro-caudal skeleton different of modern *Homo sapiens*, and it is a fact that remandibula continued to evolve with the loss of the "end to end" incisor occlusion. And there is no reason to conclude that modern *Homo sapiens*, living humankind, is the end of this internal macro-evolutionnary evolution, even large extinctions characterize its acceleration since the Pliocene.

In 1951 Pierre Teilhard de Chardin leaves New York for South Africa, a mission for the Wenner-Gren Foundation to see new fossils: *Australopithecus africanus* and *Paranthropus robustus*. He considered they belong to a phylum apart from human lineage and after thirty years of research in Asia, Teilhard concluded that human origins will be better understood in North East Africa, where the oldest artefacts have been discovered.

Conclusion

Placing evolution in a widened consideration taking into account the intra-uterine environment and integrating the embryogenesis, leads to discoveries of new evolutionary modes and tempos which are not only influenced by the environmental necessities of the species (water, fruits), but also by inner properties. This implies taking into account the homeobox genes which take a part in the ordering of the body construction, leading an internal determinism profoundly registered in the karyotype; this is a spatio-temporal order with 1) the distribution of homeogenes according to the axis of the chromosomes and 2) between chromosomes. Finally this order emerges in the phenotype expression which follows a rostro-caudal gradient: when the basi-cranial flexion stops the feet are not already formed.

The origin of both human and australopithecine head and neck postures are the dynamics of their embryonic axial development, generated by neural kinetics whose epicentre are close to the chordal apex. This morphogenetic process has nothing in common with the necessity to adopt a post-natal erected posture because of a decreasing forest cover; both are the result of the iteration of an evolutionary process observed 15 millions years ago, when the embryonic rotation of the monkey pattern changed into the great ape one, with greater verticalization of the axial skeleton and new cranio-facial contraction.

The postulate of gradual evolution created confusion between micro-evolution and macro-evolution: human origins belong to macro-evolutionary processes of increasing complexity, the first evidence is the psychomotor necessity to control the most unstable biodynamical system ever seen in all the evolutionary history of mammals. No species has its gravity centres, from the head to the pelvis, so close to their vertical axis. This is the result of a macro-evolutionnary process generating punctuated equilibria and iterative effects with amplification of their previous axial emergence: the chordo-prechordal flexions and the complexification of the central nervous system networks including the cerebellum. It is a fact that the cerebellum controls its own position in the skeleton, acquired because of embryonic verticalization, and it seems logical that the connection between neo-corticalization of the cerebellum and cerebellar hemispheres increased. *The reflexive consciousness strongly depends on these connections.*

Ecological conditions are necessary to allow the reproduction and the radiation of a new ontogenetic pattern, the environment is not be neglected, on the contrary. The reheating at the end of the Miocene (5 millions years) provoked the disappearance of the great apes from Europe to Himalayan foothills without the appearance of hominines. Introducing the mother-fetus relation calls in mind the importance of water, at least for the amniotic liquid, because pregnancy becomes longer. With such parameters, the best conditions of life for adaptation to a delay in psychomotor maturation and against a bigger vulnerability of the couple mother-infant, require the presence of environment where the rate of humidity is raised for fruits, vegetables, lakes and rivers. The females will determine the best conditions for the pregnancy and parturition, those necessary for longer feeding consumer of water. It always seemed to me natural that the pre-required ecological conditions for the success of the mutagenesis had to be the ones familiar, known and transmitted by the female, from generation to generation. It is not thus surprising that permanent bipedality appeared in the same ecological conditions as those where great apes were living, since it is not a question of locomotion. The event is a macro-evolution of the central nervous system, with an increasing complexity of its development, delay of maturation, longer period of learning and postural imbalance more complex to control. The discontinuity between the chord and the prechordal plate stigmatizes, so, not only the acquisition of the verticality, it also marks the limit between reflexes zones and the cognitive or conscious territories. *Homo sapiens* is the most recent stage of this evolution as Pierre Teilhard de Chardin had expressed "the apical bud".

The "motor" of this process can not be compared with the neo-Darwinian model, not only because of the limit of natural selection which can not explain the complete change of the embryonic development, but also because of the profound misundertstanding of the physical significance of random errors of transcription during meiosis, the loss of information or energy, entropy. Entropy is not the *natural* process able to create new embryonic patterns with the reiteration of dynamical trajectories and their amplification. Nor natural selection, neither the genetic loss of information are the *natural* evolutionary mechanisms which can explain the *biological origins* of *Homo* and Australopithecine embryogenesis (Dambricourt Malassé 1996a,b). I can not conclude that such a macro-evolutionary process is the result of increasing entropy or random loss of genetic information, when it is just the opposite which is revealed by the fossils: the iteration and the transmission of inner logics with emergent properties such as the reflexive consciousness. The neo-Darwinian model is not efficient to explain the *natural* evolutionary *origins* of *Homo sapiens* (Chaline *et al.* 2000). We do not confuse neo-Darwinian modalities between the first and the last species sharing the same body plan during millions of years and the saltationist emergence of a new body plan.

After twenty five years of research on palaeoembryology, it is clear that phylogenetic determinism had to be taken into account. Entropy, or the second principle of thermodynamics, is valid from the molecular scale of the DNA and the RNA, it is to be considered and doubtless is a part of indeterminism. But what is noticeable is the quasi-simultaneous appearance of the two lineages, 4,4 million years for Australopithecines, and at least 3 million years for *Homo*. How can we explain the quasi-simultaneity of two convergent mutageneses from the same homeogene pattern *after 15 millions years of stability*? Whatever the environmental constraints and the accidental mutations, Australopithecine and *Homo* embryogeny never emerged unless memory were integrated in the great ape genome, acquired at least from the first Simiiforms. What I mean is that without internal evolutionary logics, even ecological conditions were the best for a new species, never these lineages could emerge. The entire evolutionary curve had to be taken into account. I have come to the conclusion of a heritability of "internal logics" controlling entropy when the threshold of instability occurred in biodynamical networks of fertilized eggs. This needs new models of attractors describing the dynamic systems by adding constraints like memorization. I suppose that the probability of reaching a threshold of instability is not limited to one species. But this hypothesis does not explain why other great ape's species never evolved since the dichotomy, why the process is irreversible and transmitted only to the new fundamental ontogenesis (embryonic body plan).

This internal stability of this macro-evolutionary dynamics was discussed with René Thom and a project of formalization started in 1996 with the astrophysicist Eric Bois, to understand the difference between chaotic attractors and the macro-evolutionary process leading to modern *Homo sapiens*. Close to chaotic attractors I proposed harmonic attractors because of the conservative properties whatever the number of evolutionary thresholds (Dambricourt Malassé 1992b; 1992c; 1993b; 1996b). The introduction of mathematics to describe evolutionary trajectories of biodynamical systems is still rare (Dubois *et al.* 1992) and sometimes needs some developments to explain that attractors are neither metaphysical concepts nor "spiritual intrusion" in sciences, but the highest level of a scientific discovery (Chaline 2006).

To conclude, it is of a central necessity to understand the roots of the human embryogeny in the evolutionary continuum of fossil lineages, on the one hand, and the implications in posturology for the understanding of modern diseases, on the other. If the rostro-caudal morphogenesis becomes unstable because of the increasing complexity of homeogenes, the responsible attitude is to develop a better understanding of such evolutionary processes with a new evolutionary paradigm such as "Evo-Devo" even if the process is not neo-Darwinian. If random were the motor, it should be impossible to develop orthopaedic appliances individually when the occlusion is changing. Becoming able to predict the troubles as well as the future appliances is a responsible attitude serving humankind.

Incompletude of sciences and metaphysics in front of alterity

The consequences of these discoveries have important implications at different levels of order of reality, which invites to continue efforts of thought to unify them, integrating profound questionings. If the large evolutionary movement which leads to *Homo sapiens* is one of an evident increasing complexity, Sapiens is not the end in itself. Nothing allows us to think that the internal process ends with sapiens germinal cells. The preoccupations of Public Health indicate more the necessity of knowing how to take place in acceleration.

Of course these discoveries are non-Darwinian and seem rather a finalized macro-evolution allowing metaphysical questioning (Staune 2002); they have nothing in common with "spiritual intrusions". Neocreationism or Intelligent Design are metaphysics which integrate the non-Darwinian evolutionary process, the inner logics. But as metaphysics, they do not take into account the great number of abortive embryos (random, entropy).

The meaning of a human life is a personal research which marks the limit between the objectivity and alterity, the scientist (scientific attitude excludes the subject) and her/his human person (relations between the subjects). Understanding hominization cannot ignore the notion of person, which is the most delicate and most difficult to place in the process, because every human being develops with the same morphogenetic inheritance, which is not a person. So necessarily, the questioning and the search of understanding is outside the scientific field. Forbidding an effort of synthesis between sciences (objectivity), metaphysics (reason) and human experiences (ontology) is irrational. As a human being, as well as palaeoanthropologist face to face with the memorial of the Shoah, I can not believe that a human life is just a biochemical machine and that the crime against humanity has no ontological truth (Wiesel and de Saint Chéron 1999). There is a responsibility on evolutionary anthropologists to pursue their reflections.

Faced with emergent diseases inherited from macro-evolutionary conditions, I am convinced that *Homo sapiens* will give prior importance to rational and realistic questioning. Recognize the questioning of the evolutionary identity of *Homo sapiens* as an immaterial world heritage of the humanity becomes a real necessity, not only because of global warming, but in front of irreversible implications that the ignorance of Man and appetites of power can produce. Philosophy and metaphysics find their place in this questioning because how would the meaning of a human life be known in a process which has not reached its equilibrium state?

It is then essential to understand that the current humanity is not in itself the ontological end of the universe, of life, but the first species with sufficient capacity of abstraction to wonder about the macroevolutionary consequences of his acts, when the human mind is able to interact with the most destructive levels of energy.

It is the reflexive consciousness which is capable of objectivity with scientists, of universality with philosophers and metaphysicians. And if the intention is to give sense to the meaning of evolutionary responsibility, it is in itself an awareness of alterity. The same consciousness an artist expresses with her faith and the intelligence of her art, without breaking the mystery of her embodiment until accepting in silence the unknown fate of her bones tired by the years: "Shape and matter in bone are a slow and same thought." (Dautricourt 1986) In front of dead bones, poets feel that the silence of sense is not the echo of its absence, they feel this universal mystery is the mystery of the human condition, the mystery of love for life.

References

Bonis L. de, 2008, "African Carnivora from the Middle Miocene to the Pleistocene: new data, systematics, evolution, biogeography", *Comptes Rendus Palevol* 7: 475–86

Caro P., 2004, "Développement embryonnaire et évolution. Entretien avec Anne Dambricourt Malassé", *La Lettre de l'Académie des Sciences* 13: 14–15.

Chaline J. (ed.), 1983, *Modalités, rythmes et mécanismes de l'évolution biologique: gradualisme phylétique ou équilibres ponctués?*, Paris: CNRS.

Chaline J., Marchand D., Berge C., 1986, "L'évolution de l'homme: un modèle gradualiste ou ponctualiste?", *Bulletin de la Société Royale Belge d'Anthropologie et de Préhistoire* 97: 77–97.

Chaline J., Brunet-Lecomte P., 1990, "Modélisation des modalités de l'évolution", *Comptes Rendus de l'Académie des Sciences* 311(II): 1031–36.

Chaline J., Werdelin L., 1993, "Modes and tempo of Evolution in the Quaternary", *Quaternary International*, Oxford and New York: Pergamon Press.

Chaline J., 1996, "Origines de l'homme: la synthèse paléoanthropologique", in *Encyclopedia Universalis*, 141–156.

Chaline J., 2006, *Quoi de neuf depuis Darwin? La théorie de l'évolution dans tous ses états*, Paris: Ellipse.

Chaline J., Durand A., Marchand D., Dambricourt Malassé A., Deshayes M. J., 1996, "Chromosomes and the origins of apes and australopithecines", *Human Evolution* 11: 43–60.

Chaline J., Dambricourt Malassé A., David B., Magniez-Jannin F., Marchand D., Courant F., Millet J. J., 1998, "Quantification de l'évolution morphologique du crâne des hominidés et hétérochronies", *Comptes Rendus de l'Académie des Sciences de la Terre et des Planètes* 326: 291–98.

Chaline J., Durand A., Dambricourt Malassé A., David B., Magniez-Jannin F., Marchand D., 2000, "Were climatic changes a driving force in hominid evolution?", in Hart M. B. (ed.), *Climates: Past and Present*, London: Geological Society, Special Publication, 181, 185–98.

Coppens Y., 1996, *Résumé des cours et travaux du Collège de France, 1995–1996: Grands noms, grands moments de l'histoire des sciences.*

David B., Dommergues J. L., Chaline J., Laurin B., 1989, "Ontogenèse et évolution. Edition du Colloque international du CNRS". *Géobios*, Mem. 12: 323-335.

Dambricourt Malassé A., 1987, *Ontogenèses, paléontogenèses et phylogenèse du corps mandibulaire catarrhinien. Nouvelle interprétation de la mécanique humanisante (théorie de la foetalisation, Bolk 1926)*, Ph.D. National Muséum of Natural History, Paris.

Dambricourt Malassé A., 1988, "Hominisation et foetalisation", *Comptes Rendus de l'Académie des Sciences* 307(II): 199–204.

Dambricourt Malassé A., Deshayes M. J., 1992a, "Modeling of the craniofacial architecture during ontogenesis and phylogenesis", in Berthoz, Graf, Vidal (eds.), *The Head-Neck Sensory Motor System*, New York and Oxford: Oxford University Press, 37–46.

Dambricourt Malassé A., 1992b, "L'hominisation et la théorie des systèmes dynamiques non linéaires (Chaos)", *Revue de Biologie Mathématique*, Société française de Biologie Mathématique, nos. 117, 118, 119.

Dambricourt Malassé A., 1992c, "Comment on P. Graves 'New models and metaphors for the Neanderthal'", *Current Anthropology* 33: 49–50.

Dambricourt Malassé A., 1993a, "Continuity and discontinuity during modalities of hominization", in "Modes and Tempos in Evolution in the Quaternary", *Quaternary International* 19: 85–100.

Dambricourt Malassé A., 1993b, "Les attracteurs inédits de l'hominisation. Ontogenèse fondamentale, attracteurs chaotiques et attracteurs harmoniques", *Acta Biotheoretica* 43: 113–25.

Dambricourt Malassé A., 1996a, "Nouveau regard sur l'origine de l'Homme", *La Recherche* 286: 46–54.

Dambricourt Malassé A., 1996b, "Exemple du chaos appliqué en paléontologie humaine", Ecole thématique du CNRS, 20-24 mai 1996, *Chaos et fractales dans l'activité solaire*, Chapitre III, 214–228.

Dambricourt Malassé A., Deshayes M. J., Magnies-Jannin F., Chaline J., Marchand D., 1999, "A solution to human paradox: fundamental ontogenies and heterochronies", *Human Evolution* 14: 277–300.

Dambricourt Malassé A., Martin J. P., de Kerviler E., 2000, "Test des corrélations angulaires au cours de l'ontogenèse entre la flexion sphénoïdale et les canaux semi-circulaires de *Pan troglodytes*. Implications en paléontologie humaine", in Andrieux P., Hadjouis D., Dambricourt Malassé A. (eds.), *L'identité humaine en question. Nouvelles problématiques et nouvelles technologies en paléontologie humaine et en paléoanthropologie biologique*, Préface Yves Coppens, Paris: Editions Artcom, 24–42.

Dambricourt Malassé A., 2006a, "Evolution du chondrocrâne et de la face des grands anthropoïdes miocènes jusqu'à Homo sapiens, continuités et discontinuités", *Comptes Rendus Palevol* 5:109–17.

Dambricourt Malassé A., 2006b, "La question des origines de la lignée humaine: les modèles évolutionnistes confrontés aux fossiles, implications en orthopédie et posturologie", *Les Cahiers du Centre d'Etude de l'Ostéopathie, de la Posturologie et Sciences Annexes* 4: 38–57.

Dambricourt Malassé A., 2008, "Le peuplement humain en Eurasie: la Haute Asie et les piémonts sous-himalayens du Plio-Pléistocène à l'Holocène, origines, évolution humaine et migrations", *L'Anthropologie* 112: 370–403.

Dambricourt Malassé A., Lallouet F., 2009, "Etude au scanner des canaux semi-circulaires d'un crâne sec d'anencéphale dépourvu de flexion sphéno-occipitale; implications pour l'interprétation paléo-embryologique des hominidés fossiles" (XVème Congrès de La Société de Biométrie Humaine, Paris), *Biométrie Humaine et Anthropologie* 27: 79-86.

Dautricourt J., 1986, *Os écrits*. Catalogue d'exposition, Galerie Antoine Candau, Paris. http://jdautricourt.free.fr/

Delattre A., Fenart R., 1960, *L'hominisation du crâne étudiée par la méthode vestibulaire,* Paris: CNRS.

Deloison Y., 1999, "L'Homme ne descend pas d'un Primate arboricole! Une évidence méconnue", *Biométrie Humaine et Anthropologie* 17: 147–50.

Deshayes M. J., Dambricourt Malassé A., 1990, "Analyse des différents types architecturaux cranio-faciaux par l'approche ontogénique de l'hominisation", *Revue de Stomatologie et de Chirurgie maxillo-faciale* 4: 249–58.

Deshayes M. J., 1991, "Reconsidération de la croissance cranio-faciale au cours du développement et de l'évolution. Conséquences en orthopédie dento-maxillo-faciale" *Revue d'Orthopédie Dento-Faciale* 25: 353–65.

Deshayes M. J., 1996, "Etude dynamique et biométrique de l'architecture crânienne: application à l'enfant porteur de malocclusions dentaires", *Cahiers d'Anthropologie et Biométrie Humaine* 14: 171–89.

Devillers C., Chaline J., Laurin B., 1990, "Plaidoyer pour une embryologie évolutive", *La Recherche* 21: 802–809.

Dubois A., Chaline J., Brunet-Lecomte P., 1992, "Spéciation, extinction et attracteurs étranges", *Comptes Rendus de l'Académie des Sciences* 315: 1827–33.

Harcourt-Smith W. E. H., Aiello L. C., 2004, "Fossils, feet and the evolution of human bipedal locomotion", *Journal of Anatomy* 204: 403–16.

Jaeger J. J., 1997, *Les mondes fossiles*, Paris: Odile Jacob.

Lalauze-Pol R., Lambert S., Fellus P., Bennaceur S., Elmaleh M., 2007, "Importance du traitement précoce de la plagiocéphalie posturale, une prévention des dysmorphoses maxillo-faciales", *Revue d'Orthopédie Dento-Faciale* 41: 125–39.

McBratney-Owen B., Iseki S., Bamforth S. D., Olsen B. R., Morriss-Kay G. M., 2008, "Development and tissue origins of the mammalian cranial base", *Developmental Biology* 322: 121–32.

McKee J.K., 1999, "The autocatalytic nature of hominid evolution in african Plio-Pleistocene environments", in Bromage T., Schrenk F. (eds.), *African Biogeography, Climate Change, and Early Hominid Evolution*, New York: Oxford University Press.

Oxnard C. E., 1975, "The place of Australopithecines in human evolution: grounds for doubt", *Nature* 258: 389.

Picq P., 2007, *Lucy et l'obscurantisme*, Paris: Odile Jacob.

Rak Y., Clark Howell F., 1978, "Cranium of a juvenile *Australopithecus boisei* from the lower Omo Basin, Ethiopia", *American Journal of Physical Anthropology* 48(3): 345–64.

Richmond B. G., Jungers W. L., 2008, "*Orrorin tugenensis* femoral morphology and the evolution of hominin bipedalism", *Science* 319: 1662–65.

Schrödinger E., 1946, *What is Life?*, New York: McMillan.

Senut B., Pickford M., 2005, "Comment *Orrorin* a changé nos conceptions sur les origines des hominidés", *Anthropologie* 43: 111–19.

Spoor F., Zonneveld F., 1994, "The bony labyrinth in Homo erectus: a preliminary report", *Courier Forschungsinstitut Senckenberg* 171: 251–56.
Staune J., 2002, *Notre existence a-t-elle un sens? Une enquête scientifique et philosophique,* Paris: Presses de la Renaissance.
Wiesel E., Saint Chéron M. de, 1999, *Le Mal et l'Exil. Dix ans après,* Paris: Nouvelle Cité.
WoldeGabriel G., White T. D., Suwa G., Renne P., de Heinzelin J., Hart W. K., Heiken G., 1994, "Ecological and temporal placement of early pliocene hominids at Aramis, Ethiopia", *Nature* 371: 330–33.
White T., Suwa G., Asfaw B., 1994, "*Australopithecus ramidus,* a new species of early hominid from Aramis, Ethiopia", *Nature* 371: 306–12.
Wijnholds J., Chowdhury K., Wehr R., Gruss P.,1995, "Segment-specific expression of the *neuronatin* gene during early hindbrain development", *Developmental Biology* 171: 73–84.
Zolliköfer C., Ponce de Leon M., Lieberman D., Frank G., Pilbeam D., Likius A., Mackaye H., Vignaud P., Brunet M., 2005, "Virtual cranial reconstruction of Sahelanthropus tchadensis", *Nature* 434: 755–59.

How humans came to be so different to other monkeys and apes

Robin I. M. Dunbar

1. Introduction

We share a long evolutionary history with the great apes, in particular: the human lineage (including all the many fossil species that have existed over the last 6 million years or so since our lineage parted company with that of the chimpanzees) is firmly embedded within the African Great Ape family. Anthropologists have tended to emphasize anatomical differences between humans and other great apes (e.g. bipedalism, tool making and use) as the defining features of our species. Instead, I shall argue that the real differences lie in a much more intangible set of competences — the ability to live in the virtual world of the mind. This is the world of culture, and it is culture that, if anything, differentiates humans from other primates. To illustrate this, I will emphasize two aspects of human behavior that are, I suggest, archetypal of human culture: story-telling and religion. Both require us to be able to imagine worlds that do not physically exist. I shall argue that, apes' much-vaunted capacity for cultural learning (Whiten *et al.* 1999) notwithstanding, no other living species is even in the same league as humans in this respect. This, I will suggest, is because all other animal species lack the mental computational power needed to make it possible. The key to understanding why this is so lies in the reasons why our brains have evolved.

2. The evolution of the social brain

The social brain hypothesis (SBH) was originally proposed by Byrne and Whiten (1988) as an explanation for the fact that primates have much larger brains for body size than all other taxonomic groups (Jerison 1973). Most of this increase in brain size is the result of an enlarged neocortex (Finlay and Darlington 1990; Finlay *et al.* 2000). The essence of the SBH is that primates rely on social solutions to solve the key problems of survival and successful reproduction rather than solving these problems by individual trial and error. According to this hypothesis, primates' more complex social lives are significantly more cognitively demanding than is the case for most other non-primate species.

Considerable evidence has been adduced in support of the SBH (for recent summaries, see Whiten and Byrne 1997; Dunbar and Schultz 2007a,b). One of the core findings was a quantitative relationship between social group size and relative neocortex size in primates. On a double-log plot, mean species group size is linearly related to relative neocortex volume (indexed as the ratio of neocortex volume to the volume of the rest of the brain).

However, it is important to appreciate that it is social complexity rather than group size per se that underpins the social brain hypothesis. Attempts to test the social brain hypothesis on non-primate species (Pérez-Barbería *et al.* 2007; Shultz and Dunbar 2007; Dunbar and Shultz 2007b) have shown that, across a wide range of bird and mammal species, the social brain hypothesis takes a qualitative rather than a quantitative form. That is to say, it is pairbonded (i.e., monogamous) species that have disproportionately large brains. Anthropoid primates contrast with this pattern in that they alone exhibit a quantitative relationship between group size and brain size.

These results suggest that it might be the cognitive demands of pairbonding that underpin the social brain and were responsible for the initial impetus to evolve larger than minimally necessary for managing the somatic demands of the body in some lineages. The anthropoid primates have extended the use of these pairbond-like relationships to other group members to create what we might legitimately call "friendships" (Smuts 1985; Silk 2002) as opposed to reproductive pairbonds, thereby giving rise to the quantitative relationship that we observe in their case (Shultz and Dunbar 2007). In monkeys and apes, friendships are, in effect, alliances (coalitions whose members reliably come to each other's aid: e,g, Dunbar 1980, 1989). Coalitions buffer individuals against the stresses of group-living — those created by competition during foraging and the forms of harassment consequent on living in close proximity to others. These stresses are responsible for causing infertility in females in both captivity and the wild (Abbott *et al.* 1986; Dunbar 1980), and coalitions act to defuse and minimize these costs.

Although it is not entirely clear why pairbonds should be so demanding cognitively, my suspicion is that it has to do with the need to maintain a high level of behavioral coordination and synchrony between the pair members. This is obvious in those cases where the pair has to share rearing duties, as in many bird species. If the pair do not coordinate their behavior effectively, the individual left sitting on the eggs by a wayward mate who spends too long feeding or resting away from the nest may eventually be forced to choose between starving and abandoning the eggs. Such pairbonds are cooperative alliances designed to facilitate effective joint reproduction and the pair members have to be willing to recognize and accommodate the partner's interests otherwise the pairbond will not do its job. This kind of sensitivity to another's needs looks suspiciously like theory of mind, one of the core cognitive hallmarks of modern humans. While I do not wish to imply that monogamous birds and mammals have theory of mind, I do want to suggest that it may have been this specific need to coordinate behavior between a mated pair that provided the initial impetus for the evolution of those social cognitive skills that later underpinned full-blown theory of mind in humans.

3. Making pairbonds work

The social brain hypothesis, then, argues that maintaining and servicing the kinds of intense relationships found in pairbonds (in most birds and mammals) and friendships (among anthropoid primates) involves serious cognitive work, and that this in turn demands a proportionately large computer (the brain) to make this possible. However, cognition is only part of the story. There is a deeply emotional component to relationships, and this is especially so in respect of primate relationships.

For Old World monkeys and apes, in particular, social grooming is the principal mechanism used for social bonding. Grooming is an intense activity in which one individual leafs through the fur of another, removing bits of vegetation, dead skin and other debris. It is very much a one-on-one activity, in which the groomer is often deeply concentrated on its task. Grooming lowers the heart rate in the groomee and reduces the frequency of signs of tension and stress (yawning, scratching, etc) (Goosen 1981; Castles *et al.* 1999), to the point where the recipient of grooming can become so relaxed it actually falls asleep. Many species devote considerable proportions of their day to social grooming, and in some monkeys this can amount to as much as one fifth of the day. Grooming is physically stimulating and can be mildly painful when animals pinch and pull at hair or skin. This has the effect of triggering the release of endorphins (the brain's own painkillers) (Keverne *et al.* 1989) and it is these that are probably responsible for grooming's soporific effects.

The endorphin effects mediated by grooming play a crucial role — in ways that we do not yet understand — in the process of building the trust and reciprocity that form the basis of primate social relationships. However, to achieve these effects, animals have to invest considerable quantities of time in grooming, and there is a linear relationship between time spent grooming and the size of the social group

(Dunbar 1991; Lehmann *et al.* 2007). This investment of time in a relationship is then reflected in a corresponding willingness to come to a grooming partner's aid during conflicts (see Dunbar 1980; 1984; 1989). The role of these neuroendocrines in the bonding process seems to be critical, though as yet we have no idea why.

4. Humans and the social brain

Modern humans have much larger brains (and especially neocortices) than other primates. If we interpolate the modern human neocortex ratio into the ape equation (as the most appropriate), we obtain a predicted group size for modern humans of ~150 (Dunbar 1993). This seems to correspond quite closely to the typical size of hunter-gatherer communities (Dunbar 1993, Hamilton *et al.* 2007). More remarkably perhaps, this figure of ~150 appears frequently in a wide range of examples of human organization. It was, for example, the mean village size recorded for almost all English counties in the *Doomsday Book* as well as during the C18th, and is the typical size of the company in most modern armies, the number of recipients of a typical Christmas card distribution list in the Britain, and the size of the social network in reverse "small world" experiments, and, according to a study by the Church of England, the ideal congregation size, among many other examples (Dunbar 2008). The only substantive difference between social networks in traditional hunter-gatherer and modern post-industrial societies seems to be that, in traditional societies, everyone in the community has more or less the same network of 150 acquaintances, whereas in modern urban societies our networks are highly fragmented — one's network of 150 consists of a set of sub-networks that barely overlap. You and I may share one small set of friends, for example through a leisure activity, but there is no overlap at all in the remaining subsets: we share no relatives, we work in different places, we attend different places of worship, and so on. Networks in modern societies are fragmented and dispersed (often over considerable geographical distances), whereas in traditional societies they typically form a single integrated community.

Our social networks are highly structured, forming a series of layers that increase in size but decrease in average emotional closeness, such that successive layers include approximately 5, 15, 50, 150, 500 and 1500 members (Hill and Dunbar 2003; Zhou *et al.* 2005; Hamilton *et al.* 2007). However, the layer that encompasses the 150 individuals (that predicted by the primate group-size/brain-size relationship) seems to constitute something of a Rubicon in both social and cognitive terms. It seems to demarcate a key difference in the quality of relationships. We seem to have a qualitatively different kind of relationship to those who lie within the 150 compared to those who lie without it. One informal definition of this might be that the 150 are those whom we know as persons — everyone with whom we have a definable personal relationship which has a history. These are the individuals towards whom we feel some sense of obligation, whom we trust would help us out if we so requested, those who would reciprocate our sense of personal commitment. We know where these individuals fit into our network of relationships, and they know where we fit into theirs. It probably defines those to whom we owe personal obligations. They are the individuals who, if offended in some way by us, would cause significant problems for us elsewhere in our social world because we are all embedded within the same complex web of relationships.

In contrast, beyond this circle of 150, the nature of the relationship is very different. We can of course recognize individuals in the two outer layers (at 500 and 1500), but our relationships with them are more typological, relationships defined by rules and cues rather than personal history. We need rules of thumb to guide our interactions with them rather than being able to rely on personalized knowledge. Such rules can be formal (this is how you greet a priest) or informal (a social consensus on how one addresses one's baker or accountant). Such rules are often cued by some appropriate badge that signifies the status of the individual concerned: we dress a priest or a policeman in a particular way so that we will recognize them easily when we chance to meet them.

5. The role of cognition: the link between brain and behavior

The relationship between brain size and social group size implies that cognition imposes some kind of limit on the number of relationships an individual can maintain through time, and hence on the size of social group that can be maintained through time as a coherent social entity. However, we know very little about the kinds of cognition that might be involved. Although everyone probably agrees that this is some kind of "social cognition", quite what this actually might be is far from clear. The only thing we know much about in this respect is the phenomenon known in the developmental psychological literature as theory of mind (Leslie 1987; Perner 1991). Theory of mind is the ability to reflect on another individual's mind states. Following Grice and Dennett, philosophers of mind have long recognized that this is just one level in a potentially endless reflexive series of mind states and beliefs about mind states known as the orders of intentionality (Dennett 1989). Theory of mind (equivalent to second order intentionality) has been studied in some considerable detail because it is a major cognitive Rubicon that children pass through at about the age of 4-5 years (Leslie 1987; Perner 1991; Astington 1993).

Even though the exact processes involved in theory of mind and the higher orders of intentionality may be somewhat vague, we can nonetheless use the concept of the orders of intentionality to give us some purchase on the question as to how humans differ from other primates: the orders of intentionality form a natural scale, and seem to provide us with an index of social cognitive competence that mirrors in some way a natural hierarchy of cognitive complexity. The higher orders are just more difficult to process, and some individuals are less competent at coping with them (Kinderman *et al.* 1998; Stiller and Dunbar 2007). Even if we do not yet know what they really signify, my claim is simply that the intentionality scale provides us with a simple, useful index of social cognitive competence.

This being so, our main concern at this point is whether there is a natural limit on intentional reasoning in humans. We have tested normal adults, and it seems that this limit lies consistently at fifth order ("I *believe* that you *suppose* that I *imagine* that you *want* me to *believe* that....") across a number of studies (Kinderman *et al.* 1998; Stiller and Dunbar 2007). Approximately 65% of people have their limit at or below fifth order intentionality, and 75% have their limit at or below sixth order. There is considerable individual variation around this (Stiller and Dunbar 2007). Adult level competencies develop over a period of time between age 5 (when children first acquire theory of mind, or second order intentionality) and the early teens (when they finally acquire fifth order adult-level competencies) (Henzi *et al.* 2007).

6. Intentionality and the virtual world

The question of interest for us here is what can be achieved with different levels of intentionality. The kinds of cognitive competences represented by intentionality allow us to hold several different individuals' mind states in mind at the same time, and in this way they limit the size of social networks, and hence the size of social groups. However, it also seems likely that these competences will impose constraints on the kinds of cultural phenomena we can produce if these phenomena require us to think intentionally. Leslie (1987) noted that theory of mind may be crucial for children to be able to engage in fictive (i.e., pretend) play — in other words, play where they imagine characters or situations that do not really exist (such as dolls that can drink tea, or that the steering wheel on the back of a chair is a real car). Since this is about the world of the imagination, Leslie's point can be extended to story-telling and drama. Consider the case of Shakespeare's play *Othello*. The audience watching this play has to *believe* that Iago *intends* that Othello *imagines* that Desdemona *is in love* with Casio and that Casio reciprocates this state of *being in love* (where I highlight the successive intentional states in italics). Thus, in order to follow the story, the audience has to be able to work at fifth order intentionality, and is therefore working at the natural limits for the great majority of the population. In putting this story together, of course, Shakespeare himself has to go one level higher than his audience, to sixth order: he has to *intend* that the audience *believes*...... I have argued elsewhere (Dunbar 2005) that the difference between a competent story teller and a great one (such as Shakespeare) is their capacity to take the audience to the natural limits of their cognitive abilities. For a writer to really challenge an audience, he or she must be able to operate at a higher level still, and the fact that such skills are found in only a small proportion of the population perhaps explains why good writers are relatively rare, even though we can all enjoy a good story.

If this is a reasonable interpretation of the cognitive demands of culture, we can also use an intentional framework to explore the cognitive demands of other cultural phenomena such as religion, because this too requires us to work with an imagined world — a world that we cannot see or feel directly because it exists only in our heads. The transcendental experiences that we have during religious events are undoubtedly very real, but they are not the stuff of real physical experiences created by the external world. Rather, they belong to the inner virtual world of the imagination.

Intentionality Level	Possible statements of belief	Form of Religion
1st	I believe that god [...exists]	none
2nd	I believe that god is willing [....to intervene if you disobey his laws]	supernatural fact
3rd	I intend that you believe that god is willing [...to intervene...]	personal religion
4th	I intend that you believe that I want god to be willing [...to intervene....]	social religion
5th	I intend that you believe that god understands that I want him to be willing [...to intervene....]	communal religion

Table 1. Forms of religious belief made possible by different levels of intentionality (reproduced with permission from Dunbar 2008a).

A task analysis of the mentalising demands of religious belief suggests that it is perfectly possible to have religion with any order of intentionality from level two onwards. However, the *form* that a religion takes will depend critically on the levels of intentionality at which its adherents can work (*Table 1*). Second and third order intentionality allow one to express religious propositions, but these are purely statements of personal belief. At this level, it is difficult to incorporate the beliefs of another individual. Things improve with fourth order, but the claims I can now make about the intentions of God (or other spirits) do not commit you to anything: you can accept that I believe these things sincerely, but you do not have to accept that my beliefs are true. In effect, you have what I suggest is social religion, but it lacks the moral compulsion that truly communal religion (a religion that everyone in the community is obliged to sign up to) has. However, there appears to be a crucial phase shift at fifth order intentionality: when I accept that you believe some claim about the spirit world to be true, I *ipso facto* commit myself to the truth of your claim. We now have what I term "communal religion" — a set of beliefs that bind us into a single community whose members share the same world view. At this point, we have a very powerful mechanism for enforcing the communal will, for making us all sign up to the communal project (whatever that may be). We do not need a police force to make us behave in a religious way: we are all deeply and personally committed to it and adhere to these beliefs of our own free will.

It may be a coincidence that the number of orders of intentionality that we can typically manage is exactly the same as the number of layers to our social networks (self, 5, 15, 50, 150). If, however, this is a functional relationship (we can handle five layers to our social world as personal relationships, as opposed to the typological relationships that make up the outer layers beyond the 150, because this in some way involves handling five layers of intentionality such as friends-of-friends-of-friends, etc), then we will only be able to evolve integrated networks corresponding to the next size layer out (500 individuals) if and when we can evolve sixth order intentional competences as the norm. This would have implications for human brain size evolution, and I return to this point below.

7. Why are humans not just great apes?

I suggested at the outset that the answer to this question might lie in our ability to live in a virtual mental world. We can engage in activities that are well beyond the competences of even the great apes. Indeed, these human capacities are so far beyond the competences of other species that there is no chance that the proverbial chimpanzee sitting at a typewriter will ever produce the works of Shakespeare within any reasonable length of time (and I speak on the scale of millennia here, and not the infinite time allowed by statisticians). But why should humans and no one else have needed such extraordinary cognitive abilities?

From an evolutionary point of view, this is particularly puzzling because brain tissue is especially expensive. According to Aiello and Wheeler (1995), brain tissue is some of the most expensive matter in the body, consuming approximately 20% of total energy intake while accounting for only about 2% of total body mass (their "expensive tissue hypothesis"). This provides a very steep gradient up which natural selection has to drive brain evolution if a species is to be able to evolve a significantly larger brain than its ancestors.

There is some tentative comparative evidence (Dunbar 2003a,b), for which we now have additional support from recent as yet unpublished neuroimaging studies in humans, that intentional competences are correlated with brain (and in particular neocortex) volume. If the capacity to manage the higher orders of intentionality incurs a very significant energetic cost to the individuals concerned, then the advantages to be gained from investing in such capacities must be at least of comparable magnitude.

To understand what this selection pressure might have been, we need to consider in more detail the nature of primate sociality. At root, primate societies are implicit social contracts. Like all social contracts, their stability and functionality depends on the members trading off short and long term costs and benefits. As with pair-bonds, a group will remain stable as a coalition only providing its members are willing to compromise on some of their short term objectives in order to gain in the long run through group-level effects. For primates, these group level effects come in the form of increased survival, and hence higher reproductive rates, mainly as a result of reduced predation rates (Dunbar 1988; Shultz *et al.* 2004) though there may also be more direct benefits in terms of the survival of individual offspring (Silk *et al.* 2003; Silk 2007).

It is important to be clear here that these benefits do not constitute *group selection* in the conventional sense, by which biologists mean benefits that accrue for the benefit group at the expense of the individual members. These are *group-level* benefits that ultimately directly benefit the individual, and their costing should still be done in terms of selfish gene individual fitness. By cooperating, individuals stand to gain significant added benefits over what they could achieve on their own. But to do so, they have to manage an intermediate step, namely ensuring the group functions effectively as a group. Thus, this constitutes multi-level selection in its proper sense (i.e., some benefits accrue directly to the individual through its actions, and some accrue indirectly through the effect that its actions have on group-level functions).

The central problem for all such implicit social contracts is freeriders. There is always a significant benefit to be gained by individuals who take the benefits of sociality but do not pay all the costs (Enquist and Leimar 1993; Nettle and Dunbar 1997; Dunbar 1999). Freeriders' ability to survive in a population is directly proportional to the size of the population. Consequently, the large communities that characterise humans and their immediate ancestors face a significantly greater challenge from freeriders than those of other primates. As a result, more sophisticated mechanisms are needed to ensure that freeriders do not overwhelm the community.

Humans possess a number of cognitive mechanisms for managing freeriders, including sensitivity to social cheats, the use of social badges (e.g. clothing, shared knowledge and dialects: Dunbar 1999), costly membership rituals (Sosis and Alcorta 2003) and altruistic punishment (Orstrom *et al.* 1994; Clutton-Brock and Parker 1995; Fehr *et al.* 2002). However, without grooming we lack the endorphin-based mechanism that lies at the root of primate sociality. Since this creates a "bonding gap" (Dunbar 2009a), this has potentially serious consequences in terms of the stability of relationships, and hence the stability of the community as a whole. Something must have been needed to fill the gap, otherwise it would not have been possible to ensure the stability of the large communities that characterised humans during the later stages of our evolutionary history.

I have suggested elsewhere (Dunbar 2009a) that three specific behavioral mechanisms were brought into play to solve this problem. These were laughter, music and religion (with particular reference in this context to the rituals of religion). All three are particularly good at triggering the release of endorphins; they are also intensely social and (with the partial exception of laughter) unique to humans. And I suggest that they emerged in this order, partly because (a) we share laughter (uniquely) with the chimpanzees (Provine 2000) and (b) religion as such must have been a late innovation because it depends on language and this probably did not evolve until around 200,000 years ago (Dunbar 2009b).

The pattern of brain size evolution suggests that the freerider problem is only likely to have become a serious issue in the later stages of human evolution following the appearance of archaic humans (*Homo heidelbergensis* and allies) around half a million years ago (Dunbar, in press a). At this point, brain size began a very steep rate of increase, implying a correspondingly steep increase in the size of communities that had to be bonded. While social conformity and "good behavior" can always be imposed by punitive action on backsliders, there are limits to which such action can really enforce social cooperation. Since there are always benefits to be gained from freeriding and there is a significant chance that any one freerider will escape detection and/or punishment, there will always be a temptation to freeride, no matter how effective punishment is in reducing the absolute frequency (an identical problem occurs in relation to poaching and conservation: see Cowlishaw and Dunbar 2000). Social cooperation is far more effective when individuals act willingly because they all voluntarily sign up to the communal project (Orstrom *et al.* 1994). As a result, mechanisms that create a sense of communality are always more likely to result in the required levels of altruism than attempts to use of social control.

As has long been appreciated in the social sciences, both religion and story-telling play an important role in social bonding in all human cultures (Durkheim 2001). Religion does so through the capacity of rituals to trigger the release of endorphins, since many of these are of just the kind of mildly stressful activities that are especially good at releasing endorphins. While this may be true, we need to ask how religion differs from any number of other activities that are equally good at triggering the release of endorphins. I suggest that the important feature of religion is its capacity to provide a reason why everyone should turn up at religious ceremonies at regular intervals. In other words, theology (or at least an explanation that justifies religious rituals) plays an important role in reinforcing the regular performance of religious rituals as a communal activity, thereby ramping up the bonding effects that performing the same rituals on one's own might have. To the extent that the intellectual component of religion is a form of story-telling, story-telling itself also enters the frame as an important mechanism for community bonding.

One could argue that both of these are epiphenomena — non-adaptive by-products of the fact that we have big brains. Exactly this claim has, of course, been made for music (Pinker's [1998] "music as evolutionary cheesecake" argument), as well as for religion (e.g. Boyer 2001; Atran 2002). However, I think this claim is mistaken. From an evolutionary perspective, it rests uncomfortably with the time, money and effort that, irrespective of culture, humans invest in story-telling, music and religion. All societies value story-telling for itself as a form of entertainment, and many of these stories are intimately involved in creating a sense of community: origin stories, tribal histories, and moral tales are widespread, and all contribute directly to producing a sense of community. Similarly, every society has its own forms of music and dance, and of course it own particular form of religion.

Aside from the trivially obvious answer of language, what is it that limits these "activities of the mind" to humans alone? The answer surely has to be that only humans can cope with the fifth order intentionality that is necessary to allow these phenomena to produce the effects they do. The best that any non-human species can do seems to be second order intentionality — and even that is probably true only of great apes (O'Connell and Dunbar 2003), since the consensus is that all other animal species can only achieve first order. If mentalising skills really are an emergent property of the computational power (and hence size) of the brain (Dunbar 2003a), then the simple explanation for this striking difference between ourselves and all other species lies in the size of our brain (and, perhaps more specifically, the frontal lobes). In effect, the differences between us and all other animals comes down to the fact that the size of our brain allows us to do something computationally that is simply impossible with a smaller brain.

References

Abbott D. H., Keverne E. B., Moore G. F., Yodyinguad U., 1986, "Social suppression of reproduction in subordinate talapoin monkeys *Miopithecus talapoin*", in Else J., Lee P. C. (eds.), *Primate Ontogeny*, Cambridge: Cambridge University Press, 329–41.

Aiello L. C., Wheeler P. T., 1995, "The expensive tissue hypothesis: the brain and the digestive system in human evolution", *Current Anthropology* 36: 199–221.

Astington J. W., 1993, *The Child's Discovery of the Mind*, Cambridge, Mass.: Cambridge University Press.

Atran S., 2002, *In Gods We Trust,* Oxford: Oxford University Press.

Barrett L., Henzi S. P., Dunbar R. I. M., 2003, "Primate cognition: from 'what now?' to 'what if?'", *Trends in Cognitive Sciences* 7: 494–97.

Boyer P., 2001, *Religion Explained: The Human Instincts That Fashion Gods, Spirits and Ancestors*, London: Weidenfeld & Nicholson.

Byrne R., Whiten A. (eds.), 1988, *Machiavellian Intelligence,* Oxford: Oxford University Press.

Castles D., Whiten A., Aureli F., 1999, "Social anxiety, relationships and self-directed behaviour among wild female olive baboons", *Animal Behaviour* 58: 1207–15.

Clutton-Brock T. H., Parler G. A., 1995, "Punishment in animal societies", *Nature* 373: 209–16.

Cowlishaw G., Dunbar R. I. M., 2000, *Primate Conservation Biology,* Chicago: Chicago University Press.

Dennett D., 1987, *The Intentional Stance,* Cambridge, Mass.: MIT Press.

Dunbar R. I. M., 1980, "Determinants and evolutionary consequences of dominance among female gelada baboons", *Behavioural Ecology and Sociobiology* 7: 253–65.

Dunbar R. I. M., 1984, *Reproductive Decisions: An Economic Analysis of Gelada Baboon Social Strategies,* Princeton, N. J.: Princeton University Press.
Dunbar R. I. M., 1988, *Primate Social Systems,* London: Chapman & Hall.
Dunbar R. I. M., 1989, "Reproductive strategies of female gelada baboons", in Rasa A., Vogel C., Voland E. (eds.), *Sociobiology of Sexual and Reproductive Strategies,* London: Chapman & Hall, 74–92.
Dunbar R. I. M., 1993, "Coevolution of neocortex size, group size and language in humans", *Behavioral and Brain Sciences* 16: 681–735.
Dunbar R. I. M., 1998, "The social brain hypothesis", *Evolutionary Anthropology* 6: 178–90.
Dunbar R. I. M., 1999, "Culture, honesty and the freerider problem", in Dunbar R. I. M., Knight C., Power C. (eds.), *The Evolution of Culture,* Edinburgh: Edinburgh University Press, 194-213.
Dunbar R. I. M., 2003a, "Why are apes so smart?", in Kappeler P., Perrerira M. (eds.), *Primate Life Histories and Socioecology,* Chicago: Chicago University Press, 285–98.
Dunbar R. I. M., 2003b, "The social brain: mind, language and society in evolutionary perspective", *Annual Review of Anthropology* 32: 163–81.
Dunbar R. I. M., 2004, "Language, music and laughter in evolutionary perspective", in Oller D. K., Griebel U. (eds.), *Evolution of Communication Systems: A Comparative Approach,* Cambridge, Mass.: MIT Press, 257–74.
Dunbar R. I. M., 2005, "Why are good writers so rare? An evolutionary perspective on literature", *Journal of Cultural and Evolutionary Psychology* 3: 7–22.
Dunbar R. I. M., 2007, "Cognitive constraints on the structure and dynamics of social networks", *Group Dynamics* 12: 7-16.
Dunbar R. I. M., 2008, "Mind the gap: or why humans aren't just great apes", *Proceedings of the British Academy* 154: 403–23.
Dunbar R. I. M., 2009a, "Mind the bonding gap: constraints on the evolution of hominin societies", in Shennan S. (ed.), *Pattern and Process in Cultural Evolution,* Berkeley: University of California Press, 223-234.
Dunbar R. I. M., 2009b, "Why only humans have language", in Botha R., Knight C. (eds.), *The Prehistory of Language,* Oxford: Oxford University Press, 12-36.
Dunbar R. I. M., Shultz S., 2007a, "Understanding primate brain evolution", *Philosophical Transactions of the Royal Society of London* 362B: 649–58.
Dunbar R. I. M., Shultz S., 2007b, "Evolution in the social brain", *Science* 317: 1344–47.
Duncan J., 2001, "An adaptive coding model of neural function in prefrontal cortex", *Nature Neuroscience* 2: 820–29.
Durkheim E., 2001 [1900], *Elementary Forms of Religious Life,* Oxford: Oxford University Press.
Enquist M., Leimar O., 1993, "The evolution of cooperation in mobile organisms", *Animal Behaviour* 45: 747–57.
Fehr E., Fishbacher U., Gächter S., 2002, "Strong reciprocity, human cooperation and the enforcement of social norms", *Human Nature* 13: 1–25.

Finlay B. L., Darlington R. B., 1995, "Linked regularities in the development and evolution of mammalian brains", *Science* 268: 1578–84.

Finlay B. L., Darlington R. B., Nicastro N., 2001, "Developmental structure in brain evolution", *Behavioral and Brain Sciences* 24: 263–308.

Goosen C., 1981, "On the function of allogrooming in Old-World monkeys", in Chiarelli A. B., Corruccini R. S. (eds.), *Primate Behaviour and Sociobiology*, Berlin: Springer, 110–20.

Hamilton M. J., Milne B. T., Walker R. S., Burger O., Brown J. H., 2007, "The complex structure of hunter-gatherer social networks", *Proceedings of the Royal Society of London* 271B: 2195–202.

Henzi P., de Sousa Pereira L., Hawker-Bond D., Stiller J., Dunbar R. I. M., Barrett L., 2007, "Look who's talking: developmental trends in the size of conversational cliques", *Evolution and Human Behavior* 28: 66–74.

Hill R. A., Dunbar R. I. M., 2003, "Social network size in humans", *Human Nature* 14: 53–72.

Jerison H. J., 1973, *Evolution of the Brain and Intelligence*, London: Academic Press.

Keverne E. B., Martensz N., Tuite B., 1989, "Beta-endorphin concentrations in cerebrospinal fluid of monkeys are influenced by grooming relationships", *Psychoneuroendocrinology* 14: 155–61.

Kinderman P., Dunbar R. I. M., Bentall R. P., 1998, "Theory-of-mind deficits and causal attributions", *British Journal of Psychology* 89: 191–204.

Leslie A. M., 1987, "Pretense and representation – the origins of 'theory of mind'", *Psychological Review* 94: 412–26.

Nettle D., Dunbar R. I. M., 1997, "Social markers and the evolution of reciprocal exchange", *Current Anthropology* 38: 93–99.

O'Connell S., Dunbar R. I. M., 2003, "A test for comprehension of false belief in chimpanzees", *Evolution and Cognition* 9: 131–39.

Orstrom E., Gardner R., Walker J., 1994, *Rules, Games and Common-Pool Resources*, Ann Arbor: University of Michigan Press.

Pérez-Barbería J., Shultz S., Dunbar R. I. M., 2007, "Evidence for intense coevolution of sociality and brain size in three orders of mammals", *Evolution* 61: 2811–21.

Perner J., 1991, *Understanding the Representational Mind*, Cambridge, Mass.: MIT Press.

Pinker S., 1998, *How the Mind Works*, London: Allen Lane.

Provine R. R., 2000, *Laughter: A scientific investigation,* New York: Viking.

Shultz S., Dunbar R. I. M., 2007, "The evolution of the social brain: Anthropoid primates contrast with other vertebrates", *Proceedings of the Royal Society of London* 274B: 2429–36.

Shultz S., Noe R., McGraw S., Dunbar R. I. M., 2004, "A community-level evaluation of the impact of prey behavioural and ecological characteristics on predator diet composition", *Proceedings of the Royal Society of London* 271B: 725–32.

Silk J. B., 2002, "The 'F'-word in primatology", *Behaviour* 139: 421–46.
Silk J. B., 2007, "Social components of fitness in primate groups", *Science* 317: 1347–50.
Silk J. B., Alberts S. C., Altmann J., 2003, "Social bonds of female baboons enhance infant survival", *Science* 302: 1232–34.
Smuts B. B., 1985, *Sex and Friendship in Baboons*, New York: Aldine.
Sosis R., Alcorta C., 2003, "Signalling, solidarity, and the sacred: evolution of religious behaviour", *Evolutionary Anthropology* 12: 264–74.
Stiller J, Dunbar R. I. M., 2007, "Perspective-taking and social network size in humans", *Social Networks* 29: 93–104.
Whiten A., Byrne R. W. (eds.), 1997, *Machiavellian Intelligence II: Extensions and Evaluations*, Cambridge: Cambridge University Press.
Whiten A., Goodall J., McGew W. C., Nishida T. *et al.*, 1999, "Cultures in chimpanzees", *Nature* 399: 682–85.
Zhou W. X., Sornette D., Hill R. A., Dunbar R. I. M., 2005, "Discrete hierarchical organisation of social group sizes", *Proceedings of the Royal Society of London* 272B: 439–444.

PALEOCULTURAL APPROACH IN THE HOMINIZATION AND DIFFERENT IMPLICATIONS

Fiorenzo Facchini

Human behavior is characterized by culture. The question is defining what we exactly mean by culture. There is no general agreement about this in social and anthropological sciences. Some authors have a broad concept about the notion of culture. Culture would be each information capable of affecting individuals' behavior, something that they acquire from other individuals through teaching, imitation or other forms of social learning (see Richerson and Boyd 2005). A broad characterization of the concept of culture opens up the extension of culture to animals (birds, fish, non-human primates, etc.).

Cavalli Sforza and Feldman (1981) apply "the term 'cultural' to traits that are learnt thanks to any process of non-genetic transmission, whether by imprinting, conditioning, observation, imitation or as a result of direct teaching".

These behaviors are of great interest for the biologists, in particular the adaptive behavior, because some traditions can modify the action of the selection mechanism (see Laland, Smee and Miles 2010).

But this extension of the concept of culture to the animal world does not allow to distinguish what is specific to the human behavior.

Other authors limit the concept of culture to artistic and spiritual manifestations and to the language used by *Homo sapiens* about 100,000 years ago. This concept of culture is too restrictive. In fact, when the products of technology show some kind of planning and denote a symbolic activity, they reveal a mind capable of abstraction and therefore a self-reflection which is an indication that the human level has been reached.

Thus the application of the concept of culture is not without problems when identifying man at his origins.

Current taxonomy refers to the *genus Homo* the following species: *H. habilis/rudolfensis, ergaster, erectus, sapiens.* Some authors (Wood and Collard 1999;

2001) have recently proposed to attribute *Homo habilis* and *Homo rudolfensis* to *genus Australopithecus*. In particular these authors, on the basis of size and body shape, locomotion, relative brain size, developmental models and size of the masticatory apparatus propose two groups with different strategies: one group including the *Australopitecus species*, the *Paranthropines* and *Homo habilis* and *Homo rudolfensis*, reclassified as *Australopithecus habilis* and *A. rudolfensis*; a second group represented by *H. erectus, H. ergaster, H. heidelbergensis, H. neanderthalensis* and *H. sapiens.*

But greater consideration should be given to adaptive strategies in the attribution of the specimens to the *genus Homo*, and also to aspects related to the behavior, understood not only in functional terms (e.g. flaking of flint) but also in terms of meanings. If man is a species characterized essentially by cultural behavior, cultural capacity, in particular regarding the relation with the environment, can also be used to identify the earliest stages of the human *genus*.

In any case, the attribution of a set of skeletal remains to the *genus Homo* out of its anatomical features does not necessarily imply that specimen represents man in a philosophical sense, i.e. the thinking man.

In paleoanthropology, an identification of the species made only on the basis of morphological skeletal characters is problematic. The same difficulty appears in taxonomy, if it is based on the DNA analysis in ancient remains (so far, Neanderthals) compared to DNA in modern humans. The use of morphological or biomolecular differences in skeletal remains, if utilized to denote the human level, that is, as a criterion for the fecundity between species, is problematic.

But when we happen to meet skeletal remains that are connected with products showing a systematic and innovative work, the presence of man can be inferred, whatever morphological and evolutionary level the remains are to be attributed to.

What distinguishes human technology from non-human one (as it occurs in apes and *Australopithecus*) is the complexity of the actions by means of which this technology is performed and, even more relevant, the capacity of improving and innovating such techniques, and the significance assumed by the products in the context of everyday life.

According to Bergson (1941) "intelligence, considered in what seems to be its original feature, is the faculty of manufacturing artificial objects, especially tools to make tools and of indefinitely varying the manufacture".

According to Piveteau (1991) "The reflection that characterizes man can be defined as conscience of deferred action in making tools. It is not credible that man, since his origins, could have been *faber* earlier than *sapiens*."

Planning capacity is revealed by tool making (purpose of products, variety, progress in technology, preservation of tools) and by organization of the territory (shelters, camps). Intentionality reveals the notion of time; the subject is able to elaborate images of the past and to project them into a future that he can prefigure. In the animal world one also finds techniques (at times very complex ones), but they are regulated biologically and do not exhibit innovation and progress. They do not constitute evidence of abstract ability to project into the future. In his manifestations of intentionality, man also shows capacity of choice, self-determination and liberty, an aspect that places him on the plane of values and thus of ethics.

Another important expression of the human psychism is intentional organization of the territory, i.e. of the living spaces chosen by the hominids and devoted to various uses, related to the subsistence needs (protection, food sharing, preservation and manufactured articles).

Examples of these behaviors are recorded from around two million years ago, but tool making appears to be older, perhaps dating 2.5 million years ago.

Julien Ries (1993) observed that flint flaking implies experimentation, imagination, choice of material and form; he also attributes a symbolic meaning to the organization of the territory.

Symbolization is the other characteristic of human behavior. It consists in attributing to a sign (a sound or an object) a value that goes beyond the sign itself. Through symbolization, products of techniques are enriched with meaning and value. A symbolic value, i.e. a symbolic activity, can be recognized in systematic manufacture of tools, in the organization of territory, in the subsistence economy and in the social organization. Instrumental culture reveals a symbolism which we suggest to call *functional*, distinguishing it from the *social* symbolism expressed in language and from the *spiritual* symbolism represented by artistic and religious expressions which are not directly connected with subsistence strategies. (see Facchini 1999, 2000).

When the tool is produced for a purpose, in a variety of forms, when it is used in a given environmental context, when it is preserved (and not only used occasionally and then discarded, as in apes), when tool making is improved in time, only then can we state that it is the expression of a symbolic system of connections. *Homo symbolicus* is such because he is human, creator of tools and of art, able to communicate his internal world in various ways.

According to Paul Ricoeur (1959) "symbol leads to thinking".

We agree with Deacon (1997) who stated: "The introduction of stone tools and the ecological adaptation they indicate also mark the presence of a socioecological predicament that demanded a symbolic solution." Deacon defines man to be a symbolic species, with reference to tool making, as a system of adaptation to the environment.

In conclusion, in our opinion two behavioral characteristics denote the human cultural level: the ability to plan, expressed by tool making and domestic technologies, and symbolization, i.e. the meaning that artefacts assume. When we find signs of these abilities, we can infer that we are dealing with a human form.

Cultural behavior remains essential to identify early man and have played an important role in the evolution. Indeed, culture could have played a double role in the process of human speciation: on one side in the isolation and differentiation from other groups of hominids that did not show such behavior (e.g. the *Australopithecus* group); and on another side in the adaptation to the environment and in communication between groups that had the same cultural behavior, thus slowing down or preventing the conditions of isolation that lead to new species.

The oldest expressions of culture and early humans

Behaviors with cultural meaning are recorded around 2-2.5 million years ago. Several 2.5-to 2 million-year-old Plio-Pleistocene deposits have yielded archaeological material whose workmanship reveals rather elaborate flaking characterized by the ability to plan technical actions. This is the pebble industry (pre-Oldowan culture according to De Lumley) discovered in West Turkana (Kenya), at the Lokalalei site, and in Ethiopia at the Kada Gona (Roche *et al.* 1999; Roche 2001), Fejej and Hadar sites (De Lumley 2004). The material permits the reconstruction of the series of operations repeatedly used in many tools. The reconstruction of these operational schemes allows the recognition of technical differences in the archaeological documentation of other sites.

The industries at Olduvai (Tanzania), dating to 1.9-1.7 million years ago, also reveal definite series of operations in polyhedral and spheroidal tool working, with aspects different from those of Lokalalei. The documentation is richer between 1.7 and 1.5 million years ago, with protobiface and biface technique first observed in chopping tools found at the camps of Melka Kunturè in the Ethiopian Awash region. Examples of territorial organization are the camps at Olduvai (Tanzania), Melka Kunturè (Ethiopia) and Okote (Kenya).

The hominids associated or contemporaneous with this archaeological evidence can be attributed to Plio-Pleistocene forms of *Homo habilis* found at Olduvai (Leakey *et al.* 1964), in South Africa and Omo Valley. Thus culture is added to the morphological characteristics that separate these hominids from the Australopithecines, e.g. encephalic development and differentiation of the areas of articulate language (Falk 1983; Tobias 1983, 1996).

In the same period of time more cerebralized hominids are recorded at East Turkana, attributed to *H. rudolfensis* (Wood 1992), a name used for the first time by Alexeev (1978) for a specimen from the same area. A specimen from Malawi has also been referred to *H. rudolfensis* (Shrenk *et al.* 1993).

H. habilis and *H. rudolfensis* should be two true species. Yet, Coppens (2001) observed that they are not clearly separated by chronology, geography or ecology. Two species or two *genera*? If it is difficult to identify species throughout human evolution, it is even more difficult to do that at the dawn of humanity. Notwithstanding the taxonomic denomination, an element that could justify the attribution of both forms to the same human level is cultural behavior.

The proposal to exclude *H. habilis* and *H. rudolfensis* from the *genus Homo* (according to Wood) ignores the behavioral aspects that accompany their perfect bipedalism. These behavioral aspects should be taken into account in taxonomy because of the related adaptive strategies.

Culture in Palaeolithic man

Cultures of *Homo ergaster/erectus* in Low and Middle Palaeolithic exhibit more clearly the elaborated techniques in the bifaces and Levallois technique. Tattersall (2008) recognizes in the hand axes technique a cognitive jump.

According to Winn (2002) and other authors the symmetrical Acheulean tools are clear examples of an imposed form based on a mental template and have symbolic meanings. Even the construction of spears for hunting, as documented in the deposit of Schoeningen, dated to 400,000 years ago, must have required a conceptual activity.

The use of red ochre, to which is attributed a symbolic meaning, is documented in various deposits, some very ancient. We can quote here for example its presence in Ethiopia 1.5 million years ago, in Bed II of Olduvai (Tanzania), at Katpurin (Kenya) 285,000 years ago, at Twin Rivers (Zambia) 400,000-260,000 years ago, at Terra Amata (France) 400,000 years ago, at Becov (Czechoslovakia) 250,000 years ago, at Blombos (South Africa) 75,000 years ago, with symbolic incisions, not related to Upper Palaeolithic.

According to Marshack (1988) "rare instances of symbolizing and problem-solving in the archaeological record of the European Acheulean and Mousterian suggest highly evolved potentially variable hominid capacities… They also reveals a capacity for planning or modelling the territory and culture in time and space".

Symbolic expressions are more clearly documented in the last 100,000 years, in particular with archaic forms of *Homo sapiens*, and *Homo neanderthalensis* not only in Upper Palaeolithic.

Perforated marine gastropod shells at Oued Djebbana (Algeria) and Skhul (Israel), dating about 130,000-100,000 years ago, indicate the use of shells for personal decoration (Vanhaeren *et al.* 2006; D'Errico and Sorressi 2009).

According to Schäfer (1996), at least from the beginning of the Middle Palaeolithic there is ample evidence of non utilitarian objects (crystals, fossils, organics substances) collected and preserved by man, objects that had then the same meaning they have today.

Funerary practices, expressions of a symbolism, are documented by the burials (Tabun, Qafzeh, Skhul) in Near East 100,000 years ago, by Neanderthal man and *Homo sapiens*.

Culture as discontinuity

On the phenomenological level culture reveals discontinuity compared with non-human forms, whatever the reason or nature of this discontinuity, because it does not follow biological laws. Culture, although related to the biological sphere, is characterized as extra-biological or meta-biological, in the sense that it achieves a transcendence with respect to purely biological laws or modalities of behavior. In fact, even when it has some relationship with biological needs, culture occurs outside of any biological or behavioral determinism and, as mentioned above, is a sign of liberty and self-determination. This is particularly evident in the manifestations of spiritual and social symbolism.

According to Dobzhansky (1967), in the evolutionary history of life there have been two great moments of crisis. In consequence of these, although the organization laws and modalities of the previous phase were conserved, there was an advancement beyond the previous organization to a new level. The author proposes to call these moments "evolutionary transcendence". A first transcendence was the passage from non-life to life. The laws of chemistry are not broken, but a new organizational modality and a new relationship with the environment are established: "cosmic evolution transcended itself giving birth to life". A second moment of transcendence was the appearance of man: "biological evolution transcended itself giving rise to man". The laws regulating living beings were not cancelled, but the organizational modalities of human society are set on another level.

Ayala (2007, 2009) agrees with the previous statement and observes that moral behavior is a biological attribute of *Homo sapiens*, but moral codes are not the product of biological evolution, but of a cultural evolution. He speaks of super-organic evolution for man. "Biology precedes, culture transcends." (1998)

I would like to point out that in the terminology coined by Teilhard de Chardin (1956) the moments of transcendence according to Dobzhansky correspond to the vitalization and hominization points in the curve of complexification that he proposed.

The recognition of such discontinuity in the simplest technological manifestations makes it difficult to identify the humans in the real sense of the word. This is an open problem in paleoanthropology.

In our opinion, planning capacity and symbolism have been typically human since the origins of man, whatever the manifestations and the morphological evolutionary level may have been.

Attitude towards culture can be recognized even out of its simplest expressions, starting from pebble culture. With the passing of time, cultural manifestations become more and more meaningful and therefore human attribution turns out to be easier: pebble culture of Lower Pleistocene, Acheulean bifaces, hunt organization, leptolithic culture of the Upper Paleolithic, artistic, aesthetic, religious manifestations.

Innovation and intentional transmission finalized to learning are fundamental in this respect.

Symbolic behavior was very important for the success of humans. "The ability to communicate detailed, concrete information as well as abstract concepts allowed early humans to cooperate and plan for the future in ways unique to our species, thus enhancing their survival during rough times and boosting their reproductive success in good times." (Balter 2009)

Different implications

The implications of culture in an evolutionary perspective can be recognized at different levels, which are interconnected.

a) On paleoanthropological and evolutionary level

Cognitive capacities disclosed by culture are to be related to brain development. Cerebralization characterizes the evolutionary trend in the hominoids and among them, in the hominids and in the human line that is more and more cerebralized starting from the *Homo habilis/rudolfensis*. The process of cerebralization is a privileged direction in the evolution according to many authors (Piveteau, Dobzhansky, Teilhard de Chardin, etc.).

By culture man is able to counteract intentionally natural selection. This is an absolute novelty in the history of life. In this regard the human species represents a paradox at the evolutionary level: natural selection has produced a being able to oppose it by a modality that no longer falls within the natural game of the competition of living beings with the environment. This unique case in the world of living beings — one can say an anomaly — is explainable by the intervention of culture, a factor not to be found in other species and not directly dependent from biological properties.

Culture enters in the mechanism of the evolution of the species. We must admit a differentiation and a genetic isolation from our non-human relatives, that could be related to a cultural isolation. Moreover, cultural adaptation reduces natural selection in man. This can stop or prevent the process of isolation among the human groups which is necessary to speciation, even if morphological differentiation does not cease particularly in the adaptation to the environment. We are aware of a microevolution in the population genetics and this can be admitted in human populations as well. This can explain the difficulty in identifying species in hominization. According to some authors (e.g. Ferembach, Jelinek, Coppens, etc.), it would be more correct to speak of morphological grades or steps, rather than of species.

In paleoanthropology human species are only supposed, on the basis of morphological and ecological or biomolecular characters, not demonstrated.

b) At the ecological level

Culture realizes new strategies for survival that are accomplished through consciusness. By means of culture man adapts himself to his environment and adapts the environment to himself. Moreover culture characterises the relation between man and environment, both from the structural and from the functional viewpoint. This modality of adaptation is linked to the concept of ecological niche of the human species.

Structural adaptation implies genetic changes and functional adaptation permits a flexibility of the phenotype without genetic mutations. In this regard is to be observed that the ecological niche is not to be intended as a habitat, but as "the functional state of an organism within his community" (Elton 1927).

In the human species, culture is what characterizes the structural and functional adaptation to the environment. In some cases culture can modify, increasing or reducing, genetic adaptation and can influence the physiological adaptation. One can state that culture is emblematically the "ecological niche" of man (Facchini 1988, 2001).

This can explain the wide diffusion of man in the world. Man is an ecumenical species. With culture starts a process of humanization (Martelet 1998).

Moreover, by culture man has the possibility to oppose in some way natural selection and to manage the environment. This is a unique case in the living world. Human species shows in comparison with other species an adaptive or ecological discontinuity.

c) At the phenomenological level

Discontinuity expressed by culture not only implies a difference with respect to the rules and properties of animals, but also points out a new modality of behavior expressed by planning capacity and by symbolism. Symbolic language is peculiar to man. Freedom is a property of man. These are to be considered extra-biologic properties and they are documented by products of technology too. Man shows subjective conscience and self-determination, which contribute to determinate his behavior. They are signs of transcendence and express a phenomenological discontinuity.

The activity that man performs by means of culture (including instrumental culture) is external to fixed and constant biological schemes, is performed freely with ever innovative modalities based on individual experience.

d) At the cosmological level

Some authors (Barrow and Tipler 1986) proposed the Anthropic principle. On the basis of this principle astronomic constants turn out to be formed in order to let intelligent observers appear (strong expression of the Anthropic principle) or are such that they permit intelligent observers to be developed (weak expression).

At this regard Nicolò Dalla Porta and Secco (1991) observed that the strong principle is not demonstrable and that the weak one seems to assume more a character of ascertainment than of a proof.

Indeed, development of intelligent life is related to conditions that have actually taken place in the history of the planet, and of life on Earth, and in the whole cosmic and biologic evolution that predates man. This gives a new meaning to all reality. The Anthropic principle could be considered from another viewpoint. "Our purpose, seen in hindsight, is attached to the purpose of the whole living world, in which it seems we are allowed to consider our appearance as a particular purpose", which is made possible by the cosmological evolution (Leclerc 2008).

However, abstracting from the problems raised by the Anthropic principle, it is through the thought and conscience of man that the entire universe is thought and becomes conscious.

e) At the philosophical level

As pointed out above, Dobzhansky sees in the apparition of man the second form of transcendence in the history of life. Concerning the explanation of the nature of this transcendence and its causes it is necessary to move beyond the phenomenological aspects into a purely philosophical plane.

The phenomenological discontinuity, represented by abstractive intelligence, self-awareness, symbolic communication and language, gratuitousness and freedom, all of which cannot be reduced to purely biological activities, introduces into the picture the spiritual dimension. If matter becomes thinking (Coppens 2006), one can speak of a qualitative difference in comparison with animals, not of a difference of grade, as affirmed by Darwin, according to whom "the difference in mind between man and the higher animals, great as it is, certainly is one of degree and not of kind." (*The Descent of Man,* 1871)

The presence of the spirit is outside the realm of empirical evidences and can be dealt with only at the philosophical level whether we are talking about the human phylogenesis or about the ontogenesis. The discontinuity that can be observed on the phenomenological level may be interpreted as a transcendence at the philosophical or ontological level.

John Paul II so describes man's appearance: "With man we find ourselves in the presence of an ontological difference, one ontological leap, one could say" (John Paul II 2003). If discontinuity is observed in a long time at the phenomenological level, at the philosophical level this discontinuity must be radical, no matter what its cultural expressions are, because the spirit cannot come out of living matter, as the Pope remarked.

In the address to the Plenary Session of the Pontifical Academy of Sciences (31 October 2008) Benedict XVI reaffirmed "that every spiritual soul is created immediately by God — it is not 'produced' by parents" — and also that "it is immortal" (see *Catechism of the Catholic Church*, 366).

We can assume that the will of the Creator includes, at a certain moment of the evolutionary process, a corporeity enriched by spirit, not in the sense of an entity which is added to another one, almost placed on or beside it, but which, starting from a certain moment, exists inside the other one, as and when it is wanted by God, in a way similar to what happens in the human ontogenesis. All this for ontological reasons, not for biological reasons, if we recognize the presence of the spirit in man. The analogy with the ontogenesis is present also in the quoted address of Pope Benedict to the Pontifical Academy.

There cannot be any form of intermediate psychism which would be only partially human, as Maritain (1973) noted. But there may have been intermediate forms of animal psychism between the human one and that of the apes, as it is right to think about the *Australopithecus* (overdeveloped animals, according to Maritain 1973).

Nevertheless, the exact moment of passage into the spiritual realm, i.e. when the hominid becomes aware of himself cannot be identified by scientific methods or represented by our imagination, as remarked by John Paul II.

The obscurity in representing the appearance of the spiritual dimension is similar to what happens in the human ontogenesis.

The achievement of human level interested the first humans and since then all their descendents up to the present day humanity. After having reached human threshold in the hominization process, man's presence continues further in time with the generation of every human being who represents an ontological discontinuity with respect to animals.

Cultural manifestations cannot help in determining the exact moment of the achievement of the human threshold in hominization, but only can help in ascertaining if that threshold can be considered as reached.

Natural selection, culture and ethical behavior

Connected with the cultural attitude is the ethical behavior which requires the ability to recognize certain values and to choose freely.

Some authors derive from the Darwinian framework a concept of ethics, inspired to the socio-biology of E. Wilson, which regards culture and its manifestations as being regulated by genes. The moral characters would be evolved by animals and would have had a success by natural selection. Genes responsible for moral qualities are admitted, causal relationships between neuronal activity and behavior, self-determination and religious sense included, are looked for.

These conceptions are said to be based on the observation of emotional expressions in animals, about which Darwin recognized differences only of grade, not of quality, in comparison with man. Natural selection would act on the moral qualities in the same way it works for the physical characters. If man has had success in the evolution, it has been for natural selection. Someone speaks of coevolution for the ethic-cultural component and for the moral sense, by means of natural selection (Franceschelli 2009). This conception also inspires the Darwinian naturalism (see Franceschelli 2009; Pievani 2006; Girotto *et al.* 2009). But in this case it is postulated what instead should be demonstrated.

If some ethical behavior has had success in the society, this can not be referred necessarily to natural selection. David Wilson (2002) proposes an evolutionary approach to religions. They can be considered as adaptive units, characterized by selection of group. The natural Darwinian selection would have developed systems of belief. But this view seems ideological, because extends a biological theory beyond its field. Religion is a cultural aspect of the human society and what man realizes by means of culture has not necessarily a biological determination and an adaptive value in the Darwinian sense.

The successes of human behavior are not related to the natural processes of selection, but to culture, and cultural evolution does not follow the Darwinian scheme. Some issues, like cooperation, solidarity, altruism, familiar dimension, freedom, are conditions of success for the human species because they correspond to the biopsychic structure of man. But they have no genetic determinations, because the manifestations are different in individuals and are always freely reached. Some behaviors are connected to the fundamental or vital needs of the individual or of the species and can have favored the human groups, but the answers to the biological needs are enriched by particular significance, are perceived as values and are pursued freely and eventually transmitted by cultural selection. The experience of peoples suggests or confirms some moral behavior for the benefits to the society.

In relation with these human behaviors we must point out the responsibility of man in his choices.

Ethic behavior is exclusive of human species. According to Ayala (1987) three conditions are necessary for an ethical behavior: a) the ability to foresee the consequences of one's own actions; b) the ability to make value judgements; c) the ability to choose from a set of alternative actions. These conditions are present in man and are linked to abstractive intelligence and to freedom. They are documented by symbolic manifestations that are present from the earliest stages of mankind, as we pointed out in our previous considerations.

The freedom and capacity of values, that are at the basis of ethical behavior, are evident in modern humans, but are attested in some way also in prehistoric man.

Very important and significant have been the different forms of free cooperation in the life strategies of prehistoric man. In particular, in man, altruistic actions are undertaken in full awareness and are freely carried out (Facchini 2007).

These properties are not observed in the animal world, in which we can only find some analogies, that lead some authors to consider them as human, for a kind of anthropomorphism. But the animal behavior lacks reflected conscience and freedom, which are conditions for the moral behavior. Consequently it does not make any sense to speak of Darwinian ethic and of values derived from the animal behavior, as some authors improperly propose.

We think that the conditions of ethical behavior are present in human nature, but the moral codes are not the product of biological evolution and natural selection (as sociobiologists think). They are connected to the cognitive capacity and to cultural evolution as Ayala (2001; 2009) observed. In any case the norms of morality must be consistent with biological nature and are influenced by cultural selection.

Culture and ethical behavior appear as a transcendence from organic to human evolution, i.e. as an evolutionary event which introduces novel laws in nature, like what happened in the evolution from inorganic to organic (Dobzhansky and Boesiger 1983).

The nature of this transcendence is not a biological property and suggests a spiritual dimension.

REFERENCES

Alexeev V. P., 1978, *Paleoantropologia zemnogo shara I formirovanie chelovecheskikh ras*, Moscow: Nauka.

Ayala F. J., 1987, "The biological roots of morality", *Biology and Philosophy* 2: 235–52.

Ayala F. J., 1998, "Biology precedes, culture transcends: an evolutionarist's view on human nature", *Zygon* 33(4) 507–23.

Ayala F. J., 2007, *Darwin's Gift to Science and Religion*, Washington D.C.: Joseph Henry Press.

Ayala F. J., 2009, *L'evoluzione. Lo sguardo della Biologia*, Milano: Jaca Book.

Balter M., 2009, "On the origin of art and symbolism", *Science* 323: 709–11.

Barrow J. D., Tipler F., 1986, *The Anthropic Cosmological Principle*, New York: Oxford University Press.

Bergson H., 1941, *L'évolution créatrice*, Paris: Presses Universitaires de France.

Cavalli Sforza L., Feldman M. W, 1981, *Cultural Transmission and Evolution,* Princeton, N. J.: Princeton University Press.

Coppens Y., 2001, "Préface" to Prat S., Marchal F. (eds.), *Les premiers représentants du genre Homo en Afrique*, Paris: Artcom.

Coppens Y., 2006, *Histoire de l'homme et changements climatiques*, Paris: Fayard.

Dalla Porta N., Secco L., 1991, "Il principio antropico in fisica e cosmologia", *Il Futuro dell'uomo* 18(2) 61–110.

Darwin C. R., 1871, *The Descent of Man, and Selection in Relation to Sex*, London: John Murray.

D'Errico F., Sorressi M., 2009, "I comportamenti sibolici dei Neandertaliani e l'origine della modernità culturale", in Facchini F., Belcastro M. G., *La lunga storia di Neandertal. Biologia e comportamento*, Milan: Jaca Book, 239–56.

Deacon T. W., 1997, *The Symbolic Species. The Coevolution of Language and the Human Brain*, London: Allen Lane the Pinguin Press.
De Lumley H., 2004, "Introduction" to De Lumley H., Beyene Y. (eds.), *Les sites préhistoriques de la région de Fejej, Sud-Omo, Ethiopie, dans leur contexte stratigraphique et paléontologique*, Paris: Éditions Recherche sur les Civilisations.
Dobzhansky T., 1967, *The Biology of Ultimate Concern*, New York: New American Library.
Dobzhansky T., Boesiger E., 1983, *Human Culture. A Moment in Evolution* (edited and completed by B. Wallace), New York: Columbia University Press.
Elton C., 1927, *Animal Ecology*, New York: McMillan.
Facchini F., 1988, "Culture et spéciation dans la phylogénèse humaine", *Comptes Rendus de l'Académie des Sciences* 307(II): 1573–76.
Facchini F., 1991, "The roots of ethics: an anthropological approach", *Human evolution* 6: 461–68.
Facchini F., 1993, "La culture dans l'évolution humaine. La Vie des Sciences", *Comptes Rendus de l'Académie des Sciences* 10: 51–66.
Facchini F., 1999, "Planning and symbolism as survival strategies", in Ullrich H. (ed.), *Hominid evolution. Lifestyle and survival strategies*, Gelsen-Kirchen/Schwelm: Archaea.
Facchini F., 2000, "Symbolism in Prehistoric Man", *Coll. Anthropol.* 24: 541–53.
Facchini F., 2001, "La culture, niche écologique de l'homme", in Coppens Y. (ed.), *Origine de l'homme, Realité, Mythe, Mode,* Paris: Artcom, 209–26.
Facchini F., 2006, "Culture, speciation and the genus Homo in early humans", *Human Evolution* 21: 51–57.
Facchini F., 2007, "Anthropological roots of ethics", *Journal of Anthropological Sciences* 85: 223–25.
Facchini F., 2008, *Le sfide dell'evoluzione. In armonia tra scienza e fede*, Milan: Jaca Book.
Falk D., 1983, "Cerebral cortices of East African early Hominids", *Science* 222: 1072–74.
Ferembach D., 1986, "Conclusions" to Ferembach D. (ed.), *Les processus de l'hominisation*, Paris: CNRS, 357–66.
Franceschelli O., 2009, *Darwin e l'anima*, Roma: Donzelli Editore.
Girotto V., Pievani T., Vallortigara G., 2008, *Nati per credere,* Torino: Codice Edizioni.
Jelinek K., 1981, "Was Homo erectus already sapiens?", in Ferembach D. (ed.), *Les processus de l'hominisation*, Paris: CNRS, 85–90.
John Paul II, 2003, "Message to the Pontifical Academy of Sciences" (22 October 1996) in *Papal Addresses to the Pontifical Academy of Sciences 1917–2002*, Vatican City: Pontifical Academy of Sciences, 273
Laland K. N., Odling-Smee J., Myles S., 2010, "How culture shaped human genome: bringing genetics and the human sciences together", *Nature Genetics* 11: 137–49.
Leakey L. S. B., Tobias P. H., Napier J., 1964, "A new species of the genus *Homo* from Olduvai Gorge", *Nature* 202: 7–9.
Leclerc M., 2008, "L'unità finale del cosmo e la teleologia", in Mantovani M., Amerise M. (eds.), *Fede, cultura e scienza*, Città del Vaticano: Libreria Editrice Vaticana, 235–48.

Maritain J., 1973, *Approches sans entraves*, Paris: Fayard.

Marshack A., 1988, "Neanderthals and the human capacity for symbolic thought: cognitive and problem solving aspects of Mousterian symbol", in *L'Homme de Neandertal. Vol. 5 La pensée*, Liège: ERAUL, 57–91.

Martelet G., 1998, *Evolution et Création,* Paris-Montreal: Cerf-Mediaspaul.

Odum E. P., 1983, *Basic Ecology*, Philadelphia: CBS College Publishing [ital. trans. *Basi di Ecologia,* 1988, Padova: Piccin].

Pievani T., 2006, *Creazione senza Dio,* Torino: Einaudi.

Piveteau J., 1991, *L'Apparition de l'homme,* Paris: O.E.I.L.

Richerson P. J., Boyd R., 2005, *Not by Genes Alone: How Culture Transformed Human Evolution?,* Chicago: University of Chicago Press.

Ricoeur P., 1960, *La symbolique du mal,* Paris: Aubier.

Ricoeur P., 1959, "Le symbole donne à penser", *Esprit* Juillet-Aout, 7–8.

Ries J., 1993, *Le Religioni,* Milano: Jaca Book.

Roche H., 2001, "Comportément technique au Plio-Pleistocène", in Prat S., Marchal F. (eds.), *Les premiers représentants du genre Homo en Afrique*, Paris: Artcom, 41–44.

Roche H., Delagnes A., Brugal J. P., Feibel C. S., Kibunjia M., Mourre V., Texier P. G., 1999, "Early stone production and technical skill 2.34 Myr ago in West Turkana", *Nature* 399: 57–60.

Schrenk F., Bromage T. G., Betzler C., Ring U., Juwayeyi Y. M., 1993, "Oldest Homo and Pliocene Biogeography of the Malawi Rift", *Nature* 365: 833–36.

Schäfer J., 1996, "Die Wertschätzung außergewöhnlicher Gegenstände (non-utilitarian objects) im Alt- und Mittelpaläolithikum", *Ethnographisch-Archäologische Zeitschrift* 36: 173–90.

Tattersall I., 2008, *The world from Beginning to 4000 BCE*, Oxford: Oxford University Press. (It. trans.: *Il mondo prima della storia,* Milan: Raffaello Cortina).

Teilhard de Chardin P., 1955, *Le phénomène humain*, Paris: Seuil.

Teilhard de Chardin P., 1957, *La vision du Passé*, Paris: Seuil.

Teilhard de Chardin P., 1965, *La place de l'homme dans la nature*, Paris: Seuil.

Tobias P. H., 1983, "Recent advances in the evolution of the Hominids with especial reference to brain and speech", in Chagas C. (ed.), *Recent Advances in the Evolution of Primates*, Vatican City: Pontificia Academia Scientiarum, 85–140.

Tobias P. H., 1996, "The evolution of the brain, language and cognition", in Facchini F. (ed.), *The First Humans and Their Cultural Manifestations*, Colloquia VII and VIII, XIII Intern. Congress Prehist. Protohist. Sciences, Forlì: Abaco.

Vanhaeren M., D'Errico F., Stringer C., James S. L., Todd J., Mienis H. K., 2006, "Middle Paleolithic shell beads in Israel and Algeria", *Science* 312: 1785–88.

Winn T., 2002, "Archaeology and cognitive evolution", *Behavioral and Brain Sciences* 25: 389.

Wilson D. S., 2002, *Darwin's Cathedral. Evolution, Religion, and the Nature of Society*, Chicago: The University of Chicago Press [It. trans. 2009, *La Cattedrale di Darwin. Evoluzione, religione e natura della società*, Roma: Fioriti].

Wilson E. O., 1978, *On Human Nature*, Cambridge: Harvard University Press [It. trans. 1980 *Sulla natura umana*, Bologna: Zanichelli].
Wood B., 1992, "Early Hominid species and speciation", *Journal of Human Evolution* 22: 351–65.
Wood B., 2001, "Définition du genre *Homo*", in Prat S., Marchal F. (eds.), *Les premiers réprésentants du genre Homo en Afrique*, Paris: Artcom, 36–40.
Wood B., Collard M., 1999, "The human genus", *Science* 284: 65–71.
Wynn T., 2002, "Archaeology and cognitive evolution", *Behavioral and Brain Sciences* 25: 389.

THE CONCEPT OF EVOLUTION AS APPLIED TO THE DEVELOPMENT OF HUMAN CULTURES

Colin Renfrew

Summary

While the notion of evolution in the Darwinian sense has offered an informative framework for the evolution of species including the human species *Homo sapiens*, it has proved less effective in offering insights or understanding of the trajectories of development subsequently followed by human societies, or human "cultures", to use the prevalent archaeological term. The distinction is drawn here between the "speciation phase" and the "tectonic phase" of human development. It is argued also that the proposed evolutionary mechanism, the "meme" or "cultural virus", does not seem to offer many valid insights for cultural developments. Symbolic behavior, dependent on the formulation of culturally accepted institutional facts, has not yet been modeled successfully within a Darwinian framework. Such modeling would be highly welcome but has yet to emerge.

1. Introduction

This short paper recognizes the aspiration that the story of human cultural development over the past 100,000 years, and more especially over the past 10,000 years, be encompassed within the broad explanatory framework of Darwinian evolution. It recognizes progress towards this aspiration, exemplified by the work of Leslie White (1959) or of Cavalli Sforza and Feldman (1981), and more recently of Stephen Shennan (2002). But while considerable work has been undertaken in expressing or translating potential explanations of culture process into a broadly Darwinian terminology (e.g. Durham 1991), few insights have yet emerged that are helpful to the archaeologist or cultural anthropologist. The notion of adaptive advantage over the several million years duration of the Palaeolithic era, when evolutionary processes led from *Australopithecus* to *Homo sapiens* has been well explored (Boyd and Richerson 1985; Mithen 1996). But such arguments have been less successful when applied to the much more stroking changes visible in the human story over the past ten to fifteen thousand years (Renfrew 2007). My critique is not of the aspiration: I share the wish to deal with the human story within a broadly "scientific" framework, and admire the success of the Darwinian approach in placing the emergence of humankind within the broader framework of the evolution of species. My concern is that this approach has so far been less rewarding than one might have expected with the human story since the emergence of our species.

2. The "speciation" and "tectonic" phases of human evolution

It is now widely recognized that the emergence of our species, *Homo sapiens*, had been accomplished, in Africa, more than 100,000 years ago. The molecular genetic record supports this conclusion, and offers quite a detailed scenario for the out of Africa dispersal of our species around 60,000 years ago (Forster 2004).

The interesting observation which follows is that in evolutionary terms, thinking specifically about the genetic basis for humankind, namely the sequence of bases composing the human genome, this was established by that time. Certainly the basic Darwinian processes of variation and natural selection did not end 100,000 years ago and are still operational today. However a baby born today is not very different genetically from one born 100,000 years ago. The variation between now and then is not significantly different in scale from the variation between living members of our species today.

It is certainly the case that we do not yet understand very well the evolutionary processes which led over the past 5 million years or so from *Australopithecus* to *Homo sapiens*. That they involved the participation of human material culture — man the toolmaker — with adaptive advantages operating selectively upon the genome, is altogether likely. That is what is sometimes implied by the term coevolution. But the process was in some senses accomplished well before the out-of Africa dispersal episode of 60,000 years ago. This is what may be termed the speciation phase in the development of our species, and it was accomplished 100,000 years ago.

From then on the human story is one primarily of cultural development. Of course the processes of mutation and genetic drift did not cease. But the differences which were visible in the different cultures of the globe some 500 years ago, before the colonization processes of the second millennium AD, are not to be ascribed to differences in the DNA of their bearers.

With the human dispersal of some 60,000 years ago the picture changes. The momentum we see in cultural evolution after that point, and in particular after the agricultural and sedentary revolutions in different parts of the world, is no longer genetically based. Cultural innovation and cultural transmission have now become the dominant mechanisms. We may refer to this phase of development, where change in the genome is no longer significant, as the *tectonic* phase, laying the emphasis upon the notion of the construction of human culture (Renfrew 2007, 97). The *Oxford English Dictionary* defines tectonics as "the constructive arts in general". This phase is characterized by new forms of engagement with the material world, and the name refers to the construction of the cultural world in which we live. The distinction now is that in the tectonic phase the genotype is broadly fixed. Within the tectonic phase the evolution that is taking place is essentially cultural evolution.

3. Evolution and culture

It is worth returning for a moment at this point to reconsider what is implied by the term "evolution". Darwin's position, at its most concise, has been summarized (Szthamáry 2006):

> Units of evolution must multiply, have heredity and possess variability: and among the heritable traits, some must affect survival and/or reproduction. If these criteria are met, evolution by natural selection is possible in a population of such entities.

When we are talking of the human speciation phase, that long period of hominin development over perhaps 5 million years leading to the emergence in Africa between 200,000 and 100,000 years ago of our own species *Homo sapiens*, the processes involved may be subsumed under the term "natural selection", as so defined. It is, however, necessary to observe that "horizontal transmission" — the borrowing of cultural features between contemporaries who may not be closely related genetically — becomes increasingly important (Cavalli-Sforza and Feldman 1981).

The term coevolution, commonly used in biology to refer to the interaction of competing species, predators and prey, hosts and parasites etc. (Slatkin and Maynard Smith 1979), has been appropriated in discussions of human evolution to refer to the interacting evolutionary processes of human genes and human culture (Lumsden and Wilson 1981, 237; Mace 2005, 4). But while it is an attractive term it may carry with it problems at the more detailed explanatory level. For the mechanisms of genetic evolution are increasingly well understood, following the elucidation of the structure of DNA, and it is not clear how the horizontal transmission process implicit in this use of the term "coevolution" operates within such a framework. At the same time, however, the insights of Darwin and of Mendel have in general been upheld in the light of molecular genetics, although they could have no detailed knowledge of its mechanisms at the molecular level.

The same might be claimed for the term "meme", conceived as the cultural equivalent of the gene, and introduced into the discussion by Richard Dawkins (1976). It has been enthusiastically adopted by several authors (e.g. Mace 2005, 4). However it carries with it implications about mechanism, indeed mechanistic implications, which may be highly dubious. It is a term which Shennan (2002) generally avoids, despite the title of his book.

One criticism of the term "meme" is that it regards the units of cultural information as single atomistic particles, perhaps traveling independently. It is the case, however, that most major innovations involve complexes of related innovations, for which the structure of the recipient society must be "ready", or "pre-adapted". Single units or members do not seem an appropriate mechanism. This being said, however, the validity of the term in some cases must be respected. It is notable, for instance, how the innovation of alphabetic script was readily adopted, after its inception in the Levant of Egypt, on a widespread basis. In some areas it replaced pre-existing (usually syllabic) systems of writing. In other areas it proved apt for the introduction of writing as a novel recording system. The inception of the alphabetic, as single, coherent system, can perhaps be modeled in this way.

That human learned behavior, or "culture" frequently does behave in a gradual way, aptly described by the term "evolution", was noted by several archaeologists in the years following the publication of Darwin's *On the Origins of Species*. Among them were General Pitt Rivers and Oscar Montelius. The evolution of languages, following Darwin's insights, was modeled in tree-like form by Augustus Schleicher (1863) and has since continued to prove amenable to phylogenetic analysis (Forster and Renfrew 2006; Gray, Drummond and Greenhill 2009). That this is so may perhaps be explained by the considerable degree to which language transfer is indeed by inheritance between generations, indeed usually from the parents. It is in those cases where horizontal transfer by "borrowing" is particularly marked that the reconstruction of linguistic trees by standard phylogenetic methods becomes more dubious (Heggarty 2006).

4. Emergent properties in human culture and society

Following the out-of-Africa dispersals of *Homo sapiens* some 60,000 years go, human populations gradually became dispersed about the various continents. It is easy to characterize the subsequent histories of regional populations on these various continents as trajectories through time of the different cultures or culture systems. And it is tempting to make a comparison or analogy between the emerging cultural diversities of these groups and the emerging differences between plant or animal species as studied by Darwin. Splitting of populations, isolation, and genetic drift are processes which can lead to the differentiation into new species. So, at a very general level the differentiation into different cultures (as defined archaeologically) of the human populations along their trajectories of development might be compared with the development of new varieties and species among plant and animal populations.

These patterns, however, are not simply those of increasing diversity. There are emergent features, which might be described as of increasing complexity (Flannery 1972). Moreover, analogous events may be noted along different and apparently independent cultural trajectories. Among these may be noted:

- The use of the bow and arrow
- The construction of boats for marine travel
- The development of formalized burial practices for the ritual disposal of the dead
- The sedentary revolution, implying also the development of farming to sustain the settled villages which developed, along with a series of cultural traits including the making of pottery, the practice of weaving etc.
- The development of figurative representation (or art)
- The development of pyrotechnolgoical skills (cooking, pottery, metallurgy, glass)
- The development of systems of trade and exchange
- The development of ritual practices which often involved the veneration of postulated beings or deities
- The emergence of urban centers or cities
- The development of systems of ownership and inheritance of property
- The emergence of commodities esteemed to be of high value (gold, feathers, jade etc.)
- The emergence of writing systems
- The emergence of formalized exchange systems involving units of value (money)

Most of these are innovations which occurred independently along several trajectories of development, on several continents. And it is major developments such as these which one would like to see developed in evolutionary terms. They can of course be described in what approximate to evolutionary terms, and most narratives in the relatively new discipline of world prehistory take such a form (Clark 1961; Renfrew 2007).

Several problems arise in their analysis. And of course it must be recognized that several in the list are relatively simple technical innovations (bow and arrow, sea-going boat) while others are much more complex. In the first place, it must be questioned in what sense the different instances in various locations, as listed, may be regarded as the "same" innovation. Agriculture in different parts of the world relies upon different species, and the domestication of one plant food is not the same as that of another.

It is tempting, however, to see an analogy here with Darwinian evolution among animal species. A major cultural development, such as the inception of metallurgy or the phenomenon of urbanization might be compared with a major innovation, archived independently by some species along different lineages. Such, for instance, would be the phenomenon of flight. The flight of birds may first have been achieved by an ancestor of all bird species today. But the flight of a mammal, such as the bat, is believed to have a different ancestry. And that of flying insects is different again. The archaeologist and anthropologist may have much to learn form the evolutionary analysis of evolutionary advances along different trajectories which may in some sense be considered comparable.

5. The cognitive dimension

Many of the advances exemplified above, and others like them are firmly rooted in the cognitive dimension. They are, moreover, the result of the experience of humans with the material world. They are emergent properties which involve the simultaneous engagement of the brain and the body (Malafouris 2004; Renfrew 2004). In this respect they cannot be modeled in molecular genetic terms, and a different approach is needed which involves certain fundamental concepts.

Among these are notions of measurement, for instance the concept of weight. There is also the notion of value, as ascribed to a commodity. These come together in a nexus which is the basis of the commercial systems which form the basis for much of the social and economic life of the modern world. The notion of the "meme" here seems, at any right at first sight, too atomistic to model the phenomenon.

For many of the properties involved are examples of what the philosopher John Searle would call "institutional facts" (Searle 1995). They depend upon a consensus in understanding among humans in a group. They are not "brute facts" about properties of the material world, but rather accepted conventions operating within the social world of the society in question. It is these novel and emergent features which are difficult to model using the available tools of evolutionary theory.

REFERENCES

Boyd R., Richerson P. J., 1985, *Culture and the Evolutionary Process*, Chicago: University of Chicago Press.

Cavalli-Sforza L. L., Feldman M. W., 1981, *Cultural Transmission and Evolution, A Quantitative Approach*, Princeton: Princeton University Press.

Clark G., 1961, *World Prehistory: An Outline*, Cambridge: Cambridge University Press.

Dawkins R., 1976, *The Selfish Gene*, Oxford: Oxford University Press.

Durham W. H., 1991, *Coevolution: Genes, Culture and Human Diversity*, Stanford: Stanford University Press.

Flannery K. V., 1972, "The cultural evolution of civilisations", *Annual Review of Ecology ad Systematics* 3: 399–425.

Forster P., 2004, "Ice ages and the mitochondrial DNA chronology of human dispersals: a review", *Philosophical Transactions of the Royal Society*, Series B 359: 255– 64.

Forster P., Renfrew C. (eds.), 2006, *Phylogenetic Methods and the Prehistory of Languages*, Cambridge: McDonald Institute for Archaeological Research.

Gray R. D., Drummond A. J., Greenhill S. J., 2009, "Language phylogenies reveal expansion pulses and pauses in Pacific settlement", *Science* 323: 479–83.

Heggarty P., 2006, "Interdisciplinary indiscipline? Can phylogenetic methods meaningfully be applied to language data – and to dating languages?", in Forster P., Renfrew C. (eds.), *Phylogenetic Methods and the Prehistory of Languages*, Cambridge: McDonald Institute, 183–94.

Lumsden C. J., Wilson E. O., 1981, *Genes, Mind and Culture: The Coevolutionary Process*, Cambridge, Mass.: Harvard University Press.

Mace R., 2005, "Introduction: a phylogenetic approach to the evolution of cultural diversity", in Mace R., Holden C. J., Shennan S. (eds.), *The Evolution of Cultural Diversity: A Phylogenetic Approach*, Walnut Creek, Cal.: Left Coast Press, 1–10.

Malafouris L., 2004, "The cognitive basis of material engagement: where brain, body and culture conflate", in DeMarrais E., Gosden C., Renfrew C. (eds.), *Rethinking Materiality, the Engagement of Mind with the Material World*, Cambridge: McDonald Institute, 53–62

Mithen S., 1996, *The Prehistory of the Mind*, London: Thames & Hudson.

Renfrew C. 2004, "Towards a theory of material engagement", in DeMarrais E., Gosden C., Renfrew C. (eds.), *Rethinking Materiality: The Engagement of Mind with the Material World*, Cambridge: McDonald Institute, 23–32.

Renfrew C., 2007, *Prehistory: The Making of the Human Mind*, London: Weidenfeld & Nicolson
Schleiicher A., 1863, *Die Darwinische Theorie und die Sprachwissenschaft*, Weimar: H. Boehlau.
Searle J., 1995, *The Construction of Social Reality*, Harmondsworth: Penguin.
Shennan S., 2002, *Genes, Memes and Human History*, London: Thames & Hudson.
Slatkin M., Maynard Smith J., 1979, "Models of coevolution", *Quarterly Review of Biology* 54: 233–63.
Szthamáry E., 2006, "Darwin for all season", *Science* 313: 306–7.
White L., 1959, *The Evolution of Culture*, New York: McGraw-Hill.

Section IV
Epistemological aspects of evolution

INTRODUCTION

Since the publication of the *Origin*, an extensive debate has focused on the ontological, methodological and epistemological import of Darwinian science. In this framework the contributions collected in this section address some topical questions, starting from the evaluation of the novelty represented by the Darwinian work, mainly with respect to classical epistemology based on physics. This is strictly connected with the commitment to methodological naturalism which most authors consider implied both in Darwin himself and in his followers. One of the main problems is to establish whether it contradicts — and to what extent — or not a profitable dialogue between science and religion today.

Most of the related questions are really long standing, and today often related to the proposal of Intelligent Design (ID). However they are not questions born within the ID movement. Rather they were raised early in evolutionary thinking. Indeed, whenever the possibility to propose an alternative "biological" paradigm was prefigured, a criticism against reductionism and its methodological assumptions was formulated. Some of the contributions of this section follow this line in proposing a conceptual framework alternative to physicalism, along different paths: process metaphysics, self-organization of living systems, mathematical interpretation of the biological complexity. The section is closed by an historical reconstruction of some epistemological antecedents of the current debate.

In the starting paper, **Robert Ulanowicz** introduces a perspective that, according to him, actually mitigates many of the ostensible conflicts between science and religion. He argues that all the main assumptions of the Newtonian physics-based scientific paradigm underwent a kind of erosion in the course of the last two centuries. Natural systems' causal closure (in the materialistic and mechanist sense), their decomposability in stable minimal units, the reversibility of their dynamics, the strict determinism linking initial and successive states as well as the universality of physical laws have been put in question due to the developments in thermodynamics, in quantum mechanics, and — with particular reference to the biological domain — in the studies of ecosystems. These fields of investigation show the widespread existence of natural systems whose dynamics are the result of irreversible *processes* in which singular, unrepeatable events continuously occur. As Darwin introduced biological evolution as a process, he is seen as one of the major authors who emphasized the deep relevance of a notion which Ulanowicz here defines as "the interaction of random events upon a configuration of constraints that results in a non-random, but indeterminate outcome." The author stresses the ontological priority of processes in nature, delineating the guidelines of an "ecological metaphysics" based on a "process ecology". In this research approach radical contingency, self-influence, historical dimension, as well as the relationship between mutuality and competition play an essential role.

Stuart Kauffman maintains the unpredictability of evolution for evolutionary theory cannot be considered in any sense epistemologically closed. Criticizing the position of Simon Conway Morris and following Richard Lewontin, the author stresses that we still lack a theory fully accounting for the links binding genotypes to phenotypes, phenotypes to selection, and selection to changes in genotypes. From a purely logical viewpoint, we cannot predict the range of pre-adaptations that may occur in different conditions (indeed, we cannot even list all the possibilities). To understand the plausible occurrence of an adaptive solution that is likely to be maintained and transmitted, we should take into consideration the interplay between Darwinian natural selection and self-organization, the latter providing opportunities and constraints for the development of any biological system to be possibly selected for.

The intervention by **Dominique Lambert** deals with the epistemological problem of assessing the effectiveness of mathematics in life sciences. In particular, it concerns: a) the explanatory power of mathematical theories as to biological phenomena including major evolutionary thresholds; b) the question of whether mathematics can play in biology the same role as in physics with respect to the issue of prediction reliability. If the peculiarities of biological phenomena are not taken into account, the uncritical employment of classically intended mathematical models may bring to some misleading formal analogies and generalizations. Rather, what is needed is starting from a renewed epistemology of physics based on the "physics of universal exponents" in turn related to the "renormalization group philosophy", as this may allow "to take into account the intrinsic complexity and plurality of levels which are constitutive of the living phenomenon and, at the same time, (...) to maintain the idea of a unifying theoretical and mathematical approach to life and evolution."

In his contribution, **Elliot Sober** holds science in general and evolutionary biology in particular to rely on a principle of methodological naturalism, according to which scientific theories should be neutral with respect to the problem of whether or not God exists. As the case of biological undirected mutations shows, scientific theories or discoveries do not entail in themselves either the theistic or the atheistic worldviews. Moreover, the scientific enterprise is not confined to what is empirically testable. Science often makes use of hypothetical statements about the constitution of the natural world that may be conventional, of theoretical assumption that still cannot be tested, and of "mixed" statements that although testable do not solely comprise pieces of physical reality, as is the case when mathematical entities are involved. The problem stemming from the Creationist perspective is in placing God as the one and only valid explanation of what is observable in nature. Nevertheless, the consideration of super-natural entities or the belief in God's intervention in the natural world do not in principle prevent the making of good science, sometimes rather allowing it as happened in Newton's experience.

Finally, in his contribution, **Massimo Stanzione** adopts an historical perspective mainly focusing on epistemological issues. The analysis begins with the ascertainment of the irreducible clash between twentieth century epistemology stemming from Frege's program and the current "naturalized" epistemology which finds its roots in Quine and was differently inspired by the pragmatistic interpretation of the Darwinian research program. Despite the last century spreading of anti-naturalist philosophical positions and the deep changes occurred in science, Darwin's original program still exerts a remarkable influence, as can be seen by taking into account that by the occasion of the first Darwinian centennial celebration (1909) epistemological debates concerned many issues sharing the same conceptual core as the questions we are facing today. From the historical viewpoint, this shows not only the enduring impact of Darwin's ideas on the intellectual environment but also the continuity of an influential naturalistic tradition of thought. However, the author also points out the difference in the current confrontation between naturalism and religion as to its ideological character and invites theological thinking to a possible renewal by virtue of the challenge represented by the evolutionary worldview.

Paolo D'Ambrosio and *Massimo Stanzione*

Process and ontological priorities in evolution

Robert E. Ulanowicz

Abstract

Charles Darwin, a fervid admirer of Isaac Newton, nonetheless described evolution as a *process*, rather than as the action of laws upon objects. Against this bold initiative, the "Grand Synthesis" of Fisher and Wright and the ensuing discoveries in molecular biology ushered in the Neo-Darwinian scenario wherein ontological emphasis has reverted to material objects and mechanisms. Other life sciences, however, continue to lend themselves more naturally to description in terms of processes. The dynamics of ecosystems, for example, can be seen to rest upon a set of fundamental postulates corresponding to the attributes of processes. Mutuality stands at the ontological core of this perspective, known as "process ecology". By comparison, competition is seen to be accidental and derivative. Unlike in the Newtonian/Darwinian schema, selection in process ecology can occur internal to the system, rather than solely via the exogenous agency of "natural selection". The monist dictum of "survival of the fittest" appears to relate to only one side of a broader Heraclitean/Hegelian agonism. Such discrepancies with orthodox evolutionary theory suggest that a far richer picture of evolution (and the ethos that it informs) may be possible by reverting to Darwin's initial instinct to describe living nature primarily as process. Adopting the process perspective mitigates many of the ostensible conflicts between science and religion.

1. Evolution, process or law?

The contributions of Charles Darwin's *Origin of the Species* to scientific discourse have been monumental, and there is any number of reasons to celebrate the anniversary of its publication. For example, Darwin's work was among the first that introduced both change and history into science — considerations that were revolutionary at the time and remain controversial today. But in this essay I wish to focus upon one of Darwin's gifts to natural philosophy that is rarely emphasized — namely, the nature of Darwin's scenario for change as *process*.

Darwin's contributions undeniably were revolutionary, but Marshall McLuhan (1964) was wont to remind us how new ideas and inventions are usually perceived in the context of what went before. Now, the early nineteenth century saw an apotheosis of the consensus for a material/mechanical view of nature (Laplace 1814) that precipitated in the wake of Newton's *Principia* — although not following Newton's personal beliefs (Dellian 1992). Darwin himself was a great admirer of Newtonian methodology and aspired to become "the Newton of a blade of grass" (Depew and Weber 1995).

It is hardly surprising, therefore, that Darwin's theory for change was interpreted in terms of the prevailing Enlightenment metaphysic. Unfortunately, the phenomenon of process does not conform well to Newtonian thinking. The tendency thus was to force a fit by emphasizing and by reinterpreting those elements in Darwin's scheme that pivot upon the material and the mechanical. Signs of this drift are apparent in Fisher and Wright's "Grand Synthesis", the mathematics of which closely paralleled that used by thermodynamicists Boltzmann (1905) and Gibbs (1901) to describe the behavior of a statistical ensemble of non-interacting particles. The retro-fit was sealed by the stunning discovery and description of DNA/RNA as the material repository for biological history.

So complete has been this Newtonian revanchement that neo-Darwinian theory is considered today by many to validate the material/mechanical metaphysic. In my own country, for example, there is widespread resistance to the teaching of evolutionary theory in public schools. Some of this opposition comes from those who dispute the very facts of evolution — the "Young Earth Creationists". I would suggest, however, that an even larger number is worried more that neo-Darwinian theory is being used by some in the classroom as a Trojan Horse to inculcate youth with a hard materialism that is inimical to the values that those parents would prefer their children to adopt.

Almost forgotten by history and lost in the heat of this conflict is the reality that *process* is the very essence of the Darwinian scenario. Furthermore, upon the notion of process it becomes possible to construct a very different metaphysic as to how nature behaves. And science is desperately in need of new foundations. To wit, in my own field of ecosystem science each of the Newtonian foundations is violated in one way or another (Ulanowicz 1999). As a result, trying to apprehend living systems in terms of what Hans Jonas (1966) has called "an ontology of death" is as fruitless as pouring new wine into old wineskins.

2. Prevailing metaphysics

Before going on, I should pause to enumerate the foundations of Enlightenment science. In doing so, I will circumvent the impossible task of describing the myriad attitudes that contemporary scientists believe concerning how nature operates. Rather, I will hearken back to the early nineteenth century — the height of Enlightenment thinking — at which time there prevailed a widespread consensus on how nature behaves. I submit that virtually all shades of current beliefs retain their roots in this formulation.

David Depew and Bruce Weber in their tome, *Darwinism Evolving* (1995), conveniently enumerate the basic assumptions:

> 1. Newtonian systems are causally *closed*. That is, only mechanical or material causes are legitimate, and they always co-occur. Other forms of action are proscribed, especially any reference to Aristotle's "final", or top-down causality.
> 2. Newtonian systems are *atomistic*. They are strongly decomposable into stable least units, which can be built up and taken apart again. Atomism combined with closure gives rise to the notion of reductionism, whereby only those causes originating at the smallest scales are of any importance.
> 3. Newtonian systems are *reversible*. Laws governing behavior work the same in both temporal directions. This is a consequence of the symmetry of time in all Newtonian laws.
> 4. Newtonian systems are *deterministic*. Given precise initial conditions, the future (and past) states of a system can, in principle, be specified with arbitrary precision.
> 5. Physical laws are *universal*. They apply everywhere, at all times and all scales. The key adverb here is "everywhere". In combination with determinism, universality leads many to believe that nothing occurs except that it be elicited by a fundamental physical law.

I hasten to add that no one today believes fully in all five tenets. For example, soon after Laplace (1814) had exulted in the absolute power of Newtonian laws, Sadi Carnot (1824) demonstrated the *irreversible* nature of physical *processes*. As I mentioned, Charles Darwin (1859) himself invoked history (i.e., irreversibility and indeterminism) into his narrative. Then at the beginning of the twentieth century relativity and quantum theories surfaced to cast serious doubts upon universality and determinism.

So it would seem that after two centuries of such erosion, the classical assumptions lie fully in tatters. Nevertheless, its frayed threads continue to hold enormous sway over contemporary science. Thus it is that closure is strictly maintained in the neo-Darwinian scenario of evolution (Dennett 1995). That atomistic reductionism continues to dominate biology is evident in the contemporary prominence of molecular biology. As for determinism, a surprising fraction of scientists today continues to eschew the reality of chance, contending instead that probability simply papers over an underlying determinacy (e.g., Bohm 1989).

3. Problems with biology

I would suggest that our inability to see beyond the material and the mechanical lies in our own narcissism, or more precisely in our preoccupation with ontogeny. As a matter of fact, when it comes to organisms, matters *do* appear so tightly constrained as to be almost mechanical. The emphasis upon mechanism leads many to look for the origin of life in the appearance of just the right molecules that suddenly and magically come alive. That is, we become intent on seeing first life only in the context of a mechanical transition from complicated dead molecules to highly constrained living beings — much like the dry bones in Ezekiel's vision that took on flesh and stood up.

If we were truly evolutionary thinkers, however, we would pay greater attention to Darwinian *process* and imagine the world prior to life as one in which constraints were inchoate and flexibility was rampant — a situation not wholly unlike what exists in ecosystems. Ecologist Howard Odum (1971), for example, argued that proto-organisms could arise *only* in the context of what might be described as a proto-ecosystem — one in which complementary reactions, such as oxidation and reduction, were physically separated in space and connected via physical transport. The activities associated with this large-scale dynamics could transfer in very natural fashion to smaller "dissipative structures", or proto organisms, in analogy with how large-scale hydrodynamic eddies spawn smaller ones. Ecological dynamics, therefore, appears to afford a far more fecund theatre in which to study non-Newtonian behaviors than does our usual preoccupation with ontogeny. This is because it is easier to discern that ecological dynamics are not determined by laws, but rather by processes (Ulanowicz 2009).

My esteemed colleague and friend, Stuart Kauffman, who preceded me to this podium has argued that not all of what we see in the living world is determined by known laws. Doubtless, some will persist in labeling his assertion another "God of the Gaps" argument. For who is to say that some biological laws, akin to those of physics, are not waiting to be discovered? Well, Walter Elsasser (1981) said as much in 1981. It was he who demonstrated that it was *patently illogical* to expect "laws" in biology analogous to those in physics. Elsasser's argument centered around the obvious heterogeneity in biological systems. That biological entities all differ in at least minor ways from each other makes it impossible to define homogeneous sets of biological entities. Early last century Alfred North Whitehead and Bertrand Russell (1913) demonstrated that the operation of laws in physics is logically equivalent to operations between perfectly homogeneous sets, i.e., groupings of entities that are totally indistinguishable from each other, such as a collection of electrons or hydrogen atoms. They showed that operations between tokens of two homogeneous sets always yielded, in determinate fashion, a member of yet another homogeneous set (*Figure 1*).

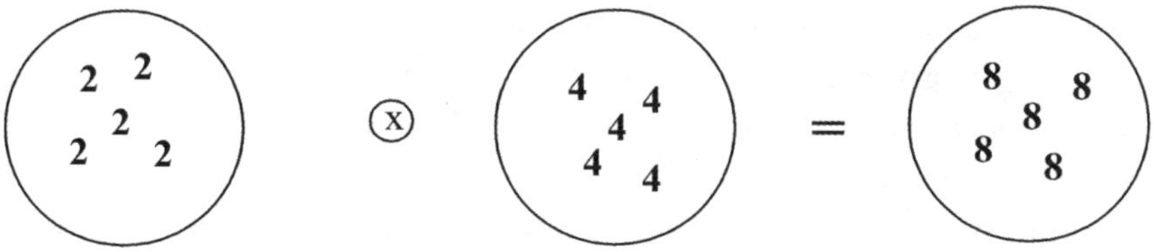

Figure 1. Operations between homogeneous sets are determinate.

By contrast, the same operations between heterogeneous groupings do not in general yield tokens of any single heterogeneous grouping (*Figure 2*). The results always exhibit some indeterminacy.

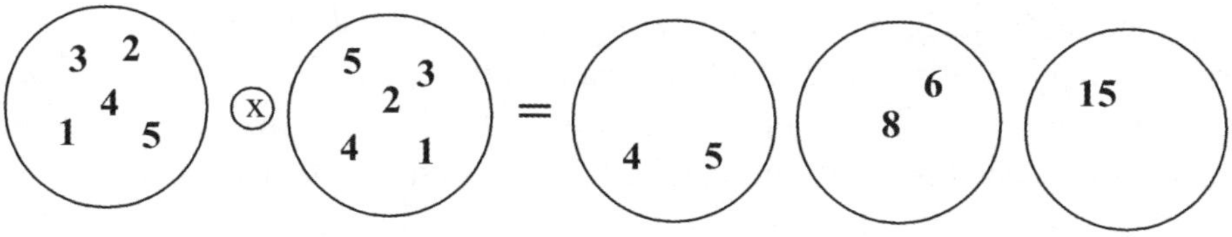

Figure 2. Operations between inhomogeneous groupings are indeterminate.

Of course, indeterminacy is hardly new. Variation is, after all, what keeps statisticians employed in biology. And, surely, one *does* observe statistical regularities throughout the living realm. Does that mean, however, that the spectre of chance is now under control — that we retain the advantage of prediction, albeit in a statistical sense? Once more, Elsasser answers rigorously in the negative by showing that probability theory does not apply universally to all chance phenomena.

Although rarely emphasized, conventional probability theory makes the tacit assumptions that chance events are simple, generic and repeatable. Elsasser (1969) demonstrated, to the contrary, that the overwhelming majority of stochastic events in biology are *totally* unique, never again to be repeated. If such an assertion sounds at first absurd, given the enormity and age of our universe, it happens to be surprisingly easy to defend. Elsasser noted that there are fewer than 10^{85} elementary particles[1] in the whole known universe, which itself is about 10^{25} nanoseconds old.[2] This means that, at the very most, 10^{110} simple events could have occurred over all physical time. It thereby follows that if any event has considerably less than 10^{-110} probability of re-occurring, it will never do so in any physically realistic time.

Now, 10^{110} is a genuinely enormous number. It might surprise some to learn, however, that it doesn't require Avogadro's Number (10^{23}) of distinguishable entities to create a number of combinations that exceeds Elsasser's limit on physical events. It doesn't require billions, millions or even thousands. A system with merely 75 or so identifiable components will suffice! It can be said with overweening confidence that any event *randomly* comprised of more than 75 distinct elements has never occurred earlier in the history of the physical universe. It follows, then, that in ecosystems comprised as they are of hundreds or thousands of distinguishable organisms, one must reckon not just with an occasional unique event, but with *legions* of them. Unique, singular events are occurring all the time, everywhere!

A necessary condition for applying probability theory to chance phenomena is that the events in question occur at least several times, so that a legitimate frequency can be estimated. Singular events, however, occur only once, never to be repeated. Any probabilities assigned to them transcend physical reality. Furthermore, these singular events constitute actual holes or gaps in the causal fabric. Akin to Heisenberg uncertainties or the Pauli Exclusion Principle, the singularities are a *necessary* part of nature, not some epistemological lacuna awaiting theoretical elaboration.

4. From laws to processes

By Elsasser's reasoning, determinism is decidedly *not* a universal characteristic of nature. I should hasten to add, however, that physical laws nonetheless remain inviolate. It's simply that no conceivable combination of the four force laws of

---

1 Today the figure is put at closer to 10^{81}.

2 A nanosecond is one-billionth of a second — the timescale of atomic reactions.

physics and the two laws of thermodynamics can possibly be stretched to cover all the possible changes amongst a complex system having, say, 35 loci for incremental change. Any particular parametric specification of laws will be satisfied by a *very large* multiplicity of possibilities. Laws continue to constrain complex biological phenomena, but they are insufficient to *determine* results. That which specifies outcomes must lie elsewhere.

As Darwin indicated, that which specifies outcomes is process. In order to maintain rigor, however, it becomes necessary to define "process" more precisely. Accordingly, we take the following as our operational definition (Ulanowicz 2009):

> *A process is the interaction of random events upon a configuration of constraints that results in a non-random, but indeterminate outcome.*

Now, the juxtaposition of "non-random" with "indeterminate" is perhaps a bit confusing, so it should prove helpful to consider a simplistic example of a process called Polya's Urn (Cohen 1976). This exercise begins with a collection of red and blue balls and an urn containing one red ball and one blue ball. The urn is shaken and a ball is blindly drawn from it. If that ball is the blue one, a blue ball from the collection is added to it and both are returned to the urn. The urn is shaken and another draw is made. If a ball drawn is red, it and another red ball are placed into the urn, etc. The first question to arise is whether a long sequence of such draws and additions would culminate in a ratio of red to blue balls that converges to a limit. It is rather easy to demonstrate that after some 1000 or so draws, the ratio indeed converges to the close

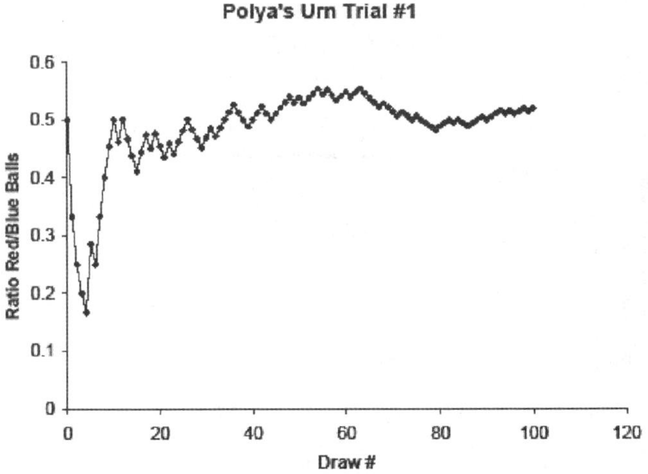

Figure 3. Polya's Urn, Trial #1 after 100 draws.

neighborhood of some constant, say 0.54591, as shown in *Figure 3*. That is, the ratio becomes progressively *non-random* as the number of draws grows.

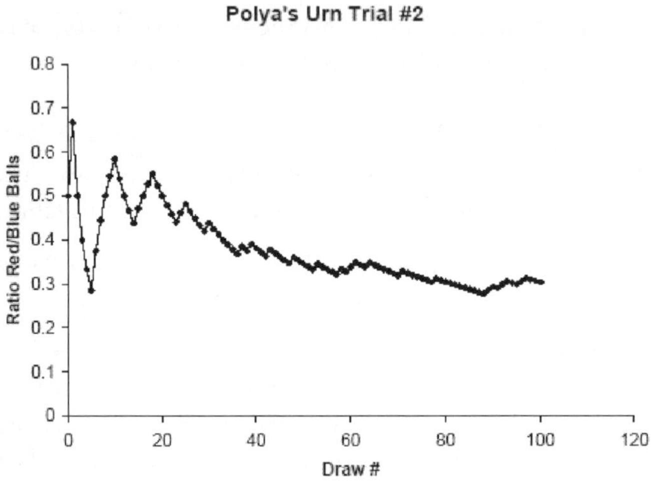

Figure 4. Polya's Urn, Trial #2 after 100 draws.

That the system does not converge closely to 0.5000 prompts a second question, namely, what would happen if the urn were emptied and the starting configuration recreated? Would the subsequent series of draws converge to the same limit as the first? It almost certainly will not. After a second 1000 draws it might approach a limit in the vicinity of 0.19561 (*Figure 4*). The Polya process is clearly *indeterminate*.

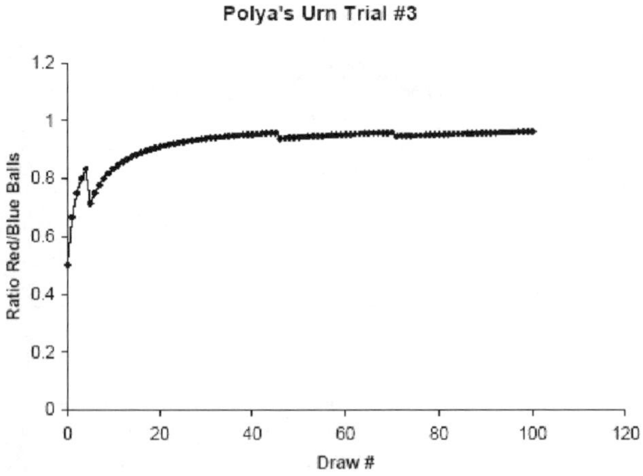

Figure 5. Polya's Urn, Trial #3 after 100 draws.

Repetition of the process many times reveals that the ratio of balls is progressively constrained by the particular series of draws (history) that have already occurred.

We note further that some histories converge to behaviors that are difficult to distinguish from mechanical, law-like dynamics interrupted by occasional noise, as in *Figure 5*. The possibility thus arises that scientific laws might be limiting forms of prior, less constraining processes (Chaisson 2001).

For later reference, we emphasize three features of this artificial, simplistic process:
(1) It involves chance.
(2) It involves self-reference.
(3) The history of draws is crucial to any particular series.

5. Order from natural processes

Of course, Polya's Urn is but an artificial process. Gregory Bateson (1972), however, provided a clue as to how natural processes could impart order to noisy affairs. He noted that the outcome of random noise acting upon a feedback circuit is generally non-random. Following his lead, I now draw your attention to a particular form of feedback — autocatalysis (Ulanowicz 1997). By "autocatalysis" I am referring to any instance of a positive feedback loop wherein the direct effect of every link on its downstream neighbor is positive (*Figure 6*).

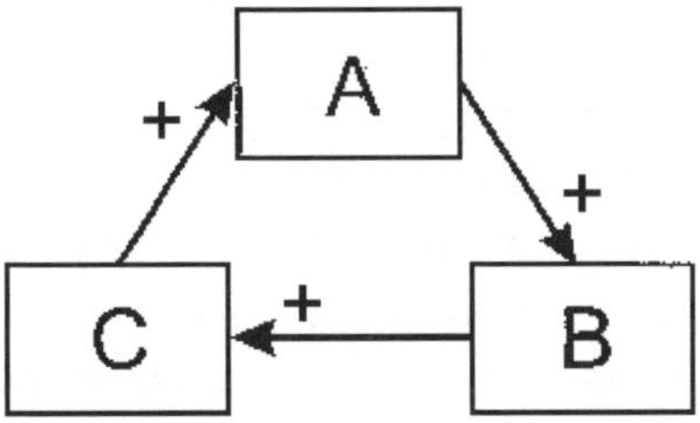

Figure 6. A three-component autocatalytic configuration of processes.

A convenient example of autocatalysis in ecology is the community that forms around the aquatic macrophyte, *Utricularia* (Ulanowicz 1995). All members of the genus *Utricularia* are carnivorous plants. Scattered along its feather-like stems and leaves are small bladders, called utricles (*Figure 7a*). Each utricle has a few hair-like triggers at its terminal end, which, when touched by a feeding zooplankter, opens the end of the bladder, and the animal is sucked into the utricle by a negative osmotic pressure maintained inside the bladder. In nature the surface of *Utricularia* plants is always host to a film of algal growth known as periphyton. This periphyton serves in turn as food for any number of species of small zooplankton. The autocatalytic cycle is closed when the *Utricularia* captures and absorbs many of the zooplankton (*Figure 7b*).

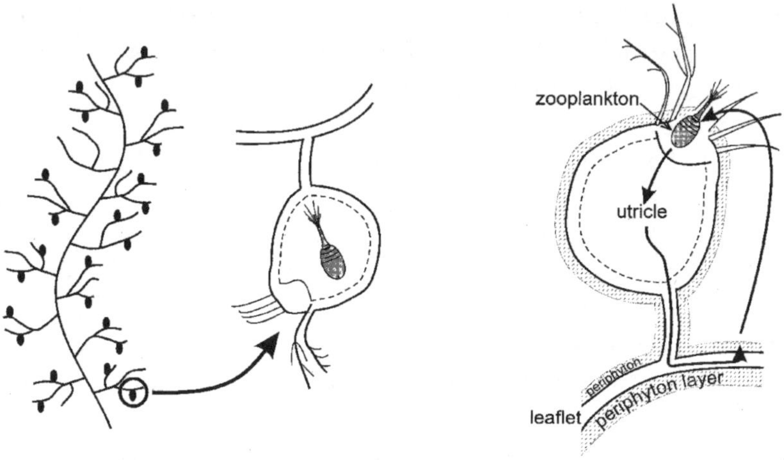

Figure 7. (a) Stem of *Utricularia* with closeup of utricle. (b) The autocatalytic processes inherent in the *Utricularia* system.

A seminal feature of autocatalysis is that it exerts selection pressure upon all of its components and any of their attendant mechanisms. Any change in a characteristic of a component that either makes it more sensitive to catalysis by the upstream member, or a better catalyst of the element that it catalyzes, will be rewarded. Other changes will at best be neutral, but more likely will be decremented by the feedback. A very important aspect of selection is that it re-enforces changes which bring more material or energy into a participating element, resulting in what can be called (in Newton's word) "centripetality" (*Figure 8*).

I believe that one cannot overstate the importance of centripetality to the phenomenon of life. Conventional Darwinism, for example, conveniently overlooks

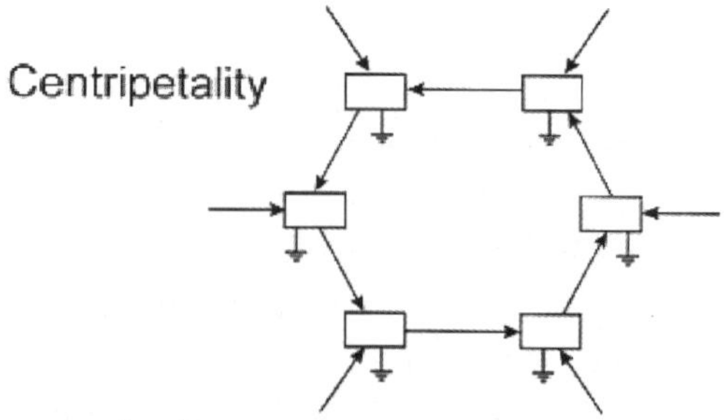

Figure 8. Autocatalysis induces centripetality.

the role of "striving" in evolution (Haught 2003). While all the various organisms are competing with one another in epic struggle, one is pressed to ask what accounts for their drive? Such striving is considered epiphenomenal to most Darwinian accounts, but here's what Bertrand Russell (1960) had to say on the topic:

> *Every living thing is a sort of imperialist, seeking to transform as much as possible of its environment into itself and its seed … We may regard* the whole of evolution *as flowing from this 'chemical imperialism' of living matter.* (Emphasis by author)

It is clear that by "chemical imperialism" Russell is identifying centripetality; and, from the perspective of systems ecology, he correctly places it at the very core of evolution.

Equally important is that centripetality is a prerequisite for competition. Without the generation of centripetality at one level, competition cannot arise at the next. Mutuality is essential; competition is an accidental consequence. To see how centripetality induces competition, we regard the sequence in *Figure 9.* In the second graph element D appears spontaneously in conjunction with A and C. If D is more sensitive to A and/or a better catalyst of C, then the ensuing dynamics of centripetality will so favor D over B, that B will either fade into the background or disappear altogether. That is, selection pressure and centripetality can guide the replacement of elements.

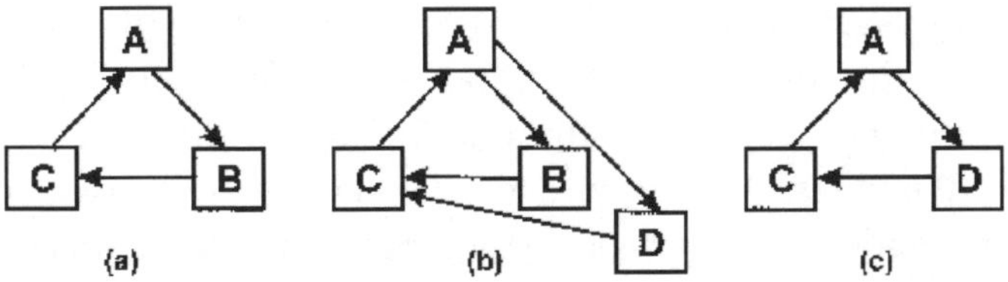

Figure 9. Centripetality induces competition.

Let us pause here to consider the ramifications of this ontological inversion. Although many feel it impossible to proceed from an existential "is" to a normative "ought", the ethos of a human community often is coupled with how it perceives nature. Thus, while emphasis upon competition is likely to promote that form of behavior, quite another prescription might follow from the recognition that mutual beneficence (the drive behind centripetality) lies at kernel of life. Corresponding moral directions are likely to be as disparate as those espoused by Thomas Huxley and Giovanni di Fidenza.

Returning to the sequence in *Figure 9*, we see how C might be replaced by E and A by F, so that in the long run, the lifetime of the autocatalytic configuration can exceed that of any of its components or their attendant mechanisms. Such supervenience by the whole over its parts explicitly contradicts the Newtonian dictum of *closure* (Clayton 2004). In fact, the other Newtonian postulates fare no better. We mentioned earlier how *determinism* is a chimera in complex systems. The asymmetric directionality in autocatalysis makes the system highly *irreversible*. The fact that each component develops in the context of its co-participants renders them all highly co-dependent over the course of time, so that the organic complex is no longer amenable to *atomistic* decomposition. Finally, the domain of any individual process is hardly *universal*, being circumscribed in time and space and subject to mitigation by processes at other levels.

6. An ecological metaphysic

As they pertain to ecosystems, all five Newtonian postulates appear ill-fitted to the description of living dynamics. The time is ripe for an entirely new, but wholly naturalistic, metaphysic — an ecological metaphysic, so to speak. In particular, the new metaphysic requires that we shift our focus from laws and objects to configurations of processes. Here we invoke the three features of the Polya process that I earmarked earlier — namely that process requires chance, self-influence and history.

Our first postulate establishes chance as a reality:

(1) *Radical Contingency*: Nature in its complexity is rife with singular events.

Organic systems are constantly being exposed to unique contingencies, but due to the self-stabilizing properties of autocatalysis, most of these events do not upset the prevailing dynamics. A miniscule few, however, may carry a system into a wholly different mode of *emergent* behavior – now perceived as an entirely *natural* phenomenon (Ulanowicz 2007).

The strictures of closure and atomism do not allow systems to maintain their integrities and grow (Ulanowicz 2009). By contrast, autocatalytic action, a particular form of self-influence, is capable of imparting form and pattern to nature. Accordingly, we replace closure and atomism by allowing for

(2) *Self-Influence*: A process in nature, via its interaction with other natural processes, can influence itself.

Thirdly, in place of reversibility we recognize, as did Darwin, that the system must retain some record of its past configurations, that is, it must possess a

(3) *History:* The effects of self-influence are usually constrained by the culmina-

tion of past such changes as recorded in the configurations of living matter.

In a scientific world bent on materialism, our conception of history tends to be dominated by DNA and similar molecular forms. But our analysis of competition suggested that the first records of organic history were more likely written into the topologies of stable, long-lived configurations of processes.

These three postulates, then, constitute a natural platform from which to cast an ecological perspective on life. In addition, the precepts spawn two corollary tenets: The first is that agency in the developmental scenario resides more with configurations of processes than with objects. Life itself is closely identified with configurations of processes. For example, Enzo Tiezzi (2006), Professor of thermodynamics and part-time hunter, asked what was different about a deer that he had just killed from the one which had been alive three minutes earlier? Its mass, form, bound energy, genomes — even its molecular configurations — all remained virtually unchanged immediately after death. What had ceased with death and was no longer present was the configuration of processes that had been coextensive with the animated deer — the very agency by which the deer was recognized as being alive.

Secondly, one can discern two opposing propensities in ecodynamics. Autocatalysis supplies the animation for systems to grow and maintain themselves. Opposing this drive is the well-known effect of the second law to degrade and dissipate existing structures. This perspective is hardly new. Diogenes reported how Heraclitus saw nature as the outcome between agonistic tendencies that build-up as opposed to those that tear-down. The direct conflict between these drives ameliorates at higher levels, however: Without the action of radical contingency, novel structures could never emerge. Conversely, larger, more constrained structures perforce dissipate more resources.

The three foregoing postulates and the two derivative observations constitute the crux of what I have called, for want of a better term, "process ecology" (Ulanowicz 2004, 2009).

7. Ontological priorities

Again it must be emphasized that these new postulates in no way abrogate any scientific laws or phenomenology that have accrued over the past three centuries. Experience with homogeneous, rarefied and weakly-interacting systems led, however, to a framework of knowledge that was found wanting whenever variety and complexity overwhelm the ability of law to specify outcomes. Most elements of the classical framework now appear as special, degenerate cases of more general processes, in analogy with how material and the laws of force are purported to have precipitated out of inchoate processes in the wake of the Big Bang (Chaisson 2001).

As with the relationship between mutuality and competition, the perspective of process ecology reorders numerous ontological priorities (Ulanowicz 2009): Process, long the stepchild in scientific discourse, now appears more fundamental than either law or material objects. Before material as we know it came to be, process was. Attributes formerly thought universal now appear circumscribed: determinism applies to a vanishingly small class of rarefied phenomena. Atomism is possible only in the purely physical realm. Monist dictums, when pursued to the extreme lead inevitably to dire consequences in a world that arises out of dual, contrasting propensities. Causalities emanating from small scales must now share the stage with supervenient influence from higher levels. Natural selection can act within a system and not just from beyond its perimeter. History is of fundamental importance. True chance quite naturally gives rise to the emergence of new phenomena. The dynamics of life are qualitatively different from that of non-living matter. Etc., etc…

8. Process ecology and theology

I wish to make it very clear that nothing I have discussed violates the constraints of methodological naturalism. There is nothing in process ecology that *in principle* would bar a metaphysical naturalist from accepting the whole of it. If what I have said happens to discomfit some, it is likely because the ecological perspective challenges cherished beliefs that some bring with them into science — especially the strict materialism that has proved such an effective weapon against transcendental convictions.

That minimalist materialism, however, excavated a gaping chasm between the transcendental and science — an abyss that swallowed the phenomenon of life. It is my hope that a process-oriented framework will allow science to address life in a direct and more rational way than hitherto was possible under the outworn metaphysic. I stress as well that bridging the chasm in no way impugns the freedom of those who wish to maintain their distance from the numinous. It simply recognizes life for what it is — a dynamic that is qualitatively different from the dead physics in which it is immersed.

It should surprise no one that any platform used to raise life out of the netherworld also mitigates several ostensible conflicts between science and believers (Ulanowicz 2004). As already noted, emergence fits comfortably into process ecology; and, by way of corollary, free will does likewise. Of course, theodicy (the problem of evil and suffering) has not been obviated, but its complexion has changed. In a world of dual dynamics, the complete elimination of petty evil and its attendant sufferings would foreclose all possibility for evolutionary change. Evil is thus seen as a problem of magnitude, rather than ontology (Ulanowicz 2004; Keller 2005; Jackelén 2009; Callahan 2003). Finally, any metaphysic that entails abundant "wiggle room" in nature (Hefner 2000) is one that cannot preclude Divine intervention. Intercessory prayer no longer defies rationality.

I would like to conclude by exhorting believers neither to fear evolution nor to try to discredit it, but rather to embrace it. Darwinian thought, freed from its procrustean box, celebrates life as distinct from death. It is too wondrous a story to be allowed to devolve into a game among little pieces of matter. Above all, I would recommend the necessity to perceive evolution as Darwin first portrayed it — as process. That perspective frees humanity from the "cosmology of despair" (Haught 2000) so fashionable in today's academic circles and allows one to entertain again a cosmology of hope.

REFERENCES

Bateson G., 1972, *Steps to an Ecology of Mind*, New York: Ballantine Books.

Bohm D., 1989, *Quantum Theory*, Mineola: Dover.

Boltzmann L., 1905, *Wissenschaftliche Abhandlungen*, Leipzig: J.A. Barth.

Callahan S., 2003, *Women who Hear Voices*, New York: Paulist Press.

Carnot S., 1824, *Reflections on the Motive Power of Heat* (translated 1943), New York: American Society of Mechanical Engineers.

Chaisson E. J., 2001, *Cosmic Evolution: The Rise of Complexity in Nature*, Cambridge, Mass.: Harvard University Press.

Clayton P., 2004, *Mind and Emergence: From Quantum to Consciousness*, New York: Oxford University Press.

Cohen J. E., 1976, "Irreproducible results and the breeding of pigs", *Bioscience* 26: 391-394.

Darwin C., 1859, *On the Origin of the Species by Means of Natural Selection*, London: J. Murray.

Dellian E., 1992, "Neues über die Erkenntnistheorie Isaac Newtons", *Zeitschrift für Philosophische Forschung* 46: 89-100.

Dennett D. C., 1995, *Darwin's Dangerous Idea: Evolution and the Meanings of Life*, New York: Simon and Schuster.

Depew D. J., Weber B. H., 1995, *Darwinism Evolving: Systems Dynamics and the Genealogy of Natural Selection*, Cambridge, Mass.: MIT Press.

Elsasser W. M., 1969, "Acausal phenomena in physics and biology: a case for reconstruction", *American Scientist* 57: 502-516.

Elsasser W. M., 1981, "A form of logic suited for biology?", in Rosen R. (ed.), *Progress in Theoretical Biology*, vol. 6, New York: Academic Press, 23-62.

Gibbs J. W., 1901 [1981], *Elementary Principles in Statistical Mechanics*, Woodbridge: Ox Bow Press.
Haught J. F., 2000, *God after Darwin : A Theology of Evolution*, Boulder: Westview Press.
Haught J. F., 2003, *Deeper than Darwin: The Prospect for Religion in the Age of Evolution*, Boulder: Westview Press.
Hefner P., 2000, "Why I don't believe in miracles." *Newsweek* 135, no. 17 (May): 61.
Jackelén A., 2009, "Creativity through emergence: a vision of nature and God", in Proctor J. (ed.), *Envisioning Nature, Science, and Religion*, West Conshohocken, Pennsylvania: Templeton Foundation Press, 180-204.
Jonas H., 1966, *The Phenomenon of Life*, New York: Harper and Row.
Keller C., 2005, *God and Power: Counter-Apocalyptic Journeys*, Minneapolis: Fortress Press.
Laplace P. S., 1814 [1951], *A Philosophical Essay on Probabilities* (Translated by Truscott F. W., and Emory F. L.), New York: Dover.
McLuhan M., 1964, *Understanding Media*, New York: Mentor.
Odum H. T., 1971, *Environment, Power and Society*, New York: John Wiley and Sons.
Russell B., 1960, *An Outline of Philosophy*, Cleveland, Ohio: Meridian Books.
Tiezzi E., 2006, *Steps towards an Evolutionary Physics*, Southampton: WIT Press.
Ulanowicz R. E., 1995, "*Utricularia's* secret: the advantage of positive feedback in oligotrophic environments", *Ecological Modelling* 79: 49-57.
Ulanowicz R. E., 1997, *Ecology, the Ascendent Perspective*, New York: Columbia University Press.
Ulanowicz R. E., 1999, "Life after Newton: an ecological metaphysic", *BioSystems* 50: 127-142.
Ulanowicz R. E., 2004, "Ecosystem dynamics: a natural middle", *Theology and Science* 2: 231-253.
Ulanowicz R. E., 2007, "Emergence, naturally!", *Zygon* 42: 945-960.
Ulanowicz R. E., 2009, *A Third Window. Natural Life beyond Newton and Darwin,* West Conshohocken: Templeton Foundation Press.
Whitehead A. N., Russell B., 1913, *Principia Mathematica*, Cambridge: Cambridge University Press.

Can evolutionary theory be epistemologically closed?

Stuart Kauffman

My contribution to the Conference centered on the implications of complexity theory to evolution. I propose to discuss this topic, but in the wider context of the Conference and its aims: What is the status of evolutionary theory, Darwinism in this Year of Darwin, and its potential theological implications? These issues fit into the framework of a set of questions bearing on whether or not evolutionary theory can be epistemologically closed such that it becomes, in principle, a predictive science. Indeed, Simon Conway Morris claims a hope for predictability when he says that, if the evolutionary tape of which Stephen Jay Gould wrote, were to be played again, all or much the same set of organisms would arise, including humans. I strongly doubt his claims, but he raises issues we must discuss. Conway Morris points, rightly, to many cases of parallelisms, or convergent evolution. A simple example would be the hydrodynamic shape of sharks and dolphins. No one doubts that selection for swimming efficiency in a water medium induces similar selection pressures for fish and marine mammals. The much larger issue is what are the "tasks" that organisms adapt to solve by natural selection, where do those tasks come from as organisms have diversified over 3.8 billion years and create niches for one another and do these tasks themselves evolve as ever new functional and selective opportunities arise in the evolution of the biosphere?

Darwin told us "descent with heritable modification, struggle for existence, hence natural selection". Is Darwinism, or Neo-Darwinism epistemologically closed such that it could be a predictive science? I will answer both "No", and more strongly, cast doubt on the possibility that the answer can ever be "Yes". This will be a radical claim.

Let me begin with an issue raised repeatedly by evolutionary biologist Richard Lewontin. He has pointed out that even at its core, evolution as envisioned by Darwin is not epistemologically closed. Lewontin has claimed that to be closed would require a theory that *i)* links genotype to phenotype, *ii)* links phenotype to selection, *iii)* links selection to change in genotype. His point is now familiar. The link from genotype to phenotype is what Scott Gilbert rightly called "the arrival of the fittest". Darwinism by itself is silent on the processes by which organisms give rise to phenotypes in the first place, and to novel phenotypes in the second place. Thus, issue *i),* the link from genotype to phenotype is barren in Darwinism itself. Much has been accomplished in the past 60 years in this area which embraces, as Gilbert and Stuart Newman rightly said, by understanding ontogeny and embryology. Below I will raise the issue of self organization and complexity theory as part of the arrival of the fittest. Here the fundamental issue is that Darwinians have thought of selection as the sole source of order in biology. A weaker version is that selection is the sole source of adaptation in biology. But this view is too weak, for beyond it is the "natural material" with which selection is able to tinker, and the ways this may both constrain and afford avenues of adaptation.

Lewontin's second issue is also fraught with difficulties: what is the link of phenotype to selection? At first glance, this issue seems rather straightforward. Darwin's finches in the Galapagos faced different foods, hence in an easily understandable way, evolved beak adaptations better enabling each species to feed on its local food stuff. An *E. coli*, deprived of beta galactosidase, and required to evolve with lactose as its only carbon source, will repeatedly evolve a new enzymatic activity from a gene now called "evolved beta galactosidase". The "task" faced by *E. coli* is clear in this controlled environment: evolve to use lactose as the sole carbon source or perish.

But the problem is far deeper. It is easiest to see the issue first in an economic context, not an evolutionary one. In the past 50,000 years, the diversity of ways humans can earn a living, and indeed the diversity of human actions, have exploded. No one could cash a check 50,000 years ago. Think of the societal innovations that had to happen to cash a check! Now let me turn to two familiar economic examples. In the early twentieth century the gasoline engine automobile replaced the horse as the major means of transportation in what is called a "Schumpeterian gales of creative destruction". When the horse disappeared as the major mode of transport and draft animal power, many other ways of making a living also perished: the blacksmith, the stable, watering troughs, the buggy, and many others. But with the automobile came its technological complements, paved roads, traffic lights, traffic police, traffic courts, bribing traffic police, an oil industry, a gasoline industry with refineries, motels, fast food restaurants, and suburbia with people living in them and now needing the automobile. These form what Robert Ulanowicz called "an autocatalytic set" of mutualisms, here in the economy. My second example is the computer, invented to calculate ballistic trajectories for naval shells in World War II. Thomas Watson Sr. of IBM said that the world might need 6 computers. He was vastly wrong. With the computer came a new opportunity to make a living by modifying computers for word processing. With word processing a new opportunity arose for sharing files. The need to share files led to the opportunity to create what is now the world wide web. In turn, the existence of the web afforded a new opportunity for electronic commerce seized by eBay. But the existence of the vast information on the web led to the opportunity to create web browsers, Yahoo and Google, and at the summit of Western civilization, Facebook.

These economic examples show us that as the economy evolves, new economic niches, and tasks to accomplish, arise in a profuse web. Below I will argue that we cannot finitely prestate these opportunities, and that no law describes this process. I will claim that the same failure of sufficient natural law arises in the evolution of the biosphere by Darwinian "exaptations". But for now, note that the same progression of ever novel opportunities to "make a living playing a 'natural game'", arises in the evolution of the biosphere. Thus Simon Conway Morris points us at a deep issue in the epistemological closure of evolution: Can we prestate how novel natural games arise and build, opportunity upon unforeseeable opportunity? If not, we cannot accomplish Lewontin's second step, the mapping of phenotype to selection.

I now turn to whether or not we can prestate the natural games, the novel niches and adaptations, that arise by Darwinian "preadaptations", or in Gould's phrase, Darwinian "exaptations". Darwin pointed out that a feature of an organism of no selective significance in the current environment might come to be of selective significance in some different environment, so be selected. I consider two biological preadaptations and one economic preadaptation. Some fish have swim bladders, sacs partially filled with air and partially filled with water. The ratio of air to water adjusts neutral buoyancy in the water column. Palaeontologists say that swim bladders arose from the lungs of lung fish. Water entered some lungs, and now there was a sac partially filled with air and water, hence poised to become a swim bladder by an exaptation. A second example are your middle ear bones, evolved from an early teleost fish by exaptations of three adjacent jaw bones. Note here that spatial relations matter. Were the three bones in the skull, spine and jaw, presumably the ear bones would not have arisen.

I now ask three questions. First, for the swim bladder and the middle ear bones, did a new functionality arise in the biosphere, neutral buoyancy in the water column and hearing? Obviously the answer is "Yes". Second, did these new functions alter the future evolution of the biosphere? Yes, new species, new proteins and so forth. Thus, these exaptations altered the becoming of the universe.

I now come to my critical third question: Do you think you could say ahead of time, or prestate, all possible Darwinian exaptations for all organisms, or, since we do not know all organisms, for humans? I have now asked thousands of people. We all agree that the answer is "No". I think I can say why we believe the answer is no. How would we list, ahead of time, all possible selective conditions? How would we know we had completed the list? How would we prespecify the one or many features in one or more organism that might constitute the preadaptation? There seem to be no ways to accomplish this.

I believe the implications are profound. Let me begin with the concept of the "Adjacent Possible". Consider a beaker with 1000 kinds of small organic molecules in it. Call these "The Actual". Now let them react by a single reaction step. Among the molecule species in the beaker may be new molecules not in the initial 1000. Call these new molecules the Adjacent Possible of the initial Actual. Over a succession of reaction cycles, the reaction system may invade ever new adjacent possibilities in the space of chemical reactions and products. Indeed the biosphere has done so for 3.8 billion years.

Now let me point to the Adjacent Possible of the biosphere. Once there were lung fish, the swim bladder was in the adjacent possible of the biosphere. Before there were multicelled organisms, the swim bladder was not in the adjacent possible of the biosphere.

Then we seem to have agreed that *we cannot list the possibilities* in the Adjacent Possible of the biosphere. I point to three deep implications, assuming this is true. First, we can never achieve the epistemological closure of evolutionary theory, for we cannot prestate the selection pressures upon phenotypes. We do not and, I claim, will never, know the full mapping from phenotype to selection, Lewontin's second step to achieve epistemological closure of evolutionary theory. Second, we cannot make probability statements about the evolution of the biosphere by exaptations. Consider flipping a fair coin 10,000 times. It will come up heads about 5000 times with a binomial distribution. But note, critically, that we knew ahead of time all the possible outcomes, all heads, all tails, and so forth. We knew all 2 to the 10,000th possible outcomes. Thus we could construct a probability measure on a frequency interpretation of probability.

Can we make probability statements for the evolution of the biosphere? Not in anything like the same sense, for we do *not* know the sample space. We not only do not know what *will* happen, we do not even know what *can* happen.

An alternative probability theory is due to Laplace: faced with N doors behind one of which is a treasure, our chance of picking the right door is 1/N. This is not a frequency based interpretation of probability and became Bayesian probabilities. But notice that we knew the number of doors, N, beforehand. We do not know N for the evolution of the biosphere. So we cannot make probability statements in anything like the same sense about the evolution of the biosphere.

My third claim is the most radical: If a natural law is a compressed description of the regularities of a process, like Newton's laws, then we can have no sufficient natural law for the evolution of the biosphere by Darwinian "exaptations". If true this says that the dream of Descartes, Galileo, Newton, Laplace, Einstein, and Schrodinger, that all that happens in the universe is describable by natural law, is false. The becoming of the biosphere is partially lawless. Yet it is also not random. The swim bladder is selected and makes selective sense in its environmental context. We confront something entirely new: the biosphere evolves in a partially lawless yet non-random way, even though we do not know how to make probability statements. We have no analogues for this is natural science, including deterministic chaos and quantum uncertainty.

I now give one example of a Darwinian preadaptation in economics. It is said to be true. Engineers were trying to invent the tractor. They needed, therefore, a huge engine block. They placed in on a succession of ever larger chasses, all of which broke. At last one of the engineers noted that the engine block was so massive and rigid that the engine block itself could be used as a chasse to hang everything else from. And that is how tractors are made and racing cars used to be made. The rigidity of the engine block put to a *new* functional use, being a chasse, is also a Darwinian preadaptation. They arise all the time in economics. Most inventions are used for purposes outside their initial inventive purpose. Thus economic and cultural evolution cannot be prestated, is partially beyond natural law, is beyond normal probability statements, yet is non-random. We must rethink our framework of ideas, for none of this exists in Newton, Einstein, or Schrodinger.

Note that in the case of these processes in evolutionary biology, these constitute the "consequential teleology" of which David Depew spoke. In the economy and culture we tend to think that we are looking at "teleology" in some simpliciter sense. But the issue is far deeper, for the opportunities, for example for web browsers, arose in some quasi-organic way as the computer allowed the web to emerge, as did the opportunity for the swim bladder in biological evolution. The coming into existence of opportunities in the economy and culture, their "ripeness" in the "world spirit" of Hegel, is itself a matter for study and understanding. Perhaps we can have no detailed law for this unfolding, yet have statistical laws.

Indeed, I want to ask whether the "opportunity" for an adaptation is a "cause" at all. I shall restate an "adaptive opportunity" as follows: "The adaptation might occur, but might not. If it occurs it will tend to be selected." Consider Aristotle's four causes: formal, final, material and efficient. Simple examples of these are: the formal cause of a house is the blueprint. The final cause of a house is, for example, my decision to build and live in it. The material cause of the house are the bricks and mortar. The efficient cause of the house are the activities of building the house. Aristotle also gave us a model of explanation in science: deduction. All men are mortals, Socrates is a a man, therefore Socrates is a mortal. With Newton, we mathematized efficient cause as deduction. Given his three laws of motion, initial and boundary conditions of balls on a billiard table, we integrate the equations of motion to find the future trajectory of the balls. But integration of his differential equations is precisely deduction.

Now consider a white tiny animal with no light sensing capacities, and an offspring with a light sensitive red spot. Let us suppose that the red spot is, in fact, an adaptation that increases the fitness of the animal bearing it. Now is the opportunity for an adaptation a cause? I think it is not a cause in any sense. First, the mutation that gave rise to the red spot may have been an acausal quantum event (on the Copenhagen and Multiple World interpretations of quantum mechanics). Second, the phrase, "tend to be selected" is a dispositional term. Think of all the possible efficient causes that might lead to the fixation of this adaptation in the population. Can we prestate the necessary and sufficient conditions for this adaptation to be selected? No. Myriad alternative efficient causes may lead to fixation and we cannot prestate them. Then we can have no law, a compact description, of the selection process. We cannot translate the dispositional "tend to" into a law. Finally, is the adaptive opportunity itself a cause? It certainly is not an efficient cause as are billiard balls hitting one another, or the sculptor chiselling the marble — a common example of an efficient cause. Nor is the adaptive opportunity a formal, final, or material cause. I conclude that adaptive opportunities in the evolution of the biosphere are not causes at all. Yet the same opportunities alter the course of evolution, hence the becoming of the universe. I think this may be fundamental, for it challenges our deepest monist belief since Empedocles and Newton, that all in the universe is what is Actual. Yet Aristotle and Alfred North Whitehead thought the "Possible" was also ontologically real. An "adaptive opportunity" is a Possibility that seems ontologically real. If so, we confront a new dualism: *Res Extensa* and *Res Potentialis*. Who knows now what this might mean?

I note that one fully consistent interpretation of quantum mechanics itself is that what is "waving" in the Schrödinger linear wave equations can be interpreted as "possibilities", which, in the famous two slit experiment, avoid the law of the excluded middle, that is, between the statements, A and not-A there is nothing. But A is true, A is possible, A is false, place the Possible exactly in the middle of the law of the excluded middle. In the two slit experiment we must say both that "It is possible that the photon did go through the first slit *and* it is possible that the photon did *not* go through the first slit." The law of the excluded middle is abrogated by quantum mechanics. Perhaps ontologically, possibilities are real. If so, the monism we have lived with since Newton in science is inadequate. This is a vast issue, warranting long consideration. And it bears on Lewontin's step from phenotype to selection, for if "adaptive opportunities" are not causes, we can, again, have no causal closure of the Actuals of evolutionary theory. Possibilities are not Actual causes.

I return to Lewontin's step I, the mapping from genotype to phenotype. It is commonly though in evolutionary biology that selection is the only source of order in biology or more weakly, of adaptations. Two of us, Stuart Newman and I, spoke against the view that natural selection is the only source of order. Newman pointed to simple reasons why many aspects of morphogenesis were natural extensions of physico-chemical processes. And why would selection and evolution not make use of these for further molding? These sources of spontaneous order were constraints and opportunities with which selection could further tinker. To what extent are these constraints and opportunities guidelines to evolution?

Consider placing phospholipids in water. They will form bilipid layers that then form hollow liposome vesicles. In essence, cell membranes are variants on liposomes. No natural selection was or is required to obtain hollow lipid vesicles given phospholipids. I now restate the point just made. To what extent do the spontaneously ordered features of organisms constrain and afford opportunities to evolution? For example, given a lipid membrane, the opportunity is in place for the establishment of membrane potentials and work cycles. Thus, cells form channels that transport Ca^{++} ions against the ion gradient from inside the cell to outside the cell using ATP to drive the process. The build-up of the Ca^{++} ions outside the cell makes that gradient steeper, hence the cell stores energy in the Ca^{++} gradient. Then, by a second channel, the Ca^{++} flows into the cell down its own gradient and pulls another solute with it. Thus, the membrane has afforded the enormous opportunity for cells to make chemiosmotic pumps that do work cycles. Is the order and adaptation here merely natural selection, or is it also the tinkered use of the natural physical-chemical properties of amphiphilic phospholipids? Obviously both self organization and selection are sources of this adaptation, hence the claim that selection is the sole source of either order or of adaptation is just wrong.

In my own presentation in the Conference, I discussed some of the roles of complexity theory — here the theory of complex non-linear dynamical systems taken as models of genetic regulatory networks. I noted that over forty years of work had shown that vast classes, or ensembles, of networks, taken as random Boolean networks, piecewise linear networks, linear ordinary differential equation networks, and most recently, continuous nonlinear differential equations with sigmoidal output functions on their inputs, spontaneously exhibit enormous order in their "ordered dynamical regime". Such networks also exhibit chaotic behavior in other parameter regimes, and most stunningly, a "critical" phase transition between order and chaos, sometimes called "the Edge of Chaos." In the ordered regime, nearby trajectories in state space flow on converging trajectories ending on dynamical attractors. It is natural to think that distinct cell types correspond to the attractors of such dynamical systems. I reported recent evidence that cell types are in fact attractors. And I discussed my efforts with my colleague, Sui Huang, to induce cancer cells to differentiate into benign, non-proliferating cells as part of cancer differentiation therapy. We have now screened 700 of 1500 small molecule drugs in an FDA approved library and found 60 "hits" that appear to induce differentiation. This is an early step, guided by the hypothesis that cell types are attractors, towards a new systematic approach to cancer therapy. Of course, very many steps are needed to realize this hope.

Examining randomly chosen members of some ensemble of networks is a new kind of statistical mechanics, averaging over a class of networks. This ensemble approach has become common in physics, where natural selection does not have to be considered. Selection plays an abundant role in evolution. Thus, the spontaneous order in non-linear dynamical models of genetic regulatory networks, like the chemiosmotic pump noted above, again suggests that we must look at a marriage between self organization which provides opportunities and constraints with which selection may tinker, and a tinkering that has brought forth the eye 11 times.

All the classes of networks noted above, random Boolean networks, piecewise linear networks, and linear ordinary differential equation models, exhibit three generic behaviors: order, chaos, and dynamical criticality. In the ordered regime, nearby states converge, the essence of homeostasis. In chaos, nearby states diverge, the butterfly effect of low dimensional chaos translated to systems with thousands of variables. In the critical phase transition regime, trajectories, on average, stay parallel. Critical networks are "magical". Critical networks maximize information transfer, maximize the diversity of organized processes they can carry out, maximize a power efficiency per unit "fuel", maximize the capacity to make reliable discrimination between past events and act reliably in the presence of noise, maximize information storage, maximize the diversity of cell types in tissue models of cells with the same regulatory network in each cell and exchanging a modest fraction of their variables, and evolve most robustly and gently. Two direct lines of experimental evidence now support the hypothesis that yeast and human cells are critical. In a recent paper, it was shown that macrophages whose Toll-like receptors were agonized one at a time, follow parallel trajectories in the gene expression space of 23,000 genes. The data are good. More, a mutated cell type is slightly chaotic, so the results are not trivial for normal cells. The second line of evidence derives from theorems showing that if a gene is deleted in a critical network, and the number of other genes that alter their behavior is studied, some genes will affect only a few genes, while a few will affect many genes. If the distribution of number of genes affected is plotted on an X axis and the number of cases affecting that many genes is plotted on the Y axis the histogram shows a fat tail to the right, to large avalanches. If the two axes are transformed to a logarithmic scale, the distribution becomes a straight line down to the right with a slope of –1.5. Stunningly, this is exactly the predicted distribution for critical networks. By contract, ordered networks would have an exponential distribution and chaotic networks would have a spike of enormous "avalanches" of genes whose activities are affected by single deletions.

Recent evidence also suggests, via arrays of extracellular electrodes, that intact awake primate brains are critical: they exhibit a power law distribution of spiking avalanches with a slope of –1.5, the signature of criticality. Perhaps cells and neural tissues are critical because such systems optimize a variety of useful properties and are attained and maintained, as a self organised feature of complex networks lying to hand for selection's sifting, by natural selection.

I remark that these two lines of experiments are entirely new in biology, for they examine the integrated behavior of thousands of genes, unlike the pathway analysis we are familiar with after 40 years of superb molecular biology. This is the true beginning of Systems Biology, in my view. We can expect such new observables to be found in abundance in the near future, examining the collective integrated behavior of the genomic regulatory system.

If cells and intact neural tissues are indeed critical, it will be another case of the marriage of self organization and selection. The spontaneous order and criticality in ensembles of genetic regulatory networks under the drift of gene duplication, rewiring of regulatory connections, and mutations to the "logic" of the regulated genes, coupled to selection for criticality as functionally optimal, with similar neural results, while not yet established with conviction, is an example of the marriage of self organization *and* selection in the mapping from genotype to phenotype. Selection is not the only source of adaptations in biology. So too is self organization. We must rethink evolutionary theory. I do not think Darwin would have minded at all.

Summary

I have argued: *i)* that evolutionary biology can never be epistemologically closed for we do not have and apparently cannot have a closed theory of the emergence of novel functionalities by Darwinian exaptations, nor, yet a theory of the opportunities afforded by the current context dependent niches of organisms with one another and the abiotic environment for ever new adaptations, hence new opportunities, in the unstatable Adjacent Possible of the biosphere. The same holds for the economy and human history. Instead we face a partially lawless yet non-random becoming where the current Actual invades its Adjacent Possible, thereby reshapes both the Actual and the next Adjacent Possible. These are new concepts in science and beyond science, for the barriers between physics and history, one with laws, the other probably without laws but with "thick descriptions" that also tell us about the real world, are both fully valid. More, if we cannot know what will happen, reason, the highest human virtue of the enlightenment, is an insufficient guide to living our lives. We need reason, emotion, intuition, imagination, story, metaphor. We need a new Enlightenment. And in place of natural law is ceaseless creativity in the biosphere, economy and culture. I believe we can use our human invented symbol, "God", to stand for this creativity and help create a shared sacred space across the globe as a global civilization of some kind, hopefully forever diverse, emerges and we tend to lapse into fundamentalisms, often religious, often hostile and even lethal.

I have also argued that *ii)* selection is not the sole source of adaptations in evolution, but that adaptation is forever a marriage of self organization and selection. All these issues require further exploration.

Some epistemological questions concerning the use of mathematics in evolution theory

Dominique Lambert

Discussing the use of theoretical physics models in biology, Evelyn Fox Keller (2007, 603) spoke about a clash of two cultures: "Physicists, she said, come from a tradition of looking for all-encompassing laws, but is it the best approach to use when probing complex biological systems?". We agree perfectly with her in the sense that no such kind of unifying equation exists, such as Dirac or Einstein-like equations, describing life as a whole or the evolution process. Many attempts to encompass life phenomena under general and a priori scheme are at best vague analogies without any predictive power and without real rooting in biological substrata. But the question now is to know if we are forced to abandon general mathematical description completely? In fact, in physics itself, the notion of universal law is broad. Compared with the description of multiscale complex systems, statistical physics has moved from a physics of fundamental laws to a physics of universal exponents related to the "renormalization group philosophy". According to this point of view what we are considering classically as "fundamental laws" are in fact only "effective laws" reflecting the properties of many connected sublevels and laws which are built only on relevant parameters characterizing in fact particular universality classes of models. The message of this paper is that we have to put aside the notion of *a priori* universal laws in order to describe life and evolution. Evelyn Fox Keller is right! But universality can be understood in the sense of multiscale physics and statistical physics. In this sense, we have of course not "the equation of life", "the theory of every evolving living being", but we could nevertheless retrieve some kind of universal and invariant properties describing life, and characterizing mathematically major regularities of the Darwinian evolution process, allowing some relevant biological testable predictions. Perhaps this is the way to overcome, or at least reduce, the clash of cultures between physics and biology.

1. How to define mathematical effectiveness in natural and human sciences?

At the beginning of the sixties, Eugen Wigner (1960) spoke about the "unreasonable effectiveness of mathematics" in physics. This effectiveness is well-known and really different from the one which can be observed in mathematical modeling of human sciences: sociology or finance. Some epistemologists, like Levy-Leblond (1982, 201), have even claimed that the role of mathematics in physics is a very specific one in the sense that it is constitutive of the physics as such. But how can we define this effectiveness? The concept is not univocal. It has several different meanings. All of them appear in biology.

When we are speaking about mathematical effectiveness, what we have first in thought, is the ability to give a representation, a formalized *description* of an empirical reality. But this meaning is very weak or poor because a description does not necessarily imply the ability to predict something interesting with respect to observations, and description is not at all an explanation (see Thom 1991). This was precisely the case with the early uses of *catastrophe theory* in ethology for example. No physiological parameters are really connected to the geometrical object used by Zeeman (1981; see Takens 1993) to model the dynamics of animal conflicts in the theory of catastrophe of René Thom (see Petitot 2006). A second example is the following. The descriptions of the diversity of biological organisms, using conformal transformations, given by d'Arcy Thompson (1992) in his beautiful book *On Growth and Form* (which influenced deeply René Thom [1997] and Conrad H. Waddington) are fascinating but at his time they were not at all connected to any reasonable biological explanation and prediction. Now, of course, it is different, if we think about the master genes and so on, which can explain some pattern changes. To take a more recent example let us consider the description of genomic correlations (see Arneodo *et al.* 2008). It is possible to describe DNA molecules as random walks on a plane. When we encounter A or T we go one step up and when we encounter G or C we go one step down. We can then check that there exist some correlations between the steps which are different for coding and non-coding parts of the DNA.[1] This can be done using some sophisticated tools such as the wavelet transform for example. But this nice description is really not an explanation. This mathematical framework cannot be biologically relevant if it can only describe the system without unveiling some

[1] If the position of the random walker is x(N), after having read N purines (A or T) or pyrimidines (C or G) the correlation function between x(0) and x(N) is $<x(N)^{2}>$ roughly proportional to $N^{-\gamma}$ with γ near 1 for coding sequences and $\gamma >> 1$ for non-coding sequences.

reasons of this difference between coding and non-coding part of the DNA molecule. A beginning of explanation of the differences occurring in genomic correlations is that the evolution pressure could be weaker on the *non-coding* parts allowing more variability and the presence of abnormal repeated sequences highly correlated. But this explanation needs another model that is not a part of the pure mathematical description. Long range correlations could also be related to the chromatin organization, but in this case we also need some additional explanatory biological and mathematical model. Thus in order to be useful in biology, mathematical models have to go far beyond a simple ability to describe an empirical situation.

Usually, the effectiveness can be measured referring to *the capacity to predict* some experimental or observational quantities. For example, the Hardy-Weinberg law can be used to compute genotype proportion in a diploid population in the absence of selection. Similarly, non-linear dynamical systems can be useful to predict the evolution of the number of predators or preys interacting in some ecological system. The effectiveness of a mathematical framework can also be *the capacity to make a retrodiction (postdiction)*. In evolution theory the reconstruction of the molecular phylogeny using algorithms and combinatorics to build evolutionary trees is something that shows the unreasonable ability of mathematics to make a retrodiction (prediction of the already flowed past).

But effectiveness of mathematical formalism can be also considered as an *ability to explain*. Indeed, to predict is not to explain. We have some nice examples with the dimensional analysis in fluid mechanics or some scale invariant laws in biology (for example Kleiber law [see Kleiber 1932; Reich *et al.* 2006] relating mass and metabolism of organisms from a cell to an elephant!). But a really useful and efficient mathematical formalism has to unveil in addition, by its own formal structure, something corresponding to the inner logic of the system. It has to shed some light on the reasons why the system is showing such and such behavior. The *backward Kolmogorov equation* (see Gardiner 1997, 55), used by Motoo Kimura for example (see Watterson 1996; Kimura 1983), not only allows us to predict the probability of fixation, after a certain number of generations, of a mutation that was initially present in the population at a given frequency. The structure of this equation derives in fact from a stochastic differential equation, which reveals something of the random way such mutation is diffusing inside the population.

A last meaning of effectiveness of mathematics can be their ability to generate some new insights. Yet false or devoid of any predictive capacity, a mathematical theory can be useful and efficient, in the sense that it is a link allowing us to think about new ideas, new formalisms which could become efficient in the future.

There are many examples in physics. Let us think of a moment of the status of several candidates for unifying quantum world and gravity (loop quantum gravity, non commutative geometry or superstrings theory). In biology we mentioned above catastrophe theory or the beautiful descriptions of d'Arcy Thompson (see Bouligand 2006) and even the Waddington (1957, 29) epigenetic landscape. While unsatisfactory at their origin, all of these have given, as fascinating analogies, some good insights to reach interesting and fruitful concepts. Effectiveness here is the ability to produce new ideas to generate potentially useful insights.

We have sketched some meanings of mathematical effectiveness. But what are the foundations of such effectiveness?

2. Some words about the origin of the (un-)effectiveness of mathematics

What are the roots of the effectiveness, or sometimes the failure of mathematics, in different empirical fields (think for example of the impossibility for financial mathematics to predict the recent crisis)? This is a deep question, too deep to be even introduced here. But we can say that this effectiveness is certainly related to the fact, as Bohr mentioned, that mathematics is a set of formal representations of conceptual relations and also to the ability of mathematics to describe *invariants* under certain class of transformations (Lambert 1997). But, precisely, you need invariants when you want to give meaning to empirical measures. Invariants are furthermore necessary conditions to recognize the presence of every element of reality. A mathematical theory, which is rich in invariants, is potentially able to be connected, somewhere, with an element of empirical reality (let us mention for example the theory of functions of one complex variable in connection with quantum mechanics or tensor analysis related to general relativity; see Penrose [2006]). Conversely, if you can extract from your empirical field some nice invariants, then the mathematization of this field is surely possible and fruitful. On the other hand, if it is difficult to extract some invariant from your domain (try to identify some invariants of the behavior of the individual financial actor, of the human psychology or even from the human history) an efficient mathematization will be difficult or impossible. This could be interesting to think about the possibility or the impossibility of modeling biological evolution using mathematical theories.

The question of the scientific and epistemological legitimacy of the use of mathematics in biology and the judgment about the abuses of mathematics in life sciences are in fact related to this criteria: "can we extract in the field some kind of invariant specific to the studied empirical domain, namely connected to some manageable pieces of reality that is relevant for experimentalists?" A second important thing is to check the ability of the framework, generating these invariants, to unveil some new explanations and to propose new observations and measurements. In fact if both things are present, mathematics will serve to give a more and more precise language to the empirical data and conversely, empirical data will shape mathematical formalism in order to adapt it progressively (like clothes on a human body) to its main features by a kind of back and forth converging process. In this sense, we will show that the nature of a biological object (which is complex, with multiple and hierarchical scales, and historical) and its specific invariants shape mathematical models forcing them to integrate multiscale and historical aspects. In fact we will do that showing the scope and the limits of several mathematical theories.

3. How mathematical modeling is constrained by biology: towards multiscale approaches

Let us try for example to describe the Indian Star Tortoise (*Geochelone elegans*)! Curiously we can approach this problem using a very recent theory that would have entertained d'Arcy Thompson. In fact the form of the shell is a very good approximation of the so-called "mono-monostatic body" the existence of which was conjectured by the famous mathematician Arnold and which was proved to exist by two Hungarian mathematicians (Várkonyi and Domokos 2006a; 2006b). This convex body also called "Gömböc" has only one stable point and one unstable point allowing the tortoise to turn easily when unfortunately placed on its shell. This nice description is not in fact a biological explanation of the emergence of the shell. Such an explanation needs to refer to descriptions of precise biochemical processes involving for example the growth factor "FGF 10"! But you also need for example some "*evo-devo*" explanation of the emergence of such shell. Then we need to know the gene network which is responsible to the building of such structure and an understanding of the physical processes leading to the emergence of an approximate Gömböc is also required. Finally we have to understand the Darwinian process leading, by mutations and selection, to this famous tortoise. Then we need descriptions at several levels of space (molecules, genes, cells, tissues) and time (evolution history). The major risk in using mathematical models in biology is to lack the multi (space and time) scales complexity. We have to respect this complexity but without complexifying to much, because then there is a symmetrical risk to get bogged down in the details.

Now we give another example. You know the famous work of Alan Turing (influenced by d'Arcy!) on the origin of coat pattern of animals. Stuart Newman and

James Dickson Murray (1989), applying this work, have done marvelous works for example on how the Leopard gets its spots. We can apply this formalism built on reaction-diffusion equations to understand for example the emergence of distinct regions in the embryo of the *Drosophila* during the early steps of its development. This reproduces apparently the good pattern but in fact it is not relevant for the biologists. "The pattern of stripes that appears during the cellular blastoderm stage of embryonic development of the fruit fly *Drosophila* looked like a pattern that could be produced by the Turing model, but experiments showed that each stripe arises independently" (Schnell, Grima and Maini 2007, 141). This is thus here a beautiful model which perfectly describes the pattern but which is not empirically adequate! We need to correct our mathematics by some deep knowledge of biochemistry. On the contrary, it was recently proved by Stephanie Sick, Stefan Reinker, Jens Timmer and Thomas Sclake, that the Turing model can be successfully used to predict the hair-follicule patterns in mice (Sick *et al.* 2006). The success of the model is related to the precise identification of the morphogens: the activator is the WNT gene and the inhibitor is the DKK gene. Thus the coupling of biochemistry, genetics (on micro-level) with reaction-diffusion (at the cell scale) has lead to some empirically adequate results. But we need also an evolutionary explanation of the emergence of these genes in life history and this implies taking into account time scales. Here too we see that a theoretical biology explanation of the pattern in mice yet non exhaustive of course must nevertheless involved multi (space and time) scales to be empirically adequate. Each integrating model will thus integrate necessarily a multiscale approach and must be shaped by the constraints of biochemistry, genetics, development, and phylogeny. Here we have to add that the Turing models can give insights to solve controversies showing for example that a change of parameters (spatial dimension and diffusion coefficient) can explain both continuous (gradualist) and discontinuous evolution (punctuated evolution à la Gould). The mathematics are here very generative (one of the facets of the unreasonable effectiveness): yet they are not completely empirically adequate, they suggest new ideas in order to find the solution of a biological controversy.

The historical feature of life, its Darwinian history can be partly modeled mathematically. At the beginning of life we have a molecular evolution that was well described by Manfred Eigen and Peter Schuster (see Eigen and Winkler-Oswatitsch 1992). The idea here is to start with a space of sequence (the combinatorial space) and to define for each sequence a fitness function leading to a "*fitness landscape*". The fitness function is in fact defined via a phenotype space (for example a secondary structure if we consider the RNA molecule). The evolution can then be seen as a jouney on a multipeak, rough, landscape. The notion of "landscape" is in fact an old one. It was firstly introduced by Sewall Wright and we know the use of such concepts for the study of the NK-models of Stuart Kauffman (models which present formal similarities with the Spin Glass models). These mathematical models of molecular evolution have been applied with success to RNA and these give some insights to understand for example the phenomenon of neutral evolution à la Kimura. Many mutations in sequence space have no effect in phenotype space.

The insights coming from the landscape concept (Kauffman 1995) are useful. In order to progress in the direction of a mathematical modeling of evolution we have to avoid being trapped by the metaphor of the static landscape. The landscape is in fact itself changing. And evolution is not a stochastic journey in a static landscape. The journey changes the landscape! "As the population moves through the fitness landscape, new peaks and valleys form, channeling its further motion" (Novak and Sigmund 2004, 793). This is a basic fact for the description of biological plasticity. In order to understand this channeling (or bottleneck) effect, we have to introduce the tools of game theory and to enter the fabulous field of the "evolutionary dynamics". The advantage of game theory is that it formulates the problem in terms of phenotypes — identified to strategies- and we know that "Darwinian fitness depends on the relative abundance of individual phenotypes within populations, and therefore game theory is the appropriate mathematical too." (Novak and Sigmund 2004, 798) But this game theory approach partly ignores the complexity of genotype-phenotype relations.

Here we realize that we have to refine mathematical models taking into account biological constraints (molecular or phenotypic ones). The choice of the static landscape theory is for example not biologically relevant. This means in fact that adding the optimization theory on a multipeak landscape to the game theory (taking into account of phenotypes-strategies) is a biological necessity that obliges us to integrate into a multiscale and complex vision many different data. We could illustrate this fact by additional examples coming from the use of graph theory in order to study biological networks in synthetic biology. Here too we could emphasize the effectiveness of this graph theoretical analysis (scale free networks, small world, random graphs…) for the understanding of biological systems and their evolutions but we could also focus on the problems and on the limits of such approaches. For example, this type of analysis becomes interesting if we couple the graph with some bio-molecular machines located at the nodes of the graph (the "real biological machines") and with a description of the interaction between such machines. We have also to couple many networks at different scales… Many interesting mathematical results come from such a perspective. For example we can quote the proof of the René Thomas conjectures concerning the multistationnarity à la Delbrück (important to understand cell differentiation for example; Thomas and Kaufman 2001a; 2001b) by Christophe Soulé (2003) of the IHESS, Paris.

The conclusion of all of this is that mathematics for biology or for evolution theory has to be shaped by real and precise constraints coming from chemistry and biology. And theses constraints oblige the mathematical framework to be a multiscale formalism and not a kind of theory of everything like the one that is awaited by theoretical physics. In a moment, we will see the consequences of this consideration for epistemology of biology.

4. How epistemology is constrained by mathematical modeling: the philosophy of multiscale systems and the end of the "clash of cultures"

Speaking about her book *Making sense of life. Explaining biological development with models, metaphors and machines*, Evelyn Fox Keller (2003, 300) said:

> The central concern of this book has been with the de facto multiplicity of explanatory styles in scientific practice, reflecting the manifest diversity of epistemological goals which researchers bring to their task. But I also want to argue that the investigation of a process as inherently complex as biological development may in fact require such diversity. Explanatory pluralism, I suggest, is now not simply a reflection of differences in epistemological cultures but a positive virtue in itself, representing our best chance of coming to terms with the world around us.

In fact the examples quoted previously could illustrate what Fox Keller said. Indeed the biological explanation cannot be considered satisfactory if we do not integrate various descriptions at multiple space and time levels. We really need a multiscale approach. This would in fact imply to renounce to some kind of unified theoretical-mathematical methods for apprehending life and its evolution confirming the "disunity of science" (p.301). But I think that we are not necessarily condemned to such an epistemological position. What is true, and we agree with Fox Keller and Michel Morange (2005), is that we are forced to renounce to a unification classically conceive starting from what Annick Lesne (CNRS, Paris VI) called a "Physics of the laws".[2] Along such lines, unification is nothing but to give general all-encompass-

[2] A. Lesne is developing a very deep and interesting "philosophy" of physics starting from the view emerging from renormalization group techniques in statistical mechanics. See for example Laguës and Lesne (2008).

ing laws or principle (for example a least action principle governing all life phenomena; or other sophisticated concepts coming from abstract mathematics: categories, such in the Robert Rosen [2008] approach).

But recent epistemological analysis of physics itself can help to escape from the trap of the so-called "physics of laws". If we followed the paths of quantum field theory, statistical mechanics and multiscale physics we discover two important elements: (i) there is a new point of view on the notion of fundamental law. In fact due to the philosophy of group renormalization what we called in the past "fundamental laws" are in fact only effective laws coming from the joint contributions of many interrelated lower levels and erasing many irrelevant micro-details; (ii) in complex systems made of many levels, we can get interesting and important information from "universal exponents" which are invariants under level changes and which characterize quantitatively and globally the systems (even if we have no access to all of their details). A nice example of this kind of exponents, also biologically relevant, is the fractal dimension of a fractal set. The transition from a "physics of laws" or a "physics of amplitudes" to a "physics of exponents", as Annick Lesne said, quoting Luciano Pietronero, allows us to overcome this irreducible disunity defended by Fox Keller. A renewed epistemology of physics can be the starting point to revisit the question of mathematization of biology and in particular of evolution. In fact this epistemology, the "philosophy of group renormalization" and on the idea of "physics of exponents", allows us to take into account the intrinsic complexity and plurality of levels which are constitutive of the living phenomenon and, at the same time, allows us to maintain the idea of a unifying theoretical and mathematical approach to life and evolution. Between a classical epistemology inherited from the myth of a physical "Theory of Everything" and the socio-epistemology of the disunity of Science, there remains a place for an epistemology of complex and multiscale systems. This can open new research fields in a philosophy of biology which excludes a kind of a priori mathematical theory of life and evolution but which enables us nevertheless to think about emerging predictable general, universal features (in the sense of "universality classes" in statistical mechanics). A task for the philosophy of science is now to build some new philosophical frameworks and concepts taking into account this multiscale nature of the life. The link with philosophy of physics will not at all be lost. We can conjecture that an epistemology of statistical physics or of "mesophysics" will successfully go, hand in hand, but without any confusion, with the new philosophy of multiscale biological systems!

References

Arneodo A., Audit B., Faivre-Moskalenko C., Moukhtar J., Vaillant C., Argoul F., d'Aubenton-Carafa Y., Thermes C., 2008, "From DNA sequence to chromatin organization: the fundamental role of genomic long-range correlations", *Mémoire de la Classe des Sciences de l'Académie royale de Belgique* (XXVIII) 2049: 7–107.

Bouligand Y., 2006, "Niveau d'organisation et morphogenèse: le point de vue de D'Arcy Thompson", in Bourgine P., Lesne A. (eds.), *Morphogenèse. L'origine des formes*, Paris: Belin, 264–85.

Eigen M., Winkler-Oswatitsch R., 1992, *Steps Towards life: A Perspective on Evolution*, Oxford: Oxford University Press.

Fox Keller E., 2003, *Making Sense of Life: Explaining Biological Development with Models, Metaphors and Machines*, Cambridge, Mass: Harvard University Press.

Fox Keller E., 2007, "Connections. A clash of two cultures", *Nature* 445: 603.

Gardiner C. W., 1997, *Handbook of Stochastic Methods for Physics, Chemistry and the Natural Sciences*, Berlin: Springer.

Kauffman S., 1995, *At Home in the Universe*, Oxford: Oxford University Press.

Kimura M., 1983, *The Neutral Theory of Molecular Evolution*, Cambridge: Cambridge University Press.

Kleiber M., 1932, "Body size and metabolism", *Hilgardia* 6: 315–51.

Laguës M., Lesne A., 2008, *Invariances d'échelle*, Paris: Belin.

Lambert D., 1997, "L'efficacité des mathématiques est-elle déraisonnable", *Comptes rendus de l'Académie royale de Belgique* (VIII) 1-6: 161–78.

Lévy-Leblond J.-M., 1982, "Physique et mathématiques", in Dieudonné J., Loi M., Thom R. (eds.), *Penser les mathématiques. Séminaire de philosophie et mathématiques de l'Ecole normale supérieure*, Paris: Seuil.

Morange M., 2005, *Les secrets du vivant. Contre la pensée unique en biologie*, Paris: Editions de la découverte.

Murray J. D., 1989, *Mathematical Biology*, Berlin: Springer-Verlag.

Novak M., Sigmund K., 2004, "Evolutionary dynamics of biological games", *Science* 303: 793–99.

Penrose R., 2006, *The Road to Reality. A Complete Guide to the Laws of the Universe*, New York: Alfred Knopf.

Petitot J., 2006, "Les modèles morphogéntiques de René Thom" in Bourgine P., Lesne A. (eds.), *Morphogenèse. L'origine des formes*, Paris: Belin, 288–97.

Reich P. B., Tjoelker M. G., Machado J.-L., Oleksyn J., 2006, "Universal scaling of respiratory metabolism, size, and nitrogen in plants", *Nature* 439: 457–61.

Rosen R., 2008, *Life Itself. A Comprehensive Inquiry onto Nature, Origin and Fabrication of Life*, New York: Columbia University Press.

Schnell S., Grima R., Maini P. K., 2007, "Multiscale modeling in biology", *American Scientist* 95: 134–42.
Sick S., Reinker S., Timmer J., Schlake T., 2006, "WNT and DKK determine hair follicule spacing through a reaction-diffusion mechanism", *Science* 314: 1447–50.
Soulé C., 2003, "Graphic requirements for multistationarity", *Complexus* 1: 123–33.
Takens F., 1993, "The work of Professor Sir Christopher Zeeman FRS", *Nieuw Arch. Wisk.* (4) 11: 251–56.
Thom R., 1977, *Stabilité structurelle et morphogenèse. Essai d'une théorie générale des modèles* (foreward by C. H. Waddington), Paris: InterEditions.
Thom R., 1991, *Prédire n'est pas expliquer* (entretiens avec Emile Noël), Paris: EsHel, La question.
Thomas R., Kaufman M., 2001a, "Multistationarity, the basis of cell differentiation and memory. I. Structural conditions of multistationarity and other nontrivial behavior", *Chaos* 11: 165–79.
Thomas R., Kaufman M., 2001b, "Multistationarity, the basis of cell differentiation and memory. II. Logical analysis of regulatory networks in terms of feedback circuits.", *Chaos* 11: 180–95.
Thompson D. W., 1992, *On Growth and Form*, Minneola: Dover.
Várkonyi P. L., Domokos G., 2006a, "Mono-monostatic bodies: the answer to Arnold's question", *Mathematical Intelligencer* (4) 28: 34–8.
Várkonyi P. L., Domokos G., 2006b, "Static equilibria of rigid bodies: dice, pebbles and the Poincare-Hopf Theorem", *J. Nonlinear Sci.* 16: 255–81.
Waddington C. H., 1957, *The Strategy of the Genes*, London: Allen & Unwin.
Watterson G. A., 1996, "Motoo Kimura's use of diffusion theory in population genetics", *Theoretical Population Biology* 49: 154–88.
Wigner E., 1960, "The unreasonable effectiveness of mathematics in the natural sciences", *Communications in Pure and Applied Mathematics* 13: 1–14.
Zeeman E. C., 1981, "Dynamics of the evolution of animal conflicts", *J. Theoret. Biol.* 89: 249–70.

Why methodological naturalism?

Elliott Sober

… any confusion between the ideas *suggested* by science and science itself must be carefully avoided.

Jacques Monod, *Chance*

1. Introduction

In the ongoing controversy between evolutionary biology and creationism, those on the science side usually maintain that evolutionary theory obeys the following principle:

Methodological Naturalism: Scientific theories should be neutral on the question of whether a supernatural God exists.

Defenders of evolutionary biology often make a second point — that this methodological principle needs to be distinguished from a thesis about what there is:

Metaphysical Naturalism: No supernatural God exists.

The thought here is that evolutionary biology has no commitment to this second form of naturalism; like scientific theories generally, evolutionary theory is committed to *ignoring* the supernatural, not to *denying* its existence.

Four clarifications are needed. The first concerns the term "supernatural." Think of nature as the totality of entities, events, and processes that have spatio-temporal location; supernatural entities, as I'll use the term, do not.[1] The second point is that methodological naturalism says nothing about the sources of inspiration that scientists may avail themselves of in developing their theories; the doctrine concerns the theories produced, not the psychological processes leading to that product. My third comment is that I have formulated these two naturalistic principles so that they are

[1] I'll avoid saying that supernatural entities are "outside" of nature, as this suggests that nature is a box and that supernatural things have a spatial location — they are outside or above the box. Nor is it part of what I mean by "supernatural" that supernatural beings are better than natural ones.

specifically about God and do not pertain to the category of supernatural entities more generally; this wider philosophical context is something to which I'll return.

My fourth clarification brings me to the quotation from Monod with which I began. Many people take evolutionary theory to suggest that there is no God.

This thought has motivated *some* atheists and *all* creationists. But here we must be careful. There is a difference between the ideas that a theory may bring to mind and what the theory *actually says*. When I hear the *Marseillaise*, I sometimes think of Humphrey Bogart, but the lyrics of the French national anthem do not entail the existence of Bogart. Methodological naturalism does not prohibit scientists from formulating theories that inspire thoughts about God, either *pro* or *con*. Methodological naturalism concerns a logical issue, not a psychological one.

What do creationists say about the relation of evolutionary biology to these two naturalisms? Sometimes they deny that evolutionary theory obeys methodological naturalism; their thought is that evolutionary theory is a materialist philosophy that embraces metaphysical naturalism and therefore violates the neutrality required by the methodological principle. However, at other times creationists agree that contemporary evolutionary biology obeys methodological naturalism, though they add that this is a defect, not as a laudable achievement. Creationists think that methodological naturalism is a shackle from which science needs to be emancipated. Refusing to consider questions about the supernatural, they maintain, is the act of an ostrich sticking his head in the sand.

My main goal in this paper is to explore the question of why science should embrace methodological naturalism. After criticizing some answers that have been suggested, I'll provide an answer that is less ambitious; I hope it is more defensible. But before I consider how methodological naturalism might be justified, I want to say something about the kind of theistic neutrality that evolutionary theory has achieved. It is obvious that the theory is not neutral with respect to *all* claims about God. For example, the theory rules out the thesis that God created life on earth about 10,000 years ago. It also is clear that the theory is inconsistent with Creationism, which is the first of the three theistic positions depicted in *Figure 1*. Creationism maintains that the complex adaptive features that we observe organisms to have could not have been produced by the evolutionary process, but were the result of God's directly intervening in nature.[2] The second position represented in *Figure 1*,

[2] Sometimes creationists describe what they take evolutionary theory to be unable to explain by distinguishing micro- from macro-evolution. For the purposes of separating the three theistic positions depicted in *Figure1*, this point does not matter.

Deism, maintains that God starts the evolutionary process and then never intervenes in what happens subsequently. Deism makes it clear that evolutionary biology does not contradict the proposition that God exists. In addition to Creationism and Deism, there is a third position to consider; it says that evolutionary theory can explain the features that organisms have, but that it is explanatorily incomplete, in that God has been at work as well. This may sound like Creationism, but it is not.

The arrows in *Figure 1* represent causal relationships. I do not assume that causality requires determinism. Smoking cigarettes causes lung cancer, but that doesn't mean that you will get lung cancer if you smoke. It doesn't even mean that smoking is part of a larger set of causal conditions that jointly suffice for lung cancer. Perhaps the universe is irreducibly chancy. As a rough approximation, we can think of causes as events that increase the probabilities of their effects. Causality can be deterministic (with complete sets of causes pushing the probabilities of their effects to a value of 1), but it need not be.

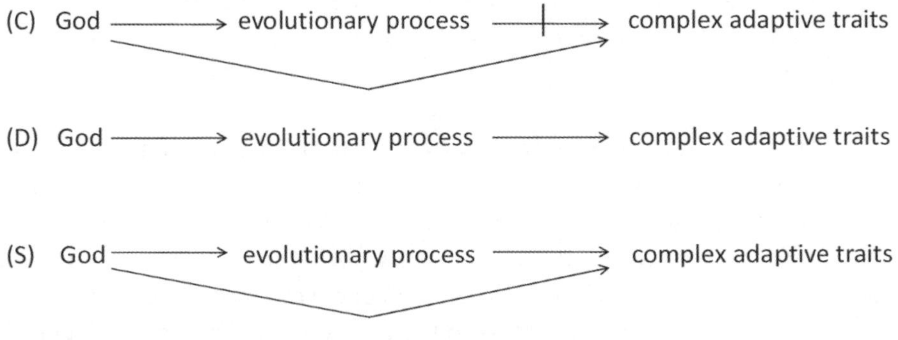

Figure 1: Three theistic positions. Creationism (C) holds that the evolutionary process is incapable of producing the complex adaptive traits of organisms that we observe, and that these are the result of God's direct intervention. Deism (D) holds that God starts the evolutionary process and never intervenes. And (S) holds that God sometimes intervenes in nature, supplementing the processes described in evolutionary theory.

Deism (as I'll understand that position) says that God builds organisms exclusively by way of the evolutionary process. This position is naturally represented in terms of a probabilistic equality:

> Pr(traits of organisms | evolutionary process & God exists) = Pr(traits of organisms | evolutionary process & God does not exist).

Once the evolutionary process is under way, what happens subsequently is unaffected by whether God exists; the process *screens off* God's existence from the traits that organisms possess. Creationism and the supplemental position (S) are both interventionist; they agree that this screening-off thesis is false.

I'll assume in what follows that the God we are discussing, if he exists, is a supernatural being. Not all conceptions of God are like this; the ancient Greeks took their gods to inhabit Mount Olympus and Spinoza thought that God is identical with nature itself. It is curious that methodological naturalism does not prohibit bringing such deities into science. Although there may be other reasons to keep them out, that is not my subject here.

Two of the three positions represented in Figure 1 involve divine intervention in the world of space and time after the universe passes beyond its first moment (if it has a first moment). But what does "divine intervention" mean? In theology, it is often understood to mean God's violating the laws of nature.[3] I do not use the term in this way. What I want to consider under the heading of (S) is the view that God *supplements* what happens in the evolutionary process without violating any laws. An intervention, as I'll understand the term, is a cause; it can trigger an event or sustain a process. Physicians do both when they intervene in the lives of their patients. Physician intervention does not entail any breakage in the laws of nature; neither does God's.[4]

2. Why evolutionary theory does not rule out an intervening God

It is not inconsistent to claim that there is more going on in the evolutionary process than is dreamt of in evolutionary biology. This is the thesis that there are "hidden variables" — causal influences on evolutionary outcomes that are not recognized in our science. If evolutionary theory were causally complete, there would be no room for this idea. However, we have no assurance that the theory covers all the facts that are causally relevant to what happens in evolution. Please note that I am not saying that there now is *evidence* that such hidden variables exist; my claim is only that they are not ruled out by current theory.

---

[3] For example, Russell's (2008) NIODA (noninterventionist objective divine action) uses the term in this way. My proposition (S) and his NIODA have much in common, despite this difference in terminology.

[4] Here I find myself in disagreement with Pennock's (1999, 195) statement that "to say that some power is supernatural is, by definition, to say that it can violate natural laws." The ability to violate natural laws might be a defining property of an *omnipotent* deity, but it isn't part of the definition of a supernatural power.

The fact that evolutionary theory does not preclude the existence of hidden variables is a simple consequence of the fact that the theory is probabilistic. To see why, let's begin with a simple analogy — the tossing of a coin. Suppose you toss a coin repeatedly and obtain evidence that justifies the following conclusion:

(1) Pr(the coin lands heads at t_2 | the coin is tossed at t_1) = 0.5.

Someone who believes that determinism is true can accept proposition (1), but will maintain that there is more involved in the process of tossing the coin than is described in this probability statement. The determinist will maintain that:

(2) Pr(the coin lands heads at t_2 | a complete description of the initial conditions at t_1) = 0 or 1.

Although propositions (1) and (2) may seem to disagree about what the coin's probability of landing heads is, in fact they do not. Both describe the coin's probability of landing heads, but they relate that event to different propositions that describe what came before. Since statements (1) and (2) conditionalize on different propositions, both can be true. It makes no sense to ask what an event's true probability is. This is like asking what the true distance to Madison is. There is no such thing. There is the distance from Chicago to Madison and the distance from New York to Madison. Distance is inherently relational. Similarly for probabilities.

To further illustrate this point, consider the Newtonian model of coin tossing that Diaconis (1998) describes. The initial conditions for a toss determine whether the coin lands heads or tails. The reason a sequence of tosses exhibits a mixture of heads and tails is that the initial conditions vary from toss to toss. To simplify matters, we assume that there is no air resistance, that the tossed coin spins around a line through its plane, and that the coin lands without bouncing (perhaps in sand). The relevant initial conditions are then fixed by specifying the values of V (the upward velocity of the tossed coin) and ω (the angular velocity, specified in revolutions per second). If V is very low, the coin doesn't go up much when it leaves the tosser's hand; if ω is very low, the coin, as Diaconis says, "rises like a pizza without turning over." Depending on the values of V and ω, the coin will turn over 0, 1, 2, 3, … times before it lands. Suppose the coin we are considering starts each tossing session by being heads up in the tosser's hand. Then, if the coin turns over 0 or an even number of times, it lands heads; if it turns over an odd number of times, it lands tails. These different possibilities correspond to the regions of parameter space shown in *Figure 2*. Starting at the origin and moving Northeast, the different stripes correspond to 0 turns, 1 turn, 2 turns, etc.

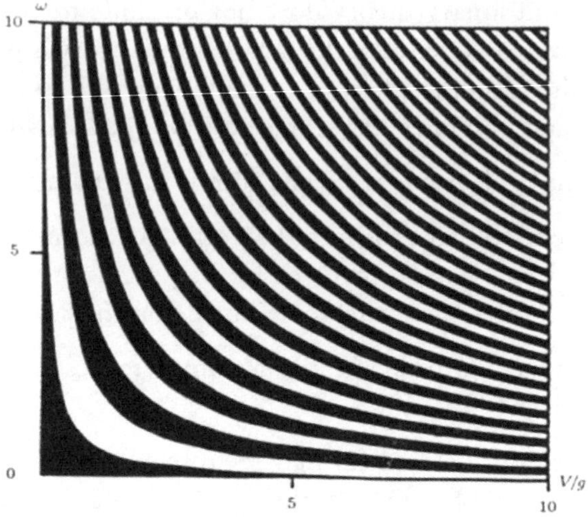

Figure 2: Coins in black regions of parameter space land heads; coins in white regions land tails (from Diaconis 1998).

In this Newtonian model, each outcome of tossing the coin is a deterministic consequence of the initial conditions, as proposition (2) asserts. However, this does not mean that proposition (1) is mistaken in its claim that the coin has a probability of landing heads of 0.5, given just the fact that it is tossed. A probabilistic model of coin tossing is consistent with the thesis that the system is deterministic. If determinism is true, there are hidden variables, not represented in the probability model (1).

Let's consider the parallel situation with respect to the claim that biologists are making when they say that mutations are "undirected" or "unguided." By this they simply mean that mutations do not occur because they would be useful to the organism. Mutations do have their causes — for example, radiation — but that is a different matter. Biologists perform experiments to test the hypothesis that mutations are undirected. Let's consider a very simple experiment that brings out some of the important features that more complex and sophisticated experiments have. Consider a species of blue organisms. Suppose these organisms would benefit from protective coloration if they were placed in a green or a red environment; being green is better than being red if the environment is green, but the reverse is true if the environment is red. Our experiment is to place some of these blue organisms in a red environment and some in a green environment and then record the frequencies with which red and green mutations occur, as shown in *Figure 3*.

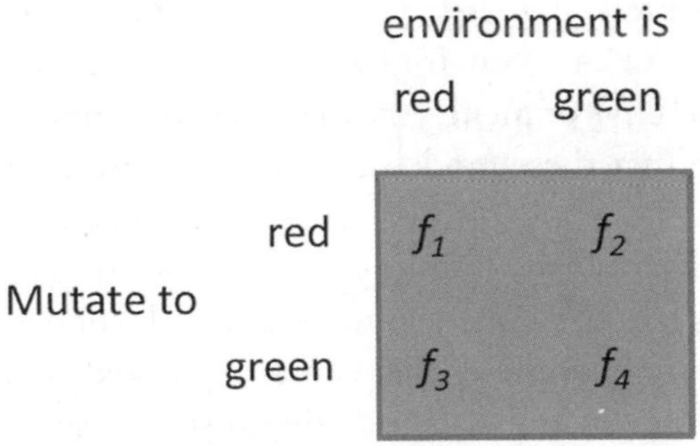

Figure 3: The frequencies with which blue organisms mutate to red or to green in different environments.

Suppose the observed frequencies are nearly identical. The conclusion a scientist will draw is that mutation probabilities are not influenced by what would be good for the organism. This should be understood for what it is — a good hypothesis whose justification comes from the frequencies of events in a sample. My point here is the same as the one about coin tosses. The hypothesis that the different mutations have the same probabilities in different environments does not rule out the possibility that there are hidden variables; perhaps each mutation that occurs is the result of its own suite of deterministic causes. If the data do not rule out hidden variables, they also do not rule out *supernatural* hidden variables.[5]

Our view of whether mutations are guided by unseen forces should be shaped by the same considerations that govern our view of whether coin tosses are influenced by unseen forces. Experts on coin tossing will tell you that coins do not land heads because this would be good for gamblers. Geneticists will tell you that mutations do not occur because they would be good for the organisms that have them. We should accept what the experts are saying, but should realize that their task is the interpretation of frequency data.

In the simple experiment that I described in which we track the mutations that blue organisms experience, the competing hypotheses describe the probabilities of

---

[5] Darwin's premier North American advocate, Asa Gray, wanted to supplement Darwin's theory with the claim that mutations are directed by a deity. Gray (1888) advised Darwin "to assume, in the philosophy of his hypothesis, that variation has been led along certain beneficial lines." I like the fact that Gray suggested that this addition be placed in the *philosophy* of the hypothesis, not in the *hypothesis* itself. In any event, Darwin declined to make the suggested addition.

different mutations in different environments. One hypothesis says that the probability of a red mutation in a red environment is greater than the probability of a red mutation in a green environment. Another says that these probabilities have the same value. Notice that these models say nothing about whether there have ever been mutations in the whole history of life that God made sure would happen. Scientists do not have a way of testing this theistic assertion. However, that does not show that it is false. Scientists sometimes use the derisive comment "not even false" to characterize hypotheses that cannot be tested. The derision can be separated from a point on which theists and atheists should agree: there is a difference between hypotheses that the evidence tells us are false and hypotheses that our data do not permit us to test.

The thesis that mutations are undirected is sometimes presented as a rock-bottom "philosophical" commitment of evolutionary biology, akin to materialism and just as central to the scientific worldview. This is a mistake. True, the thesis *is* central to biology, but it has nothing to do with materialism or theism. It is like the thesis that an organism can't synthesize vitamin *D* from sunlight or can't regenerate lost limbs. It is an empirical question whether mutation probabilities have the shape they do because of what would be good for the organism. The fact that many organisms do not experience guided mutations does not rule out the possibility that some do, in some environments. If it turns out that some organisms have the capacity to favorably adjust their mutation probabilities in the light of environmental change, then evolutionary biology will have the task of explaining why this is so. Mutation rates vary from species to species, across environments for a given species, and they have different values for different parts of an organism's genome. It is a good scientific question why this is so.

The fact that mutations are undirected should not be a problem for theists. Maybe God arranged for mutations to be undirected. And if some mutations in some organisms in some environments turn out to be directed in the sense I have described, that is no threat to atheism. Atheism has no more of a stake in mutations being undirected than it has in organisms being unable to synthesize vitamin *D* from sunlight.

Theists can of course be deists, holding that God starts the universe in motion and then forever after declines to intervene. But there is no contradiction in their embracing a more active God whose interventions into nature fly under the radar of evolutionary biology.[6]

---

[6] The probabilistic character of evolutionary theory's modeling of mutations is one reason that the theory does not rule out hidden variables, but the theory is probabilistic in another way, and this provides a second context in which hidden variables are not precluded. Here I

Divine intervention isn't part of science, but the theory of evolution does not entail that none occur.[7]

3. Should science talk only about what exists in nature?

What reasons have been offered for accepting methodological naturalism? Sometimes the claim is advanced that science, by definition, eschews discussion of the supernatural. If the point is put by saying that *natural* science says nothing about the *super*natural, the claim may sound like a tautology. But this definitional ploy accomplishes nothing. Even if "science" is defined as an activity for which methodological naturalism is true, the question remains of why time and energy and money should be devoted to doing science rather than to doing *shmience.*[8] Shmience is just like science, except that it doesn't definitionally require a commitment to methodological naturalism. As mentioned earlier, intelligent design theorists want shmience to replace science as a project to which society devotes itself. A substantive reason is needed for thinking that methodological naturalism provides good advice for inquiry. The victory that the definitional argument achieves is empty.

It is interesting to note, in this connection, that science does not, as a matter of fact, avoid postulating supernatural entities. Here I am not talking about what science says concerning the existence of God. Rather, I have in mind a different sort of supernatural entity — numbers. Evolutionary theory entails that numbers exist, and numbers are supernatural entities. Or, at least, that's the view of numbers endorsed by a certain philosophy of mathematics. Mathematical Platonism says that numbers and other mathematical objects do not have spatio-temporal location. Mathematical Platonism is not universally accepted among philosophers. However, it does have a lot to be said for it, and many philosophers think it is correct.

Here is a brief sketch of why Platonism looks plausible. Consider the claim that there are infinitely many prime numbers. This is a true statement, as any number theorist will tell you. But what are these things called numbers? What must they

---

mean the fact that biological populations are finite and so natural selection is represented as a probabilistic cause.

[7] Although evolutionary theory does not rule out hidden variables, there are results in quantum mechanics that rule out certain types of hidden variables. Theists may want to consider whether they should take issue with physics about this; they have no similar bone to pick with evolutionary biology.

[8] My "schmience" is what Plantinga (1997) calls "sience"; the argument I give against the definitional ploy is his.

be like to make the statement true? First, it is important not to confuse numbers and numerals; numerals are names for numbers. The statement about primes isn't about names; it's about the things those names name. The statement would still be true if there were no language users, and hence no names for the numbers. Indeed, the statement would still be true if there were no matter in the universe. This is what leads Platonists to claim that numbers are supernatural entities.

This may explain why many philosophers think that Platonism is a plausible account of pure mathematics, but what has this to do with evolutionary theory? The answer is that many statements in mathematized evolutionary theory entail that numbers exist. Scientists hardly notice that their models have this implication, but such models are everywhere. Consider, for example, the claim that the rates of molecular evolution in two lineages are different. The Platonistic commitments of this statement become visible when it is stated a bit more formally:

> There exists a number $d \neq 0$, such that $R_1 - R_2 = d$, where R_1 is the rate of evolution in the first lineage and R_2 is the rate in the second.

Or consider the claim that the fitness of a certain trait (*T*) in a population is frequency dependent and is a linear function of its frequency. In other words:

> There exist numbers m $\neq 0$ and b such that the fitness of trait $T = mx + b$, where x is the frequency of the trait.

These claims are subject to empirical test, and it may turn out that the evidence leads us to reject them both. Perhaps there is a molecular clock in the two lineages; perhaps trait *T*'s fitness is frequency *in*dependent. The important point is that scientists do not recoil from these two models because they entail that numbers exist. And if these two models are rejected, there are other models, with other Platonistic commitments, that evolutionary biologists embrace.

I mentioned before that the usual picture of the relationship of methodological naturalism and metaphysical naturalism is that the former does not entail the latter. Now let's ask the converse question: does the metaphysical thesis entail the methodological thesis? There is a simple argument for saying *yes*: if there are no supernatural entities, a true scientific theory cannot claim that such things exist. If the goal of science is to find true theories, the entailment relation is established. The problem with this argument is that science needs mathematics, whether or not Platonistic entities exist. If numbers do not exist, then mathematics is a useful fiction; indeed, it is an *indispensible* fiction. Scientists shouldn't worry about whether numbers exist; they should just help themselves to the assumption that they do. Whether numbers exist is something for philosophers to puzzle over.[9]

4. Are claims about the supernatural always untestable?

Another way to defend methodological naturalism is to claim that statements about the supernatural are untestable. This is an old saw, and several philosophers (e.g., Quinn 1984, Laudan 1988) have cited counterexamples. I have already mentioned the following statement:

> A supernatural deity created life on earth 10,000 years ago.

Scientists have abundant evidence for life's being much older; if so, this statement about the supernatural is testable. The same point pertains to many statements that are about numbers:

> The number of apples in the basket is prime.

Both of these example statements are about supernatural entities, but they are not *solely* about the supernatural. The former is about life on earth *and* God; the second is about apples *and* numbers. Both of these "mixed" statements are testable.

Pennock (2009, 550–551), elaborating on ideas he developed earlier (Pennock 1999), disagrees:

> ... both Laudan and Quinn cite the young-earth creationist view that God created the earth 6,000 to 10,000 years ago as a hypothesis that is testable and found to be false. But this and other examples that are offered to show the possibility of tests of the supernatural invariably build in naturalistic assumptions that creationists do not share. Confronted with the empirical evidence for an ancient earth, creation scientists dismiss the relevance of any such ob-

[9] I defined an entity's being supernatural to mean that it lacks spatio-temporal location; from this it follows that numbers are supernatural entities, at least according to Platonism. But what if we adopt a different definition of "supernatural?" Draper (2004), for example, says that a supernatural entity is one that can affect the natural world without being part of it. If numbers aren't causes, this entails that numbers aren't supernatural. I guess it follows that they aren't natural, either, if being natural requires spatio-temporal location. According to Draper's definition, numbers would be neither natural nor supernatural. In any event, the question I am addressing does not depend on whether you use Draper's definition of "supernatural" or mine. Why is methodological naturalism right to council silence with respect to supernatural deities if science is entitled to construct theories that entail the existence of mathematical objects?

> servations on the ground that God simply made the earth *appear* to be old (or "mature"). Some think of this as a test of faith so that one learns to accept the authority of the Bible over that of one's (mere) senses. The point here is that we cannot overlook or ignore, as Laudan and company regularly do, the fact that creationists have a fundamentally different notion from science of what constitutes proper evidential grounds for warranted belief. The young-earth view is certainly disconfirmed if we are considering matters under MN [methodological naturalism], but if one takes the supernatural aspect of the claim seriously, then one loses any ground upon which to test the claim.

Pennock is describing an exasperating style of argument that creationists deploy. Let us see how it applies to a simpler example. Consider the statement:

(Purple-ID) A supernatural deity caused everything in the world to be purple.

This statement is about the supernatural and it makes an observational prediction. Of course, a defender of Purple-ID might reply that things only *seem* to have colors other than purple. Friends of Purple-ID can do for their theory what friends of young-earth creationism have done for theirs. However, this is a fact about *people*, not about *propositions*. Notice the shift from propositions to people in the passage I quoted from Pennock. He begins by discussing a *proposition* (that God created the earth 6,000 to 10,000 years ago) and then shifts to a fact about how creationists defend this proposition, pointing out that "*creationists* have a fundamentally different notion from science of what constitutes proper evidential grounds for warranted belief [emphasis mine]." It is true that creationists have been unscientific, but this is a fact about them; nothing follows about the character of the theory they wish to defend. Consider a dogmatic Darwinian or a dogmatic Newtonian who argues unscientifically; this fact about them does not show that their pet theories are unscientific.

Pennock (2009, 552) thinks that the words that come out of the mouths of creationists have a radically different meaning from the same words that come out of the mouths of noncreationists. He thinks that when scientists consider "the earth is 10,000 years old" or "everything is purple," they assign the relevant terms a "naturalistic meaning," but when young-earth creationists or defenders of purple-ID utter these sentences, the sentence has a "supernaturalistic meaning." This claim of Pennock's is a claim in the philosophy of language, one that I find implausible. Creationists and anti-creationists disagree about a great deal, but that doesn't show that they assign different meanings to statements like "everything is purple" and "the earth is 10,000 years old." It is true but irrelevant that young-earth creationists and friends of Purple-ID might adopt assumptions that make it impossible to test their theory. Being dogmatic about your pet theory does not entail that the theory you are being dogmatic about takes on a meaning that differs from the meaning the theory has when contemplated by your less dogmatic associates.

So I stand by my claim that there are mixed statements that are testable; mixed statements, recall, are ones that are about both supernatural entities and things that have spatio-temporal location. But what about statements that are *purely* about the supernatural? Consider another example that Pennock (1999, 196) discusses, namely Philip Johnson's (1990) claim that "God creates for some purpose". Does this statement have implications about what occurs in nature? Perhaps not. Pennock asserts that Johnson's claim is untestable, and maybe he is right.[10] However, it does not follow that *all* statements about the supernatural are untestable.

This suggests the following thesis: even if some of the mixed statements that creationists make are testable, their core propositions are not, and this is what makes their theory untestable. One problem with this thesis is that it is hard to tell what the core propositions are. Is "God exists" the only core proposition in young-earth creationism, or is "God created the earth between 6,000 and 10,000 years ago" also part of the core? The other problem is that even if some of the core propositions in a theory are untestable, it does not follow that the theory as a whole is untestable.[11]

Let us apply this question about creationism to science itself. Is *every* statement in a scientific theory testable? Mathematized theories in biology and in other sciences entail that numbers exist. Is the existence of numbers empirically testable?

There is an irony here. The idea that scientific statements must be testable is familiar to us in part because of the large influence that Popper's (1959) views about falsifiability have had. Popper thought that testability, understood in terms of his idea of falsifiability, was the solution to the demarcation problem, which is the problem of separating scientific statements from nonscientific statements. Philoso-

---

[10] The account of testability that I develop in chapter 2 of Sober (2008) takes account of an idea that philosophers call Duhem's thesis, which says that scientific theories rarely make predictions on their own, but need to be supplemented by auxiliary propositions to do so. Perhaps "God exists" does not, all by itself, make predictions, but that is not enough to show that it can't make predictions when suitably supplemented.

[11] Consider the following two adjacent claims from Pennock (1999, 195): "... supernatural hypotheses remain immune from disconfirmation" and "creation-science does include supernatural views at its core that are not testable." The latter does not entail the former.

phers have recognized for a long time that falsifiability is a flawed account of testability (Sober 2007; 2008), but that is not the irony I have in mind. Rather, the irony is that some of Popper's contemporaries, the logical positivists, also thought that testability is a concept of central importance to science, but they denied that *every* statement in a scientific theory must be testable. Carnap (1950) and Reichenbach (1938), among other positivists, held that scientific theories often contain *conventional* elements. These are statements that are in a theory because they are useful, not because we can offer evidence that they are true. Carnap held that statements like "physical objects exist" and "numbers exist" are included in scientific theories for this reason. Physicists do not run tests to see whether physical objects exist; rather, they *assume* that there are physical objects, and then test statements like "electrons exist." And the mathematician's belief that numbers exist is not based on a mathematical proof that they do; rather, mathematicians *assume* that numbers exist, and then construct proofs of statements like "there are infinitely many primes." Contemporary discussion of methodological naturalism in connection with the debate between evolutionary biology and creationism owes a lot to Popper; the positivist heritage has largely been forgotten. This is too bad, because the idea that science uses framework assumptions that are untestable has a lot to be said for it.[12]

The positivists were not the only ones who saw that a whole theory can be testable even if some of its parts are not. This difference between whole and part is something that Popper also saw; he did so by exploring the logic of his concept of falsifiability. Popper (1959, 249) notes that a theory can be falsifiable even though some of its consequences are not. Although there is nothing paradoxical about this logical fact about falsifiability,[13] it does pose a problem for Popper's criterion of demarcation. It is hard to see how a theory can be scientific if some of its parts are not. Inspired by Popper, you may be tempted to criticize a theory by focusing on an element in the theory, declaring that element unfalsifiable, and then concluding that the whole theory is unfalsifiable. Unfortunately for those who seek

[12] Popper (1959) regarded methodological rules, like the requirement of falsifiability, as conventions. But given his methodology, a genuinely scientific statement must be falsifiable.

[13] It is standard in philosophy of science to think of theories as "closed under logical implication." This means that if a theory *T* has *C* as one of its logical consequences, then *C* is "part" of *T*. In classical logic, every theory entails a tautology. However, tautologies are not falsifiable. And even with regard to *T*'s nontautologous consequences, some of these can fail to be falsifiable even if T is falsifiable.

such silver bullets, this is not an analysis that can withstand logical scrutiny.

It is easy to miss the fact that an empirically well-confirmed scientific theory may contain assumptions that are not empirically confirmed at all. This is because there is something intuitively attractive about the following principle:

> *The Special Consequence Condition of Confirmation*: If observation *O* confirms theory *T*, and theory *T* entails that *C* is true, then *O* confirms *C*.

Confirmation does not mean that the observation *proves* that the theory is true; observational evidence hardly ever has such power in science. Rather, think of confirmation as the process of making the theory more plausible than it was before.

The Special Consequence Condition, so named by Hempel (1965, 31), may seem right, but it is not. Here's a simple example that illustrates why. You are playing poker and would dearly like to know whether the card you are about to be dealt will be the Jack of Hearts. The dealer is a bit careless and so you catch a glimpse of the card on top of the deck before it is dealt to you. You see that it is red. The fact that it is red confirms the hypothesis that the card is the Jack of Hearts, and the hypothesis that it is the Jack of Hearts entails that the card will be a Jack. However, the fact that the card is red does not confirm the hypothesis that the card will be a Jack.[14] There is strong evidence for the existence of electrons, and the existence of electrons entails that there are physical objects; however, it does not follow that there is a strong evidence for the existence of physical objects.

5. Is violating methodological naturalism a science-stopper?

Another popular way to defend methodological naturalism is to shift from a thesis about *propositions* to one about *people*. Instead of discussing the propositions that creationism entails, we are asked to consider the effect that abandoning methodological naturalism would have on people. The claim is that violating methodological naturalism is a "science stopper".[15] If we allow ourselves to talk about supernatural entities, we will stop doing serious science; the methodological principle is said to be a necessary prophylactic. To evaluate this argument, let's begin with a simple historical point. Many central figures in the Scientific Revolution thought that science needs God to explain some natural phenomena. For

[14] See Sober (2010a) for a simple explanation of why the Special Consequence Condition is mistaken within a Bayesian framework.

[15] The label for this argument is from Plantinga (1996); the argument is advanced in Pennock (1999, 292) and in Miller (2007).

example, Newton thought that the solar system would collapse without divine intervention. Newton did not always follow the precepts of methodological naturalism in his work, but this did not prevent him from doing good (or great) science. What is true about the science-stopper argument is that if you use "God wanted things to be so" as your *one and only* response to all the observations you make, you'll never do serious science. But the same is true of an obsessive devotion to the proposition "all events happen because carrots are orange." If you believe that this is the *one and only* explanation of what occurs, *that* will stop you from doing serious science. However, that is no reason to prohibit scientists from mentioning the color of carrots. Introducing God into science does not *necessarily* shut down the whole show, although it will do so if it is carried to monomaniacal extremes (as will any *idée fixe*).

Newton's belief that the stability of the solar system is due to divine intervention did not stop him from doing great scientific work on the law of gravitation. This shows that you can introduce God to explain *X* and still do good science about *Y*. But doesn't saying that God explains *X* rule out doing serious scientific work on why *X* is true? It need not. A scientist can believe that everything that happens in nature happens because it is God's will and still try to discover naturalistic explanations for natural phenomena. It may be replied that introducing the existence of God in this way is scientifically idle. Maybe so, but that does not show that invoking the existence of God must bring science to a stop.

My argument depends on a distinction — between saying "God did it" as *part* of an explanation and saying this as one's *entire* explanation. The latter practice, if generally adopted, would spell the end of productive science. But that does not show that the former would also be lethal. This is why the science-stopper argument fails to justify methodological naturalism.

6. If numbers, why not God?

Mathematics is a necessary framework for science. Many of the scientific theories we prize presuppose that numbers exist. No such argument has been produced for including mention of God in scientific theories. Indeed, there have been many predictive scientific theories, developed over several centuries, that are silent on the question of whether God exists; these provide ample evidence that *science does not need the God-postulate*. These "silent theories" can be supplemented; one can add to them the claim that God exists, or the claim that he does not, or the claim that we don't know whether there is a God, but in each case the supplements aren't consequences of the science; rather, they are philosophical add-ons.

Arguments against introducing the claim that God exists into scientific theories have often been *in-principle*; they attempt to show that this postulate *necessarily* prevents science from reaching one of its goals. For example, it is claimed that the resulting theories cannot be tested or that introducing the God-postulate precludes the development of naturalistic explanations. The argument I would offer is more modest. Naturalistic science has been a success. Nature has presented us with scores of problems that have been addressed successfully by theories that are theistically neutral. We know, by inspecting the history of science, that theism is not needed. The modest defense I would offer of methodological naturalism is simply this: *if it isn't broken, don't fix it.*

One glib retort to this suggestion is that science is *already* broken, and so it *does* stand in need of fixing. This response stems from expecting science to do more than it is capable of doing. If you want scientific theories to tell you what is right and what is wrong, or what the meaning of life is, you will be disappointed. But this does not mean that *science* has failed. Science is not in that line of work; it is a misplaced *scientism* that holds that all meaningful questions can be answered by science and by science alone. This is what Huxley (1892) meant by "scientific naturalism" and what the positivists aimed to defend by way of their testability theory of meaning. It is worlds away from the methodological naturalism that I am defending.

Could a new form of inquiry be pursued in which theories are constructed that are both empirically successful and also are committed to the existence of God? There is a trivial way to do this. Just take a theory that we now admire, one that is theistically neutral, and append to it the statement "and this is God's will." The result is "theistic science," but the theistic add-on is doing no real scientific work. It is idle. The same trick can be used to construct an atheistic science, and the atheistic addition is also scientifically idle. There is a second way of creating a theistically committed scientific theory that also is trivial, but in another way. Consider a theistically neutral theory *T* that happens to say nothing about the empirical proposition *E*. Let's add to *T* the postulate "God wanted *T* to be true, but ensured that *E* would be false." Now the theistic add-on *does* say something about nature that the initial theory *T* failed to address. Even so, the new, augmented theory is not interesting. If you want to find out whether *E* is true, why not just investigate this directly? Why do so by appending the hypothesis "God wants *E* to the false" to a theory *T* that says nothing whatever about *E*? It is obscure how a theistically committed science could proceed in which the theism is not idle. Is it reasonable to expect that a theistically committed science, or shmience, will be better than the science we now have? It certainly could be a lot worse.

To modern scientists, it may seem obvious that science needs mathematics, but does not need theology. But like all marriages and divorces, these two have their histories, and what seems inevitable after the fact need not have seemed inevitable before. One reason that methodological naturalism became more and more persuasive over hundreds of years was that naturalistic explanations increased in number and power. But the triumph of methodological naturalism had another source; theists developed a theological picture according to which a benevolent deity made the world in such a way that methodological naturalism would turn out to be a successful research tool. This is the idea to which Darwin (1859) appealed when he began the *Origin* with a quotation from his mentor William Whewell:

> But with regard to the material world, we can at least go so far as this – we can perceive that events are brought about not by insulated interpositions of Divine power, exerted in each particular case, but by the establishment of general laws.

Methodological naturalism was promoted by scientists *and* religionists; in fact, the same people often played both roles at once (Numbers 2003).

When he heard Laplace's exposition of the nebular hypothesis (a Newtonian explanation of the origin of the solar system), Napoleon was taken aback. "Where is God in your theory?" he asked, and Laplace is said to have replied that he had no need of that hypothesis. Many of Darwin's contemporaries were shocked that he did not appeal to the direct activity of a designing deity to explain organic diversity. Darwin could have said what Laplace said; in fact, he did say as much. Atheists may want to reach for Ockham's razor at this point and argue that these theories are evidence against the existence of God precisely because these theories show that the God hypothesis is not needed in science. But the inference from "science can explain phenomenon *X* without invoking the existence of God" to "God does not exist" is shaky (Sober 2010b). It also is true that science can explain why gold melts at a certain temperature without postulating the existence of dinosaurs, but that isn't evidence that there were no dinosaurs. We need to attend to Monod's good advice: "any confusion between the ideas *suggested* by science and science itself must be carefully avoided." For some people, Newtonian theory and Darwinian theory *suggest* that there is no God. However, this is not what these theories say; it is a philosophical interpretation that requires additional premises. Molière's M. Jourdain was astonished to learn that he had been speaking in prose for so many years. We should not be astonished, when we discuss science, to find that we are actually doing philosophy.

Acknowledgments
My thanks to Glenn Branch, Joshua Filler, Gerald McKenna, David Miller, Bradley Monton, Ronald Numbers, Robert Russell, and William Stoeger for useful discussion.

REFERENCES

Carnap R., 1950, "Empiricism, semantics, and ontology", *Revue Internationale de Philosophie* 4: 20–40. Reprinted in Carnap R., 1956, *Meaning and Necessity*, Chicago: University of Chicago Press.
Darwin C., 1859, *On the Origin of Species by Means of Natural Selection,* London: John Murray.
Diaconis P., 1998, "A place for philosophy? The rise of modeling in statistical science", *Quarterly of Applied Mathematics* 56: 797–805.
Draper P., 2004, "God, science, and naturalism", in Wainwright W. (ed.), *The Oxford Handbook of Philosophy of Religion,* Oxford: Oxford University Press, 272–303.
Gray A., 1888, "Natural Selection not inconsistent with Natural Theology", in *Darwiniana,* Cambridge, Mass.: Harvard University Press, 1963, pp. 72–145.
Hempel C., 1965, "Studies in the logic of confirmation ", in *Aspects of Scientific Explanation and Other Essays in the Philosophy of Science,* New York: Free Press, 3–47.
Huxley T., 1892, *Essays upon Some Controverted Questions*, London: Macmillan.
Johnson P., 1990, *Evolution as Dogma. The Establishment of Naturalism*, Dallas: Haughton Publishing Company.
Laudan L., 1988, "The demise of the demarcation problem", in Ruse M. (ed.), *But Is It Science? The Philosophical Question in the Creation/Evolution Controversy*, Buffalo: Prometheus Books, 337–51.
Miller K., 2007, *Finding Darwin's God. A Scientist's Search for Common Ground between God and Evolution*, New York: Harper Collins.
Numbers R., 2003, "Science without God. Natural law and Christian beliefs, an essay on methodological naturalism", in Lindberg D., Numbers R. (eds.), *When Science and Christianity Meet*, Chicago: University of Chicago Press, 265–86.
Pennock R., 1999, *Tower of Babel. The Evidence against the New Creationism*, Cambridge, Mass.: MIT Press.
Pennock R., 2009, "Can't philosophers tell the difference between science and religion? Demarcation revisited", in Pennock R., Ruse M. (eds.), *But is it Science? The Philosophical Question in the Creation/Evolution Controversy,* New York: Prometheus Books, 536–77.
Plantinga A., 1996, "Methodological naturalism?", in van der Meer J. (ed.), *Facets of Faith and Science*, vol. 1, Lanham: University Press of America, 177–221. Reprinted in

Pennock R. (ed.), 2001, *Intelligent Design Creationism and its Critics,* Cambridge, Mass.: MIT Press.
Plantinga A., 1997, "Methodological naturalism? Part 2", *Origins and Design* 18, no.2 (an expanded version of Plantinga 1996)
http://www.arn.org/docs/odesign/od182/methnat182.htm
Popper K., 1959, *Logic of Scientific Discovery,* London: Hutchinson. References to the revised edition of 1968.
Quinn P., 1984, "The philosopher of science as expert witness", in Cushing J., Delaney C., Gutting G. (eds.), *Science and Reality. Recent Work in the Philosophy of Science,* Notre Dame: Notre Dame University Press.
Reichenbach H.,1938, *Experience and Prediction,* Chicago: University of Chicago Press.
Russell R., 2008, *Cosmology from Alpha to Omega. The Creative Mutual Interaction of Theology and Science,* Minneapolis: Fortress Press.
Sober E., 2007, "What is wrong with Intelligent Design?", *Quarterly Review of Biology* 82: 3–8.
Sober E., 2008, *Evidence and Evolution. The Logic Behind the Science,* Cambridge: Cambridge University Press.
Sober E., 2009, "Parsimony arguments in science and philosophy", *Proceedings and Addresses of the American Philosophical Association* 83/2: 117–55.
Sober E., 2010a, "Evolution without naturalism", in Kvanvig J. (ed.), *Oxford Studies in Philosophy of Religion,* Oxford: Oxford University Press.
Sober E., 2010b, "Parsimony Arguments in Science and Philosophy", *Proceedings and Addresses of the American Philosophical Association* 83: 118-155.

The Recurrent Debate: Darwinism and Philosophy

Massimo Stanzione

Let me start from some epigraphic — and contradictory — quotations:

- My theory would give zest to recent and fossil comparative anatomy; it would lead to the study of instinct, heredity, and mind-heredity, *whole (of) metaphysics.* (Darwin 1887, vol. II, cap. 1, 8; my italics)
- Plato ... says in *Phaedo* that our imaginary ideas arise from the preexistence of the soul, are not derivable from experience, *read monkeys for preexistence.* (Darwin 1838, 291; my italics).
- *4.111* Philosophy is not one of the natural sciences. [...] *4.112* Philosophy aims at the logical clarification of thoughts. [...] *4.1121* Psychology is no more closely related to philosophy than any other natural science. [...]

  4.1122 Darwin's theory has no more to do with philosophy than any other hypothesis in natural science. (Wittgenstein 1918)
- *It is difficult to draw a sharp line between philosophy and natural science. The naturalist* who introduces a new principle, or demonstrates a fact which throws a new light on existence, not only renders an important service to philosophy but *is himself a philosopher in the broader sense of the word.* [...]The name of Charles Darwin stands with those of Galileo, Newton, and Robert Mayer, names which denote new problems and great alterations in our conception of the universe. (Hoffding 1909, my italics)

My main theses will be the following:

1) The impact of Darwinism on philosophy, especially on epistemology and ethics, was immediately deep and very large. It would be an interesting but impossible task to analyze and explicate the historical reasons of this intellectual trend. Therefore I'll focus my paper on some case studies. For the same, practical reasons, the ethical dimension will be left out of my paper.

2) In the first Darwinian centennial celebration (1909), the scientific and epistemological debates was already focused on the same sensible questions we are confronted with today. That should be a point of interest, to my mind, because this demonstrates that Darwinian naturalism was a scientific research program with a substantial core, not a metaphysics of evolution.

3) Even if the scientific framework is today dramatically changed, and in spite of anti-naturalist paradigms in philosophy (the phenomenological and the analytical ones) during the last century, the original Darwinian project which intended to naturalistically explain the origin of the human mind, human knowledge and human morality in terms of evolutionarily based acquisitions, never ceased. Indeed it is today stronger than in the past.

4) Anyway, looking deeper in the history of ideas, we find many evidence supporting that, also in general philosophy, there was a long standing (albeit underestimated) development of naturalist trends, both in the professional field and in the wider intellectual climate, animated by leading scientific thinkers.

5) To justify my theses, I'll briefly compare two influential positions in naturalism, which I denominate "the analytical program" and "the substantial program" respectively. The first program is now well known and debated in the Anglo-Saxon analytic tradition as a consequence of the so-called Naturalized Epistemology proposed by Quine. The latter program coincides with the uninterrupted strain of intellectual works inspired by naturalistic thinking (ranging from the genetic to the evolutionary epistemologies of Baldwin, Piaget, Lorenz, Campbell and Riedl, up to the evolutionary, even if *non-naturalistic* theory of Popper).

6) I will argue that both these programs do not immediately justify the contemporary *ideological* confrontation between naturalism and religion.

7) Finally, although theology is not my concern, I would like to suggest that the evolutionary worldview has been received — and should be received also today — as an urge to renovate theological thinking.

1. Fundamental preliminaries about the many meanings of "Naturalism"

Let me preliminarily clarify how the *ontological, methodological, epistemological* and *metaphysical* naturalism will be here intended, under the background assumption that these dimensions are reciprocally interrelated (and sometimes confused).

Naturalistic ontology (NO) affirms that the space-temporal states, events and processes which, on their whole, constitute the totality of physical universe (or *Nature*, or *World*), exist as substantially prior to and/or independent from our ways of perceiving (parts of) them, conceptualizing (parts of) them and knowing (parts of) them.

This definition implies two *further metaphysical* options (in a sense which will be soon clarified).

You can say either:

- that nothing exist over and above the Nature or World (*Deus sive Natura*); that the mind is nothing but a natural process and that no supernatural reality exists;
- or that the natural World, *no matter whether you consider the mind as a natural process or not, depends upon a* (personal or impersonal) different entity: a Supernatural Being, its *Creator*.

Methodological Naturalism (MN) is a different thesis, often called "the principle of the causal closure of the physical world" (CCP).

If positively stated, it assures that every physical effect has a sufficient physical cause. The physical world constitutes a causally closed domain: if a physical event has a cause at *t*, then it has a physical cause at *t*.

Obviously, if true, this thesis distinguishes physics from all other subject domains. The biological realm, for example, is not causally closed in this sense, since biological effects (say the extinction of the dinosaurs) could have non-biological causes (say the impact of a meteorite). Importantly, the mental realm is not causally closed, either: a mental pain can be caused by sitting on a physical drawing pin, or a train of thought can be interrupted by a loud noise.

Viewed from the negative side, the principle states that, in explicating natural events, any present or future physical theory should be completely independent of religious or metaphysical views or commitments.

Anyway, logically the CCP principle *does not assert that everything is physical*, but only that those things that *are* physical have a physical cause.

Even so the CCP does give rise to a powerful argument for reducing many *prima facie* non-physical realms to physics: for it indicates that anything that has a causal impact on the physical realm must itself be physical. That's the reason why the MN is so often fused with the strongest version of the NO.

In fact, many *prima facie* non-physical realms — such as the biological, meteorological, and mental realms — certainly do seem to be both causally affected by physical events, and to have a causal influence on the physical realm.

Following this path of reasoning, *the MN comes very near to the reductive physicalism program*: the *prima facie* non-physical realms *ought* to be *a posteriori* reduced to the physical one.

Moreover, this very path is not a mandatory path. Indeed, CCP essentially requires that any scientific explanation concerning the empirical realm should be in accordance with the closure principle, and not that scientific explanations are necessarily reducible to physical ones (a point that the famous biologist Ernst Mayr continuously vindicated).

Is the CCP an a-priori principle? It seems implausible. The CCP is only a highly empirical claim, whose acceptance derives from detailed empirical evidence about the causes of physical effects. But in the MN framework it implies a more substantial consequence: the CCP is a prime example of the way that philosophical results, just like scientific results, characteristically depend on empirical findings.

The next step we are now driven to do is introducing and defining epistemological naturalism (EN), which is about the nature and the justification of our beliefs.

"With Dewey I hold that knowledge, *mind and meaning* are part of the same world that they have to do with, and that they *are to be studied in the same empirical spirit that animates natural science*. There is no place for a prior philosophy." (Quine 1969, 26; my italics)

In other terms, if you reject the speculative approaches to cognition (based on a merely logical analysis of language and on the presumed existence of some a priori truths) you will be brought to a broader vision of our cognitive possibilities. "Drawing on the deliverances of the sciences, naturalists view members of our species as highly fallible cognitive systems, products of a lengthy evolutionary process. How could our psychological and biological capacities and limitations *fail* to be relevant to the study of human knowledge? "(Kitcher, 1992, 58). Briefly speaking: "Epistemology, or something like it, simply falls into place as a chapter of psychology and hence of natural science." (Quine 1969, 82)

Now, what could we add about *metaphysical naturalism (METN)*? Should it not be considered simply *scientism*?

According to Ruse: "[metaphysical naturalism] is an *epistemological and ontological* commitment to the nature of knowledge *and* of the real world. The metaphysical naturalist denies that there is anything beyond scientific knowledge. *Such a person I take to be an atheist (not just an agnostic) and a materialist* (inasmuch as materialism makes sense today)." (Ruse, 2005, 84–85, my italics)

Most post-analytic philosophers manifest today a very similar orientation: the EN (the original field of their "naturalistic return" in analytic philosophy) ought to be identified or strictly related, thanks to CCP, with metaphysical naturalism (METN).

Now, it's true that both EN and METN imply the non existence of anything which goes over the scientific knowledge, but *in principle they are different*, because one could be a provisional atheist in doing physics, a supporter of an instrumentalist epistemology, a supporter of NO and a theist in metaphysically interpreting the tenets of NO!

The disputable conflation between EN and METN is based on a preceding move: the implicit or explicit tendency to identify METN with the strong version of NO, starting from the idea — contrary both to the original Quinean statement and to the soundest philosophical tradition — that ontology *is* metaphysics (not only a substantial branch of metaphysics). Metaphysics? But what does it mean?

It would be foolish to offer here a simple definition of "metaphysics", a term which sounds as a coarse word to the ears of so many scientists and philosophers. A satisfactory definition could be offered *only* in the framework of Ancient and Medieval philosophy, where "metaphysics" had its subject matters: the "being as such" or "the first causes of things" or the study of "things that do not change". But after the seventeenth century, "metaphysics" began to be a catch-all category, a repository of philosophical problems that could not be otherwise classified: not epistemology, not logic, not ethics.

Some problems that Aristotle and the Medievals would have classified as belonging to physics, because in no way related to first causes or unchanging things (e.g. free will, purposiveness, mind-body, personal identity across time) became to be assigned (at least partly) to metaphysics. It was at about that time that the word "ontology" was invented to be a name for science of being as such, an office that the word metaphysics could no longer fill.

So Christian Wolff famously distinguished between "general metaphysics" (Ontology), and the various branches of "special metaphysics", which study the being of objects of various special sorts, such as souls and material bodies. Note that, anyway, Wolff does not assign "first causes" to general metaphysics: *the study of first causes belongs to a branch of special metaphysics* (i.e., natural theology).

Wolff's definition is objectionable because it is not evident that the problems of "special metaphysics" do, in any non trivial sense, really involve being. It is highly debatable whether "minds" (mental properties and/or states or processes), free will, the unity of consciousness and the personal identity could be considered "beings"!

But until now, no attempt to do better than Wolff seems really successful. The same holds for the many (strong and weak) attempts made, at least since the time of Hume, in order to demonstrate the *impossibility* of metaphysics.

The strong version of this thesis was construed by logical-positivists: "all metaphysical statements are meaningless, because the meaning of a statement consists entirely in the predictions it makes about possible experience." But this thesis is a non analytic and non-predictive statement: therefore, if it is true it is meaningless; or, if it is meaningful, it is false.

The weak form of the thesis is: metaphysical statements are meaningful, but human beings can never discover whether any metaphysical statement is true or false (or probable or improbable or warranted or unwarranted).

This idea is at least as old as Kant, but versions of it that are much more adherent to the naturalistic framework have been presented by some authors, which in the introduction of De Caro and Macarthur (2004) are defined "liberal naturalists" (John McDowell, Hilary Putnam, Donald Davidson, Barry Stroud, John Dupré, Colin McGinn).

But McGinn's (1999) dissolution of metaphysics reaches the conclusion that the human mind is (as a matter of evolutionary contingency, and not simply because it is "a mind") incapable of a satisfactory treatment of a large range of philosophical questions (a range that includes *all* metaphysical questions).They are… *mysteries*! In other words: science is not enough to fill the "gaps" between physical basis and its derivative realms, and philosophy is a futile endeavor!

Rebus sic stantibus, I honestly prefer another move. From now to the end of my paper, I'll use the term "naturalistic metaphysics" (at least as metaphysics is to be understood in the METN) simply as a compendium of philosophical problems which are, in fact, historically and culturally present in the intellectual Darwinian and post-Darwinian naturalistic framework (professional systematic and non systematic philosophy included) but cannot be clearly defined as intrinsically different from other parts of philosophy.

2. Back to the recurrent debate: Darwinism and philosophy in the past and today

According to Hoffding (1909), there are four main points in which Darwin's investigations possess philosophical and scientific importance.

1) Darwin's energetic renewal of the old idea of evolution had his chief importance in strengthening the conviction of this real continuity in the world, that is a continuity in the series of form and events. It helped to produce the great realistic/naturalistic movement which characterizes the last third of the nineteenth century.

2) Struggle for life and natural selection are principles which have been applied, more or less, in every department of thought. The philosophical importance of these ideas does not stand or fall with the answer to the question, whether natural selection is a sufficient explanation of the origin of species or not: it has an independent positive value.

3) Human thought itself is, then, a variation (or a mutation) which has been able to persist and to survive. Is not, then, the problem of knowledge solved by the evolutionary hypothesis? Many authors gave an affirmative answer to this question before and after the appearance of the *Origin of Species.*

4) The Darwinian theory of natural selection seemed to change the whole conception of life, and particularly all the conditions on which the validity of ethical ideas depends. Darwinism, it was said, has proclaimed brutality. On the contrary, Darwin has, indeed, rendered a great service to ethics in making the difference between the life of nature and the ethical life appear in a strong light.

There is a conceptual core in Darwin's thought which has both a scientific and an epistemological-metaphysical relevance (in the previously stated meaning of METN). This core was almost forgotten in the 1959 Darwin centennial celebration published by Sol Tax (1960). In the famous Whewell's terminology it appears to be a "consilience of inductions" resulting in many principles. They are:

Two *ontological principles*:

- the common descent principle (or continuity of life principle);
- the principle of species as space-temporally contingent clusters (transformative principle);

Some *explicative principles* of the sufficiency of certain non-teleological (in Darwin's terminology "mechanical") processes in order to account for the adaptation and apparent purposiveness of living beings. In Darwin's thought, these processes embrace:

- the struggle for life (natural selection) and further forms of selection (sexual and group selection), together with embryological laws (included the so-called recapitulation law);
- the related principle of the growing complexity of life (progressive principle);
- the analogical principle concerning the evolutionary process pertaining to the biological realm on one hand, and the cultural processes pertaining to language, ethics and the higher faculties of human mind on the other.

As you can see, this core isn't merely scientific, but philosophical and … (in a legitimately not bad sense) metaphysical.

3. Long standing naturalism: supporting evidences from the history of philosophy

To paraphrase a saying which gained considerable notoriety a decade ago, "… we are all naturalists now. But, even so, this common naturalism is of a very vague and general sort: is an admission of a direction more than a clearly formulated belief. It is less a philosophical system than a recognition of the impressive implications of the physical and the biological sciences. And, not to be outdone, psychology has swelled the chorus by pointing out the organic roots of behavior and of consciousness." These are not words of mine; they constitute the very *incipit* of Roy Wood Sellars (1922) *Evolutionary Naturalism*!

Therefore, even if Kitcher (1992) is right in asserting that the naturalistic stance in philosophy was crucial both *before* the emergence and *after* the overthrow of Frege's program (which launched the "linguistic turn" in epistemology), *pace* Kitcher *the naturalistic stance was also alive during the so-called linguistic turn*, especially in the realistic and pragmatistic traditions. With a relevant proviso: if you do not consider the adoption of an analytic framework as a progress on its own, the history of ideas teaches us that the core of the naturalistic program has been in fact "progressive" only on the *scientific* side of it, *not* on the *philosophical* side. Starting from Darwin's own work, in the variegate cluster (or family) of naturalistic theories all the most relevant epistemological and philosophical themes of contemporary naturalism were debated. Recently there has been a renewed appreciation of this historical continuity (see e.g. Kim 2003, Madden 2000), which I hope will confirm and enrich the basis of some of my own works (Stanzione 1981, 1984, 2004). In the epistemological field, it is today customary to oversimplify the story, identifying this core, without any deeper historical analysis, with the tenets of the post-Quinean "naturalized epistemologies".

To my better knowledge, their radical formulation runs as follows:

- The cognitive assertions that go beyond immediate experience have *neither deductive nor inductive validity*; to justify inductive procedures it is therefore necessary to assume the provisional validity of the results of some of these procedures, thanks to the circular reference to the hypothetical ontology of our present-day sciences. Fortunately "our innate standards show a gratifying tendency to run with the grain of nature. This concurrency is accountable, surely, to natural selection." (Quine 1974, p.19) Unfortunately "the appeal to a law of natural selection would be unwarranted, since the law rests in turn on induction. *What natural selection contributes, rather... is a reason why induction works, granted that it does.*" (Quine 1974, p. 20)

- Truth has to do with the fit between our beliefs and the reality they purport to describe. Yet, in practice, all we can do is to compare beliefs with beliefs, never a belief with a direct known reality. Therefore it is possible to accept the *correspondence meaning of truth and the regulative role of truth in science* and, at the same time, *to admit coherence as the major but still fallible symptom of truth.*
- *Epistemological naturalism is therefore an (almost) descriptive theory*. But it is also a viable way out of the skepticism (i.e.: the inescapable result of the ultimate failure of any foundationalist program).

Eighteen years before Kitcher's essays, D.T. Campbell wrote an influential essay on evolutionary epistemology (Campbell 1974). In reconstructing the realistic and naturalistic trends which, in Kitcher's terminology, we could define "anterior to the linguistic turn", Campbell quotes Peirce, James, Dewey, Huxley, Mach, Simmel and many others. Now, let me integrate his work and quote some relevant naturalist thinkers living in Darwin's time (F. A. Lange, J. Fiske, E. Montgomery, C. Wright, F. E. Abbot; see Stanzione 1981) in order to strengthen my thesis about the old-standing naturalist trends in philosophy — and their multifarious ideological aspects.

4. The German background of American evolutionary thinkers: Lange on philosophy and biology

As the modern ones, the first proponents of naturalism aimed to escape skepticism about human perceptive and categorical apparatus and to surmount the traditional opposition between the empiristic and the Kantian ways of thinking by virtue of the evolutionary theories (and especially the Darwinian selective one).

In a relevant sense, the outstanding F.A. Lange's treatise *Geschichte des Materialismus und Kritik seiner Bedeutung in der Gegenwart* (voll. 1-3, 1865-1874) is to be considered a real masterpiece, which was very influential on both European and American naturalistic trends in philosophy. Lange expressed in it some really prophetic ideas on many relevant topics of the present philosophical discussion on the better way to interpret the Darwinian naturalistic heritage. These topics are the following.

The problem of chance, necessity (universal natural laws) and the concept of un-intentional adaptations in respect to teleology.

Lange says:

> Is it quite true, [...] that in place of mysteriously working causality [hidden variables] we have only the "possibility" of happy accidents ? What we see is not possibility but actuality. The single case seems to us only possible (seems to us "fortuitous") because it is regulated by the activity of natural laws, which in

our human apprehension have nothing to do with this special result of their reciprocal action. In the great whole, however, we can see the necessity. Amongst the countless cases the favourable ones too must happen for they are actually there ... In fact, this does not so much refute all teleology as afford an insight into the objective nature of the adaptiveness of the phenomenal world. We see clearly that this adaptiveness in the individual case has nothing human about it; nay, that, so far as we have yet observed, it is not brought about by higher wisdom, but by means which, in their logical value, are clearly and distinctly the lowest that we know. This estimate, however, is again only based on human nature, and so there remains for the metaphysical, the religious conception of things, which overpasses these limits in its imaginations, room for the setting up of a teleology which must be simply and definitively rejected from physical research and the critical philosophy of nature (Lange 1865, II, section 2, cap. 4, 36).

Gradualism, "saltationism" and the disturbance of equilibrium (in a really Piagetian flavor).

We may also assume that the changes in organisms which have been left to themselves have not been completed with quite such imperceptible slowness as Darwin's own view seems to require, but that after every important change in the conditions of existence there has resulted, as it were by starts, a rapid development of some forms and a retrogression of others. We may very well assume also that every such disturbance of the natural equilibrium produces a tendency to variation, and thus gives opportunity for the origin of new forms, which rapidly establish and perfect themselves when the conditions are favourable to them. All the various principles which modern inquirers have introduced into the doctrine of descent in order to complement the principle of natural selection, as, e.g., migration, the isolation of species, etc., are only more or less happily apprehended special aspects of the decisive main principle of the *disturbance of equilibrium*, which must necessarily produce the stability of species where the conditions long remain identical (Lange 1865-1874, II, section 2, cap. 4, 45).

"Laws of forms" and evolution through natural selection.

[E.g.] Imagine the ready-hewn stones for a Gothic and a Byzantine cathedral so piled up, on two sites of like form and most limited dimensions, as to use every inch of space, and that the two heaps attain the same external shape. Then it is very easily conceivable that these masses of material at some distance appear like two exactly similar structures. But if the stones are separated and properly put together, from the one of these piles there can only result the Gothic, and from the other the Byzantine cathedral.

If this is once recognised, we must also draw the consequences, partly by recognising that chemical relations have their rule, and, as it were, their plan of development, but partly, too, by appreciating the whole attitude of morphology to the genesis of organisms. *We must, that is, admit the principle that unknown peculiarities of matter, probably chemical, may exercise a decisive influence on the development of beings, on their future form and their modes of life, although these very peculiarities are already present in the first elementary forms without producing any difference cognisable by us* (Lange 1865-1874, II, section 2, cap. 4, 62; my italics).

The "origin of life" problem (and its possible solutions, including F. Crick's "panspermia")

A priori, it is, of course, very much more probable that from the beginning of life there was a greater number of germs not completely alike and not equally capable of development, whether these germs came from the *meteoric dust of cosmical space*, or whether life developed itself from the monera of the ocean depths (Lange 1865-1874, II, section 2, cap. 4, 63; my italics).

Teleology, Theology and Causality

If, therefore, the opposition to Darwin proceeds in part openly, in part half unconsciously, from predilection for the old teleological view of the world, sound criticism can only in answer draw the line, *that no opposition is scientifically justified which does not just as much as Darwinism itself start from the principle of the intelligibleness of the world, joined with the thorough application of the principle of causality.* [...]

Meanwhile there is also a teleology which is not only compatible with Darwinism, but is almost identical with it [...] If Darwinism, as compared with the gross anthropomorphic teleology, appears as a theory of chance, this is only its thoroughly justified, negative side. Adaptations proceed from the conservation of relatively fortuitous formations, but these formations can only be called fortuitous in so far as we can assign no reason why this particular form appears at this moment. [The necessary and eternal] laws indeed do not immediately produce what is adapted, but they produce a multitude of variations, a multitude of germs, in which the special case of what is adapted, of the persistent, is perhaps relatively very rare. [...] It might be easily demonstrated that even in the highest actions of man this principle of the conservation of what is relatively best adapted still plays its part, everywhere co-operating with the most subtle apparatus of a specific reaction. Even the great discoveries and inventions, which form the basis of higher civilisation and intellectual progress, are still subordinate to that universal law of the conservation of the strongest, while they are at the same time tested by the most delicate methods of science and art. The whole question of correct teleology may be reduced to this. (Lange 1865, II, section 2, cap. 4, 65-67; my italics)

In the Kantian philosophy, therefore, which has sounded these questions deeper than any other, the first stage of teleology is directly identified with the principle which we have repeatedly spoken of as the axiom of the intelligibleness of the world — and Darwinism in the wider sense of the word, i.e., the doctrine of a scientifically intelligible theory of descent, not only does not stand in contradiction with this teleology, but, on the contrary, is its necessary presupposition. The 'formal' finality of the world is nothing else than its adaptation to our understanding, and this adaptation just as necessarily demands the unconditional dominion of the law of causality without mystical interferences of any kind, as, on the other hand, it presupposes the comprehensibility of things by their ordering into definite forms (Lange 1865, II, section 2, cap. 4, 69).

5. Philosophy and biology in the first Anglo-Saxon evolutionary thinkers

In discussing the principle of natural selection as the principal agent of cerebral development, John Fiske wrote in his *Outlines of Cosmic Philosophy*:

> The view of cerebral action here adopted settles the long-vexed question between the Lockian and Kantian schools as to the sources of knowledge … The strength of Locke's position lay in the assertion that all knowledge is ultimately derived from experience — that is, from the intercourse between the organism and the environment. The strength of Kant's position lay in the recognition of the fact that the brain has definite tendencies, even at birth. The Doctrine of Evolution harmonizes with the two seemingly-opposite views, by showing us that in learning we are merely acquiring latent capacities of *reproducing* ideas (Fiske 1874, 160–61; italics mine).

In the same years, also Edmund Montgomery (a philosopher-scientist born in Scotland but educated in Germany), Chauncey Wright and Francis Ellingwood Abbot were trying to answer these same question.

Considered by some as a pioneer of organicism, Montgomery did research on protoplasm and extended his conclusions to an explanation of the emergence of conscience. Philosophically interested in Kant's gnoseology, he criticized it from a realistic point of view in his Montgomery (1871). According to Montgomery: "only a pre-established harmony, developed trough evolution, can render a definite perception that rises in the perceiver, tightly and minutely representative of the correspondent perceivable entity that exists outside of the perceiver".

Francis Ellingwood Abbot was, as John Fiske, one of Wright's friends and a proponent of scientific realism. The philosophy of science maintained by Abbot was devoted to correcting the crudities of naive realism, but in a radically *antinominalistic* way. "Relationism or scientific realism (of which *universalia inter res* may be adopted as an apt formula) teaches that universals, or genera and species, are, first, objective relations of resemblance among objectively existing things; secondly subjective concepts of these relations, determined in the mind by the relations themselves; and thirdly, names representative both of the relations and the concepts." (Abbot 1885, 60)

The evolutionary approach was used by the first naturalistic philosophers to surmount the traditional problematic of natural kinds also in language and in consciousness.

Chauncey Wright, one of the founders of American pragmatism, went considerably on in developing the naturalistic program. Here are, in short, Wright's most relevant assertions *from an epistemological and philosophical viewpoint.*

Wright, in his interpretation of Darwinism, succeeded in avoiding the most serious conceptual mistakes that were very common at his time. He had no progressivistic (or, as Lewontin and Gould would put it, "Panglossian") view of evolution, according to which all biological functions, including cognition, are constantly improving, due to their increasing adaptation to the outer world. Stimulated by Darwin himself, he had a polemical confrontation with both Wallace and Mivart. Against Wallace he succeeded in arguing that it is not necessary to state a supernatural intervention in order to justify the peculiar characteristics of human mind (language and morals). Against Mivart's argument (the impossibility to explain trough natural selection the independent but convergent formation of complex organs) Wright produced a line of reasoning very similar to the line used by Gould and Vrba (1982) in introducing the "exaptation" concept. Two are, in Wright's opinion, the general hypotheses prefiguring "exaptation" and "bricolage".

> The one, which is so conspicuous in the principles of comparative anatomy, *how few the fundamental structures are, which have been turned to such numerous uses*; that is, how meager have been the resources of Natural Selection, so far as it has depended on the occurrence of structures which were of no previous use, or were not already partially useful in directions in which they have been modified by the selection and inheritance of 'individual differences'.

The other, how important to Natural Selection have been the principles of indirect utility and "correlated acquisition", dependent as they are on "ultimate physical laws":

> ...in a long course of development, even in cases of highly specialized faculties, existing uses have risen in succession or alternately to the place of first importance, as in the various uses of the hand. This principle of a plurality of

> existing uses involves a very important influence in secondary uses, whether these are incidental or correlative acquisitions, or are the more or less surpassed and superseded ones (quoted from the last Wright's letter sent to Darwin, 1875, February 24).

That is to say — going back to Darwin's problem of the relationship between efficient and final causes in the emergence of the mind — that *there exists a direct relationship* between the emergence of self-consciousness and the faculties that preceded it in genetic development, *but there is no necessary causal link* in this relationship. In short, according to Wright the new usages are only accidentally related to the old faculties: they are involved in a continuous process of natural selection of "exaptations".

In fact, Wright was already aware that organisms actively solve problems thanks to the structure of their internal organization. Therefore he did not conceive cognitive innovation — viewed as the fundamental step in the transition from perceptive conscience to reflective self-consciousness — as the *direct* result of selective pressures from the outer world. He conceived the emergence of human mind not only as a result of a balance between necessity and fortuitousness, but in terms of a new concept: the *unconditionality*. In Wright's opinion, this concept could assure the, in principle, possible coexistence of determinism and "casual" changes.

The notion of unconditionality came out of his criticism of the identification of the scientific notions of law and causal relationship with the notion of efficient cause alone. *Unconditionality is an hypothetical logical link which can explain a "creative" anomaly in causal structure.* There is no deterministic reason why new orders in reality are actually produced by natural selection without violating any natural physical cause. The process can be explained only ex post, thanks to historical reconstruction efforts. This spontaneous and non-directional process explains the finalistic behavior of organisms without making recourse to final causes.

In this framework, Wright also faces the issue of categorizing the elements of sensitive experience putting together Kantism, Darwinism and associationism and redefining the relationship between the knowing being and what is known in terms of *constructive conceptualization*. This is the distinguishing feature of his naturalism as opposed to naive realism. Conceptualization means organizing phenomena within possible inferential domains, translating them into signs which at first are exclusively natural, and then accessible to being named. Therefore, even if, obviously, not all natural knowledge is rational knowledge, *it is legitimate to view also the implicit knowledge of organisms as a special kind of inference based on meaningless signs.*

In his criticism of the Kantian concepts of the pure a priori forms of sensitive perception, Wright assumes that the a priori element that actively organizes the perceptions of data and of their relationship is a sort of *inductive generalization*, which depends on *phylogenetic memory*. Memory actively organizes chains of mnestic hints*: it does not so by way of mere association but by selection.*

Associations form an "order which, once established to some degree, keeps the senses constantly under control and governs our attention in perception". In other terms, some of these associations have been selected, established and transformed in "regulative elements of mnemonic origin", that is to say in a gnoseological a priori.

To explain this, Wright makes recourse to the Aristotelian concept of entimematic reasoning which can be defined as a sort of imperfect syllogism, which draws its conclusions from only implicit or probable assertions or from signs. *Animals are unaware of the general characteristics of signs*, they can only perceive the *systematic link* between "a" sign and the (inferred) existence of "the" thing that corresponds to it, without understanding at all the connection between the sign and the thing and "without classifying the sign under the general category of signs".

To sum up, the theory of signs proposed by Wright is particularly interesting essentially because it links the origin of natural signs to the *joint* action of *two* mechanisms that nowadays seem to be of paramount importance in the naturalistic explanations of the mind: *memory* and *selection*. In other terms, those structures which ensure that men can re-categorize continuously their memories (Edelman and Tononi 2001), are absent or insufficiently developed in animals. Therefore, at least on a speculative level, Wright explicitly maintains that categorization processes, memory and continuous selection of possible matching between external and internal signs are interconnected in order to give rise to the human self-consciousness and language.

In 1873 he wrote, after a personal Darwinian invitation, a long article entitled *Evolution of Self-Consciousness*. This article was devoted to explaining the genesis of self-consciousness by denying all the rigid traditional barriers (instinct/intelligence, perception/intellect, consciousness/self consciousness) and by showing the new functions that consciousness assumes in the human species and, above all, in language (the ability to think and communicate in linguistic forms).

According to Wright, language first begins as a set of instinctive and unintentional acts which is socially elaborated upon to serve the purposes of communication without its creators having this purpose in mind. When it reaches the mnemonic and reflexive phase, language becomes a system of inventions which, due to its own freedom and creativity, helps in the thinking self-clarification of the thinker, and therefore not only in communicating.

Though inadequate, Wright's theory still serves our purposes because it demonstrates the crucial role of the epigenetic developmental dimension. *Consciousness is an epigenetic phenomenon which has to be explained by means of a modified Darwinian theory: in Wright's terms, a selective theory of progressive, experiential construction of natural signs.*

6. The naturalistic research program and its continuity in the non-analytical trends of the nineteenth century

Of course, in the philosophical field, "the naturalistic program" embraces not only epistemology, but many other dimensions: anthropology; a unified research framework in human and natural sciences; ethics and aesthetics. They were already included in the original Darwinian framework, which is therefore to be viewed as a paradigmatic "research program". That is demonstrated by an historical event: the gathering, in June 1909, of scientists from 167 different countries in Cambridge to celebrate the hundredth anniversary of Charles Darwin's birth and the fiftieth anniversary of the publication of *Origin of Species* (see Seward 1909). Not only: "the 1909 Darwin celebration ... represents a significant watershed in the history of modern biology that allows historians to assess the status of evolution prior to the advent of the chromosome theory of genetics." (Richmond 2006, 447). It is also a proof of the persistence of the naturalistic "substantial" program even when the mechanism of natural selection was increasingly under attack, and the evolutionary theory was in disarray.

In Seward (1909) you will find three chapters on naturalistic anthropology ("The Descent of Man", "Charles Darwin as an Anthropologist" and "Some Primitive Theories of the Origin of Man"), respectively written by the anatomist G. Schwalbe, the well known zoologist and evolutionary monist Ernst Haeckel, and by the social anthropologist and mythologist J.G. Frazer, who was notoriously very influential in the early stages of the modern studies of comparative religion (and of Freudian metapsychology). But you will find also five chapters using a naturalistic methodology in a very large range of socio-cultural, ethical and religious domains. They are entitled: "Darwinism and sociology", "The Influence of Darwin upon Religious Thought" (and "on the study of religions"), "Evolution and the Science of Language", "Darwinism and History".

Now, do you remember my former Quine's quotation: "With Dewey I hold that…"? Well. In reasserting and vindicating the American origins of contemporary philosophical naturalism, Kim (2003) quotes the views of philosophers like Roy Wood Sellars, John Dewey, John Herman Randall, Sydney Hook, and… of the neo-empiricist Ernest Nagel (!). And in the same paper some of the central tenets of naturalism, such as an adherence to "scientific method" as the sole source of knowledge and the causal/explanatory closure of the natural world, are taken in account and considered. But, in my opinion, if we take in account also the "substantial program", the scope of this historical reconstruction should be enlarged, in order to embrace all sides of the exceptional flourishing of evolutionary and naturalistic thinkers during the so-called "eclipse of Darwinism". I am referring here to authors commonly classified as supporters of emergentist, holistis (or systemic), or dialectic theories, but often philosophically well informed and scientifically active before the rise of the Synthetic Theory. Their naturalism was really multifarious, especially in evaluating the relationship between the evolutionary paradigm and the quest for a "spiritual life".

Some of them are today very famous, as the Russian inventor of bio and noosphere Vladimir Ivanovich Vernadsky (1863-1945), which influenced both Theilard de Chardin and Jean Piaget; the holist and emergentist thinker Samuel Alexander, the founder of philosophical (and theological) process naturalism (Whitehead). Some, on the contrary, are almost forgotten as e.g.: the Russian physician and economist Alexander Bogdanov (1873-1928), who wrote a monumental treatise on a general science of organization based on both selective and cooperative processes (see Bogdanov 1922; see also Stanzione 2004); the South-African Jan C. Smuts, a supporter of a strong ontological holism (see Smuts 1926; see also Procacci 2001); the Italian "synthetic theorist" Eugenio Rignano (1870-1930), who was, together with Federigo Enriques, the founder of *Scientia*, the most important Italian review of philosophy of science. Rignano was also acquainted with Henry Berr, the French founder of *La Revue de Synthèse*, aimed to integrate historical and scientific cultures in the framework of a vast project of scientific synthesis. Today remembered mostly for its socio-political works (e.g. his proposal about the social significance of the inheritance tax), he was in fact an influential figure also in psychology and theoretical biology: e.g. Dewey 1938 approvingly quotes the biological theory of reasoning formerly stated in Rignano 1920; on the contrary, Needham 1927 is a polemical response to Rignano's 1924, considered as a romantic defense of the vitalistic and teleological positions.)

7. Darwinian naturalism and theology: some provisional indications

With a great logical jump, let me turn to the open question of naturalism and religious faith. After our historical survey and on the basis of our best knowledge, we are driven to subscribe, as our favored theses: 1) The opportunity to encourage in theology of that evolutionary method of observation and study, which has shaped all modern research; 2) The recoil of Christian apologetics towards the ground of religious experience, a recoil produced by the pressure of scientific criticism upon other supports of faith. But there are at least six problems that remain open. In fact, if you admit that:

- The great advance of our knowledge of organisms has been an important element in the general advance of science. It has acted upon all other branches of natural inquiry, and it held for a long time that leading place in public attention which is now occupied by speculative physics. Consequently it contributed largely to our present estimation of science as the supreme judge in all matters of inquiry; and that
- It contributed largely to the supposed destruction of mystery and the disparagement of metaphysics which marked the last age, as well as to the recommendation of scientific method in branches of learning where the direct acquisitions of natural science had no place....

The logical consequence will be that:

1) There really exists what has been described as a conflict between Science and Faith, between Theism (abstract Theism) and Natural Selection. (Note that there is no difficulty at all between Theism in the abstract and *Development* in the abstract.... but there does arise a difficulty... because natural selection seems to be a rival explanation of Special Creation as we know it). The argument from design (it is not the argument *for* design) seems to be shaken by the presence of an explanation which is an alternative. This explanation, described in the largest outline, states that a thing comes to be what it is because, unless it were what it is, it could not be at all.

2) In the second place, there is a conflict between Bible cosmogony and *Development* in the abstract. Abstract Theism must limit Natural Selection. But Positive Theology, Revealed Theism, has a battle yet to fight with (what may be called) Abstract Development.

3) In the third place, there is the difficulty about the origin and dignity of man, about which I will here say nothing — as already stated in the preliminaries of my paper.

4) In the fourth place, there is the new Psychology, a physiological psychology (today better known as "cognitive neurosciences") which attempts to displace the psychology which is more properly so called because it believes in the existence of a spiritual principle, a Psyche. That is a serious and difficult branch of the conflict.

5) In the fifth place, certain parts of biological speculation seem to give a new force to the generalized attack on the doctrine of Free Will (as demonstrated e.g. by Dennett 1995).

6) Last, and more practically important than any, there exists a tendency in natural studies in some way to obscure the authority of tradition.

To distinguish the elements of debate is useful only in *some degree.* But the real problem is how to deal with a Society which has reached some such state. In order to attain this aim, you could adopt five distinct methods.

1) First, there is the method which might be described as *the method of segregation* (e.g. S.J. Gould's NOMA). This method seeks to emphasize a division of the camps, and I must be content here to say briefly that it is obviously absurd and hopeless.

2) *More attractive seems the method of detailed conciliation*, the persevering attempt to make the Bible come into line, or to express the same thing with the teaching of Science. That kind of line of conciliation seems to me frankly hopeless.

3) *Then there is the line of detailed criticism*; ... the policy of trying to show here and there *that Science is not sure of its ground* — according to the line of reasoning exemplified by Plantinga (1997) —, of drawing the line where the sea of ignorance lies around this province of light. It is a hopeless method, because those who depend upon the failure of a vigorous body of thought, surely are putting their hopes in a very shaky place. *It is like a policy which depends upon the weakness of other nations.*

4) Apparently near to this last, but quite distinct from it in my mind, is the method of rival agnosticism. This is the method of meeting the skepticism of Science by a rival skepticism with which Religion shall fortify itself against Science. This method relies rather upon disparaging experience *en bloc*. It is a well-known philosophical attitude, of course, and has a great deal to be said for it; but what does it result in? You will not turn *misologists* because the tide of argument sets against you.

5) Then there is, after all these, the metaphysical criticism of Materialism as such. Its object is to recall men to the true pivot of thought — i.e. our human special condition: contingence, imperfect rationality and the need to obtain a collective satisfaction of our (individual and collective) existential needs. But it is not the work which can help the layman when he has crossed the bar and lost his faith. It is meaningless to him. It is no good, when a man is showing you that this or that experiment is fatal to your view, to suggest a revision of the scheme of experiment as such, to disparage or put in its proper place a whole region of information, a whole world of means of learning.

I conclude that the only way is to take a definite account of all aspects of religious experience in a descriptive and non normative way (A naturalistic conclusion!)

Should this conclusion appear theologically superficial and unsatisfying to you, that would be no problem for me. What I just wrote is simply a summing up of a lecture. It was entitled "Science and faith". Its author was the Anglican reverend P. N. Wagged, who wrote it in 1901, eight years before his already quoted "The influence of Darwin upon Religious Thought" (1909, first Darwin's Centennial). A century after I can only completely agree with him!

References

Abbot F. E., 1885, *Scientific Theism*, Boston: Little, Brown & Co.

Bogdanov A., 1922, *Tektologiya: Vseobschaya Organizatsionnaya Nauka*, 3 vols., Berlin and Petrograd-Moscow; translated in English as *Essays in Tektology: The General Science of Organization*, Seaside, CA: Intersystems Publications, 1980

Campbell D. T., 1974, "Evolutionary Epistemology", in Schilpp P.A. (ed.), *The Philosophy of Karl Popper*, La Salle: Open Court, 413–63.

Darwin C., 1887, *Life and Letters of Charles Darwin*, vols. I–III, London: Murray; notebook written in the year 1837.

Darwin C., 1838, *The Notebooks on Man, Mind and Materialism,* in Gruber H. E., *Darwin on man. A psychological study of scientific creativity; together with Darwin's early and unpublished notebooks. Transcribed and annotated by Paul H. Barrett, commentary by Howard E. Gruber. Foreword by Jean Piaget,* London: Wildwood House, 1974.

De Caro M., Macarthur D. (eds.), 2004, *Naturalism in Question*, Cambridge, Mass.: Harvard University Press.

Dennett D., 1995, *Darwin's Dangerous Ideas*, New York: Simon & Schuster.

Dewey J., 1938, *Logic, the Theory of Inquiry*, New York: Henry Holton & Co.

Edelman G. M., Tononi G., 2001, *Consciousness: How Matter Becomes Imagination*, London: Penguin Books.

Fiske J., 1874, *Outlines of Cosmic Philosophy*, vol. II, London.

Gould S. J., Vrba E. S., 1982, "Exaptation: a missing term in the science of form", *Paleobiology* 8: 4–15.

Hoffding H., 1909, "The influence of the conception of evolution on modern philosophy", in Seward A. C. (ed.), *Darwin and Modern Science*, Cambridge: Cambridge University Press, 446–64.

Kim J., 2003, "The American origins of philosophical naturalism", *Journal of Philosophical Research,* special supplement on "Philosophy in America at the Turn of the Century", Charlottesville: Philosophy Documentation Center, 83–98.

Kitcher P., 1992, "The Naturalistic Return", *The Philosophical Review* 101: 53–114.

Lange F. A., 1865, *Geschichte des Materialismus und Kritik seiner Bedeutung in der Gegenwart* (2nd rev. edn., 2 vols., 1873/1875, Leipzig); English quoted edition in a single volume *The History of Materialism and Criticism of Its Present Importance*, 2nd edn., New York: The Humanities Press, 1950.

Madden E., 2000, "Introduction", in Ryan F. X. (ed.), *The Evolutionary Philosophy of Chauncey Wright*, vols. 1–3, Bristol: Thoemme Press.

McGinn C., 1999, *The Mysterious Flame: Conscious Minds in a Material World*, New York: Basic Books.

Montgomery E., 1871, *Kant's Erkenntnisslehre wiederlegt vom Standpunkt der Empirie. Ein vorbereitender Beitrag zur Begrundung einer physiologischen Naturauffassung*, München.

Needham J., 1927, *Man a Machine. In Answer to a Romantical and Unscientific Treatise Written by Sig. Eugenio Rignano & Entitled "Man not A Machine"*, London: Kegan Paul.

Plantinga A., 1997, "Methodological Naturalism?", Part I in *Philosophical Analysis: Origins & Design*, 18, 1, http://www.arn.org/docs/odesign/od181/methnat181; Part II in *Philosophical Analysis: Origins & Design*, 18, 2;
http://www.arn.org/docs/odesign/od182/methnat182.htm

Procacci S., 2001, *Alle radici dell'olismo. La filosofia della natura di J.C. Smuts*, Napoli: Edizioni Scientifiche Italiane.

Quine W. v. O., 1969, "Ontological Relativity", in Quine W. v. O., *Ontological Relativity and Other Essays*, New York: Columbia University Press.

Quine W. v. O., 1974, *The Roots of Reference*, La Salle: Open Court Publ.

Richmond M. L., 2006, "The 1909 Darwin Celebration. Reexamining evolution in the light of Mendel, mutation, and meiosis", *Isis* 97: 447–84.

Rignano E., 1920, *Psicologia del ragionamento*, Bologna: Zanichelli; translated as *The Psychology of Reasoning*, New York: Harcourt, Brace & Co., 1923.

Rignano E., 1922, *La memoria biologica*, Bologna: Zanichelli; translated as *Biological Memory*, London: Kegan Paul, 1926.

Rignano E., 1926, *Man Not a Machine. With a Foreword of Prof. Hans Driesch*, London: Kegan Paul.

Ruse M., 2005, "Darwinism and Naturalism. Identical twins or just good friends?", in Hösle V., Illies C. (eds.), *Darwinism and Philosophy*, Notre Dame: University of Notre Dame Press, 83–91.

Sellars R. W., 1922, *Evolutionary Naturalism*, Chicago: Open Court Publ. Company.

Seward A.C. (ed.), 1909, *Darwin and Modern Science*, New York/Cambridge: Putnam's Son & Cambridge University Press.

Smuts J. C., 1926, *Holism and Evolution*, London/New York: MacMillan.

Stanzione M., 1981, "Introduzione", in Campbell D. T., *Epistemologia evoluzionistica*, Roma: Armando, 7–62 (Italian translation of Campbell D. T. 1974).

Stanzione M., 1984, "Epistemologia evoluzionistica: confronti e critiche", in Continenza B. *et al.*, *Evoluzione e modelli*, Roma: Editori Riuniti, 193–300.

Stanzione M., 2004, "Selezione, organizzazione e metodo scientifico in A. A. Bogdanov", in Bogdanov A. A., *Quattro dialoghi su scienza e filosofia*, Roma: Odradek, 63–94.

Tax S. (ed.), 1960, *Evolution after Darwin: The University of Chicago Centennial*, Chicago: University of Chicago Press.

Wittgenstein L., 1918, *Logisch-philosophische Abhandlung*; translated in English as *Tractatus logico-philosophicus*, London: Kegan Paul, 1922.

Wright C., 1973, "The Evolution of Self-Consciousness", in North American Review, April 1873; rep. in Wright C., *Philosophical Discussions*, New York: Henry Holt & Co., 1877, 199–266.

Wright C., 1975, "Letter to Charles Darwin, February 24, 1875", in *Letters of Chauncey Wright, with some account of his life by James Bradley Thayer*, Cambridge, Mass.,1878, 330–38.

Section V
Evolution, metaphysics and teleology

Introduction

This section is dedicated to a philosophical assessment of biological evolution. The gathered papers show the variety of positions and approaches typical of philosophical examinations about so complex a topic.

One of the most difficult issues is the significance, scope, and role of anthropology. Indeed, when we speak of man, we have the feeling of entering a domain that does not pertain to science, philosophy, or theology alone. However, each of these disciplines claims anthropology to be its proper and exclusive object. This raises the problem of whether or not a reconciliation between scientific naturalism and the idea of man as a cultural being is possible. Such a peculiarity of anthropology, moreover, is precisely the reason why this domain is also the source of almost all the major conflicts between science, philosophy and theology. The first paper of this section by **Jürgen Mittelstrass** deals precisely with this problem and with the paramount difficulties that emerge in the contemporary research in bioethics.

Cardinal **Georges Cottier** takes the dynamism of nature as displayed by evolutionary theory into account and stresses the necessity of avoiding any form of concordism. After having recalled that when potential conflicts between science and religion arise, the easiest solutions are not often the best ones, the author points out the need to distinguish in any theory statements that are more certain from those that are less so. The main aim of the contribution is to show the radical difference between creation on the one hand, and mutations occurring during the evolution process on the other. Another relevant issue is how to understand the human soul: Cardinal Cottier points out the traditional Catholic understanding of the rational soul as form of the human body. This raises the important problem concerning the relations between soul and body, considerably increased when the Cartesian dualistic view became widespread in Modern Ages.

The contribution by **Vittorio Hösle** on "Why teleological principles are inevitable for reason. Natural theology after Darwin" takes us for a journey across Darwin's religious ideas. Thereafter, the author focuses on the theistic perspective of the botanist Asa Gray, one of the earliest and most accurate interpreters of Darwin's theory, being among the few who knew about it even before publication. Looking at Darwin's religious ideas through Gray's eyes also provides an opportunity to critically evaluate the influence of the positions maintained by Spinoza, Leibniz, Hume and Kant on the involved debate about teleology in the natural world. Darwinian evolution and theism are finally shown to be logically compatible. As the author puts it, theism regards the universe as created by an intelligent mind, thus entailing a teleological order, for minds act according to ends; but this also implies both the existence of intelligent beings able to grasp such a finality and a world intrinsically intelligible. As one of the most powerful scientific theories, Darwinian evolution may testify the intelligibility of nature and, therefore, support the rational belief in the "formal purposiveness" of nature as a whole.

David J. Depew addresses the issue of teleology in biological systems, taking into account remarkable positions in the history of philosophical thought, and with punctual reference to Darwin's original view of contingency, natural selection, and adaptation. In particular, adaptation is seen as a process in which fitness-related characters arise and are diffused precisely for the sake of the advantages they provide ("consequence teleology"). As the author suggests, we may clearly see the point when considering the range of natural selection as the interactive processes through which organisms progressively develop in highly coordinated and articulated systems, as supported by modern Evo-Devo research trends in biology. Aristotle's view of internal teleology in organisms and the related notion of *epigenesis* as well as Darwin's idea of co-adaptation throughout the biological domain, may still help in conceiving the *post facto* directionality in evolution "in richer ways than the crude idea that 'evolutionary arms races' favor ever more elaborate 'vehicles' for passing genes across generations."

In his insightful contribution, **William R. Stoeger** deals with the problem of ascribing essential characters like directionality, finality and relationality to the natural world on the basis of established scientific knowledge. Drawing our attention to the processes that brought to the formation of the universe, the author proceeds to explain that investigations in modern cosmology show us a universe in constant evolution: it unfolded in more and more differentiated and complex aggregates, which in turn set the stage for further emergent structures, eventually allowing the rise of biological systems and human consciousness. In this whole dynamic process of growing complexification, new functionalities and relationships among natural systems also arise, so that we may speak of an overall directionality towards higher forms of interconnectedness, not deterministically leading to particular results, rather constraining natural processes within a range of possible outcomes. The author's considerations, in which issues such as information, emergence and "functional teleology" play a major role, may be crucial for any attempt at elaborating the key notions of an updated philosophy of nature consistent with what science has presently to teach.

In the paper by **Ludovico Galleni** an attempt is provided to interpret the very complex and sometimes difficult to understand thought of the Jesuit Teilhard de Chardin. The author both presents the scientific production of Teilhard and advances a personal interpretation according to which the French Jesuit's understanding of evolution could be considered a synthesis of the contributions of Lamarck, Darwin, and Mivart. In particular he finds the contribution of Lamarck as providing a study of the issue of directionality or teleology in evolution, while that of Mivart is about the widespread convergences observed in nature. The latter point is also relevant from a scientific point of view since it points out the necessity to integrate the traditional Darwinian understanding in a wider synthesis.

The historical reconstruction of **Ronald Numbers** is very helpful to understand the roots of the Intelligent Design movement that can be found in American creationism. The phenomenon is very complex and had different stages. A first anti-evolutionist movement appeared in the 1920s. A few years after, the self-educated "flood" geologist G. Price gave rise to several partial initiatives. It is only in 1941 that an American Scientific Affiliation is created that evolved from initial fundamentalist positions to more updated ones. As a reaction, in the 1960s a more conservative Creation Research Society was established, claiming to be a true research society. In 1968 the U.S. Supreme Court struck down as unconstitutional the last of the laws from the 1920s outlawing the teaching of evolution. This forced creationists to abandon any thought of making the teaching of evolution illegal, and turned their attention to writing legislation that would allow the teaching of "creation science" alongside that of "evolution science." The immediate consequence was the replacing of the words "creation" and "creationists" with the euphemisms "intelligent design" and "design proponents." In this way, the Intelligent Design movement was born. As their proponents defined it, intelligent design provided a frame of reference that "locates the origin of new organisms in an immaterial cause: in a blue-print, a plan, a pattern, devised by an intelligent agent."

In conclusion, I would like to draw attention to the remarkable papers by **S. Gilbert, W. Hurlbut** and **M. Buiatti** included in the Appendix to the volume: they may well further enrich reflections on this section. These papers were not delivered as contributions to the Conference but were offered during a preparatory Workshop organized few months earlier.

Gennaro Auletta

Evolution and the conditio humana: A philosophical introduction

Jürgen Mittelstrass

1. Belief and science

Maybe it is due to the Darwin year and its thrilling advertising that we witness today, even if it is to the surprise of many, a new eagerness both on the scientific as well as on the theological side to win the game about the explanation of Man — Where is he from? Where will he go to? What is his nature? For many, mostly those of a fundamentalist orientation, the aim is to reach an explanatory sovereignty in matters scientific as well as theological, as if the point was to turn belief into science and science into believing. Philosophically speaking, we are dealing with a new form of naturalism on the one hand, and a new form of theologism on the other.

Naturalism is understood as the claim to the universal explanatory competence of science, or, respectively, as the thesis that the natural world (including Man) and the sciences explaining it are the sole basis for the explanation of all things. One variety of such a naturalism is what I shall here call *biologism*. That signifies, on the one hand, the idea of an absolute hegemony of biological explanations over others, including philosophical and theological ways of explaining, and, on the other hand, the subordination of the (scientific and philosophical) concept of validity under the concept of evolution (its philosophical representative being the so-called evolutionary epistemology). According to this, biology explains everything, even knowledge itself. Today one may occasionally find both varieties, the methodological as well as the epistemological, in evolutionary biology as well as in brain research. Biologism is, just like naturalism in general, universalist as well as reductionist.

Let us use *theologism* here to stand for the claim to the universal explanatory competence of theology, insofar as this understands itself as a competitor to scientific, and in particular natural scientific explanatory competence. One may conceive of pantheism — the doctrine that God exists in all objects of the world, or that God and the world are identical, respectively — and theosophy — including the assumption of a capacity of understanding of a higher kind (than the scientific one) — as older varieties of this (though in a philosophical rather than theological guise), and all fundamentalist views as younger varieties, among which are *creationism,* for instance in the form of the so-called scientific or *Young Earth creationism* (employing a literal exegesis of *Genesis*), or the so-called *evolutionary creationism* (which interprets natural selection as an intervention in evolution by God). The doctrine of *intelligent design* is the most recent form of creationism (neocreationism).

Naturalism as well as theologism of the forms described here bear witness to a relapse into views which seemed to have been long superseded. Indeed, epistemologically speaking, any kind of *reductionism*, scientific as well as theological — but philosophical too, of course — makes the world a simple place, a too simple place, and thus carries to excess reasonable explanatory aims, it also loses the capacity to think in a variety of dimensions, for instance those of science and those of the life-world. Already the Greeks had understood that these are not reducible to each other; here, the discovery of the possibility of science does not entail a replacement of the life-world, and with this I mean guiding principles for the life-world, but rather to their enlargement.

It is the same in theology as well as in philosophy. It is important to take notice of scientific insights and to bear them in mind, but to always take into account the philosophical insight that science is not everything, also and in particular in the matters that concern Man in his very own condition, in his thinking, feeling and perceiving and in his self-conception. Once again: it would be a fundamental error to suppose that theology was under the obligation, like philosophy, to compete with scientific knowledge, and it would equally be a fundamental error to suppose that science explains everything, and that it will always have the last word, even in matters of belief. If that were the case, science (here in the form of a biologism) and theology (here in the form of a theologism, for instance of the creationist variety) would irreconcilably be opposed to each other. The fact that this is occasionally the case today bears witness not to a loyalty to principles, scientific or theological, but to a fundamental misunderstanding — on both sides.

In other words and relating to how the current discussion about evolution and belief is structured: just as science, once it has gone beyond an ill-understood reductionism, may take a neutral stance towards religious belief, religious belief can also live with the sciences without losing itself in their results. That also holds for (Darwinian) evolutionary theory. Why should God, theologically speaking, not have created a universe that is developing and following Darwinian laws? One may be creationist (in the best theological sense) without endorsing creationism (in the pseudo-scientific sense). George Coyne, the former Director of the Vatican Observatory, even goes one step further: "If they respect the results of modern science and, indeed, the best of modern biblical research, religious believers must move away from the notion of a dictator God or a designer God, a Newtonian God who made the universe as a watch that ticks along regularly." (Coyne 2005, 7) There is a bridge to God from Darwin as well; one just has to know how to walk on it, respecting the autonomy of belief as well as that of science.

But the old theme of belief and science was not supposed to be the topic here. The question is rather how the *biological nature* of Man, with which biology deals in the framework of a theory of evolution, and his *cultural nature,* with which the humanities and theology are dealing, relate to each other, and in which way the *conditio humana*, the human condition, may bring itself to bear vis-à-vis scientific explanations of itself. The perspective I will adopt in what follows is neither a scientific one in the strict sense (for instance in the form of a science-theoretic treatment of evolutionary theory) nor a theological one in the strict sense (for instance in the form of a further examination of the theme of science and belief), but rather a philosophical one, more precisely: an anthropological and ethical one. I intend to deal with what may be considered the right criteria for looking at our biological and cultural nature.[1]

2. Anthropological distinctions

Philosophical anthropology has always distinguished between the biological and the cultural nature of Man, in other words between what is natural to him in a physical and biological sense, and what pertains to him culturally, what his cultural essence is. This, however, does not mean that both essences, the physical and the cultural, fall apart, and that therefore, as Descartes for example holds, Man disintegrates into two essences.

Descriptively, within the context of biological systematics, Man is a sub-species of the species *homo sapiens*, namely *homo sapiens sapiens*, and is the only recent member of the genus *Homo*. But this definition includes only the empirico-physical side of Man, not that which makes up the essence of humanity *ascriptively*, namely its form of self-description and (not conclusively established) self-determination. The latter was described classically as the *animal rationale*, a being en-

[1] For a more detailed analysis of what follows see Mittelstrass (2001 and 2007).

dowed with and determined by reason, or as a being lying between animal and God. Here, it would be a category error to interpret what we do and what we think as the products of natural processes, whereby even interpreting becomes part of nature, a "natural fact". But we fall into a new form of naiveté if we oppose this interpretation with a claim that scientifically discovered facts have no influence, or at least ought to have no influence, on the self-determination of Man. Thus it is a matter of adopting a scientifically informed and philosophically considered position, one which is beyond mere *biologism*, as described before, and *culturalism* — with this I mean (as distinguished from biologism and naturalism in general) the claim to universal explanatory sovereignty of the humanities —, which, in other words, is beyond an absolute distinction between biological and cultural explanations, and which refers to both the lives and the laws that shape our lives. Such a position should neither reduce Man to (pure) nature, nor to the (absolute) spirit he aspires to be.

Modern philosophical anthropology mirrors this situation. It takes its point of departure from two opposing conceptions: that attributed to Max Scheler and that of Helmut Plessner. According to Scheler, "Man" is the "X that can behave in a world-open manner to an unlimited extent." (Scheler 1927, 49) According to Plessner, "Man" is characterised by an "eccentric positionality" (Plessner 1928, 362ff.), whereby his eccentric existence, which does not possess a fixed centre, is described as the unity of mediated immediacy and natural artificiality. Accordingly, Plessner formulates three fundamental laws of anthropology: (1) the law of natural artificiality, (2) the law of mediated immediacy, and (3) the law of the utopian standpoint (Plessner 1928, 309-346; see Lorenz 1990, 102f.). Similarly, Arnold Gehlen states the thesis that Man is by nature a cultural being (Gehlen 1961, 78), and in doing so, his cultural achievements are seen as compensation for missing organs and "Man" is defined as a creature of defect (Gehlen 1972, 37).

Stipulations of a similar kind can also be found in the history of philosophical anthropology. Thus Man is called the creature without an archetype by the Italian Renaissance philosopher Giovanni Pico della Mirandola: he himself, according to the will of his Creator, is to determine the "form," that is, the cultural form in which he wishes to live (Pico della Mirandola 1942, 106). According to Immanuel Kant, the question "What is Man?" can only be answered if we already have answers to the questions, "What can I know?," "What ought I to do?" and "What may I hope?" (Kant 1958, 448) The attempt to determine "1. the source of human knowledge, 2. the extent of the possible and profitable use of knowledge, and finally 3. the limits of reason" (Kant 1958, 448), is itself an anthropological research programme, and on the background of the critical philosophy of Kant it is an open research programme that defines Man according to what he can achieve in theory and practice. For Friedrich Nietzsche, finally, Man is the not-yet-determined animal (Nietzsche 1968, 79), and thus science too is seen as the expression of human endeavour "to determine himself" (Nietzsche 1968, 533). Furthermore, one of the reasons for the difficulty of saying what Man is lies in the fact that Man is the (only) creature that possesses a reflective relation to itself — that Man, as Heidegger says, is the creature "that in its being relates understandingly to its being" (Heidegger 1977, 52f.) or that it is "concerned in its being with this being itself." (Heidegger 1977, 12) This opens up a broad horizon of possible self-interpretations of Man, and, at the same time, a broad horizon for an answer to the question, what a human being, what Man, is. The only thing that is clear is what can be called, with regard to the essential openness of Man, the *anthropologically basic situation*.

This openness affects all phases of human development, both from an ontogenetic and from a phylogenetic point of view. There is no "natural" fate in the becoming of Man, as an individual or as a species that might be conclusively determined by biological laws, even though of course the "schema" of this development is proscribed by biological regularities. Thus, there is no adulthood before childhood, no reverse ageing, no Achilles who is young until he dies. In psychological terminology: the architecture of human ontogeny is incomplete (Baltes 1997), and not merely in earlier stages, but throughout a lifetime. "Incompleteness" is here understood as the result of the "constructive" nature of Man, and therefore as an anthropological constant.

3. Evolution and ethics

Today we are promised great advances above all from the developments of modern biology, for example in medicine. But there are great risks as well, for instance in the thoughtless or irresponsible application of bioengineering. There is nothing fundamentally new in this. Ground-breaking inventions have become saddled with dangers and risks of abuse of a new and usually unimagined order throughout human history. What may be new in the case of modern biology is the fact that developments in biological knowledge now appear to place Man in the unique position of being able to change his own nature, and that this development has implications for ethics (in the conventional sense). Man intervenes ever more powerfully in evolution, even in his own, and he changes the measures with which he previously described and regulated his situation, that is to say, the human condition.

While we have known since Darwin that Man, not only from the point of view of philosophy and culture, but also biologically, has no fixed essence, he is nevertheless, even though this is imperceptible to the individual and only recognisable to science over long periods of time, subject to evolutionary changes. It has only recently become clear that Man can intervene in these changes himself — an ability to deliberately change his own genetic constitution, and that of his progeny. In fact, the *conditio humana* itself is changing: in the sense that now even Man's biological foundations are at his disposal. This creates a completely new and momentous situation in the domain of anthropology as well as in the domain of ethics.

There are various consequences that have been drawn from this situation. One is the call for a *bioethics* code, where this is understood as a code of applied ethics that deals specifically with biological states of affairs. Such a code would prescribe watchfulness and particular measures in certain fields, as well as certain applications that could be formulated as rules for an ethics of responsibility. Such rules, if applied to developments in genetic technology, might include a rule of considering the consequences through a careful checking for possible undesirable results and also a rule of caution, thus enabling one to choose the option that offers the greatest security of prognosis and the least expected harm. However, the debate on the ethical problems of biology extends far beyond bioethics into the direction of *ecological ethics*, which attempts to change the foundations of ethics itself.

The point of departure of such a concept of ethics is often an argument about *going against nature*. According to this, genetic engineering and interventions in human reproductive processes do something that is the business of nature alone; they intervene in a regulatory manner in a self-regulating nature. Gene transfer may cross species boundaries, and thus infringe on the "identity of species" (Altner 1991, 214) and disturb the (relative) stability of ecological balances (Altner 1991, 217). In arguments of this kind — which also include, by the way, the notion that animals, plants and the elements are like us and we are like them "in the whole of nature" (Meyer-Abich 1984, 24) —, we encounter biological uncertainty — what is the "identity of species" supposed to be? — coupled with ethical unclarity — what does ethics have to say about the order of species, that is, about biological classifications, or even about nature as a whole, however that is imagined? Those who think (and write) this way are confusing the empirical (biological states of affairs) with the normative and commit the naturalistic fallacy, that is, they infer what ought to be from what is, they derive norms from facts.

Precisely this is the case in the well-known arguments of Hans Jonas. He declares the natural to be the highest norm and views any intervention into these natural processes as an offence against "naturally" given norms. For instance, for Jonas, the technology of cloning is in "contradiction to the dominant strategy of nature" (Jonas 1985, 179) and thus cannot be justified. The natural — here in the form of a natural reproduction — consequently appears here as something not to be interfered with, and as something that pursues its own goals, with strategic means and that by these means makes itself the highest normative authority. As a matter of fact, the attempt is made repeatedly to construct an ecological ethics on the basis of an inference from facts to norms (which usually reveals a concealed naturalism) and to then oppose this new ethics in the form of *physiocentrism* to the *anthropocentrism* that has long dominated ethics and which is now (in many respects erroneously) declared to have been fundamentally mistaken.

A central role in such arguments is played by differing concepts of nature, and thus a bioethics that has been expanded into an ecological ethics is also called an ethics of nature. For the anthropocentric position — both in questions of ethics and of nature — Man is the point of departure of all arguments and nature has no intrinsic moral value. For the physiocentric position, nature is characterised by its own (absolute) intrinsic value, which at the same time implies duties of Man toward nature. To be more precise, we can distinguish between *pathocentrism* (all sensible creatures have a moral value), *biocentrism* (all living creatures have a moral value) and *radical physiocentrism* which, as just explained, makes all of nature the bearer of moral value. Common to all these variants is that values, which in fact are always the result of valuations, are declared to be a part of nature itself.

The expansion of a bioethics (a sub-area of applied ethics), to biological ethics in the form of, or against the background of physiocentrism, is thus based on a misunderstanding. This expansion not only makes ethics dependent on a particular view of the world, but also leads by its naturalistic premises to a new *ethical biologism*. Biologism as described before obtains entry to ethics. Biology here is expected to be an advisor and also a legislator in ethical affairs. And this in turn involves both a philosophical and a biological misunderstanding, since modern biology teaches us how permeable the boundaries are between the natural and the artificial, that is, those processes determined by Man. The appeal to nature in ethical questions, which made sense in archaic cultures, no longer makes sense here.

And one more thing: the notion that moral conduct as a particular form of social behavior is itself the product of evolution or could be given an evolutionary explanation leads one astray if it is understood in an absolute sense as a foundation of ethics. Whereas in the first case of a biologistic ethics, natural relations are to be taken as the standard of ethics, in the second case ethics would be a product of these relations, and thus our ethical deficits would not be due to the failings of reason but to an evolution that was unfinished and unable to cope adequately with Man (see Wolters 1998; Vogt 1998). An *evolutionary ethics* in this sense would be a convenient excuse for tasks unaccomplished in Man's dealing with himself and with nature.

But these tasks are what have to be addressed — which is why it is only a rational ethics, that is, an anthropocentrism in ethics rightly understood, which is able to solve them. Nature gives no ethical lessons, either in the form of physiocentrism or in the form of evolutionary ethics. Nature only reminds us when harm is caused — think of environmental problems — of the unfinished tasks of rational ethics.

4. Homo faber

In 1488, Giovanni Pico della Mirandola wrote the following about God's intentions towards Man: "We gave you neither a fixed dwelling, Adam, nor a particular appearance, nor any special talent, in order that you might have and own the dwelling, the appearance and the talents that you desire for yourself. (...) We made you neither heavenly nor earthly, neither mortal nor immortal, so that you might form yourself as your own, worthy, free and creative sculptor." (Pico della Mirandola 1990, 5-7) One hundred years later (1596) Johannes Kepler writes in the dedication letter of his *Mysterium cosmographicum*: "We perceive how God, like one of our own architects, approached the task of constructing the universe with order and pattern, and laid out the individual parts accordingly, as if it were not art which imitated nature, but as if God himself had looked to the mode of building of Man who was to be." (Kepler 1938, 6; trans. Duncan 1981, 53-55)

What Pico della Mirandola and Kepler still affirm in a pious and expressive language is nothing other than the extension of the concept of Man as *homo sapiens* to include that of *homo faber*, both with regard to himself and to his world. Pico della Mirandola's characterization of Man as "his own sculptor" corresponds to Nietzsche's and Plessner's definition of Man as the not-yet-determined animal, or indeed to Plessner's characterization of Man by means of his eccentric positionality (which is juxtaposed to the undistanced centricity of the animal). Similarly, Kepler's characterization of a *homo faber* competing with God paradigmatically corresponds to the modern notion of scientifically supported technical cultures, in which Man creates and encounters — both in and by means of his productions — not only the world, but indeed himself.

Is Man his own work, in the way that the (modern) world is his work? Certainly not in the sense that Man is an artefact that created itself. For even in his role as *homo faber*, Man remains bound to what always has been called the *conditio humana*, and what is meant by the work-like character of Man is above all his self-determining ("cultural") essence, not his natural or biological essence. Nonetheless, such distinctions, which are also boundaries, are beginning to give. Against the background of modern scientific and technical developments, the possibility has raised its head that along with the rational nature of Man (that which makes him *homo sapiens*) we might change not only his external (physical and social) nature but also his internal (biological) nature.

Biologism on the one hand, which reduces Man to his biological nature, is here joined by setting absolute the *homo faber* on the other side. In the so-called "Posthumanism" or "Transhumanism", which is up to no good in the latest discussions about the future of Man (see Silver 1998; Bostrom 2005), a perfecting of Man, made possible by technological and medical advances, is endorsed, as well as the overcoming of the limitations of the species Man which have been taken as natural till now. Apparently, the idea is to play God (see Midgley 2000; Kass 1997 and 2002). In both cases, in being handed over to biological or to technical evolution, Man would lose his nature, he would either step back into nature or become a machine.

Once again: Will Man put at his own disposal all the "parts" that make up his essence — body, soul and reason? Has he become master of his own nature in a sense which would have been unimaginable even for Pico della Mirandola or Kepler? I think that we must accustom ourselves to the fact that this power of Man over himself will increase, driven as it is by scientific and technical development. But we must at the same time preserve, in opposition to this development, those indispensable things which are experienced in love and in happiness, in sickness and in death, and in which, despite the threat of the triumph of *homo faber* over *homo sapiens*, an essential part of our humanity is contained. Might this be what Pico della Mirandola meant when he had God say to Man that the latter was created neither heavenly nor earthly, neither mortal nor immortal?

Conclusion

Casting a concerned glance at the world, the German philosopher and poet Georg Christoph Lichtenberg once remarked: "Why should there not be levels of souls reaching up towards God, such that our world was the work of one of them who did not quite know his business, an attempt?" (Lichtenberg 1971, 410) One cannot very well follow Lichtenberg's line, either in the tracks of scientific biologism or in the tracks of religious fundamentalism. The responsibility for the state of the world is ours, and not that of nature or a beginner god. But it is true that the modern world is the work of Man, and that he possibly did not quite know his business. Let us direct our self-understanding and our actions, including our scientific and technological understanding and our scientific and technological actions, in such a way that we aim at a *humane* world, and do not conduct experiments without thinking through the effects either on nature or on ourselves. This also means that Man, in a certain sense, cannot be optimised, at least not in so far as with such an optimization he would step out of his own nature. This is what we call *conditio humana*. In it the finiteness of Man in all its facets is mirrored — in happiness and misery, hope and despair, belief and knowledge. Knowledge of the laws of natural evolution will do nothing to change that.

References

Altner G., 1991, *Naturvergessenheit: Grundlagen einer umfassenden Bioethik*, Darmstadt: Wissenschaftliche Buchgesellschaft.

Baltes P. B., 1997, "On the incomplete architecture of human ontogeny: Selection, optimization, and compensation as foundation of developmental theory", *American Psychologist* 52: 366–80.

Bostrom N., 2005, "In defence of Posthuman Dignity", *Bioethics* 19: 202–214.

Coyne G., 2005, "God's Chance Creation", *The Tablet* 259, 6: 6–7.

Gehlen A., 1961, *Anthropologische Forschung: Zur Selbstbegegnung und Selbstentdeckung des Menschen*, Reinbek: Rowohlt Taschenbuch Verlag.

Gehlen A., 1972, *Der Mensch: Seine Natur und seine Stellung in der Welt* [1940], 9th edn., Wiesbaden: Akademische Verlagsgesellschaft Athenaion.

Heidegger M., 1977, *Sein und Zeit* [1927], 14th edn., Tübingen: Max Niemeyer.

Jonas H., 1985, "Laßt uns einen Menschen klonieren: Von der Eugenik zur Gentechnologie", in Jonas H., *Technik, Medizin und Ethik: Zur Praxis des Prinzips Verantwortung*, Frankfurt: Insel-Verlag, 162–203.

Kant I., 1958, "*Logik A 25, Werke in sechs Bänden*", in Weischedel W. (ed.), vol. III, Darmstadt: Wissenschaftliche Buchgesellschaft, 417–582.

Kass L. R., 1997, "The Wisdom of Repugnance", *The New Republic* 216, 22: 17–26.

Kass L. R., 2002, *Life, Liberty and the Defense of Dignity: The Challenge for Bioethics*, San Francisco: Encounter Books.

Kepler J., 1938, *Prodromus dissertationum cosmographicarum continens Mysterium cosmographicum, Gesammelte Werke*, vol. I, Dyck W. V., Caspar M., Hammer F. (eds.), Munich: C. H. Beck'sche Verlagsbuchhandlung (translation by Duncan A. M., 1981, *Johannes Kepler. Mysterium Cosmographicum / The Secret of the Universe*, New York: Abaris Books).

Lichtenberg G. C., 1971, *Sudelbücher K 69, Schriften und Briefe*, vol. II, Promies W. (ed.), Munich: Carl Hanser.

Lorenz K., 1990, *Einführung in die philosophische Anthropologie*, Darmstadt: Wissenschaftliche Buchgesellschaft.

Meyer-Abich K. M., 1984, *Wege zum Frieden mit der Natur: Praktische Naturphilosophie für die Umweltpolitik*, Munich: Hanser.

Midgley M., 2000, "Biotechnology and monstrosity: why we should pay attention to the 'Yuk Factor'", *Hastings Center Report* 30, 5: 7–15.

Mittelstrass J., 2001, "The Anthropocentric Revolution and Our Common Future", in Raff W.-K. *et al.* (eds.), *New Pharmacological Approaches to Reproductive Health and Healthy Ageing (Symposium on the Occasion of the 80th Birthday of Professor Egon Diczfalusy)*, Berlin, Heidelberg, New York: Springer (*Ernst Schering Research Foundation. Workshop Supplement* 8), 57–67.

Mittelstrass J., 2007, "Science and the search for a new anthropology", in Sánchez Sorondo M. (ed.), *What Is Our Real Knowledge about the Human Being?*, Vatican City: The Pontifical Academy of Sciences (*Pontificiae Academiae Scientiarum Scripta Varia* 109), 101–109.

Nietzsche F., 1968, *Jenseits von Gut und Böse* [1886], in Colli G., Montinari M. (eds.), *Werke: Kritische Gesamtausgabe*, , vol. VI/2, Berlin: de Gruyter, 1–268.

Nietzsche F., 1973, *Nachgelassene Fragmente Frühjahr 1881 bis Sommer 1882*, in Colli G., Montinari M. (eds.), *Werke: Kritische Gesamtausgabe*, vol. V/2, Berlin: de Gruyter.

Pico della Mirandola G., 1942, *De hominis dignitate. Heptaplus. De ente et uno, e scritti vari*, Garin E. (ed.), Florence: Vallecchi.

Pico della Mirandola G., 1990, *De hominis dignitate/Über die Würde des Menschen*, Buck A. (ed.), Hamburg: Meiner.

Plessner H., 1928, *Die Stufen des Organischen und der Mensch: Einleitung in die philosophische Anthropologie*, Berlin and Leipzig: de Gruyter.

Scheler M., 1927, *Die Stellung des Menschen im Kosmos*, Darmstadt: Otto Reichl.

Silver L. M., 1997, *Remaking Eden: Cloning and Beyond in a Brave New World*, New York: Avon Books.

Vogt M., 1998, "Evolution/Evolutionstheorie(n) 2.3", in Korff W. *et al.* (eds.), *Lexikon der Bioethik*, vol. I, Gütersloh: Gütersloher Verlagshaus, 717–21.

Wolters G., 1998, "Evolution/Evolutionstheorie(n) 2.2", in Korff W. *et al.* (eds.), *Lexikon der Bioethik*, vol. I, Gütersloh: Gütersloher Verlagshaus, 714–17.

The Metaphysical Significance of Creation

Card. Georges Cottier op

1. An epistemological problem

1. Raising the question of man's intrinsic constitution, his destiny, and his place in the universe, evolution addresses the philosopher and the theologian — and they can scarcely fail to incorporate a consideration of it into their conclusions. Based on the convergence of a multiplicity of facts as evidenced by various experimental and systematic disciplines on the scientific level, evolution is put forward in theories having quite a high level of probability, even if they still remain open problems.

My purpose here is to offer for your consideration a series of free-ranging reflections on some problems relative to this subject — problems both theological and philosophical. It is the latter that will chiefly concern me.

In this connection I would like to emphasize the importance of epistemological questions; not a few difficulties are created by the confusion of spheres of research, resulting in extrapolations and contradictions with regard to the meaning of certain claims.

2. It is very widely believed that the problem of the compatibility of the statements of Holy Scripture with our conception of the world emerged with the advent of modern science.

In reality however, the Fathers of the Church had encountered a problem similar to our own. The cosmological systems prevalent in Hellenistic culture were the object of a consensus founded on rational arguments, even if they had their point of departure in data of immediate experience that was non-critical, merely approximate, and for this reason often mistaken.

St Thomas refers to the position of the Fathers with regard to the specific question of the anteriority in time of *materia prima*, as pure potentiality, over form. He denies the possibility of such an anteriority. The question was important for its impact on the interpretation of the biblical account of the creation in six days (see Aquinas 1965, IV, a. 2). There is a clear distinction between Augustine, who discussed the topic many times, and the "other Fathers", including Basil.

After having summarized the various explanations, Thomas gives a theological conclusion: of the positions concerned, he writes, Augustine's is the most subtle and it defends the Scripture best against the contempt of the unbelievers; by contrast that of the "other Saints" is easier (*planior*) and at first sight more consonant with the letter (*magis verbis litterae quantum ad superficiem consona*). But neither the one nor the other of these positions is in disaccord with the truth of faith; for the text (*circumstantia litterae*) is compatible with the sense of both. Given this compatibility with the truth of faith, discussion of each must bear on the intrinsic validity of the arguments put forward.

This judgment gives rise to a few reflections. It should be noted that neither for Thomas nor for the Fathers cited by him is the question of evolution raised in the modern sense of the term. On the other hand the time dimension comes in through the simple fact that the biblical account speaks of six days. But the interpretation of these six days put forward by various commentators is very free: they have recourse to metaphor and some go in the direction of concordism, but it is clear that *creationism* cannot appeal to the patristic tradition.

Moreover, if Thomas contents himself with affirming the compatibility of the various theories with the teaching of the faith, the reason is of an epistemological order. Considered in themselves, these theories do not impose themselves as necessary, they do not have the demonstrative force of science in the Aristotelian sense of the word. They are plausible constructions that do not go beyond the likely or the probable. The lesson needs to be remembered: assertions and theories must be evaluated according to their truth value, i.e. according to the nature of the arguments.

I have referred to this issue because it incites us to epistemological rigor. Not all the claims that compose the theories of evolution are equally certain. Interpretation features extensively here. While there is a starting-point in well-established facts, the resources of the imagination come into play as well. The theologian must however offer a correct hermeneutic of the biblical text — and even more so today than the Fathers or the Scholastics could do.

Creation

3. We have to make a critical effort and liberate our minds from the influence of the positivist mentality so widely dominant in contemporary culture and in philosophical form too. We need fully to recognize the necessity for metaphysics and the capacity of the human intellect for knowing truth beyond the phenomenon, in its profundity and ontological diversity, however imperfectly.

The concept of creation is moreover a metaphysical concept. It is true - as Etienne Gilson demonstrated — that we owe knowledge of the creation historically to the Judaeo-Christian revelation. The encounter with this revelation led to something of a liberation for philosophical reason, which thus took a step away from the conception of emanationism, as current in classical and Hellenistic thought.

In its first and proper sense then, creation signifies the mode according to which all that which is, under whatever form, comes from God, who is the first and universal cause. God is the being subsistent of itself, *Ipsum esse per se subsistens*. The other beings are not of themselves, they receive being, they are by participation.

There exist various levels of participation, so that from the ontological point of view beings are more or less perfect, they have their cause in the first Being, which possesses perfection in fullness, it is perfection itself.

By creation is understood then the production of the totality of beings by the first cause, which is God. Prior to this production nothing is presupposed. Nothing means: non-being. Hence the definition: *creare est aliquid ex nihilo facere*, to make something out of nothing. The word refers to the creating cause. *Something (aliquid)* and *nothing (nihil)* are two metaphysical notions. Both the *something* and the *nothing*, but particularly the latter, need to be explained. When therefore some authors talk about nothing, we need to ascertain whether they understand *nothing* in a radical sense, or as a species of *something* (see Cottier 2002, 131–43; 2008).

The mode of knowing

4. Reflection on the problem of creation encounters the obstacle of our human knowledge, of which the direct object is the material world, and, consequently, the world of the imagination. It is the reason for which our knowledge needs imagination.

Our world is characterized by the phenomenon that the Ancients called the process of "generation and corruption", that is of mutation. This phenomenon, in addition to the process of being born and dying, involves the continuous and multiple transformations, substantial or accidental, of beings. Consequently our primary experiences always include succession and time.

This is reflected in language. *Facere* (active sense) or *fieri* (passive sense) *ex nihilo* seem to indicate an anteriority in time and to suppose that time embraces in its comprehension nothing and being together. In reality this is an illusion, since time, as measure of movement, supposes the existence of things that move.

St Thomas notices this. The spontaneous prereflexive tendency of our mind is to think creation as a mutation. We will have to distinguish then, through a reflexive gaze of the intellect at itself and at its expressions, between the object that we know and the *mode* according to which we know it and express it.

Creation is not a mutation. In a mutation in fact the same thing is presented differently before and after, whether from the substantial point of view or the qualitative or quantitative. But in creation it is the totality of the substance of the things that is produced. Consequently there is not something identical that is presented differently before and after, except in the manner of our representation. In fact, we conceive that a certain thing that did not actually exist before, exists now. It is necessary then that the intellect distinguish critically the content which is thought from its mode of proceeding. In other words, it is a matter of taking away the movement or mutation connected to our mode of conceiving: there remains then the actual content of the concept of creation, which is a relation of causality. Creation places in created reality a relation, so that creation in the creature is the relation to the Creator God, principle of his being (Aquinas 1888–1889, I, q. 45, a. 3). But all relations are not of the same nature. Some are given in reality: the eye is referred to the object seen, vision modifies it. But reason can establish a correlative relation between the object seen and sight, which does not modify the object in fact. We must therefore distinguish between the "real relation" and the "relation of reason". On the basis of the real relation, reason can always establish a relation of correlative reason.

In the active sense, creation expresses the divine essence, with the connotation of a relation to the creature. This relation is a relation of reason. In fact, God, pure act, does not undergo modifications, for that would suppose potentiality and imperfection in him. The relation of the creature to God, however, is a real relation. For all that which is, the creature is oriented to God, on whom he depends in what he is in his existing; he is carried and sustained by Him, he receives everything from Him (Aquinas 1888–1889, I, q. 45, a.3, ad 1).

From this analysis there emerges a conclusion of crucial significance. Movement, becoming, and time have no place in the definition of creation. Creation signifies the radical dependence of the creature on the cause of the totality of being, which is God. Creation is the relation that the creature, in all that he is, has with his Creator.

Consequently, in many cases recourse to the concept of creation in the debates on evolution is inappropriate.

The explanations we have given so far allow us to clarify the meaning of the definition: *creare est aliquid ex nihilo facere.* The *ex nihilo* is the language of mutation proportioned to our habitual mode of perceiving reality. Since *nihil,* Nothing, is the negation of being, there cannot be a something *anterior.* Moreover the actual concept of nothing, which signifies non-being, is a logical instrument that our intellect, in its natural dynamism and tending as it does towards the real — being — develops to be able to enter into the ontological intelligence of reality. The concept of nothing, i.e. of non-being, is produced by the negation of the affirmation — which comes first — of being. To think nothing is to think the absence or radical and absolute negation of being.

The hypothesis of the eternity of the world

5. For Aristotle the world is eternal. This thesis had found a new actuality in the thought of the great Arab commentators. It seemed at first sight to be opposed to the doctrine of creation. Thomas examines the arguments of Aristotle carefully, and his conclusions can be helpful for our discussion (see Aquinas 1888–1889, I, q. 46; 1961, II, c. 31-38; 1965, q. III, c. 13-17).

Between the absolute of God, who is being *per se* and necessary and the beings that receive from Him their being and that can in fact not be, the distinction is radical. To create is an act of the sovereign freedom of God, it proceeds from His will and from His wisdom. This supposes that God is personal. Moreover, the perfection of God is such that if He had not created, nothing would be lacking to his perfection, to which consequently the creation does not add anything.

These elements of doctrine are presupposed in Thomas's critique of the arguments of Aristotle, which are founded on the concept of necessity.

Since the will of God is the cause of beings, for a being to exist necessarily, God must will it, as the necessity of the effect depends on the necessity of the cause. But God wants of necessity and solely his own infinitely perfect good. It is not therefore necessary that God should have willed that the world exist for ever. The world will be eternal if God wants it to be. But an eternal world would itself be created - creation signifying the radical dependence of the created being on the will of the Creator.

Thomas studies the arguments of Aristotle carefully and points out that none is apodictic; they are probable rather than compelling.

The conclusion is not then non-eternity, but the non-necessity of his eternity supposing that he is eternal. In other words philosophy cannot by itself decide whether the world is or is not eternal. That it began in time we know thanks to revelation.

Therefore Thomas opposes theologians, including St Albert, his Mentor, and St Bonaventure who claim to prove the beginning of the world in time. Erroneous claims provoke the contempt of unbelievers.

6. Let us add two observations. The first concerns the hypothesis of an eternal world. Its "eternity" would not be of the same nature as the divine eternity, which,

according to the definition of Boethius, is "tota simul". Rather, an eternal time would involve succession (see Aquinas 1888–1889, I, q. 46, a. 2, ad 5).

Second observation: unlike ancient physics, modern physics studies the origin of the universe in time. Thus it may be thought that there exists a certain convergence between the theory of the big bang and the doctrine of faith. We are not however talking here about a proof in the rigorous sense of the term.

Moreover the claim is not to be considered as an argument in favor of creation. The big bang event does not signify the emergence of *our* universe from nothing. The principle is applicable here: *ex nihilo nihil fit.* The event presupposes the pre-existence of the factors that have produced it. The big bang itself, at the origin of the universe and of time, also supposes creation; it is not creation in the metaphysical sense of the term.

The philosopher must attend to what the experimental sciences are saying. He must moreover propose an interpretation of them that is properly philosophical and alert to the difference in the fields of research, alert as well to the specificity of the questions that pertain to the one or the other field and of the methods proper to them. What pertains to one discipline must not be introduced into the fabric of another. Moreover the lexicological issue is important, for the same words take on a different meaning in one field of knowledge from the one they have in another.

2. The origin of man

7. It is obvious that the problem of the origin of man constitutes a specific and crucial problem for the philosophical interpretation of evolution. It is hardly surprising that the great materialist ideologies of the past century and again the materialist philosophies of today have expected to find confirmation in evolutionism. They look to the origin to explain the nature of the fully constituted result. The questionableness of their demonstration can be seen if we consider the Marxist concept of the "qualitative leap". In my opinion, the explanation does not clarify anything but on the contrary betrays the presence of ignorance and confusion.

On this question then philosophy has different things to say that directly concern the doctrine of faith, which here exercises a regulatory function.

Down through history, the Magisterium of the Church has had to pronounce on diverse aspects of a fairly complex question. The first interventions concerned the humanity of Christ. But they also directly concerned the essence of man, his unity, his spirituality and the immortality of his soul, and the meaning of the resurrection. The Magisterium has also intervened on the transmission of human life and on the vocation of parents as collaborators with God. Our contemporary discourse on the dignity of man, as person willed by God for himself, supposes the biblical revelation of the *imago Dei.* The teaching can be summarized in the two affirmations, and they have the status of principles.

The rational soul is the form of the human body. Pius IX, who put forward the traditional doctrine, used a developed formula: *anima rationalis vera per se et immediate corporis forma* (Denzinger and Schönmetzer 1976, 2828).

In his Encyclical *Humani Generis* (12 August 1950), Pius XII spoke about evolution. Assuming that it is scientifically attested, this concerns the origin of the human body. In fact, *anima enim a Deo immediate creari cattolica fides nos retinere iubet*, the immediate creation of the human soul is a truth of faith (Denzinger and Schönmetzer 1976, 3896).

The human soul

8. The concept of the soul is a key concept in Greek thought. Its meaning is broad and it varies moreover from one author to another. There is the well-known metaphor of Plato on the body as tomb or prison of the soul. It may happen that a plurality of souls is attributed to a single individual. In addition to this there is the added problematic of the *anima mundi.*

Aristotle defines the soul as form of the living body. But his commentators down to the Arabic ones diverge on the nature of the intellect, viewed as a single one for all individuals. First the Western Middle Ages and then again the Renaissance devoted themselves to this issue, and it was the object of a crucial debate.

It was this debate that would necessitate various interventions by the Magisterium in its concern to eliminate all ambiguity on the meaning of the *imago Dei.* It is the intellective soul that is *per se* and immediately the form of the human body. The affirmation is equivalent to positing unequivocally the nature as person of each individual of the human species.

St Thomas perspicaciously establishes the compatibility of intellectuality, that is of spirituality, with the fact of informing matter. This bipolarity of a single soul constitutes the paradox of man. Consideration of evolution cannot prescind from an appraisal of the problem.

The human soul being by its essence intellective or spiritual, it cannot emerge or arise from the potentiality of matter. It is directly created by God.[1] The task of philosophy is to indicate the reasons for this claim.

---

[1] John Paul II (2003, 373) referred to this in his discourse to the Pontifical Academy of Sciences in 1996.

The concept of the soul, something not often referred to today, in particular in the study of the questions that concern us here, seems to be numbered among those outdated ideas that barely elicit a smile of ironic condescension. A philosophical approach to the question of evolution cannot however ignore it. The fact that the Magisterium continues to speak of the soul is a sign that the concept must be valid for the philosopher too.

The difficulty is derived from the profound change undergone by the meaning of the notion. Though the use of the term has persisted, in reality the reference is to something else. In fact, the Aristotelian conception has been replaced by the Cartesian schema of the thought-extension duality, *pensée et étendue*: on the one hand the body which obeys mechanical laws, on the other, the purely spiritual soul. According to this way of seeing, evolution would be reduced to a species of history of bodies.

In parenthesis it should be noted that a conception like the Cartesian one can have a certain heuristic value, as an instrument for research into certain aspects of the living, but not as an image of the living in its specificity and complexity.

The Aristotelian conception is different. Aristotle starts from biology. The soul is the substantial form of the living, the inner principle of its unity and organization and of the coordination of its activities. Living beings are distinguished according to different levels, recognizable by their specific activities, of a vegetative, sensitive, and intellective order.

To each level corresponds a type of soul, the higher form assuming the activities of the lower.

Only God, giving being in its totality, creates. As first cause, his general motion leads and guides the activity of secondary causes, which proceed from nature and from its energies. Evolution is the development in time of the immense activity of these secondary causes that belong to nature, it presupposes the general motion of the first cause.

This development is the actuation of the energies of nature. The appearance of the intellective or spiritual soul as such requires a direct intervention of God. The spiritual soul of each human being does not proceed from the energies of matter, but is created immediately by God. The philosophical reading of evolution recognizes in the progressive perfecting of the sensitive soul of the higher primates a *preparation* for the infusion of the spiritual soul directly by God the Creator. With this infusion man appears in the world.

Purified of erroneous ideas and prejudices that attribute the principal role in generation to the male, Aristotle's position on the growth of the embryo as taken up by Thomas constitutes a valid analogy for the comprehension of evolution. For an understanding of these processes, as Maritain notes, the concept of *causa dispositiva* is essential (Maritain 1992, 573–648).

Among the various issues here to which philosophical reflection must pay attention is the vast world of animal psychology, with the problem of the properly cognitive capacity of the sensitive soul. In man, intellective knowledge has need of the knowledge of the senses — from which it is specifically distinct.

We have seen that in the materialist perspective the concept of a qualitative leap was without any meaning, since everything, *ex hypothesi*, proceeds from the energies of matter. Referring to the first Cause and to its power of innovation, it is legitimate to speak of such a leap, to indicate the irreducibility of thought. But the human intellect is not superimposed on the sensibility of an animal, the intellect and the sensibility are in symbiosis. The problem of the interpenetration, in the distinction, of intellect and sense, is essential for a philosophical doctrine of evolution.

St. Thomas

9. Evolution, attested by scientifically established facts, cannot fail to be of concern to philosophy.

This interest can consist neither in concordism nor in mere transcription. The question must be discussed at the level of the consideration of the being of nature, object of the philosophy of nature. At this level legitimate questions arise that do not necessarily have significance for scientific research. Such a one is teleology. To want at all costs to introduce philosophical concepts into the very fabric of the discourse of the empirical sciences is a methodological error that generates mistakes and nonsense.

There is in general a great readiness to speak of the confrontation between science and faith or science and theology. But theology as such cannot do without the support and mediation of philosophy.

In this perspective, it is worth pondering the potential value of the thought of St Thomas with respect to the subject of evolution. A major reason for pursuing such an investigation is that Thomas the theologian is also a metaphysician of genius. Let me offer some examples.

In the fourth question *De Potentia*, which I referred to at the beginning of this paper (Aquinas 1965, q. IV, a. 2), Thomas emphasizes the advantages of the interpretations of the six days of the creation as proposed in the thinking inspired by St Basil. This interpretation allows us to reject those philosophies for which matter is uncreated or which attribute the formation of lower things to causes other than God, positions that belong to the Greek tradition. It is an approach based on the thesis that the world did not come to its perfection straight out of nothing, but gradually.[2] Of course we are talking about a representation here, but it has the merit of highlighting succession in time.

Thomas's observation must not be forced. In his vision, the important thing is not succession, but gradation: "*In rebus naturalibus gradatim species ordinatae esse videntur: sicut mixta perfectiora sunt elementis, et plantae corporibus minerablibus, et animalia plantis, et homine aliis animalibus*, in the things of nature the species seem to be ordered according to a gradation: as the mixed are the most perfect of the elements, the plants more perfect than mineral bodies, the animals more perfect than the plants and men more perfect than the other animals." (Aquinas 1888–1889, I, q. 47, a. 2)

Man, because he is also an animal — a rational animal — is included in the gradation. The *Summa contra Gentiles* contains reflections on the same subject which have both great anthropological and great metaphysical significance. I am thinking of two texts that complement each other.

In chapter 68 of Book 2, Thomas establishes how an intellective substance, such as the human soul, can be the form of the body. He has recourse to a principle of Dionysius the Areopagite (*De Div. Nom.* C. VII): the divine wisdom has united

---

[2] "quod autem opera sex dierum, secundum alios Sanctos, non fuerint simul sed successive producta, non fuit ex defectu potentiae Creatoris, qui simul, potuisset omnia producere; sed ut ordo divineae sapientae in rerum institutione demonstraretur, qui res ex nihilo in esse producens non statim post nihilum in ultima perfectione naturae eas instituit, sed primo fecit eas in quodam esse imperfecto, et postea eas ad perfectum adduxit, ut sic gradatim ex nihilo ad ultimam perfectionem mundus perveniret; et sic diversis gradibus perfectionis diversi dies deservirent; ut sic et eorum esse ostenderetur a Deo procedere, contra illos qui ponunt materiam increatam, et nihilominus perfectionis rerum ipse Deus auctor appareret, contra illos qui rerum inferiorum formationem causis aliis adscribunt"

"That the works of the six days according to the other saints were produced not simultaneously but by degrees, was not owing to lack of power in the Creator who could have produced all things at once, but was directed to the manifestation of God's wisdom in the production of things, in that when he made things out of nothing he did not at once bring them from nothingness to their ultimate natural perfection, but conferred on them at first an imperfect being, and afterwards perfected them, so that the world was brought gradually from nothingness to its ultimate perfection. Thus different days corresponded to the various degrees of perfection, and it was shown that things derived their being from God, against those who contended that matter was uncreated, and that moreover he is the author of their perfection, against those who ascribed, the formation of the lower world to other causes" (Aquinas 1965, IV, a. 2).

the lowest level (*fines*) of the higher beings to those that are the highest of the lower beings. Thus the supreme in the order of bodies, the human body, attains to the lowest of the higher order, that is to the human soul, which occupies the last level among the intellectual substances. And he cites another Neoplatonic text, the *Liber de Causis* (§ 2, 8) on the human soul as being like a horizon and a boundary between corporeal and incorporeal beings.

> Potest autem obici quod substantia intellectualis esse suum materiae corporali comunicare non possit, ut sit unum esse substantiae intellectualis et materiae corporalis: diversorum enim generum est diversus modus essendi; et nobilioris substantiae nobilius esse. Hoc autem convenienter diceretur si eodem modo illud esse materiae esset sicut est substantiae intellectualis. Non est autem ita. Est enim materiae corporalis ut recipientis et subiecti ad aliquid altius elevati: substantiae autem intellectualis ut principii, et secundum propriae naturae congruentiam. Nihil igitur prohibit substantiam intellectualem esse formam corporis humani, quae est anima huamana. Hoc autem modo mirabilis rerum connexio considerari potest. Semper enim invenitur infimum supremi generic contingere supremum inferioris generis: sicut quaedam infima in genere animalium parum excedunt vitam plantarum, sicut ostrea, quae sunt immobilia, et solum tactum habent, et terrae in modum plantarum adstringuntur; unde et beatus Dionysius dicit, in VII cap. *de Div. Nom.*, quod divina sapientia coniungit fines superiomm principiis inferiorum. Est igitur accipere aliquid supremum in genere corporum, scilicet corpus humanum aequaliter complexionatum, quod attingit ad infimum superioris generis, scilicet ad animam humanam, quae tenet ultimum gradum in genere intellectualium substantiarum, ut ex modo intelligendi percipi potest. Et inde est quod anima intellectualis dicitur esse quasi quidam horizon et confinium corporeorurn et incorporeorum, inquantum est substantia incorporea, corporis tarnen forma. — Non autem minus est aliquid unum ex substantia intellectuali et materia corporali quam ex forma ignis et eius materia, sed forte magis: quia quanto forma magis vincit materiam, ex ea et materia efficitur magis unum. Quamvis autem sit unum esse formae et materiae, non tamen oportet quod materia semper adaequet esse formae. Immo, quanto forma est nobilior, tanto in suo esse superexcedit materiam. Quod patet inspicienti operationes formarum, ex quarum consideratione earum naturas cognoscimus: unumquodque enim operatur secudum quod est. Unde forma cuius operatio superexcedit condtitionem materiae, et ipsa secudum dignitatem sui esse superexcedit materiam. Invenimus enim aliquas infimas formas, quae in nullam operationem possunt nisi ad quam se extendunt qualitates quae sunt dispositiones materiae, ut calidum, fiigidum, humidum et siccum, rarum, densum, grave et leve, et his

> similia: sicunt formae elementorum. Unde istae sunt formae omnino materiales, et totaliter immersae materiae (Aquinas 1961, II, c. 68).[3]

In chapter 22 of Book 3, Thomas discusses the various forms of ordering of things to their ends. He puts forward the doctrine, complementary to the above mentioned one on the ordering of forms, of the aspiration and tendency of matter to an actuality always more perfect, that is to a higher form. Thus the *materia prima* is firstly in potency to the element, then to the mixed, and so on, it desires (*appetit*) the most perfect actualization, the ultimate level of forms, that is to say the human soul. Consequently, all the forms of the cosmos tend to man.

---

[3] "Nevertheless, it may be objected that an intellectual substance cannot communicate its being to corporeal matter in such fashion that the two will be united in the same act of being, because diverse genera have diverse modes of being, and to the nobler substance belongs a loftier being.

Now, this argument would be relevant if that single act of being belonged in the same way to the matter as to the intellectual substance. But it does not. For that act of being appertains to the corporeal matter as its recipient and its subject, raised to a higher level; it belongs to the intellectual substance as its principle, and in keeping with its very own nature. Nothing, therefore, prevents an intellectual substance from being the human body's form, which is the human soul.

Thus are we able to contemplate the marvelous connection of things. For it is always found that the lowest in the higher genus touches the highest of the lower species. Some of the lowest members of the animal kingdom, for instance, enjoy a form of life scarcely superior to that of plants; oysters, which are motionless, have only the sense of touch and are fixed to the earth like plants. That is why Blessed Dionysius says in his work On the Divine Names that "divine wisdom has united the ends of higher things with the beginnings of the lower." We have, therefore, to consider the existence of something supreme in the genus of bodies, namely, the human body harmoniously tempered, which is in contact with the lowest of the higher genus, namely, the human soul, which holds the lowest rank in the genus of intellectual substances, as can be seen from its mode of understanding; so that the intellectual soul is said to be on the horizon and confines of things corporeal and incorporeal, in that it is an incorporeal substance and yet the form of a body. Nor is a thing composed of an intellectual substance and corporeal matter less one than a thing made up of the form of fire and its matter, but perhaps it is more one; because the greater the mastery of form over matter, the greater is the unity of that which is made from it and matter.

But, though the form and the matter are united in the one act of being, the matter need not always be commensurate with the form. Indeed, the higher the form, the more it surpasses matter in its being. This fact is clearly apparent to one who observes the operations of forms, from the study of which we know their natures; for, as a thing is, so does it act. That is why a form whose operation transcends the condition of matter, itself also surpasses matter in the rank of its being.

For we find certain lowest-grade forms whose operations are limited to the class of those proper to the qualities which are dispositions of matter; qualities such as heat, cold, moisture and dryness, rarity and density, gravity and levity, etc. And those forms are the forms of the elements: forms which therefore are altogether material and wholly embedded in matter."

> Cum vero, ut dictum est, quaelibet res mota, inquantum movetur, tendat in divinam similitudinem ut sit in se perfects; perfectum autem sit unumquodque inquantum fit actu: oportet quod intentio cuiuslibet in potentia existentis sit ut per motum tendat in actum. Quanto igitur aliquis actus est posterior et magis perfectus, tanto principalius in ipsum appetitus materiae fertur. Unde oportet quod in ultimum et pefectissimum actum quem materia consequi potest, tendat appetitus materiae quo appetit formam, sicut in ultimum finem generationis. In actibus autem formarum gradus quidam inveniuntur. Nam materia prima est in potentia primo ad formam elementi. Sub forma vero elementi existens est in potentia ad formam mixti: propter quod elementa sunt materia mixti. Sub forma autem mixti considerata, est in potential ad animam vegetabilem: nam talis corporis anima actus est. Itemque anima vegetabilis est in potentia ad sensitivam; sensitiva vero ad intellectivam. Quod processus generationis ostendit: primo enim in generatione est fetus vivens vita plantae, postrnodum vero vita animalis, demum vero vita hominis. Post hanc autem formam non invenitur in generabilibus et corruptibilibus posterior fonna et dignior. Ultimus igitur finis generationis totius est anima humana, et in hanc tendit materia sicut in ultimam formam. Sunt ergo elementa propter corpora mista; haec vero propter viventia; in quibus plantae sunt propter animalia; animalia vero propter hominem. Homo igitur est finis totius generationis (Aquinas 1961, III, c. 22).[4]

In my opinion these reflections are precious and fruitful for a philosophical consideration of the human significance of evolution.

A metaphysics of the act of being (*actus essendi*) is in essence attentive to the dynamism of being that is evident also at the level of nature. It offers pointers for a philosophical reflection on evolution, whose mediation is indispensable if fideism is to be avoided.

---

[4] "As we said, since any moved thing, inasmuch as it is moved, tends to the divine likeness so that it may be perfected in itself, and since a thing is perfect in so far as it is actualized, the intention of everything existing in potency must be to tend through motion toward actuality. And so, the more posterior and more perfect an act is, the more fundamentally is the inclination of matter directed toward it. Hence. in regard to the last and most perfect act that matter can attain, the inclination of matter whereby it desires form must be inclined as toward the ultimate end of generation. Now, among the acts pertaining to forms, certain gradations are found. Thus, prime matter is in potency, first of all, to the form of an element. When it is existing under the form of an element it is in potency to the form of a mixed body; that is why the elements are matter for the mixed body. Considered under the form of a mixed body, it is in potency to a vegetative soul, for this sort of soul is the act of a body. In turn, the vegetative

References

Aquinas Th., 1961, *Liber de veritate catholicae Fidei contra errores infidelium seu Summa contra Gentiles*, Marc P., Pera C., Caramello, P. (eds.), Taurini-Romae: Marietti.

Aquinas Th., 1965, *Quaestiones disputatae de potentia*, Pession P. M. (ed.), 10th edn., Taurini-Romae: Marietti.

Aquinas Th., 1888–1889, *Opera omnia iussu impensaque Leonis XIII P. M. edita, t. 4-5: Pars prima Summae theologiae*, Romae: Ex Typographia Polyglotta S. C. de Propaganda Fide.

Cottier G. M., 2002, "La doctrine de la création et le concept de néant", in Cottier G. M., *Le désir de Dieu*, chap. VII, Paris: Parole et Silence.

Cottier G. M., 2008, "La doctrine philosophique et théologique de la création chez Thomas d'Aquin", contribution to the Plenary Session of the Pontifical Academy for the Sciences on Evolution.

Denzinger H., Schönmetzer A., 1976, *Enchiridion symbolorum definitionum et declarationum de rebus fidei et morum*, Fribourg: Herder.

John Paul II, 2003, "Address to the Pontifical Academy of Sciences", October 22 1996, in *Papal Addresses to the Pontifical Acadmy of Sciences 1917–2002 and to the Pontifical Academy of Social Sciences 1991–2002*, Vatican City: Ex Aedibus Academicis, 370–374.

Maritain J., 1992, "Vers une idée thomiste de l'évolution", in *Approches sans entraves*, in Maritain J. and Maritain R., *Oeuvres completes*, XIII, Fribourg-Paris: St Paul.

soul is in potency to a sensitive soul, and a sensitive one to an intellectual one. This the process of generation shows: at the start of generation there is the embryo living with plant life, later with animal life, and finally with human life. After this last type of form, no later and more noble form is found in the order of generable and corruptible things. Therefore, the ultimate end of the whole process of generation is the human soul, and matter tends toward it as toward an ultimate form. So, elements exist for the sake of mixed bodies; these latter exist for the sake of living bodies, among which plants exist for animals, and animals for men. Therefore, man is the end of the whole order of generation." Maritain (1992) takes inspiration on this text.

Why teleological principles are inevitable for reason. Natural theology after Darwin

Vittorio Hösle

In grateful memory of the eminent historian of science Matthias Schramm (1928–2005)

One of the most amiable characteristics of Charles Darwin is doubtless his modesty. In his *Autobiography*, which was begun in 1876 and continued to receive additions till 1882, he himself sincerely avers that his works has been "over and over again greatly overpraised" (1993a, 126) and never claims to be particularly talented in metaphysical matters: "My power to follow a long and purely abstract train of thought is very limited; I should, moreover, never have succeeded with metaphysics or mathematics." (140) Needless to say, he understood that his revolution in biology had transformed the concept of species and lessened the gap between humans and other animals; for he showed in detail how the principle of natural selection could explain innumerable facts in the organic realm that had been previously regarded as irreducible, including the existence of species and human behavior in all his facets. Still, the enormous impact that *On the origin of species* of 1859 had immediately outside of the scientific community of biologists on the public at large and its quick spill-over to religious and philosophical issues seem quite surprising.[1] Only one sentence at the end of the book hints to the origin of humankind (1968, 458), dealt with in his later works of 1871 and 1872, and there are no explicitly antireligious consequences drawn in this or in any of Darwin's published books. Only at the end of *Variation of animals and plants under domestication* of 1868 did Darwin ask, fully aware of transgressing the limits of his own discipline, some critical questions regarding the compatibility of the mechanism

[1] A good overview of classic texts of the debate can be found in Young and Largent (2007).

of natural selection with traditional theism. Darwin's intellectual self-restraint, which distinguishes him so strongly from some of the contemporary neo-Darwinists as, e.g., Richard Dawkins,[2] was not simply a cautionary move, due to concern for his wife or fear of social ostracism: he had thought seriously about religious issues in the years immediately after his return from his voyage around the world and had come to the conclusion that his theory did not entail atheism. It is thus proper to start the following reflections with an analysis of his quite subtle religious ideas, which are certainly not inferior to those of most contemporary scientists addressing religious issues (I). I will then put them into the context of a more general development of the concept of God and nature and discuss the most important attempt of the nineteenth century to render Darwinism compatible with theism, namely that by Asa Gray. The focus on Gray is due to the simple reason that progress in philosophy is, if it takes place at all, of a very different nature than in science;[3] therefore, we can learn from philosophical reflections of past centuries far more than we can learn from scientific theories of the same time (II). Finally, I will discuss in which sense even after the Darwinian revolution a teleological interpretation of nature remains not only a possibility, but a necessity (III).

I.

In the chapter of his *Autobiography* dedicated to "Religious belief" — pages published in complete form not earlier than 1958, since the first edition of 1887 had been subjected to family censorship — Darwin describes the reasons that led to the fading of the Anglican belief in which he had been educated. After all, he had for some time contemplated becoming an Anglican minister himself. He claims, and as far as we know his development, truthfully, that he had never had strongly developed religious sentiments (91); certainly he had never been an Evangelical. But at the same time he had imbibed traditional Anglican theology, and at the beginning of his voyage on the "Beagle," during which his favorite book was Milton's *Paradise Lost*, he was sometimes scorned by orthodox officers "for quoting the Bible as an unanswerable authority on some point of morality." (85) How did the slow evaporation of these beliefs, which he in vain tried to resist, come about? The reasons and causes mentioned by Darwin have relatively little to do with his great scientific discovery. What are they? We can subdivide the chapter in question into three main sections. In the first (A), Darwin analyzes the arguments against

[2] See particularly (2006).

[3] See the defense of a cyclical theory of the history of philosophy in Hösle (1984).

the Bible being based on divine revelation; in the second (B), he moves from revealed to natural theology. Here he discusses both the arguments in favor of (B 1) and those speaking against (B 2) the existence of God, among which the theodicy problem figures; and he mentions also the logically independent issue of the immortality of the soul (B 3). Thirdly (C), he reflects on his own development — one might say, on the evolution of his own religious beliefs and on their epistemological status. Such reflections are found at the end of the chapter, but also B is interspersed with them; I will discuss them in the context of (C).

(A) With regard to the first point, he deals separately with the Old and the New Testament. Concerning the former, he mentions on the one hand historically false stories narrated in it, on the other hand the primitive feelings ascribed to God (like those "of a revengeful tyrant"). Concerning the second, he is disturbed by the connection of the New with the discredited Old Testament. Additionally, the intrinsic incredibility of miracle stories, the credulity of the people living in those times, the impossibility of excluding that the Gospels were written later than the events narrated, as well as the contradictions among the evangelists, led him "to disbelieve in Christianity as a divine revelation."(86) He continues to recognize the beauty of the morality of the New Testament, but sees that it partly depends on later interpretations that did not capture the originally intended meaning. This original meaning comprehends, e.g., the doctrine of everlasting punishments of non-believers, and he declares this doctrine "damnable." (87) Clearly, there is nothing in these arguments connected to natural science: They are based on ethical ideas and historical research; and even the general skepticism with regard to miracles is not so much a result of the scientific development as its presupposition. Most of the arguments mentioned as well as others directed against the *formal* idea of revelation had already gained currency in the eighteenth, some in the seventeenth century (think only of Spinoza and Lessing), even if they had had more impact on the Continent than in England. But David Friedrich Strauss' *Das Leben Jesu, kritisch bearbeitet* of 1835/36 came out as *The Life of Jesus Critically Examined* already in 1846 in the translation of Marian Evans, later better known under her pen-name George Eliot, and clearly this book changed our relation to the New Testament as no other before or afterwards.[4] Darwin himself recognizes that the arguments proposed do not have "the least novelty or value". By "value" he can only mean "value as original contributions", hardly "validity", for he insists that he has "never since doubted even for a single second that my conclusion was cor-

[4] On the philosophical struggles to make sense of the Bible see Hösle (1999).

rect" (87) — one of the passages whose publication was prevented by Emma Darwin. The collapse of the naïve belief in revelation, by the way, does not exclude an understanding of the core Christian message, insofar as it is justifiable by reason, as a manifestation of the divine and of the whole of the Bible as a progressive revelation; and it certainly did not prevent the growth of natural theology in early modernity, but on the contrary stimulated it. It was only the latter's crisis that hastened the decline of theism. Let us see what Darwin contributed to this crisis.

(B 1) In the second section he begins by stating in a very concise manner the contribution of his theory to natural theology. "The old argument of design in nature, as given by Paley, which formerly seemed to me so conclusive, fails, now that the law of natural selection has been discovered." The sentence claims neither that the theory of natural selection[5] is *incompatible* with theism nor that it has *destroyed all proofs of God.* In the most natural, even if admittedly not the only grammatically possible, interpretation, it does not even state that the argument of design has been overthrown by the new theory; but it limits this claim to the argument of design *as exposed by Paley.* Darwin had thoroughly studied Paley in Cambridge; he mentions that in order to pass his B.A. exams, he had to read his works *A View of the Evidences of Christianity* as well as *The Principles of Moral and Political Philosophy*, but obviously here he has in mind his last work, the *Natural Theology* of 1802 with the famous watchmaker comparison,[6] a book whose logic "gave me as much delight as did Euclid. The careful study of these works, without attempting to learn any part by rote, was the only part of the Academical Course which, as I then felt and as I still believe, was of the least use to me in the education of my mind." (59) Still, one could understand Darwin's earlier quoted sentence as implying that Paley's special creationism is the only valid form of the argument of design. But that this is not the case is proven by the context. For Darwin continues: "Everything in nature is the result of fixed laws" and then: "But passing over the endless beautiful adaptations which we everywhere meet with, it may be asked how can the generally beneficent arrangement of the world be accounted for?" (87 f.) Here Darwin seems to recognize the point, both quite obvious and insisted upon by a friend like Asa Gray, that the set of the laws of nature within which natural selection alone can operate is not itself the result of natural selection. The mechanism of natural selection

---

[5] The intentionalist meaning of "selection" has to be rejected when thinking about natural selection. "In the literal sense of the word, no doubt, Natural Selection is a false term" (Darwin 1988a, 66)

[6] In the twelfth edition, alone accessible to me, see (1809 1 ff., 416 ff.)

can explain the adaptation of single organisms and species only under the presupposition of a system of laws that render such an evolution possible. If no planet had water, e.g., there would be no life, given the set of laws of our world; and it is easy to imagine other worlds in which natural selection could not lead to forms of life more complex than prokaryotes. Thus, the correct interpretation of Darwin's claim must be that it is directed only against a form of the physicotheological argument using special creationism. The causes for the development of the various species are immanent. But as long as we see ends *in the world as a whole*, not in the single species, we should still be allowed to ask why the world is such that in it awe-inspiring entities like organisms exist. In this form, so one could say, even if Darwin does not use this language, the cosmological and the physicotheological proof merge: we are searching for a cause not simply of the world, as in the cosmological proof, nor of the adaptation of the single species, as in the traditional physicotheology, but of a world that allows for the evolution of organisms as we know them. And indeed Darwin does not reject this type of argument: "This follows from the extreme difficulty or rather impossibility of conceiving this immense and wonderful universe, including man with his capacity of looking far backwards and far into futurity, as the result of blind chance and necessity. When thus reflecting I feel compelled to look to a First Cause having an intelligent mind in some degree analogous to that of man; and I deserve to be called a Theist." (93) We will return to this passage, since in a later addition to it Darwin weakened the statement.

(B 2) Darwin then addresses the classical theodicy issue, i.e. the existence of evil as an argument *against* the existence of God. Of the various types of evil he mentions only suffering. At the beginning of his reflections, he does not seem to presuppose that the world has to be the best possible, or at least such that it could not be better, in order to justify the ways of God; he seems (but the appearance is deceiving) to be satisfied with the balance being in favor of happiness.[7] And this the sanguine Darwin clearly affirms, even if he is aware that the weighing of pleasure and suffering is very difficult.[8] But he sees in natural selection, which presupposes an overproduction, an argument for his belief. "According to my judgment hap-

---

[7] Leibniz sometimes uses the balancing argument, even if it cannot prove that the actual world is the best possible; see *Essais de théodicée* § 258 (1978, VI 269).

[8] A modern version of this belief is forcefully articulated by J. Balcombe (2006). The contrary position has been urged most forcefully by the pessimist philosopher Arthur Schopenhauer, who sarcastically asks us to compare the pleasure of an animal devouring another with the pain of the devoured one (Parerga and Paralipomena § 149; 1977, 9.317). But the comparison is

piness decidedly prevails, though this would be very difficult to prove. If the truth of this conclusion be granted, it harmonizes well with the effects which we might expect from natural selection. If all the individuals of any species were habitually to suffer to an extreme degree they would neglect to propagate their kind; but we have no reason to believe that this has ever or at least often occurred. Some other considerations, moreover, lead to the belief that all sentient beings have been formed so as to enjoy, as a general rule, happiness." (88) Darwin then adds that if suffering lasts long it diminishes the power of action, while pleasurable sensations do not have this depressing effect; "on the contrary they stimulate the whole system to increased action." Thus it is reasonable that natural selection uses pleasure rather than pain as stimulus for action, even if the latter "is well adapted to make a creature guard itself against any great or sudden evil." (89) Darwin mentions various types of pleasure, as those from exertion of the body and mind and from sociability, and sums up: "The sum of such pleasures as these, which are habitual or frequently recurrent, give, as I can hardly doubt, to most sentient beings an excess of happiness over misery, although many occasionally suffer much. Such suffering is quite compatible with the belief in Natural Selection, which is not perfect in its action." (89 f.) At the end of the third chapter of *The Origin* he will console himself with the reflection "that the war of nature is not incessant, that no fear is felt, that death is generally prompt, and that the vigorous, the healthy, and the happy survive and multiply." (1968, 129) Nevertheless, the existence of any amount of suffering remains a disturbing problem, and this even more among animals than among humans, since the latter, and only they, may benefit from suffering and improve morally.[9] "This very old argument from the existence of suffering against the existence of an intelligent first cause seems to me a strong one." (90) This sentence proves that Darwin did regard the problem of suffering as a serious challenge to theism. However, he did not believe that his own theory had increased the difficulty of the problem. On the contrary, he thought that it had strengthened the idea that the balance is in favor of pleasure — an idea which is certainly not a sufficient, but a necessary presupposition of any solution to the problem.

---

unfair, for what should be looked at is the pleasure that the animal now devoured enjoyed earlier in its whole life, not the momentary pleasure of the predator. Schopenhauer furthermore believes that, although a violent end is not as common among humankind as among other animals, still human life is more miserable, because the pain is not simply transient, but both anticipated and remembered (§ 153; 1977, 9.318 ff.).

[9] With regard to this issue, Darwin agrees with Schopenhauer (Parerga and Paralipomena §173; 1977, 9.351).

However, there is one feature of the Darwinian theory that in his eyes did not simply destroy one of the *proofs* of God, but which might be directly *inconsistent* with God's existence. In our chapter of the *Autobiography*, he merely alludes to it and refers to the last pages of *Variation of animals and plants under domestication*, affirming that "the argument there given has never … been answered". What is his argument? There Darwin compares the beneficial effect of natural selection based on variation to the use of stone fragments whose shape is the result of various accidents by an architect. As he generally does in his work, Darwin embraces determinism; that is, he insists on the existence of sufficient causes for the shape of the stones, even if they are not better known than the causes of variation were at his time. "And here we are led to face a grave difficulty, in alluding to which I am aware that I am travelling beyond my proper province. An omniscient Creator must have foreseen every consequence which results from the laws imposed by Him." (1988b, 371) But is it plausible to assume that God has really intended all variations, even those that lead to particularly repellent behavior of animals? And if we deny this in one case, how can we maintain it in another? "However much we may wish it, we can hardly follow Professor Gray in his belief 'that variation has been led along certain beneficial lines,' like a stream 'along definite and useful lines of irrigation'." Two things particularly worry Darwin: the plasticity of organization, which leads also to harmful deviations, and the redundant power of reproduction, which, according to the Malthusian basis of his theory, entails the struggle for existence. These "must appear to us superfluous laws of nature. On the other hand, an omnipotent and omniscient Creator ordains everything and foresees everything. Thus we are brought face to face with a difficulty as insoluble as is that of free will and predestination." (372) We will come back both to Gray's theory and to Darwin's argument which seems to challenge most directly theism, since it states that the mechanism of natural selection, whose existence cannot be doubted, is somehow "superfluous" for an omnipotent and omniscient Creator.

(B 3) In the *Autobiography*, Darwin only shortly addresses the problem of immortality. He thinks that the belief in it is "strong and almost instinctive", since the thought seems intolerable that life will become extinct when the sun will grow too cold for life. "To those who fully admit the immortality of the human soul, the destruction of our world will not appear so dreadful." (92) Still, the reader does not have the impression that Darwin himself shares that belief.

(C) The passage just quoted shows that in Darwin's view the fact that a belief is almost instinctive does not prove that it is true. Indeed one of the most amazing features of the chapter is the capacity of observing one's own beliefs, as it were, from outside. Darwin offers, so to speak, a "natural history" of his own religious views, as he had done with those of humankind in the second chapter of *The Descent of Man, and Selection in Relation to Sex* (1981, I 65 ff.). This stance is partly based on his work as a biologist, which included, particularly in *The expression of the emotions in man and animals*, unbiased observation not only of his children, but also of himself. Partly it presupposes the rejection of an intuitionist epistemology, i.e. the position that there are indemonstrable and nevertheless legitimate basic convictions concerning religious matters. From at least Friedrich Heinrich Jacobi to Alvin Plantinga, philosophers of religion have used such an epistemology, which is indeed tempting; for it is not easy to see how without basic beliefs the whole idea of justification can begin to make any sense. And once this is granted, it is hard to deny our religious basic beliefs the same status that we grant to our non-religious basic beliefs.

Darwin, however, pokes fun at this position by quoting the remark of an old lady directed against Darwin's notoriously skeptical father: "I know that sugar is sweet in my mouth, and I know that my Redeemer liveth." (96) His main objection against the epistemological position that such "inward convictions and feelings are of any weight as evidence of what really exists" is that different people often vary in their basic convictions, and that sometimes these are logically incompatible with each other. Among religions, e.g., there is no consensus. "But it cannot be doubted that Hindoos, Mahomadans and others might argue in the same manner and with equal force in favor of the existence of one God, or of many Gods, or as with the Buddists of no God." (90 f.; analogously with regard to moral beliefs, 99) What is even more, our own convictions change over time. Darwin describes the almost religious awe he felt as a young naturalist in the Brazilian forest, "but now the grandest scenes would not cause any such convictions and feelings to rise in my mind. It may be truly said that I am like a man who has become color-blind." Darwin even thinks that his earlier quasi-religious feelings were in fact not of a religious nature, but more kindred to the sense of sublimity (92). He seems to suggest that the religious dimension was an *interpretation* of raw qualia that in themselves were not at all necessarily connected with the idea of God. In general, he reports an increase of skepticism in the development of his mind: "Nothing is more remarkable than the spread of skepticism or rationalism during the latter half of my life." (95) However, this self-distancing is not applied to the moral feelings. "The highest satisfaction is derived from following certain impulses, namely the social instincts." (94) Sometimes we are allowed, even obliged, to act against the expectations of other people; then a person "will still have the solid satisfaction of knowing that he has followed his innermost guide or conscience." (95) But it is not clear why we should regard ourselves more bound by our conscience than by our religious feelings; after all, also the demands of conscience vary among different people.

This self-distancing plays a central role in the already mentioned later addition to the passage in which Darwin declares himself to be a theist. He mentions that his conviction of the existence of a First Cause is based on a process of inference: "But then arises the doubt — can the mind of man, which has, as I fully believe, been developed from a mind as low as that possessed by the lowest animal, be trusted when it draws such grand conclusions?" (93) The belief in God was after all inculcated into children, and it could well be that the connection between cause and effect is not necessary, "but probably depends merely on inherited experience." (93) Therefore, Darwin declares himself no longer a theist, but an agnostic.[10]

II.

Darwin's religious ideas are not revolutionary, but are based on the development of science and philosophy in early modernity. There is little doubt nowadays among historians of science that the miracle of modern science was profoundly rooted in a religious vision of the world.[11] The latter offered both a moral justification to the novel enterprise of easing human life through the development of science and technology and a metaphysical foundation to the basic concept of universal natural laws that do not tolerate any exception, a concept still alien to the ancient world, which lacked its theological presuppositions.[12] It is true that several of the fathers of modern science were not orthodox Christians, i.e., they rejected the doctrines of Trinity and/or of the divinity of Christ, as Isaac Newton did; but most of them were committed theists or deists. The distinction between the latter two concepts, by the way, is fuzzy and often misleading. This is partly

---

[10] Already on February 17th, 1861he had written to Asa Gray: "With respect to Design & … I have no real objection, nor any real foundation, nor any clear view. — As I before said I flounder hopelessly in the mud." (1994, 30) Darwin refers to various earlier letters, as, e.g., from November 26, 1860 (1993b, 496). The correspondence with Gray encompasses in a fascinating mixture botanical, theological, and political issues.

[11] See Yolton (1990).

[12] Suffice it to point to the end of the second book of Aristotle's *Physics*, where the author rejects the primitive pre-Darwinian theory of Empedocles, who does not yet use differential reproductive rates (198b10 ff.). The whole argument works only because Aristotle ignores the concept of natural law, as Hans Wagner rightly observes (1979, 479).

due to the fact that the various elements used to define "deism" are not equivalent logically and are not instantiated simultaneously empirically. Hegel, e.g., like Tindal or Toland, does not accept a revelation beyond reason, but takes the doctrine of Trinity more seriously than any other philosopher of modernity; Dickens does not believe in the divinity of Christ, but regards the morals of the Gospel as unsurpassed and unsurpassable. For historians of theology, a multidimensional continuum between traditional theism and natural or rational theology is more helpful than the strict opposition of the concepts.[13]

Another reason why the distinction between deism and theism creates more problems than it solves is that one of the main dividing issues — whether God only created the world or continues to intervene in it — is amenable to an intermediate solution: God operates according to the laws created, but in the development of the world according to His laws, He himself is present, in certain axiologically relevant events more than in others. Immediately after the publication of *The Origin*, a famous and bestselling volume by liberal theologians, *Essays and Reviews*, triggered disciplinary measures by the Church of England. (This did not prevent, however, one of its authors, Frederick Temple, from becoming later the Archbishop of Canterbury.) In this book, the Reverend Baden Powell, who had rejected on theological grounds special creationism even before Darwin's book, argued that the belief in a world order without interruptions, i.e., without miracles, shows more respect to God than the assumption that he has periodically to fix his creation by special interventions (1860, 114).

Where does this conception originate? The idea that divine action is necessarily channeled according to the laws of nature was forcefully articulated by Spinoza. But the laws are not sufficient; every event must have also a secondary cause, none can be explained only by the eternal properties of God, i.e., his natural laws.[14] The attractiveness of this idea is not limited to the pantheist context, within which Spinoza develops it; nor does the idea as such entail a theory of the substantial

[13] The decline of deism in the nineteenth century has various reasons — one, no doubt, that it had penetrated and transformed orthodox Christianity. It became superfluous. — I use "natural theology" as equivalent of "rational theology", even if one may argue that natural theology is the subset of rational theology that uses only a posteriori arguments for God's existence. Rational theology is broader, since it recognizes also a priori arguments such as the ontological and the moral ones.

[14] See *Ethica* I 28. Curley (1969) recognized that by doing so Spinoza anticipated the Hempel-Oppenheim scheme of explanation.

identity of body and mind or a rejection of any kind of physico-theology, even if Spinoza himself has only scorn for it as for any realist theory of value.[15] As is well-known, the search for the ends of God in the creation of various natural objects was an important part of early science — think of the Boyle Lectures, which start in 1692. In England, this tradition lasted much longer than on the Continent, namely till the Bridgewater Treatises (1833–1840). But already in the eighteenth century, the various "Theologies of Insects" etc. were widely made fun of, for example by Voltaire. In his masterful book *Natur ohne Sinn?*, Matthias Schramm has shown how the research by Leibniz, Euler, and Maupertuis that finally led to extreme principles, particularly the principle of least action, was already a reaction against the theological absurdities that found God in the most limited means-end-relations, either between organs and their organisms or, worse, between other species and humans. A formal teleology was regarded as more promising than the traditional one oriented towards concrete contents (Schramm 1985, 71).[16] Leibniz is the most prominent among those thinkers who mediate between the Spinozian postulate that God necessarily acts through laws and a finalistic interpretation of the world, which no theist can give up. Leibniz accepts the principle of sufficient reason, and thus an all-pervasive system of efficient causes, but at the same time teaches that they do not logically exclude final causes. The world of consciousness cannot be grasped without finality, and even the system of the laws of nature, particularly the laws of movement, are not simple consequences from logic, but presuppose an axiological principle.[17] God chooses among all possible worlds the world with the maximal value: His wisdom is manifested in the choice of the whole, not in that of single ends.

Darwin's rejection of special creationism was nothing else than the application to species of Spinoza's idea that God acts only through antecedent conditions and general laws: Species are not eternal features of the world, and thus also their existence has to be explained by antecedent conditions (earlier species) and general

[15] *Ethica* I Appendix.

[16] See, e.g., Kant, *The Only Possible Argument in Support of a Demonstration of the Existence of God*, A 63 f. Kant speaks of the parsimony of nature (A 63, 141).

[17] See *Principes de la Nature et de la Grace, fondés en raison* (1978, VI 598-606; 603). One of the great insights of Kant's aforementioned work is that even the existence of logically necessary laws of nature can be interpreted as a manifestation of divine reason (A 56, 101, 118); thus, he combines Spinoza and Leibniz. Interesting are, e.g., his reflections on the geometric properties of the hexagon and its importance in nature (A 95, 141, 152). Compare Darwin (1968, 247 ff.) on the bees' comb.

laws (among which the mechanism of natural selection is pre-eminent). Darwin had indeed a profound awe for universal laws, and this awe was one of the forces behind his discoveries. His son William wrote shortly after his death about Darwin: "As regards his respect for the laws of nature it might be called reverence if not a religious feeling. No man could feel more intensely the vastness or the inviolability of the laws of nature…"[18]

The rationalist frame of 17th and 18th century philosophy was communicated to the early Darwin not by Spinoza or Leibniz, but by William Whewell[19] and John Frederick William Herschel, both of whom he personally met and neither of whom ever accepted his theory.[20] Among the great philosophers the most important influence was David Hume, which may explain the increasing skepticism of the aging Darwin with regard to religious issues. I already alluded to Hume's influence when I spoke about Darwin's offering a "natural history" of his own religious views, and indeed the early notebooks show a far-reaching impact of the Scottish thinker.[21] The works quoted are *A Treatise of Human Nature*, *An Enquiry Concerning Human Understanding*, *A Dissertation on the Passions*, *The Natural History of Religion*, and the *Dialogues Concerning Natural Religion*. The latter book must have been particularly fascinating for Darwin, because it develops a powerful critique of the argument from design, even if long before Darwin's own theory — which, interestingly, is adumbrated in Part VIII of the book (Hume 1947, 185). Hume's influence is also present in Darwin's rejection of free will;[22] and in the late *Autobiography* the main argument against the validity of the cosmological proof reflects Hume's criticism of the concept of cause. Also Hume's idea of explaining the evolution of religion according to universal laws is mirrored in Darwin's criticism of a philosopher "who says the innate knowledge of creator <is> <<has been>>

[18] This unpublished letter of January 4, 1883 to Francis Darwin is quoted according to Sloan (2005, 143).

[19] Whewell's third Bridgewater Treatise *Astronomy and General Physics Considered with Reference to Natural Theology* is quoted already in the notebooks (C 72; 1987, 262 and C 91; 1987, 266) and will offer the first of the two citations placed at the beginning of *The Origin*. The passage can be found at the beginning of the eighth chapter of the third book (1833, 267).

[20] An intermediate step between the general theory and its application to biology was the uniformitarianism of Hutton and Lyell. See Gray (1884, 109) and Huxley, who called "the 'Origin of Species' the logical sequence of the 'Principles of Geology'" (1896, 232).

[21] See C 270 (1987, 321), C 267 (1987, 326), M 104 (1987, 545), M 155 (1987, 559), N 101 (1987, 591), N 184 (1987, 596).

[22] M 27 (1987, 527), M 30 f. (1987, 526 f.), N 49 (1987, 576 f.).

implanted in us ... by a separate act of God, & not as a necessary integrant part of his most magnificent laws. of which we profane <<degnen>> in thinking not capable to <do> produce every effect, of every kind which surrounds us." (M 136; 1987, 553). This passage is significant because it proves that the rejection of special creation or revelation in the early Darwin went together with a commitment to theism or deism. Indeed, Darwinism as such is compatible with a both a Spinozian and a Leibnizian interpretation, and it is not science, but philosophy itself that has to decide which of the two interpretations is the more plausible one.

Most prominent among those of Darwin's contemporaries who understood this point and even tried to integrate the idea of natural selection into a theistic worldview was his friend and admirer, the great American botanist Asa Gray, one of the few persons to whom Darwin had communicated his theory before it was published.[23] (In fact, it was a letter written by Darwin to Gray in 1857 that was read on July 1st, 1858 in the Linnean Society in order to prove that Darwin had developed his theory before he knew about Wallace's analogous one.) Gray, who was an orthodox Protestant and had a remarkable philosophical intelligence, defends the central tenets of Darwinism with acumen against its critics (e.g., Louis Agassiz), but maintains that it is nothing other than an hypothesis — however, the best hypothesis available, which theologians are well advised to respect (1880, 458 ff.; 1884, 260 f.). His reflections on the philosophical consequences of Darwinism impress the reader because of his complete grasp of the theory,[24] his awareness of the methodological and epistemological issues at stake, his elegant style, his sober religiosity, his impartiality and even respect for different opinions and, last not least, his subtle humor.[25] It is thus not surprising that his various essays on Darwinism and religion were soon, namely in 1876, collected in a volume (reissued in 1884) and that in 1880 he was invited to deliver two lectures on *Natural Science and Religion* at the Theo-

[23] Another important Christian Darwinist was St. George Mivart, whose relations with Darwin, however, quickly worsened after the publication of his *On the Genesis of Species* of 1871.

[24] "I declare that you know my Book as well as I do myself," Darwin wrote him on July 22, 1860 (1993b, 298).

[25] I may mention that the copy of *Natural Science and Religion* I am using (from the Libraries of the University of Notre Dame) is graced by an autographed dedication in Latin by the author to the "amicissimo" B. Peirce — no doubt his Harvard colleague Benjamin Peirce, the noted mathematician, devout Christian and father of the greatest American philosopher, Charles Sanders Peirce, whose late philosophy of evolutionary love has some similarities with Gray's theory.

logical School of Yale College.[26] His first text, published already in March 1860 in the *American Journal of Science and Arts*, is a review of *The Origin*, which at the end addresses the philosophical and theological consequences of Darwin's intellectual breakthrough. He rightly states that the author is silent about how he "harmonizes his scientific theory with his philosophy and theology" (1884, 56) and tries to reconstruct its implicit philosophy. Alluding to Paley's watchmaker analogy, he wittily asks: "What is to hinder Mr. Darwin from giving Paley's argument a further *a-fortiori* extension to the supposed case of a watch which sometimes produces better watches, and contrivances adapted to successive conditions…?" (57) Gray insists on the fact that God has to be conceived as non-temporal; and he says that every philosophical theist must adopt either the idea that the Creator's intervention is done from all time or through all time.[27] He sees dangers connected to both views, the first leading to atheism, the second to pantheism, and clearly prefers the second. "Natural law, upon this view, is the human conception of continued and orderly Divine action." (58) In order to undermine any distinction between initial creation and later interventions, Gray appeals, in my eyes ironically, to "profounder minds to establish, if they can, a rational distinction in kind between his <sc. God's> working in Nature carrying on operations, and in initiating those operations." (59) Gray insists on a "continued directing intelligence",

[26] Already September 26th, 1860 Darwin reacted with the following words to the 1860 articles by Gray: "I do not pretend to be a good judge, as I have never attended to Logic, Philosophy &c; but it is my opinion that you are best reasoner, of any man, let him who he may, that I ever read. … The two last essays are *far* the best Theistic essays I ever read." (1993b, 388) October 24, 1860 Darwin wrote Gray that Lyell had affirmed: "It would be well worth while if a little Book could be got up by Asa Gray for the theological part is so admirable." (1993b, 443) Gray's two books deserve indeed to be reprinted today. This great American intellectual might even posthumously contribute to an end of the destructive culture war in the USA about evolution. For he was, as Darwin called him, "a hybrid, a complex cross of Lawyer, Poet, Naturalist, & Theologian! — Was there ever such a monster seen before?" (1993b, 350) In letters to others, however, Darwin was not so generous. Alluding to Comte's theory of the three stages, Darwin criticizes in a letter to Charles Lyell of August 1st, 1861 Herschel and Gray for being still caught up in the theological state of science (1994, 226 f.).

[27] In another essay, the theory of occasional direct action is mentioned as third possibility. But while it is called the most popular, Gray claims that it is the least attractive to thoughtful people (1884, 159). He calls the idea that each organ was executed by God "an idea which has been set up as the orthodox doctrine, but which to St. Augustine and other learned Christian fathers would have savored of heterodoxy." (357) In (1880, 83) also Aquinas, Leibniz, and Malebranche are mentioned.

but this is compatible with a general plan. While at the end he grants that there may have been an independent origination of certain types and thus accepts "as much intervention as may be required", he insists "that Natural Selection, in explaining the facts, explains also many classes of facts which thousand-fold repeated independent acts of creation do not explain, but leave more mysterious than ever." (61) Only a few months later, he published in the same journal a dialogue "Design versus Necessity" on different ways of assessing the value of Darwin's book for natural theology.[28] Interesting is the note added at the end that the opponent to the theistic interpretation might have used the concept of contingency instead of necessity (86).

The third essay was also published in 1860 and has the programmatic title "Natural Selection not Inconsistent with Natural Theology". Gray mentions that Darwin's gradual evolutionary theory fits well with an old principle of natural philosophy — it "answers in a general way to the Law of Continuity in the inorganic world, or rather is so analogous to it that both may fairly be expressed by the Leibnitzian axiom, *Natura non agit saltatim.*" Gray insists that this principle cannot be presupposed a priori, but "naturalists of enlarged views will not fail to infer the principle from the phenomena they investigate — to perceive that the rule holds, under due qualifications and altered forms, throughout the realm of Nature." (123) (Needless to say, the later theory of punctuated equilibrium is not a form of saltationism, but a form of gradualism in the broad sense of the term.) Gray is aware that "the principle of gradation throughout organic Nature may … be interpreted upon other assumptions than those of Darwin's hypothesis" (126), but certainly the Darwinian theory offers an explanation of the phenomena that we observe. Gray exemplifies his theory with regard to individuality, which is connected to the loss of vegetative reproduction and was only slowly acquired in the course of evolution, even if it is the "very ground of *being* as distinguished from *thing*". Only unicellular plants are called by Gray real units, not so the more complex ones. "In the ascending gradation of the vegetable kingdom individuality is, so to say, striven after, but never attained; in the lower animals it is striven after with greater though incomplete success; it is realized only in animals of so high a rank that vegetative multiplication of offshoots are out of the question, where all parts are strictly members and nothing else, and all subordinated to a common nervous center — is fully realized only in a conscious person." (125) In the last section of the essay on "Darwin and his Reviewers", Gray insists on the very dif-

[28] I have not been able to find out for whom "D.T." stands, which is the name of the interlocutor of "A.G.", which stands for Asa Gray.

ferent quality of the reviews of *The Origin*; and he is particularly irritated by those who insinuate that Darwin's book is atheistic. Gray recognizes that Darwin has purposely been silent on theological issues, and in a witty application of the issue at stake to Darwin's behavior in proposing the issue comments: "This reticence, under the circumstances, argues design, and raises inquiry as to the final cause and reason why. Here, as in higher instances, confident as we are that there is a final cause, we must not be over-confident that we can infer the particular or true one." (144) One possible explanation may be that he is not familiar with philosophical inquiries and thus focuses only on secondary causes; another may be that he enjoys being attacked as atheist by inconsiderate people. In general, Gray writes, a theist is only committed to the idea that there are ends in nature, but not at all to a specific claim that a particular thing or event is intended by God. "Most people believe that some were designed and others were not, although they fall into a hopeless maze whenever they undertake to define their position." (138) Every intelligent person accepts general besides particular providence, i.e. that many events are willed by God only insofar as they are the consequence of general laws, and no thoughtful theist ascribes design only to supernatural events (149). In one point, however, Gray treads upon dubious terrain, and that is his claim, forcefully denied by Darwin, that "variation has been led along certain beneficial lines." (148) It is true that the causes of variation were unknown at that time (and even today we do not know all of them), and it is also true, as he says, that Darwin's theory does not necessarily entail that there are more harmful deviations than were already known to exist. "Good-for-nothing monstrosities, failures of purpose rather than purposeless, indeed, sometimes occur; but these are just as anomalous and unlikely upon Darwin's theory as upon any other." (147) But in this passage he tends to play down the negative variations, and he somehow insinuates the direct presence of God when the secondary causes are not yet known. In any case, Gray does not claim that Darwinism leads to natural theology; in fact, he recognizes that the arguments for design may not convince everybody. "But we may insist, upon grounds already intimated, that, whatever they were good for before Darwin's book appeared, they are good for now." (152) Even if he does not use this language, Gray points to the fact that the *genesis* of an organism does not contribute at all to the question whether its astonishingly adapted structure is due to design (151 f.; cp. 259).[29]

[29] An analogous argument can be found in Kant's *Universal Natural History and Theory of Heaven*: The mechanical explanation of the genesis of the solar system does not exclude its teleological interpretation (A XX ff., 71).

Certainly, Darwinism can be interpreted in an atheistic way (159); but this interpretation does not follow from the theory. "If you import atheism into your conception of variation and natural selection, you can readily exhibit it in the result." (154) Gray recognizes the chance moment (which he rightly locates not in selection, but in variation) as the major stumbling block. A large number of these variations "are not improvements, but perhaps the contrary, and therefore useless or purposeless, and born to perish." (156) But Gray retorts by saying that nature abounds with analogous instances, not only in the inorganic world, but even among humans. "Some of our race are useless, or worse, as regards the improvement of mankind; yet the race may be designed to improve, and may be actually improving." (157)

Of particular interest is Gray's claim that Darwin brought back teleology to natural science, "so that, instead of Morphology *versus* Teleology, we shall have Morphology wedded to Teleology." (288) Teleology (the term was coined by Christian Wolff) here means that one species is useful to another. The reason for this resurgence is that we owe to the idea of the mutability of species ecology proper.[30] If species are not eternal, but survive only due to a constant struggle for existence, then "the structure of every organic being is related, in the most essential yet often hidden manner, to that of all other organic beings, with which it comes into competition for food or residence, or from which it has to escape, or on which it preys." (Darwin 1968, 127) It is this holistic approach to nature that justifies analyses of causal interdependences between different species and leads to a "restoration of teleology" (357), as Gray writes in the final essay of the volume "Evolutionary Teleology". But this is not only important for the natural science of biology; it makes the argument from design more plausible to the objectors who always pointed to the dysteleological structures in nature. "In the comprehensive and far-reaching teleology which may take the place of the former narrow conceptions, organs and even faculties, useless to the individual, find their explanation and reason of being. Either they have done service in the past, or they may do service in the future." (375) The imperfections have the function of keeping the evolutionary process going on and of producing better adaptations. "In this system the forms and species, in all their variety, are not mere ends in themselves, but the whole a series of means and ends, in the contemplation of which we may obtain higher and more comprehensive, and perhaps worthier, as well as more consistent, views of design in Nature than heretofore." (378) Again, Gray does not deny that the argument from design may be rejected: he is familiar with both

[30] The term "ecology" was coined in 1866 by the Darwinian Ernst Haeckel (see Leps 2004, 601).

Hume's and John Stuart Mill's well-known objections (not, however, with Kant's). But he reiterates his point that, while the question of design in the world will be discussed by philosophers so long as that world lasts, there is nothing in Darwinism inimical to that assumption. It only teaches us to look at Nature as a whole and find, or reject, design in her as a whole (379). But does not natural selection render final causes completely superfluous? Gray points to the fact that selection operates only if there occurs variation towards new forms; thus it cannot explain what in fact it presupposes (385 f.).

In his Yale lectures Gray makes this point more clearly. For him, not the whole theory of trans-specific evolution, but natural selection itself is not a simple hypothesis, but a truth (1880, 46), while the idea that variation is all-directional and accidental does not enjoy the same status. But whatever the exact causes of variation, at that time still unknown, natural selection cannot be one of them and thus it cannot *on its own* explain the slow evolution towards more and more complex forms (49). At the same time, together with variation, "the principle of natural selection, taken in its fullest sense, is the only one known to me which can be termed a real cause in the scientific sense of the term." (70 f.) However, it is a cause only together with the laws of the living world that remain as astonishing after Darwin as they were before him "Does it <sc. natural selection> scientifically account for the formation of any organ, show that under given conditions sensitive eye-spot, initial hand or brain, or even a different hue or texture, must then and there be developed as the consequence of assignable conditions? Does it explain how and why so much, or any, sensitiveness, faculty of response by movement, perception, consciousness, intellect, is correlated with such and such an organism? I answer, Not at all! The hypothesis does none of these things. For my own part I can hardly conceive that any one should think that natural selection scientifically accounts for these phenomena." (73) And thus the argument from design can be used as well, *if* one is already a theist (84 f.). The conflict is "not between Darwinism and direct Creationism, but between design and fortuity, between any intention or intellectual cause and no intention nor predicable first cause." (89) Gray at the end mentions that explicit arguments for theism do not come from the study of nature — it only never bars them. His own reason for believing in a divine principle of the world is that it "gives us a workable conception of how 'the world of forms and means' is related to 'the world of worths and ends.' The negative hypothesis gives no mental or ethical satisfaction whatever." (91) Gray also gestures towards an epistemological argument by quoting a sentence ascribed to James Clerk Maxwell "that he had scrutinized all the agnostic hypotheses he knew of, and found that they one and all needed a God to make them workable." (91) In any case, Gray interprets evolution as the slow instantiation of the values willed by God. "As the forms and kinds rise gradually out of that which was well-nigh formless into a consummate form, so do biological ends rise and assert themselves in increasing distinctness, variety, and dignity. Vegetables and animals have paved the earth with intentions." (93) He then adds some (unsatisfying) reflections on the "insoluble" problem of the free will (97) and ends with some arguments for the immortality of the human soul.

We have seen that Darwin at the end of *Variation* remained skeptical with regard to Gray's attempt to reconcile natural theology with Darwinism. Partly this has to do with the fact that Gray looks at Darwinism from a theistic perspective, which for Darwin was not as evident.[31] But Darwin did not even regard as obvious the much weaker thesis that his theory was *compatible* with theism. As far as I can see, there are two main arguments against the compatibility claim. The one is not used by Darwin himself, but plays an important role in the usual accounts of the decline of physicotheology. According to it, the whole concept of evolution destroyed the idea of design.[32] For an absolute mind must be able to immediately get at its ends; why should there be a long way towards them? However, already Leibniz discusses this issue and accepts that progress in perfection in finite beings may be compatible with the best possible world, since change towards the better may be itself a perfection.[33] Indeed, one might argue, such a world may be more interesting and challenging than one that is absolutely stable.

The second argument is related, but more specific: Darwin points to the redundant power of reproduction and the harmful deviations that follow from the plasticity and extreme variability of organisms. Even if he does not say it, it seems to me that for him the Malthusian overproduction is somehow the opposite of the principle of the least action that so attracted natural theologians in the course of the eighteenth century. The negative variations and the overproduction with subsequent selection do not seem the quickest way to the results God might have in mind and thus incompatible with providence.[34] How can these doubts be answered? With regard to the harmful variations, one can claim with Gray that the

---

[31] Cp. Miles (2001).

[32] See Schramm (1985, 54 and 164 f.), who follows Hans Freudenthal.

[33] Cf. his text *An mundus perfectione crescat* (1965, 368 ff.).

[34] In a letter to Gray from June 5th, 1861, Darwin rejects the idea of designed variation and writes: "what an enormous field of undesigned variability there is ready for natural selection to appropriate for any purpose useful to each creature." (1994, 162; see also 1993b, 275, 389).

occurrence of such deviations was already known before Darwin; so his theory does not add anything negative to the old arguments of theodicy. On the contrary, natural selection *prevents* these harmful deviations from spreading; it thus limits the negative effects of chance and can even claim to be purposive with regard to adaptation.[35] Darwin himself indeed thought that one reason against special creationism was strictly theological: The latter ascribes quite ugly results to the direct will of God (1968, 263). As he wrote to Gray on May 2, 1860: "I cannot persuade myself that a beneficent & omnipotent God would have designedly created the Ichneumonidae with the express intention of their feeding within the living bodies of caterpillars, or that a cat should play with mice." (1993b, 224) The conceptual tool to deal with such ugly phenomena (which are easily surpassed by human suffering and cruelty) is old. Whoever interprets, in my eyes reasonably, divine omnipotence as God being the cause of everything, is well advised not to give up the normative difference between various states of affairs. Suffering is not willed by God in the same degree as pleasure, even if it is impossible to deny that both are willed by Him, if they occur and if He is supposed to be omnipotent. The intellectual tool used by a theist such as Leibniz is to say that God accepts certain intrinsically negative states because they are necessarily connected with the positive states He aims at with his antecedent will, i.e., in a strong sense of the term, yet independently of the consequences connected with them.[36]

But why did God accept a world in which suffering occurs at all, such as the mechanism of natural selection certainly entails? One classical answer is to say that one decisive criterion in the choice of the world that God created was the simplicity of the laws of nature, and that this again and again entails suffering, including innocent humans. Even if one rejects any "direction" of the variations, one may argue within a theistic worldview that the mechanism of natural selection has been chosen by God for four reasons: first, natural selection is an extremely simple mechanism with enormous causal power — Wallace in 1859 compared it with "the centrifugal governor of the steam engine." (1991, 300) Second, the overproduction, which, together with the scarcity of the resources, variation, and inheritability of traits, entails natural selection as a logical consequence, is an expression of the principle of plenitude: all possible life-forms are tried, even if only the compossible ones will last. The value of life is the reason why it tries to spread as much as it can, even if the price it has to pay is the struggle for existence, which then

[35] See McMullin (2008) and already Darwin (1994, 226).

[36] See *Essais de théodicée* § 114 (1978; VI 166).

leads to the development of more complex life forms. Third, paradoxically the overproduction is the reason for the creation of scarcity, which obliges organisms to behave economically and to try to optimize resources: thus something analogue to the principle of the least action is imposed upon organisms by the redundant reproduction. And fourth, within a determinist universe, natural selection together with other natural laws and given antecedent conditions had to lead to the results that God wanted, such as the existence of morally responsible beings.

III.

Asa Gray's reflections, supported by the final considerations, have hopefully shown that Darwinism and theism are not logically incompatible. But, as Gray himself wrote, this does not prove that theism is true. Theism, the assumption that the world has been created by an absolute mind, certainly entails a teleological interpretation of the world as a whole, for minds act according to ends. The inversion, however, does not seem to hold; for both Aristotle and Schopenhauer defend a teleological interpretation of nature even without belief in a Creator God: the ends are immanent in nature. What are the arguments for theism? The argument from design is indeed far from being cogent. *If* there is a divine principle of the world, then we have to interpret the world in terms of ends, but it is not the world as such that forces upon us a teleological interpretation. There is too much apparent dystelelogy in it. We may explain it away as necessary for desirable ends, but only if we already have independent arguments for a transcendent principle. Furthermore, Hume in the posthumous *Dialogues Concerning Natural Religion* of 1779 and Kant in *The Only Possible Argument in Support of a Demonstration of the Existence of God* of 1763 as well as in the *Critique of Pure Reason* of 1781 (B 648-658/A 620-630) have rightly individuated the main weaknesses of the argument.[37] Their lasting insight is that the argument cannot be grounded on an empiricist basis. To begin with Hume, he writes that we can infer causes from effects, when we have observed the two together in other cases; "but how this argument can have place, where the objects ... are single, individual, without parallel, or specific resemblance, may be difficult to explain." (1947, 149) Furthermore, it is not clear

[37] It remains an astonishing fact in comparative philosophy that the thinker who came closest to Hume's criticism is the Indian Ramanuja from the 11/12th century in the *Sri Bhashya*, an author, of course, unknown to Hume. See Yandell (1999, 205 ff.). The fact proves a strange convergence of the development of the human mind, one is tempted to say: some teleology in the history of philosophy.

why the cause of the world does not itself need a cause (161). Since according to empiricism all our knowledge is based on experiences of the world, we can only say that the cause *of* the world must be analogous to one of the causes of order that we find *within* the world, and all minds we know are finite. "This world … is very faulty and imperfect, compared to a superior standard; and was only the first rude essay of some infant Deity, who afterwards abandoned it, ashamed of his lame performance…" (169); it could also be the common work of various deities (167). Beside reason as cause of artworks, also instinct, vegetative and sexual reproduction are causes of remarkable objects. Why then should we assume that one of these causes is more likely to be the origin of the world than another? "Any one of these four principles above mentioned (and a hundred others which lie open to our conjecture) may afford us a theory, by which to judge of the origin of the world; and it is a palpable and egregious partiality, to confine our view entirely to that principle, by which our own minds operate." (178) One may deny the last sentence by embracing a Cartesian theory of the epistemological priority of the First-Person-Perspective; but if one does not, then Hume's supposition "that the world arose by vegetation from a seed shed by another world" is no less rational than the belief in the creation of the world by a mind. Hume is aware of the fact that there is also an a priori argument for theism, namely the combination of the cosmological and ontological prove proposed by Demea, but his interlocutors Cleanthes and Philo are unanimous in rejecting it (188 ff.).[38]

Already in the *Universal Natural History and Theory of Heaven* of 1755 (A XIV f.), and particularly in *The Only Possible Argument in Support of a Demonstration of the Existence of God* of 1763, Kant had argued that the argument from design could only prove an architect, not a creator of the world (A 116) and that it could not show that the cause of the world is most perfect or even a single one (A 199 ff.). The first criticism is repeated in the *First Critique* (B 655/A 627), where the general insight is expressed that the cause inferred can only be proportional to the effect. The empirical way can thus never lead to absolute totality (B 656/A 628).[39] Only

---

[38] Cleanthes's counterarguments partly beg the question, partly make wrong presuppositions (thus he confuses causes with reasons, 190), and since Philo points to the very different problem that the necessity of God may entail the necessity of the world (191) and thus a Spinozian conception, we must here leave it open whether Hume shared all the arguments by Cleanthes, who in the other parts of the dialogue argues embarrassingly badly.

[39] For fairness's sake one has to mention that, e.g., Paley is aware that the attributes ascribable to the cause of the world are only "beyond all comparison" (1809, 443 f.).

by a surreptitious use of the cosmological argument, which wants to demonstrate a necessary being and which itself presupposes the ontological argument, can the physicotheological argument lead to an absolute being. According to the later Kant, however, the cosmological and the ontological proof are also irredeemably flawed, while the early Kant had still defended a new version of the ontological proof, which he later gave up. Still, as is well known, Kant's contribution to natural theology is not limited to the destructive work done in *The Only Possible Argument* and in the *First Critique*. The *Critique of Practical Reason* introduces God as a postulate of practical reason (A 223 ff.), and even if the epistemological status of this postulate is unclear and controversial, clearly Kant can claim to have given the moral argument for the existence of God a new foundation. This is linked to Kant's radical break with eudemonist ethics: the question of what our duty is cannot be reduced to the problem what makes us happy. But if the moral fact cannot be reduced to natural desires like that for happiness, then the moral Ought is not part of this world. But how can it operate in this world? How can we have a duty to act according to the moral law if the natural world is not in principle malleable by the demands of the Ought? These questions are dealt with in the *Critique of Judgment* of 1790, whose second part, the "Critique of teleological judgment", is the most impressive defense of teleological thinking in modernity after the collapse of traditional physicotheology. Kant's defense, however, insists on teleology being a regulative, not a constitutive principle (B 270): it is a subjective principle of judging the phenomena. Kant reiterates his opinion that physicotheology does not achieve what it promises. It does not have any potential to address the question of the ultimate end of nature (B 401); and it cannot infer a cause with absolute qualities: if it does so, practical reason is secretly supplementing what theoretical reason on its own cannot do (B 404).[40] Indeed, based on purely physicotheological arguments, the cause of the world might as well act with instinct as with reason (B 409). Only ethicotheology can lead to a true concept of God, physicotheology on its own might well lead to some form of demonology (B 414). The ultimate end that we, always aware of the subjective nature of this ascription, have to ascribe to God within ethicotheology are human beings under moral laws (B 415 f.). On this ground, however, a teleological interpretation of nature, with which we humans remain inevitably connected, is legitimate (B 419).

---

[40] See also Kant's 1788 essay *On the Use of Teleological Principles in Philosophy*, A 36 f.

Certainly no contemporary attempt to make sense of teleology in nature can ignore Kant.[41] The main reasons why the naturalist worldview proposed nowadays by neo-darwinians such as Dawkins is so unappealing are the two realms of normativity and of mentality. Naturalism does not render justice to the fact that as persons we inevitably have ultimate intellectual and moral ends that cannot be reduced to the execution of biological programs. One might try to explain causally, e.g. by sociobiological means, how our moral concepts developed, but even if these attempts at explanation have indeed shed light on the *genesis* of some of our ideas, e.g. in the field of sexual morality, they are in principle unable to explain where the *validity* of our moral convictions comes from. This is the general problem of any *merely* evolutionary account: the space of reasons, both in the theoretical and in the practical sphere, is irreducible to the realm of causes. Since thinking and arguing are actions, and action is inevitably oriented towards ends, we need a theory of finality in order to make sense of what we as thinking and acting beings are. This is, as Kant rightly understood, the starting point of any teleology. But why is it not sufficient to assume nature as mere realm of causes on the one hand and rationality as an independent sphere on the other?

The answer is that the unity of nature that evolutionary biology has so powerfully pointed out is indeed inimical to any dualist theory of being. The only rational and moral agents that we know are, after all, complex animals, and if it is not plausible to assume that the categorical moral demands imposed upon humans are simply the result of a contingent evolution, then the idea is tempting that the development of moral beings was an end of nature. But are agents with moral ends the only end of nature? It seems a crude anthropocentrism to dismiss prehuman nature as having only instrumental value for humans. It is more plausible to see in the slow evolution of mentality and the capacity to have purposes another end of nature. Even if there are causal mechanisms for the development of species, and probably also for the origin of life, one can hardly deny that life with its marvelous adaptation of the organs to each other and the whole organism is characterized by a form of teleonomy distinct both from the inorganic realm and from the rational finality of humans. And there is nothing incompatible with modern science (although it is not entailed by it) if one interprets the development from inorganic

[41] This is not to say that all of his ideas in the field of philosophy of religion are valid. Thus, Kant's criticism does not overturn the ontological proof in the version proposed by Plantinga (1982, 197 ff.). Its validity would eliminate the subjective limitations to which Kant subjects teleological arguments.

objects to organisms and self-conscious minds themselves as an unfolding of ever more complex ends: bringing forth entities that have increasingly complex ends is, so to speak, an end either of nature or of its creator; and the highest end is the generation of a being that can ask the question what an ultimate end is.[42]

But not only are there practical ends for rational agents, such as justice; one theoretical end must be the recognition of truth, without which neither science nor philosophy makes sense. Now this end presupposes two things: first, there must be rational beings able to follow inferences, and, second, nature must be intelligible. This has implications for the way in which nature must be structured. If one rejects interactionism, i.e. the idea that mental states can cause physical states, the argument from natural selection will fail to *explain* the development of mental states, for they are of no use. No doubt those err who aver that epiphenomenalism is inconsistent with Darwinism. For one may assume that there are laws of supervenience guaranteeing that there are certain mental states that accompany determinate physical states. But, as we already know, the existence of such laws is presupposed, and not explained, by Darwinism. However, there is an additional issue. If we want to take our own thoughts seriously, we have to assume that some of the physical states of brain are causally so interconnected that the corresponding mental states, e.g. in a logical deduction, are also logically connected. Such a presupposition is not simply wishful thinking; it is the transcendental presupposition of any rational investigation.[43] An analogous teleological explanation might be given for those constant parameters of nature without which life or intelligent life could not have developed. (Of course, it is not possible to justify *all* natural laws with such considerations.) It might be objected that it is a brute fact that such fine-tuning holds, but stopping at brute facts rarely helps either science or philosophy. Our philosophical curiosity is more satisfied if we can reduce laws to a principle of the world that is itself a mind and wants to be recognized by inner-worldly finite minds. Such a reduction is kindred in spirit to the satisfaction of the scientific curiosity, as when Darwinism no longer had to accept the different species as mere facts but could explain their existence. But, of course, in the case of metaphysical

[42] For a concrete elaboration of the way how transcendental values are realized in evolution, see Hösle (2005); on a possible teleological interpretation of evolution towards humans, see Wandschneider (2005); on the overcoming of the dualism between nature and Ought, see Illies (2006).

[43] See Hösle (2007). The dialogue discusses the arguments against interactionism; and one of the interlocutors defends a form of parallelism inspired by Leibniz.

explanations *reasons*, and not *causes* as in scientific explanations, are at stake. Alternative, non-teleological explanations of those constant parameters can hardly convince. It is extraordinarily implausible that they and all the laws of nature are determined by logic, as Spinoza seems to have believed. The multiverse model, the idea that our world is only one of many other worlds that do not have such teleological features, on the other hand, is *per definitionem* empirically unverifiable; and there is nothing metaphysically attractive in the multiplication of actual worlds.[44] Furthermore, the argument can easily be mirrored to lessen the burden of theodicy: if we cherish the idea that there are many actual worlds, it could be that God has created all possible worlds in which good balances evil; we are in one such world, but not in one with maximal value.

But not only must nature according to this account be structured so as to bring forth minds; it must also be intelligible. And this means that we have reason to look for theories that are both simple and fertile. Needless to say, any good scientific theory must match the empirical facts; but matching them is a necessary, not a sufficient condition for the quality of a scientific theory. Already in the *Critique of Pure Reason* Kant insisted on such regulative principles as the continuous *scala naturae* of creatures (B 670ff/A 642 ff., particularly B 696/A 668); in the *Critique of Judgment* (B XXVIII ff.), they are interpreted as an expression of the formal purposiveness of nature.

Darwin's theory is certainly one of the most powerful scientific theories. It enhances the consilience of the various biological disciplines, but also of biology and the sciences of humans. Based on a very simple argument, it allows us to find causes for entities such as species that before him had resisted any explanation. It does not at all deny the beauty and complexity of the living world, but teaches us to see it as a complex and fragile web. It can easily be destroyed and for its preservation we humans maintain a moral responsibility. It increases our faith in the intelligibility, and thus in the formal purposiveness, of nature at the very modest price that we can no longer jump directly to the First Cause whenever we do not yet understand the relevant secondary causes.

---

[44] We have to distinguish the multiverse conception in a strict modal sense from Lee Smolin's idea of "cosmological natural selection". Since according to it new "universes" are created by collapsing black holes, they all form a single world even if some constant parameters may differ in the various "universes". The theory is scientific, not metaphysic, and does not add anything relevant to the well-known fact that only small parts of the world are places for life and mind. Furthermore, the theory has very little to do with natural selection.

REFERENCES

Balcombe J., 2006, *Pleasurable Kingdom. Animals and the Nature of Feeling Good*, New York: Macmillan.

Curley E., 1969, *Spinoza's Metaphysics. An Essay in Interpretation*, Cambridge, Mass.: Harvard University Press.

Dawkins R., 2006, *The God Delusion*, New York: Bantam Press.

Darwin C., 1968, *The Origin of Species by Means of Natural Selection*, Harmondsworth: Penguin.

Darwin C., 1981, *The Descent of Man, and Selection in Relation to Sex*, Princeton: Princeton University Press.

Darwin C., 1987, *Notebooks, 1836-1844*, Ithaca: Cornell University Press.

Darwin C., 1988a, *The Origin of Species 1876*, in Barrett P. H, Freeman R. B. (eds.), *The Works of Charles Darwin,* vol. 16, London: William Pickering.

Darwin C., 1988b, *Variation of Animals and Plants under Domestication. Vol. II*, in Barrett P. H., Freeman R. B. (eds.), *The Works of Charles Darwin,* vol. 20, London: William Pickering.

Darwin C., 1993a, *The Autobiography of Charles Darwin 1809-1882*, New York/London: W.W. Norton & Company.

Darwin C., 1993b, *The Correspondence of Charles Darwin. Vol. 8. 1860*, Cambridge: Cambridge University Press.

Darwin C., 1994, *The Correspondence of Charles Darwin. Vol. 9. 1861*, Cambridge: Cambridge University Press.

Gray A., 1880, *Natural Science and Religion. Two Lectures delivered to the Theological School of Yale College*, New York: Charles Scribner's Sons.

Gray A., 1884, *Darwiniana. Essays and Reviews pertaining to Darwinism*, New York: D. Appleton and Company.

Hösle V., 1984, *Wahrheit und Geschichte. Studien zur Struktur der Philosophiegeschichte unter paradigmatischer Analyse der Entwicklung von Parmenides bis Platon*, Stuttgart-Bad Cannstatt: Frommann-Holzboog.

Hösle V., 1999, "Philosophy and the interpretation of the Bible", *Internationale Zeitschrift für Philosophie* 2:181-210.

Hösle V., 2005, "Objective Idealism and Darwinism", in Hösle V., Illies C. (eds.), *Darwinism and Philosophy*, Notre Dame: Notre Dame Press, 216-242.

Hösle V., 2007, "Encephalius. A conversation about the mind-body-problem", *Mind and Matter* 5.2: 135-165.

Hume D., 1947, *Dialogues Concerning Natural Religion*, Indianapolis: Bobbs-Merrill.

Huxley T. H., 1896, *Darwiniana. Essays*, New York: D. Appleton and Company.

Illies C., 2006, *Philosophische Anthropologie im biologischen Zeitalter*, Frankfurt: Suhrkamp.

Leibniz G. W., 1965, *Kleine Schriften zur Metaphysik*, Darmstadt: Wissenschaftliche Buchgesellschaft.

Leibniz G. W., 1978, *Die philosophischen Schriften*, Hildesheim/New York: Georg Olms Verlag.

Leps G., 2004, "Ökologie und Ökossystemforschung.", in Jahn I. (ed.), *Geschichte der Biologie*, Hamburg: Nikol-Verlag, 601-619.

McMullin E., 2008, "Could natural selection be purposive?", in Seckbach J., Gordon R. (eds.), *Divine Action and Natural Selection*, Singapore: World Scientific, 115-125.

Miles S. J., 2001, "Charles Darwin and Asa Gray discuss teleology and design", *Perspectives on Science and Christian Faith* 53: 196-201.

Paley W., 1809, *Natural Theology; or, Evidences of the Existence and Attributes of the Deity, Collected from the Appearances of Nature*, London: J. Faulder.

Plantinga A., 1982, *The Nature of Necessity*, Oxford: Clarendon.

Powell B., 1860, "On the study of the evidences of Christianity", in Temple F. (ed.) *Essays and Reviews*, London: John W. Parker & Son, 94-144.

Schopenhauer A., 1977, *Zürcher Ausgabe. Werke in zehn Bänden*, Zürich: Diogenes.

Schramm M.,1985, *Natur ohne Sinn? Das Ende des teleologischen Weltbildes*, Graz/Vienna/Cologne: Styria.

Sloan P. R., 2005," 'It might be called Reverence'", in Hösle V., Illies C. (eds.), *Darwinism and Philosophy*, Notre Dame: Notre Dame Press, 143-165.

Wagner H., 1979, *Aristoteles. Physikvorlesung. Übersetzt von Hans Wagner*, Darmstadt: Wissenschaftliche Buchgesellschaft.

Wallace A. R., 1991, *An Anthology of His Shorter Writings*, Oxford: Oxford University Press.

Wandschneider D., 2005, "On the problem of direction and goal in biological evolution" in Hösle, V., Illies C. (eds.), *Darwinism and Philosophy*, Notre Dame: Notre Dame Press, 196-215.

Whewell W., 1833, *Astronomy and General Physics Considered with Reference to Natural Theology*, Philadelphia: Carey, Lea & Blanchard.

Yandell K. E., 1999, *Philosophy of Religion. A Contemporary Introduction*, London: Routledge.

Yolton J. W. (ed.), 1990, *Philosophy, Religion and Science in the Seventeenth and Eighteenth Centuries*, Rochester: Rochester University Press.

Young C. C., Largent M. A., 2007, *Evolution and Creationism. A Documentary and Reference Guide*, Westport: Greenwood Press.

Accident, Adaptation, and Teleology in Aristotle, Empedocles, and Darwinism

David J. Depew

1. Chance in *The Origin of Species*

Charles Darwin framed the *Origin of Species* to meet criteria for inductive science that had been set out by John Herschel in his *Preliminary Discourse on the Study of Natural Philosophy* (Herschel 1830; Hodge 1977). Accordingly, he was distraught when he learned that Herschel, to whom he had sent a copy of his newly published book, was not persuaded. "I have heard by a round about channel," Darwin wrote to Charles Lyell, whose methodology in the science of geology Herschel had earlier praised, that "Herschel says my book 'is the law of higgledly-pigglety.' What exactly this means I do not know, but it is evidently very contemptuous. If true, this is a great blow and discouragement".[1]

It got worse. An incoherent interpretation of the *Origin* soon took hold in quarters hostile to Darwin that natural selection is no better than the pounding of a large number of monkeys on a large number of typewriters over a period of time long enough to produce, by pure chance, *Hamlet, King Lear, Othello,* or the *Principia Mathematica* of Newton. When Francis Bowen, a professor of natural theology at Harvard University, took this line, Darwin protested in a letter to his botanist friend Asa Gray, who was managing his case in America:

> It is monstrous … that [Bowen] should argue against the possibility of accumulative variation and actually leave out entirely selection … The chance that an … improved pouter-pigeon should be produced by accumulative variation without man's selection is almost infinity to nothing; and so with natural species without selection.[2]

---

1 Darwin to Lyell, December 10, 1859, in (Burkhardt et al. 1985- [henceforth CCD], vol. 7, 421).

2 Darwin to Gray, 26 November, 1860 (CCD , vol. 8, 496).

Darwin had reason to feel discouraged. His theory did not equate selection with the mere persistence of coincidentally variations that happen to be useful. He took variation to be a necessary, not a sufficient condition of natural selection. He postulated the gradual shaping of heritable variations into adaptations by a process that necessarily extends across generations. At first, these variations merely happen to elevate the probability of survival and reproduction. But as they spread through an interbreeding population they do so precisely *because* they amplify and elevate these probabilities. They *become* adaptations.[3]

To see how Darwinism construes adaptations, we should distinguish between the multi-generational *process* of adaptation, *states* of (relative) adaptedness or fitness that make a trait adaptive in a certain population, and *traits* that are full-blown adaptations. These distinctions became clearer, more precise, and more tractable when they were mathematically reformulated in terms of "the probability revolution" that was only beginning when Darwin wrote, but that subsequently transformed twentieth-century science (Hodge 1987; Gayon 1992; Depew and Weber 1995). Probabilistic reframing has allowed us to see that overall fitness is a dispositional property. Like the solubility of sugar in water, it is undergirded by myriad small physical differences in the development of organisms that enhance their chances of reproductive success (Mills and Beatty 1979; Brandon 1981; Brandon 1990; Sober 2000). For his part, Darwin was aware that there is an element of chance in whether, even if an organism has a useful variation, it will result in enhanced survival and reproduction (Darwin 1859, 82). But he was especially insistent on the chanciness of whether useful variations will be forthcoming in the first place. What he meant by calling variation 'chance' in relation to natural selection is that variation, however caused, must arise independently of utility if it is subsequently to be shaped into adaptations by the trans-generationally selective process of adaptation. It was on this point that Darwin took himself to differ from Lamarck, especially Lamarck as interpreted by Lyell (1830-33, vol. II).

Formulations of Darwin's theory in which chance occurrences are presented more as causes than conditions of organic structure and function persist to this

[3] How quickly this process can take place is not my topic today. My concern is not with how many generations it takes, but with natural selection's necessarily multi-generational nature. Darwin was not always as clear as he could be on either topic. His gradualism assumed many generations. Moreover, he did not always distinguish clearly between natural selection within a single generation — what Robert Brandon called "environmental selection"—and natural selection operating in a concerted way across generations. Only the latter can be the proper locus of adaptations (Darwin 1859, 81; Brandon 1990).

day. They are especially common between two sorts of critics who are otherwise often at odds with each other: defenders of intelligent design and some mathematical physicists. Intelligent design advocates exaggerate the elements of chance in Darwinism because conscious design is especially persuasive when it is presented as the only reasonable alternative to a succession of highly improbable coincidences. That was Bowen's tactic. It was not, however, the source of Herschel's complaint. He objected because he was a physicist. He could not recognize in the process Darwin described any laws of nature of the sort with which he was professionally familiar. The laws of physics apply to homogeneously described quantities, not chaotically heterogeneous ones.

There remains, however, a more challenging factor at work in characterizations of Darwinism as a theory of "frozen accidents." Even when we have acquired a more accurate understanding of natural selection, Darwinism's appeal to chance denies that the history of life "from monad to man," as the phrase has it, consists of a process in which earlier stages unfold for the sake of later stages, with its usual implication that properties of lower organisms come to be for the sake of the uses to which higher organisms put them. Accordingly, those who see Darwinism through the lens of overall teleology often extend the role of chance in Darwinism from the micro-evolutionary relation between variation and utility to the macro-evolutionary relation between earlier and later moments in the history of life. Darwinians who oppose large-scale teleology of this sort sometimes embrace the same extension and so claim that evolutionary history is indeed a history of accidents (Gould 1989).

Darwin's own view on this subject was roughly the following. Among the growing variety of evolutionary ideas that had been advanced since about 1800, the *Origin* was novel in proposing that environmental exigencies impinging on developing organisms in what Darwin called "the struggle for existence" are strong enough to belie an intrinsic causal bond between ontogeny (the development of individual organisms) and phylogeny (the successive emergence of higher forms of life). Rather than fine-tuning good fit with environments, as Lamarckians believed, Darwin held that internal directional tendencies toward phylogenetic complexity by strong analogy with individual development cannot explain adaptive fit:

It is preposterous ... to say that, after a certain unknown number of generations, some bird had given birth to a woodpecker and some plant to the mistletoe ... This assumption seems to me to be no explanation, for it leaves the case of the coadaptations of organic beings to each other and to their physical conditions of life untouched and unexplained (Darwin 1859, 1).

To be sure, Darwin thought that spasms of undirected variation are set off when an interbreeding population meets difficulties (Darwin 1859, 43, 82; Hodge 1987). In this he differed from his twentieth-century successors, who think the generation of variation is in the last analysis a spontaneous but inevitable consequence of the physical structure of the genetic material. It results from the probability of copying errors. Still, for Darwin, as well as his successors, which, if any, of these variations will be amplified in a concerted way by the process of adaptive natural selection is antecedently unpredictable and initially unrelated to whatever good effects result. To be sure, Darwin did not deny a parallel between ontogeny and phylogeny. Indeed, he recognized that changes in ontogeny are the source of the variation on which natural selection works and accepted something like the same progressive phylogenetic order postulated by other advocates of "the doctrine of development" (Richards 1992; Nyhart 2008). But Darwin's shift in the locus of causality to contingent matches between what natural variation generates and what adaptation requires accounts for this order as an *ex post facto* result of the fact that as life ramifies under the imperatives of struggle the difficulty of accessing new resources and so of opening new niches will bias evolution's direction toward kinds that possess increasingly powerful means of exerting agency on their environments and, in supporting these abilities, can on the whole be expected to be more structurally complex. The superiority of Darwin's theory over Lamarck's is that he was able to explain extinctions. In such cases, the required variation is simply not forthcoming. Accordingly, progressive though it might be, the history of life forms for Darwin resembles a bush more than a tree, as has been frequently noted.

In Darwin's day it was difficult even for his converts, not to speak of his opponents, to separate the etiology of adapted traits from the larger issue of evolutionary direction and purpose. Darwin's confession to his American correspondent Gray that he was "in a muddle" on this point suggests that the problem was no less puzzling to him.[4] Even today the absence of an overall goal seems to many to make Darwinism's implied history of life random, mindless, blind, and accidental. In John Burrow's introduction to the Penguin edition of the *Origin,* for example, we read that "Nature according to Darwin is the product of blind chance and blind struggle, and man is a lonely intelligent mutation." (Darwin 1969, 43) A no less striking misformulation is Steve Fuller's one. In Darwinism, he claims, "chance mutations are the driving force of evolution" and natural selection consists of "compounded historical accidents." (Fuller 2007, 31, 48)

[4] Darwin to Gray, 3 July 1860 (CCD, vol. 8, 273).

Fuller might have been misled by the seemingly innocent formula "random variation plus natural selection," which often appears in evolutionary textbooks as well as in the polemical writings of the intelligent design theorists who suspect these books of secret materialism. The formula is misleading because it fails to distinguish between the spontaneity of variation, which is grounded in the properties of replicating macromolecules, and randomness. The randomness of the variation on which natural selection works comes into view only in and through its relation to selection. So it makes no sense to speak of random or chance variation apart from and prior to natural selection. Moreover, the spontaneous changes in DNA that are at the root of organic variability are, statistically speaking, regular enough to be used as an evolutionary clock. Thus if they were evolution's driving force, as Fuller claims, the history of life would not look at all like the series of accidents he takes it to be.[5] The point is moot in any case because variation, whether it is random or regular, is not in fact evolution's driver. While the process of adaptation is affected by many constraints and contingencies, it is always more causal than the variation it works on. If this were not so, Simon Conway Morris could not even begin to lay the stress he does on evolutionary convergence (Conway Morris 2008).

On the basis of just such misrepresentations, however, one is often invited to view Darwinian evolution as a whole as the result of a series of accidents, with the implication that the human species itself is an accident. Theists who protest this supposed implication can find support in the proclamations of physicists like Steven Weinberg, who remarks that evolutionary biology differs from physics insofar as it "depends on historical accidents" (Weinberg 1993, 42) rather than on laws of nature of the sort with which he, like Herschel, claims familiarity: universal laws that deductively subsume a homogeneously described array of instances and are in principle reducible to the laws of more basic sciences. Weinberg parts company with intelligent design proponents only in factually accepting that human beings are, as Barrow and Fuller assert, accidents of history (Weinberg 1993, 245). Both sides in this debate labor, however, under the same false assumptions.

I diagnose the fallacy as follows. In interpretations of Darwinism that speak of "random variation *plus* natural selection" the element of contingency in the overall

---

5 Gray proposed just such a view to Darwin in order to save overall teleology. Natural selection, he conceded, is not directional, but might it not be possible for God to rig the flow of the variation on which it works in an overall direction? Darwin rejected this argument as too strong to preserve the very concept of natural selection (Gray 1963; Darwin to Gray, November 26, 1860, CCD, vol. 8, 496).

course of evolution is first exaggerated and then projected back into the core of the theory of natural selection. The resulting notion that natural selection ranges over a sea of purely chance events that drive it here and there is then re-projected outward onto evolution as a whole. This picture has the effect of exaggerating the accidental quality of both natural selection and evolutionary history. In what follows, I will separate the two topics. I will first identify a limited but objective degree of what Kant called natural purposiveness, and even of natural teleology, in the process by which natural selection brings forth adaptations. I will then suggest how this kind of teleology affects large-scale teleology. To make my points clearer I will at times compare Darwin to Empedocles and Aristotle.

2. Adaptations and biological teleology

Thomas Henry Huxley was devoted to spreading the scientific worldview through society. He championed Darwin to that end. In consequence, he opposed "fallacies ... loudly expressed in the early days of the Darwinian controversy ... that charge Mr. Darwin with having attempted to reinstate the old pagan goddess Chance." (Huxley 1888, 552-53) We have already encountered the tendencies to which he was referring. Huxley responded by taking a 180° turn. For Darwinism, he says, chance is merely ignorance of deterministic causes. As science progresses, it will be found that

> there is not a curve of the waves, not a note in the howling chorus, not a rainbow-glint on a bubble, which is other than a necessary consequence of the ascertained laws of nature; and that with a sufficient knowledge of the conditions competent physico-mathematical skill could account for, and indeed, predict every one of these so-called 'chance' events (Huxley 1888, 554).

In our own time, this dream has been revived by molecular geneticists, who sometimes believe that in the genetic code they have at last found ways to "compute" the organism in ways comparable to physicists' ability to compute orbits (Gilbert, 1993). Reframing Darwinism in terms offered by the probability revolution allows us to see why this dream is illusory. As Elliott Sober has shown, the properties over which natural selection and other determinants of evolutionary trajectories range, such as fitness or relative adaptedness, belong to or at the very least are only visible in organisms considered as members of populations whose reproductive rates are being compared. A Laplacean or Maxwellian demon whose eyes were trained only on the computable paths of individual atoms, molecules, genes, or even single organisms would fail to notice the phenomena that evolutionary theory attempts to explain, not to speak of the causes that do the explaining (Sober 2000). Even philosophers of biology who subscribe to reductionistic ideals of science have given up on the idea of reducing evolutionary to molecular biology (Rosenberg 2007).

Just as dubious is Huxley's concession that theists might "without risk of contradiction" believe that an intelligence that computes the deterministic history of the universe might have an overall purpose in mind (Huxley 1888, 554). His point is that if the world were governed by chance, as Darwin's detractors took him to imply, this "wider teleology," as Huxley calls it, would be impossible. The living world would be simply too chaotic to be viewed even by God as purposive. Determinism removes this impediment. Huxley commends his "wider teleology" as better than the "old" teleology, which "supposes that the eye … was made with the precise structure it exhibits for the purpose of enabling the animal which possesses it to see." He says that the old teleology "has undoubtedly received its death blow" at Darwin's hands (Huxley 1888, 554).

But this is as weak as Huxley's computationalism. Teleology has several meanings. In all of them, however, it is an inherently explanatory concept, not a post-explanatory interpretive one. It cites ends, purposes, goals, functions, intentions and other such items as answers to questions about *why* a sequence of events unfolds as it does. As such, teleology has links to causal processes that rival law-governed deterministic accounts more than Huxley admits. If anything, Huxley had it backwards. While a sound account of adaptation by natural selection will avoid exaggerating the role of chance, it will also evade reductionistic determinism enough for a Darwinian to affirm without equivocation that in point of fact the eye *has* come to be "for the purpose of enabling the animal which possesses it to see."

Let me support this claim. It was probably Huxley's widely circulated deterministic and reductionistic argument that accounts at least in part for Darwinism's reputation among Oxford and Cambridge intellectuals of the 1870s and 1880s as having revived ancient Greek materialism. In sending him a copy of his translation of Aristotle's *On the Parts of Animals*, for example, the Oxford classicist William Ogle playfully imagined a meeting between Darwin and "the father of naturalists." "I can fancy the old teleologist," Ogle wrote, "much astounded to find that … Democritus, whom he thought to have effectually and everlastingly squashed, had come alive again in the man he saw before him." (Ogle to Darwin, 17 January, 1882; see Gotthelf 1999)

Ogle might more aptly have cited Empedocles than Democritus. Empedocles was no less a materialist than Democritus, but he introduced more objectivity into our inescapable impression of the natural purposiveness of organic traits. By allowing spontaneity (*to automaton*) into the universe, he was able to regard claims that organic traits "save" (*esothê)* the organisms that have them as objectively true, even if only accidentally caused. The story he tells is this. When cosmic history is more under the influence of love than of strife, various body parts are independently formed by the chemical affinities of the elements — so much air, so much water, so much fire, so much earth. This much is the work of necessity (*anankê*). These parts then spontaneously (*automatou*) conjoin and remain conjoined whenever in combination they just happen to be stable in, and so fit for (*epitêdeios*), the environments in which they spring up. All the organisms we observe, Empedocles claims, are coalescences of parts "saved" in this chancy way. Other possible combinations never got off the ground or quickly died out (Empedocles, fr. 57 DK; Aristotle, *Physics* II.8.198b29-32).[6]

This is indeed a selection process. To that extent, Empedocles is closer to Darwin than Democritus is. Beyond this similarity, however, there are large differences between Darwinian natural selection and Empedocles's appeal to what his critic Aristotle called incidental (*kata symbêbêkos, per accidens*) final (*hou heneka*) causality: causality in which an event that arises independently of a good outcome just happens to bring about that outcome.

One difference is that there is no appeal in Empedocles to differential reproductive success within kinds. This rules out the trans-generational etiology required for Darwinian adaptations, leaving Empedocles with the "preserved accident" view of the selection process from which I have been distancing Darwin and Darwinism. On any neo-Empedoclean reconstruction of Darwinism, the first, accidental instance of a variant that is causally responsible for comparative reproductive success already counts as an adaptation. We have already seen, however, that on any coherent formulation of Darwinian theory adaptations lie at the end of a process of adaptation. In fact, the necessarily gradual emergence of adaptations, which enhance the capacities of populations of interbreeding organisms for metabolizing, growing, and reproducing, would count in Aristotle's terminology as proper (*kata auta, per se*), not incidental (*kata symbêkêkos, per accidens*), final causality. Proper final causality is causality that runs through a process whose constituent moments, to the extent that something

[6] For Empedocles, apt combinations of body parts are inherited in male and female seeds, each preserving only part of an apt organic structure. The separate halves combine in reproduction to yield a whole organism. Empedocles probably regarded organisms as ontologically identical with these seeds, which expand in the course of growth, contract in the generative cycle, and then expand again. In addition to holding that this process is too marvelous to be an accident, Aristotle objected that Empedocles' doctrine denied that organisms actually come to be at all; they are not independent substances, but merely modifications of the elemental substances.

does not interfere, emerge as they do because they have a *telos*, as Aristotle puts it, or, in Kant's terms, are naturally purposive (*Zweckmessig*). To be sure, Aristotle would not agree that this ontogenetic process finds a counterpart in large-scale evolution. He did not think that large-scale evolution is even possible. Still, in Aristotle's technical terms Darwinian adaptations have properly final causes. They reliably have certain effects and come to be precisely *because* they have these effects (Depew 2008).[7]

For these and others reasons, most analytic philosophers of biology have opted for what is called the consequence-etiological, historical, or selected-effects analysis of what counts as an adaptation (Ayala 1970; Wright 1976; Brandon 1981; Brandon 1990; Sterelny and Griffiths 1991).[8] We have seen that for Darwinism the effects of initially chance variations gradually increase the relative adaptedness of the organisms that possess, transmit, and amplify these variations. The adaptive effects of past selective events are embedded to various degrees in highly heritable developmental processes. Over phylogenetic time this results in an ordered set of adaptations. The consequence-etiological analysis states that the adaptations that come to be entrenched in each ontogenetic cycle are explained by the concerted series of past effects on relative reproduction rates, all driven by ordinary causality, that propel traits through populations. This account neither exaggerates nor underestimates the element of chance in Darwinism. It gets it "just right."

Francisco Ayala may have been the first, but he was not the last, to argue that consequence-etiological explanations of adaptation by natural selection also count as teleological explanations in the sense Aristotle urged against Empedocles (Ayala 1970; Dobzhansky *et al.* 1977; Wright 1976; Brandon 1981; Lennox 1993; Depew 2008). Adaptations not only have certain reliable effects in the efficient conduct of organic life, but they come to be just because they have these effects. Having evolved for the sake of realizing the ends that they regularly bring into existence, they are properly finally caused in Aristotle's sense.

This approach to biological teleology relieves the antinomy that Kant, who could think of natural teleology only on the model of design, thought he saw in

---

[7] Aristotle retains the merely incidental final bond between the origin of a trait and its life-enhancing effect that he ascribes to Empedocles in the etiology of the simplest organic kinds that he calls spontaneously generated. He calls them spontaneously generated for this very reason.

[8] Consequence etiological analyses of adaptation are often conflated with analyses of what counts as a biological *function*. The second seems to me a different and more difficult claim than a consequence etiological analysis of *adaptations*.

natural purposes. Specifically, it alleviates the problem of reverse causality. Kant could readily imagine how conscious representations of intended or designed future events could lead to their achievement. But he could not imagine how the end-states of the embryological process, for example, could influence the sequence of earlier states that bring functional traits into existence by natural, not intentional, means. The consequence-etiological approach enables us to see that adaptations come to be from selected past effects that feed forward into new generations by a combination of positive and negative feedback loops. This makes it possible to say that adapted traits come to be and to exist for the sake of their adaptive effects with as much conviction as Aristotle. Reductionists, who tie explanation to underlying efficient-material causes, will reject this conclusion. But we have already seen that evolutionary biology is deeply resistant to their demands.

Ayala's approach to consequence-etiological teleology treats it as an explication of the apparent design of organisms and their adaptations. This analogy was inherent in the natural theology that Darwin opposed, which has tended to saddle Darwinism with the assumption that its *explananda* are identical with or very like those of intelligent design (Lewens 2007).[9] As the consequence-etiological approach to biological teleology has been further developed, however, it has increasingly presented itself as an alternative to thinking of organisms as analogous to designed artifacts. The latter requires an "engineering" conception of fitness that is at best a useful heuristic for suggesting adaptationist hypotheses, but should not be too deeply projected into the ontology or etiology of organisms. The consequence-etiological analysis of adaptation is certainly applicable to organisms when they are thought of as assemblies of functionally adapted modules that, in the absence of a presiding intelligence, are under the control of evolved information-processing programs that run on replicating macromolecules. Yet this "design-without-a-designer" conceit carries with it an aggregative conception of organisms as collections of independently optimized adaptations that not only echoes the intelligent design paradigm that it overtly opposes, but also bears traces of Empedocles's modularized account of the agglomeration of already individuated hands, legs, heads, and other body parts. To that extent it flirts with assigning accidents too large a role in explaining the origin of organic traits to account for the fact that adaptations properly so-called arise for the sake of the advantages they procure. The basic difficulty is that design-without-a-designer talk is parasitic on the intelligent design paradigm. The first instance of an intelligently designed trait obviously

---

[9] See Behe (1996) for an extreme example of this tendency. For Behe, an advocate of intelligent design, adaptations are not only like machines; they are machines.

exists for the sake of its intended task as much as, or even more than, later instances. Thus unless great care is exercised the distinction between initial, chance variations and their subsequent shaping into adaptations by the process of natural selection can easily be blurred. The stress on discrete adaptations as products magnifies the risk of inscribing endings into mere beginnings. It downplays the process of adaptive natural selection, especially when it analogizes this process to computer search programs and other such tropes (Dennett 1995). It can be replied that critics of engineering conceptions of adaptation are reading too much into a mere metaphor. But this is a perilous rebuttal. It amounts to tacitly confessing that design-without-a-designer advocates either do not have an analysis of adaptation at all, as distinct from an adaptationist rhetoric suited to creationist or formerly creationist audiences, or are tacitly relying on a more literal analysis such as consequence etiology.

By contrast, the consequence-teleological view of adaptation *is* well displayed when the beings over which natural selection ranges are conceived as epigenetically self-organizing developmental processes (Oyama, Griffiths, and Gray 2001). Aristotle already knew from close observation that organisms and their parts come to be by a process in which an originally indeterminate, but highly dynamic proximate matter differentiates itself into a progressively more articulated whole. Every point in the ontogenetic process is causally determined by its immediately prior state. Having reached its *telos,* the formative process begins over again by renewed reproduction (*Generation of Animals* II.2.1.735a10-25; 2.4.740a1-24). As a result, the organized and differentiated wholes whose stage-by-stage emergence Aristotle describes in *The Generation of Animals* come to possess the kind of substantial integrity that is required if organisms are to exhibit, in degrees that vary with psychic (*psychikos*) capacities proper to their kinds, behavioral plasticity and resilience in the face of the environmental pressures (*On the Soul.* II.1.412all-413a10, IV.416a6-9; *Generation of Animals* I.18.722b18-24). Part of Aristotle's case against Empedocles is to insist that if organic kinds were not substantially organized in this way not just some, but all of them would have perished or never come to be in this strife-ridden world. The very fact that we obviously share the world with a plentitude of kinds of living beings is proof enough for Aristotle, then, that organisms are not and cannot be aggregations of ontologically separate parts.

Through the concept of consequence teleology, more of the epigenetic Aristotelian ontology of organisms can be transferred to an evolutionary, even Darwinian, worldview than one might initially have suspected. This is most easily seen when we think of natural selection as a phenomenon that arises only in certain sub-classes of self-organizing autocatalytic dissipative structures (Depew and

Weber 1995; Weber and Depew 2001). These are, in Aristotle's terms, life's "proximate matter." The enhanced heritability presupposed by natural selection properly so-called does not lose diachronic features that are already visible in the chemical and biochemical forms of selection from which it emerges. On the contrary, the ontogenetic trajectories of organisms show in intensified form the same directional patterns of bifurcation and differentiation that are observable in the less tightly bounded and informed systems to which they are coupled as nodes of energy, work, and dissipative flow through ecological systems (Depew and Weber 1995).[10] This point enables us to understand development as the repository, through common descent, of generatively entrenched adaptations and as the site of evolutionary novelties that can be tested by selection. Happily, as Scott Gilbert has recognized,[11] the epigenetic-developmental perspective is once again becoming the focal point of evolutionary biology, making analogies between organisms and designed entities less and less appropriate and making design-based analyses of adaptation, whether they take a presiding intelligence or natural selection to be the designer in question, less and less convincing.

3. Is etiological causality *really* teleological?

I hear an objection. Consequence teleology, you may say, "isn't really teleology" (Walsh 2007). Whether this is true depends on what one thinks teleology 'really' is and how it is related to final causes. The term itself goes back only to 1728. It was devised by the philosopher Christian Wolff to refer to the providential nature of the pervasive system of mutual uses that connects various parts of the *scala naturae* (Wolff 1983). It originally referred, that is, to inferences of the sort satirized by Voltaire in *Candide,* such as that Providence has given us nose bridges in order to support eyeglasses. Treatises on natural theology, which reached their most elaborate development in eighteenth and early nineteenth-century Great Britain, are full of arguments of this "prevision and provision" sort (Hodge 1982). We learn, for example, that coal was put into the secondary strata so that much later we would have fuel to burn (Buckland 1836, 497-98). This was originally called "the doctrine of final causes" because Wolffians ascribed other aspects of biology, including reproduction, to mechanical efficient causes. The point of teleology was to show that the hierarchical system of mutual uses among parts of the *scala naturae*, culminating in the way inanimate stuffs and lower kinds serve the higher

[10] See also the contributions to this volume by Kauffman, Newman, and Ulanowitz.

[11] See his contribution to this volume.

purposes of human beings, is so finely tuned that it must be referred directly to a benevolent intelligence. In this way final causality and teleology became associated with intelligent design on a cosmic scale. It was evidenced by phenomena that resisted natural efficient causal explanation. This is, I believe, what teleology commonly connotes even today. It is the source of biology's continued entanglement with theology, especially "god-of-the-gaps" natural theology.

There is, however, more to be said. Wolff's privileging of intentionally-designed external use over the end-orientation of ontogenetic development bothered Kant. He responded not by rejecting the doctrine of teleology, but by turning it upside down. By adopting a finally-caused epigenetic rather than an efficiently-caused preformationist view of reproduction, he, like Aristotle, made organic self-formation the primary locus of teleology, reduced mutual uses to mere effects of efficient causal chains, and treated their pervasive linkages in the system of nature as a whole as a matter of reflective, not determinate, judgment. Darwin, too, rejected Wolffian teleology. "If it could be proved that any part of its structure had been formed for the exclusive good of another species," we read in the *Origin,* "it would annihilate my theory." (Darwin 1859, 201) Darwin himself sometimes traced the etiology of adaptations to what he called final causes (Lennox 1993). Moreover, on at least one occasion he was willing to approve using the word 'teleology' to refer to explanations of finally caused adaptations in this sense. In rebutting his Harvard colleague Bowen's accident-based misinterpretation, Asa Gray claimed that "Darwin's great service to natural science is bringing back to it teleology; so that instead of morphology versus teleology, we have morphology wedded to teleology." (Gray 1963, 237) Darwin replied, "What you say about teleology pleases me especially."[12]

Darwin was comfortable speaking this way because in this context he was using the term teleology not in the Wolffian-Voltairean sense, but as referring to the pervasive internal purposiveness and end-orientation that the great comparative anatomist Cuvier, who was himself influenced by Kant, saw in the relation between each species' integrated structure and its unique "conditions of existence" (Grene and Depew 2004). This Aristotelian-Kantian-Cuvierian sense of teleology had been brought to Britain by William Whewell, whom Gray was actually quoting when he spoke of "morphology wedded to teleology."

If William Ogle, the classical philologist who had sent Darwin his translation of *The Parts of Animals* had known as much as we now know about Aristotle's biological

[12] Darwin to Gray, June 5, 1874 (Darwin F. 1888, vol. 3, 189). See also Hösle's contribution to this volume.

treatises he would have seen that "the father of teleology" himself was not much less insistent than Darwin that species cannot acquire traits oriented to the good of other species by any form of properly final, that is teleological, causality (Lennox 1993; Lennox 2001). Aristotle certainly recognized that members of each species play systematically supportive roles in one another's flourishing. But for him this system of mutual uses is an effect, not a cause, of each organism's coming to be for the sake of realizing its distinctive range of species-specific psychic capacities. To be sure, this system is good for all concerned. This good, moreover, is all the more realizable because it unfolds against a metaphysical backdrop in which Being is inherently a plentitude that stretches from pure mind to pure matter. For Aristotle emergence is unnecessary for the same reason that reductionism is impossible. Nonetheless, Aristotle distinguished between coming to be for the sake of a good proper to oneself (*heneka* + genitive) and, having done so, existing for the sake of the good of another (*heneka* + dative) (Kullmann 1998). It would seem that by Wolff's standards, far from being the "father of teleology," Aristotle was not a teleologist at all!

What had happened is that by Darwin's time Wolff's original sense of this term was no longer monopolizing its meaning. Indeed, except where the argument from design continued to play an ideological role in *public* discourse, as in Great Britain, internal teleology of the Aristotelian-Kantian-Cuvierian sort had become its preferred meaning among experts, including Darwin. It referred to the internal final causality involved in epigenetic self-formation. "Design-without-a-designer" Darwinians have overlooked this fact. Writing in opposition to direct descendents of British natural theology, they treat intelligent design as the fixed, unproblematic meaning of teleology and in consequence as Darwinism's whipping boy (Ghiselin 1984; Dennett 1995; Ruse 2003). By treating teleology in this univocal way they have ignored important historical, and highly contingent, aspects of the term. I can say why by summarizing my argument thus far. If any conception of teleology is useful for understanding adaptation it is the consequence-etiological account. Adaptations so analyzed are properly finally caused and teleologically explained. This is not an arbitrary or polemical commandeering of an old term. It analytically expresses insights that have been deepening and broadening ever since the epigenetic revolution of the late eighteenth century. From this revolution, in which Kant and Darwin both participated, emerged the modern discipline of biology. As this discipline advances the design and use teleology with which it was entangled becomes less and less relevant.[13]

---

[13] Ernst Mayr has suggested that the term "teleology" be left to its Wolffian misery and replaced by "teleonomy" (Mayr 1976, 384-404). But this notion, as Mayr defined it, is restricted to the

What remains to ask is whether consequence teleology can or should be extended to the course of evolution as a whole, ontogeny giving rise to phylogenic progress that recapitulates itself in each new generation even as it pushes further. As Peter Bowler has shown, post-Darwinian elites, including many clerics, made themselves comfortable with macroevolution by convincing themselves that recapitulationism preserves much of the pre-modern, anthropocentric *scala naturae* (Bowler 1988). They simply regarded the notion that lower beings emerge for the sake of human beings as an unfolding diachronic rather than an eternally synchronic affair. Upon inspection, this popular representation comforts because it tacitly retains the external Wolffian teleology to which its advocates were already accustomed. It simply disguises use teleology by dressing it up as internal teleological ontogeny on a grand scale.[14] I agree with Bowler that this picture of teleology is inherently non-Darwinian (Bowler 1988). Contrary to core Darwinian principles, it assumes inner tendencies that pull ontogeny toward greater complexity, autonomy, mentality, or other such goods. Small wonder that twentieth century Darwinians have been hostile to it—so hostile that at times they have overreacted by abandoning the developmental perspective altogether or by asserting that evolution exhibits no directional tendencies at all. In recent decades, Darwin's own view of *post facto* directionality has been restored. It might be articulated, however, in richer ways than the crude idea that "evolutionary arms races" favor ever more elaborate "vehicles" for passing genes across generations.

One way of enriching the directional thrust of life's history would be to build on the ontological status of organisms as energy-using, entropy-dissipating developmental processes. By their very nature communities of organisms so described are deeply embedded in the ecological systems of which they are constituent parts. These systems are under the proximate influence of the Second Law of Thermodynamics. Accordingly, they exhibit temporal, irreversible, and directional aspects.[15]

end-directedness of cybernetic systems as they achieve homeostatic states. In addition, Mayr conceives of development as "read out only" from an evolved program, thereby underestimating the interactive nature of development and evolution that has come to the fore in "evo-devo."

[14] This way of thinking goes back not to Aristotle, but to the Stoics, whose sense of the world as a single living substance denied the distinction between internal and external teleology. All uses are as internal as external in this view. In the Judaeo-Christian-Islamic creationism that supervened on the Stoic picture, however, this account undermined internal teleology altogether by its dominant appeal to a divine architect whose primary object is use teleology. Wolff's term 'teleology' arose from this background.

[15] See Ulanowicz's contribution to this volume.

It thus becomes easier to understand in-principle directional tendencies in the history of life, rendering it unnecessary to invoke either self-interested competition or benevolent "prevision and provision" to beat back entropy and cause directional change.

It also becomes easier in this light to discern the foundations of what Darwin called "the co-adaptation of organic beings to each other and to their physical conditions of life." Darwin, we recall, saw ecological mutualism precisely when he let go of inner drives toward phylogenesis. In this spirit, I see no reason why the consequence-teleological approach to Darwinian adaptations is not applicable to well-integrated and differentiated communities of social insects, to certain forms or degrees of symbiosis,[16] to group phenomena, and even perhaps to the regulation of the atmosphere by expansion and contraction of biomass. The properties of such systems have certain effects. The systems that have these properties have come to be just because they have these effects. Some of these properties might be adaptations precisely defined. Others might be consequence-teleological in a different way. A macro-ecological view of this sort provides a site for evolutionary consequence teleology that has nothing to do with late nineteenth century popular pseudo-teleology. It also has the advantage of replacing the anthropocentrism that lies at the heart of that defective view, which gave rise to modern racism, with an ethic of care in which, with Darwin, we can discern and appreciate the kinship of all living things.

In sum, consequence teleology allows us to contemplate the biosphere with far greater theological equanimity than we can look upon the barren Laplacean scene conjured up by Huxley. Even if nature is not inherently teleological, there are local, but plentiful teleological phenomena in it. Moreover, natural science itself provides teleological explanations for these phenomena, even from a strict Darwinian perspective. Theology does not have to supply the teleology. It is already there for theologians to contemplate.

References

Ayala F., 1970, "Teleological explanations in biology", *Philosophy of Science* 37: 1–15.

Behe M., 1996, *Darwin's Black Box: The Biochemical Challenge to Evolution*, New York: Free Press.

Bowler P., 1988, *The Non-Darwinian Revolution,* Baltimore: Johns Hopkins University Press.

[16] See Margulis' contribution to this volume.

Brandon R., 1981, "Biological teleology: questions and explanations", *Studies in the History and Philosophy of Science* 12: 91–105.
Brandon R., 1990, *Adaptation and Environment*, Princeton: Princeton University Press.
Buckland W., 1836, *Geology and Mineralogy with Reference to Natural Theology*, London: Pickering.
Burkhardt F. *et al.* (eds.), 1985- [CCD], *The Correspondence of Charles Darwin*, Cambridge: Cambridge University Press.
Conway Morris S., 2008, *The Deep Structure of Biology: Is Convergence Sufficiently Ubiquitous to Give a Directional Signal?*, West Conshohocken: Templeton Foundation Press.
Darwin C., 1859, *On the Origin of Species*, London: John Murray. Facsimile 1st edition: 1964, Cambridge, Mass.: Harvard University Press.
Darwin C., 1969, *The Origin of Species,* Burrow J. (ed.), Harmondsworth: Penguin (6tth edition).
Darwin F. (ed.), 1888, *Life and Letters of Charles Darwin*, London: John Murray. Reprint: 1959, New York: Basic Books.
Dennett D., 1995, *Darwin's Dangerous Idea*, New York: Simon and Schuster.
Depew D., 2008, "Consequence etiology and biological teleology in Aristotle and Darwin, *Studies in the History and Philosophy of the Biological and Biomedical Sciences* 39: 379–90.
Depew D., Weber B., 1995, *Darwinism Evolving: Systems Dynamics and the Genealogy of Natural Selection*, Cambridge, Mass.: MIT Press.
Dobzhansky T., Ayala F., Stebbins G. L., Valentine J., 1977, *Evolution*, San Francisco: Freeman.
Fuller S., 2007, *Science vs. Religion? Intelligent Design and the Problem of Evolution,* Cambridge: Polity Press.
Gayon J., 1992, *Darwin et l'après Darwin: Une histoire de l'hypothèse de sélection naturelle*, Paris: Editions Kine. English trans.: 1998, *Darwinism's Struggle for Survival*, Cambridge: Cambridge University Press.
Ghiselin M., 1984, "Introduction to Darwin's *The Various Contrivances by Which Orchids are Fertilized by Insects*", Chicago: University of Chicago Press.
Gilbert W., 1993, "A vision of the grail," in Kevles D., Hood L. (eds.), *The Code of Codes*, Cambridge: Cambridge University Press, 83–97.
Gould S. J., 1989, *Wonderful Life*, New York: Norton.
Gould S. J., Lewontin R. C., 1979, "The spandrels of San Marco and the Panglossian paradigm: a critique of the adaptationist programme," *Proceedings of the Royal Society of London B* 205 (1161): 581–85.
Gotthelf A., 1999, "Darwin on Aristotle," *Journal of the History of Biology* 32: 3–30.
Gray A., 1963, *Darwiniana*, Dupree A. (ed.), Cambridge, Mass.: Harvard University Press (Originally published 1876, New York: Appleton).
Grene M., Depew D., 2004, *Philosophy of Biology: An Episodic History*, Cambridge: Cambridge University Press.
Herschel J., 1830, *Preliminary Discourse on the Study of Natural Philosophy*, London: Longmans, Rees, Orme, Brown and Green.

Hodge M. J. S., 1977, "The structure and strategy of Darwin's 'long argument'", *British Journal for the History of Science* 10: 207–46.

Hodge M. J. S., 1987, "Natural selection as a causal, empirical, and probabilistic theory", in Krüger L., Gigerenzer G., Morgan M. (eds.), *The Probabilistic Revolution,* vol. II, Cambridge: Cambridge University Press, 233–70.

Huxley T., 1888, "Teleology", in Darwin F. (ed.), *Life and Letters of Charles Darwin*, vol. II, London: John Murray. Reprint: New York: Basic Books, 1959.

Kullmann W., 1998, *Aristoteles und die Moderne Wissenschaft*, Stuttgart: FranzSteiner Verlag.

Lennox J., 1993, "Darwin *was* a teleologist," *Biology and Philosophy* 8: 405–21.

Lennox J., 2001, *Aristotle's Philosophy of Biology*, Cambridge: Cambridge University Press.

Lewens T., 2007, "Adaptation", in Hull D., Ruse M. (eds.), *The Cambridge Companion to the Philosophy of Biology*, Cambridge: Cambridge University Press, 1–21.

Lyell C., 1830-33, *Principles of Geology*, London: John Murray.

Mills S., Beatty J., 1979, "The propensity interpretation of fitness," *Philosophy of Science* 46: 263–86.

Nyhart L., 2008, "Embryology and morphology," in Ruse M., Richards R. (eds.), *The Cambridge Companion to the Origin of Species*, Cambridge: Cambridge University Press, 194–215.

Oyama S., Griffiths P., Gray R. (eds.), 2001, *Cycles of Contingency: Developmental Systems and Evolution*, Cambridge, Mass.: MIT Press.

Richards R., 1992, *The Meaning of Evolution*, Chicago: University of Chicago Press.

Rosenberg A., 2007, "Reductionism (and anti-reductionism) in biology", in Hull D., Ruse M. (eds.), *The Cambridge Companion to the Philosophy of Biology*, Cambridge: Cambridge University Press, 120–38.

Ruse M., 2003, *Darwin and Design*, Cambridge, Mass.: Harvard University Press.

Sober E., 2000, *The Nature of Selection*, Cambridge, Mass.: MIT Press (1st and 2nd editions 1984).

Sterelny K., Griffiths P., 1999, *Sex and Death: An Introduction to the Philosophy of Biology*, Cambridge, Mass.: MIT Press.

Walsh D. M., 2006, "Organisms as natural purposes: the contemporary evolutionary perspective." *Studies in the History and Philosophy of the Biological and Biomedical Sciences* 37: 771–91.

Weber B., Depew D., 2001, "Developmental systems, Darwinian evolution, and the unity of science", in Oyama S., Griffiths P., Gray R. (eds.), *Cycles of Contingency: Developmental Systems and Evolution*, Cambridge, Mass.: MIT Press, 239–54.

Weinberg S., 1992, *Dreams of a Final Theory*, New York: Pantheon.

Wolff C., 1983, *Philosophia rationalis sive logica, pars I,* in Ecole J. (ed.), *Gesamelte Werke*, Band I, Hildesheim: Georg Olms Verlag.

Wright L., 1976, *Teleological Explanation*, Berkeley and Los Angeles: University of California Press.

EMERGENCE, DIRECTIONALITY AND FINALITY IN AN EVOLUTIONARY UNIVERSE

William R. Stoeger sj

Though this paper was given in a section of the Conference devoted to theology, it will consist of focused reflections on a mixture of science and philosophy oriented towards revealing the natural sciences' — including evolutionary biology's — consonance with theology. In order to do this effectively I shall first situate evolutionary biology within a broader cosmic context — within the unfolding of the universe itself from the Big Bang onwards — beginning from a very hot, dense, simple, and undifferentiated state, and gradually expanding and cooling. As it cooled, new things became possible — both on macroscopic scales (superclusters of galaxies, clusters of galaxies, galaxies themselves, clusters of stars, stars, and planetary systems) and on microscopic scales (a large number of chemical elements, followed by simple molecules and in some places with the right conditions very complex molecules, leading to life). That is, as the universe expanded and cooled, it gradually became ever more lumpy and complex. There was a time — the first 300,000 years after the Big Bang — when there were no stars and galaxies, nor any other elements besides hydrogen, helium and a little bit the lightest metal, lithium. All the other elements were manufactured within stars, or as some of them exploded in supernovae. Without stars, we simply would not be here. Thus, cosmic evolution sets the stage for biological evolution — puts in place the essential building blocks for life and the environments within which it can emerge.

Secondly, I shall stand back and distill from the sciences some of the more important general and philosophically and theologically relevant characteristics of nature. Emergence is certainly one of those. But so are directionality and finality, as we shall show. Other such broad key characteristics include evolution itself — the fact that the universe and nature itself are always changing and adjusting to new conditions and environments — with old things disappearing and completely new systems and organisms emerging from earlier ones through the self-organizing tendencies of natural agents, which is complemented by *bona fide* natural selection at the biological level. Then there is the fragility of nature — nothing last forever — but previous stages and systems provide the material and more importantly the stable plateaus from which new systems are generated. And then there is relationality — a key characteristic of nature and the universe. All novelty and emergence is really due to the constitutive relationships at lower levels which enable and effect the emergence of novel systems and organisms at higher levels. Along with the importance of these relationships, are several other key features: the nested hierarchies of organization at hundreds — if not thousands — of different levels on this planet and possibly on others; the relative formational and functional autonomy of nature — it does not need to be micromanaged; its inherent dynamism; and finally its deep interconnectedness — the same laws of physics and chemistry function throughout the universe, and everything is related to everything else, often in highly differentiated ways.

There is no doubt that, as the universe has evolved from the Big Bang, roughly 13.7 billion years ago, it has undergone continual differentiation and complexification.

In the course of that cosmic history — and more recently in the history of our planet, the Earth — an amazing array of intricately related, and interdependent systems and networks of systems has emerged at all levels, and then evolved further into new systems and networks. Living organisms, and among them conscious, rational and freely choosing organisms, are the most notable of these. They exhibit capabilities and behaviors far beyond those of their basic components. This appearance of new objects and systems with qualitatively different and often more advanced capabilities and behaviors from networks of more simple ones is often referred to as "emergence" — for example the emergence of life, the emergence of consciousness, the emergence of intelligence. This does not mean that these emergent phenomena are not explainable in terms of the processes and relationships among lower-level systems and entities — these obviously provide the necessary conditions for life, consciousness and intelligence — but simply that such emergents and their properties and operations are not causally reducible to or determinable by their individual operations.

This continually unfolding emergence of new and intricately organized systems and organisms strongly suggests a directionality in the history of the universe, and in the history of the Earth and of life on it — at least in the general movement from the very simple to increasing levels of complexity. And from that suggestion, which seems to reflect similar ideas of various philosophical and theological traditions, many recent interdisciplinary pundits postulate an overarching finality or teleology — a purposefulness — to the unfolding universe, and to nature itself as it evolves on Earth, and possibly elsewhere. This has been spurred by the controversial but very popular so-called "anthropic coincidences," or "anthropic principle" (see Stoeger 2007, 445, and references therein), the recognition by many cosmologists and physicists, that the universe appears to be fine-tuned for complexity — and therefore for life and consciousness. In other words, if any one of a number of key physical or cosmological parameters — e. g., the strength of the gravity, the strength of the strong nuclear force, of electromagnetism, or the initial rate of expansion of the universe as it emerged from the Planck era (the Big Bang) — were just slightly different, the universe would be completely sterile — without the possibility of complexity and life.

Though it seems impossible to either confirm or deny such an overarching cosmic purpose on the basis of the natural sciences alone, it is clear that within systems and organisms themselves, a certain local, focused teleology has emerged — as differentiated functionality. Each component of a complex system or organism has a particular function within it — a function which is often essential to its survival and integrity. We have for instance in the bodies of mammals the life-giving functions of the heart, the kidneys, the lungs, the brain and its key components. And as any system or organism is always a part of some larger system, organism or ecology, it in turn fulfils a certain function, or set of functions — which is often interpreted as having a certain "purpose" within that larger system. And natural selection itself supplies the preference for the organisms which are more fit and functionally adapted relative to a given environment. This itself implies a certain directionality, even finality.

It is far from controversial to recognize this pervasive and amazing pattern of the emergence of novelty and incredible variety throughout the history of the cosmos and of the Earth. In this presentation I intend to reflect briefly but carefully on the general conditions and processes underlying this continuing emergence. It is somewhat controversial to go further and maintain that there is a general directionality to the unfolding cosmos — and to the overall evolution of systems within it. Among most biologists there is strong resistance to asserting that. However, there are some, along with a number of biophysicists, cosmologists, astrobiologists, and complex systems specialists, who strongly support this conclusion on scientific grounds. In order to do so, we need to understand what is meant — and what is not meant — by this directionality. I intend to explore this briefly, including a sketch of its scientifically accessible basis. By "directionality" in this sense — as a scientifically accessible or discernible movement — I do not mean one with a single unique, or even definite, goal, but simply one which proceeds towards a definite range of possible outcomes — which become more focused and delimited as evolution continues. It is not necessarily goal determined — though it may be — but is primarily process driven. Specific processes naturally imply a limited range of outcomes.

This continual emergence of novelty in nature — which, as I've already hinted, is really due to the rich differentiation of lower-level components and their internal and external synchronic and diachronic relationships (combinations and interactions) with one another and with their immediate environments — along with directionalities and functional teleologies they induce, reflects the deep consonance and compatibility of cosmological and biological conclusions about origins with the best that Jewish, Christian, and Muslim theologies of creation have to offer. This is not at all surprising, given that, from a theological perspective, Nature is the ongoing "work" of the Creator. Oftentimes, however, either our concepts of Creator and creation are so inadequate, or our interpretations of scientific conclusions so philosophically distorted or shallow, that an authentic and careful *rapprochment* between the two becomes nearly impossible.

And finally I shall reflect briefly, and I hope carefully, on the scientifically controversial issue of teleology, both the teleological aspects stemming from functional differentiation within systems and organisms and the suggestion of an overarching teleology — or finality — for which some find support in contemporary physics, cosmology, chemistry and biology. Certainly the major theological traditions strongly affirm such a teleology — and major philosophical traditions have argued to that. But, to what extent can strictly scientific conclusions of natural sciences as such validly support a definite purpose to our universe? And whether or not they can, what are the scientifically accessible relationships, processes and regularities which effect the implementation of that overall teleological movement? These are the questions I shall attempt to answer provisionally here.

1. Emergence and directionality in the early universe

Under the dominant influence of gravity, proto-galaxies condensed from the expanding cosmic hydrogen and helium (and a little bit of lithium, the lightest metal) — there were no other chemical elements — and then very soon the first stars appeared. As we have already stressed, only with stars were the elements beyond hydrogen, helium and lithium synthesized. Eventually, the stars and their remnants enriched the cosmos with carbon, oxygen, phosphorous, iron, silicon, and all the remaining 92 natural elements. But, in order for stars to form, a number of very delicate conditions had to be fulfilled. It is only because the universe has been expanding and cooling since the Big Bang and that gravity of a certain strength continues to operate — along with the precise strengths of the other basic physical interactions — that stars were able to form at all. In fact, if the universe had been expanding a little more quickly than it was at the beginning — or a little more slowly — stars would not have formed. In that case there would have been no chemistry — and the building blocks necessary for the eventual emergence of life would never have been produced. The universe would have remained sterile and boring forever!

Thus, the expanding, cooling, complexifying history of the universe since the Big Bang 13.7 billion years ago provides the overarching, all-embracing physical directionality — and conditions — for cosmic, chemical and biological evolution (Stoeger 1998, 163; Davies 1998, 151). Although we cannot say precisely or in detail exactly what had to emerge from that process, it is clear — just from the way gravity operates and local matter condensations heat up on collapse — that the universe had to become lumpy on a number of scales, that those lumps would continue to interact gravitationally, and that within their internal stellar furnaces new — heavier — elements, with completely new properties would be manufactured. This is the basic directionality inherent in the cosmic system itself — given the laws of physics and certain "initial conditions" at any early stage we consider. Given those conditions then gravity and other physical regularities and relationships inevitably lead to a definite rather narrow range of outcomes. Possible future configurations and possibilities are tightly — though only rarely uniquely — restricted.

In this early stage of cosmic evolution, there are two examples of lower level emergence worth examining briefly. First, there are the large, intermediate and smaller size gravitationally bound systems — galaxies, clusters of galaxies, stars and their planetary systems, and clusters of stars (e.g. globular clusters). Secondly, on the microscopic level, there is the formation of elements heavier than hydrogen and helium in cores of stars, and in supernova explosions.

In the first case — astronomical systems — gravity and angular momentum conspired to act on density fluctuations in the cosmic plasma — once matter decoupled from radiation about 300,000 years after the Big Bang — to form this vast array of hierarchically ordered — diverse — systems of various sizes. And each of those continued — and continue — to interact and evolve in various ways, again because of gravity and angular momentum, among other lesser influences.

In the second case, all the heavier elements and chemical compounds are gradually synthesized. Nuclear physics along with electromagnetism, acting within the evolving but specific conditions of temperature, pressure and chemical composition in stellar cores — and in the case of compounds in the cooler atmospheres of some stars or on planets — enable their formation. These elements and compounds have completely different properties and potentialities than the components from which they are formed. Why? Precisely because of the particular relationships between the particles (protons, neutrons and electrons) that make up the atoms, or between the atoms that make up compounds like salt and water. Salt, for instance, has completely different properties than the sodium and chlorine which constitute it. And the amazing substance water is very different from the hydrogen and oxygen from which it is synthesized. These are examples of emergence on the chemical level — and illustrative of the constitutive nuclear and chemical relationships which endow these substances with their novel characteristics and capabilities — their unique and very specific or focused ways of interacting with other substances in particular environments. Again, in systems involving given elements and compounds — under specific conditions — there is always a directionality. There is a directionality from the past to the present situation — dictating that this outcome, or one similar to it, will occur. And there is a directionality towards the future specifying a certain relatively narrow range of types of subsequent outcomes (Stoeger 1998, 163) — unless very different conditions and influences from outside the system under consideration intervene.

2. Pre-biology and the emergence of life and consciousness

It is not possible in this very short and sketchy talk to summarize the various stages of evolution and the processes which are involved. I have spent time briefly describing the formation of simple physical systems in our universe, and that of the elements and simple compounds — because they are basic to everything else that follows and because already at those levels, it is clear that there is at least a type of emergence — depending upon networks of relationships and leading to differentiation, complexification and diversity — and, of course, a certain internal directionality.

It is worth pointing out, too, that all this depends heavily on the fundamental order with which the universe is endowed — at the level of physics, and the chemical order which emerges from that as the elements are produced. This is connected, once more, to the "anthropic principle." What is the origin of that order? And what enforces it, in such a way that higher levels of order emerge, as more complex systems and entities arise?

I shall now very briefly speak of the processes that lead to and characterize the emergence and directionality at higher levels of complexity leading to life and consciousness, and then describe in outline the three general categories (levels) of emergent phenomena first given by Terrence Deacon (2006, 111; 2007, 88).

Once we have diverse localized environments — say on a planet — characterized by very particular stationary conditions — or cycles of conditions — of temperature, pressure, alkalinity or acidity, chemical composition, available free energy, etc., various new chemical species are produced. And not only that but there will be chemical cycles involving a number of different but interrelated chemical species which manifest certain regularities within a given environment or ecology — some species, for instance, catalyzing the production of other species — and being catalyzed in turn by yet a third category. The evolution of such chemical systems, though self-sustaining for a time, will certainly change the environment — leading to other conditions, providing new niches for slightly or very different systems. These constitute new plateaus of organization stability, from which further evolutionary changes can take their departure. Or there may be an intrinsic or extrinsic alteration in the system — on the macroscopic or on the microscopic level — which drives its development in a different direction. In all cases, what is critical are the effective intrinsic and extrinsic relationships which effect the production — the emergence — of new and different systems, and the selective reinforcement of some types over others. This *is* already a primitive type of natural selection!

This general pattern of system-building and networking develops even more potential when differentiated species of molecules (e.g. DNA or RNA) begin to carry detailed information which is implemented by other molecules (RNA or proteins), and reliably reproduce themselves within their chemical systems. This obviously allows for a detailed propagation of successful characteristics for flourishing in a given environment to later generations. With some mechanism of variation (e.g. genetic mutation or symbiotic incorporation) among the information carrying and information implementing molecules, new and more successful characteristics can develop and eventually dominate in the system. This is basically natural selection — considered to be the principal "process" responsible for biological evolution. But it is considered by many researchers in the field to be already important in advanced pre-biotic chemical networks — e. g. Manfred Eigen's autocatalytic hypercycles (Eigen and Schuster 1978; Küppers 1983, 125). This is, as Bernd Olaf Küppers (1983, 2; 1990, 136) mentions, the crucial stage of *instructed synthesis* of molecules like nucleic acids and proteins, following the earlier more primitive stage of *non-instructed* synthesis of pre-biotic molecules, which provides the immediate building blocks for the emergence of reproductive chemical systems. As I have already implied, however, the very existence and effectiveness of natural selection depends strongly on the entire scaffolding of physical, chemical and biological processes which underpin the almost innumerable internested levels of order and dynamic organization represented by living systems and their supporting environments. Natural selection can only work on what has already emerged and prospered — and that presupposes the continuing successful operation of all the interacting networks of systems to which we have been referring.

Once such sophisticated non-living systems have developed, a further advance involves their strategic separation from one another — and encapsulation — while at the same time remaining in continual mutually beneficial interaction with one another and with their environment — for energy, material resources, and waste disposal. An essential general feature of these systems is the prevalence of information-controlled feedback loops, which enable them to adjust continually to internal and external conditions, to maintain homeostasis and to assure their successful operation. Thus, even at this level there is what can be called teleonomic or goal-like behavior, which relies on selection and effective transmission of information crucial to the survival of the system. As a result the way these systems behave and evolve is radically *context dependent* (Küppers 1990, passim; Stoeger 1998, 177, and references therein). Thus, they manifest not only bottom-up and same-level causality, but also top-down causality — particularly because of the effective relationships their subsystems have with one another, and the system itself has with its environment.

As these separated but interacting systems evolve, they form higher level systems with one another — manifesting co-operative and symbiotic behavior. This enables modularity and redundancy to develop — both in those systems that are living, and in those which are not. This is clear in eukaryotic cells and in multicellular animals, for instance. Each module of the system — e.g. the heart, the lungs, etc. — has its internal workings shielded from the rest of the organism. The organism itself does not need to monitor all those details, as long as the lungs are functioning properly. However, each module interacts with the organism in very specific ways, receiving nourishment and information from it for its successful operation, but also providing certain essential and very specific products or functions for the overall health and survival of the entire organism — in the case of the lung oxygen for the entire body. Natural selection continues to act on populations of these organisms or systems — enabling those which have emerged that are more fit for a given environment to flourish.

We have briefly indicated here some of the key features which are involved in the emergence of more and more complex systems — and eventually living organisms. What is absolutely necessary for this to occur are: 1. a basic underlying order (that is, physics and all that it presupposes), which as systems develop, leads to higher levels of order (chemistry and biology); 2. differentiation and diversity of systems — and therefore of their properties and capabilities; 3. evolving highly differentiated relationships among these entities and systems; 4. the effective generation, selection, preservation, reproduction, and co-ordinated implementation of information within and among systems and organisms. All these — particularly the internal and external constitutive relationships among lower-level components of systems and organisms — are at the basis of the phenomenon of emergence. They also obviously manifest a certain directionality — both the overall orientation towards certain outcomes, based on the conditions which obtain now and on the processes and relationships which dominate their continued operation and evolution, and the continuing drive to maintain themselves and their specific functioning within the larger systems of which they are a part.

Thus, this directionality and functional teleology or teleonomy is deeply based in the laws of nature themselves — the regularities, processes and relationships which characterize our world. This is something I have always insisted upon, and which other specialists in this area have also stressed — for instance Paul Davies (1998, 151, and references therein) in his many writings. It is helpful to spend a few more minutes exploring how this functional teleology within systems can be explained. All the components needed have already been pointed out — and our brief discussion of the different levels of emergence in the next section will shed more light on what I am about to say.

As we have already seen, as complexification proceeds — even long before living organisms emerge — there is a differentiation of structures (e.g. stars and planets, and the atoms and molecules of the different chemical elements and compounds) on many different levels. This automatically involves a differentiation of functions or capabilities — of the ways specific systems or structures interact or relate to others. Then these diverse systems — with different functions or specializations — evolve in their relationships and interactions, often forming symbiotic — mutually beneficial — more highly complex systems. This means that the specific function of each component eventually becomes not only beneficial to but also essential to the survival and successful operation of the larger more complex system. Each component thus fulfils a definite "purpose" within the larger system. In fact, each component has been selected because of what it gives the larger system. Natural selection is continually reinforcing and enhancing this choice — modifying it in light of what leads to more successful organisms (Ayala 1970, 1; 1998, 101). Thus, the particular focused *outcome* of each component or subsystem is why that component is present to begin with, and why it continues to play an essential role in the organism (Auletta, Ellis, and Jaeger 2007, 1159). Presupposing chemical, biological and physical compatibility and accessibility, as well as functional appropriateness, the "purpose" of each component emerges within the overall successful operation of the larger system of which it is a part. It fulfills a particular role within that system, which is essential to its life, development, or successful operation in its environment. This is how functional teleology arises — and how it flows from the underlying laws of nature which are operative at different levels in a complex system or an organism.

3. The different levels of emergence

Not all types of emergence are equal — or even similar. Looking in more detail at the various general categories of emergence will aid us in appreciating and understanding how it occurs, and the role the important relationships among the components and within their environment play in constituting emergent systems as complex and novel. It will also enable us to understand how the distinctive selection and amplification processes that are often involved function, as well as the origin of the directionalities and functional teleologies which are implicit in them. Terrence Deacon (2006, 111; 2007, 88) has given us three general levels or orders of emergence, which are well worth considering.

Deacon (2007, 93) defines emergent phenomena as those "having novel properties not contained in their constituent components and exhibiting regularities that cannot be deduced from laws affecting their constituents." He goes on to stress, however, that at the same time the behavior and properties of emergent systems do not violate or replace the physical and chemical regularities and relationships governing the components. What is always critically important in emergent phenomena are, according to Deacon, the dynamical configurational regularities of the whole system which strongly affect — select, channel, reinforce and amplify — the constituent interactions. As Deacon (2007, 95) further comments: "What needs explaining is how some systems come to be dominated by higher-order causal properties such that they appear to 'drag along' component constituent dynamics, even though these higher-order regularities are constituted by lower-order interactions." As already mentioned, the nonlinearities, positive and negative feedback loops, and selective amplification of certain lower-level and intermediate-level outcomes all play a central role in emergent systems and organisms.

Deacon's first-order emergence is exemplified by the pressure and temperature of a gas, or the surface tension and turbulent flow of liquids. These properties are not those of the individual molecules making up the gas or the liquid but are statistically based properties of large collections of such molecules. These emergent properties, however, can be understood and explained by the relationships among the molecules making up the gas or liquid — the statistical regularity of the relationships among the molecules leads to these emergent phenomena. The same is true of the emergence of galaxies or planetary systems, and at a higher level to emergence of the different chemical elements from more basic ones via fusion, or the production of chemical compounds like salt and water from their underlying chemical constituents.

A helpful representative of second-order emergence is the growth of a snowflake. In this and other similar cases of emergence — leading to very unique and beautiful structures, or intricately organized, but non-living behaviors — the overall macroscopic configuration which begins to develop plays an essential role in constraining and guiding the microscopic processes which underlie the system, so that they are preferentially activated at some positions and not in others. This can be looked upon as selection of a few from a very large number of possibilities, and the amplification of the selected alternatives. In the case of the snowflake, prior growth sets the pattern for later growth. More advanced second-order emergent systems exhibit "autopoiesis" — dynamic self-organizing behavior. Chemical systems for instance can have several — or even many — different kinds of components which therefore interact very differently with one another but at the same time as part of interconnected cooperative network — as in the autocatalytic hypercycles (Eigen and Schuster 1978; Küppers 1983) to which we briefly referred above, or in the much simpler chemical oscillators (see, e.g. Peacocke 1983, 40 ff.) which so many have investigated over the past 50 years.

Living systems — organisms — are the primary examples of third-order emergence. In this case, not only are "the higher-ordered states repeatedly re-entered into the lower-order dynamics," as Deacon (2007, 106) is fond of saying, leading to reinforcement and amplification of those higher-order states, but there is also an ongoing cumulative memory and evolution of the living system across time, as the detailed information concerning the successful microscopic and macroscopic structures of the past are reintroduced in descendants of those systems. There is here very complex selection, reinforcement and in some cases amplification of earlier self-organizing systems and their slow or rapid evolution in slightly or even very different organisms via natural selection, symbiosis, etc.

These are the principal levels of emergence Deacon has identified. George Ellis, a well-known physicist and cosmologist, has added others (Ellis 2006, 99) — such as those illustrated by learning in animals, and at even a higher level, by rational behavior and the use of symbolic language in human beings. Again, all of them rely heavily on the complex, highly differentiated and dynamic relationalities among their constituents at many different levels of organization, and on innumerable deeply nested feedback control loops. These characteristics allow them to manifest the directionalities and functional finalities which we have already emphasized in our discussion.

4. Overarching teleology in nature and in the universe

At this point we can briefly explore whether or not the natural sciences themselves provide evidence for an overall teleology — or finality — guiding or directing nature towards certain outcomes — or goals (Stoeger 1998, 182). From our discussions so far, is there any definite scientifically accessible way of reaching a conclusion on this issue? What about the anthropic principle, and the apparent fine-tuning of the universe it reveals? I am convinced that on the basis of our scientific findings so far, we have to conclude on strictly scientific grounds that there is no secure evidence of such a global finality. But, at the same time, there is also no scientific evidence ruling it out. On the one hand, it is clear that the physics and chemistry of our universe are open to the possibility of life and consciousness: We are here! Furthermore, as we have mentioned, slight changes in the underlying physics and chemistry would have rendered the universe completely sterile.

So one might be led to suspect some deep cosmic purpose. But on the other hand, we know that there are vast numbers of cosmic venues where these possibilities have not been realized. In fact it may be — we shall probably never know — that we are the only location where intelligent life has emerged. This could lead to assuming that we are a happy accident — one of very few stellar systems, and one a very few universes where life and consciousness have arisen. Furthermore, there is the startling bleakness of the cosmic destiny which is provided by cosmology — it's either freeze or fry. It now looks very much like it will be "freeze"! Science has no way of exploring the intentionality of agents which fall outside its investigative capabilities. Thus, we really cannot say one way or the other about an overarching teleology.

On some philosophical grounds, and on theological grounds, of course, we can affirm such a teleology, or finality. The underlying order of the universe — not to mention its existence — requires an ultimate ground or basis, and almost certainly a radically relational basis. That ground or basis — the Creator — can be argued to have a definite purpose for ordering the universe is this way, rather than some other way. From Christian, Jewish and Islamic revelations intentional creation is central, and is motivated by the Creator's love and desire to share the Creator's existence and goodness with others, inviting them to share in full, free and loving communion with the Creator. We can compellingly argue this theologically. Though we cannot argue it scientifically — the Creator as Creator is inaccessible to the natural sciences — we can certainly say that all that we have found in the sciences supports a deep compatibility and consonance with our less inadequate understandings of the Divine. And, certainly, relationality-based emergence, along with the directionalities and focuses of local functional finalities at every level and the relative autonomy of nature, is at the core of this profound consonance.

REFERENCES

Auletta G., Ellis G. F. R., Jaeger L., 2008, "Top-down causation by information-control: from a philosophical problem to a scientific research program", *J. R. Soc. Interface* 5: 1159–72.

Ayala F. J., 1970, "Teleological explanations in evolutionary biology", *Philosophy of Science* 37:1–15.

Ayala F. J., 1998, "Darwin's devolution: design without a designer", in Russell R. J., Stoeger W. R., Ayala F. J. (eds.), *Evolutionary and Molecular Biology. Scientific Perspectives on Divine Action,* Vatican City State and Berkeley: Vatican Observatory Publications and The Center for Theology and the Natural Sciences, 101–16.

Davies P., 1998, "Teleology without teleology: Purpose through emergent complexity", in Russell R. J., Stoeger W. R., Ayala F. J. (eds.), *Evolutionary and Molecular Biology: Scientific Perspectives on Divine Action,* Vatican City State and Berkeley: Vatican Observatory Publications and The Center for Theology and the Natural Sciences, 151–62.

Deacon T. W., 2006, "Emergence: the hole at the wheel's hub", in Clayton P., Davies P. (eds.), *The Re-emergence of Emergence: The Emergentist Hypothesis from Science to Religion,* Oxford: Oxford University Press, 111-150.

Deacon T. W., 2007, "Three levels of emergent phenomena", in Murphy N., Stoeger W. R., S. J. (eds.), *Evolution and Emergence: Systems, Organisms, Persons,* Oxford: Oxford University Press, 88–110.

Eigen M., Schuster P., 1978, *The Hypercycle: A Principle of Natural Self Organization,* New York: Springer-Verlag.

Ellis G. F. R. , 2006, "On the nature of emergent reality", in Clayton P., Davies P. (eds.), *The Re-emergence of Emergence: The Emergentist Hypothesis from Science toReligion,* Oxford: Oxford University Press, 79–107.

Küppers B.-O., 1983, *The Molecular Theory of Evolution,* New York: Springer-Verlag.

Küppers B.-O., 1990, *Information and the Origin of Life,* Cambridge, Mass.: The MIT Press.

Peacocke A. R., 1983, *An Introduction to The Physical Chemistry of Biological Organization,* Oxford: Oxford University Press.

Stoeger W. R., 1998, "The immanent directionality of the evolutionary process, and its relationship to teleology", in Russell R. J., Stoeger W. R., Ayala F. J. (eds.), *Evolutionary and Molecular Biology: Scientific Perspectives on Divine Action,* Vatican City State and Berkeley: Vatican Observatory Publications and The Center for Theology and the Natural Sciences, pp. 163–90.

Stoeger W. R., 2007, "Are anthropic arguments, involving multiverses and beyond legitimate?", in Carr B. (ed.), *Universe or Multiverse?,* Cambridge: Cambridge University Press, 445–57.

Teilhard de Chardin: *Moving towards* humankind?

Lodovico Galleni

1. Introduction: the epistemological bases

This paper focuses on the scientific work of Pierre Teilhard de Chardin, on his contribution to the debate on evolutionary theories, and on his impact on the relations between science and theology. In earlier publications in connection with the fiftieth anniversary of the death of the French Jesuit, I suggested that a careful analysis of Teilhard's work will show how a *specific* model of the interactions between science and theology — which I call "the model of the third millennium" — is an underlying part of what has created this impact (Galleni 2005).

This model highlights not only the influences of science on theology, but also those of theology on science, though there is an asymmetry involved that has become clear after a careful reconstruction of the scientific and intellectual careers both of Teilhard de Chardin (Galleni and Groessens-Van Dyck 2007) and of other authors who in various ways have tackled the same problem.[1]

It seems to me that this asymmetry is necessary for the science/theology interactions to be fruitful — and it is to be understood thus: theological and religious reflection can offer hints to the scientist that can help him in the construction of theories; but it can never make any difference to the results reached by the method (or methods) proper to every science. On the other hand, theological research, and in particular that on God the Creator, cannot avoid having a vision of nature as emerging from the various sciences "in their legitimate autonomy", to cite the doctrinal statements of the Second Vatican Council. Here "legitimate autonomy of science" clearly means that science responds only to its proper method in the development of its work. There is however a further Council statement that is of importance from the point of view of the doctrine of truth, to wit, that the results of science come together to form the truth about man.[2]

---

[1] The asymmetry is already evident in the models of interaction and exclusion between science and theology discussed in Galleni (1992a; 2007, 3–33).

[2] Cf. Concilio Ecumenico Vaticano II, Costituzioni, Decreti, Dichiarazioni, AVE, Roma, 1966.

At this point, a question quite naturally arises: in the economy of the universe, is the appearance of the free and thinking creature a purely random event, a fortunate number coming up in the lottery of the universe, or is it also the result of some kind of necessity — so that it can therefore be reconstructed through laws of nature that are at least in part deterministic?

In the specific case of life sciences — admitting that they respond only to their proper method as well — Teilhard de Chardin not only belongs among the most important evolutionary biologists of the twentieth century, but undoubtedly looks like the most innovative among them, thanks to his scientific contributions in the areas of complexity and the Biosphere (Galleni 2010).

However, in my view these same Teilhardian arguments with regard to the theory of biological evolution derive from a new approach to the relations between science and theology. Without investigating this issue, it will not be possible then to discuss Teilhard's scientific work.

In order to do so, I will begin by summarizing the three principal alternative theoretical models for the interpretation of biological evolution in relation to the position of man in nature.

The first hypothesis is the one so potently argued and reaffirmed by J. Monod (1970): that man is the lucky number that emerged from the "tombola" of the Universe. The second sees Man as the result of rigidly deterministic laws that science describes; the third sees him as the possible, if not probable, result of laws that are not rigidly deterministic — where the stochastic aspects also play a role that is non-secondary. In this latter case, we could understand the "processuality" of nature metaphorically as a *moving towards* which may under the right conditions attain to consciousness. It is this interpretation of the third model that I aim to affirm here on the basis of a study of the scientific work of Pierre Teilhard de Chardin.

2. Theoretical models and scientific theories of evolution

Each one of the three theoretical models I have listed has favored and directed the development of a different specific evolutionary theory. As contemporary epistemology has now demonstrated, such models do not in fact just come out of observation and/or experiment, but out of more complex interactions that involve philosophy and theology as well, not to mention the dense network of relationships that each scientist has with the social milieu in which he was trained and in which he operates (Galleni 2010, 11–20).

To make matters simpler, I will be looking to the epistemological work of Imre Lakatos (1978) for guidelines in understanding the nature and role of theoretical models with respect to scientific theories properly so called.

For Lakatos a scientific theory is an intellectual construction with an extremely complex structure. It has two parts: the first part represents what Lakatos calls the hard core, which contains the fundamental part of the theory; if this part were to be falsified by observation and/or experiment, that would of necessity involve the abandonment of the theory itself. Around this hard core there is then a second part, which Lakatos calls the protective belt: this contains lines of research connected to the hard core by a relation of subordination. As working hypotheses, these serve on the one hand to extend ever further the explanatory capacity of the theory, demonstrating its fertility; on the other, to keep the hard core protected from anomalies, from unresolved problems that are the origin of conflicts, thus allowing the theory to survive for longer.

Now let us take a closer look at the structure of the hard core. According to Lakatos, in addition to the principles and axioms typical of the theory, this also contains meta-physical orientations, an expression we are to understand literally: concerning not only what is beyond observation and experiment, but also what comes from social, political, philosophical, religious and even theological ideas of the author of the theory.

This part of the theory is fundamental, but it is often suppressed by the constructor of the theory, perhaps because in many cases he is either not fully aware of the metaphysical nature of certain of his assumptions, or because he doesn't really understand it. The far from easy task of reconstructing and clarifying these aspects will fall then to the historian and to the epistemologist.[3]

In this connection, the case of Monod can act as a useful example: his chance-based vision of evolutionary mechanisms is a modern version of the analogy of the architect used in earlier times by Darwin: evolution being determined by two disconnected chains of causes that meet up by chance. Up to this point we are on factual terrain — and indeed the thesis refers to a series of observations and experiments that Monod correctly judges to be scientifically objective.

But the philosophical theses on the place of man in nature and on chance chosen as the tool for the interpretation of the universe do not derive from scientific objectivity, but from the hard core of the whole theory of Jacques Monod, that is from existentialist metaphysics — something which is furthermore clarified by the reference to the *Myth of Sisyphus* (the title of a collection of essays by Albert Camus [1942, 161–66]), which we find at the beginning of *Chance and Necessity* (Monod 1970).

[3] For the importance of historical research in epistemology and for a clear reference to the structure of scientific theories, see Feyerabend (1975, Italian edition 1985, 21–25).

But first, to avoid any kind of misunderstanding, it will be as well to explain that I shall be using the Popperian and Lakatosian complex of epistemological instruments exemplified up to now without espousing the skeptical vision of scientific explanation and its bearing on truth that other epistemologists have in fact drawn from it.

Popper himself (1969) declared that he was fully aware of the danger that falsifications would encourage contemporary epistemology to move closer to the nominalism of Bellarmine than to the realism of Galileo. Precisely because they are falsifiable, scientific theories will never give us an absolute definitive certainty about how things really are: they will always remain no more than instruments that help us to understand better how the Universe functions and perhaps to have some control over its processes. In the case of evolutionary biology, this conclusion would be devastating, since it could serve — as has in fact happened — to create doubt over the fact of evolution itself! Fortunately, Popper himself suggests a crucial way out of this impasse, one which I consider valid and to be shared: while it is true that human knowledge of itself is constituted by theories that are by nature necessarily falsifiable, this does not mean that within it some certain and definitive acquisitions cannot be also attained: for example, says Popper, that the earth turns around the sun, or I would add, that living beings have evolved and present *common kinship connections that result from a divergent descent.*

These we can consider henceforth to be *acquired facts* secured by human knowledge (Popper 1969).

When speaking of evolution it is necessary in point of fact to distinguish on the one hand between evolution as a process to be reconstructed by means of research of a historical nature, a process whose factual existence is then as well assured as that of the Roman Empire; and on the other, the theories put forward to explain this fact (Galleni 2002b). These latter are and will be by their very nature always falsifiable — even though when dealing with an order of phenomena as complex as those of evolution, it will be necessary to anticipate recourse to the concept of *local falsification.* To understand this better let us take as example a theory like natural selection: it functions well in many cases, but it can be falsified in others. This does not mean that it has to be abandoned or integrally revised as a theory; but merely that it could coexist alongside other theories (Galleni 2000, 89–99; 2010a, 111–15).

We are now basically ready to begin our historical-critical survey of Teilhard de Chardin's work. We shall discover thus the extent to which, in his case more than in that of other scientists, it is easier to grasp the connections between the structure of the hard core of his theory and the evolutionary hypotheses formulated by him, precisely because they are explicit and evident. But first, I want to refer to the key episodes in his life story, drawing on it to find a definitive explanation of the asymmetrical character of the relation between science and faith, perfectly exemplified by the case of Teilhard.

Pierre Teilhard de Chardin was born in France at Sacernat in 1881 (one year before the death of Charles Darwin), and died in New York in 1955.

He lived out his religious vocation in the Society of Jesus and his scientific vocation first in Paris, in the laboratory of the palaeontologist and palaeoanthropologist Marcellin Boule, then in China. There he took part in the group working on the fossil remains of what has become known as Peking man. For this and for other work he is commonly regarded as the founder of the modern palaeontology and geology of the Chinese subcontinent. But he is also remembered for his fascinating synthesis of evolution with Catholic theology, a synthesis that alongside the rich harvest of French philosophy and theology before and after the Second World War, was to open up one of the paths that would lead the way to the perspectives of the Second Vatican Council.[4]

Here however we are not so much interested in the strong impulse that Teilhard's ideas on Christian philosophy and theology received from his research into evolution,[5] as in the influence of the former on the construction of the Teilhardian scientific theory and on the orientation of his research, in a relationship that was

---

[4] A very short note on Teilhardian theology: Cardinal Henri de Lubac observed how extremely robust this theology is both from the point of view of the coherence of its lines of research and from that of its biblical references. The problem is that because he was not able to publish his theological writings in his lifetime, a large number of texts that are almost all posthumous have to be analyzed patiently and lovingly for this high theological quality to be revealed. Once this has been done the quality of Teilhardian theology stands out clearly. Yet, as Cardinal de Lubac again remarked, all too often Teilhard de Chardin has been attacked by superficial and prejudiced readers on the basis of scanty and hurried readings: see de Lubac (2008).

On the other hand Tanzella-Nitti devotes more than one page of the Christological entry written by him for the *Dizionario Interdisciplinare di Scienza e Fede* to Teilhard de Chardin, emphasizing the pioneering ideas to be found in the Teilhardian Christology and expressing reservations only on the ambiguities of the language. But in this same dictionary in my entry on Teilhard de Chardin I emphasized how the ambiguities of the language were fundamentally down to the fact that he was not allowed to publish his philosophical and theological writings in his own lifetime. See Tanzella-Nitti (2002) and Galleni (2002a).

[5] As I have already written with reference to the work of the Reformed theologian Karl Heim as an illustration, no reflection on God as Creator can evade the vision of the Creation that comes from science (see Galleni 1992a, 11–14).

in fact reciprocal, although "asymmetrical" in a sense that I shall now clarify (See Galleni 2003). No theological research on the subject of God as Creator can evade what science, in its "legitimate autonomy" (an expression taken, let me emphasize, from the Constitution *Gaudium et Spes* of the Second Vatican Council), tells us of the structure of Creation. On the other hand, as we have already said, following the reading of Imre Lakatos, the construction of a scientific research programme cannot evade an important part of the hard core, constructed not on observation and experiment but on a different base that I would call "more argumentative". Nothing strange then about the fact that a scientist like Teilhard de Chardin, who was at the same time also a theologian — or at any rate someone capable of being receptive to the proper questions of theology — should use these questions to construct his research programme.

The important thing is to maintain the asymmetry: theology must in fact accord great importance to the description that science gives of the Creation and it can also suggest research pathways, but in the end it should also take into account of the fact that science reaches conclusions (where of course it is actually able to reach definitive conclusions) on the basis of its method or methods. On the other hand, science must live with the awareness that theories are not at all the pure unfiltered results of observation and experiment but are constructed in a more complex and structured manner, so that hints coming from more "metaphysical" paths of research[6] are fundamental — even if ultimately on the questions that concern the description of nature, the last word is left to those natural conclusions drawn (to paraphrase Galileo) from sense experience and necessary demonstrations.

At this point we can take up our investigation again and see that the best examples of these interactions are precisely the models and theories originally put forward to explain the evolution of living beings and then often used consequently to generate general conclusions about the place of man in nature.

Research into the structure of these models as found in the work of Lamarck, Darwin and Wallace, and Mivart, will clarify the relationships between their theories and the philosophical part of the hard core, laying the basis for discussion of the Teilhardian model.

---

[6] Let me re-emphasize here for the sake of clarity and precision and to avoid misunderstandings, that I use the word metaphysical in the literal sense of the word, i.e. of all that goes beyond the physical, meaning all that goes outside the important but restricted sphere of observation and experiment. See Galleni (2005, 162–63).

Jean Baptiste Monet, Monsieur de Lamarck, put forward the first complete and coherent theory on the transformation of living beings. We can take as our point of departure, chronologically speaking at least, the opening lecture in his course of Year X, held in Paris in 1802, from which emerged the first sketch of the theory of evolution — or transformism as Lamarck called it. The fundamental mechanism was a dialectical relation between the living being and the environment, through which a change of environment provoked an adaptive response and therefore a change in the living being that was then transmitted to any offspring.

The use and disuse of organs and the heredity of acquired characters are then the carrying structure of the Lamarckian theory (Lamarck 1969). Moreover, the evolutionary vision transformed the great Aristotelian chain of beings, which placed groups of living beings on different steps according to their growing structural complexity, into a true and proper genealogical tree. There are varying interpretations as to the existence of a more or less progressivist vision in the transformism of Lamarck. In my opinion however, this directional movement that develops along an ascendant chain is analogous to the *moving towards* to which we referred earlier, and it has a clear philosophical reference, i.e. to the Enlightenment concept of progress (see Ruse 1996). Furthermore, the mechanism which makes possible its realization has a particular cultural antecedent that is not so much philosophical as theological. Lamarck belongs in fact with the so-called mechanistic Deists who rejected the personal God of the revealed religions and in particular the Biblical God. The God of Enlightenment reason is only the guarantee of the good functioning of natural mechanisms because he is their origin. In the particular case of evolution, he sets in motion the chain of causes and they then develop automatically as in the case of a clock which, once assembled and wound up, works by itself. Just as with the clock, it will then be necessary to have an uninterrupted chain of causes to ensure continuity of connections between the mechanisms involved: we are then looking at a deterministic model. On the other hand, the vital forces that spur the animal to adapt are also always physical forces, even if specific to life and all yet to be researched and discovered. In principle, the living being is only a machine whose components can be dismantled and then put together again in a different way. The important thing is to discover the physical force that actually provides the living glue for the new being.[7]

[7] Likewise, the forces which, in Mary Shelley's *Frankenstein*, transform a series of assembled pieces into a living being are also physical, in an excellent literary example of Cartesian reductionism and mechanism. See Shelley (2002).

With Lamarck we have the first evolutionary theory that is complete and coherent (the use and the disuse of the organs connected to the dialectical animal-environment relationship and to the heredity of acquired characters), in addition to the precise indication of a scientific research programme, which presupposes a chain of interconnected causes in a continuous relationship of cause and effect *connections*. All this then in a perspective that moves towards progress.

Keeping our focus firmly on our central topic of evolution, the mechanism proposed to explain it by Alfred Russel Wallace and Charles Robert Darwin is different: it was put forward first in two separate articles in a pamphlet published by the Linnean Society of London[8] and then in a volume published in 1859 by Darwin on his own (Darwin 1859). The mechanism is the one universally known under the name of "natural selection", whose theoretical model we could call "the *unconnectedness* model".

The structure of the Darwin-Wallace theory was excellently synthesized by J. Huxley (1942). According to the latter, the theory can be presented schematically as three facts and two deductions. The first fact consists in the observation that offspring are more numerous than their parents. The second fact is born out of the observation of the demographic stability of a species: notwithstanding a level of fecundity that is more or less elevated, the number of individuals remains more or less constant from generation to generation.

These two facts suffice to take us to the first deduction: that there exists a competition between individuals which translates into nothing less than a "struggle for survival". This first part of the theory is derived directly from the work of Thomas Robert Malthus, who wrote the first version of his own book as far back as 1798, with reference however only to human populations (Malthus 1993).

A third fact served to extend the hypothesis of Malthus and translate it into a mechanism with a universal bearing (the second deduction) — the fact of hereditary variability: individuals of a species vary appreciably one from the other and this variability is inherited. At this point, if individuals vary and the environment varies as well, this latter will favor the survival from generation to generation of those individuals who bear the characters that are best adapted to a certain environmental situation. Here then the choice of individuals (selection) is completed by the (natural) environment.

---

[8] Darwin and Wallace (1858). The articles were published in the same number of the *Journal of the proceedings of the Linnean Society: Zoology*, London, in 1858. The paternity of the theory goes therefore to both authors.

The aspect that concerns us most here is that natural selection is the result of two *unconnected* chains of causes: on the one hand the mechanisms of hereditary variability (which neither Darwin nor Wallace knew) and on the other the selective function that the environment exercises with regard to variability.

The relations between these two mechanisms are illustrated by Darwin with the analogy of the architect.

Natural selection acts like an architect who must build a house at the foot of a landslide, making use of the stones from the landslide itself. Can we say, asks Darwin, that in such a case the form of the stones can be ascribed to chance? Not insofar as the form is down to the rock from which the stones come, from the forces of erosion that have acted on them, and from the knocks that the stones have undergone in falling. Nor can it be said that the architect for his part uses the stones randomly: for in setting them in place he has criteria on which he orders them so as to compose a wall or an arch, he chooses them and uses them for a purpose and aims to exploit their forms to the maximum. But the actual form of the stones is not of course shaped with a view to the use that the architect will make of them. By this theory (that of two unconnected causal chains which encounter one another *by chance* in a particular situation) Darwin intended to emphasize the presence of chance factors within the evolutionary mechanisms,[9] which Monod (1970) would affirm to be valid on the terrain of molecular biology as well.

An omniscient Creator, wrote Darwin, ought to have foreseen every consequence of the laws imposed by him on nature, but can we reasonably think that he has foreseen the form of every rock or fragment of rock in terms of the use that the architect will make of it? If we consider it nonsensical to suppose that the various fragments have been organized for the good of the architect, can we think that he has organized the innumerable variations of animals and plants subjected to domestication for the good of the breeder? (Mivart 1871, 242–88).

No reasonable person can consider that the variations were formed with a view to the choice that the environment would make of them and that they are in some way guided by intention. Ultimately, if the variations had been providentially foreseen, the possibility of negative variations or the reproductive excess that has led to the struggle for existence, to the survival of the fittest, and to natural selection would be superfluous laws of nature. On the other hand, since an omnipotent and omniscient Creator orders everything and foresees everything, why then

[9] For a synthesis of the architect analogy and a discussion of the concept of chance in evolutionary biology, see Galleni (2000, 114–24).

would he have to use dramatic mechanisms like the struggle for survival, the presence of variations that are in part also negative or at any rate highly inefficient? The plasticity of variation, according to Darwin, could lead to negative deviations of the structure just as reproductive excess led to the struggle for survival. Could a Creator who foresees everything have accepted mechanisms of this kind?

As can be seen, there is a very precise relation between the aspect of unconnectedness, i.e. a mechanism based on unconnected chains of causes, and the demise of a vision of natural theology in which Darwin himself had been trained when studying at Cambridge to become a Church of England cleric.

The key point was the philosophical weight to be given to this chance encounter of unconnected chains of causes. In fact the two authors would present different philosophical visions of the theory of natural selection. For Darwin it seems to preclude any kind of possibility of a *moving towards* thinking being, given the profound arbitrariness of the mechanisms, for Wallace on the contrary it fits into the various ways in which purpose is manifested in the universe, and this purpose is that of preparing the universe to receive thinking being (Wallace 1895, 141–66). It is therefore particularly interesting to observe how the same theory allows and divides into different philosophical interpretations.

But there is a third perspective that soon comes on the scene. In 1871 the volume by St. George Jackson Mivart *On the Genesis of Species* was published in London (Mivart 1871).

The author, a zoologist of the school of T. Huxley, whose entry into the Royal Society was supported by Darwin himself, wrote this book to confirm the importance of the evolution of living beings, thenceforth to be regarded as beyond doubt, but emphasizing the limits of natural selection. All the examples and problems that from then down to our days would be cited to present the limits of natural selection were already set out in Mivart's lucid book: they are the origin of complex structures, parallelisms, threshold effects, periods of stability and rapid changes, self-organization. All these examples, set out in detail and with precision, would incite Darwin to add a chapter in the sixth English edition of his book to reply to Mivart's objections, a sign of the brilliance of this latter author. It need not concern us here to enter into detail on the objections to natural selection or the solutions to them put forward by Mivart. For Mivart there must exist a mechanism of evolution internal to living beings that ensures that many of the principal transitions develop for a programme that can be described in terms of general deterministic laws that are similar to those of chemistry. Ultimately, writes Mivart, once the laws of chemistry were known, we became capable of synthesizing urea, and when we have understood the laws of evolution we shall be capable of synthesizing a fish in the laboratory.

Basically we are talking about the most Galilean scientific programme of the three that we have briefly summarized. Galileo spoke of the laws of nature that God had written in the language of mathematics; thus, for Mivart, God had written with the deterministic laws typical of the language of chemistry. The philosophical reasons were clear: Mivart was a Catholic of the school of J. H. Newman and therefore he sought general laws that would explain the process of *moving towards* humanity and overcome in some way certain random aspects of natural selection with arguments and examples that would have an important influence outside the Anglosaxon world.[10] But the time has now come to turn to Pierre Teilhard de Chardin.

3. The synthesis of the three models in Pierre Teilhard de Chardin and evolution as a *moving towards*

I have devoted a good part of my theoretical work to the study of this author's scientific activity and to his impact both on contemporary theories of evolution and on the relation between science and theology. The conferences that were held on the fiftieth anniversary of his death have made it possible to take stock of the ongoing research on this important figure of a man of faith and science who has had an extremely positive influence on Catholic theology. But in my papers I have basically focused on his scientific work (see Galleni 1992b, 153–65; 1995; Galleni and Groessens-Van Dyck 2001, 5–104).

There is a common denominator in all of the Teilhardian scientific work, that of finding the experimental proofs that, reintroducing a preferential element within the random fact, show with the force of scientific research that the principal characteristic of evolution is in fact the *moving towards* inherent in it: of matter

[10] Mivart's perspective would have interesting developments both in the Russian context with the geneticist N. Vavilov who, working on parallelisms, would study in the genera and species those deterministic regularities that Mendeleiev had described for chemistry (Vavilov 1922), and in the Latin context with the Italian and Spanish school (see Leonardi 1950). We cannot however pursue this particular avenue, leading to the debate between the Latin and Russian schools and to the links — clear or yet to be researched — with the Catholic and Orthodox traditions. For now it will suffice to refer to two volumes that record the acts of two colloquia held in Saint Petersburg and Pisa and published in bilingual Italian-Russian editions (Reati and Galleni 2004; Grib, Galleni and Reati 2007).

towards complexity and of life *towards* complexity and consciousness. This is the closest connection between Teilhard de Chardin and the work of Lamarck, but revisited at this point within a perspective of Christian theology and philosophy.[11]

This line of research would lead Teilhard, as we shall see, to develop two important contributions to the theories that attempt to explain the mechanisms of evolution: biology as science of the infinitely complex and geobiology as science of the Biosphere and a necessary instrument for the application of the techniques of complexity to evolution.

His long and laborious life would be lived then in the period in which Darwin and Wallace's hypothesis of natural selection was connected to Gregor Mendel's laws of genetics, thus bringing about the general revision of the mechanisms of evolution that goes under the name of the modern synthesis. But while the synthesis would basically be achieved in the Anglo-American cultural environment, Teilhard de Chardin would be trained in the scientific milieu of French palaeontology and in particular in Marcellin Boule's laboratory (Galleni and Groessens-Van Dyck 2001; Vialet and Hurel 2004) and then he would work for twenty years in China. Only in the period after World War II would he find his way to the United States for the last period of his active life.

Man of science and man of faith, he would live his religious vocation in the Company of Jesus and his scientific vocation, as we have just said, in the capacity of a trained palaeontologist and palaeoanthropologist.

In the course of the preparation of his doctoral thesis he became acquainted with a geologist, Jean Boussac, with whom he was to have an important exchange of letters during World War I — which both men would spend in the trenches. From these letters there emerges very clearly the shared idea of finding a different ap-

---

[11] The reference to Lamarck is essential: in fact this is probably the most important influence of the French zoologist on Teilhard de Chardin, i.e. evolution as a *moving towards* which, as we have already seen, can be traced back in Lamarck — as an instance of influential metaphysics — to the Enlightenment's cherished idea of *moving towards* progress. This is the most important heritage of Lamarck — even if we shall clearly then see in the Teilhardian work an inheritance of acquired characters that is actually linked to cultural inheritance. On the other hand, the actual concept of *moving towards* represents one of the points at which, like the meridians in the neighborhood of the pole, to use Teilhard's own terminology again, science, philosophy, and theology meet. There are then in this case evident points of contact that we have to know how to "manage". Here we see the necessity of a new discipline, science-and-theology, which must show us the paths and the tools we need if we are to work on the points of contact (see Jackelen 2001).

proach to geology by looking to go and research possibilities of a global vision that would go beyond purely reductionistic mechanisms. To go beyond the study of this or that stratum or this or that fossil and elaborate tools that make it possible to work on overall views. The correspondence would be interrupted by Jean Boussac's death in the trenches.[12]

When the War was over, Teilhard de Chardin would go back to Marcellin Boule's laboratory and complete and publish his doctoral thesis. And already in the publication of the material in this thesis we can see emerging what would be a guiding thread of his research: the quest for parallelisms and convergences, as experimental examples that would confirm his theory of evolution as a *moving towards*. Let us try to penetrate further into this part of Teilhard's work. According to the mechanisms illustrated by the analogy of the architect, hereditary variability was formed in fact by its own mechanisms that were without any connection to the selection that acted on them. In evolutionary lines that were henceforth separate, variability put forward ever new and diverse raw material to selection, and the case where an evolutionary sequence was run through in a parallel manner by evolutionary branches now separated had to be extremely rare. Therefore to find parallelisms meant to suggest that evolution had taken place thanks also to systems and mechanisms different from the typical Darwinian ones. Basically, parallelisms served to indicate that once a determinate evolutionary stage had been reached, the following steps were already at least in part determined and so, even if the basic trunk subdivided, the branches would follow parallel trajectories. Fundamentally this meant an end to a groping quest for possible solutions that would have to lead to a separation and continuous distancing of the branches of phylogenesis.

A particular but crucial aspect for Teilhard de Chardin was the parallelism towards the formation of ever more complex cerebral structures, an evident proof of animal evolution as a *moving towards* complexity and consciousness.

In presenting the Prosimians of the Quercy phosphorites, Teilhard de Chardin would describe the presence of a *moving towards* a growth in height and cerebral structures demonstrated by parallelisms with the phylogenetic branch that would lead to Man. We are at the beginning of the Nineteen twenties, we are still in France and the Teilhardian research project is starting to become clearer. Parallelisms are an important pathway of research to show the basic idea of evolution

12 Teilhard de Chardin and Boussac (1986). Fascinatingly enough, it was precisely the geologist Boussac who suggested to the priest Teilhard de Chardin the idea of reading a mystic, Angela da Foligno, and the idea of a universe that is in relation to its Creator in its globality.

as a *moving towards*. But there is also another far from negligible aspect: taking up the discussion initiated with Jean Boussac and tragically interrupted by war, and looking for a better understanding of how to develop a global approach to evolution. Here there is a first important encounter, with the Russian geochemist Vladimir Vernadsky who was in Paris as a guest of the Sorbonne. Together, Teilhard, Vernadsky and the philosopher Edward Le Roy would coin the term Noosphere that basically points to an attempt at a global approach applied to the totality of thinking beings. The theory of the Noosphere is not relevant to this paper and in any case I have already studied it in depth in connection with the Biosphere and with the possibility that the two spheres may be connected by a relation of symbiosis (Galleni and Scalfari 2005; Galleni 2010b). Here I am concerned to emphasize the way that the two themes (that of the global approach and that of parallelisms) proceed together and how furthermore there emerged an interesting theoretical connection with Russian thought. In addition to Vernadsky, at the same period the Russian geneticist N. Vavilov wrote his contribution on the theory of parallel series in genetics, to which I have already briefly referred.

But the 1920s produced a crucial development in the life of Pierre Teilhard de Chardin: his collaboration with Father Emil Licent sj, another Jesuit and palaeontologist who was conducting excavations in China and setting up a Museum of Natural History in Tien Tsin, grew closer and Teilhard moved to China where he remained for about twenty years until the end of the Second World War. The study of the geology, the palaeontology, and the palaeanthropology of the Chinese subcontinent represented in fact a great opportunity for the scientist Teilhard, who for the first time found himself facing the fundamental challenge for any evolutionist: studying evolution on the continental level, i.e. over vast areas and through long durations.

The Chinese experience brought together various strands in Teilhard's work. He took part in numerous scientific expeditions in the vast spaces of the Chinese subcontinent, among them the so-called "yellow cruise" organised by Citroën. Teilhard represented one of the last of the great scientist-explorers. But there is another important aspect: in the course of one of the first expeditions to the Ordos Desert in 1923, Teilhard had an experience typical of many mystics, the experience of the desert, and he then wrote the *Mass on the World*, one of the most important texts of twentieth-century mysticism, in which the experience of totality was the subject of forceful consideration (Teilhard de Chardin 1996).

It is fascinating to reflect that it was precisely from this moment that Teilhard's thinking came to center on the problem of the global approach. A few months after the expedition, he in fact wrote a letter that is a mini-manifesto for the proposal of a new discipline: one that will consider biology as the science that studies complexity. We are talking then about one of the first calls — if not in fact simply *the* first — for scientific research into complexity.

The global approach previously discussed with Jean Boussac here begins to take on its definitive structure. The theory that was emerging can be considered as an ensemble of important claims and research programmes. The first claim was that only in geometry do objects maintain the same characteristics while changing scale. In all the other sciences a change of scale calls for research into mechanisms that are lost on lower scales. It is an uncompromising critique of reductionism: in process of reduction, information is lost. This is interesting since ultimately the great force of the Synthetic Theory was precisely in the application, necessary at that time, of the reductionist method. Incidentally, we need to remember how the Synthetic Theory, erroneously confused with Neodarwinism — which is quite another matter — unifies genetics (Mendelian and post-Mendelian) and the Darwin-Wallace theory of natural selection (see Galleni 2010c).

Now emerges the problem of Biology as a science that studies the infinitely complex, understanding by complexity precisely the fact that an object of a higher hierarchical order has new characteristics that cannot be studied by reducing it to its parts. It is as we have said a strong critique of the reductionist paradigm.

On the other hand a very specific instrument is recommended — i.e. a close relation between evolutionary biology and Biosphere – in order to investigate evolution on the continental and global level.

A letter written from Tsien Tsin on 23 March 1924 ends with a strong emphasis on the need for a science of the Biosphere: evolution should be studied on the level of more or less broad portions of Biosphere, writes Teilhard, and so we need to begin to think of a biology of the Biosphere, just as there is beginning to be a chemistry of the Lithosphere.

A few years ago a new review was launched called *Geobiology*, with the aim of studying evolutionary mechanisms on the Biosphere level. The opening editorial cited various authors who in different ways could be considered precursors of the idea of geobiology, but there is no reference to Teilhard.[13] And yet, as we shall see, as early as the beginning of the Forties Teilhard, giving shape to the idea that had started to form in his mind almost twenty years earlier, had founded a review in Peking whose title, *Geobiologia*, is the Latin word for *Geobiology*. This clearly

[13] See "Editorial", *Geobiology* 1 (2003): 1.

shows that it would be quite right to consider him the founder of the discipline of that name (Teilhard de Chardin 1967).

In that same year, in a letter written on his return from a geology congress in Peking to Christophe Gaudefroy, another important priest-scientist figure, he gave vent to all his dissatisfaction with the purely reductionist method of geology. The concern that he emphasizes is very clear: when we have described all the strata and the types of rock, what shall we do? We need to reintroduce the possibility of tools for an overall vision (Teilhard de Chardin 1988).

So we see the renewed emergence of the idea of a more general vision to be concretised in an initial proposition, that of continental evolution. Only by studying evolution on the broad continental level in fact, and thus eliminating or reducing the distortions connected to small numbers, could the true basic rhythm of evolutionary mechanisms be reconstituted.[14]

Continental evolution was then the tool for the study of evolution at the level of more or less broad portions of Biosphere. At this level, important evolutionary aspects could emerge that were lost in reductionist research. Moreover, to show that the principal characteristic of evolution was not the random dispersion of types, but the *moving towards*, what must emerge according to the Teilhardian hypothesis was the presence of phenomena of parallel evolution.

At this point the programme of Teilhardian research was complete.

Let us attempt a closer analysis: the hard core was composed of the hypothesis of evolution as a *moving towards* complexity and, in animal life, towards consciousness. This hard core clearly has a "metaphysical" part that is the part connected to the need to find a particular position for the place of man in nature. Moreover Teilhard, discussing the positions of his palaeontologist colleague G. G. Simpson, an exponent of the Synthetic Theory, emphasised how the negation of the presence of phenomena of *moving towards* in animal evolution was more closely connected to what he called the metaphysics of the radical Darwinists (well exemplified by the case of Monod mentioned earlier), than to a precise scientific objectivity (Galleni 1988).

To take our analysis further: the hard core points us to some definite pathways for research. The fundamental pathway is connected to evolutionary parallelisms: life tends to follow paths that evolve in a parallel way. The parallelisms had already been illustrated by St. George Jackson Mivart, but also analyzed at length by the Russian scientist N. Vavilov at the very beginning of the Twenties.

[14] For a clear analysis of the relations between the evolutionary theory of Teilhard de Chardin and contemporary debate, see Galleni (1995, 41–42).

The basic point was that of a deterministic model linked to models of chemistry; today it can be exemplified by the term *evolutionary constraints.* Once an evolutionary structure has been attained, this determines subsequent transitions more or less rigidly; therefore, if a phyletic branch divides, its branches evolve not by diverging, but by staying parallel — or with some shared original characters, that remain so, even if the phyletic branches are now separated.

Mivart's idea was one that has reemerged many times in the history of evolutionary thought, of an evolution which takes place through internal mechanisms and which, as we have seen, represents a third hypothesis with respect to that of Lamarck and Darwin-Wallace.

Teilhard goes beyond this hypothesis, first of all since he points to a precise experimental pathway, connected to research into evolutionary mechanisms studied in greater or lesser portions of Biosphere: the method of research of continental evolution, as Teilhard called it. Moreover, he was fully aware of the difficulty of a rigidly deterministic vision: he would in fact write that evolution gropes its way forward between the play of large numbers and randomness. Only in part, then, were the reasons for the parallelisms to be sought in internal mechanisms; it was necessary to find others too, connected, as Teilhard would write later, to the general laws of the evolution of the Biosphere. The pathway of research that would become part of the hard core of Teilhardian evolutionism would then be that of the general laws of evolution of the Biosphere, studied of course by geobiology – laws which, as we shall see, could be deterministic only in part.

Here then we have a true and proper research programme which gives us exact paths to follow, the most important of which is research into parallelisms through the technique of research into continental evolution.

Teilhard became increasingly involved in the work of palaeontological and palaeoanthropological research in the Chinese subcontinent. Furthermore: taking up the great didactic tradition of the Jesuits, he sent some Chinese scientists to train in Paris and create a school of Chinese palaeontology. A participant in the working group on the remains of *Homo erectus*, more commonly known as Peking Man, he would basically devote himself to the remains of Stone Age industry and to other signs of culture and to dating their remains. This is of great interest for my work. Then in fact, to wit at the end of the Twenties, in the absence of a method of dating that was wholly reliable, the remains became associated with a certain fauna and the attempt was made to connect this with other fauna contained in well-known strata, extracting from the succession of the strata chronological evidence useful for finding a relative dating. In order to assign a date to

Peking Man, Teilhard hit on the so-called reddish rodent earths, which he studied closely, finding there one of the best examples of evolutionary parallelisms. This was the topic of the evolution of the Sifneidi, a small group of mole rats that he followed over a long duration (about twenty million years) and over wide areas (i.e. the whole Chinese subcontinent), applying the geobiological method.

In the Sifneidi, an initial trunk divides into three branches, clearly distinguishable by the different form of the occipital bone (flat, concave, or convex). But the three branches that are now separated likewise develop common characteristics: all move towards a growth in size, the cervical vertebrae fuse and the molars pass from limited to continuous growth. This is a case, as Teilhard writes, of a group orthogenesis. Here it needs to be emphasized that the term orthogenesis, often used in a very unclear manner by palaeontologists themselves to indicate — and these are the words of Teilhard himself — an almost magical linearity of the evolutionary *phila*, is used here as a synonym of parallel evolution: the directionality of evolution is demonstrated by the fact that phyletic branches which are now separated follow the same pathways, evolving therefore in a parallel — or directional manner. (It is no accident, as we have already noticed, that Mivart too, in his critique of Darwinian selectionism, had devoted a chapter to parallelisms.) The evolution of the mole rat of the Chinese Pleistocene is a splendid example of that evolution of parallel series that a few years earlier Vavilov had described in plants. On the basis of such examples, Teilhard would end up justifying the birth of a new discipline of evolutionary biology: geobiology (Galleni 1992b) — a very important point for our subject. Let me illustrate this by an analogy: the trees of phylogenesis are generally represented as trees at rest and with branches that constantly diverge. The ones envisaged by those who study parallel evolution (Mivart, Vavilov, Teilhard) are like trees subjected to a strong wind: the wind takes the branches and arranges them in parallel. But the wind is a physical force, I must emphasize for clarity, which imposes itself and integrates what is allowed by growth and continuous divergence: a force that in some way integrates that of the unconnected play of mutation-selection.

In a paper completed in the course of the last period of his life, that of his American exile, Teilhard would come to speak of the possibility that there may exist evolutionary morphogens, thus transferring a term taken from embryology into the schema of evolutionary biology with a particular technical meaning.

If morphogens are important, it is because they operate in a complex system like that of the embryo. What is the complex system to which we are to refer for evolution? Here we come to the last two original points in Teilhard: the definition of complexity and the attention to the Biosphere.

Teilhard explains very clearly that complexity is not the mere aggregation of parts, nor is it the mere indefinite geometrical repetition of unity, but:

> *combination,* i.e. the particular and superior form of grouping whose distinctive characteristic is that of correlating a certain fixed number of elements, few or many it matters not, with or without phenomena of aggregation and repetition, in a closed system of determinate range. Such are the atom, the molecule, the cell, the pluricellular organism etc.
>
> (...) with combination there emerges a type of group that is structurally complete at every moment (...) a unity that is truly and doubly "natural" in the sense that, organically limited in its contours with respect to itself, it allows to appear, furthermore, specific phenomena of autonomy, at determinate higher levels of internal complication. A certain complexity that progressively liberates a certain "centreness", not of symmetry, but of action. "Centre-complexity" we might say, to abbreviate and clarify (Teilhard de Chardin 1970, 31–32).

The relation with the concept of system is clearly visible, an object made of parts and relations between parts that is here clarified with the stress on the correlation between the parts within a clearly-defined object, specified to be a closed system of determinate range. But the breakthrough in Teilhard is to focus on the centrality of movement: the complex object is stable, not because it is stable from the geometrical point of view, but because it remains stable while constantly changing, therefore around certain parameters that remain fixed, a concept that Teilhard designates by the expression centreness of action. Here we clearly see the relation with the work of Waddington, his theory of the epigenetic landscape and the application to living systems of the general theory of systems.

But as we said, in Teilhard the system is no longer the single individual, which from an embryo becomes adult, but the Biosphere itself.

In fact the Teilhardian definition of complexity harmonizes well with the Teilhardian definition of Biosphere. For Teilhard begins by reminding us how living beings, considered as a single entity, form a single system connected to the surface of the earth. In the system, the elements are not simply aggregated together like grains of sand (and here we turn to the definition of complexity), but organically interdependent. Furthermore the organic husk is not physically separable from the inorganic and hence arises the importance of the Biosphere seen as a system of interacting parts.

The Biosphere ceases at this point to be a mere metaphorical entity, a tool that defines an ensemble of objects and relations that are not easy to describe or comprehend, to become a physical reality, as objective and essential as the other spheres that surround and compose the surface of the earth (Teilhard de Chardin 1943).

I cannot at this point run through the theory of the Biosphere in its developments within the work of Teilhard: at most I can make brief reference to subsequent developments.

Here I am concerned only to note how the two fundamental themes of biology as science of complexity and the designation of the Biosphere as the complex system that constitutes the ultimate object for the application of the techniques of complexity to evolution represent the two fundamental contributions of Pierre Teilhard de Chardin, palaeontologist geologist, and palaeoanthropologist to theories of evolution (Galleni 2010a).

What needs to be emphasized here is connected to the fact that these two points are fundamental today for the study of evolutionary biology and are necessary if we are to recognize, describe, and understand those phenomena of parallelisms and canalizations that account for the moving towards of evolution: a *moving towards* complexity and consciousness that Teilhard would summarize in the law of complexity-consciousness.

In sum, in an article of a few years ago, I ventured to put forward an analogy (Galleni 1997).

The *moving towards* of the Biosphere can be compared to animal morphogenesis. At bottom embryonal development concerns a system, the embryo that changes in time and diversifies, constantly keeping the parameters that allow survival stable, and it moves along precise lines of canalization. Thus the evolution of the Biosphere too *moves towards* an evolutionary line that leads to complexity and consciousness while keeping stable the parameters that allow the survival of life itself (this is the idea of an Italian geologist of the end of the nineteenth century, also a priest: Antonio Stoppani, which would then be taken up by James Lovelock; see Galleni 1996). There is then at the Biosphere level too a change in stability and this change in stability is verifiable if there are phenomena of canalizations and parallelisms.

It is then curious and fascinating that in one of the last letters written by Teilhard to his fellow scientist Pierre Leroy, with whom he had shared the adventure of geobiology in China, there emerged a clear reference to the work of Waddington, whose theory of the epigenetic landscape did in fact describe the canalizations of development (Leroy 1976, 177–79).

4. Evolution, which model?

We have seen that with the publication of Mivart's book, a debate began on the models of evolution: firstly that of Lamarck, linked to the dialectical relationship of living being-environment and to connected chains of causes. The second, that of Darwin and Wallace linked to the unconnected play of variation-selection based on the random encounter of unconnected chains of causes. It is from this mechanism that arises the interpretation which attributes to evolution the most random character as well as the denial by some authors of the possibility of finding the experimental traces of a *moving towards.*

The third model, Mivart's, is the most strictly deterministic and compares the laws of evolution with those of chemistry. It is an evolution that develops with chains of rigidly deterministic causes and by mechanisms internal to the living being, based on phenomena of self-organization, threshold effects, parallelisms, phases of stability and of rapid changes.

The Teilhardian thesis opens up a pathway of synthesis: the canalizations and the parallelisms are the result of mechanisms that are researched at the Biosphere level. At the population level the mechanisms are fairly close to the Darwinian ones.

As we have already noted, Teilhard wrote in fact that evolution gropes its way forward, between the play of large numbers and randomness; but, changing scale, here we see the intervention of relations at the Biosphere level and the emergence of clear lines of directionality demonstrated by the parallelisms which may be described as the result of mechanisms of *moving towards*: The Biosphere *moves towards* the maintenance of stability. But the *moving towards* is the result of mechanisms that are not strictly deterministic, and the fascinating developments of the modern theory of complexity show how the *moving towards*, represented for example by deterministic chaos and by so-called strange attractors, is not a *moving towards* that is down to mechanisms that are rigidly deterministic (see Benci and Galleni 1998; Kauffman 2000; Morowitz 2002). Teilhard's scientific work is the starting-point for a project of synthesis which will probably give better results with the application of modern techniques of complexity to the theory of the Biosphere, but which in the meantime allows us one final reflection.

The *moving towards* identified by Teilhard and extended by him to become a tool for the interpretation of the universe, takes place with mechanisms which can be described by studying the general laws of evolution of the Biosphere, which are however in part probabilistic and therefore classifiable in the category of possibility; a universe, in short, whose fundamental characteristic is (in philosophical and theological terms) freedom: that of *moving towards* the reception of the free and thinking creature.

Acknowledgments

The author wishes to thank Massimo Stanzione (University of Cassino) and Silvana Procacci (University of Perugia) for their careful and benevolently critical reading of this text.

References

Benci V., Galleni L., 1998, "Stability and instability in evolution", *J. Theor. Biol.* 194: 51– 549.

Camus A., 1942, *Le mythe de Sisyphe*, Paris: Gallimard.

Darwin C. R., 1859, *On the Origin of Species by Means of Natural Selection*, London: John Murray.

Darwin C. R., Wallace A. R., 1858, "On the tendency of species to form varieties; and on the perpetuation of varieties and species by natural means of selection", *Journal of the Proceeding of the Linnean Society of London. Zoology* 3: 46–50.

de Lubac H., 2008, *Teilhard postume*, Paris: Cerf.

Feyerabend P. K., 1975, *Against Method: Outline of an Anarchistic Theory of Knowledge*, London: Verso (Italian edition, 1985, *Contro il metodo,* Milano: Feltrinelli).

Galleni L., 1988, "Monismo e pluralismo teorico nella teoria dell'evoluzione", *Metamorfosi* 9: 83–90.

Galleni L., 1992a, *Scienza e Teologia, proposte per una sintesi feconda*, Brescia: Queriniana.

Galleni L., 1992b, "Relationships between scientific analysis and the world view of Pierre Teilhard de Chardin", *Zygon* 2: 153–65.

Galleni L., 1995, "How does the Teilhardian vision of evolution compare with contemporary theories?", *Zygon* 30: 25–45.

Galleni L., 1996, "Prospettive per un approccio multidisciplinare al problema della Biosfera che si evolve come un'unica entità complessa", in Cerrai P., Galleni L. (eds.), *Da Pisa a Como e ritorno*, Pisa: SEU, 17–32.

Galleni L., 1997, "From Teilhard de Chardin to Lovelock: a metaphorical approach to Gaia hypothesis and evolution", *Annales Biotheoretici* 1: 68–101.

Galleni L., 2000, *Biologia*, Brescia: La Scuola.

Galleni L., 2002a, "Teilhard de Chardin, Pierre", in Tanzella Nitti G., Strumia A. (eds.), *Dizionario Interdisciplinare di Scienza e Fede*, Roma: Urbaniana University Press-Città Nuova, 2121–23.

Galleni L., 2002b, "Evoluzione", in Tanzella Nitti G., Strumia A. (eds.), *Dizionario Interdisciplinare di Scienza e Fede*, Roma: Urbaniana University Press-Città Nuova, 575–90.

Galleni L., 2003, "Il rapporto scienza-fede secondo il modello di Pierre Teilhard de Chardin", in Maraldi V. (ed.), *Teologia della creazione e scienze della natura*, Bologna: Edizioni Dehoniane, 15–33.

Galleni L., 2005, "Pierre Teilhard de Chardin, scienza e teologia nella prospettiva del terzo millennio" *Revista Portuguesa de Filosofia* 61: 159–84.

Galleni L., 2007, *Ciencia y Teologia*, Buenos Aires: Editorial Epifania.

Galleni L., 2010a, *Darwin, Teilhard de Chardin e gli altri…le tre teorie dell'evoluzione*, Pisa: Felici.

Galleni L., 2010b, "Teilhard de Chardin and the Science of the Biosphere", in Duffy K. (ed.), *Rediscovery Teilhard's fire*, Philadelphia: Saint Joseph's University Press, 197–206.

Galleni L., 2010c, "Pierre Teilhard de Chardin e l'evoluzione come "muoversi verso" l'essere pensante", in Calabi L. (ed.), *Il Futuro di Darwin*, Torino: UTET, 77–98.

Galleni L., Groessens-Van Dyck M. C., 2001, "Lettres d'un paléontologue, Neuf letters inédites de Pierre Teilhard de Chardin à Marcellin Boule", *Revue des questions scientifiques* 172: 5–104.

Galleni L., Groessens-Van Dyck M. C., 2007, "A model of interaction between science and theology based on the scientific papers of Pierre Teilhard de Chardin", in Sweet W., Feist R. (eds.), *Religion and the Challenge of Science*, Aldershot: Ashgate, 55–72.

Galleni L., Scalfari F., 2005, "Teilhard de Chardin's engagement with relationship between science and theology in light of discussions about environmental ethics, *Ecotheology* 10 (2): 196–214.

Grib A. A., Galleni L., Reati F. (eds.), 2007, *Scienza e Fede in dialogo: P. Teilhard de Chardin e Pavel Florenskij*, Saint Petersburg: Saint Petersburg University Press.

Huxley J., 1942, *Evolution: The Modern Synthesis*, London: Allen and Unwin.

Jackelen A., 2001, "Introduzione", in Galleni L. (ed.), *Scienza e Teologia, un nuovo campo di ricerca e insegnamento per antichi problemi*, Pisa: Quaderni Stenoniani, 3–9.

Kauffman S., 2000, *Investigations*, Oxford: Oxford University Press.

Lakatos I., 1978, *The Methodology of Scientific Research Programs*, Cambridge: Cambridge University Press (Italian edition, 1985, *La metodologia dei programmi di ricerca scientifici*, Milano: Il Saggiatore).

Lamarck J. B., 1969, *Opere*, Torino: UTET.

Leonardi P., 1950, *L'evoluzione del vivente*, Brescia: Morcelliana.

Leroy P., 1976, *Lettres familières de Pierre Teilhard de Chardin mon ami, 1948-1955*, Paris: Le Centurion.

Malthus T. R., 1993, *An Essay on the Principle of Population*, Oxford: Oxford University Press.

Mivart G. J., 1871, *On the Genesis of the Species*, New York: D. Appleton.

Monod J., 1970, *Le hasard et la nécessité*, Paris: Seuil (Italian edition, 1971, *Il caso e la necessità*, Milano: Mondadori).

Morowitz H. J., 2002, *The Emergence of Everything*, Oxford: Oxford University Press.

Popper K. R., 1969, "Tre punti di vista a proposito della conoscenza umana", in Trinchero M. (ed.), *Scienza e Filosofia*, Torino: Einaudi, 11–47.

Reati F., Galleni L. (eds.), 2004, *Teoria dell'evoluzione: lo sguardo della scienza e della fede cristiana*, Gargnano: Centro per il dialogo italo-russo.

Ruse M., 1996, *Monad to Man: The Concept of Progress in Evolutionary Biology*, Cambridge, Mass.: Harvard University Press.

Shelley M., 2002, *Frankenstein ovvero il Prometeo moderno*, Milano: RCS

Tanzella Nitti G., 2002, "Gesù Cristo: Rivelazione e Incarnazione del Logos", in Tanzella Nitti G., Strumia A. (eds.), *Dizionario Interdisciplinare di Scienza e Fede*, Roma: Urbaniana University Press-Città Nuova, 703–704.

Teilhard de Chardin P., 1943, "Géobiologie et Geobiologia", *Geobiologia* 1: 1–8.

Teilhard de Chardin P., 1967, "Lettres inedites à un savant de ses amis", *Christus* 54: 238–58.

Teilhard de Chardin P., 1970, *Il posto dell'uomo nella natura*, Milano: Il Saggiatore.

Teilhard de Chardin P., 1988, *Lettres inédites*, Monaco: Le Rocher.

Teilhard de Chardin P., 1996, *L'Inno dell'Universo*, Brescia: Queriniana.

Teilhard de Chardin P., Boussac J., 1986, *Lettres de Guerre Inédites*, Paris: O.E.I.L.

Vavilov N. I., 1922, "The law of homologous series in variation", *Journal of Genetics* 12: 46–89.

Vialet A., Hurel A., 2004, *Teilhard de Chardin en Chine correspondance inédite (1923-1940)*, Paris: Editions du Muséum-Edisud.

Wallace A. R., 1895, *Natural Selection and Tropical Nature. Essays on Descriptive and Theoretical Biology*, London: Macmillan.

Scientific creationism and intelligent design

Ronald L. Numbers

For the past century and a half no issue has dominated discussions of science and religion more than evolution. Indeed, many people see the creation-evolution debates as the central issue in the continuing controversy. And for good reason. Over a century after the scientific community had embraced organic evolution, many laypersons continued to scorn the notion of common descent. In the United States, where polls since the early 1980s have shown a steady 44–47 percent of Americans subscribing to the statement that "God created human being pretty much in their present for at one time within the last 10,000 years or so," nearly two-thirds (65.5 percent), including 63 percent of college graduates, according to a 2005 Gallup poll, regarded "creationism" as definitely or probably true (Jones 2005; Moore 2005).[1] As we shall see, such ideas have been spreading around the world.

1. Creation and creationism

In 1929 an obscure biology teacher at a small church-college in northern California self-published a book titled *Back to Creationism*. This brief work, appearing just as the American antievolution movement of the 1920s was winding down, attracted little attention. And it would deserve scant mention today except for the fact that it was one of the first books to use the term "creationism" in its title. Until well into the twentieth century critics of evolution tended to identify themselves as antievolutionists rather than creationists (Clark 1929).[2]

Three factors help to explain this practice. First, the word already possessed a well-known meaning unrelated to the creation-evolution debate. Since early Christianity theologians had attached "creationism" to the doctrine that God had

---

[1] The figure of 65.5 percent appears in the profile for Question 30c, surveyed August 5–7, 2005.

[2] The only earlier book with which I am familiar is Judson D. Burns (1908). Burns, a small-town physician in Iowa, defended the special creation of the first humans but ignored the meaning of "creationism".

specially created each human soul — as opposed to the traducianist teaching that God had created only Adam's soul and that children inherited their souls from their parents. Second, even the most prominent scientific opponents of organic evolution differed widely in their views of origins. Some adopted the biblical view that all organisms had descended from the "kinds" divinely created in the Garden of Eden and preserved on Noah's ark. Others, such as the British geologist Charles Lyell (1797–1875), advocated the spontaneous but nonsupernatural appearance of species in regional "*centres* or *foci* of creation." Still others followed the leading American antievolutionist, the Harvard zoologist Louis Agassiz (1807–1873), in arguing for repeated plenary creations, during which "species did not originate in single pairs, but were created in large numbers." (Numbers 1998, Ch. 2: "Creating creationism: meanings and uses since the stage of Agassiz"; see also Rupke 2005, 143–72) Third, even Bible-believing fundamentalists could not agree on the correct interpretation of the first chapter of Genesis. A majority probably adopted the ruin-and-restoration view endorsed by the immensely popular *Scofield Reference Bible* (1909), which identified two creations (the first "in the beginning," the second associated with the Garden of Eden) and slipped the fossil record into the vast "gap" between the two events. Another popular reading of Genesis 1, advocated by William Jennings Bryan (1860–1925), the leading antievolutionist of the time, held that the "days" mentioned in Genesis 1 represented immense ages, each corresponding to a section of the geological column or perhaps to a period in the history of the cosmos. Only a handful of those writing against evolution insisted on what later came to be known as young-earth creationism but was then called "flood geology": a recent special creation of all "kinds" in six twenty-four-hour periods and a geologically significant flood at the time of Noah that buried most of the fossils (Numbers 1998, 52–53).

Flood geology was the brainchild of the scientifically self-educated George McCready Price (1870–1963). A Canadian by birth, Price converted to Seventh-day Adventism as a youth and had accepted the writings of the Adventist prophetess, Ellen G. White (1827–1915), as divinely inspired. Throughout her life White had experienced religious dreams and trance-like visions, which she and her followers saw as divine. During one episode she claimed to have been "carried back to the creation and … shown that the first week, in which God performed the work of creation in six days and rested on the seventh day, was just like every other week." She also endorsed a 6,000-year-old earth and a worldwide catastrophe at the time of Noah that had buried the fossils and reshaped the earth's surface.[3] There was nothing novel about White's history, except its timing. By the middle of the nineteenth century, when she began writing, almost all evangelical expositors on Genesis and geology had conceded the antiquity of life on earth and the geological insignificance of Noah's flood (Stiling 1991).

As a young man full of religious zeal, Price dedicated himself to providing a scientific defense of White's outline of earth history. Although he could scarcely tell one rock from another, he read the scientific literature voraciously — and critically. Early on it struck him that the argument for evolution "*all turned on its view of geology*," which provided the strongest evidence for both the antiquity of life and its progressive development. But the more he read, the more he became convinced that the vaunted geological evidence for evolution was "a most gigantic hoax." Guided by Mrs. White's "revealing word pictures of the Edenic beginning of the world, of the fall and the world apostasy, and of the flood," he concluded that "the actual facts of the rocks and fossils, *stripped of mere theories*, splendidly refute this evolutionary theory of the invariable order of the fossils, *which is the very backbone of the evolution doctrine*." (Numbers 2006, 91–92)

In 1906 Price published a booklet titled *Illogical geology: The weakest point in the evolution theory*, in which he offered a $1,000 reward "to any who will, in the face of the facts here presented, show me how to prove that one kind of fossil is older than another." Before his death in 1963 he would author some two dozen books, the most systematic and comprehensive being *The new geology* (1923). In it, he restated his "great 'law of conformable stratigraphic sequences,' " which he modestly described as "by all odds the most important law ever formulated with reference to the order in which the strata occur." According to this law, "Any kind of fossiliferous beds whatever, 'young' or 'old,' may be found occurring conformably on any other fossiliferous beds, 'older' or 'younger.' " To Price, so-called deceptive conformatives (where strata seem to be missing) and thrust faults (where the strata are apparently in the wrong order) proved that there was no natural order to the fossil-bearing rocks, all of which he attributed to Noah's Flood (Numbers 2006, 95–97; McCready Price 1923, 637–38).

Despite repeated attacks from the scientific establishment, Price's influence among non-Adventist fundamentalists grew rapidly. By the mid-1920s the editor of *Science* could accurately describe Price as "the principal scientific authority of the Fundamentalists,"[4] and Price's byline was appearing with increasing frequency in a broad

[3] Much of this essay is based on Numbers (2006). On White's views, see p. 90. See also Numbers (2008).

[4] *Science* (5 March 1926): 259.

spectrum of religious periodicals. Nevertheless, few fundamentalist leaders, despite their appreciation for Price's critique of evolution and defense of a biblical flood, gave up their allegiance to the "gap" and "day-age" theories for his flood geology.

In *Back to creationism*, the book with which we began this essay, one of Price's former students, Harold W. Clark (1891-1986) tried to establish Price's Bible-based theory as "the science of creationism." This "new science," he wrote optimistically,

> will interpret the records of the rocks, the lives of plants and animals, and human history, in the light of the creation story. ... As men go deeper into the science of creationism, the inmost secrets of the cell and the atom will display the power of the Creator in ways that have never been understood; and in the degeneracy and evil that biology and sociology bring to light will be seen the activity of the counter-power [i.e., Satan] that has been trying to mar the beautiful creation. ... The time is ripe for a rebellion against the domination of evolution, and for a return to the fundamentals of true science, BACK TO CREATIONISM. (Clark 1929, 138–39, italics added; see Numbers 2006, 142–48)

2. Organized creationism

As the American antievolution movement petered out in the late 1920s, a few diehards tried to keep the protest alive by organizing a new society. Their efforts, however, immediately ran into two obstacles: a paucity of trained scientists and the continuing disagreement over the meaning of Genesis 1. Price had never finished college nor even taken an advanced course in science, though Clark in the early 1930s world earn a master's degree in biology at the University of California in Berkeley. Other antievolution activists with some exposure to science were Harry Rimmer (1890–1952), a Presbyterian evangelist and self-described "research scientist" who had briefly attended a homeopathic medical school; Arthur I. Brown (1875–1947), a Canadian surgeon whose handbills described him as "one of the best informed scientists on the American continent"; S. James Bole (1875–1956), a professor of biology at Wheaton College, who had earned a master's degree in education and would in 1934 receive a Ph.D. in horticulture from Iowa State College; and Bole's colleague on the Wheaton faculty, L. Allen Higley (1871–1955), a chemist (Numbers 2006, Ch. 4: "Scientific creationists in the age of Bryan").

In 1935 Price, Clark, Rimmer, and Higley joined with a few others to create "a united front against the theory of evolution." The resulting society, the Religion and Science Association, quickly dissolved, however, when the members fell to squabbling about the age of the earth, with Price and Clark supporting flood geology, Rimmer and Higley pushing for the gap theory, and still others arguing for the day-age interpretation. As one frustrated antievolutionist observed in the 1930s, fundamentalists were "all mixed up between geological ages, Flood geology and ruin, believing all at once, endorsing all at once." How, he wondered, could evangelical Christians possibly turn the world against evolution if they themselves could not even agree on the meaning of Genesis 1? (Numbers 2006, Ch. 6: "The Religion and Science Association")

A few years after the demise of the Religion and Science Association Price and a small number of mostly Adventist colleagues in southern California, where he had retired, organized a Deluge Geology Society, which for several years in the early 1940s published a *Bulletin of Deluge Geology and Related Science*. The group consisted of "a very eminent set of men," bragged Price. "In no other part of this round globe could anything like the number of scientifically educated believers in Creation and opponents of evolution be assembled, as here in Southern California." By far the best-trained scientist in the society was a Missouri Synod Lutheran, Walter E. Lammerts (1904–1996), who had earned a Ph.D. degree in genetics at the University of California, Berkeley, and was teaching horticulture at its southern branch in Los Angeles. The society's most exciting moment came in the early 1940s, when it announced the discovery of giant fossil footprints, believed to be human, in geologically ancient rocks. This find, one member predicted, would demolish the theory of evolution "at a single stroke" and "*astound the scientific world!*" But even this group of flood geologists, who all agreed on the recent appearance of life on earth, divided bitterly over the issue of "*pre-Genesis time for the earth*," that is, whether the inorganic matter of the earth antedated the Edenic creation. About 1947 the society died (Numbers 2006, Ch. 7: "The Deluge Geology Society").

By this time a more ecumenical society of evangelical scientists had appeared on the scene: the American Scientific Affiliation (ASA). Created in 1941 by associates of the Moody Bible Institute, the association at first took a dim view of evolution. By the end of the decade, however, the presence of a number of well-trained young scientists who embraced theistic evolution (or its intellectual sibling "progressive creationism") was dividing the association. The most influential of the insurgents were J. Laurence Kulp (1921–2006) and Russell L. Mixter (1906–2007). Kulp, a Wheaton alumnus who had earned a doctorate in physical chemistry from Princeton University and then completed the course work for a second Ph.D. in geology, had established himself at Columbia University as an early authority on radioisotope dating. As one of the first evangelicals with advanced training in geology, he spoke with unique authority. Worried that Price's flood geology had "infiltrated the greater portion of fundamental Christianity in America primarily due to the absence of trained Christian geologists," he set about exposing its abundant scientific flaws. In an influential paper first read to ASA members in 1949, he concluded that the "major propositions of the theory are contraindicated by established physical and chemical laws." Mixter, meanwhile, was pushing for greater acceptance of the evidence for limited organic evolution. While teaching biology at Wheaton College, he earned a doctorate in anatomy from the University of Illinois School of Medicine in Chicago in 1939. Before long he was nudging creationists to accept evolution "within the order" and assuring them that they could "believe in the origin of species at different times, separated by millions of years, and in places continent apart." (Numbers 2006, Ch. 9: "Evangelicals and evolution in North America")

3. The creationist revival

In 1954 Bernard Ramm (1916–1992), a theologian-philosopher associated with the leadership of the ASA, brought out a book audaciously called *The Christian View of Science and Scripture.* Damning "hyperorthodox" Christians for their "narrow bibliolatry" and "ignoble" attitude toward science, this avatar of neo-evangelicalism urged Christians to quit getting their science from Genesis and adopt the progressive creationism so popular within the ASA. He dedicated his book to one of the founders of the ASA and thanked Kulp for vetting the book for "technical accuracy." Ramm aimed his harshest rhetoric at the flood geology of Price, whose growing influence among fundamentalists he regarded as "one of the strangest developments of the early part of the twentieth century." Despite Price's manifest ignorance, his brand of creationism had come, at least in Ramm's imagination, to form "the backbone of much of Fundamentalist thought about geology, creation, and the flood." (Numbers 2006, 208–11)

Many evangelicals, including Billy Graham (b. 1918), hailed Ramm's book, but fundamentalists tended to respond angrily to what they regarded as an arrogant and heterodox attempt to equate progressive creationism with *the* Christian view. Ramm's attack provoked one young fundamentalist, John C. Whitcomb, Jr. (b. 1924), a Princeton-educated Old Testament scholar teaching (and working on a doctorate) at the fundamentalist Grace Theological Seminary, into turning his dissertation into a spirited response to Ramm and a defense of "the position of George M. Price." When Whitcomb approached the Moody Press about publishing his study, the editor recommended that the biblical scholar recruit a trained scientist as coauthor. He eventually found an acceptable, if not perfect, partner: Henry M. Morris (1918–2006), a fundamentalist Baptist who had earned a Ph.D. in hydraulics from the University of Minnesota and had just taken over as head of the large civil-engineering program at Virginia Polytechnic Institute (Numbers 2006, Ch. 10: "John C. Whitcomb, Jr., Henry M. Morris, and The Genesis Flood").

As defenders of Price's flood geology, Whitcomb and Morris faced the difficult — perhaps impossible — task of not being dismissed as "crackpots" for trying to promulgate his theory. Early on Morris suggested to Whitcomb that it might be best "simply to point out Price's arguments as a matter of historical record, and then leave your main emphasis on the Scriptural framework and the geological implications thereof." Later, as he and Morris neared the end of their project, Whitcomb shared his own concerns about being identified with the disreputable Price and his strange church:

> I am becoming more and more persuaded that my chapter on "Flood Geology in the Twentieth Century" will hinder rather than help our book, at least in its present form. Here is what I mean. For many people, our position would be somewhat discredited by the fact that "Price and Seventh- Day Adventism" (the title of one of the sections in that chapter) play such a prominent role in its support. My suggestion would be to supply for the book a fairly complete *annotated bibliography* of twentieth-century works advocating Flood-geology, without so much as a mention of the denominational affiliation of the various authors. After all, what *real* difference does the denominational aspect make?

In the end the authors camouflaged their intellectual debt to Price by deleting all but a few incidental references to him and all mention of his Adventist connections.

In 1961, after Moody declined to publish their book, the Orthodox Presbyterian Rousas J. Rushdoony (1916–2001), founder of the ultra-right-wing Christian Reconstruction movement, guided them to a small fundamentalist press in Philadelphia, which finally brought out *The Genesis Flood.* Although one critic accurately described the book as "a reissue of G. M. Price's views brought up to date," it created a sensation within the evangelical community.

Two years after the appearance of *The Genesis Flood* a small group of Christian scientists energized by Whitcomb and Morris's book — and increasingly annoyed by the ASA's drift toward evolution — walked out of the ASA and founded their own hyperorthodox society, the Creation Research Society (CRS). Leading this effort, both administratively and financially, was the Lutheran geneticist Lammerts, who until this time had maintained a low creationist profile. The initial eighteen-man CRS steering committee imprecisely reflected the theological composition of the emerging young-earth creationism movement: six Missouri Synod Lutherans, six Baptists (four Southern, one Regular, and one independent), two Seventh-day Adventists, and one each from the Reformed Presbyterian church, the Christian Reformed church, the Methodist church, and the Church of the Brethren. The committee included five biologists with Ph.D.'s earned at major universities, two more biologists with master's degrees, and one biochemist with a doctorate in biochemistry. There were no physicians in the group and only one engineer, Morris. Twelve of the eighteen lived in the Midwest, four in the Southwest, one in California, and one in Virginia (Numbers 2006, Ch. 11: "The Creation Research Society").

The CRS claimed to be a "research society," but it conducted few investigations outside of libraries. The chairman of the committee on research, Larry G. Butler (1933-1997), a Baptist biochemist at Purdue University and one of the few active members of the CRS with a major academic appointment, grew increasingly frustrated with the proposals he received. Hoping to "present an image of scientific respectability as much as possible without Biblical compromise," he diligently tried "to exclude authentic psychopaths, cranks, & kooks" looking for a forum for their farfetched ideas. As he quickly discovered, too many fellow creationists suffered from a fondness for the sensational: "We make astonishing observations (human footprints contemporary with dinosaurs); we postulate dramatic upheavals (sudden deposits of masses of ice from a planetary visitor); we propose sweeping scientific generalizations (negation of the entire system of 14C dating." Although some colleagues in the society pushed him for the presidency, he found himself increasingly impatient with what he called "the lunatic fringe" of creationism. Discouraged by the failure of his efforts to raise the scientific standards of creationist research, he resigned from the board of directors in 1975 and later allowed his membership (in both the CRS and his church) to lapse (Numbers 2006, 283–85).

Despite a common commitment to young-earth creationism, disagreements soon arose. One of the most significant was over the issue of speciation. As biologists discovered more and more species, it became clear to creationists that Noah's ark could not have accommodated representatives of each one. Thus many of them adopted the solution of a former student of Price's, Frank Lewis Marsh (1899-1992), who argued that the Genesis "kinds" should not be equated with species but with families or what he called *baramins.* This solved the problem of space on the ark but created another one: how had the kinds preserved on the ark produced so many genera and species, and in only 4,300 years. It seemed likely, for example, that the *Canidae* family — including domestic and wild dogs, wolves, foxes, coyotes, jackals, and dingoes — had descended from a single kind. Morris and most of his colleagues embraced rapid "microevolution". However, as a geneticist, Lammerts knew that that was scientifically impossible, that there must have been a second creation to repopulate the earth after the deluge. Unfortunately for him, the Bible never mentioned such an event; so his supernatural solution never caught on (Numbers 2004, 84–100).

For a young-earth creationist organization, the CRS grew rapidly. On the occasion of its tenth anniversary it boasted a membership of 1,999, with 412 of them holding advanced degrees in science. By this time society leaders were switching from "flood geology" as the name of choice for their model of earth history and substituting the labels "creation science" and "scientific creationism." In truth, there was little difference between the old and the new, except that scientific creationism made no mention of biblical events and persons, such as the Garden of Eden, Adam and Eve, and Noah's flood. However, the focus on the flood remained the same. Morris made this clear in a book titled *Scientific Creationism* (1974):

> The Genesis Flood is the real crux of the conflict between the evolutionist and creationist cosmologies. If the system of flood geology can be established on a sound scientific basis, and be effectively promoted and publicized, then the entire evolutionary cosmology, at least in its present neo-Darwinian form, will collapse. This, in turn, would mean that every anti-Christian system and movement (communism, racism, humanism, libertinism, behaviorism, and all the rest) would be deprived of their pseudo-intellectual foundation.

Driving the switch in labels was a desire to have a product acceptable for use in public schools, especially in California, which was revising its guidelines for teaching science. Tellingly, *Scientific Creationism* appeared in two almost identical versions: one for public schools, stripped of all references to the Bible, and another for church schools, which retained biblical references and added a chapter on "Creation according to Scripture" (Numbers 2006, Chap. 12: "Creation science and scientific creationism").[5]

Scientific creationists liked to contrast the "creation model" of origins with the "evolution model" — and to insist that the former was just as scientific as the latter. In their own minds — and as revealed in their published writings — they loved science and simply wanted to protect its good name. In selling their two-model approach to school boards and state legislatures, they repeatedly appealed not only to their scientific credentials but to their desire to promote science. "Stress that creationists are not proposing to teach the "creation story of Genesis" in the schools," advised Morris, "but only to show that the facts of science can be explained in terms of the scientific model of creation." (Numbers 2006, 276–77)

In 1968 the U.S. Supreme Court struck down as unconstitutional the last of the laws from the 1920s outlawing the teaching of evolution. This forced creationists

[5] On changing interpretations of the Constitution, see Larson (2003, 93–94).

to abandon any thought of making the teaching of evolution illegal and turned their attention to writing legislation that would allow the teaching of "creation science" alongside that of "evolution science." The creationists sought scientific status for their views in order to circumvent the constitutional separation of church and state, which had implications for the teaching of religion in schools. The Bill of Rights in the U.S. Constitution forbade Congress from passing any "laws respecting an establishment of religion, or prohibiting the free exercise thereof." Before World War II the Supreme Court had interpreted this narrowly in its literal sense; in the late 1940s, however, it held that the Constitution had erected "a wall of separation" between church and state. At a time when public-opinion polls were revealing that "half of the adults in the U.S. believe God created Adam and Eve to start the human race," the movement for "balanced treatment" enjoyed a large reservoir of popular support (Gallup 1979, 14).[6] In the end only two states, Arkansas and Louisiana, adopted the two-model approach. In 1982 a federal judge in Arkansas, having been tutored by the philosopher Michael Ruse (b. 1940) on the demarcation criteria that allegedly distinguished science from nonscience, declared the Arkansas law to be infringement of the constitutional requirement to keep church and state separate; three years later a court in Louisiana reached a similar decision. The US Supreme Court ratified these judgments in 1987, while allowing, in the words of one justice, that "teaching a variety of scientific theories about the origins of humankind to schoolchildren might be validly done with the clear secular intent of enhancing the effectiveness of science instruction."[7]

4. Intelligent design

The Supreme Court's decision dashed the hopes of creation scientists who had expected their stripped-down version of creationism to pass constitutional muster, but it did little to dampen the widespread antipathy toward evolution in America. Few found the decision more disappointing than two creationist authors, Dean H. Kenyon and Percival Davis, who had drafted a manuscript tentatively titled *Biology and Creation* in anticipation of the demand for a high-school textbook when the court ruled for creationism. Their optimistic publisher calculated a financial bo-

[6] Beginning in 1982 Gallup began assessing what percentage of Americans subscribed to "Creationism, that is, the idea that God created human beings pretty much in their present form at one time within the last 10,000 years." Between 1982 and 2008 the number varied between 43 and 47 percent. See www.gallup.com/poll/21814/Evolution-Creationism-Intelligent-Design.aspx.

[7] The best introduction to the legal debates over creationism is Larson (2003).

nanza of "over 6.5 million in five years." When the court virtually wiped out the market for creationist texts, Kenyon and Davis quickly sanitized their manuscript by substituting *Of Pandas and People* for the original title and replacing the words "creation" and "creationists" with the euphemisms "intelligent design" and "design proponents." As they defined it, intelligent design (ID) provided a frame of reference that "locates the origin of new organisms in an immaterial cause: in a blueprint, a plan, a pattern, devised by an intelligent agent."[8]

Of Pandas and People may have begun as a conventional creationist work, but it put into play a new slogan in the ongoing campaign against evolution: intelligent design. The intelligent-design *movement* began in the early 1990s with the publication of an antievolution tract, *Darwin on trial* (1991) by a Berkeley law professor, Phillip E. Johnson (b. 1940). Upset by the anti-Christian stridency of some Darwinists — such as Richard Dawkins — the Presbyterian layman set out to expose what he saw as the logical weaknesses of the case for evolution, particularly the assumption made by its advocates that naturalism is the only legitimate way of doing science. Ever since investigators of nature in the early nineteenth century had shifted from natural philosophy (which allowed for appeals to the supernatural) to science (which did not), practitioners, regardless of religious persuasion, had refrained from invoking divine or diabolical forces when explaining the workings of nature. In short order, explaining nature naturally became the defining characteristic of science, for Christians as well as for atheists. In contrast to metaphysical naturalism, which denied the existence of a transcendent God, this methodological naturalism supposedly implied nothing about God's existence. Johnson vehemently disagreed. Professing to see little difference between methodological naturalism and scientific materialism, he set out to resacralize science or, as one admirer put it, "to reclaim science in the name of God." If the evidence warranted a supernatural explanation, Johnson argued, then invoking intelligent design should count as a legitimate scientific response. Intelligent design, as one insider conceded, was simply a politically correct way to refer to God.[9]

Johnson aspired to pitch a tent big enough to accommodate all antievolutionists who were willing to set Genesis aside (at least temporarily) and focus on the purported scientific evidence against evolution. Although a few young-earth creationists sought shelter in the tent, Morris and other Bible-based creationists resented

[8] This section is based on Numbers (2006, Ch. 17: "Intelligent design").

[9] On methodological naturalism, see Numbers (2003, 265-85). The evangelical philosopher-theologian Paul de Vries (1986) coined the term.

the effort of the intelligent designers to marginalize their views and to avoid "having to confront the Genesis record of a young earth and global flood." In the mid-1990s the founder of the right-of-center Discovery Institute in Seattle invited ID theorists to establish an institutional home within the institute called the Center for the Renewal of Science and Culture. With a year or so they had raised "nearly a million dollar in grants." The most generous donor was Howard Fieldstead Ahmanson, Jr. (b. 1950), heir to a fortune made in the savings-and-loan business. An intimate of Rousas J. Rushdoony, the theocrat who had found a publisher for Whitcomb and Morris's *Genesis Flood*, Ahmanson, like his mentor, sought "the total integration of biblical law into our lives."[10]

By this time several younger men had joined Johnson as the public face of the movement, among them Michael J. Behe (b. 1952), a Catholic biochemist at Lehigh University. In 1996 the Free Press of New York released Behe's *Darwin's Black Box: the Biochemical Challenge to Evolution*, the first antievolution book in seven decades published by a mainstream publisher.[11] In his book Behe argued that biochemistry had "pushed Darwin's theory to the limit ... by opening the ultimate black box, the cell, thereby making possible our understanding of how life works." The "astonishing complexity of subcellular organic structure" — its "irreducible complexity" — led him to conclude that intelligent design had been at work. "The result is so unambiguous and so significant that it must be ranked as one of the greatest achievements in the history of science," he concluded grandiosely. "The discovery of [intelligent design] rivals those of Newton and Einstein, Lavoisier and Schroedinger, Pasteur and Darwin." The tip of the hat to Darwin was no slip. In contrast to most of his colleagues in the movement, Behe did not rule out the possibility of divinely guided evolution.

More typical of attitudes toward theistic evolution within the ID camp was that of another rising star, the mathematician-philosopher William A. Dembski (b. 1960). "*Design theorists are no friends of theistic evolution*," he declared.

> As far as design theorists are concerned, theistic evolution is American evangelicalism's ill-conceived accommodation to Darwinism. What theistic evolution does is take the Darwinian picture of the biological world and baptize it, identifying this picture with the way God created life. When boiled down to its scientific content, theistic evolution is no different from atheistic evolution.

---

[10] For a scholarly appraisal of Reconstructionism, see Molly Worthen (2008, 399-437).

[11] In 1925 Macmillan of New York published George Barry O'Toole, *The case against evolution*.

On the origin of organic forms his position did not vary much from that of the scientific creationists. While acknowledging that organisms had "undergone some change in the course of natural history," he believed that such changes had "occurred within strict limits and that human beings were specially created." As an expert in probability theory, Dembski focused on the unlikelihood of organisms arising by accident, and especially on a method for detecting intelligence, his much-maligned "explanatory filter." Like Johnson, Dembski attacked evolution as part of a much larger strategy to revolutionize the way science was practiced. "*The ground rules of science have to be changed*," he declared quixotically. "We need to realize that methodological naturalism is the functional equivalent of a full blown metaphysical naturalism." For a brief period at the turn of the millennium the prolific Dembski headed an ID center at Baylor University, described as the "first intelligent design think-tank at a research university."

Intelligent design emerged as front-page news in 2005, after a group of parents in the Dover, Pennsylvania, filed suit against the school board for promoting ID in ninth-grade biology classes. The religiously conservative board had instructed teachers to tell their students about the weaknesses in Darwin's theory and direct them to *Of Pandas and People*. The case, like the creation-science trials of the 1980s, hinged on whether the recommendation of ID theory constituted the teaching of religion and therefore violated the US Constitution. Behe appeared as the star witness for the defense but scarcely helped his side when he lamely, but honestly, conceded that ID "does not propose a mechanism in the sense of a step by step description of how these structures arose." In the end the judge condemned the school board for its actions — memorably declaring it a "breathtaking inanity" — and ruled that ID was "not science" because it invoked "supernatural causation" and failed "to meet the essential ground rules that limit science to testable, natural explanations." A conservative Christian himself, the judge rejected as "utterly false" the assumption "that evolutionary theory is antithetical to a belief in the existence of a supreme being and to religion in general."

5. Into all the world

Although scattered critics of evolution could be found around the globe throughout the twentieth century, organized antievolutionism rarely appeared outside the United States before the late twentieth century. When Price lived in England for four years in the mid-1920s he found little interest in fighting evolution, even among conservative Christians. In the early 1930s, however, a band of British antievolutionists, led by the barrister and amateur ornithologist Douglas Dewar (1875-1957), formed the Evolution Protest Movement (EPM) — after the Zoological Society of London had rejected a paper of his on mammalian fossils, leading him to conclude that evolution had become "a scientific creed." During its first quarter-century the EPM reached a membership of about 200 and established tiny outposts in Australia, New Zealand, Canada, and South Africa (Numbers 2006, Ch. 8: "Evangelicals and evolution in Great Britain").

The creationist awakening in the United States in the 1960s sparked a number of brush fires around the world. The head of the somnolent EPM predicted that Whitcomb and Morris's "revolutionary re-interpretation" of earth history would usher in "a new era." Indeed, it did. "More than any other single factor," explained one British creationist, "this scholarly but highly controversial volume lifted creationism from the Gospel Hall tract-rack to the College seminar room." By 1980 young-earthers had largely captured the EPM; that year they changed its name to the Creation Science Movement. Just as American creationists dreamed of getting into the public-school curriculum, British creationists aspired to air time on BBC (Numbers 2006, 355-62).

In the wake of visits from Morris and his irrepressible sidekick Duane Gish (b. 1921), sometimes joined by other colleagues at the Institute for Creation Research (founded in 1972), antievolutionists around the world began rallying around young-earth creationism. Still, as late as 2000, the American paleontologist Stephen Jay Gould (1941-2002) confidently assured non-Americans that they had nothing to fear. "As insidious as it may seem, at least it's not a worldwide movement," he said. "I hope everyone realizes the extent to which this is a local, indigenous, American bizarrity."[12]

Gould, a great scientist, proved to be a false prophet. Even as he spoke, creationism was becoming a truly global phenomenon, successfully overcoming its "Made in America" label and flourishing not only among conservative Protestants but also among pockets of Catholics, Eastern Orthodox believers, Muslims, and Jews. Conservative Protestants, however, continued to lead. In Australia, for example, young-earth creationists in 1980 established an energetic Creation Science Foundation.[13] After seven years one of its co-founders, the charismatic former high-school biology teacher Kenneth A. Ham (b. 1951), moved to the United

[12] This section is based largely on Numbers (2006, Ch. 18: "Creationism goes global").

[13] On creationism in Australasia, see Numbers (2002) and Numbers and Stenhouse (2000); both of which are reprinted in Coleman and Carlin (2004).

States to work with Morris at the Institute for Creation Research. In 1994 he launched his own creationist ministry, Answers in Genesis (AiG), headquartered in northern Kentucky, just south of Cincinnati. Within a decade AiG had emerged as the most dynamic creationist organization worldwide, with Ham alone speaking to over 100,000 people a year. In 2007, to great fanfare, AiG opened an impressive $27-million-dollar Creation Museum, which attracted hundreds of thousands of visitors annually. South Korea became another major center of Christian creationism. Since its founding in the winter of 1980-1981 the Korea Association of Creation Research has established branches throughout the land, published a successful bimonthly magazine, and held thousands of seminars. In 2000 it began a program of sending creationist missionaries to other countries, the first going to Indonesia, a predominantly Muslim country.

The spread of organized creationism from Christianity to Islam began in the mid-1980s, when the Muslim minister of education in Turkey contacted the Institute for Creation Research with a request to help promote a two-model curriculum that would teach both creation and evolution. In 1990 a small group of young Turks in Istanbul formed the Science Research Foundation (BAV in Turkish), headed by the shadowy Adnan Oktar (b. 1956), who had adopted the pen name Harun Yahya. A student first of interior design and then of philosophy, young Oktar had grown increasingly distressed with the materialism that flourished in Turkish universities, a philosophy he linked to Darwinism and Zionism. The activities of the cult-like BAV repeatedly brought him to the attention of the police, and earned him a jail sentence on at least three occasions. As part of their "great intellectual campaign against Darwinism," Oktar and his circle produced scores of books, including *The Evolution Deceit: The Collapse of Darwinism and Its Ideological Background* (1997), millions of copies of which circulated in many languages. Although the Qur'an did not require belief in a young earth, twenty-four-hour creation days, or a global flood, Oktar for years drew heavily on the writings of young-earth creationists for his critique of evolution. By the early twenty-first century, however, he seemed to be moving into the more intellectually compatible ID camp — so much so that the Discovery Institute listed Harun Yahya's Web site as "An Islamic Intelligent Design Site." But in a pique over the ascendancy of a former disciple in the ID world, Oktar dismissed ID as just "another of Satan's snares" because of its failure to recognize Allah. However branded, his antievolutionist crusade prompted a widespread debate among conservative Muslims.

On a much smaller scale creationism also acquired a foothold among Orthodox Jews, who, despite believing that God had created the world no more than 6,000 years ago, had typically paid little attention to Christian efforts to stop the spread of evolution. Occasionally an individual Jew had spoken up, but it was not until 2000 that Jewish creationists organized the Torah Science Foundation, a largely Israeli-American group inspired by the Lubavitcher rebbe Menachem Mendel Schneerson (1902-1994). Eager to maintain an identity separate from Christian fundamentalists, these Jewish antievolutionists meshed the teachings of the Torah and the Kabbalah with off-the-rack creationism to create a uniquely Jewish product.

Continental Europe, perhaps the most secular region on earth, at first proved resistant to American-style creationism. But conditions changed rapidly. With the demise of the Soviet Union in 1991, evangelical Christianity boomed in Russia, and along with it creationism. Before long bureaucrats in the Russian ministry of education were cosponsoring creationist conferences, collaborating with American creationists on the writing of textbooks, and urging that creationism be taught to help restore academic freedom in Russia after years of state-enforced scientific orthodoxy. As one academician put it, "no theory should be discounted after the long Communist censure." Other former Soviet-block countries — Poland, Hungary, Romania, and Serbia — also witnessed the spread of creationism. In 2004 the Serbian minister of education, an Orthodox Christian, instructed primary-school teachers that they should no longer have students read a "dogmatic" chapter on Darwinism in the commonly used eighth-grade biology textbook, and the following year the Romanian ministry of education granted permission for teachers in both public and Christian schools to elect to use a creationist alternative to the standard biology textbook.

Sporadic outbreaks of antievolutionism also occurred in Western Europe. In 2004, for instance, the Italian minister of education announced her intention to eliminate the teaching of evolution for students ages eleven to fourteen, which prompted mass protests and a quick retreat. The following year the Dutch science and education minister triggered a fierce debate in the Netherlands by suggesting that the teaching of intelligent design might help to heal religious rifts because Christians, Jews, and Muslims all believe in creation. The furor prompted one alarmed observer to ask "Is Holland Becoming the Kansas of Europe?"

Assessments of the depth of antievolution sentiment in Europe in the early twenty-first century must rely on public-opinion surveys. One of the earliest polls of European attitudes toward creation and evolution, in 2002, found that forty percent favored naturalistic evolution, twenty-one percent endorsed theistic evolution, twenty percent (with the Swiss leading the way) believed that "God created all organisms at one time within the last 10,000 years," and nineteen percent remained undecided (Kutschera 2003). Four years later the BBC shocked many when it announced the results of a poll showing that "four out of 10 people in the UK think that religious alternatives to Darwin's theory of evolution should be taught as science in schools." The survey indicated that only forty-eight percent of Britons believed that the theory of evolution "best described their view of the origin and development of life." Twenty-two percent said that "creationism" best described their views, seventeen percent favored "intelligent design," while thirteen percent were undecided (Numbers 2006, 408).[14]

According to a Gallup poll in 2005 almost twice as many Americans preferred "creationism" to "intelligent design," with 58 percent of the respondents regarding creationism as definitely or probably true compared with 31 percent for intelligent design and 55 percent for evolution (such figures hint at a lack of clarity on the issues). More than a quarter (28 percent) reported being unfamiliar with intelligent design; 11 percent, with creationism; 8 percent, with evolution.[15]

Well into the new century creationism and ID continued to roil American politics at the local, state, and federal levels. The 2008 US presidential election was no exception. The Republican nominee for president, Senator John McCain, a Southern Baptist, advocated teaching students "all points of view" about the origins of humans, as did his Pentecostal running mate, Governor Sarah Palin of Alaska. "Teach both," she said. "You know, don't be afraid of information. Healthy debate is so important, and it's so valuable in our schools. I am a proponent of teaching both." The Democratic candidates, though also religious, unequivocally supported science. Senator Barack Obama, a member of the United Church of Christ, dismissed ID as "not science." As he explained to the York, Pennsylvania, newspaper:

> I'm a Christian … I believe in evolution, and I believe there's a difference between science and faith. That doesn't make faith any less important than science. It just means they're two different things. And I think it's a mistake to try to cloud the teaching of science with theories that frankly don't hold up to scientific inquiry.

His pick for the vice presidency, Senator Joe Biden, a Catholic, dismissed ID as "malarkey".[16]

The big question looming over this entire discussion is why so many people reject evolution. Unfortunately, there is no simple answer, such as lack of education or hatred of science. Most antievolutionists profess a love of science; they refer to

---

14 www.Mori.com/polls/2006/bbc-horizon.shtml. See also Allgaier and Holliman (2006).

15 See www.gallup.com/poll/21814/Evolution-Creationism-Intelligent-Design.aspx ; see also Moore (2005).

16 Karamartin (2005); Kizzia (2006); Murphy (2006); Joyce (2008); "Real time with Bill Maher", Episode 74, 7 April 2006, interviewing Joe Biden.

young-earth creationism as creation science and regard intelligent design as a scientific theory. Some, such as the Seventh-day Adventists, reject evolution largely because their founding prophet told them that evolution was Satanic. Others, such as the Christian fundamentalists, believe that evolution contradicts the plain meaning of God's word in Genesis. Some critics have linked "Darwinism' with unsavory social and political movements, such as German militarism after World War I, to communism after World War II, to atheism and materialism today. Virtually all take the view, promoted by anti-evolutionists and scientific materialists alike, that evolutionary thought is incompatible with genuine religious belief. Faced with an apparently stark choice, they elect to maintain their religious faith. All of this suggests that these movements will not succumb to evolutionary orthodoxy any time soon.

References

Allgaier J., Holliman R., 2006, "The emergence of the controversy around the theory of evolution and creationism in UK newspaper reports", *Curriculum Journal* 17: 263–79.

Burns J. D., 1908, *What is man? or, Creationism vs. Evolutionism*, New York: Cochrane Publishing Co.

Clark H. W., 1929, *Back to Creationism: A Defense of the Scientific Accuracy of the Doctrine of Special Creation, and a Plea for a Return to Faith in the Literal Interpretation of the Genesis Record of Creation as Opposed to the Theory of Evolution,* Angwin: Pacific Union College Press.

Coleman S., Carlin L. (eds.), 2004, *The Cultures of Creationism: Anti-evolutionism in English-speaking Countries*, Aldershot: Ashgate.

de Vries P., 1986, "Naturalism in the natural sciences,' *Christian Scholar's Review* 15: 388–96.

Gallup 1979, "The Christianity Today-Gallup poll: an overview", *Christianity Today*, 21 December 1979.

Jones J. M., 2005, "Most Americans engaged in debate about evolution, creation" (13 October), http://poll.gallup.com

Joyce T., 2008, "Obama talks about York", *York Sunday News*, 30 March 2008, 1, 7.

Karamartin C. J., 2005, "McCain sounds like presidential hopeful", *Arizona Daily Star*, 24 August 2005, www.azstarnet.com/sn/politics/90069

Kizzia T., 2006, "'Creation Science' enters the race", *Anchorage Daily News*, 27 October 2006, dwb.adn.com/news/politics/elections/v-printer/story/8347904p-8243554c.html

Kutschera U., 2003, "Darwinism and Intelligent Design: The new Anti-Evolutionism spreads in Europe", *NCSE Reports* 25 (September-December): 17-18.

Larson E. J., 2003, *Trial and Error: The American Controversy over Creation and Evolution*, 3rd edition, New York: Oxford University Press.

McCready Price G., 1923, *The New Geology*, Mountain View: Pacific Press.

Moore D. W., 2005, "Most Americans tentative about origin-of-life explanations" (23 September 2005), http://poll.gallup.com

Murphy F., 2006, "Obana townhall: math, science add up" *Naperville Sun*, 18 January 2006, obama.senate.gov/news/060118-obama_townhall/ (not science)

Numbers R. L., 1998, *Darwinism Comes to America*, Cambridge, Mass.: Harvard University Press.

Numbers R. L., 2002, "Creationists and their critics in Australia: an autonomous culture or 'the USA with kangaroos?' ", *Historical Records of Australian Science* 14 (June 2002): 1–12.

Numbers R. L., 2003, "Science without God: natural laws and Christian beliefs," in Lindberg D. C., Numbers R. L. (eds.), *When Science and Christianity Meet*, Chicago: University of Chicago Press, 265–85.

Numbers R. L., 2004, "Ironic heresy: how young-earth creationists came to embrace rapid microevolution by means of natural selection", in Lustig A. J., Richards R. J., Ruse M. (eds.), 2004, *Darwinian Heresies*, Cambridge: Cambridge University Press., 84–100.

Numbers R. L., 2006, *The Creationists: From Scientific Creationism to Intelligent Design*, expanded edition, Cambridge, Mass.: Harvard University Press.

Numbers R. L., 2008, *Prophetess of Health: A Study of Ellen G. White*, 3rd edition., Grand Rapids: Eerdmans.

Numbers R. L., Stenhouse J., 2000, "Antievolutionism in the antipodes: from protesting evolution to promoting creationism in New Zealand", *British Journal for the History of Science* 33: 335–50.

O'Toole G. B., 1925, *The Case against Evolution,* New York: Macmillan.

Rupke N. A., 2005, "Neither creation nor evolution: the third way in mid-nineteenth century thinking about the origin of species", *Annals of the History and Philosophy of Biology* 10: 143–72.

Stiling R. L., 1991, "The diminishing deluge: Noah's flood in nineteenth-century American thought," unpublished PhD thesis, University of Wisconsin-Madison.

Worthen F. M., 2008, "The Chalcedon problem: Rousas John Rushdoony and the origins of Christian Reconstructionism", *Church History* 77: 399–437.

Section VI
The theological debate

Introduction

The relationship between Darwin's theory of evolution and religious doctrines has been presented as highly problematic at least since the public encounter between Thomas Henry Huxley and Samuel Wilberforce in the Oxford Theological Association, in 1860. The "theological debate" has originated a wide variety of positions, from the plain acceptation of the accordance between evolution and religion, to their full rejection: religion should be definitively abandoned, as a result of the new evolutionary world view, according to some people; evolutionary theories cannot be accepted, as plainly contrary to revealed doctrine, according to others. American creationism was, already in the twentieth century, the religious movement whose social and public consequences reached a higher level of confrontation.

A certain number of Christian scientists, theologians and philosophers received *The Origin of Species* with great interest, seeing "grandeur in this view of life, with its several powers, having been originally breathed by the Creator." Some of them, at the same time, were conscious of the fact that the new theory required a deep reflection on the traditional exposition of the theological doctrine of creation, and the Biblical interpretation of *Genesis*. New theological insights have been developed as a consequence of the new comprehension of life offered by biological evolution. The six contributions of the last part of this volume aim to sketch a general panorama of some of the main questions involved in the theological debate around evolution that followed the publication of Darwin's theory.

In the first contribution, "Creation in the Old Testament. Some pointers", **André Wénin** offers a reflection on the meaning of the Biblical texts on which the doctrine of creation is based. The Biblical narration of the creation has been often considered as opposed to the scientific account of life evolution. However, Wénin stresses how the Biblical account does not really deal with creation as such, but with human beings and their relation with the world. Moreover, the Old Testament shows an essential tension in this relationship: the human being is presented, specially in the book of Genesis, as the center of creation, endowed with the mission of dominating and governing the world. In the book of Job, however, the human being seems to be "decentered": the world escapes his control, making evident the condition of creatureliness and its limits. It is a sort of "binocular vision", in which the position of human being, image of God, is at the same time central, and yet also relative, as part of the world.

Jean-Michel Maldamé considers in his contribution "The various meanings of the word evolution in science, philosophy and theology." After distinguishing the "scientific facts" from the explanatory principles of the theory of evolution, Maldamé turns his attention towards the philosophical notion of "creation", offering an important distinction between "beginning" and "origin". The root of many of the present difficulties, leading either to rejecting creation on a materialistic basis, or to rejecting evolution in order to save the doctrine of creation can be seen in the false identification of creation with the act of production of the "first instant" of reality. On the contrary, the aforementioned distinction makes possible the existence and distinction of science and metaphysics. In this way, theology opens the way to better understanding the relationship between God and the world.

Creationism has been traditionally the most radical anti-evolutionist movement grounded on a religious basis, although it seems to have been replaced, in the last decades, by intelligent design. Although it was initially an American phenomenon, creationism and intelligent design have recently been extended also to Europe and the other continents. **Jacques Arnould** examines from the theological point of view the reasons and the impact of their proposals. According to Arnould, this phenomenon should be perceived as an opportunity to better define the content and limits of science and theology. Science cannot became dogmatic, but in the same way religious doctrine need to submit to a careful theological analysis. The role of natural theology, in particular, seems to require today a new attitude, perhaps less confident, that would not be confused with the demonstrative methods of science.

The contribution of **Rafael A. Martínez** presents the first response of the Vatican authorities to Darwin's theories, which was poorly known until the opening of the Archives of the Congregation for the Doctrine of the Faith, in 1998. Although no explicit condemnation of evolution was issued, theology handbooks spread the idea that the Holy See opposed evolutionary theories. But the actual circumstances were not clearly known. The new documents show how a general anti-evolutionist attitude, common in theologians, was balanced with Catholic evolutionist proposals, giving place to a somehow "pragmatic policy" avoiding that evolution could became "another Galileo case".

Pierre Teilhard de Chardin has been undoubtedly the author that more deeply tried to offer an integration of evolutionary biology and theological doctrines, in particular with the revelation of Christ. **Georges Chantraine** offers a detailed account of Teilhard's perspective, in with evolution was seen as the law of complexity of growth, from biosphere through "Noosphere" to "Christo-sphere". Following Henri de Lubac's reading of Teilhard, Chantraine presents Teilhard's attempt to establish a relationship between the "physical phenomenology" and Christian significance, trying to clarify the methodological problems of his proposal.

Robert J. Russell summarizes and assesses some of the recent theological debates around evolutionary, molecular, and developmental biology. Starting with Vatican II documents and key John Paul II statements on Science and Religion, mainly from the Message to the Director of the Vatican Observatory, Russell offers a constructive theological interpretation of evolution that includes theologies of creation and redemption and draws on the writings of Karl Rahner, Celia Deane-Drummond, Denis Edwards, Jurgen Moltmann, Ted Peters and Martinez Hewlett, and Joseph Życiński. According to Russell, the future calls for a continued and robust engagement between Christian theology and evolutionary biological sciences.

Rafael A. Martínez

The creation in the Old Testament. Some pointers

André Wénin

It is hardly a secret that the subject of evolution is of great interest today. It is obvious however that the Old Testament does not refer to evolution. It limits itself to a discourse on creation that is essentially evocative and poetic, a discourse that does not stand out as significantly different from the thinking in the surrounding cultures of the ancient Near East. In actual fact Genesis looks here more as though it is derivative from the culture of that environment. However, when the Old Testament texts that speak of the creation are studied more closely, it is quite surprising to see that they do not really deal with the creation as such. The biblical text is quick to shift the focus onto the human being and humanity, as though what mattered was to allow the reader to situate himself in the world around him that is his world. At least, this is what emerges from the first two chapters of Genesis.

In fact the so-called "accounts of the creation" in Genesis 1–2 — a quasi-liturgical poem followed by a mythical-style narrative — place the accent on the complex relationship between God and that which is created and on the relatively central position of the human being in the terrestrial sphere over which he receives the responsibility to exercise dominion. This centrality of humanity and of its role in that which is created, which is echoed in Psalm 8, is however relativized by other texts of a sapiential type, like the divine discourses at the end of the book of Job (chap. 38–41) or Psalm 104: these two texts, while clearly affirming the sovereignty of God, emphasize that human beings are fully at home in the created, with which they are in solidarity to such an extent that any idea of dominion is in fact contested. My paper aims to go more deeply into this paradoxical vision of the creation that we find in the Old Testament.

1. Genesis 1–2: the human being at the center of the created world

Perhaps precisely because they initiate an account in which the subject is essentially human beings, the first two chapters of Genesis have a fundamentally anthropocentric intention. This is of course expressed in a different way in the two chapters, but as we shall see, the two ways are convergent and complementary.

The long poem that recounts the creation of the universe in seven days (Gen. 1: 1–2: 4) is composed in such a way that the work of creation culminates, so to speak, in the human being. Everything contributes to highlighting it. It comes at the end of the six days and it is the object of three discourses by God. The creation of the human being is preceded by a deliberation of the Creator with himself: this makes the moment solemn while preparing point by point what is to follow. Moreover, for the first time, God addresses himself to one of the creatures ("and God said to them", v. 28), and does so in order to entrust to humans a responsibility that presupposes the exercise of a certain power over the world and over its other inhabitants — a purpose already expressed when God formulated the project of making humans (v. 26). This dominion will take on all its significance when the Creator ceases to display his own dominion in entering the repose of the seventh day, distancing himself in this way from the created world henceforth entrusted to the responsibility of humans (Gen. 2: 1–3).

What goes before in the account of Genesis 1 ultimately constitutes the arrangement of the framework in which will evolve those unique beings that are humans. Composed with great meticulousness, the evocation of the creative act manifests a will to classification and ordering that, although not completely alien to a scientific mentality, proceeds first of all from the desire to show the coherence of the action of the Creator and to underline the harmony of the totality of his work. God begins logically by establishing a spatio-temporal framework: the heavens, the seas and the earth with its plants for space, the alternation of days and nights and the calendar regulated by the stars for time (Gen. 1: 3–19: days 1 to 4; a very regular section in 3 groups of 69 words, or 207 in all). God then populates the three spaces that have been organized, always proceeding methodically: the seas and the heavens are filled first with fishes and birds respectively; next, the earth welcomes the terrestrial animals, and then the humans, who receive power over the terrestrial space and over the animals (Gen. 1: 20–31: days 5 and 6; 206 words in irregular groups — perhaps intended to suggest an image of swarming life). This is the global framework of the story that will then be told by Genesis and the books that follow it.

We need to dwell for a moment on what is said of the human being in this section, since it obviously constitutes the culmination of the divine work while also receiving a specific role in the created world. Other creatures are complete when they acquire a body following the divine words, but the human being — or rather humans — is not finished, as some details of verses 26–27 suggest (see Balmary 1993, 109–116; Wénin 2007, 37–46). The image of God that the human bears within himself is not yet "like": the disappearance in v. 27 of the term "likeness" present in v. 26 shows that. With regard to *'adam*, the narrator states, in fact: "in the image of God he created him, male and female he created them": created in the image of God, the human is not like Him, since he is also in a certain proximity to animals, as is suggested by the use of the worlds *zakar* and *neqevâ*, "male and female", employed in Hebrew for animals as for humans. In this sense, the change of pronoun which goes from the singular ("created him") to the plural ("created them") is probably significant: in the image of God, humanity is one, but like the animals, he is also plural.

It is against this backdrop that we need to understand the word of blessing addressed then to humans at v. 28: "Be fruitful, multiply, fill the earth and subdue it. Be masters of the fish of the sea, the birds of heaven and all the living creatures that move on earth". Like the animals of the sea and air blessed at verse 22, humans receive a blessing that invites them to be fruitful, to multiply and to fill the space inhabited by them. But they alone are commanded by God to subjugate the earth and dominate the animals, as he announced in forming his project at verse 26. Commanding them in this way to exercise their dominion, God calls them as it were to act like him who, from the beginning, clearly exercises a sovereign dominion over the elements of the universe. In short, he shows humans a way to fulfill the image of God in themselves by resembling him in the exercise of their dominion over the world itself. The words utilized to evoke this power both connote a certain potency. *Radah* (v. 26. 28) means "to oppress, to crush" (see Gal. 4: 13), "to dominate" (Lev. 26: 17) like a king who exercises his own sovereign power (1 Kings 5: 4 ; Ps. 110: 1). Where *kabash* is concerned (v. 28), it has as its meaning "to subject" conquered enemies (2 Sam. 8: 11) or " bring" slaves "to submission" (Jer. 34: 11). The power which God commands humans to exercise therefore presupposes the use of force, in the image of the power displayed by the Creator in his action as described in Genesis 1.

This order is immediately followed by another word with which God gives a vegetable food first to humans and then to animals: the former receive cereals and fruit, the latter "the foliage of the plants" (v. 29–30). This last seems to be anecdotal. And yet we are talking here about the last word of the Creator and also the longest. And, in the context, its significance is very far from superficial. In fact, after having enjoined humans to dominate the animals, to give humans a solely vegetable diet different from that of the animals is tantamount to suggesting to them that it is possible to dominate the animals without killing them, thus setting a limit to their own power. And just as animals do not receive the same diet, food will not become an object of conflict with them. This means that human dominion over the animal can be free of all violence, in the image and likeness of the power that God displays in the creation from the beginning of this first page of the Book, a power that does not destroy anything and imposes limits on itself, as is clearly indicated for example, by the rest of the seventh day. The duty of dominating the animals can therefore be fulfilled with mildness, as a dominated dominion (see Beauchamp 2005, 105–144). It is in this way that humanity in general and humans in particular will realize their vocation of resembling the image of God. If they succeed, the animals, guided by those pastors full of mildness, will not even eat one another, as God suggests at verse 30 when he gives them for food all the foliage of the plants. In this way, humanity will be able to preside over a universal harmony, a fulfillment of the hidden desire of God in his work of creation (see Isa. 11: 6–9). What is at stake is the choice that humanity must make between a dominion that is violent and boundless, on the one hand, and a dominion that is gentle and measured, on the other — the only kind that will enable humanity to be fulfilled in the image of God.

The next page of Genesis turns on the same subject and clarifies it further. As Jan Fokkelman writes, "the 'account of paradise' is not, in reality, a second account of the creation, but a fuller study of man as creature" (Fokkelman 2002, 134). Thus, in the account of Eden as in the poem of the seven days, with much logic and without apparent effort, Yahweh God arranges a framework for life appropriate to the human, who is here the first to be created (2: 7). As in Genesis 1: 28, God entrusts to *'adam* a precise responsibility with regard to creation (2: 15); as in 1: 29, he gives him vegetable food while addressing him to ask him to accept that something is withheld: the human is in this way placed before a limit with respect to which he will have to make a choice. There is then a basic analogy between these two pages that is indicative of a profound thematic continuity, and the narrator clarifies this by "zooming" in from the whole of humanity, which is introduced on the sixth day (Gen. 1: 26–29), to one individual human being who will be the protagonist of the narrative of Eden (2: 7). But let us take a closer look at what is related in this way about the creation of the human being.

From the beginning of the second page (2: 5.7), the play on words between *ha'adam* and *ha'adamah* makes clear the close relationship that unites the human to the humus: the earth will be cultivated by one who has just been shaped from the soil of the ground. But the human being is not the only one to be "taken out of the soil": the trees that the Creator causes to grow (2: 9) and the animals that he forms (2: 19) are made in the same way, as is underlined by the repetition of the espression *min-ha'adamâ*. The human being is therefore profoundly bonded to the mineral world, and like the plants and the animals, he is wholly part of the physical universe. The narrator, however, also shows us what the human has that is unique. Vegetables grow from the earth, and human beings and animals are "formed" out of this same matter; living beings therefore have an additional feature in common. Having said this, there are two elements in the account that distinguish the human from the animals. These are related in a rather surprising way that calls for interpretation on the part of the reader.

On the one hand, only the human is said to be formed our of "dust from the ground" (2: 7 – RSV Translation), dust being a sign of death in the first Testament (see for example Gen. 18: 27; Ps. 22: 16; Job 7: 21; Dan. 12: 2). In this way it seems that from the beginning the text is implying the mortal destiny of the human being which is referred to explicitly at the end of the account, in 3,19: "until you return to the ground … for dust you are…", are the words of God to speak of natural death. But is not this return to dust also perhaps true for the animals? So why say it only of the human being? Maybe the aim is to point to one of his peculiarities, namely the fact that he knows he is dust, that he is mortal, and that this knowledge sets him apart from the condition of the other living beings of creation. On the other hand, the narrator relates that, to turn the being he has formed into a living being, Yahweh God "blew the breath of life into his nostrils" (*nishmat ḫayyîm*, 2: 7), something that will not happen for the animals. This too is a surprising fact, given that many animals obviously breathe as well. This "breath" is not therefore breath pure and simple, and all the more so in that the narrator explicitly states that it comes from God. As early as Gen. 1: 3, God uses his own breath to speak, to give names (see Wénin 2007, 59–60), and it is precisely this that will make the human when Yahweh God presents the animals to him: he will call out their names (*shemôt*), thereby demonstrating that they cannot provide him with a "face to face" encounter, given that they do not answer him (2: 19–20). Thus, the human stands out among living beings in that, like God, he uses his own breath to name things and beings. It should be noted moreover that in acting in this way he exercises over the animals that mild dominion which is referred to at the end of chapter 1.

To summarize then, in this account *'adam* is connected to the *'adamah* with his whole corporeal being and connected to God through the word that allows him to emerge from nature and that confers on him a possibility of dominating it — in the image of God, we might say. This subject is moreover covered again and at greater depth in another passage in the same account in connection with the garden — a protected enclosure according to the etymology of the word *gan* — and with regard to the role that God assigns there to the human being: "Yahweh God planted a garden in Eden, which is in the east, and there he put the man he had fashioned" (Gen. 2: 8); from the soil, Yahweh God caused to grow every kind of tree enticing to look at and good to eat (v. 9 – RSV translation). Then, "Yahweh God took the man and settled him in the garden of Eden to cultivate and take care of it" (v. 15 – RSV translation). The task and the responsibility of the human are indicated by two verbs: *'avad*, "to cultivate" since we are talking about a garden, but more generally "to work" or even "to serve", and *shamar*, "to take care of" this enclosed place and "to watch over" it. The polysemy of the verb *'avad* suggests that the activity of dominion that consists in cultivating and working the garden supposes that it is treated with respect as a servant does with his master. Along the lines of what is outlined in Gen. 1, the power exercised will then remain measured. In fact the human does not live in the garden as owner or boss, but as gardener, as servant. That is what is emphasized in its own way by the verb *shamar* which indicates that the human ought, like any good custodian, to take care of the garden. Again, the dominion in question here needs to be dominated or self-limiting.

The relationship between the human and the garden, however, does not go in one direction only. In fact, just as soon as Yahweh God has placed the human in the garden, he turns to him to give him the trees for food, all the trees (2: 16). In this way, he inaugurates a form of exchange: the human "will work" the garden which God has arranged; in exchange, the garden will offer him its trees "enticing to look at and good to eat" and he will feed off them. But there is more: if the human "takes care of" the garden and "watches over" it, the closed garden will for its part "protect" (*ganan*) him. What is outlined in this way, is so to speak a "covenant" between the human being and the garden, between humanity and nature which welcomes him, a relation of exchange in which the human can enjoy goods offered by the garden that he will cultivate and care for. But an attentive study of the account reveals that this covenant depends in reality on the human: the better the human works and takes care of the garden according to the divine plan, the better the garden will protect him and feed him. The future of this relationship of exchange depends then on human responsibility — as the end of chapter 1 has already made clear in its way.

The limit to the exercise of dominion over the garden is then suggested in the two verbs that describe the role that the human must take up in it. This role is moreover explained subsequently by Yahweh God when he sets a limit in enjoining the gardener not to eat from a tree (v. 17). I cannot enter here into the details of the analysis of this injunction, whose meaning is as rich as it is complex, still less into the explanation of the fact that it should have been the tree of "the knowledge of good and evil" that forms the object of the limit set by God; I have given my view on this topic elsewhere (see Wénin 2007, 62–69; 2008, 303–318); suffice to say here that it is not a matter of forbidding humans to acquire knowledge or pursue science. Taking the narrative at its most superficial level, I will content myself with this observation: the fact that without any explicit reason the human being does not have access to all the trees shows at least that he is not the owner of them. If he was the actual owner of the garden, we would not be able to understand why the enjoyment of one of the trees would be forbidden him. He cannot then exploit it limitlessly. If he does so, he will sign his own death warrant. Not because the legitimate owner will come and punish him — nowhere is it stated that Yahweh God is the owner of the garden — but because in rejecting the logic of the covenant by allowing himself to be guided by greed, that is, by an uncontrolled and all-encroaching desire, he would be adopting an attitude that is of itself the bringer of death. As at the end of chapter 1, to consent to a limited dominion is the choice that will allow life to develop fully.

To conclude this overview of Gen. 1: 1–2: 17: each in its own way, these two pages set the human being at the center of the created world, thus developing the theme of the creation not for itself, but from a perspective of theological anthropology with a definite ethical slant. Each page also suggests that this unique place of humans, connected with their privileged relationship with God, involves them in a particular responsibility: that of exercising dominion over the world. This dominion will however contribute to the harmony of the created only to the extent to which it accepts the proper limit imposed by respect for life, puts itself at the service of nature, and is exercised with the mildness of the God who creates without destroying. It is not primarily the future of the world that depends on this, but the future of each human and the future of humanity itself.

2. Some poetic pages: the human being decentered

This conception of the human being as center of the creation finds a potent echo in Psalm 8: 4–6: "What are human beings that you spare a thought for them, or the child of Adam that you care for him? Yet you have made him little less than a god, you have crowned him with glory and beauty, made him lord of the works of your hands, put all things under his feet..." — This is not however the only discourse in the Hebrew Bible on the subject. There is in fact a great poetic text that refers to the creation while explicitly moving humanity away from the central position attributed to it in the texts we have looked at so far: I am referring to two long discourses that Yahweh addresses to Job from out of the heart of the tempest in answer to his repeated appeals (Job 38–39; 40–41; on what follows, see Lévêque 1987, 261–299). In magnificent poetic tones, God develops a vision of the world in which the place of the human being is largely relativized. The plain truth seems to be that he chooses to take up a contradictory position to that of the discourse in Genesis 1 in which the human is situated at the center and set in a position of dominion over the terrestrial universe and the animals.

In the first response that he addresses to Job, Yahweh refers to the creation, going through a series of questions highlighting how limited is the dominion of each human as of humanity itself over the world and even over his own existence. A first limit has to do with temporality. The universe, in fact, existed long before humanity and man: "Where were you when I laid the earth's foundations?", asks the Creator (38: 4a). A second limit, which derives in part from the first, has to do with his knowledge. God goes on to say: "Tell me, since you are so well-informed!". And he adds (v. 5–8): "Who decided its dimensions [of the earth], do you know? Or who stretched the measuring line across it? What supports its pillars at their bases? Who laid its cornerstone to the joyful concert of the morning stars, and unanimous acclaim of the sons of God? Who pent up the sea behind closed doors, when it leapt tumultuous from the womb?…" Whatever the state of our current progress in the sciences, any such progress always raises more questions than it gives answers, so the question remains: "Have you an inkling of the extent of the earth? Tell me about it, if you have!" (38: 18). A third limit has to do again with the power of the human being. He does not have any purchase on the passing of time: "Have you ever in your life given orders to the morning?" (38: 12); nor does he have any purchase on the climate either: "Can your voice carry as far as the clouds, and make the pent-up waters do your bidding? Will lightning flashes come at your command and answer: here we are?" (38: 34–35). The stars completely elude him: "Can you fasten the harness of the Pleiads, or untie Orion's bands? Can you guide the Crown season by season and show the Bear and its cubs which way to go?" (38: 31–32). His power does not extend to the immensity of space: "Have you been right down to the sources of the sea and walked about at the bottom of the Abyss?" (38: 16); death, for its part, remains a mystery for him: "Have you been shown the gates of Death, have you seen the janitors of the Shadow dark as death?" (38: 17). All in all, the power of humanity over nature — except to destroy it, we might say today — is absolutely relative in comparison with the forces that preside over the equilibrium of the world and which, according to the book of Job, depend on the Creator.

After having thus relativized the power of humans over the nature that surrounds them, God addresses the animal world. He dwells on certain animals whose way of life he takes the time to describe, noting that these animals have no need of humans. Where humans are mentioned, their position is relativized and doubt is cast on their capacity for dominion. Thus, wild beasts like the lioness or the raven find their food without any necessity for the intervention of the human being (38: 39–41). The same applies to the mountain goats or the deer when they give birth to their young ones and accompany them for a while until they gain their freedom (39: 1–4). Worse: the creation is also full of animals that are quite useless for man because they are too untameable to be useful. Thus the wild donkey and the "brayer" live in inhospitable steppes far from the hubbub of the cities and the noise of the traffic, enjoying a freedom that is unknown to those animals whom their drivers have "domesticated" (39: 5–8). In the same way, the wild ox will not allow itself to be tamed as a beast of burden and man cannot exploit it to work: "Is the wild ox willing to serve you? … will he harrow the furrows for you? … Can you depend on him to come home and pile your grain on your threshing-floor?" (39: 9–12). The ostrich, for its part, seems stupid, negligent, and harsh on its young; she would not be of any utility to humans, but "if she bestirs herself to use her height, she can make fools of horse and rider too" (39: 13–18). By contrast, the horse performs wonderful services for the human and the warrior in particular; but this is not what confers on it the quality that makes it strong, proud and intrepid in battle (39: 19–25). Finally, for humans, the hawk and the eagle remain out of reach; they do not take orders from humans and live their life very well without humans (v. 26–30). Therefore, whatever their relationships with the human being, these animals always elude him in one way or another. The human must not therefore believe that it is he who is their boss.

After a pause in which, at the request of Yahweh, Job replies that he has nothing to say in answer (40: 1–5), a second divine discourse takes up the same subject again to cast doubt once more on human dominion by questioning the extent of the power that man is able to exercise. One of the ways of affirming this power and doing so with pride would consist in resisting the arrogant and the wicked who disfigure humanity: "Bring down all the proud, strike down the wicked where they stand. Bury the lot of them in the ground, shut them, every one, in the Dungeon. And I (The Lord) shall be the first to pay you homage, since your own right hand is strong enough to save you", says God to Job 40: 12–14). But after all, is humanity really capable of overcoming and dominating the animality that lurks in its own breast? Then he cannot put on a display of pride and robe himself "in splendour and glory" (40: 10). What are we to say moreover of the prodigious animals in the face of which the human is completely impotent? Their mere existence suffices to put a question mark against his alleged centrality in creation. In the first place there is the *behemôt*, described as a monster of power having the features of a hippopotamus — a living armoured vehicle. Yahweh says of it that is "the first of the works of God" (40: 19). The following passage in the poem shows how all of nature puts itself at the service of this herbivore to feed and protect it (40: 20–22). Powerful though they are, the floods do not frighten it (40: 23). For all the weak points that it does have, the man who manages to overpower it is a brave man indeed (40: 24). The leviathan, for its part, is described as being like a crocodile. Along the lines of what has just been said of the uncatchable *behemot*, the description of it is preceded by a series of questions that point to the smallness of the human being by comparison with it. He cannot catch it, enslave it, domesticate it, hunt it, or even simply face up to it (40: 25–32). Everything in the leviathan suggests the power, the brutality, and the invulnerability of a terrifying and invincible dragon. And the poet concludes his long description: "He has no equal on earth … of all the lordly beasts he is king" (this last probably refers to the proudest animals like the wild beasts (41: 25.26). Here again we see a definitive relativization of the central place that humanity might think it holds in the world. For if dominion depends on force, these animals would have more of a right to claim it than the human.

As Jean Lévêque writes with reference to Job, humans have a tendency to take into consideration only a world that is commensurate with their power, "without measuring how far the real is capable of resisting them". But like Job, they must "get used to the idea that certain forces of their world are neither tameable nor disturbable at will." (J. Lévêque 1987, 295) Among these forces, there is also that which wounds the human being, that which causes him suffering, that which escapes his control, like the two monsters mentioned. More generally however, the condition of creatureliness implies limits, even if only those imposed by the fact of living in space and time.

Along with this magnificent poem of the book of Job which strongly relativizes the centrality of the human in the creation of which Gen. 1–2 speaks, there are also poems of praise or rather of gratitude towards God the Creator. In these poems, the world is described as a gift that humans receive with life. The wonderment in the face of the creation sung by these Psalms makes the human being not the master of the world, but rather the witness to a generosity and a goodness that precede him and that belong to the God from whom he receives his life. The most beautiful and longest of these poems is Psalm 104 in which the created world is described at length. In the universe that he contemplates in this way, the Psalmist sees the fruit of the wisdom of God, a wisdom that surpasses man, who can nonetheless admire it: "How countless are your works, Yahweh, all of them made so wisely! The earth is full of your creatures" (Ps. 104: 24) — this subject is a common one in the Old Testament and it is sometimes the object of specific treatments (see for example. Prov. 8: 22–31; Sir. 24: 2–6). In its proper place, each thing contributes to the harmony of the whole, even the chaotic abyss of the primordial waters, the darkness of the night and the Leviathan that God has created to make fun of it (104: 6–9.20.26). Even the return to the dust, natural end of life, adds its own touch to this universe in which everything sings the praise of the God who is the lord of life: "Turn away your face and they panic; take back their breath and they die and revert to dust. Send out your breath and life begins; you renew the face of the earth" (104: 29–30).

The place that this poem reserves for the human being is notable. It begins by describing the heavens, the clouds, the wind, and the lightning (104: 3–4), before referring to the creation of the earth (v. 5). Thus we see appear from time to time the mountains, the precipices, the rivers, and the sea (v. 6–9), and this water makes animal life possible: "you opened up springs … supplying water for all the wild beasts, the wild asses quench their thirst…" (v. 10–12). The water also allows the species of greenery to grow on which animals and humans feed; the latter can therefore cultivate the soil to make from it their own bread and festal vines and oil (v. 13–15). After this brief reference to humanity, the poet continues with his description: the trees, the sparrows and the storks, the animals of the mountain (v. 16–18). Then comes the turn of the stars that regulate the rhythm of time to which human work is subjected (v. 19–23), before the sea is evoked with its luxuriant population and also its shipping traffic (v. 25–26). Finally, as in Genesis 1, the Psalmist speaks of the gift of the food that assures the life of the living until, returning to the dust, they give place to others (v. 27–30). This quick survey of the poem suffices to demonstrate the modest position of the human being in the framework. He is one of the living beings who populate the earth like many others and he does not occupy any particular place. His unique characteristic is the work that the generosity of nature makes possible. Moreover, the work referred to by the poet, as if in passing, has nothing to do with some kind of dominion of humans over the universe. It serves only to assure their sustenance and give them joy (v. 14–15 and 23). The point is here then that man is not in any way the measure of things. Of course he has his own place in the harmonious world that eludes him to a great extent, a world he does not dominate, a world that he can however admire in finding in it the trace of a God worthy of his praise. For as the only creature endowed with speech, he can praise the Creator, as is underlined by the ending of the poem: "I shall sing to Yahweh all my life, make music for my God as long as I live. May my musings be pleasing to him, for Yahweh gives me joy" (v. 33–34).

Conclusion

While seemingly so contradictory, these two conceptions actually provide us with a binocular vision onto the reality that they contemplate, and this puts everything into perspective. The most important point they have in common is that they make the God of speech the sole origin — in the strong theological sense of the word — of creation, as if to underline the fact that it is his Word — or his wisdom (Prov. 3: 19–20; 8: 22–31) — that establishes everything in being and attributes to everything its proper place; and, as Psalm 33: 4–6 says, "The word of Yahweh is straightforward, all he does springs from his constancy, ... the faithful love of Yahweh fills the earth. By the word of Yahweh the heavens were made, by the breath of his mouth all their array". Therefore, Paul will say that what can be known of God has been evident from the creation of the world (Rom. 1: 19–20) while John for his part will state that in the beginning it was by the divine logos — at one and the same time life and light — that everything was made (John 1: 1–5).

As to what concerns humanity and the human being, created in the image of God and called to perfect themselves in his likeness, they have a place and a role that are at the same time central and crucial and yet also relative, to the extent that humans are an integral part of the created world in which they exercise their limited dominion that constitutes a task and a responsibility for them. Relativized they may be, but their choices involve consequences that can have death-bringing repercussions on the created world. This underlines the radical ethical impact of their particular status in creation.

Hence, even if the discourse of the first Testament remains partial and is not without internal tensions, it offers significant material for a theology of creation insofar as we do not however forget that its discourse has nothing of the dogmatic. It is rather to be read as mythical and poetic and can serve for theological development only if it is part of a hermeneutic project that takes account of the specific language of the texts and puts their particular contribution into perspective.

References

Balmary M., 1993, *La Divine origine. Dieu n'a pas créé l'homme,* Paris: Grasset.
Beauchamp P., 2005, "Création et fondation de la loi en Gn 1, 1 – 2, 4. Le don de la nourriture végétale en Gn 1, 29s.", in Id., *Pages exégétiques,* coll. Lectio divina n° 202, Paris: Cerf, 105–44.
Fokkelman J. P., 2002, *Comment lire le récit biblique. Une introduction pratique*, coll. Le livre et le rouleau n° 13, Bruxelles: Lessius.
Lévêque J., 1987, "L'argument de la création dans le livre de Job", in ACFEB, *La création dans l'Orient ancien,* coll. Lectio Divina n° 127, Paris: Cerf, 261–99.
Wénin A., 2007, *D'Adam à Abraham ou les errances de l'humain. Lecture de Genèse 1, 1 – 12, 4,* coll. Lire la Bible n° 14, Paris: Cerf.
Wénin A., 2008, "Le précepte d'Adonaï Dieu en Genèse 2,16–17. Narration et anthropologie", *Revue des Sciences religieuses* 82: 303–18.

The various meanings of the word evolution in science, philosophy and theology

Jean-Michel Maldamé op

1. Introduction

The word Evolution, in its general sense, refers to a continual process of change. In the philosophy of nature, however, it refers to what occurs when an original seed germinates and develops into rich maturity, or when in due course, what was hidden, or in a latent state, becomes manifest. In a more restricted sense, the word is used to characterize a scientific theory whose object is to render life intelligible. The usual phrase is then, "the theory of evolution", which not only gives a description or a classification of living objects, but proposes to explain their unity and diversity. The word evolution can be used with a fourth meaning when it refers to the active subject which presides over phenomena: it is then written with a capital E. Because of these several different meanings, the concept of evolution is at the center of many sharp debates, often occasioned by misunderstandings.

I shall first attempt a few methodological clarifications, in order to better determine what the "theory of evolution" really is. Then I shall confront this expression with the theological concept of creation. We shall see how God's action and natural processes described by science can fit together, before advancing a theological proposition in order to renew the dialogue between science and faith.

2. The scientific theory of evolution

In order to avoid mistakes, one should be aware of the fact that the word "theory" refers to a general interpretation of certain facts. One should therefore make a list of those facts, before examining how they are combined into a theory, by means of principles.

2.1 Scientific facts

The theory of evolution rests on numerous independent facts concerning various fields and specific approaches: they are not raw facts, but appear in the light of specific sciences. Here is a list of seven of them:

1. First, there is no spontaneous generation. Successive generations are linked together by a genealogy. Scientists must study lineages. Research then opens on to a temporal dimension and a history of vital phenomenon.

2. Researchers take into account the living objects of the past, through the study of fossils. They inscribe them into a structure where forms survive through time, while allowing changes to appear. In spite of the scarcity of fossils — since the conservation of very old remains of living objects is uncertain — naturalists have gathered enough of them to start comparing and classifying them.

3. The inscription of living objects, when their fossilized remains are found, occurs within a chronological frame. It has been possible to construct such a frame thanks to the progress of geological sciences, which allow us to say that the history of the Earth is over four billion years old. Such a protracted stretch of time allows us to classify artifacts by arranging them chronologically, in such a way as to allow us to situate transformations which have been elsewhere observed.

4. We now know better than ever before that the conditions of life fall within the scope of precise environmental conditions. The study of this fact — ecology as science — has shown how the environment offers all sorts of possibilities, and imposes all sorts of constraints. When the environment is modified, living objects must adapt, change or disappear.

5. Investigations into lineage and the observation of variations in forms has led to the study of plasticity in living objects. It is accounted for by the role played by genes. Genes carry the information which presides over the constitution of living organisms, and which are transmitted from one generation to the next. Since the transmission from one generation to the next does not occur in an absolutely identical way, that leaves room for changes, for mutations which provoke variations in the shapes of living objects. The study of embryology allows us to observe internal mechanisms of regulation and programming of living objects.

6. The living objects under observation all belong to a population, understood as a body of beings sharing a common fate.

7. Finally, one notices that precise rules preside over the transformations in populations. Rules can be written down in mathematical formulas, whose development enables us to construct behavioral models of populations dealing with random facts. One should also mention thermodynamics, whose processes show how an open system can organize itself out of disorder, and how stability can be maintained in spite of the law of entropy.

Such facts are taken into account by scientists, who want to explain away the diversity and the unity of all living objects. For that, they must interpret facts; their interpretations are based on principles of a different order, different from description and observation. So that, in order to understand the theory of evolution, one must explain its principles. This is a necessary step since the current controversies bear more often than not on the principles of interpretation, rather than on the facts as such.

2.2 The principles of the theory of evolution

The principles have the status of concepts. They belong to science, since they contribute to its elaboration.

1. The first principle is that the world is intelligible according to an exacting science. This implies that the explanation must remain within the realm of scientific explanation and should not appeal to interventions external to science, breaking with the natural course of transformations. This is contained within the notion of nature.

2. The second principle is the universal character of the laws and rules which preside over the phenomenon of nature. Past facts can be analyzed in compliance with laws which have been established today and verified in current experiments. One should not refer to "another world", structurally different from the world observed today.

3. The universality of the laws of the universe is linked to a principle of homogeneity which requires that biology should define rules for, and impose constraints to the living objects of yesterday, of today and of tomorrow.

4. A fourth principle is that the diversity of living objects falls within the scope of a common identity: so that when our mind wants to reduce the multiplicity of living forms to unity, and is in search of a general explanation which would allow to link all living objects one with the other, it should respect the nature of life.

5. The notion of "animal species" can no longer be seen as an absolute notion. One should rather make use of the notion of "population", which is more flexible, to describe observations.

6. Besides, living beings can be studied through the concepts of mathematical sociology: which leads us to favor the notion of population, understood as a geographical unity where beings are inter-fertile.

7. The last principle is that human beings must not be excluded from observation or from scientific explanation.

It appears, from the bare enumeration of these principles, that they do not have the same status as facts. Facts are unquestionable, whereas principles can be challenged for philosophical reasons which evade scientific methods. Those principles are the basis of what we call a philosophy of nature, which, as such, is open to a philosophical debate.

The theory of evolution then appears as a unified vision where exacting experimental verifications take place, an exacting urge for logical coherence and for natural philosophy.

2.3 The scientific theory of evolution

Thanks to the principles which preside over the interpretation of the facts mentioned, the theory of evolution presents itself as an explanation of the diversity and the unity of all living beings. The theory does so in a manner which is specific, according to the rules of a scientific construction which utilizes concepts of different orders, operational concepts for facts, and general concepts for purposes of synthesis. The general concepts (general ideas, paradigms or themes) have a universal value. So that, even if the work of the founders is for a great part obsolete, the horizon opened by Lamarck and Darwin is still being explored, and their intuitions have not ceased to be substantiated by observations and analyses.

1. The theory of evolution sets forth a general history of living beings. It even inserts it into a vaster continuum, that of the biosphere and even more comprehensively, that of the solar system and of the cosmogeny.

So, in order to present the living objects and link them together, the theory of evolution places the variation of species or transformism within the frame of a history which classifies beings in the shape of a genealogical tree, starting from an original cell, through differentiation.

The theory of evolution comes under the heading of historical inquiry: the facts which are studied cannot be exactly repeated (we cannot return to the time when life first appeared on earth…). The theory accounts for what has been. It has therefore, a status of a particular nature: it does not correspond to the epistemological criteria defined by Karl Popper, who took fundamental physics as his model.

2. The theory of evolution does not just retrace a history. It proposes a research program, and opens itself to new vistas. The layout of the genealogical trees is an anticipation, which calls for a confrontation between results which have been obtained in different fields. Therefore, the theory of evolution plays its role as a theory, by proposing research programs. It suggests hypotheses, which must be verified. It establishes research programs aimed at verifying, asserting, correcting, even invalidating what is accepted as truth. It is an open theory: if it never ceases to modify the branches of the tree as is made necessary by the observation of new

facts, such modifications, far from invalidating the theory, confirms its ability to account for new elements.

Ever since Charles Darwin, the theory has never ceased to renew itself by appropriating new analyses. Even if there is more to be discovered than what is known, and even if results are under continual transformation, the theory of evolution does not cease to show its fecundity, because it enables us to understand the unity and the diversity of living objects.

Within this framework, scientific explanations ought to promulgate laws of transformation. They are often called the "mechanisms of evolution", by analogy with the physics of movement (cinematic or dynamic), for what is at stake here is to explain how transformations operate in the course of time.

On this point, sharp debates take place among scientists about the mechanisms at play, particularly on the following questions: is change continual? Does it occur in stages? The debate develops inside the theory of evolution; it does not invalidate it, it invites new developments.

As far as the dialogue between science and religion is concerned, the problem of continuity plays a major role, for it leads to the following question: is the grandeur of mankind not obliterated by its continued insertion on the tree of the living objects? Should one appeal to God's intervention? If so, does it clash with the scientific explanation?

3. Creation

The opposition to the theory of evolution illustrated by certain believers bears on this point: do we have to renounce continuity, in the name of the theology of creation? For unbelievers, it is symmetrically the opposite: is it possible to believe in creation, when one sees the continuity which exists in the tree of life? In order to be able to give an adequate answer, we must consider the concept of creation, which, unfortunately, is ambiguous: it is a source of misunderstandings, which originate in the vision of the world of those who use it.

3.1 The mechanistic vision of creation

Today, the usual meaning of the creation refers to the beginning of the world. For ordinary people, the creative act consists in the production of the various elements which constitute the world and their arrangement into an optimal, well ordered system, like a well conceived and well regulated machine. God has given an initial impulse to this mechanism and set it in motion, and ever since, everything happens according to the laws of nature. The act of creation proper is limited to this single first instant. It is in this sense that the first words of Genesis are understood, when translated by "In the beginning", whereas the Hebrew term *berechit* and the Greek term *archè* have a richer metaphysical dimension.

Such a conception of creation seems to me to be at the root of all the present difficulties, either to base materialism, or to motivate the rejection of the theory of evolution by certain believers. In order to find a way out, we must try to draw a line between "beginning" and "origin".

3.2 Beginning and origin

1. The term: "beginning" designates the very first instant in the duration of a phenomenon, contrary to the word "origin", which refers to ontological or metaphysical realities. The origin deals with the constituent condition of all that appears in the course of an event. It refers to an act which cannot be isolated inside the series of events which form the web of transformations. It is not objectively verifiable through a measuring system or through a physical-chemical representation.

To speak about the origin, is to speak about an act which is not limited to one specific moment in time. The origin must not be mistaken for the beginning. To speak about the beginning is to acknowledge something new and therefore to distinguish between what was before and what is now. It means establishing a difference in time, conceived as an ongoing, continuous experience.

The origin does not confine us to one single instant in time. It is not one occurrence among others, but it is the constituting condition of all that is, at every moment, in the course of events which have happened in the time-space continuum. The origin cannot be adequately circumscribed to a scientific theory.

This distinction between origin and beginning is quite clear, on the conceptual level. Unfortunately, such is not the case for many people, because everyday language confuses the two terms. There is a reason for this: the origin seems to be clearer in the beginning, since it is the moment when "being" begins. The use of the word is ambiguous, and therefore one should be very careful, and aware of the fact that people often use one word for the other.

2. The search for the origin is beyond the scope of the scientific method. It rather pertains to the philosopher's field. The philosopher wonders about new occurrences in a different framework of thought. Science only takes into consideration the transformations of energy. The philosopher wonders about the reason why such transformations exist. He asks questions about the very existence of life and the universe. His approach is of a different order than that of the scientist.

a. Is such a question legitimate? It could be said that it reaches beyond the limits of the human mind and therefore is unable to access a reliable form of knowledge. One could also consider that it does make sense, and that trying to find an answer to it is legitimate.

b. In such a case, the interrogation can go on: what is the reason for the existence of living objects? Not such or such a one — but the sheer reason for just being there? One takes then a metaphysical road.

Such a question could be answered by saying that life is self-sufficient and that it is its own *raison d'être*. It exists by itself, is in itself its own justification. The universe then becomes an absolute and words like Life and Nature are written with capital initial letters.

One could observe, on the other hand, that the course of the history of life is of such a nature, that its subject cannot be made into an absolute: it must therefore be referred to something other than itself, which alone is an absolute, which lies beyond and outside the space-time continuum and is transcendent. Philosophers call him God, using a language common to the philosophical heritage as illustrated by Plato and Aristotle, and to monotheistic religions.

c. In such a perspective, one acknowledges that the existence of the universe is due to the action of God. Philosophers also use the term "creation" to designate the act through which God calls into being the universe, which is other than himself: such a gift is coextensive to the duration of the course of ages, as experienced by us in the present time.

d. The word "intervention" seems inappropriate. Its use is responsible for the current difficulties, which are three in number. First, the awkwardness of the discourse held by many Christians (among whom many priests) who use this word without explaining it; secondly, the agnostic or materialistic rejection of all references to a causality which unbelievers place at the same ontological level; lastly, the deficiencies of apologetic, which resorts to ignorance in order to appeal to a special action of God.

To get round those ambiguities, one should now try to clarify the way in which the orders of action assemble themselves in a process which implies several actors. This point is essential to show not only that creation and evolution can be reconciled, but that it is imperative to conciliate them.

3.3. Cooperation and synergy

The distinction I have introduced between beginning and origin enables us to understand why science and metaphysics, here complying with monotheistic theology, can allow for their differences and, at the same time, not ignore each other.

1. Creation lies at a different level of causality than the act of transformation. It is therefore important to link up the orders of causality. The reference to the action of God during the process should not be understood as an interaction. In an interaction in physics, chemistry or biology, the active elements are of the same order and work together, for just as in classical mechanics, forces combine with one another. Only the action of God is able to join some other action without modifying it and, *a fortiori*, without altering it.

The classical image which makes it possible to understand this is that of music. In a piece of music, everything is produced by the instrument and everything is produced by the musician who plays that instrument. All that is heard, is the result of both the instrument and the musician, without it being possible to separate what belongs to one or the other. Such a union, which in no way alters the principles of action (the musician and his instrument), is made possible because there is a difference of nature between those who act as musicians, and the instruments. This image enables us to understand what takes place in life, which is, at the same time, the result of the action of factors studied by the sciences of nature and anthropology, and of a transcendent principle. Because this principle is of a different order, and because it does not intervene as an element of the world, one must acknowledge that everything is of God, and everything is of nature. By reason of his transcendence, God can act without distorting the laws of nature. The usual difficulties that arise on the subject are caused by the fact that we think of God as an architect who transforms a given reality. God's action consists in giving to beings the ability to be what they are.

2. The notion of creation then appears in its metaphysical dimension, as a reply to the question of the origin. It is in no way a means of getting round the inadequacies of scientific knowledge. On the contrary, it receives support from that knowledge. It witnesses the temporal development of an action carried by a being who is distinct from others. Thus supported by the theory of evolution, by a judgment which has to do with philosophy, the human mind acknowledges in the course of life the action of a principle which addresses the question of the origin. The notion of creation, thus analyzed, does not come under the heading of the search for a beginning, but is a matter for appraisal for the present. Creation is not an act of the past; it is the current relation between the Creator with all that exists. Natural theology had fostered a conception of God as one factor among others — the first one.

To say that everything is of nature, and everything is of God, is a liberating experience, since the two species of causality are not exclusive one of the other. The action of God consists in giving to beings the ability of being what they are. By bestowing being upon them, God gives them the ability to act according to their own nature — let us remember that the word nature is not static.

3. The assertion of God's action in the course of evolution no longer looks like it may have looked in the positivistic age. It was looked upon as a violence, or as a trick played by the Almighty, apparently bent on the realization of some purpose outside nature. What has been said shows, on the contrary, that the action of God is the realization of nature's wishes. The creating action takes everything to its culminating point. The theory of evolution shows in what area the creative action develops. It opens on the question of meaning.

4. The meaning of evolution

The question of finality is at the heart of the debates between the theology of creation and the theory of evolution. Is it possible to address it? In order to do so, one must draw a clear line between teleonomy and teleology. As a matter of fact, ever since mathematics has been used in science, the Aristotelian notion of finality has needeed to be made more accurate by differentiating between "finality" in the full sense of the word, i.e "teleology", and "finality" used in a restricted sense, i.e "teleonomy".

4.1. The levels of being

The term "teleonomy" is used in biology; it refers to the construction of life in its autonomy and its unity. For clarity's sake, it is important to specify that the use of this word does not have the same scope, depending on how scientists look at their work. In order to make things easier to understand, one should underline the difference that exists between various conceptions of life — starting with what is privileged in scientific studies: the gene, the individual, the group or life considered as a whole.

1. The first option consists in laying stress on the gene — it is the thesis defended by Richard Dawkins. Since the individual who reproduces himself is defined and characterized by his genes, that author considers genes as the main actors in the history of life. It follows that organisms are there in order to allow genes to carry on their journey through time, by passing from one generation to another. The aim of each generation is to optimize the abilities of genes which gather together, in order to have the best chances to deal with the difficulties and constraints caused by the environment. The organism is the instrument of this struggle for life, whose subject is the gene.

2. Another option, diametrically opposite, considers a group, or a population. In which case, the behavior of the individual is subservient to the group. He takes advantage of his aptitudes for the benefit of the group. As a matter of fact, closely-knit societies, or those which are best integrated, are more likely to resist aggressions, more likely to adapt and therefore evolve through ages. It may happen that the individual does not develop all his possibilities, abdicating them for the advantage of the group. This can be observed in the behavior of animals which live in collectivities, like the colonies of bees, where the workers curb their potential activities.

3. Between these extreme positions, others favor the individual, considered as a strong unit integrating elements which are the constituents of life at all levels, starting with the biochemical elements, down to behavioral dispositions. Selection deals with individuals, not their genes or the group to which they belong.

This diversity of views originates in an option which is philosophical in nature: if science places itself in the perspective of a study of functions, and if it is legitimate to favor such or such function, it is important to acknowledge that such a choice is linked to a philosophical approach, therefore to political, social, or moral options. The Christian notion of "person" can then legitimately come into the debate.

4.2. A transcendent Creator

Within the scope of the metaphysical option, which acknowledges the existence of a transcendent creative principle, the use of the concept of "teleology" is welcome. The notion of creation is in harmony with the results of science to suggest a divine design. Such an expression has the merit of acknowledging the importance of evolution in the history of life on planet Earth, and opens up other perspectives, like extra-terrestrial life, biogenesis coextensive to cosmogenesis, etc.

It is important to observe that acknowledging the transcendence of the creative principle implies two assertions: in the first place, because this principle is transcendent, it does not distort the laws of nature which it posits. In the second place, the acknowledgement of the action of the Creator is conveyed in a philosophical language, which transposes what is understood by human action. This conceptualization when it occurs does not leave the concepts in the same register of language. The use of terms coming from human experience needs to be amended. The words are legitimate, but their use must not lead to ignore the fact that language must host moments of negation. To say that God is the Creator implies that one leans on the human experience of production, but at the same time one admits that such a production is of a different order: the "starting from nothing" (*ex nihilo*) of the traditional formula is that negative moment. Similarly, to speak of a divine design, according to the image of an architect's work, needs to be corrected by negations bearing on the modalities of the action. The meaning of transcendence necessarily implies the reference to a negation, which is present at the very heart of the assertion.

The apologetics which uses terms describing a divine intervention to make up for the inadequacies of the scientific theory, and the intense religious considerations share in the same misconception: they misjudge the status of the theological language.

5. The design of God

At the end of this critical survey, we have reached a crucial point: the road to reason is adventurous, because the field of human knowledge is endlessly progressing.

5.1. The starting point for reflection

Experiencing the world through the senses does not simply consist in an immediate capture of the world. It opens up to a comprehensive perception of events and individual facts, which are part of a global vision. The observation of life does not consist in just storing up curiosities, it encourages the observer to build up the image of the great tree of life where branches, orders, families, species and varieties spread out, according to the structural order of their aptitudes. A history then appears, where mankind occupies a singular place. Thus life appears like a successful combination, arousing a feeling of admiration, perhaps even of wonder. This is the point where our reflection takes its root.

Each individual endeavors to reach perfection by giving to his potential an optimum actualization. This could be called "first love". Or one could more simply use the word "dynamism", in accordance with an analogy borrowed from the world of physics, but also present at every ontological level. For a scientific observer, life is a tension towards the best realization of each and every one. The tree of evolution illustrates the ascent towards a greater internal unity, and its widening out emphasizes the spreading display of varied forms.

Such a dynamism is uppermost in the minds of human beings. It then takes on a new name, and is called "desire". The word has been used by medical doctors and psychologists, bent on the study of the physical and psychological reality of human beings. The roots of desire reach far beyond the level of clear consciousness. Such a desire opens up and flourishes in language, which is a structuring element of the human species, since it is at the same time the reflection, and the engine of the process of humanization. Language is a reflection, because it establishes the irreducible difference from other animals; it is an engine, because it produces the culture where human nature reaches its full realization. Language is then the engine of thought, where what is expressed becomes ideas.

Mankind thus belongs to a long history. Its apparition is not, as traditionalists believe, the infusion of an immortal soul into a pre-human body, but rather a global development, since the ancestors of mankind — whose history is told through an extension of the customs of the populations close to *homo sapiens* — also have their own cultural elements. Mankind is thus placed inside a long history which encompasses it and, by comparison, its specific nature reveals itself as what should strictly be described as "the emergence of the soul". Evolution, therefore, is also a "history of the soul", it does not deal only with matter — as the spiritual dualism used by the apologetic unfortunately claims.

5.2. The history of salvation and evolution

Theologians illustrate these views by resorting to two essential categories. The first category lies at the heart of the Gospel, it deals with the concept of realization; the second one is right at the center of the theological renewal consecrated by the Second Vatican Council, the history of salvation.

1. One can speak of "realization", when what was latent and concealed, in a rough state or in a state of great dispersion, becomes unified. It is, in the full meaning of the word, a recapitulation. The word, as a matter of fact, comes from the art of rhetoric. There, the orator, after having set out the various elements of the case, or of the issue, he is addressing, proceeds with giving a synthesis of his talk. The audience have in mind the elements, they have followed the rigour of the presentation. Recapitulation consists in repeating the argument in a global formula, which is all the clearer as it is brief. It can therefore be said that the human being, positioned as he is on the great tree of life, recapitulates it; which means that all that is found among other living objects is found in him, but that these elements have all changed, because they are part of his unity. The human being is not only at the top of the tree of life, as was claimed in Darwin's days; he is not only the ultimate point, as Bergson or Teilhard de Chardin believed; we see him as the recapitulation of all that can be found in the diversity of living objects. Such a recapitulation is not only a juxtaposition: it is an accomplishment, in the sense that all that is taken up has a dimension which verges on infinity.

2. The theological notion of "history of salvation" can be useful here. It takes root in the Biblical Texts. In the elaboration of such a vast corpus, one can see how the history of a kingdom centered around Jerusalem has widened to include first

the cultural journey from Egypt to Mesopotamia revolving on the Law, then the history of a large crowd of people obsessed with the figures of the patriarchs Abraham or Jacob, and also the destiny of mankind as a whole, captured in its patriarchs or emblematic figures, like Adam, Cain, Noah, thus laying the foundations of a universal history of mankind. This unity operates through the acknowledgement of an action of God, the unique, transcendent master of time and worlds. Can this notion of "history of salvation" — which is not, in any way, a sacred history, but the history of mankind falling the prey to every possible distress as the reverse side of its grandeur — can this notion be used in the perspective of science enlightened by the theory of evolution? Yes, it can! Because every beginning is linked to what has come before, to what has conditioned or caused it to happen. So that, the millions of years of the history of life can be seen in the eyes of faith, and be made into a history led by God, realizing His design.

3. In this perspective, it is possible to consider the place occupied by human beings as eminent, a place which can be described as pivotal. It is not static, as in the spiritual tradition, which locates the spirit outside time, but dynamic. The word "pivotal" refers to a direction, the determination of which depends on intellectual choices. If one favors such or such other principle, the direction of the axis will vary. The Christian faith leads one to opt for the pivotal direction of mankind. The presence of mankind has in times past been shown in biblical texts written some three thousand years ago, in a context completely different from today's context; it is obviously linked to the justification of Neolithic culture which arose in the Near East. It is impossible to adopt it in a literal sense as the fundamentalists insist on doing. But it can be revisited. On the one hand, it is quite clear that human beings are the most complex creatures among all others; on the other hand, they are capable of reflection and their nature is a cultural nature. Thus, scientific knowledge is not denied; on the contrary, it is made legitimate, since it is a decisive instrument in answering the questions emblematically inscribed by Gauguin at the end of his quest for absolute happiness: "Who are we? Whence do we come? Whither are we going?" or, in philosophical terms, what are the origin, and the end.

5.3. The grandeur of man

This global perspective invites us to re-examine our understanding of God's design. By placing the human being at the center of language and communication in human societies, speech has become the unquestionable objective criterion to state what the essence of man is. Speech is then given to be heard in its full richness. It is not only an instrument to convey or transfer information — like animal languages: it establishes a relationship with the other, it is a call, an act of listening, which structures the subject. A specific dimension opens up. The human being is not only a force in the process of evolution according to its own dynamism. He is a responding being. He is a responsible being, not only in the moral or legal sense of the word, but, more radically, at the ontological level.

It then appears, for the believer, that being human means answering the call of the other. The "other" can be him who manifests himself through the Alliance, according to the famous expression used by Pascal in his *Memorial*, "the God of Abraham, Isaac and Jacob, not the God of the philosophers and scientists", the God who calls and to whom the believer answers. To enter this perspective is not substantiating a fixed vision of nature, but on the contrary realizing that it is better understood when inscribed in a relation where the call is the decisive moment. The action of God is not seen as that of a dominating master sitting at the top of his eternity, seeing and foreseeing everything in an absolute, infallible manner, but on the contrary as that of one who respects the order he has created and which establishes a nature. God is not one who uses cunning in order to reign over the world without the world knowing. God is the one who calls. He attracts us, appeals to us and his appeal creates an area of liberty. The better we reply to his appeal, the larger the area of liberty becomes. To be human, is to answer the call: the call of mankind, where solidarity, communication, responsibility are concerned. But also, the call of the Creator conveyed by human mediations in the course of history, therefore in a temporality which has a founding quality. All this enables us to acknowledge the reference to "a special action of God" (as Pope Benedict XVI recalled, in his address to the Pontifical Academy of Sciences, November 1st, 2008) without using the word "intervention", in order to respect the wisdom of God, who uses the laws of nature to do his work.

The notion of revelation is not thought of, as it is in the fixist theology, as an opposition, or exception, to the natural processes discovered by scientific reasoning, but as the realization of the fundamental urge for truth inscribed in human nature. The call of God is not part of a series of other calls, but the root of all calls. Fanaticism, sectarianism and dogmatism are the consequences of our forgetting this dimension.

Theological conclusion: The Holy Ghost as Creator

The notion of creation conveys those riches in the theological tradition, which asks questions about the whys and wherefores of the world. It is not enough for the believer to say that the world was created by God: he must know why God has created the world. The traditional answer runs against a tradition according to which creation is a necessity.

The monotheistic tradition challenges this perspective, which implies an ontological poverty of the divine being inscribed in the necessity of its evolution. The reason for the creating act is not the result of a need, but the product of his generosity. God does not need to create: He does so out of love. The verb to love is here understood as meaning to wish other people's welfare. Love is then the key word which illuminates life. It designates the energy which lies at the root of the great tree of life. With an even stronger energy, the Christian tradition challenges the gnostic, or Manichaean temptation, which sees in the material world the result of a fall.

The God of the Christian faith is not that of deism, not even that of theism. The traditional reading of the Bible has been able to find these words in the first verse of Genesis, where God acts through His word, and creates a world over which hovers His Spirit. The verb which conveys the action of the Spirit (the breath of God) designates the movement of a bird fluttering about (*merahephet*) over its nest (root: *rhp*): it is a protecting presence, not weighing heavily over it; it brings nourishment and is an incitement to flight. Thus, our vision of the tree of life is not one where we would try to introduce divine interventions, but on the contrary to look for its trace in the dynamism of life. The Spirit of God, who is God himself, is at work in a spiritual manner. He animates, stimulates, educates, protects, guards, vivifies.

The attention paid to the experience of the Spirit by Christians since Pentecost is usually reserved for mystical studies. But the personal action of the Spirit is not limited to mystics. I think it can be extended to the whole creation, and more particularly to the living. It is thus appropriate to develop an important element of Christian theology: the personal role played by the Holy Ghost in creation, and in leading the history of salvation towards its full realization.

Creationism, intelligent design and evolution. A theological perspective

Jacques Arnould op

1. Theologians wanted!

The Creationist Crisis has crossed the Atlantic and is now threatening Europe. Although principally an American phenomenon that sprang up more than a century ago as a reaction to the work of Charles Darwin and the dissemination of his ideas, in Europe too it has now become more than just a problem discussed by a few Bible groups, based on American texts translated into local languages. The issue has many ramifications, whether religious, political or ideological, and seems already to be of sufficient importance to worry the political, moral and religious authorities of contemporary Europe while continuing to gain support every day. The first volume of *The Atlas of Creation* caused quite a stir when it was widely distributed to European schools, as well as to scientists, the media, and other influential figures in 2007. It had the effect of an ideological bombshell, leaving no one in any doubt about the existence of a crisis and (why shy away from the word?) a crusade, or indeed several crusades right here in Europe.

Epistemologists, philosophers of science and sociologists (Numbers 2006; Lecourt 1992) all warn against the danger of failing to distinguish between the movements that once existed, still exist today or are emerging, and of tarring them all with the same creationist brush; the different attitudes they each display to studies, findings and the scientific method should help us avoid any such confusion. Without wanting to descend into caricature, and without making the slightest judgement or still less condemning anyone, I think we can distinguish between three separate stages representing three distinct types of movement opposing the theory of evolution as it has developed from the work of Darwin (Arnould 2007). First there are old-fashioned creationists who adamantly refuse to consider any science that does not agree with the Biblical or Koranic canon; "Nothing but The Bible" or "Nothing but The Koran" might be their slogans. Then there are modern creationists who are no more tolerant of evolution or Darwinian and Neo-Darwinian theory than the former but who have invented what they call "creation science" that they would like to see taught on an equal footing (under what is known as "balanced treatment") with the science of evolution; to me this is simply a *falsification* of science. Finally there is *Intelligent Design* that accepts the idea of evolution but explains that "certain features of the universe and of living things are best explained by an intelligent cause, not an undirected process such as natural selection." Some think this could or even should be called neo-creationism, a position that merits debate as the movement does not insist on the existence of a God as creator; personally, I regard intelligent design as an erroneous and unjustified alternative to current theories of evolution.

One of the arguments most often used by opponents of creationism and intelligent design is the scientific weakness of these positions: the *impact factor* (by which is understood the number of citations or references made to an article) seems to support this; an epistemological analysis shows it clearly. They also point out the number of court rulings that have judged these movements to be religious and not scientific, after hearing philosophers, historians or sociologists of science; and in our secular societies, religion is deemed to have no place in biology teaching.

However, just as Francisco J. Ayala can speak of "Darwin's gift to science and religion", I think it is only fair to talk of the gift made by creationism and intelligent design to science and religion. This does not mean that I intend to defend these movements blindly, but to stress the way they have given science and religion an opportunity, an invitation, indeed an obligation to define their content better, as well as the limits to their scope and their authority. No one in the scientific community would deny that notions of causality and finality, chance and necessity, hierarchy and progress, are complicated notions to manipulate, especially where studying and teaching the life sciences are concerned. Science must also be careful to avoid becoming so dogmatic as to exacerbate the reaction of certain religious bodies. A symposium in Paris, organised by the French Ministry of Education in November 2008, clearly identified the challenges that must be accepted today if science teaching is to progress as it should.

In the same way, I am convinced that the movements supporting creationism and intelligent design can be a gift to Jewish, Christian and Muslim thinkers (I shall only deal with Christianity in this paper), as long as they accept the need to submit their ideas to cool, careful theological analysis. Although, as a Christian, I must first see most creationists, or believers in intelligent design, as brothers or sisters in Christ, I also have a duty not to let myself be blinded by facile or erroneous statements or arguments concerning theology. I must therefore plead here in favor of more frequent recourse to and input from theologians, in debates about the Creationist Crisis, not only in order to develop new aspects of theology concerning creation, as proposed by the other contributors to this session, but also to undertake a real theological diagnosis of this crisis.

2. What is theology?

It is not widely known (or is easily forgotten) that the origins of theology (etymologically: the discussion of the divine) go back to the Greek philosophical tradition; Plato mentions it in his *Republic*, when discoursing on the value of mythology in education. Even after the arrival of Christianity, theology retained its link with those great stories that are in fact the myths by which Mankind addresses and tries to answer the great mysteries of his condition on Earth; and creation theology is a very good illustration of this. While it must proclaim God as Creator, as the Apostle says "ready always to give answer to every man that asketh you a reason concerning the hope that is in you" (1 Pet. 3: 15), theology must also provide answers to men and women in our own time who wonder about the origin, meaning and purpose of human existence, in a cultural context where myths about creation from a Cosmic Egg or the Genesis story have been replaced by the Big Bang or Lucy, the Australopithecine.

The Holy Scriptures are our primary source on which to build a coherent structure for a theology whose purpose is to explain Christian faith, to speak of God and to link all things to God; there can therefore be no question of ignoring or rejecting them — we should rather be constantly trying to make them more accessible. Over the centuries, the faithful have always been encouraged to read the Scriptures not only with their eyes and their hearts but also with their intelligence. In his *Address to the World of Culture* in Paris on 12 September 2008, Pope Benedict XVI quoted Augustine of Dacia: "The letter indicates the facts; what you have to believe is indicated by allegory." Although the Magisterium of the Catholic Church refused to allow Père Lagrange, the founder of the Jerusalem Bible School, to publish his exegesis on the Book of Genesis, and, during the Modernist Crisis, denounced everything that seemed too closely associated with science as the "poison of error", the Second Vatican Council and its Constitution *Dei Verbum* finally brought the Catholic Church round to acknowledging, not only that literature could exist in a variety of *genres*, but also that it was important to record accurately the popular ways in which hagiography was experienced, spoken of or expressed in earlier times, as well as the modes of expression used in different places in normal human relationships in those days. Knowledge of text and context (for example, through historical criticism), literary and stylistic analysis, archaeological and historical research, semiotics, narrative analysis: contemporary exegesis uses a multitude of techniques to interpret texts that are venerable in every sense, and which themselves instruct the faithful to "open thy mouth and eat that which I give thee" (Ezek. 2: 8–10), thus recognising the importance of thoroughly digesting their contents. Even if the Biblical accounts of the Creation continue today as they have done in the past to be a source of friction, disagreement and even conflict between religious and scientific visions of the world, its origin and its evolution, it would be wrong to try and solve the problem by ignoring all references to the Scriptures or, on the contrary, by insisting on strict literalism and concordism. If non-believers are invited to respect these writings because of their roots in *mythology* (even if a believer sees much more in them than that), believers should undertake an intelligent and reasoned analysis of what they consider to be holy writings, in the very name of *theology*. Christians need not go so far as to abandon the Genesis story, Adam and Eve, the accounts of the Fall or the Flood, as a way of relating the creation of the world and of mankind; it is simply necessary that they maintain a theological perspective (and therefore to some degree a mythological one), without attempting to be in the least scientific. The words of Augustine of Hippo remain relevant today, when he wrote in his *On Genesis*: "Since divine revelation leaves many things obscure, we can arrive at various conclusions by interpreting them and adopting an opinion on an obscure issue; we must therefore be prepared to abandon them if discussion shows them to be contrary to truth ... Those who are not Christian can acquire certain notions about the natural sciences through reason and experience ... If non-believers can therefore demonstrate, with sufficient proof, the truth of a fact, it is our duty to show that such a fact is not contrary to Scripture; we must use reason, discussion and experience, insisting only on convincing proofs and so cling to our faith that we will not be led astray by the glib talk of false philosophy or frightened by the superstition of false religion." (Augustine 2002, chaps. 18, 19 and 21) It is true that the intellectual position called for by St Augustine is not easy to adopt, but it is nonetheless the condition for a healthy relationship between the different branches of knowledge that are science, philosophy and theology.

Although they remain the most important, the Scriptures are not our only theological reference; tradition is another. Nowadays, we and our contemporaries have a somewhat ambiguous relationship with the idea and reality of tradition: this can vary from rejection, pure and simple, to a blind acceptance of authority with the argument that whatever has been believed, thought and practised for centuries must necessarily be true. A more nuanced approach is needed. Tradition offers a way of correlating the past with the present, providing theologian and believer alike with both a reality principle touchstone and an experience of freedom within a set of possibilities and restrictions supplied by scriptural references, elements of dogma and the everyday experience of the faithful. It is in this respect, and this respect alone, that traditions may be called "living", not when they are merely used as the basis or justification for a particular position or set of rules. In this field a heavy responsibility is borne by the Magisterium, the teaching authority of the Church, which means different things to different Christian denominations. A particularly delicate subject, calling on the Magisterium's dual tasks: "to ensure that the faith of the Christian community remains ever fresh and relevant, while defending the freedom of the faith against the institution's own logic and justification", in the words of Christian Duquoc, a French theologian. Duquoc also adds: "the Magisterium is not the primary motor of the faith's vitality, which remains the intensity of the Christian communities." (Duquoc 1996, 1061) Along with the theologians and the Magisterium, the faithful are indeed the third side of a triangular dynamic that is essential to the vitality of Christianity; their common sense (sometimes called *sensus fidelium*) and experience are an essential ingredient of the theological process. For example, there is nothing to stop the theologian from developing a theology of original sin setting up a meta-historical Adam existing in a sort of parallel world until the moment of his first and primordial disobedience. But what would be the likelihood of such a construction, close to Gnosticism, finding favor with Christians who live in an imperfect world, confronted with an everyday struggle against evil, whose experience of sin and desire for salvation are a million miles from that image of Adam, so distant from their condition as humans?

This shows more clearly the challenges and even the risks faced by theology and theologians. Their task is not simply to write catechisms for children, or *Summae* (like St Thomas), as was the case at a time when these activities were an important part of ecclesiastical duties in a culture that had a high regard for compendia of all kinds; it is much more to do with incompleteness, light and shade. When the theologian Karl Rahner was asked what dogma was, he answered along the lines of: "Dogma is like a streetlamp at night. Some will use it to explore the shadows and the dark, to go into the unknown without fear of getting lost; others, like the drunkard, will use it as support to hold onto and stop them falling!" Theologians are in the first group: they base their work on the Scriptures and on tradition and submit their conclusions to be judged and validated by the Magisterium and the faithful, the better to navigate the obscurity of the questions and situations of our modern world, the jungle of opinions and knowledge, while practising the virtues of audacity and modesty. Which is why their conclusions can be so different: the theological tradition is anything but monolithic; it is carried out in a historically-determined multiplicity of methods, languages and practices; theologies can be Biblical, patristic, scholastic, monastic, mystical, narrative, political, liberationist, feminist, and so on. The only thing that makes theology a coherent whole is the articulated plurality of all these theological expressions.

3. The age of secular theologians

Since they deal in light and shade, does this mean that theologians do not seek evidence? Of course not. They are descendants of "Doubting Thomas", the Apostle who asked out loud for the proof of the Resurrection of Christ that so many believers after him would have liked: "Except I shall see in his hands the print of the nails, and put my finger into the print of the nails, and thrust my hand into his side, I will not believe!" (John 20: 25) Humans need to test facts, to experience reality, to handle the evidence, whether they be theologians or research scientists.

Indeed, history is full of examples illustrating this very human characteristic, moments when scientists came to think of themselves as wise men, over-confident about the extent to which their theories matched reality. This was the case at the end of the nineteenth century, just before the revolution in physics that would mark the beginning of the age of relativity; it was also the case in the seventeenth century, when Galileo sparked a revolution that was no less important for science. From that time on, the laws of nature would not necessarily concur with our immediate experience: we might see the Sun rising and setting, as though it were going round the Earth, but it was really the Earth that was going round the Sun.

Amos Funkenstein, a professor at Berkeley who died in 1995, wrote a major study about "secular theology in the Modern Age" referring to those who did not intend their work for theology departments, nor get involved with questions of holy doctrine, but still managed to deal with classical theological issues: God, spirits and demons, salvation and the Eucharist. Funkenstein considered that Leibniz was an archetypal secular theologian (Funkenstein 1986) and asked what role such thinkers played in the emergence of modern science. The answer he gave was that they borrowed the methods and results of the theological imagination and applied them to the scientific imagination, thus opening up new horizons. He concluded that it was with a view to broadening the logical horizon of divine omnipotence that the theologians of the Middle Ages sought to distinguish between the possible and the impossible, between the idea of Almighty God and His ordinary, "reasonable" powers; they thus developed orders and states of nature different from ours, sometimes contrary to the facts but nonetheless logically possible. The secular theologians subsequently used these mediaeval, purely theological speculations to formulate laws of nature, in the sense used by modern science, as hypotheses that do not necessarily concur with our immediate experience of reality.

I should like to dwell a moment on one of the other examples studied by Funkenstein, Isaac de La Peyrère, who published two works in 1655 in which he invented and defended the concept of pre-Adamites, the idea that Adam and his descendants were late-comers in the history of mankind and not at all the first men (La Peyrère 1655a; followed in the same year by a smaller volume 1655b). This idea not only contradicts the usual reading of the Biblical story as told in the Book of Genesis, but threatens to overturn the providential vision of history and of an ordered world, compatible with the idea of an all-powerful God and strict determinism. There might be other histories than that of Israel and also other accounts of creation than the story of six days, so that a fossil would not need to be explained away as a *lusus naturae* or joke of nature. In short, La Peyrère referred to the Bible in such a way as to develop a counter-history, by which I mean a history that differed from the theological-scientific *vulgate* that was the official version at that time, based primarily on appearance or evidence.

Let us now leave it to the philosophers of science to continue down the path cleared by Funkenstein and elucidate the question of the relationship between science and reality, while we return to religion and theology: might not those who work in these fields be just as fascinated, indeed tempted, by the force of proof and the evidence of reality? I have already suggested an answer with the example of Thomas; I must now demonstrate what I mean.

4. The achievements and limitations of natural theology

Believers are just as fascinated by evidence drawn from reality as are scientists; how else can we explain the fascination with miracles or the never-ending quest for a demonstration of the existence of God — even if this is still a work-in-progress today? How else can we explain the shifting alliances as well as the walls and barriers between theologians, philosophers and scientists, erected and dismantled over the ages in accordance with intellectual and cultural movements and the swings in strength from one group to another?

Since the seventeenth century, scientists have kept their distance from the religious and theological approach; for some of them, it was a matter of life or death.

In his History of the Royal Society of London, published in 1667, Thomas Sprat explained that the members "meddle no otherwise with divine things, than only as the Power and Wisdom and Goodness of the Creator is displayed in the admirable Order and Workmanship of the Creatures. It cannot be denied, but it lies in the Natural Philosopher's hands best to advance that Part of Divinity; which, though it fills not the Mind with such tender and powerful Contemplations as that which shows us Man's Redemption by a Mediator, yet it is by no means to be passed by unregarded, but is an excellent Ground to establish the other … These two subjects, God and the Soul, being only forborn, in all the rest they wander at their Pleasure." (Sprat 1667, 81–83)

Theologians have not always demonstrated the same degree of prudence, nor maintained the same distance between their intellectual territory and that belonging to science: even after the seventeenth century, some continued to practise "natural theology", right up to our own time with its enthusiasm for intelligent design. Since Richard Dawkins' "Blind Watchmaker" became a best-seller, everyone has learned about the book entitled *Natural Theology* published by William Paley in 1802 that strongly influenced Charles Darwin as a young man. In it, Paley wrote that if a man found a watch on the beach and concluded it was the work of a designer but denied that there was design behind the rest of creation, then: "This is atheism: for every indication of contrivance, every manifestation of design, which existed in the watch, exists in the works of nature; with the difference, on the side of nature, of being greater and more, and that in a degree which exceeds all computation. I mean that the contrivances of nature surpass the contrivances of art, in the complexity, subtlety, and curiosity of the mechanism; and still more, if possible, do they go beyond them in number and variety; yet, in a multitude of cases, are not less evidently mechanical, not less evidently contrivances, not less evidently accommodated to their end, or suited to their office, than are the most perfect productions of human ingenuity." (Paley 1802) The biologist Michael Denton refers specifically to Paley in the introduction to his book *Nature's Destiny*: he proposes to write his own natural theology, proving that everything in evolution has happened and continues to happen as though the Creator had prepared a set of prefabricated parts, ready for assembly (Denton 1998). Denton has no doubt that the cosmos is suited to support life, not just up to a point but ideally, which he claims as proof of the existence of design. What should we think of this type of approach, which seems so close to the arguments for intelligent design?

To stay within the bounds of Christian theology, I find it instructive to examine the way Denton refers to the works of Paley and other natural theologians of the seventeenth, eighteenth and nineteenth centuries; he either seems or pretends to be ignorant of the modern and perhaps also the ancient history of this philosophical position and its associated literature. By mentioning and referring to Paley, Bridgewater and Bentley, while forgetting Thomas Aquinas, Luther and Barth he seems to be seeking a respectable theological cover for a position that does not deserve any such thing.

First Paley and later Darwin knew better than Denton where to set the limits of natural theology. In the passage quoted previously, Paley underlines the difference between the works of nature and those produced by the hand of man. Darwin too wonders whether it is possible to compare the works of nature with the works of man, but he seeks neither to support (as we might indeed expect) nor to disapprove of the possibility of believing in God (he leaves his readers free to choose). He starts by denouncing the tendency to attribute absolute perfection to the works of nature; he recalls "what Helmholtz, the highest authority in Europe, has said about the human eye; that if an optician had sold him an instrument so carelessly made, he would have thought himself fully justified in returning it." (Darwin 1871) He then stresses the limitations, indeed the dangers, involved in comparing too closely the products of nature and those produced by human skill. You wish to compare the eye and the telescope? By all means, he says, such curiosity is perfectly natural. But it would surely be presumptuous to conclude that the former could have taken shape in the same way as the latter, the product of "the continuous and lengthy application of the highest human intelligence". Such an analogy may hold up when it is a question of function, but not for its construction or development. This seems to be a sensible remark; it reminds me of the observation by Etienne Tempier, Bishop of Paris, who, to defend the idea of a possible plurality of worlds (including worlds like ours), explained that it was not for the human spirit to guess at the possible limits to the omnipotence of God the Creator, even if theology might try and persuade us that we are the only intelligent beings and that our world is at the centre of the universe; this was in 1277, when the geo-centric cosmology seemed unassailable. It would be a shame if the recent revival of interest in natural theology, handed down to us by the Greeks and Thomas Aquinas, should be hijacked by the supporters of intelligent design as a way of defending their position.

The reserve expressed here does not mean that the Christian and theological outlook can afford to abandon the methods of natural theology; history simply shows us that we need to be vigilant about the conditions in which it is exercised. We particularly need to ask ourselves (and perhaps ask those who defend creationist theories such as intelligent design) how natural theology can accommodate revelation. Believers can choose between two approaches, the one developed by Thomas Aquinas and adopted by Vatican I, or the one developed by Luther during the Heidelberg Disputation in 1518. "If anyone shall have said that the one true God, our Creator and our Lord, cannot be known with certitude by those things which have been made, by the natural light of human reason: let him be anathema", declared the Council in 1870. But Luther had said "That person does not deserve to be called a theologian who looks upon the invisible things of God as though they were clearly perceptible in those things which have actually happened. This is apparent in the example of those who were 'theologians' and still were called fools by the Apostle in Romans 1: 22. Furthermore, the invisible things of God are virtue, godliness, wisdom, justice, goodness, and so forth. The recognition of all these things does not make one worthy or wise." (Luther 1518, thes. 19) Two radically opposed positions which history has shown not to be the exclusive characteristics of either Catholicism on the one hand nor Protestantism on the other; since Luther's time, there have been many changes of position. I shall only consider the following two cases.

First of all, Luther's vehemence should not be taken as a condemnation of philosophy and the role of reason in belief. He may have been highly aware of the imperfections of Man the sinner, but he seems to have understood, or to have sensed intuitively, that natural theology as it had been practised until then was about to collapse under the influence of modern scientific methods, which is more important for our purpose. After Galileo, and also later, after Darwin, nature fell silent; it could no longer occupy the same favored position in the journey to finding God. It was not as though the link between philosophy (to which natural theology belongs) and Christian theology had actually been broken; it had simply become more fragile: the belief in nature as something stable, illustrated and defended in contemporary notions of the cosmos and of unchanging species[1], had to give way to the idea of a nature immersed in history. Christians had to surrender the symbolism of a cosmos that provided a reassuring framework for the only meaningful "history", involving salvation, and face up to the idea of a much more dramatic

[1] As Darwin saw it, those who held creation as a dogma, whom he sometimes called creationists, believed that species were immutable, each created separately. He wrote: "if I have erred in giving to natural selection great power, which I am far from admitting, or in having exaggerated its power, which is in itself probable, I have at least, as I hope, done good service in aiding to overthrow the dogma of separate creations." (Darwin 1871, vol. 1, 153)

reality. "Letting go of the harmonious spectacle of cosmic regularity as a mirror of divinity" writes Christian Duquoc, "means accepting the insecurity of the historic process as a hypothetical consequence of the will of God." (Duquoc 1983, 79) The theologian Wolfhart Pannenberg argues that contemplating the cosmos must give way to experiencing reality as history — without forgetting the importance in Christian thinking of the Word of God and the Scriptures, and while remembering all the precautions and restrictions mentioned previously. Luther's warning, repeated in the twentieth century by the Protestant theologian Karl Barth, should not lead to the total rejection of natural theology, but rather invite us to reconsider it in the context of the world as we experience it today, in other words in the light of so much new scientific knowledge and its constant progress.

The second case I propose to consider here may be difficult for believers to admit, whether Christian, Jewish or Muslim. Natural theology as practised by Denton (as the various intelligent design theorists say they do, either implicitly or explicitly) is not about developing a vision of God but a vision of Man. "It speaks of something missing at the centre of the human condition, not with a view to filling it but to indicating what it means about the existence of some Other that we are unable to identify", explains Duquoc again (1983, 84). Even so, natural theology is not reduced to the mere level of anthropology. In an essay inspired by the work of Eric Weil on Kant, Pierre Fruchon shows how anthropology and theology are related in the question of meaning: "for Mankind, meaning can only ultimately be found in God, but God is not meaning in Himself; He only provides meaning for Mankind, for the finite being who seeks and wonders, who finds meaning in what he perceives as the reality around him and man himself in search of an explanation: the fact that man is inherently a seeker-after-truth." (Fruchon 1976, 43) The lesson is obvious: meaning raises questions, it cannot be imposed; it can be tested *a posteriori*, but it cannot be proved. Contingency is not a threat (because it would be the opposite of meaning), but a favor, bringing together the conditions which allowed the appearance of life and therefore of mankind.

Reconsidered like this, in the light of the work of twentieth century theologians and philosophers, natural theology seems very different from the position of creationists who remain convinced that the latest scientific evidence, in both astronomy and biology, will help them convert non-believers or persuade doubters. The conclusion of Stephen Hawking's *A Brief History of Time*, published in 1988, seems just as inappropriate and pretentious: "However, if we do discover a complete theory, it should in time be understandable in broad principle by everyone, not just a few scientists. Then we shall all, philosophers, scientists and just ordinary people, be able to take part in the discussion of the question of why it is that we and the universe exist. If we find the answer to that, it would be the ultimate triumph of human reason — for then we would know the mind of God." (Hawking 1988, 175) Is Hawking one of those "extreme rationalists" about whom Albert Einstein spoke, those who claim that science can do anything, including develop a Grand Unified Theory or G-U-T? Such thinkers are to science what the partisans of intelligent design are to theology, like those who practiced natural theology before the modern era or those who think they can *prove* the existence of G-O-D (God). Hawking at least recognized, twenty years later, that he had been mistaken and admitted that the hope of one day finding a unified theory to explain our world was a delusion: "I am now content that our attempts to understand will never be complete and that we shall always need to make new discoveries" (Küng 2008, 30ss); scientific research is condemned to watch the limit of what is known constantly recede as it advances towards it, sometimes faster even than the progress of science itself.

In the end, there will be no battle between G-U-T and G-O-D. So much the better. No one will ever prove the existence of God. And in the same way, no one can hope to attain the uttermost shores of knowledge, and claim that an all-embracing science has defined nature.

5. Are there any scientific discoveries that are negative for theology?

I may have given the impression that, for the subject that interests us here (the relationship between science and religion) I am in favor of keeping these two areas of study completely apart and that I approve of the notion of *NonOverlapping MAgisteria* (NOMA) proposed by Stephen Jay Gould (1999). I certainly believe that, in crisis or conflict situations, the first thing to do is to separate the belligerents, set clear boundaries to their specializations and the extent of their authority, whatever the principle behind this choice (the distinction between the why and the how, secularism, NOMA or some other) as long as the objective is to facilitate subsequent dialogue: in their constant curiosity and searching-after-truth, humans cannot be content with a position that seems like schizophrenia, with answers that never coincide.

Although I strongly believe in dialogue, I am still fully conscious of a possible imbalance between the opposing camps of science and theology and remain doubtful about certain positions held mostly, indeed almost exclusively, by fellow-theologians (or at least by believers) and much more rarely by scientists —

about the possible convergence of the two points of view.[2] Believers may wish for it and may indeed try and achieve it, but scientists need to meet them half-way. In other words, it is no use trying to stake a claim to the exclusivity of scientific approval, as the creationists want to do; to me, it seems preferable to adopt the approach described at the end of the nineteenth century by Dalmace Leroy, a French Dominican, who wrote in the Preface to his work *L'évolution des espèces organiques* ("The Evolution of Organic Species"): "I believe that the same will happen with evolutionism as happened with Galileo's theory; it may frighten the orthodox at first, but when feelings have calmed down the truth will emerge, once shorn of all exaggeration on both sides. It may then not seem such a bad thing that a man of the church should have spoken well of it in advance. Let us render unto Caesar the things that are Caesar's so that we may then invite Caesar in turn to render unto God the things that are God's." (Leroy 1887, 10–11) It is worth noting that Dalmace Leroy speaks of inviting, not of obliging: he does not believe for one moment in the efficacy of militant proselytising, of appropriation for ideological ends, of forced convergence. Not all scientists are willing to hear a speech such as that given by Benedict XVI, at the *Collège des Bernardins* in Paris. The Pope concluded with the following words: "A purely positivist culture, relegating the subject of God to the purely private and non-scientific domain, would mark the capitulation of reason and the renouncement of its highest possibilities and would therefore be a blow to humanism, and would have serious consequences." I can understand the surprise and annoyance of fellow-scientists on hearing or reading these words: why and how are they expected to bring God into the field of science? Unless science is to be defined differently and given a different mission and other methods than those it has today, it should not be expected to take God into account nor to explain God, although scientists are clearly free to believe in God themselves. It is regrettable that Benedict XVI's magnificent initiative of extending an open hand to the world of culture should have caused such bitterness in the world of science, partly through disagreement over the terms used.

I am obliged to paint an even blacker picture (!) and abandon the ideological territory inhabited (or dreamt) by the creationists, the partisans of intelligent design and no doubt others besides. Let me explain: I do not believe that it is wise to expect *a priori* anything positive for theology from science, its inquiries and discov-

[2] The different kinds of relationship between science and faith can be classified into four types: conflict, independence, dialogue and integration (Barbour 1990; Karakash and Schäfer-Guignier 1992; Euvé 2000).

eries, its theories and pronouncements. Please note that I say *a priori*; positive aspects may be discovered *a posteriori*, bringing joy to believers and scientists alike, but that is not what interests me here. I am referring here to an idea proposed by Merleau-Ponty, in his lectures at the *Collège de France*. After first affirming (1968, 118) that, "by looking to science for answers, philosophy can acquire knowledge of how creatures actually function that it might never be able to understand any other way" Merleau-Ponty continues "Science tends to make, as certain physicists have said, 'negative philosophical discoveries' (London and Bauer)." (Merleau-Ponty 1968, 119) As an example, he cites "the scientific assessment of space-time forms in non-Euclidian geometries and the physics of relativity" which "teaches us to reject the common-sense appreciation of space and time which takes no account of the situation of the observer" (for we now know that such things are influenced by the fact of being observed) and which "prepares us to give their full ontological meaning to certain descriptions of space and time as perceived — space and time that are polymorphous, but for which neither science nor common sense can define more than a few traits." (Merleau-Ponty 1968, 120) There are other examples as well, in quantum mechanics or dynamics, in which one cannot describe nature from the outside, as a mere spectator, but in which the observer inevitably becomes a participant. In other words, Merleau-Ponty states that scientific discoveries with a philosophical dimension, those that bring about fundamental conceptual changes, often arise through negative discoveries, opportunities and starting points for reversals of perspective.

I believe that the same thing applies to theology: theologians need to admit that there can be such things as theologically negative discoveries, scientific discoveries that threaten to overturn their usual way of looking at things. Without directly challenging the message of the Scriptures nor that of the tradition to which they belong, they find themselves obliged to change the way in which they understand the world (that is to say the part that natural theology plays in their studies), as well as the way they express their thoughts about God, Mankind and the world (and, primarily, the way they incorporate mythology, narrative and history in their work).

6. Apocalypse

In October 2005, after a court case in Dover, Pennsylvania, brought by parents who refused to allow intelligent design to be taught to their children in science lessons, the Evangelical Pastor Pat Robertson broadcast the following warning on the *Christian Broadcasting Network*: "if there is a disaster in your area, don't turn to God, you just rejected Him from your city!" He later wrote: "God is tolerant and loving, but we can't keep sticking our finger in His eye forever. If they have future problems in Dover, I recommend they call on Charles Darwin. Maybe he can help them." The Pastor seemed almost to be calling down apocalyptic fire and brimstone on the recalcitrant citizens of Dover! It is a shame that the word apocalypse should have come to suggest a cataclysmic end to the world, thus losing its etymological meaning of "revelation". (In French, the Book of Revelations, the last Book of the Bible, is known as the *Apocalypse*.) It seems to me that the work of theologians should have apocalypse as its aim, in the sense of revelation. It is the honourable task of theologians to strive to answer questions that have not yet been asked, to try and formulate notions that did not even exist previously; to paraphrase the title of a famous lecture by Hans Jonas, it is and always will be vital to develop new notions of God after every event, fortunate or unfortunate, creative or destructive, predicted or unpredicted, beneficial or malign, that Mankind will undergo in the course of its history. These strong words from a German-born Jewish philosopher should inspire theologians of any religion that recognizes a "Creator God": "That God cares for his creatures is one of the most familiar tenets of the Jewish faith"; but "this caring God is not a magician who, by simply caring instantaneously causes the consequence that he wishes for: on the contrary, he has left room for action by other players in the scenario, so that his caring depends on their own actions. He is therefore also a god in danger, a god running a risk of his own." (Jonas 1994, 26–27) Such a God, concludes Jonas, cannot be all-powerful!

I think that the theological perspective, applied to what at the beginning of this paper I called the Creationist Crisis, must extend to this point or, better still, start out from this point. It must question the evidence for these statements, even if they come from the Scriptures and from tradition, not for the purpose of rejecting them, but to make the meaning more accessible where it has not yet been fully revealed. In both their research and their pronouncements, theologians must resign themselves to remaining in a state of incompletion. They must heed Rahner and his illustration of theology as being like a streetlamp, quoted above, and disprove Diderot when he says "Wandering in a vast forest at night, I have only a faint light to guide me. A stranger appears and says to me: 'My friend, you should blow out your candle in order to find your way more clearly.' This stranger is a theologian."

References

Arnould J., 2007, *Dieu versus Darwin. Les créationnistes vont-ils triompher de la science?*, Paris: Albin Michel.

Augustine of Hippo, 2002, *The Works of Saint Augustine. On Genesis*, New York: New City.

Barbour I., 1990, *Religion in an Age of Science*, San Francisco: Harper.

Denton M., 1998, *Nature's Destiny. How the Laws of Biology Reveal Purpose in the Universe*, New York: Free Press.

Darwin C., 1871, *The Descent of Man and Selection in Relation to Sex*, 2 vols., London: John Murray.

Duquoc C., 1983, "La théologie naturelle. Son enjeu dans le débat ouvert par la Réforme", *Lumière et Vie* 32, No. 161.

Duquoc C., 1996, "Théologie", Catholicisme, Paris: Letouzey et Ané.

Euvé F., 2000, *Penser la création comme jeu*, Paris: Cerf.

Fruchon P., 1976, *Existence humaine et révélation. Essais d'herméneutique*, Paris: Cerf.

Funkenstein A., 1986, *Theology and the Scientific Imagination from the Late Middle Ages to the XVIIth Century*, Princeton: Princeton University Press.

Gould S. J., 1999, *Rocks of Ages. Science and Religion in the Fullness of Life*, New York: Ballantine.

Hawking S. W., 1988, *A Brief History of Time. From the Big Bang to Black Holes*, London: Bantam Press.

Jonas H., 1994, *Le concept de dieu après Auschwitz. Une voix juive*, Paris: Editions Payot & Rivages.

Karakash C., Schäfer-Guignier O., 1992, "Typologie des articulations entre science et foi religieuse", in Bühler P., Karakash C. (eds.), 1992, *Science et foi font système. Une approche herméneutique*, Geneva: Labor et Fides.

Küng H., 2008, *Petit traité du commencement de toutes choses*, Paris: Seuil.

La Peyrère I. de, 1655a, *Præadamitæ sive exercitatio super versibus 12°, 13° et 14° capitis V Epistolæ D. Pauli ad Romanos.*

La Peyrère I. de, 1655b, *Systema theologicum ex præadamitarum hypothesi.*

Lecourt D., 1992, *L'Amérique entre la Bible et Darwin*, Paris: PUF, 2007.

Leroy D., 1887, *L'évolution des espèces organiques*, Paris: Perrin & Cie.

Luther M., 1518, *The Heidelberg Disputation.*

Merleau-Ponty M., 1968, "Le concept de nature" (1956-1957), in *Résumé de cours. Collège de France, 1952-1960*, Paris: Gallimard, 1968.

Numbers R. L., 2006, *The Creationists. From Scientific Creationism to Intelligent Design*, Cambridge/London: Harvard University Press.

Paley W., 1802, *Natural Theology or Evidences of the Existence and Attributes of the Deity, Collected from the Appearance of Nature*, London: Faulder.

Sprat T., 1667, *The History of the Royal Society of London, for the Improving of Natural Knowledge*, London.

The reception of evolutionary theories in the Church

Rafael A. Martínez

1. Catholic attitude on evolution

The study of the Catholic reaction to Darwinism was focused, until very recently, on a series of local controversies, most of them in the latter part of the nineteenth century (Stebbins 1988, Glick 1988a, Paul 1988, Engels and Glick 2009). One might guess, considering the centralized structure of the Catholic Church, that these local conflicts would announce an approaching crisis, similar to that of Galileo's time, or to the crisis in the Evangelical world in the 1920s. But the crisis never arrived. The last part of the nineteenth century was a period marked by strong doctrinal controversies, but issues related to evolution never appeared in official statements made by Roman authorities.

On the contrary, from the second or third decade of the twentieth century evolution appeared as more and more openly accepted by Catholic theologians. In 1932, Ernest Messenger published a long account of the state of the matter. It was centered on the evolution of human beings, which was in fact the main doctrinal point. Messenger's conclusion was that no opposition existed between Christianity and the scientific theory of evolution (Messenger 1932).

More recently, Pius XII and John Paul II publicly declared evolution as compatible with the Catholic faith.

In his Encyclical *Humani Generis*, Pius XII stated that

> the Authority of the Church does not forbid that ... research and discussions ... take place with regard to the doctrine of evolution (Pius XII 1995).

Evolution was, in his views, an acceptable hypothesis in the biological level. Almost fifty years after, on October 22, 1996, Pope John Paul II declared that the theory of evolution was considered today as more than a hypothesis .

> Today, almost half a century after the publication of the Encyclical, new knowledge has led to recognize that the theory of evolution is more than a hypothesis ... The convergence, neither sought nor fabricated, of the results

of work that was conducted independently is in itself a significant argument in favor of this theory (John Paul II 2003).

But, what has happened a century before?

1.1 The cultural context: the "conflict thesis"

The encounter between Darwinism and Catholic theology was marked by the tension between science and Christianity prevailing during the second half of the nineteenth century. The advances of the natural sciences together with archaeological discoveries of ancient cultures, were perceived as a threat to Christianity. Theological positions were attacked in the name of science. Publications hostile to Christianity in general and the Catholic Church in particular multiplied everywhere. This is the period when John William Draper published his *History of the Conflict between Religion and Science* (Draper 1874), and Andrew Dickson White, his more ambitious work titled *A History of the Warfare of Science with Theology in Christendom* (White 1896). These two books enjoyed great success, not only in the United States, but also within Europe and in Catholic countries. Both authors held that there was a permanent and inevitable conflict between science and theology.

In France, Louis Jacolliot argued that Christianity was no more than a variant, lacking any historical basis, of Indian myths, and presented his ideas as a consequence of scientific rigor: "A new world is emerging. Science, with its rigorous methods, has dealt a mortal blow to religious poetry and historical legends, and the day is approaching when only sensate, rational, and human phenomena will be believed" (Jacolliot 1913, 1).

In such an environment, the enthusiasm that evolutionism sparked, was frequently mixed with attacks on religion and, in the name of evolution, defense of agnosticism, atheism, materialism, and free thought.

An additional factor was that the arguments for the theory of evolution were at that time presented according to several competing theories. Julian Huxley later characterized this period as the "eclipse of Darwinism" because of the uncertainty surrounding natural selection. According to Huxley, the reaction against natural selection set in during the 1890s as Darwinism became more and more theoretical, with no apparent support from experimentation — "merely case books of real or supposed adaptations" (Huxley 1943, 23). Several anti-Darwinian theories were in open competition (Bowler 1983). The absence of a consensus created a climate propitious for Catholics to advance their own hypotheses, or to deny evolution.

All this helps to understand why Catholics frequently assumed a stance hostile to evolutionism, which they could picture as an instrument used by materialists to attack religion. The ideological baggage that evolution had acquired in some authors, made it an obvious target for believers to attack.

1.2 The reaction of Catholic theologians

Catholic theologians applied a severe critique to evolutionism, especially as it treated the origin of Adam's body. Some of them asserted that the direct divine creation of Adam's body was Catholic doctrine. Matthias Joseph Scheeben (1835–1888) considered an heresy to pretend that man, insofar as concerns his body, is descended from monkeys, even if accepting that God created the soul (Scheeben 1961, 160–61). The Jesuit theologian, Camillo Mazzella (1833–1900) affirmed that the direct creation of Adam's body by God, even though might not officially be a Church doctrine, could not be denied by Catholics, as the contrary would constitute a "rash" doctrine lacking a solid basis (Mazzella 1896, 343–74). There were, among Catholic writers, many that considered it impossible to reconcile evolution and doctrine.

Opposition to evolution was commonly grounded in a too literal interpretation of Sacred Scripture and Church Fathers, and on theological reflection, based mainly in Aquinas' philosophical principles. But it was not easy to find direct support in authoritative declarations of the Church. It was possible to quote some local decisions, such as a provincial Council held in Cologne in 1860, but these interventions were not enough to make evolution a thesis opposed to Catholic doctrine. A declaration of Roman authorities, the Pope or at least some authoritative Roman Congregation, would be necessary. But the only references to actions of the Holy See against evolutionism were very brief: a few Catholic authors were mentioned as a proof that the authorities of the Holy See were contrary to evolution.

A typical case is that of the Jesuit Christian Pesch who wrote:

> The opinion rejected here had already been reproached several times by the Roman authorities. In 1891 Leroy published a book [he cites it] in which he defended the opinion of Doctor Mivart. But he went to Rome in 1895 "to receive a warning" [*ad audiendum verbum*], was ordered to retract his views, and he did so ... Some years later Zahm wrote a book in which again he defended Doctor Mivart's opinion as probable. But in 1899, he too was ordered by the Congregation of the Holy Office to withhold his book from sale. Therefore, it is clear that the Congregation of the Holy Office opposes this opinion (Pesch 1908, 82).

Very similar reports were offered by other theologians (Tanquerey 1911, Beraza 1921, Boyer 1940, Rahner 1953, Parente 1959, Alszeghy 1967, Juste 1965, Ruiz de la Peña 1996), but in this account there were factual errors. It was known that Rome had intervened on some occasions, but the exact picture was enveloped in darkness. There were no public documents, but almost all the available data came from a series of commentaries by Salvatore Brandi in *La Civiltà Cattolica*, a journal of the Roman Jesuits which, without being an official publication of the Vatican, has always had a special relationship with the Holy See.

2. Evolution at the Index

Further clarification has been impossible until very recently. Only when in 1998 the Vatican opened the archives of the CDF (Congregation for the Doctrine of Faith), which contains the archives of the old Congregations of the Holy Office and of the Index, it became possible to gain access to new information. Along with Mariano Artigas (deceased in December 2006) and Thomas Glick, we have tried to reconstruct in detail the attitude of the Vatican authorities (Artigas, Glick and Martínez 2006). Here we will try to give a general view of the main conclusions.

The research covers roughly the pontificate of Leo XIII (1878–1903). A few books dealing with evolution, written by non-catholic or anti-catholic authors, had already been condemned by the Congregation of the Index. These were the cases of Louis Jacolliot's *La genèse de l'humanité*, condemned by a Decree of the Sacred Congregation of the Index on March 7, 1876, and John William Draper's *History of the Conflict between Religion and Science*, on September 4, 1876. In the next few years the Decrees of the Congregation would also condemn, among others, Nicola Marselli's *The Origin of Humankind* (June 27, 1881), Pietro Siciliani's works (April 3, 1882), Émile Ferrière's *Le Darwinisme* and *The Scientific Errors of the Bible* (April 7, 1892), and Odón de Buen treatises on Geology and Zoology (June 1, 1895). Only the last case had to do properly with scientific texts (Odón de Buen attitude was, however, very anti religious). In all these cases the reason of that decision was not just evolution, but the general materialistic and sometimes anti-catholic views, and both questions are not easily distinguishable.

A Vatican attitude about evolution could much more easily be perceived when Catholic authors, trying to combine Christian doctrine with evolutionism, were involved. The three main cases were those of Raffaello Caverni in 1878, Dalmace Leroy in 1895, and John Zahm in 1898. Some of these were the source of the accounts provided by theologians and historians on the attitude of the Roman authorities, but they were poorly known before the opening of the Archives.

The first case is that of Raffaello Caverni (1837–1900), an Italian priest from Florence interested in science and its history. Towards the end of his life he wrote a monumental history of the experimental method in Italy. In a 1877 book, he proposed the reconciliation of evolution with Catholic doctrine. This book was denounced to the Congregation of the Index by his own archbishop, was condemned by the Congregation in 1878, and listed in all the later editions of the Index. Nevertheless, the case has been practically ignored.

The second figure is Marie-Dalmace Leroy (1828–1905), a French Dominican. The second edition of *The Evolution of Organic Species*, his book favorable to the reconciliation of evolution and Catholicism, was published in 1891. Four years later, *Le Monde* published a letter written by Leroy from Rome, in which he retracted his position, explaining that his hypothesis had been judged untenable, having been examined in Rome by the relevant authority. No decree was published.

The protagonist of the third case is John Augustine Zahm (1851–1921), an American priest of the Congregation of Holy Cross, professor of physics at the University of Notre Dame. Zahm published a number of volumes on science and religion, one of which, published in 1896 was condemned by the Congregation of the Index in 1898. But the corresponding decree was never published.

We have analyzed also the case of three other Catholic authors: Geremia Bonomelli, John C. Hedley, and St. George J. Mivart. They have been presented as having been condemned by their evolutionist opinions, although no action at all took place.

Before presenting how the Index dealt with these three cases it would be convenient to consider the nature and the procedures of the Congregation of the Index.

2.1 The Index of Prohibited Books

The *Index of Prohibited Books* was a publication that listed the books whose reading, possession or publication was prohibited for Catholics.[1] We have notice of such lists of books from antiquity, for example, a list produced by a Roman Council in 494 A.D. This activity acquired new importance in the sixteenth century as a consequence of the Lutheran Reform and the spread of printing. Printed books and pamphlets were one of the principal means of diffusion of Lutheran doctrine. From 1517 and 1520 more than 300,000 copies of Luther's printed works were sold, and around 3,700 edition of his works were published from 1517 to his death in 1546 (Bujanda 1985, 12). That provoked a reaction by the Catholic Church aimed at impeding the printing, sale, possession or reading of these publications. As a result, the ecclesiastical censorship was organized in a systematic way.

---

[1] On the history of the *Indices* of prohibited books, see the Introduction to Bujanda's edition of the first Roman *Indices* (Bujanda 1990, 27–44).

The Index did not start, however, with that centralized character. At first, books were condemned by civil and ecclesiastic authorities. The first *Index of Prohibited Books* was published in 1544 by the Faculty of Theology of the University of Paris. Five different editions were published until 1556, and more than 500 titles were included. The University of Louvain published three *Indices* in 1546, 1550 and 1558. Louvain attempted to publish also an "index of permitted books" that would have ruled the academic activity, but this kind of catalog was not adopted by the other *Indices*. In Venice the first *Index* was prepared in 1549 by the Inquisition, the Apostolic Nuncio and the "Tre Savii sopra Eresia" appointed by the Serenissima Republic of Venice. This *Index* was never officially published. Five years later a different *Index* was issued; however, it was also suspended after only one year.

Several more *Indices* were published in Milan and Florence (1554), Portugal (1547, 1551, 1561) and Spain (1551, 1554, 1559). In 1557 the first Roman *Index* was printed under Paul IV (the ancient General Inquisitor, Gian Pietro Carafa); however, it was not published, and a different edition was finally approved at the end of 1558. It was extremely intolerant: all works written by heretics were forbidden, also if not dealing with religious questions, and it should be modified soon through two decrees: *Instructio circa indicem* (February 1559) and *Moderatio indicis* (June 1561).

The publication of the Decrees of the Council of Trent led to the preparation of a new *Index* in 1564, under Pius IV, and to the formulation of ten general rules that remained in force for more than 300 years. The rules determine the works that had to be condemned: all the writings of the chief heretics (Luther, Calvin...), other heretics' writings dealing with religion, obscene books, books on astrology, divination or magic; the rules establish also the conditions to read and retain condemned books. The rules establish also the authority of local bishops and the Inquisition, in particular to "expurgate" book that contain errors, to give authorization to print books, and to forbid new books in their region. In fact, several *Indices* were published afterwards in some Catholic countries: Anvers (1569, 1570, 1571), Portugal (1581), Munich (1582), Spain (1583, 1584).[2]

In 1571, Pope Pius V created the Congregation of the Index as a permanent institution in the Vatican to scrutinize publications. After several projects, the Con-

---

[2] For an edition containing all the books listed on the Roman *Index* from 1600 to 1966, with a comprehensive introduction, see Bujanda and Richter (2002). Bujanda has also published other studies on the *Indices of Prohibited Books*, in Anvers, Spain, Louvain, Milan, Paris, Portugal and Venice (Bujanda 1985).

gregation published a new *Index of Prohibited books* in 1596. From time to time new editions would include the books prohibited since the preceding edition.

The operations of this institution would later be more precisely defined by Sixtus V's 1588 reform of the Roman Curia. In 1753, Benedict XIV established with greater precision the procedures to be followed in the examination and censorship of books and set by the constitution *Sollicita ac provida* the very norms which continued in force during the period here examined. This constitution stressed also the consultors required attitude: in the examination of denounced books, consultors should act without any prejudice (country, school, institution…)[3] and guided only by the common Church doctrine taught by the approved Councils, the Roman Pontiff and the Church Fathers; opinions expressed by the authors should be understood in their full context; when possible, ambiguous sentences should be interpreted in the most favorable sense.

In 1897 Leo XIII initiated a comprehensive reform of the *Index*, simplifying the body of rules that had accumulated over the centuries and revising their content. In 1917 Benedict XV abolished the Congregation of the Index and assigned to the Holy Office the tasks related to the *Index of Prohibited Books*. Finally, in 1965 Paul VI left the *Index* with no status in ecclesiastical law (Paul VI 1965), while preserving its spirit.[4] The final edition of the *Index* was published in 1948.

2.2 The Index procedures

The Congregation of the Index examined publications that had been denounced. It could also carry out such an inquiry on its own initiative, but that was not the common case. If the result of the inquiry were negative, a *decree* was published whereby the book was added to the *Index*.

The term "Congregation" refers mainly to the meeting (*congregatio*) of a group of Cardinals which had the mission of deciding about the prohibition of denounced books. But it refers also to the permanent office that carried the necessary work related to this task. As was the case with the other Congregations, that of the Index was headed by a Cardinal Prefect, aided by the Master of the Sacred Apostolic Palace (*Maestro di Sacro Palazzo*) as his permanent assistant, an office equivalent to the current theologian of the pontifical household. The Congregation

[3] "De variis opinionibus atque sententiis in unoquoque Libro contentis, animo a praejudiciis omnibus vacuo, judicandum sibi esse sciant" (Benedict XIV 1753, 15).

[4] "Eundem tamen non amplius vim legis ecclesiasticae habere cum adiectis censuris" (Congregation for the Doctrine of the Faith 1966, 445).

had a Secretary; with the exception of the first, who was a Franciscan, the rest were always Dominicans. In addition, there were two or three subalterns. A group of theologians (mostly from Roman academic institutions) collaborated with them as consultors. Their mission was to study each denounced book, report on it and propose a resolution.

The Congregation of the Index reached its decisions in three phases. In the first place, the Prefect, assisted by the Secretary, charged one or more consultors with the examination of the work denounced. The consultor submitted his report in writing. In the period under consideration, these verdicts, with a few exceptions, were printed for distribution in the meetings of the Congregation.

The second step was a meeting called the Preparatory or Particular Congregation, in which the consultors, chaired by the Secretary and assisted by the Master of the Sacred Palace, examined and discussed the reports that the same consultors had prepared. This Preparatory Congregation had no power to make decisions. Its function was to prepare the cardinals for their deliberations, examining and discussing the reports that the same consultors had prepared. After their discussion, the consultants were to draw up a recommendation for each work examined, usually including a vote tally. The proposal was transmitted to the Cardinals, together with the reports of each consultor.

Several days later the General Congregation would meet, in which the Cardinals who were members of the Congregation of the Index participated. The task of the Cardinals was to judge the works submitted for examination, taking into account the reports of the consultors and the deliberations of the Preparatory Congregation, and to decide what kind of punishment should be accorded to each one of them. The Secretary of the Congregation and the Master of the Sacred Palace also attended the meetings of the General Congregation.

After the General Congregation, the Secretary was received in audience by the Pope, for him to confirm the decisions taken. The Secretary would explain the cases studied and the decisions taken and the Pope would then order the publication of the Decree that converted such decisions into Church law. After being informed of the decisions, the Pope customarily gave his approval. But this was not a simple bureaucratic action, because the Pope, with some frequency, ordered other works added to the decree, or otherwise bypassed the regular Index procedure.

The decree was then printed in large format and posted in the appropriate places in Rome (the Vatican, the Palace of the Chancery, and a few more). All decrees had the same structure. At the top of the page was the Pope's escutcheon, flanked on either side by the figures of St. Peter and St. Paul. The text began with a formulaic paragraph, always the same, stipulating that the Cardinal-members of the Congregation had met on a specific date; nothing ever changed in this opening paragraph except for the date. Next came the list of books that the decree prohibited. Only the author and publication data of the book were given, and nothing more. This is extraordinarily important, because nothing was ever said about the reason for prohibiting the book, which is why some cases have been unexplained until the present. When the author was a Catholic and had accepted the Congregation's decision, the following sentence was added: "The author, in a praiseworthy manner, has submitted [to the decree] and has rebuked his book". At the end of the document came the date the decree was ordered, its publication date in Rome, together with the signatures of the cardinal Prefect and the Secretary of the Congregation.

2.3 How decisions were reached

The decisions of the Congregation of the Index were based on already existing doctrine. The Congregation could not, of its own accord, decide whether a doctrine was acceptable or not: it could only apply already existing doctrine to specific books. Clearly, when the consultor examined books, they used their own arguments, but the decisions themselves had to be based on already existing decisions of Popes, Church Councils or of the Congregation of the Holy Office: although theological arguments might be adduced, these did not have the value of public doctrinal authority, because they were never published and were only known to those who participated in the activities of the Congregation. Still they are a valuable source of information. The reports preserved in the archive reveal the arguments used in each case.

In the specific case of evolutionism, the Congregation of the Index continually found itself in an area where no doctrinal judgment had been clearly defined by the relevant authorities, and it had therefore to base its decisions on the arguments that surfaced within the Congregation itself. Perhaps this explains why the decisions of the Congregation of the Index, although on several occasions contrary to evolutionism, were always put forth moderately. As we will see, such moderation also is explained by the considerate way in which Catholic authors were treated, as well as the Orders or religious Congregations to which certain authors belonged: on some occasions the public condemnation of the book was replaced by a brief retraction by the author, or by some phrasing that fell short of a retraction.

The particular and general Congregations were convened with varying frequency, generally twice or three times yearly. Attendance was not very regular.

The number of cardinal members of the Index, although the total varied, was always quite high: between 1870 and 1900 it varied from 18 to 38. A total of 105 different cardinals were members of the Index in those years. Many resided outside of Rome and never appeared at the General Congregation, or very infrequently, if it coincided with their presence in Rome for other reasons. This was the case of the Cardinals of Rhodes, Ferrara and Naples, who attended the Congregation only once. Moreover, many of the Curia Cardinals who resided in Rome were members of several Congregations, and so could not always attend every meeting. If, as happened in some cases, they were also Prefects of another Congregation, it could occur that they might never participate in the sessions of the Index. Such, for example, was the case of Cardinal Miecisłas Ledóchowski (1822–1902) after becoming Prefect of the Holy Congregation for the Propagation of the Faith.[5] In the years mentioned (1870–1900) only 53 cardinals took part in the meetings (including the three just mentioned, who participated in one meeting only), with attendance varying between 5 and 13 cardinals at each meeting.

Attendance at the Preparatory Congregation was not much more regular. A total of 109 consultors are mentioned in the annual Pontifical volumes (*Annuario Pontificio*) for the period 1870–1900, but only 78 of these actually participated in any of the Congregations. The number of participants to the congregations, including the Secretary and Master of the Sacred Palace, varied between 7 and 14 on each occasion.

These numbers are quite high, and they could give the impression of a "huge working machine", but that would be far from reality. In many cases cardinals — and sometimes consultors — were named in old age, lasting in their office only a few years. More than two thirds of the 105 cardinals mentioned died before the turn of the century. The average time at the Index was less than 9 years, and the average participation (excluding those who never assisted) was slightly over 9 meetings.

It could be useful to consider a shorter period: 1894–1899, that correspond to the Secretariat of Marcolino Cicognani op (1835–1899). His annotations were quite regular, and that will permit to follow the life of the Congregation with some detail. In these six years, twelve General Congregations took place. The total number of different Cardinals member of the Index was 46 (around 20-30 each year); only 17 participated in a meeting, and only 9 were present in more than five Congregations. We can identify a "hard core" of around eight or nine cardinals that were regularly present: Serafino Vannutelli (1834–1915; Index Prefect from 1894 to 1896), Andreas

[5] To "Propaganda Fide" were entrusted all the affairs related with countries without stable Catholic Hierarchy, or "mission territories".

Steinhuber (1825–1907; Index Prefect from 1896), Camillo Mazzella (1833–1900; former Prefect of the Index from 1889 to 1993, and afterwards Prefect of the Congregation of Studies), Gaetano Aloisi-Masella (1826–1902; not so regular), Lucido Maria Parocchi (1833–1903; Vicar of Rome), and, from their nomination around 1895-1896, Girolamo Maria Gotti (1834–1916), Raffaele Pierotti (1836–1905; former Master of Sacred Apostolic Palace), and Francesco Segna (1836–1911; former "Assessore" of the Holy Office). Giuseppe Maria Graniello (1834–1896) was also quite regular until his death in January 1896. When in 1896 Pope Leo XIII decide to prepare a series of new rules for the Index, the aforementioned cardinals (plus Luigi Galimberti, until his death on May that same year) were called to participate in a series of Extraordinary General Congregations, in order to prepare the new rules.

The main weight of the Index work fell on the Secretary and his staff, which was very small. Until 1896 it was formed only by a "writer", who took care also of the archive, and a doorkeeper. In January 1896 Cicognani wrote in the Diary that he presented a request to the Pope that a second writer were added. The request was granted, but, as the budget was not enough, it was decided to add a tax to the requests for permission to read and retain prohibited books. This decision was very soon discarded, and the Index kept the custom of not asking any fee.[6]

There is another curious annotation in the Diary, that gives some idea about how the Congregation of the Index carried its work. In January 1899, Cicognani presented a new request to the Pope, this time about the offices assigned to the Congregation, at the *Palazzo della Cancelleria*. His words (this time he wrote in Italian) need no comment:

> few and small rooms, unadorned, dim, and moist, with the entrance at the middle of the fourth stair, over a step, with danger of falling for those coming

[6] ACDF, Index, Diari, vol. XXII, fol 15*r* (January 9, 1896): "Cum Secretaria huius nostrae S: Congregationis ex tribus tantum membris constituta sit, nempe ex Secretario, ex Amanuense, qui munere etiam Archivistae fungitur, et ex clavario (vulgo Bidello, o Portiere); unde non parvum emolumentum ei esset si quartum adiungeretur: de consilio et mandato Emi. Card. Praefecto, Pater a Secretis in audientia extraordinaria habita supplicem libellum SS. D. N. Papae, ad hunc effectum porrescit. Deficiente vero pecunia pro novi officialis stipendio, Sacrae enim Congregationi suppeditat omnes sumtus ex bonis S. Sedis Administratio Vaticana: Ssmus. benigne annuit, et iuxta petita, perpensis quoque oeconomicis conditionibus eiusdem S. Sedis, mandavit, ut quartus addatur officialis sub nomine *Protocollistae et Scriptoris*, atque pro eius stipendio ad unumquodque Rescriptum legendi libros prohibitos applicetur in posterum taxa duarum libellarum titulo 'pro exercitio' non obstante consuetudine contraria huius S. Congregationis concedente gratis facultatem legendi ac retinendi libros prohibitos".

in and out; from the ceiling of Fr. Secretary's room a shower of dust keeps falling covering books and papers.[7]

3. Raffaello Caverni

Raffaello Caverni (1837-1900) was an Italian priest from Florence, professor of physics and mathematics, author of a monumental *History of the Experimental Method in Italy* (Caverni 1972). He was a good friend of Antonio Favaro, the editor of the Edizione Nazionale of Galileo's complete works, although afterwards they became estranged. Caverni was a well-known but nonconformist parish priest near Florence. His book, *New Studies of Philosophy: Lectures to a Young Student* (Caverni 1877) had as its central theme the possibility of harmonizing evolution and Christianity, but it was also very critical of some aspects of ecclesiastic education. *La Civiltà Cattolica*, which at the time was published at Florence, issued a long very negative review of the book.

Caverni's book was denounced to the Index by the Archbishop of Florence, Eugenio Cecconi, after a brief local investigation. After a regular examination, the General Congregation decided to condemn the book. Caverni accepted the decision, and the book was placed on the *Index of Prohibited Books*. But, curiously enough, nobody ever knew about that. But lets go by stages.

We know, from the Archive documents, that this was the first occasion in which evolution was directly under the examination of the Roman Congregations. A report of the General Congregation states that "until now the Holy See has rendered no decision on the system mentioned", that is, Darwinism.[8] The book was examined by the Dominican fr. Tommaso Maria Zigliara (1833–1893). He would become one of the main leaders of the "Neo-Thomist" movement. A year later Zigliara was named cardinal and became Prefect of the Congregation of Studies.

Zigliara's report (nineteen pages long, dated May 25, 1878) was very negative. He had already presented his views in his *Summa philosophica* (Zigliara 1891): transformism was a materialistic doctrine, as it presupposes a spontaneous evolution guided only by the forces of nature. So, his judgement could not be very positive.

[7] ACDF, Index, Diari, vol. XXII, fol 46*r* (January 11, 1899): "...poche e piccole stanze, disadorne, oscure, umide, coll'accesso a metà della quarta scala sopra un gradino con pericolo di cadere per chi entra ed esce; dal soffitto della stanza del P. Segretario cade di continuo una pioggia di terriccio che copre libri e carte..."

[8] Archive of the Congregation for the Doctrine of Faith (ACDF), Index, Protocolli, 1878–1881, fol. 73.

According to Zigliara, the book "merits inclusion in the *Index of Prohibited Books*".[9]

The proposal was accepted by the Preparatory Congregation on June 27, 1878.[10] On July 1, 1878, the General Congregation unanimously concluded for the prohibition of the book.[11]

It could be useful to note that Zigliara's proposal was not only grounded on the opposition between evolution and Catholic doctrine. He had in mind a particular philosophical interpretation of Darwinism, that he considers as incompatible with faith: "Darwinian evolution … is nothing more than the material part of total evolutionism, which is the same as Hegelian pantheism".[12]

The intention of the Congregation of the Index was to "indirectly condemn Darwin", but that intention was not really reached. The decree was published without any mention of evolution or Darwinism, as the title of the book made no mention of either. And years later, when the polemic about evolutionism reached its peak, in the internal documents of the Congregation Caverni's case was remembered only once. Even *La Civiltà Cattolica* was not aware of that "indirect condemnation".

Caverni himself attributed the condemnation to "the ignorant fanaticism of those who felt attacked by him". He also criticized several aspects of Church life, such as the education of seminarians in Italy and the scholastic method applied in these programs, his criticisms having been leveled against the Jesuits in particular. One might well have thought that those critiques were what provoked the prohibition of the book. And that interpretation has been maintained until very recent studies. A study of Caverni published in 2001 states: "The reasons behind prohibition of Caverni's book cannot be found in the theory of evolution as he presented it, or at least not only in it. In that book Caverni leveled a harsh critique at various aspects of the ecclesiastical world, its culture in particular" (Pagnini 2001, 43).

Caverni's book was denounced and condemned in 1878. During the next fifteen years no other book defending the harmonization between evolution and Catholic doctrine was deferred to the Index. Darwin was not included in the *Index* — his grandfather, Erasmus Darwin, had been included for his *Zoonomy*, a pre-evolutionary and quite materialistic essay — and neither were Thomas H. Huxley, Herbert Spencer or Ernst Haeckel. This fact should suggest that there was no systematic

[9] ACDF, Index, Protocolli, 1878–1881, fol. 71, p. 18.

[10] ACDF, Index, Protocolli, 1878–1881, fol. 66. See also ACDF, Index, Diari, vol XX, p. 202.

[11] ACDF, Index, Diari, vol XX, p. 203.

[12] ACDF, Index, Protocolli, 1878–1881, fol. 71, p. 18.

investigation of publications. Whether a book was examined by the Congregation or not depended, in great part, on someone formally denouncing it. The Holy See granted special importance to books written by Catholic authors and in Catholic countries, because they were more likely to disturb the life of the Church.

4. Marie-Dalmace Leroy op

The two other cases were very different. In both cases the Congregation of the Index decided to condemn the book, but the prohibition was never published. In consequence, Leroy and Zahm's books were never included in the *Index.*

Marie-Dalmace Leroy (1828–1905) was a French Dominican. After the French Revolution, the Dominican Order was suppressed in France. Henri Lacordaire (1802–1861) undertook the task of restoring it. Leroy was one of the first to take the Dominican habit after the restoration of the Province of France, in 1850.

In 1887, he published a book entitled *The Evolution of Organic Species* (Leroy 1887). The reviews it received induced him to prepare a new edition, corrected and expanded, in which he attempted to better explain certain controversial points. This new edition was published in 1891 under the title *The Evolution Limited to Organic Species* (Leroy 1891). His intention was to demonstrate that evolution is compatible with Christianity. However, polemics were inevitable. Joseph Brucker criticized in the Theological journal *Études* both editions, mainly on the basis of biblical arguments (Brucker 1889, Brucker 1891).

Three years later, Leroy was denounced to the Congregation of the Index. The Leroy affair is very well documented in the Vatican archives (Artigas, Glick and Martínez 2006, 65–100). The book was denounced by a French layman, "Mr. Ch. Chalmel, Officier d'Académie" (an honorary title in the field of teaching and education). In a highly respectful tone, he submitted two questions for the consideration of the Congregation. The second refers to Galileo in a very naive way, and it was never considered: the Cardinal Prefect (probably) crossed it with two strong lines of red pencil. The Cardinal drew attention to Leroy's book, and the examination started.

In the Congregation of the Index, Leroy's book passed through a very careful study: until four written reports were produced, some very long. The first report proposed that no action be taken against Leroy's book. In this report, written a short time after the publication of the encyclical *Providentissimus Deus*, the Franciscan Teofilo Domenichelli admitted that it was correct to say that in the first chapters of *Genesis* the Bible uses figurative language in a human mode, to condescend (as Dante would say) to our understanding, and that the question of whether the Biblical Creation narrative ought to be interpreted literally or allegorically is open to free debate among theologians. Domenichelli recalls the warnings of Saint Augustine in *De Genesis ad litteram*, lib. I, 18–19, not to be facile in presenting a literal interpretation of the first pages of Genesis "in order not to expose oneself to ridicule in the presence of unbelievers." Domenichelli proposes that no measures be taken against Leroy.[13]

At this point in time, the Secretary of the Congregation simply assumed that no measure would be adopted, but Domenichelli's report was not persuasive enough, and two new reports were ordered in the General Congregation. The second report, written by Ernesto Fontana (1830–1910), recently appointed as bishop of Crema, near Milan, proposed that the book not be prohibited, but that the author be warned.

The author of the third report, Luigi Tripepi (1836–1910), was harsher. Although he too did not propose to condemn the book in a decisive way, his arguments tend to demonstrate the complete incompatibility of evolution and Catholic doctrine. His main argument was, behind the scenes, that evolutionism "is nowadays abandoned and refuted as false and absurd by the same rationalists and unbelievers who previously supported it".[14] From the theological and philosophical point of view, Tripepi's report was based on the arguments of Cardinal Mazzella, who he follow closely (sometimes literally), and were mainly grounded on a very literal interpretation of the Genesis, according to which God created Adam's body immediately, that is, through a "special action".

Tripepi proposed to prohibit the book or to warn the author, inviting him to retract, but he personally favored the second, more benign option, out of consideration of the author and his Order.

The Preparatory Congregation in which Fontana and Tripepi report were discussed took place on January 17, 1895. The recommendation of the consultors was, curiously, harder than the proposals made by Fontana and Tripepi: "the doctrine, as it is found in the book, should be proscribed. Let the author be invited, via the Master General, to retract it publicly, as his own initiative".[15] The reason of this new decision can be found in two other documents.

In a memorandum, the Secretary Cicognani wrote "There was a heated debate … It was first proposed to conclude with a *moneatur* [warning] to the author.

---

[13] ACDF, Index, Protocolli, 1894–1896, fol. 86, p. 13

[14] ACDF, Index, Protocolli, 1894–1896, fol. 125, p. 9.

[15] ACDF, Index, Diari, vol. XXII, fol. 9r. The Master General of the Dominicans then was Andreas Frühwirth (1845–1933).

The current resolution was because the doctrine of the book (or better, evolutionism) is opposed by science and faith".[16]

That new resolution seems to be the effect of the intervention of another consultor, Enrico Buonpensiere, op. The Archives conserve a fourth report, written by Buonpensiere four days after the Preparatory Congregation. The eight-pages handwritten report contains, as Buonpensiere says, "all that I said orally about the work of Father Leroy" in the Preparatory Congregation.

> Evolutionism, as all Catholic philosophers teach, is resolutely condemned by *ontological* and *empirical* science. In *ontology*, the essence of any object is an *immutable type,* that is, *incapable of any change* [evoluzione,] *whether towards the greater or towards the lesser*. In empirical science there is an inexorable law of *hybridization* which maintains living species distinct, in such a way that from the pairing of two organisms belonging to *different species* no fruit will be obtained, or else such fruit is totally *infertile*.[17]

This is close to the wording that Cicognani used in the memorandum for the Cardinal Prefect, and reflects the importance he now attached to Buonpensiere's opinion. It may well be that Cicognani asked Buonpensiere to commit the arguments he originally expressed in the debate, to writing.

According to Buonpensiere, evolutionism cannot even be called a *hypothesis*: it is a simple *desideratum* of materialism. From this assertion Buonpensiere derives a second point, as a logical consequence: it would be "rash and anti-Christian" to try to support evolutionism on Revealed Doctrine. "Therefore, the presumptuousness of those who seek to make evolution equivalent to Christian dogma is foolhardy and anti-Christian". It is important to note that the rashness or temerity of the opinion is linked with the *falsity*, from the scientific point of view, of the evolutionist theories. Yet there is a third point. Leroy's attempt would be also "anti-Catholic", because opposite to the way in which Revelation has been interpreted by the Tradition of the Church.

In the second General Congregation, a hardening of the position finally crystallized. All but one Cardinal formulated the decision that won the day: the book would be proscribed, but without publishing the decree and a public retraction would be sought from Leroy.

After the retraction, Leroy tried to get permission to publish a revised version of the book, which triggered two additional written reports. It was not permitted. Again in 1902 Leroy asked a new permission, that was refused again. He accepted

---

[16] ACDF, Index, Protocolli, 1894–1896, fol. 127.

[17] ACDF, Index, Protocolli, 1894–1896, fol. 117–18, p. 3; underlined in the original.

the decision, but asked again if the reason was that evolution theory has to be considered as contrary to faith, or if it was his own exposition, that was incorrect. Leroy did not receive any answer.

5. John A. Zahm csc

Things were slightly different in Zahm's case. In 1896, the American priest John Augustine Zahm (1851–1921), professor of Physics at the University of Notre Dame, published *Evolution & Dogma*, a book in which he argued for the compatibility of evolutionism and Catholic doctrine (Zahm 1896). Zahm extended his thesis to the origin of man (that is, of the human body). The Congregation of the Index decided to condemn the book, but the decree was never published. Both Zahm's supporters and his adversaries waged a long struggle with the Roman authorities, all documented in numerous letters.

Evolution & Dogma had been quickly translated into French and Italian, as he was a very well known author. Zahm denunciation was probably part of a much more extended plot. The Italian edition of Zahm's book was denounced to the Congregation of the Index by Otto Zardetti, a Swiss-born Archbishop that had been Bishop of Saint Cloud, Minnesota, from 1889 to 1894. In 1897 Zardetti was living in Rome, after less than a year as Archbishop of Bucharest, Romania. Zardetti was related to the so-called "conservative" group of American Catholic Hierarchy. Zahm case, in fact, can be seen as part of the "Americanist" crisis.

In the nineteenth century, many American Catholics were European immigrants, and it was often difficult for the established population to view them as authentic Americans. Catholics of German origin wanted to have their own schools and hierarchy, while the Irish supported a complete *Americanization* of the Catholic Church. The Americanists also had their own positions with respect to the social question (most Irish immigrants were workers; German immigrants were often middle-class), Catholic and public education, some religious issues, and they placed special emphasis on everything related to democracy and the separation of Church and State. At the end of the nineteenth century there was tension between the Americanists, sometimes called "liberals", and those Catholics described as "conservative". Zahm was closely associated to the Americanist party, and battle against evolution was in some way part of the anti-Americanism battle.

When Zahm's denunciation reached the Congregation, the report was asked again to Enrico Buonpensiere. Of course, his position had not changed. From the scientific point of view, he always considered evolution as a false theory. The proofs presented by Zahm are, in Buonpensiere's opinion "simple assertions, and arbitrary explanations of biological phenomena that, with the same ease and certainty that Zahm proposes them, can be perfectly refuted by others, inasmuch much as they lack both philosophical and traditional theological arguments".[18]

Towards the end of his report, Buonpensiere presents a most unusual proposal: he formulates his own theological thesis, attempts to demonstrate it, and proposes that the Congregation condemn the contrary doctrine, that is: "God formed Adam's body not immediately from the mud of the earth, but out of the body of an anthropomorphic brute, whom he had prepared in order to be produced by the forces of natural evolution from lower matter".[19]

Of course, this proposal was not approved: a condemnation of this type did not fall within the competence of the Index. Rather a notable disparity of opinion emerged in the Preparatory Congregation. Six of the consultors proposed that the work be condemned; three proposed that the author only be warned, as "this doctrine [evolution] is not opposed to any defined dogma nor does it impugn the faith." Finally, one said "that it should not be condemned now, while he Holy Office is asked to resolve the doubt: Does transformism contradict divine revelation? Afterwards let this Congregation proceed against the book".[20]

With respect to the General Congregation, Cicognani wrote that "the most Eminent Cardinals, after a long and serious discussion, decreed the prohibition" of Zahm's book. Again we encounter a serious debate, which is only logical bearing in mind that there was no unanimity as to the theological standing of evolution. Domenichelli proposed that such standing be determined. Buonpensiere proposed a concrete definition. An anonymous consultor proposed that the question be clarified by appeal to the Holy Office. But the harder line finally prevailed.

In the meantime Zahm was not inactive. He had very good connections in Rome and among American bishops. For almost two years (1896–1897), he had been in Rome as General Procurator of his Congregation. He was also very close to the leading people of Americanism, the so called "Liberal Wing" of the American Hierarchy. John Ireland, Archbishop of St. Paul, Minnesota, was considered the leader; John Joseph Keane, former Bishop of Richmond and first President of the Catholic University of America, was at that time in Rome; Dennis J. O'Connell, the best friend of Zahm in Rome, was considered "the agent" of the American

---

18 ACDF, Index, Protocolli, 1897–1899, fol. 180, p. 8.

19 ACDF, Index, Protocolli, 1897–1899, fol. 180, p. 45.

20 ACDF, Index, Diari, vol. XXII, fol. 39r (August 5, 1898).

Hierarchy in Rome. In a tightly-knit group, and counting on the support of Card. Serafino Vannutelli, former Prefect of the Index and now Prefect of the Congregation for the Bishops, they tried by all means to stop the publication of the Decree. Monday, November 7, in an audience with the Pope, Cardinal Vannutelli put the question directly before Leo XIII and asked him not to publish the decree. The Pope accepted. News of the Pope's decision soon reached Zahm. Keane wrote first, right after Vannutelli came to visit him on Wednesday: "Card. Serafino has just been to see me. Last Monday he asked the Pope that your sentence sh'd not be published. The Pope agreed willingly. So it is settled".[21]

However, early in 1899 a new campaign to publish the decree against Zahm started again. On January 2, Salvatore Brandi, the Director of *La Civiltà Cattolica* wrote to Archbishop Corrigan of New York, known as the leader of the "Conservative Wing", to bring him up to date on the situation in Rome, mentioning both *Evolution & Dogma* and Americanism. He announces that the encyclical against Americanism is finally ready and would be made public in a few days, in spite of opposition to its publication. He adds: "In the current number of the Civiltà, Your Grace will read an article of mine on *Evolution and Dogma*. It preludes, I am pretty sure, to a decree against Dr. Zahm's work".[22]

The document about Americanism, a Letter from the Pope to Cardinal Gibbons, archbishop of Baltimore, titled *Testem Benevolentiae* ("Witness of Benevolence") was in fact published on February 22, 1899. No decision about Zahm followed, however, although the news about *Evolution & Dogma* being put into the Index continued to spread. In April a new crisis seemed to open, when a review of the French edition of *Evolution & Dogma* (Zahm 1897) appeared in *Annales de Philosophie Chrétienne*. The Congregation decided to warn Zahm again, in an official letter to Zahm's Superior.[23]

This time Zahm acted quickly. On May 16 he wrote to the French translator, asking him to do everything he could to withdraw the book from sale. The wording is carefully studied so as not to explicitly mention the prohibition, nor give the slightest hint of a retraction:

> I learn from unquestionable authority that the Holy See is opposed to the further distribution of your translation of my work "Evolution & Dogma," & I,

[21] Keane to Zahm, November 9, 1898: University of Notre Dame Archives (UNDA), J. A. Zahm Collection, Box 1, Folder 12.

[22] Brandi to Corrigan, January 2, 1899: UNDA, CANY I-1-I [2].

[23] Cicognani to Français, April 25, 1899: UNDA, J. A. Zahm Collection, Box 1, Folder 12.

> therefore, beg of you to use all your influence to have the work withdrawn from sale.[24]

The same day Zahm wrote also, in very similar terms, to his Italian translator, Alfons M. Galea, a well known man of affairs and philanthropist form Malta. What Zahm had not foreseen was that this letter would be made public several days later. Perhaps Galea himself decided that a quick way to comply with the wishes of the Holy See was to publish Zahm's letter. Perhaps someone intervened. What is clear is that on May 31, the newspaper *Gazzetta di Malta* published Zahm's letter along with a statement from Galea, associating himself with Zahm's position. After a few weeks, *La Civiltà Cattolica* published these documents with minimal commentary. It only recalled that in two previous issues, it had warmed of the "many errors" in Zahm's book. Just a simple factual note, but it was evident that for Brandi this was a victory. Indeed all this material, just as *La Civiltà* published it, has been used for more than a century as an authoritative source for the assertion that the Holy See opposed Zahm's book. Nothing more was required. After that, the case was considered closed, even though Zahm neither went to Rome nor retracted.

6. The Vatican policy towards evolutionism

It is not an easy task to present the conclusions of these facts. There are some similarities among them, but each case was the product of very different circumstances. In the three cases the Vatican intervened in response to denunciations that reached the Congregation of the Index from the outside. The Holy Office played no role in any of these cases. The conflicts were not generated by any policy of the Roman authorities with respect to evolutionism.

In the three main cases in which evolution was examined in the Congregation of the Index, the authorities and the experts that examined the cases had actually a wide diversity of opinions. Evolutionism provoked severe tensions. Discussions were sometimes strong, particularly in some Catholic media, such as *La Civiltà Cattolica*. From the outside, one might well think that the Vatican adopted a careful policy towards evolutionism. The surprise was to realize that, in a sense, there was no policy at all. The actions of the authorities responded to particular circumstances, not to any carefully designed plan.

---

[24] Zahm to Flageollet, May 16, 1899: ACDF, Index, Protocolli, 1897–1899, fol. 273; copy sent to Sacred Congregation of the Index.

The Vatican authorities were aware of the fact that no condemnation of evolutionism had been issued, and apparently they were not anxious to provoke one. They examined the writings of our protagonists when each was denounced. The analysis of those writings was done on the basis of the existing doctrine. The Congregation could not, of its own accord, decide whether a doctrine was acceptable or not: it could only apply already existing doctrine to specific books. But there was no official doctrine regarding evolutionism. This explains why the reports of the experts followed no uniform pattern. Nor was there any fixed pattern that could predict the decision of the Cardinals, or even of the Pope. We will see that in one of the major cases, the Pope prevented the publication of a prohibition decided by the Cardinals.

In the specific case of evolutionism, the Congregation of the Index continually found itself in an area where no doctrinal judgment had been clearly defined by the relevant authorities, and it had therefore to base its decisions on the arguments that surfaced within the Congregation itself. Perhaps this explains why the decisions of the Congregation of the Index, although on several occasions contrary to evolutionism, were always put forth moderately. On some occasions the public condemnation of the book was replaced by a brief retraction by the author, or by some phrasing that fell short of a retraction.

Perhaps it can be said that the Holy See adopted a pragmatic policy with respect to evolution. No energetic intervention was considered necessary. One might even think that Roman Authorities preferred to avoid condemnations by using less controversial measures, such as personal retractions, without directly compromising the Vatican. Perhaps they hoped that this approach would be sufficient to hinder the diffusion of evolutionary ideas.

There certainly existed in the Catholic tradition elements sufficient to reconcile evolution with the Bible, and there was no lack of Catholic theologians who proposed to resolve the difficulties. Evolution could be reconciled with Creation and divine activity in the world taking seriously the traditional distinction between God as "Primary Cause" and created "secondary causes" that act in accordance with the laws that God impressed upon nature. This was the means adopted by those who said that evolution could be harmonized with Christianity. Still, the more severe attitude generally prevailed among theologians in the nineteenth century. Special creation of species was perceived as more in line with Christian doctrine and with tradition. This was especially clear with respect to the origin of the human body.

The Vatican authorities seem to have shared the same attitude that prevailed, at that time, among theologians. The archival documents make clear that, in the internal deliberations of the Congregation of the Index there was an almost general consensus for rejecting evolutionism, above all when it is applied to the origin of the human body. It is patent that in some cases uncertainties existed. Several times, the consultors proposed to withdraw the accusation. Other times the suggestion was that the question be clarified by appeal to the Holy Office to resolve the doubt: "does transformism contradict divine revelation?" But the Congregation of the Index did not consider it necessary to take any of these steps, and made the decisions that seemed most prudent.

References

Alszeghy Z., 1967, "El evolucionismo y el Magisterio de la Iglesia," *Concilium* 26: 366–73.

Artigas M., Glick T. F., Martínez R. A., 2006, *Negotiating Darwin. The Vatican Confronts Evolution, 1877-1902*, Baltimore: The Johns Hopkins University Press.

Benedict XIV, 1753, *Apostolic Constitution Sollicita ac provida*.

Beraza B., 1921, *Tractatus de Deo creante*, Bilbao: Elexpuru.

Bowler P. J., 1983, *The Eclipse of Darwinism. Anti-Darwinian Evolution Theories in the Decades around 1900*, Baltimore & London: Johns Hopkins University Press.

Boyer C., 1940, *Tractatus de Deo creante et elevante*, 3rd edn., Rome: Gregorian University.

Brucker J., 1889, "L'Origine de l'homme d'après la Bible et le transformisme," *Études* 26: 28–50.

Brucker J., 1891, "Bulletin Scripturaire," *Études* 28: 488–97.

Bujanda J. M. de, 1985, *Index de l'Université de Paris, 1544, 1546, 1547, 1549, 1551, 1556*, Sherbrooke: Centre d'Études de la Renaissance, Université de Sherbrooke.

Bujanda J. M. de, 1990, *Index de Rome, 1557, 1559, 1564. Les premiers index romains et l'index du Concile de Trente*, Sherbrooke: Centre d'Études de la Renaissance, Université de Sherbrooke.

Bujanda J. M. de, Richter M., 2002, *Index librorum prohibitorum 1600–1966*, Sherbrooke: Centre d'Études de la Renaissance, Université de Sherbrooke.

Caverni R., 1877, *De' nuovi studi della Filosofia. Discorsi a un giovane studente*, Florence: Carnesecchi.

Caverni R., 1972, *Storia del metodo sperimentale in Italia*, New York: Johnson Reprint Corporation, Reprint from the original edition, Florence: Civelli, 1891–1910.

Congregation for the Doctrine of the Faith, 1966, "Notificatio (June 14, 1966)," *Acta Apostolicae Sedis* 58: 445.

Draper J. W., 1874, *History of the Conflict between Religion and Science*, New York: D. Appleton & Co.

Engels E.-M., Glick T. F. (eds.), 2009, *The Reception of Charles Darwin in Europe*, London: Continuum.
Glick T. F., 1988*a*, "Spain," in Glick Th. F. (ed.), *The Comparative Reception of Darwinism*, 2nd edn., Chicago: University of Chicago Press, 307–345.
Glick T. F. (ed.), 1988*b*, *The Comparative Reception of Darwinism*, 2nd edn., Chicago: University of Chicago Press.
Huxley J., 1943, *Evolution. The Modern Synthesis*, New York and London: Harper & Brothers.
Jacolliot L., 1913, *Christna et le Christ*, 2nd edn., Paris: Flammarion.
John Paul II, 2003, "Discourse to the Pontifical Academy of Sciences (October 22, 1996)", in Sánchez Sorondo M. (ed.), *Papal Adresses to the Pontifical Academy of Sciences 1917–2002*, Vatican City: Pontifical Academy of Sciences, 370–374.
Juste R., 1965, "La teología católica y el problema de la evolución humana. Un siglo de historia eclesiástica," *Revista española de teología* 25: 393–414.
Leroy M.-D., 1887, *L'évolution des espèces organiques*, Paris: Perrin.
Leroy M.-D., 1891, *L'évolution restreinte aux espèces organiques*, Paris and Lyons: Delhomme et Briguet.
Mazzella C., 1896, *De Deo creante*, 4th edn., Rome: Forzani.
Messenger E. C., 1932, *Evolution and Theology. The Problem of Man's Origin*, New York: MacMillan.
Pagnini S., 2001, *Profilo di Raffaello Caverni, 1837–1900, Con appendice documentaria*, Florence: Pagnini e Martinelli.
Parente P., 1959, *De creatione universali de angelorum hominisque elevatione et lapsu,* 4 ed., Turin: Marietti.
Paul H. W., 1988, "Religion and Darwinism", in Glick Th. F. (ed.), *The Comparative Reception of Darwinism*, 2nd edn., Chicago: University of Chicago Press, 403–436.
Paul VI, 1965, "Apostolic Letter Integrae Servandae," *Acta Apostolicae Sedis* 57: 952–55.
Pesch C., 1908, *Praelectiones dogmaticae quas in Collegio Ditton-Hall habebat. Vol. III, De Deo creante et elevante, De Deo fine ultimo*, 3rd edn., Friburg: Herder.
Pius XII, 1995, Encyclical *Humani Generis*, in *Enchiridium delle Encicliche*, Vol. 6 Bologna: Edizioni Dehoniane, 628–60.
Rahner K., 1953, *De Deo creante et elevante et de peccato originali*, Innsbruck: Typescript source notes.
Ruiz de la Peña J. L., 1996, *Imagen de Dios. Antropología teológica fundamental*, 3rd edn., Santander: Sal Terrae.
Scheeben M. J., 1961, *Handbuch der katholischen Dogmatik. Book III Schöpfungslehre*, 3rd edn., Friburg: Herder.
Stebbins R. E., 1988, "France", in Glick Th. F. (ed.), *The Comparative Reception of Darwinism*, 2nd edn., Chicago: University of Chicago Press, 116–163.

Tanquerey A., 1911, *Synopsis Theologiae Dogmaticae Specialis, Vol. II: de Fide, de Deo Uno et trino, de Deo Creante et Elevante, de Verbo Incarnato*, 13th edn., Rome-Tournai-Paris: Desclée.

White A. D., 1896, *A History of the Warfare of Science with Theology in Christendom*, New York: D. Appleton & Co.

Zahm J. A., 1896, *Evolution & Dogma*, Chicago: D. H. McBride,

Zahm J. A., 1897, *L'évolution et le dogme*, Paris: Lethielleux.

Zigliara T., 1891, *Summa philosophica in usum scholarum*, 8th edn., Paris and Lyons: Delhomme et Briguet.

Evolution according to Teilhard de Chardin

Georges Chantraine sj†

This paper will make a connection between science (the first part of this colloquium) and theology (the third part), for Pierre Teilhard de Chardin wanted to connect palaeontology, his field, with the revelation of Christ. I. We will consider first of all Teihard's ideas on evolution. II. We will next try to see Teilhard's own perspective: evolution considered according to the law of the complexity of growth; this law passes from biosphere through Noosphere to "Christo-sphere". III. Such an idea of evolution falls within the field of science, but also links this field with what is outside it, a relationship between the "physical phenomenology" that Teilhard describes and a possible Christian significance. IV. Is this relationship between what is inside and what is outside the field of science tenable? Is it possible to cross between what is a scientifically sophisticated phenomenon and Christology? These are two questions of method.

I. The idea of evolution

Fr Pierre Teilhard de Chardin was aware of the antithesis between the concept of a universe in genesis, held since the sixteenth century and that of a static universe formerly conceived by the ancients, following Biblical Revelation. He also knew of the antithesis between "a kind of divinity, completely immanent in the world" and "the transcendent Christian God" (OT9f, 238[188]). He set himself to resolve this double antithesis. On the other hand, there was "the idea of Evolution germinated in me, like a seed: whence it came I cannot say" (OT13a,

† Fr George Chantraine passed away on May 23, 2010, without reviewing the edition of his article. Texts from the *Oeuvres de Pierre Teilhard de Chardin* (Paris: Seuil 1955–1976) will be quoted as OT followed by the volume, a letter identifying the different writings included in the volume, and the page number; when possible, the English edition page number will be added in square brackets. All Teilhard's texts quoted in the article will be listed in the References according to the French edition of Teilhard works and its English translation. Other references will follow the usual Author-Date system (*Editor's note*).

32[24])[1] that for Teilhard ruled out the idea of pantheism, as responding to "an innate, ungovernable need to attain self-fulfilment by accession not, indeed, to *others*, but to become *the Other*" (OT13a, 32[24]).[2] In the end, Teilhard "began to live his faith in the God of Jesus-Christ within the dynamism of evolution" (Schellenbaum 1971, 447).

From his first writings[3], Teilhard affirmed evolution — which he first called transformism —, not according to a theory, that of Lamarck or of Darwin[4], but according to that which it is: the historic and organic dimension of life and the universe, and according to two new sciences, Palaeontology and Prehistory (OT3j, 249–50[177]).[5] Evolution, at the experimental level, "is now an *indubitable* fact for science" (OT10c, 164[139]). Understood as "*organicity* of the stuff

[1] Without doubt, the idea came to him from Fr Joseph Maréchal, whom he first knew in 1910, and from Marcellin Boule meeting him in 1912 (Arnould 2004, 52).

[2] The notion of evolution by concentration is presented in the *Journal* in November 1917 (Arnould 2004, 52–53). It was in February 1917 that Teilhard spoke in his Journal of "ontological Evolution" (Schellenbaum 1971, 76).

[3] P. Schellenbaum (1971, 73–77) has considered the "first steps of Teilhard towards a law of evolution".

[4] Included under the word 'evolution' are theories vastly different from one another (To C. Cuénot, 11.04.53, Cuénot 1958, 311). Teilhard places the human Phenomenon in comparison with the Lamarckian anti-chance and with "the Darwinian play of external forces and … chance" (OT1, 163[149]). Teilhard thus answers a question and notes, from a personal standpoint, that there are "functional complementarities, — one could say 'symbioses'" between Lamarckian and Darwinian factors, that apply differently depending upon whether it is a question of micro-biology or biology on a greater scale. We will not say with J. Arnould (2004, 290) that the thought of Teilhard is Lamarckian. In fact, Teilhard tries to reconcile these contraries in respect of a "deeper insight that reveals the psychological depth of Teilhard" (Mooney 1966, 48–49) and also the third Infinity, the Gulf of complexity, as we will see. — In 1953, Teilhard wrote: " 'Darwinism' and 'transformism' are words that already have only an historical interest" (OT10f, 286[238]). We shall see below a critique of "Darwinian transformism". It is already implicit in *The Natural History of the World*: "the biological sciences will increasingly stress the views of the physical and organic interdependence of living forms that Lamarck and Darwin translated, knowing no better, into terms of simple generation, adaptation and heredity" (OT3e, 147[104]). On Darwinism, Neo-Darwinism and Christian theories concerning evolution see A. Dulles (2009, 89–99). — The perspective of Teilhard on evolutionism is different in this regard to that of Cardinal Christoph Schönborn (2007, 75–96).

[5] Teilhard addressed palaeontology in 1923 (OT2a, 51–81[33–57]). In 1913 he presented the first pages of a publication on prehistory which was in process (Teilhard de Chardin 1913; 1913a). He did not mention there comparative anatomy, molecular biology, etc.

of the universe (a temporal, combined with a spatial, organicity)", it is "evidencial fact" (OT9g, 246[193]).[6] This is "the great intellectual event which will in the future mark the nineteenth century" (OT3j, 248[175]). Teilhard frees the "often masterly insights" of the "first generations of transformists" from association with "a great deal of defective explanation and false philosophy" (OT3a, 17[7])[7] and more generally from a "two-fold intellectual and emotional isolation" where we live religiously in relation to the world (OT10a, 113[94]). Transformism is "a very general relationship of physical dependence and continuity between organized forms" (18[8]). The development of life is subject to an "*internal* law" (31[18]), a "law of perfectly clear distribution" (32[18]), which shows "unity" of life "beneath an extraordinary variety" (OT3d, 116[81]). Further, there is a "transformist paradox": "Life perceived by the first advances of science as a fluid continuity, resolves itself, as our researches progress further, into independent and discontinuous terms" (OT3d, 118[82]).[8] This follows from the fact that "*palaeontology* (like all long-distance vision) only reveals *maxima*" (OT3d, 132[94]; see OT3g 177[126]).[9] But, this paradox doesn't affect the distribution law (OT3d, 120[84]; 123[87]). Our living universe (as our material universe) "*has a structure*,

[6] In his Address to the Pontifical Academy of Science, on 22 October 1996, John-Paul II declared: "The theory of evolution is more than a hypothesis" (*L'Osservatore Romano*, French edition, 29 October 1996). [The English version published by *L'Osservatore Romano*, "new knowledge has led to the recognition of more than one hypothesis in the theory of evolution," seems to be a misunderstanding of the original French text: "de nouvelles connnaissances conduisent à reconaître dans la theorie de l'evolution plus qu'une hypothèse". *Editor's note*]. He thus distanced himself from the Encyclical of Pius XII *Humani generis* (Pius XII 1950, 3895) according to Benedict XVI (Horn and Wiedenhofer 2007, 154–55). The Pope remarks at the same time that "the theory of evolution is still not a complete and scientifically verifiable theory". This is not what Teilhard thinks. According to Teilhard, the only hypothesis is the continuing convergence of the complexity-consciousness underway for six hundred million years (OT9g, 247[194]).

[7] We will refer to this text, "How the Transformist Question Presents itself Today" (OT3a, 15–40[7–25]) by giving only the page number in parentheses.

[8] He also refutes the objection of "fixists": "Far therefore from proving that the animal world forms an exceptional realm, refractory to history, the discontinuity of the phylectic series is a positive indication of the reality of a biological evolution, as easily recognizable as that of the Roman Empire" (OT3f, 162[114–15]).

[9] "In reality all that we must conclude, in zoology, from the absence of intermediate forms, is that since the biosphere reacts in exactly the same manner to the methods of our historical analysis as everything that we most certainty know to be evolutionary, it is itself evolutionary by nature" (OT3g, 178[127]).

and this structure can only be due to a phenomenon of growth" (33[20]; see OT3d, 123–126[87–89]). These "successive growths of life may be *the substance of a history*" (36[22]).[10] "Understood in the beginning principally as a need for change, evolution has become principally a law of birth" (OT3d, 141; OT3j 249[101]).[11] Such is "the essential" of transformism (OT3g, 191[137]). It is not in this sense a "hypothesis"[12] (39[25]), but "a general method of research, accepted in practice by all scholars" (OT3h, 215[152]);[13] it is "the particular expression, applied to the case of life, of the law which conditions our whole knowledge of the sensible universe" (39[25]).[14] In 1950, Teilhard clarified: evolutionism is not "a simple method. It is in fact a new and common dimension of the universe, and consequently affects the totality of elements and relations of the universe". It is "the expression for our minds of the world's passage from the state of 'cosmos' to the state of 'cosmogenesis'" (OT3m, 348[246]). It comes out of the discovery of time (OT3g, 180[127]), or more precisely "Organic-time", that is to say, "time whose total unrolling corresponds to the gradual, progressive and irreversible elaboration of a collection of organically linked elements" (OT3g, 183[130]). It is a science, having as its object "empirical succession in time", that is distinguished from the philosophical view that looks for "underlying causality" (OT3h, 216[153–54]). Further, Teilhard rejects the objection that declares the theory of evolution condemnable "since it affirms that greater springs from the less": "In so far as it is possible to define in the natural movement of evolution an absolute direction for man, all that modern evolutionary theory declares is that in the spa-

---

[10] "*Transformist is, at bottom, no more than an admission that we can plot the history of life* as we plot the history of human civilizations or of matter" (OT3h, 214[152]).

[11] A "law of *birth*" is not a "law of determinism" (OT3e, 147[104]). "Every birth corresponds to a *discontinuity* or jump (*quantum*), of nature and extent that varies with each case" (OT2b, 192[136]).

[12] Teilhard adopts, according to the formulation of Dominique Lambert (1999, 32–33; 41; 68), a critical and realist epistemology: "the synthetic theory of evolution is not only a sort of convenient hypothesis to confirm observations on fossils, it says something correct on the reality of the historical development of life". Christopher F. Mooney (1966, 49) speaks of an "experimental affirmation of the coherence of being".

[13] "Understanding", is "in terms of modern science", "integrating in the cosmic evolutionary" (OT2b, 189[133]).

[14] Or again: "the extension to zoology and botany of a form of science (historical science) which is continually extending its rule over all human sciences (physics and chemistry, the study of religions, institutions, etc.)" (OT3h, 215[152]).

tio-temporal reality of the cosmos the great *succeeds* the less. And this is both indisputable, and un-condemnable. A *process* is not a philosophical *explanation*" (OT3m, 347–48[245]).

This law of transformism supposes the law of irreversibility, distinct from "orthogenesis", which is "development in the same direction" (OT3b, 73–74[49–50]).[15] It also supposes "the 'plasmatic' role of the living psyche", what is natural is not identified with what is "mechanical" (OT3d, 134[96]); provided that "life is conceived (as all spiritual causes should be) as a synthetic force of a higher order than that of the physico-chemical forces, capable of co-ordinating them and acting on then without ever destroying or falsifying their determinisms" (OT3d, 135[96]). It is a question of a "autonomous power of organization and differentiation, in no way localized in individuals, but diffused in any large portion of animated matter" (OT3d, 138[99]), of a "mysterious but not metaphorical biosphere" (OT3d, 139[100]). There may also be "a certain physical 'immanence' (forgive the philosophical term) in life" (OT3d, 140[101]). In the present distribution of life, "we can at least distinguish some general characteristics expressive, either of an astounding power of expansion and malleability, or of a general ascent towards greater consciousness and freedom" (OT3g, 168[118]).

Transformism evidently postulates "the absence of all empirical beginnings" (OT3g, 184[131]). It implies a choice: "Would you insist on placing the primacy of being in the plural and unconscious at the risk of making the world unthinkable and incapable of life? Then everything reduced to its lowest terms becomes matter. — Do you not understand, we ask, that on the contrary only union and synthesis make the universe blessed and stable? In order to be, would you not choose to seek the absolute sense of all growth in the direction of this upper pole? In that case, by virtue of the link between things established by evolution, everything is related to the highest. Everything becomes, if not spirit, at least distant preparation, spiritized 'matter'" (OT3g, 187[133–34]). It is the choice of "spiritual evolutionism", that "which places the true future of the world's development and being on the side of the spirit, that is to say of synthesis. This evolutionary theory, which depends on a belief in a higher pole of the universe might be called an evolutionism by convergence" (OT3g, 193[138]; see Ratzinger 1969, 240 s.). Teilhard approaches evolution by climbing from below to above by synthesis, not descending

[15] Orthogenesis is "the manifest property of living matter to form a system in which 'terms *succeed each other* experimentally, following the constantly increasing values of centro-complexity' " (OT1, 115[108]).

from above to below through analysis (OT3h, 221[158]). On several occasions (see OT2b, 185–234[132–171]; OT7j, 333–53[319–37]), he suggested "that such a relationship is furnished by the visible process of complexity/consciousness[16] which 'at right angles' to entropy,[17] irresistibly by the play of great numbers, draws a fraction of matter simultaneously to arrange and interiorize itself on itself; human socialization being, from this point of view, merely the highest phase of the 'complexification' and 'growth into consciousness' of the biosphere" (OT3o, 365–66[256–57]).[18] It is above all a third infinite, along with the Immense and the Minute,[19] the two Infinite of Pascal. By this, Teilhard departs from "Darwinian transformism" for he replaces this theory's formula, "survival of the fittest", with survival of the "most complex" (OT7i, 316[302]), before saying that evolution is not only "conscious" in man (J. Huxley), but also "convergent" (OT7i, 322[308]; on J. Huxley, see OT3k, 320[228]; OT3l 330[234]). He regards Humanity therefore as "the key to and direction of evolution" (OT3h, 221–22[158]).

---

[16] "Complexity — or … the degree of 'centration' of the corpuscles in themselves" (OT3k, 316[225]). "As I understand it here, complexity is an *organised*, and consequently *centred heterogeneity*" (OT3k, 313[222]). "The term 'consciousness' is taken in its widest sense to indicate every kind of psychicism, from the most rudimentary forms of interior perception imaginable to the human phenomenon of reflective knowledge" (OT1, 53[57]). Teilhard formulated this law around 1920 thanks to a change in perspective: he went from the "spiritual power of matter" to the "spiritualisation of matter" (Schellenbaum 1971, 116).

[17] See the definition of entropy in *The Human Phenomenon* (OT3i, 235–36[167–68]). There are two physical functions: "Entropy has been replaced for me by 'the highest consciousness' as the essential physical function of the universe. The world seems to me to 'tumble,' if I can so put it, forward and upward upon the spiritual; and this inversion of cosmogony has the consequence of giving a cosmic consistency to the centers of consciousness — to the monads: this personal treasure, the center within every soul, is imperishable, and the supreme Center must be both lovable and loving" (Letter 19.01.29, Teilhard de Chardin 1956, 118).

[18] Teilhard, always being sure of the foundations, proceeds more cautiously in *The Phenomenon of Man* (OT3i, 237–38[167–69]).

[19] In a universe of three infinites, superior living beings (man, for example) "take their place at the summit of a special branch; and in his terminal position (*in which they directly prolong the line of the atoms and molecules*) they form an *extreme*, by the same right as a galaxy or an electron" in the infinitely great and the infinitely small (OT3k, 315[224]). The "law of complexity/consciousness" is a "cosmic tendency" "as fundamental as entropy or gravitation" (OT2b, 195[139]). — In the formula "complexity-consciousness", consciousness is the correlative of complexity (OT3k, 318–19[225–26]) — The universe of two infinites is that of Einstein's formulae (OT3k, 324[232]).

"Energies of a psychic nature everywhere control the development if live and man by his thought has renewed the face of the earth" (OT3h, 222[159]). Thought, "a reality of cosmic and evolutionary nature" constitutes, "in good Evolution", an "axial and superior form of Life" it forms a "Noosphere" (OT3j, 254[181]; see also Martelet 2005, 31–32).Three successive steps characterize human ultra-speciation: the breakthrough in reflective thought, culture and convergence (OT3o, 370–376[260–65]). This happens as a return to scientific research: "if man is (as Julian Huxley has written) only evolution which has become 'reflexively' conscious of itself: then it logically follows that by way of introspection we should be in the position to understand directly, in the position to understand directly, in the very modalities of our action, at least certain of the factors that formerly presided over the transformation of life. And one idea particularly sizes our mind, that an important role in the appearance of new zoological characteristics might perhaps be ascribed to the 'forces of invention'" (OT3l, 330[224–25]; see OT7k, 387[366]). It is important, Teilhard considered in 1955, that Palaeontology "attempt … fully to integrate the so-called forces or factors of 'orthogenesis' in its theories" (OT3p, 384[269]), contrary to what is being done by the neo-Darwinists (particularly from the United States) (OT3p, 385[270]).[20] Such an integration, wrote Teilhard in 1950, produces a "Neo-anthropocentrism, no longer of position, but of direction in evolution": man is no longer the centre, as he was before Galilee and Darwin, but he is "leading shoot of the world in growth" (OT3m, 349[247]; see also OT3i 232–33[165–66]; OT3k 305[216]). Despite objections, Teilhard seriously doubted whether "the essential vital process of complexity-consciousness has reached its ceiling" (OT3n, 358[252]). "Is not the noosphere[21] implacably condemned from birth to wither and disappear on account of the limited and precarious basis afforded it by our planet? *What is the future of the spirit of the earth*?" (OT6a, 48[39]). "But one point at least seems capable of being established by *analysis of the present facts*; and this is that unless we make up our minds to admit that the cosmos is intrinsically an absurdity, the growth of the spirit must be taken as irreversible" (OT6a, 49[40]). "From the moment that it admits thought, a universe can no longer be simply temporary or of limited evolution: it must by its very structure emerge into the absolute" (OT6a, 50[41]).[22] "*Since man*

[20] There is a brief presentation of neo-Darwinism by Marc Leclerc (1998, 443).

[21] On the Noosphere, see (OT3c, 89ff.[61ff.]). The word comes from Édouard Le Roy (OT3j, 254[181]).

[22] It is also the object of a choice between Death (and rebellion) and the absolute (OT3e, 109–111[77–79]).

and in man, simple evolution tends gradually to mutate into auto- (or self-) evolution" (OT3n, 360[254]). The consequence of which is that in a "*régime of auto-evolution*, this 'arranging' form of energy (energy of invention and combination) appears and (despite its incredible smallness in terms of 'ergs' or 'calories'), begins to play an increasingly decisive role in the progress of the World. This is quite certain" (OT3n, 361[254]; OT3l, 330[228]).[23] In speaking of energy, Teilhard similarly writes in 1953: "what is holding, paradoxically 'cantilevered' on entropy, this vast mass of improbable arrangements constituted by twentieth-century man, is ultimately the awareness (still vague, but already infinitely more developed than in the Neolithic) that 'something' is waiting for us in the depths of time to come" (OT7k, 388[367]). According to this evolutionism, the world is not only charged with "a formidable responsibility", but also "infused from the humblest level of belief in God, with an irresistible attraction" (OT3g, 193[139]; OT3k, 323[232]; OT6a, 52–57[43–47]). It "accompanies and supports the rise of universalists views and aspirations which are even more characteristic of our modern age than any surge of individualism" (OT3g, 196[141]) and which are "the natural climate of Christian faith" (OT3h, 223[159]). So emerges an "integral Energetics of the Universe",[24] which is "the fundamental problem of action" (OT6d, 156[124])[25] that, inspired by Blondel, does not have the "critical rigour" (Leclerc 1998, 444) and that is rooted in "Christian spirituality" (Lubac 1971, 14).

In *How I believe* (OT10b, 115–52[96–132]), Pierre Teilhard justifies or explains the movement of the Spirit described in *The Spirit of the Earth* as the movement of his faith in the world and in the Universal Christ. Faith in the world is presented as "faith in the unity of the world, faith in the existence and faith in the immortality of the spirit which is born from the synthesis of the world — these three faiths, summed up in the worship of a personal and personalising centre of universal convergence" (OT10b, 141[121]). What allowed Fr Teilhard to keep his

---

[23] From here he poses the question: "Life, undoubtedly, sets up automatisms that we must understand scientifically. But does it set them purely *automatically*? This is the whole question" (OT3l, 331[235]).

[24] Teilhard shows the same approach without speaking directly of evolution in *The Spirit of the Earth*; the principle of this approach is the human phenomenon: "The first thing I saw was that man alone can help man to decipher the world" (OT6a, 26[20]). It is a question of Man being the arrowhead of evolution.

[25] "At the very moment when life reflects on itself — by virtue of the same movement — it finds itself facing *the problem of action*" (OT6a, 53[43]; see Schellenbaum 1971, 435, and 40, 45 n.116).

Catholic faith; after having broken with oriental religions and humanist pantheisms, as not satisfying his faith in the world, was the universal Christ, "a synthesis of Christ and the universe", the "inevitable deployment of the mystery in which Christianity is summed up, the mystery of the Incarnation"[26] (OT10b, 146[126]): "ahead of us, *a universal cosmic centre* is taking on definition, in which everything reaches its term, in which everything is explained, is felt, and is ordered", a "physical pole of universal evolution" (OT10b, 147[127]).[27] Its relationship to the world is twofold: "By disclosing a world-peak, evolution makes Christ possible, just as Christ, by giving meaning and direction to the world, makes evolution possible" (OT10b, 148[128]).[28] The Universal Christ transposes "into terms of physical reality the juridical expressions in which the Church has clothed her faith" (OT10b, 148[128])[29], "no longer simply as a fragile creation of my individual thought, but as the product of a long historical revelation, in which even those whose faith is weakest must inevitable recognize one of the principal lines of human progress" (OT10b, 149[129]). It is at the same time "the authentic expression of the Christ of the gospel". Christ become "even greater in order still to remain the same Christ" (OT10b, 149[129]).

---

[26] Gustave Martelet (2005, 218–33) notes the relationship between such a Christ and "the new face of God".

[27] This is the object of "Christology and Evolution" (OT10a, 93–113[76–95]). Teilhard clarifies there: "Since, in concrete fact, only one single process of synthesis is going on *from top to bottom* of the whole universe, no element, and no movement, can exist at any level of the world outside the 'informing' action of the principal centre of things" (OT10a, 107[88]) Compare with *Hominization* (OT3c, 103[72]): "From top to bottom of the series of beings, everything is in motion, everything is raising itself, organizing itself in a single direction, which is that of the greatest consciousness".

[28] In *Christology and Evolution* (OT10a, 108[89–90]), Teilhard expresses the same idea in an historical manner: "In all truth, we will be perfectly justified in saying that evolution has preserved our God for us if, through evolution, our religion is forced to recognize the existence of the universal Christ and the fullness of his efflorescence. But to balance this (and this is even more true) we shall have to add that the universal Christ have appeared just at the right time to protect the idea of evolution from itself". For me, this expression is more imaginative and less rigid. Christ is the "saviour of evolution", because the Universal Christ is "personal" (109[90]). In the same way, in *The God of Evolution,* Teilhard wrote: "Christ, on the other hand, and Christ alone, as time passes, becomes a more and more real being for a particularly vigorous section of mankind; an this is a result of a twofold process which, paradoxically, continually both personalizes and universalizes him more fully as the years go by" (OT10f, 289[241]).

[29] See also *Christianity and Evolution* (OT10e, 210[180]). Juridical signifies declared as such (for example, "King").

II. The rise of history and the descent of God

In a still unpublished letter of the 9 April 1967, Fr Henri de Lubac proposed to Étienne Gilson, a key for reading — the word is his — the works of Fr Pierre Teilhard de Chardin: "If, according to him, everything rises in the interior of the history of the world, it must first of all, in each case 'descend' from God: each time the context suggests this, he says it with a clarity, even an insistence, without fail".[30] Evolution ascends, we know, according to the law of complexity-consciousness; the descent of God works in the cosmic Christ. According to the interpretation of Fr de Lubac (2002), that we will follow, of which truth was the scholar and the mystic Teilhard the initiator and in what way can this truth enlighten science and theology? (Rabut 2003, 209, 189, 157–58)[31] Such are the two questions posed.

1. Teilhardian phenomenology

The Human Phenomenon is a "scientific treatise". It only considers "the phenomenon" but considers "the *whole* phenomenon" (OT1, 21[29]).[32] Let us try to understand what Teilhard understood by this and understand the question of the author: are we in a position "without overstepping the limits of strictly positive science, to find room for such an idea as the 'within of things'?" (Lubac 2002, 97[72]; see OT3c, 78[52]; OT1, 49–64[53–66]; Daleux 2001). This is the result of the fact that, in the Universe, "the world is completely transformed from one circle to another" (OT9c, 122[91]). To see the human Phenomenon is "to appreciate and effect the transposition that must be applied to the organico-physical laws and concepts established first for the world of inorganic matter and then for that of vitalised matter, when we enter the world that has been given a new form by the specifically human power of reflection, in other words, *the hominised world*" (OT9c, 123[92]). The "inside" is the result, according to Teilhard, of the inverse of entropy, which is life (OT3i, 237–39[167–70]); on a moral level, resulting in a "appetite for living," (OT3i, 240[171]) "a sort of internal gravitation," an attraction "upwards, from within" (OT3k, 322[230]).

---

[30] Henri de Lubac to Étienne Gilson, Lyons, 9 April 1967, Archives Gilson, Toronto; see Lubac 1986, 111 n. 5. See also the study of this double movement in Lubac (1965, 195–227; 2002, 14–15[13–14]).

[31] The relationship between Pierre Teilhard and Henri de Lubac has been considered by Xavier Tilliette (unpublished work), Jörg Disse (2001, 94–110) and Jean-Pierre Wagner (2003, 199–214).

[32] The same thought is in *Hominization*: "we will try to look at man as pure natural scientists, adding nothing to him and, at the same time, taking nothing from him" (OT3c, 78[52]).

1.1 The entire phenomenon

The phenomenon in its roots[33] is a collection of "experimental relationships" and of "laws of recurrence", registered at the interior of a certain sphere. "I have chosen man as the centre, and around him I have tried to establish a coherent order between antecedents and consequences" and "to discover … between the elements of the universe … an experimental law of recurrence which would express their successive appearance in time" (OT1, 21[29]).

According to this law, the *Human Phenomenon* and similar works present an "anthropocentrism of movement" (OT2c, 297[209]). "Man is no longer merely at the centre of nature, as he was in the fixed geocentric system … but at the growing-point or peak, man is seen to be 'a specific summit'" (Lubac 2002, 110[81]). Since the appearance of man, the experimental reality has adopted "a new method" (Lubac 2002, 126[93]). "Cosmogenesis is being continued in an anthropogenesis" tending "towards a divine focus of mid which draws it onward" (OT1, 301–302, cited by Lubac 2002, 127[93]).

"When it reflects upon itself, the consciousness of the individual acquires the formidable property of being able to foresee the future, in other words, to foresee death. And at the same time it recognizes that to collaborate any longer with life's efforts would become psychologically impossible, if something — that something being what is best in its work — could not succeed in escaping total destruction … As applied to the individual the idea of *total* death may not, when first entertained, prove too great a shock; but extended it to the whole of mankind in its unity, and our minds revolt".[34] "From the objection itself, Père Teilhard extract an initial, though not a complete answer, when he distinguishes within this power of despairing reflection a 'demand for irreversibility'", and therefore, a solution (Lubac 2002, 57[43]; see also 142[104]; see OT7b, 115–116[97–127]; OT3k, 323[232]).

"Supremely *complex*", man is "supremely *free and conscious*" (OT7b, 106[101]; Lubac 2002, 150[109]; see also OT3c, 106ff.[74ff.]). "Similarly, the Noosphere, which in its initial significance is a reality with which we are already presented, is also, in a second sense, and predominantly, an ideal to be realized morally and by the use of our freedom … Hominization means 'the appearance of morality and

[33] To see the "human Phenomenon" as Fr. Teilhard understands it, "*there is ultimately no more simple, nor more radical way of doing so than to become conscious of the unity of cosmic evolution*" (OT9c, 121[90]).

[34] Quoted by de Lubac (2002, 56-57[42]), we do not find it in the given reference (OT7e, 203–210[197–203]).

the crisis it involves'; and it is for that reason that Anthropogenesis, 'surrounded as it is by dangers,' is a drama" (Lubac 2002, 150–51[109]).[35]

How do we return? "What had been until then a blind urge towards an 'infallible progress' became 'the power of reflective organization' and 'a progress that may be accepted or rejected'; and the effective achievement of that progress can consist only in a 'rise into the improbable through the triumph of freedom'" (Lubac 2002, 151[110]).

In the order of liberty, "in Père Teilhard's view, a double victory is assured. The first is the victory of Spirit over matter" (Lubac 2002, 149–50[108]). The Spirit represents "the indestructible part of the Universe" (OT6a, 51[41]; cited by Lubac 2002, 150[109]). " 'The ebb of entropy' will be overcome by 'the rising tide' of Noogenesis" (OT5c, 103; cited by Lubac 2002, 150[109]; see Ratzinger 1969, 240). "A second victory, involved in and conditional upon the first, is equally assured. This is the victory of the risen Christ, whose Pleroma will one day be accomplished … For the Christian, man's 'biological success' must be much more than a probability: it is an absolute certainty" (Lubac 2002, 150[109]; see OT5h, 305[246–47]). We shall return to the second victory.

Let us consider the victory of the Spirit. It is fulfilled by "the fundamental choice of the individual", then by "the *common choice* of the mass of Mankind" (OT5b, 78[59]). In effect, a day must come, thinks Teilhard, when men "having finally become aware of their unity to all things and their intimate link with all the rest, will have between their hands the fullness of their soul to throw freely into the divine Source". It is thus that "the final choice" will take place: "whether a world is to revolt or to adore" (OT5a, 31[19]). It is most probable, according to Teilhard that the abyss between rebellion and adoration would deepen along with the advance of hominization towards the final option (Lubac 2002, 157–58[113–14]). For, the same choice between rebellion and adoration is imposed upon each generation.

Thus Pierre Teilhard de Chardin saw the *entire* phenomenon. But he saw *only* the phenomenon; he saw it with fresh eyes according to the parameter of duration (see among other texts OT10a, 97[78]), one of irreversibility and of a synthetic fashion.

1.2 Nothing but the phenomenon

This is the duration he so sharply perceived: time "measured by the development of global Reality" (Lubac 2002, 27[23], see OT5d, 111[86]; OT10b, 127–28[106–107]; 135–36[114–15]; OT9d, 129–46[98–112]).

[35] The references to Teilhard are found there. This tragedy is described for the current era, the end of the Neolithic, in (OT3c, 108–109[75–76]).

How did Fr Teilhard de Chardin try to establish his human phenomenology as science? He considered "the time has come to realise that an interpretation of the universe — even a positivist one — remains unsatisfying unless it covers the interior as well as the exterior of things; mind as well as matter. The true physics is that which will, one day, achieve the inclusion of man in his wholeness in an coherent picture of the world" (OT1, 30[35–36]; 51[55]). Unification, obtained by complexification and organization, is accompanied, in effect, "by a growth of inner consciousness and freedom". Also, "the principal movement of reality is a synthesis" (OT6b, 72–73[56]).[36]

Which is the result of such a synthesis? An "inversion of the Cosmogony".[37] By his synthesis, Teilhard bring about, also, the pursuit of an "extension of concepts" or their "distortion" (d'Armagnac 1957, 29–39). "This applies to the concepts of orthogenesis, energy, and consciousness" (Lubac 2002, 237[167]). "Analogy works out perfectly, and the philosophers may take comfort. At the same time, it is in fact precisely the analogical character of the concept that raises difficulties for the scientists" (Lubac 2002, 239[168]) since positive science requires univocity of concepts.

A positive science needs to be based in addition upon the measure of quantity. And yet, "Fr Teilhard ... sets out to determine 'the qualitative laws of growth' " (OT1, 54[58]; Lubac 2002, 239[168]). He measures duration by "the rise of reflection" (OT7h, 301–302[289]). "Finally, he puts forward the view that in the phenomenal 'without' can be distinguished a 'transphenomenal' element which is something 'invisible', and which nevertheless science cannot entirely leave out of account; he goes so far, too, as to postulate the reality of 'souls'" (Lubac 2002, 239[168]).[38]

In the space of "chaos" which lies between science and philosophy, between the knowledge of man and the sciences of nature (To P. Valensin 04.07.20, Teilhard de Chardin 1974, 63), Teilhard wants to introduce his "third abyss", the abyss of complexity.[39] While the first two, the two "pascalian" abysses of infinity and nothingness, all abstract and mathematical, are "infinities of dispersion", this abyss of

[36] "Man, the center of perspective, is at the same time the *center of construction* of the universe" (OT1, 27[33]). "In the human spirit, as in a single irreplaceable fruit the whole life of earth — that is to say, in brief, its whole cosmic value — is gathered and sublimated" (OT6a, 35[28] ; quoted by Lubac 2002, 236[166]).

[37] Letter (to Max and Simone Begouën?), 1929 (Teilhard de Chardin 1956, 118). Teilhard confirmed this in other places in his work.

[38] "Being is time, it does not only have time" (Ratzinger 1969, 240 s.).

[39] Teilhard wants it first of all in science, namely in the "hierarchical law of increasing complexity in unity" (OT9a, 56[29–30]).

complexity is "the abyss of synthesis" (Baruzi 1907, 237). And that establishes a "kind of phenomenology or generalised physics" (OT1, 49[53]; Lubac 2002, 240–41[168–69]; see also OT11c, 231–36[212–15]; d'Armagnac 1957).[40]

Such a scientific understanding rests upon three conditions (Galleni 2010). We are already familiar with the first two: the irreversibility of the Spirit and evolutionary personalization, and both in the end suppose another, supreme condition, which ensures in advance the solidity of the group. This is called Omega Point.[41] These three conditions presuppose faith. Still, according to Teilhard, to believe, is "to effect an intellectual synthesis" (OT10b, 119[98]; 128[107]). The irreversibility of the Spirit is postulated, let us insist, by means of the absurdity of absolute death (OT6a, 23–57[19–47]).[42]

2. The cosmic Christ

So the rise of cosmic evolution comes from above, from the descent of God. If we accept this affirmation, Christ, who comes from God, takes on a cosmic dimension. He is the Omega point. From this comes a double question: what is the link between Teilhardian phenomenology and Christ?[43] What is the link between the cosmic Christ and Jesus of Nazareth?

Respecting the distinction between scientific and religious disciplines, Teilhard proposed to combine these two questions in a twofold fashion: on the one hand, by means of hypothesis, he suggests it is clarified in the "structure and the workings of the Universe" thanks to the "Christian phenomenon"; such is the point of the important essay from 1948 *My Fundamental Vision* (OT11b, 177–223[163–208]); on the other hand, he shows to the Christian (Lubac 2002, 100–101[74–

---

40 "My 'phenomenology' is not that of Husserl or Merleau-Ponty" (To C. Cuénot, 11.04.53, Cuénot 1958, 311). After having discovered space following Copernicus and Galilee, after having discovered the double abyss of the Minute and the Immense, we have, wrote Teilhard, discovered time, the third dimension and the third infinite "Through the notion of 'third infinite' (the 'infinity of complexity')", Fr Teilhard wanted us to learn "the 'gesture' — as he put it — that will enable us to 'overcome the illusion of Quantity' " (Lubac 2002, 90[66], quoting OT10b, 126[106]); he wanted to "dissipate the mirage of matter" (OT11b, 177–223[163–208]).

41 God is also called Omega by Teilhard (OT10f, 288[240]), as is Christ. It is that Christ the "ultimate summit (that is, the only possible God) of an evolution definitively recognized as a movement of convergence" (OT10f, 291[243]).

42 However, it remains that men after the Neolithic era recognize this absurdity: it is a vital choice (OT3c, 110[76]).

43 The role of Christ in the evolution of the world is studied by Christopher F. Mooney (1968). See also Peter Schellenbaum (1971).

75]) and also to the non-believer, what is renewed in the Christian phenomenon when he takes into account the human Phenomenon: it is in this perspective that *The Divine Milieu* appears. In the two cases, but according to two inverse perspectives, the Christian, "drawing upon an added source of knowledge, may advance yet another step" (Lubac 2002, 12[12]; citing OT5h, 285[231]). *The Religion of Teilhard de Chardin*[44] sheds light upon the internal link between these two perspectives: who does allow science[45] to see the Christian phenomenon by means of hypothesis? What does the human phenomenon ask of the Christian phenomenon in accepting being entirely itself and for the Christian phenomenon to manifest its true dimension? We shall try to answer these two questions in the following two sections.

Once we decide the act of faith, we can understand in two ways "the deep structure" of Omega Point. Without blurring the truths of faith and science, Teilhard presents several alternatives concerning Omega Point: as a unity of pure convergence or as Being that already attracts us; as being dependent on the forces of evolution or independent of them; some Thing or Someone, a centre of unification or transcendent reality, that may be the home of immanence and height of transcendence. He wonders which of the two terms " 'has the weight of logic in its side', his science does not force personal assent; all it does is constantly to offer 'rational invitations to an act of faith', and in the final act in that series, it 'emerges' itself into 'worship'" (Lubac 2002, 262–63[182]). Here the supernatural act of faith is only prepared; it strengthens the scientific reasoning of Fr Teilhard (Lubac 2002, 265–66[183–84]) and will bring it new light. But this reasoning is content to say nothing but what is "christianisable" (Letter 16.03.31, Teilhard de Chardin 1956, 140[172]).[46]

---

[44] Gaston Fessard (1964, 224) appreciates the value of this book: "Fr H. de Lubac has given, in *La Pensée religieuse du père Teilhard de Chardin*, in a way a full commentary and a masterly illustration because it is exactly balanced. In effect, he doesn't conceal the 'insufficiencies of language and thought' of the Teilhardian synthesis, concerning theology and even more metaphysics, which justify, in the life of his author, the prudence of his Superiors and explain today the reservations of the Magisterium. But at the same time, he brings out the deep correctness of the aim, the intrinsic conformity to the tradition, indicating in the end which 'great paths of research' it opened and which 'accurate maternal intuitions in it are capable of furnishing Christian thought with new blood'".

[45] Science has two aspects: "1) Because the scientific study of the world is analytical, at first it makes us follow a direction that leads away from divine realities. 2) On the other hand, since this same scientific insight into things show us the *synthetic structure* of the world, by its natural extension, turns us back to the unique center of things, namely, which is God our Lord" (OT9a, 47–48[21–22]).

[46] On the distinction between science, philosophy and revelation, see Henri de Lubac (1965, 171–94).

III. Christ and evolution

We shall now concern ourselves with the second perspective, that of *The Divine Milieu*. After having tried to establish that there is in man an "impulse towards *Some Thing* ahead", he concludes in the perspective of *The Divine Milieu* that this impulse "cannot achieve its full fruition except by combining with another and still more fundamental aspiration — one from above, urging him towards *Some One*" (OT5i, 374[302]). Teilhard thus distinguishes between nature and grace and brings them together "in the interplay (so wonderfully contrived) of what I call the two lines of curvature (or convergence) — the cosmic ('natural') and the Christic ('super-natural')" (Letter 07.08.50, Teilhard de Chardin 1957, 100[303]). From 1920 onwards, for him it is impossible "to read St Paul without seeing, beneath his words, with dazzling clarity, the universal and cosmic dominance of the Incarnate Word" (Lubac 2002, 144–45[105], quoting OT9b, 84[56]). He sees break forth "the astonishing, emancipating, harmony between a religion that is Christic, and an evolution that is convergent, in type" (Lubac 2002, 146[106], quoting OT13b 108[93]). Such is the purpose of *Science and Christ* (OT9a, 45–62[21–36]).

How are science and Christ brought together? "If there are without any doubt exact physical laws for the development of spirit within history, why should there not be, *a fortiori*, laws for its further expansion and ultimate culmination?" (OT11a, 169[155]; Lubac 2002, 176[126]). There is thus a phenomenology of "Christ the evolver". Only, to describe the relationships of one to another, two words take on two correlative aspects under which their relations should be seen: "transformation" or "integration" and "critical point" (Lubac 2002, 178[127]).[47] The "critical point" is "a point of annihilation", a point of death (Lubac 2002, 185[132]). Correlatively, the concept of transformation is always double faceted. It poses simultaneously continuity and discontinuity, permanence and change, the same and the other (Lubac 2002, 186[132]; 197–98[141–42]). By reason of this "critical point", all concordism is eliminated and, by reason of transfiguration, all "figurative anticipation" is also eliminated. The transfigured is unrepresentable. "Everything, accordingly, will perhaps end 'through a turning about, psychic rather than sidereal, which will have the appearance of death but will be in fact a liberation from the material plane of history, and an Ecstasis in God.' In the dis-

[47] Other terms: "transfiguration", "reversal", "metamorphasis" or again "excentration" and "inversion". See also Henri de Lubac (1971, 12). He speaks there also of "segregation", of "creative union", of "point Omega" and of "energetic".

appearance of what must perish, this will be 'the consummation of the divine Milieu' " (Lubac 2002, 192[137], quoting OT6a, 57[47]). Thus in one way, without gnosis, "the Universe … follows man's destiny" (Montcheuil 1949, 175; Lubac 2002, 193[136]) and correlatively, "the 'cosmic convergence', brought about through 'the critical point of co-reflexion', does not by itself entail 'the Christic emergence'" (Lubac 2002, 193[137]; see OT13b 97[82]). On earth, among man, certain natural conditions matured. It is the "point of human maturity" that will correspond to the "Parousia-point". We must expect this event in total detachment. It is a "total transformation", "continuity in discontinuity" (Lubac 2002, 193[138], 200[142]).

Teilhard was tirelessly concerned to show in "the sudden revival of human evolution" an "*irresistible* rise towards the *Personal*" (OT5f, 265[215]). "A Universe of personal substance" and a Universe "of converging curvatures" — which amount to the same thing — is in the eyes of Fr Teilhard de Chardin "the only milieu capable at once of receiving the Christ we adore, and of interesting the man of whom we dream" (Letter 22.11.36, Teilhard de Chardin 1974, 323). It is faith in Jesus that satisfies for Teilhard his new "propulsive faith in the Ultrahuman" and his "ascensional faith in God" (Lubac 2002, 221[156]).

1. What the cosmic Christ brings to the human Phenomenon

The solution science expects, without finding it, has been shown through divine Revelation. The disciples of Christ "believe in the vivifying power of death. They … maintain that through Faith man can conquer what is inevitably determined… Then, perhaps, if we again reflect on our human condition, we might be in a better position to find this issue that 'opens up for consciousness in the hearth of things'; hope of 'an escape in death' might take root in us and flourish. And then, 'as through magic' our 'terror of Matter and Man' will be 'transformed, reversed, into peace and trust'; we shall feel that we are already, in principle, 'freed,' for there is indeed a World-heart" (Lubac 2002, 71[54]). As rational as Teilhard might want it, this requirement remains in his eyes "faith in a supreme centre of personalization, of convergence, and of cohesion, in which alone salvation for the Universe can be conceived" (Lubac 2002, 57–58[43], citing OT3g, 195–96). God, who universally brings events to life, is such a centre (Letter 27.05.23, Cuénot 1958, 65). In addition, "no belief that is of the purely natural order, however firmly based it may be in reason, can be sufficient at this juncture. It could never, Père Teilhard believed, succeed in completely overcoming practical pessimism in real life. For it is not enough to understand the necessity of suffering: if we are willing to accept suffering, nothing can take the place of the 'astonishing Christian revelation' which shows us that, 'provided it be accepted *in the right* spirit, it can be transformed into an expression of love and a principle of union'" (Lubac 2002, 57–58[43], citing OT7g, 256[248]). "The believer must still pray, so that he may not succumb 'to the temptation to curse the Universe and its Maker'" (Lubac 2002, 58–59[44], quoting OT4). This holds up to the point where death accompanies every change for good or for ill. "While, however, with some (whom it leaves permanently broken up and diminished) it is purely *ad mortem*, with others (whom, as it breaks them up, it compensates) it is a transition — the unique issue — that leads to the new life" (OT12b, 221[173]).

It is thus that death is "the door that opens on to transfiguration". It is perfect passivity, if it is "an actual interior state and as an attitude of obedience and love proposed to our freedom" (Lubac 2002, 75[56]). The apparent annihilation is "the supreme gateway of Life", of the Divine Milieu, which "present in, and exercising its attractive force in the inaccessible depth of every creature" nevertheless always evades its embrace, withdrawing "always further bearing us along with it towards the common centre of all consummation": always by the cross and by its mystery (Lubac 2002, 79[57–58], citing OT4, 135[113]).

2. Jesus of Nazareth and the cosmic Christ

From *The Universal Element*, Teilhard considered "as insufficient any explanations concerning the unity of the world through the 'will of God' or by his 'creative Action' and suggested as the principle of unity 'the cosmic influence of Christ'" (OT12e, 401–409[OT13, 221–18]). "It was this doctrine that replaced for him, after some searching, the idea of a 'natural soul of the world'" (Lubac 1965; see OT12c, 248–50[180–82]).[48] "The universal Christ, is the Christ influencing everything *physically*". For all that, it did not take the place of created freedoms (Lubac 1965; see OT9b, 85–86[57]).

From a slightly different perspective, that of *The Divine Milieu*, Teilhard suggests that in his own humanity, the historical Christ is "not only the perfect Man, the ideal Man, but the entire Man, one who gathers, from the depths of his consciousness the consciousness of every man" (OT6b, 72ff.[53ff.]). Teilhard knows and proclaims that "even today abandonment of the historical character of Christ (that is, the divinity of the historic Christ) would mean the instant dismissal into

[48] It has not been possible to find some of the references given by the author to Lubac 1965 (*Editor's note*).

the unreal of all the mystical energy accumulated in the Christian phylum during the last two thousand years. Christ born of the Virgin, and Christ risen from the dead: the two are one inseparable whole" (OT10d, 187[159]; see also Lubac 1971, 17).

So Teilhard can still affirm: "The mystical Christ, the universal Christ of St Paul, has neither meaning nor value in our eyes except as an expansion of the Christ who was born of Mary and who died on the cross. The former essentially draws His fundamental quality of undeniability and concreteness from the latter. However far we may be drawn into the divine spaces opened up to us by Christian mysticism, we never depart from the Jesus of the Gospels" (OT4, 141[105]). In other words, "without historic and traditional Revelation, Our Lord evaporates" (OT10e, 201–216).[49] "Is in the mystery of the Eucharist that Père Teilhard found the vital synthesis of these aspects of the Christ who was to him always one and the same Christ" (Lubac 2002, 87[64]). And also in the Church: the Church, "Christic pole of the earth; through Rome, I mean, runs the ascending axis of hominization" (To M. Teillard-Chambon, 07.10.48, Teilhard de Chardin 1957, 95[299]). For "without the Church Christ evaporates, or crumbles, or is annulled!" (Lubac 1965, citing a note from 4 November 1916).

The "universal Christ", is the Christ that transforms in Him the human Phenomenon. So Teilhard can speak of "panchristism", originally a Blondélian concept.[50] He affirms that in this "there is no false pantheism" (OT9b, 87[59]). Henri de Lubac shows this amply (Lubac 1965). This panchristism is the means of a "pantheism of differentiation" (OT7f, 234[225]; cited by Lubac 2002, 223[157]) where God is all in all.[51]

---

[49] I did not find this citation there but I found the equivalent: "If it is indeed true that it is through Christ-Omega that the universe in movement holds together, then, correspondingly, it is from his concrete germ, the Man of Nazareth, that Christ-Omega (both theoretically and historically) derives his whole consistence, as a hard experiential fact. The two terms are intrinsically one whole, and they cannot vary, in a truly total Christ, except simultaneously" (OT10e, 211[181]). This passage was underlined by Henri de Lubac in his copy.

[50] On the Blondelian original, see in addition Lubac (2002), *Lettres intimes* (Teilhard de Chardin 1974, 455; 457): "T had contemplated the problem of action following Maurice Blondel, whose influence on him, we noted, was profound. Even though he transferred the Blondelian themes into a totally different framework of thought, they remained determinants for him". See Virgoulay (1990, 201-209).

[51] 1 Cor. 15: 28 is cited thirteen times in Greek in the works of Teilhard (Lubac 1965).

IV. Thinking of evolution and Christ today. Questions of method[52]

How should we think of evolution and Christ fifty years after the death of Pierre Teilhard de Chardin? Is it legitimate to pose such a question? Nothing prevents us from wondering if this axis of life passes Christ by. In an inverse sense, what relationship exists between the Pauline vision of Christ and the cosmic role of Christ?[53] These questions suppose links between science, philosophy and theology (Lubac 1971, 13). In order to pass from the science of Life to theology the mediation of philosophy is necessary, for it is by philosophy that man reflects upon that which is and upon his destiny. Let us examine these questions of method, leaving aside the current vision of evolution.[54]

Teilhard de Chardin showed that the science of Life itself passes from its outside to an inside: the Spirit that appears in terms of the biosphere is not just a result that is noticed; it is also a mover, which gives direction. Once we have noticed it, we cannot but notice that we have put ourselves in motion by the object. Henceforth we must yield to the origin of the Spirit's movement. No only does the most come from the least — this is the outside —, but the most gives rise to the least according to the law of complexity-consciousness and according to the third Infinite — this is the inside. It is by the necessity of structure that the science of life is worked out according to the inside.

Fr Teilhard de Chardin developed this point of method. Organically, duration is irreversible (see OT9c, 125[94]). "Side by side with the measurable current of entropy, or running across it, there is another current, impatient of measurement; it is disguised in the material, comes to the surface of the organic, but is most clearly visible in the human. This is the imponderable current of Spirit" (OT9c, 126[95]). The world "is irreversible, which means that, considered in its evolutionary essence, it is imperishable" (OT9e, 226[179]). A key is there in order to "explore from within the universe that physics has, until now, tried to apprehend from without" (OT9c, 126–27[95]). "If physics is ever to be the scientific cosmogony it hopes to be, it will become progressively more impossible

[52] We have not considered the risks of Teilhardian thought; on this subject see Mooney (1966, 229-242).

[53] These are the questions of Teilhard: "What form must our Christology take if it is to remain itself in a new world?" (OT10a, 95[76]). And he also aims to solve this question: "Christianity and evolution: not two irreconcilable points of view, but two ways of looking at things that are designed to dovetail together, each completing the other" (OT9f, 240[189–90]).

[54] See Reisse *et al.* 2006; Baudry 2002; Ladrière 1976; Facchini 2002. On evolution according to Pierre Teilhard de Chardin, see Arnould 1996; Birx 1991; Brechtken 1983; Glässer 1984.

for it to confine itself to the study, from outside, of the phenomena governed by the laws of great numbers and subject to entropy. It will have to have recourse to complex symbols or functions in which the second aspect and the second current of things can be expressed: by those I mean the individual spontaneity and progressive organicity of the elements, seen from within" (OT9c, 128[96–97]). Teilhard knows that he will have against him, because of irreversibility, affirmed within *The Human Phenomenon*, "the pure scientists ... as well as the experts in pure metaphysics" (To Max and Simone Bégouën, Peking, 09.05.40, Teilhard de Chardin 1957, 44[263]), who "following the prophet Einstein, many physicists (or speakers) begin to proclaim the existence of a finite and reversible universe where duration must give way before space, the only definitive and fundamental reality", while according to him, every true science conceives " an irreversible march of the spirit" (To Henri de Lubac, 31.07.30, Teilhard de Chardin 1974, 219).[55]

Besides, is only scientific analysis sufficient for knowing the real? By means of analysis, we can only, thinks Teilhard, "reach the extreme lower limits of the real" and "obtain the unity of impoverishment in homogeneity" (Lubac 2002, 235[165]; see OT6d, 141–200[113–62]). According to him, the moment has come to overturn the perspective boldly, to oppose "a true science, more comprehensive in its object and more positive in its results, against a negatory illusion that has become a parasite on a stunted science" (Lubac 2002, 235[165]). This negative illusion is one of "a sort of inside-out metaphysics or mysticism, to accept a materialist or nihilist dogma: materialist, if he fails into the illusion of 'regarding the elements of the analysis as more real than the terms of the synthesis'; nihilist, if he sees the universe disintegrating and eluding him even as his analysis proceeds" (Lubac 2002, 233[164], quoting OT10b, 125[105]).[56] "Such, according to Père Teilhard, was the mistake or the inevitable fate of nineteenth-century science" (Lubac 2002, 233[164]). "And in spite of many disappointments", the spirit which produced such an illusion "has by no means disappeared" (Lubac 2002, 234[165]).

Thus Teilhard sought to found a science of the inside. One philosophical question remains to be treated: can the results of such a science of life be clearly read? From a general manner, is there an induction other than scientific induction? Is there a philosophical induction? Gaston Isaye established it by calling the result of this induction:

---

[55] The connexion with Bergson seems evident (Worms 2004).

[56] Josef Ratzinger (1969, 240) noted this consideration in Teilhard.

"philosophy of nature" (Isaye 1987, 170–201).[57] Jean Ladrière approved such a method because of the autonomy of the sciences maintained by Isaye and because of the reciprocal relationships between sciences and the philosophy of nature (Isaye 1987, 21).[58] So Teilhard showed why science implies an inside, but did he show the link between the science of life and philosophy and "their mutual interaction"?

So in theory the link between science and philosophy is established. It can also be in practice. While science has been changing — and remains changing — no longer only thanks to Greek *theory*, by discursive reason, inherited from Descartes and from Kant, or by a positivism derived from Auguste Comte, but thanks to the capacity received by man from God, his Creator, to know and to subdue the entire universe, it puts in place — *volens nolens* — a history oriented towards an end that it can watch as a whole in a synthetic glance. Bioethical committees show the practical necessity of uniting science and morals or ethics. How does such a union work? Could the link take shape between a science limited to space, to reversibility and to analysis and an ethic that cannot miss at least suggesting the duration, the irreversibility and the synthesis that brings about the free act? Is it not important to clarify the freedom of each person and of the whole society, that science accepts the theoretical base which founds it, even at the risk of turning against God the Creator the freedom of man which it exercises? In terms more philosophical than Teilhardian, a philosophy of nature sheds light on the sense of scientifically known phenomena. Could it abstract from this sense so that it is given to man, that is to say, abstract from the final destiny of man? If the universe has a sense for man, how can man put this into action?

A second question of method bears upon the link between the science of life, such as Teilhard conceived, and theology. Without dealing with the relationship between the sciences of nature and theology (See Puddefoot 1998, 1076–1077), we shall suggest a limit to the thought of Teilhard, which is of a more scientific order. Fr H. de Lubac mentions this as a "daring extrapolation" (Lubac 2002, ch. 19).[59]

[57] Christoph Schönborn explains that the philosophy of nature is the mediation between faith and science (Schönborn 2007, 81ss).

[58] Marc Leclerc presented the philosophy of Fr Isaye by considering man. He also explains the role of language in philosophical induction: "the mastery of language that allows the appearance of culture came to *humanise* the hominid formed by this evolution, without being adequately explained by physiological transformations what renders them possible." (Isaye 1987, 444)

[59] Simply by critiquing the imprecisions of language, Fr de Lubac in this book does not implicate the doctrinal rectitude of Teilhardian thought. Reminding the *Monitum* of the Holy Office on 30 June 1962, Cardinal Paul Poupard (2004, 23) mentioned "some ambiguities and grave doctrinal errors" contained in Teilhard's work. In his *Mémoire sur l'occasion de mes écrits*,

We know that the phenomenology of Teilhard and the cosmic Christ have a twofold link between them. Nevertheless, does Teilhard not make a "daring extrapolation" in respect of a second critical point of Man? There exists, in effect, according to him, a "second critical point of reflection, collective and superior": a "critical point of socialisation" or of "co-reflection", that will finally mark the full "humanity" (OT1, 279[251]; 320[287]; OT5b, 81[63]; see Martelet 2005, 179–93).

And yet, the "threshold of reflection," by which man was constituted (OT1, 180[164]), is it necessarily unique? Teilhard doesn't deny that "the threshold of reflection bears in itself something definitive" (OT1, 90[88]). And, concerning the new critical point, he thought of a promotion, not extinction. The Being towards which humanity tends, would be endowed with the same essential attributes, but with an increased intensity; beings would be personal, of a reinforced personality.

Yet, "after telling us that he proposed to study Spirit exclusively as a phenomenon, in its objective manifestations that are patient of scientific observation, he later proceeds to say, not so happily, that 'Spirit is not a meta- nor an epi-phenomenon; but *the* phenomenon.' In short, the 'phenomenon of man' would appear to him to have embraced, at least in man's eyes, the whole of the 'human paradox.' In other words, he may not have been sufficiently on his guard against the trap laid for him by his phenomenological point of view. That point of view, which in principle should have been adopted simply as a matter of method, would seem to have ended by becoming exclusive — at least in some stages of the systematic construction Père Teilhard attempted. That is why it would not be possible to fully recognize in that construction the originality of the paradoxical reality represented by human thought, or reflective consciousness: a reality that is born in time, grows in time, is conditioned in countless ways by time, and yet, as soon as it appears, by its very being, dominates time. *Quasi in horizonte existens aeternitatis et temporis*"[60] (Lubac 2002, 306[212–13]; citing OT6c, 118[94]; OT1, 179[163]).

"Is Père Teilhard much more successful when, in order to establish his concept or to make it more probable, he has recourse to a factual argument"? (Lubac 2002, 306[213]), that is, "the fantastic spectacle, staring us in the face, of a rapidly rising *collective Reflection,* moving in step with a increasingly unitary organization"

Fr de Lubac (2006, 106) mentions the *Monitum* "by quite vague concepts, the importance of which was sometimes exaggerated". The *Monitum* was equally aiming at his book on Teilhard. A text that had to be presented to the Council demanded the condemnation of Teilhard (Lubac 2006, 118). In fact, the Council pronounced nothing.

[60] "As though it existed on the horizon of eternity and time" (Thomas Aquinas, *Summa Contra Gentiles* II, 80, 13 — *Editor's note*).

(OT13a, 48[38]). Here we find the observation of a fact and its interpretation curiously run together. If we accept the fact, does it follow "that we are travelling through an 'ultra-evolution' towards the 'return' of a 'new critical point' beyond which the 'ultra-human' would come into its own"? (Lubac 2002, 308[214]). However, Fr Teilhard does not forget that this future Humanity is made up of persons who are each responsible for their last decision.

He does not confuse the two processes of personalization, that of socialization and that which comes from God and from Christ. But, "Père Teilhard is anxious to bring out their reciprocal relationship and, for a dialectical purpose we might say, he first adopts a point of view that looks at it from his side: in some passages, then, he gives the impression, though it is no more than an impression, of developing a philosophy — one might almost call it a mythology — that is questionable, whereas in reality the inspiration behind him is already the indisputable truths of his faith" (Lubac 2002, 310[215]). The "critical and *final* point" (OT7h, 302[290]) "represents our passage, by translation or dematerialization, to another 'sphere of the Universe'[61] … The point of 'ultra-hominisation', it will mark, as we have just seen, the moment without extension that is to follow immediately upon our access to the trans-human. This will be the final condition and instantaneous sign of the Parousia" (Lubac 2002, 311–12[217]).

Another question: how are the universal Christ of St John and St Paul and the Christ Omega articulated? "It is the merit of the works of Jean-Michel Maldamé, from the Pontifical Academy of the Sciences, to have collected the positive aspects of the Teilhardian vision, by going beyond them and correcting that which, among them, could lead us towards pantheism" (Lambert 1999, 53; Maldamé 1993, 181–98; 1998, 202–203; 2000; 2006). However, it is only a "risk" (Lambert 1999, 158) and Dominique Lambert absolves Teilhard of pantheism, when Teilhard speaks of "God all in all" (Col. 3: 11; Lambert 1999, 55).[62]

Yet, is the inside thought of as that of human history? This is the question of Gaston Fessard. Certainly Teilhard recognizes this: "never forget that between the world and God there is Man" (Fessard 1964, 232). But, in order to understand him, thinks Fessard, it is important to understand that the affirmations of Teilhard "concern essentially *historicity* and a properly *supernatural* historicity, even if they do not reach

[61] Henri de Lubac (2002, 185–200[132–42]) has explained it in a chapter on the *Transfiguration of the Cosmos.*

[62] Jean-François Thomas (1998, 103-120) shows how Teilhard distances himself from the temptation of "pagan" pantheism or of "diffusion", opposing "convergence" to pantheism.

natural historicity — or "the evolution of the Cosmos" ... — that *by the intermediary* of *human* history, so of our historical actuality" (Fessard 1964, 239).

Conclusion

Thus are the fundamental intuitions of Teilhard philosophically and theologically better placed.[63] However, a question might seem to remain: "What is properly "nature or 'evolution' as subject? In fact they don't exist! When we say that nature does this or that, this can only be an attempt to regroup a series of events as a subject, which nevertheless doesn't exist as such. It seems to me evident that this verbal expedient — perhaps inevitable — is comprised in itself of claims of a certain weight" (Ratzinger 1969, 155). We can note, however, two things with Dominique Lambert: "if we want to correctly appreciate the type of ontology underlying contemporary science, we must recognize that it *effectively* reaches reality, but that it only grasps it with a view that decomposes it, reducing it by following a unifying hierarchy assembling one without the other" (Lambert 1999, 25).

Pierre Teilhard de Chardin looked for, even in science, the concrete universal and the person. If it stays uniquely positive, science ignores the human phenomenon; it even risks misunderstanding it, by reducing it to the elementary, to matter and, on the other hand, to nothing, because it doesn't see the essence of the human phenomenon. From the side of the Catholic religion, if it was not concrete, as it is mystic, it risked forgetting, being more rooted in the human phenomenon than the Cosmo genesis and could not instead communicate to the anthropogenesis the momentum in the cosmic Christ that only it possessed. The thought of Teilhard de Chardin is bold for these two reasons. It is so because it deploys a first gift, perceptible in the *Hymn to Matter*: "Blessed are you, powerful matter, irresistible march of evolution ... Blessed are you universal matter, immeasurable time..." (OT12f, 478[OT13a, 75]). This first gift is equally that of *The mystic Milieu* (OT12a, 153–92[115–149]). The thought of Teilhard is bold for this other reason: it "reconciles opposites without allowing one to be destroyed by the other, and

---

[63] However, according to Cardinal Ratzinger (1969, 240) these intuitions "cannot be considered as a definitive response, because their foundations, established within the natural sciences, are limited to the anatomical and anthropological field (with the exception of genetic processes) and that the philosophical and theological development of these concepts is insufficient". Let us note, however, that geneticists participated in the colloquy of Palaeontology of April 1947 (Arnould 1996, 288); "And all is connected because all is genetically built". Solages here (1967, 81) explains Teilhard de Chardin.

that bring together extremes by pursuing both union and distinction to their limits" (Lubac 2002, 226[160]). It uses paradox[64] and synthesis.

So, it seems to us, the ways are methodologically open for a scholar who takes up the heritage of Teilhard and for a theologian who thinks, perhaps, for a future council, of a Christology, which would have a cosmic dimension.

References

Arnould J., 1996, *Darwin, Teilhard de Chardin et Compagnie: l'Église et l'évolution*, Paris: Desclée de Brouwer.

Arnould J., 2004, *Pierre Teilhard de Chardin*, Paris: Perrin.

Baruzi J., 1907, *Leibniz et l'organisation religieuse de la terre*, Paris: F. Alcan.

Baudry G.-H., 2002, *Le sens de l'évolution en question*, Saint-Étienne: Aubin.

Birx H. J., 1991, *Interpreting Evolution: Darwin and Teilhard de Chardin*, Buffalo, N.Y.: Prometheus.

Brechtken J., 1983, *Evolution und Transzendenz: über unser wissenschaftliches Weltbild von heute und die Frage nach Gott, unter besonderer Berücksichtigung der evolutionstheoretischen Theodizee bei Pierre Teilhard de Chardin und Hoimar von Ditfurth*, Berne: P. Lang.

Cuénot C., 1958, *Pierre Teilhard de Chardin*, Paris: Plon.

d'Armagnac C., 1957, "Philosophie de la nature et méthode chez Teilhard de Chardin", *Archives de Philosophie* 20: 5–41.

Daleux A., 2001, *Teilhard de Chardin: Science et foi réconciliées?*, Paris: Gabriandre.

Denzinger H., Schönmetzer A., 1976, *Enchiridion symbolorum definitionum et declarationum de rebus fidei et morum*, Friburgo: Herder.

Disse J., 2001, "Henri de Lubac und Pierre Teilhard de Chardin. Eine Annäherung", Reifenberg and van Hooff 2001, 94–110.

Dulles A., 2009, "God en evolutie", in *Communio. Internationaal Katholiek Tijdscrift*, 34/2: 89–99.

Facchini F., 2002, *Origini dell'Uomo ed evoluzione culturale. Profili scientifici, filosofici, religiosi*, Milan: Jaca Book.

[64] Teilhard also uses the paradox; we have already seen it (see Lubac 2002, 273[189]). He describes its use: "my paradoxes may serve to make felt with a more immediate impact points of which classical philosophy calls for qualifications in wider and more flexible terms" (Lubac 2002, 267[185]; cited without reference, perhaps in a letter to Henri de Lubac). — So it is described by Henri de Lubac in the works of Teilhard — and so he practiced himself, the paradox differs of the *Aufhebung* of the Spirit in its negativity or beyond the survival, which is neither life nor death according to the deconstruction of Derrida.

Fessard G., 1964, *La vision religieuse de Teilhard de Chardin*, in *L'Homme devant Dieu. Mélanges offerts au P. Henri de Lubac*, vol. III, Paris: Aubier-Montaigne.
Folscheid D. (ed.), 1990, *Maurice Blondel. Une dramatique de la modernité*: Paris: Editions Universitaires.
Galleni L., 2010, "Teilhard and the Science of the Biosphere", in Duffy K. s.s.j. (ed.), *Rediscovering Teilhard's Fire*, Philadelphia: Saint Joseph's, University Press, 197-206.
Glässer A., 1984, *Evolutive Welt und christlicher Glaube : Pierre Teilhard de Chardin, 1881–1955*, Ratisbonne: F. Pustet.
Horn S. O., Wiedenhofer S. (eds.), 2007, *Creazione ed Evoluzione. Un convegno con Papa Benedetto XVI a Castel Gandolfo*, Bologna: Dehoniane.
Isaye G., 1987, *L'affirmation de l'être et les sciences positives*, Leclerc M. (ed.), Paris-Namur: Lethielleux-Presse universitaire.
John Paul II, 1996, "Discours du 22 octobre 1996 à l'Académie pontificale des sciences", *Osservatore Romano*, éd. franç (29 octobre).
Lacoste J.-Y. (ed.), 1998, *Dictionnaire critique de théologie*, Paris: PUF.
Ladrière J., 1976,"Penser la création", in *Communio. Revue catholique internationale* 1/3 53–63.
Lambert D., 1999, *Sciences et théologie. Les figures d'un dialogue*, Namur: Presses universitaires de Namur.
Leclerc M., 1998, "Évolution", Lacoste 1998, 442–45.
Lubac H. de, 1965, *Teilhard posthume. Réflexions et souvenirs, précédé de Blondel - Teilhard de Chardin, correspondance 1919* (Œuvres complètes, vol. 26), Paris: Cerf.
Lubac H. de, 1971, "Préface", Schellenbaum 1971, 9–18.
Lubac H. de, 1986, *Lettres de Monsieur Étienne Gilson*, Paris: Cerf.
Lubac H. de, 2002, *La Pensée religieuse du père Teilhard de Chardin* (Œuvres complètes, vol. 23), Paris: Cerf [*The Religion of Teilhard de Chardin*, trad. R. Hague, New York: Desclee: 1967].
Lubac H. de, 2006, *Mémoire sur l'occasion de mes écrits*, Paris: Cerf.
Maldamé J.-M., 1993, *Le Christ et le cosmos. Incidence de la cosmologie moderne sur la théologie*, Paris: Desclée.
Maldamé J.-M., 1998, *Le Christ pour l'Univers*, Paris: Desclée
Maldamé J.-M., 2000, *En travail d'enfantement création et évolution*, Saint-Etienne: Aubin.
Maldamé J.-M., 2006, *Création et Providence. Bible, science et philosophie*, Paris: Cerf.
Martelet G., 2005, *Teilhard de Chardin. Prophète d'un Christ toujours plus grand*, Bruxelles: Lessius.
Montcheuil Y. de, 1949, *Leçons sur le Christ*, Paris: Éd. de l'Épi.
Mooney C. F., 1966, *Teilhard de Chardin et le Mystère du Christ. La révélation chrétienne dans un système évolutionniste de pensée*, Paris: Aubier-Montaigne.
Pius XII, 1950, *Humani generis*, Denzinger and Schönmetzer 1976, n. 3895.
Poupard P., 2004, "Teilhard invite à reconnaître l'homme au coeur de l'univers", *La Croix* (21 octobre): 23.

Puddefoot J. C., 1998, "Sciences de la nature", Lacoste 1998, 1076–1077.
Rabut O. A., 2003, "Dialogue avec Teilhard de Chardin", *Henri de Lubac ou l'hospitalité de la théologie, Revue des sciences religieuses* 77/2: 199–214.
Ratzinger J., 1969, *Wer ist dans eigentlich – Gott?*, Schutz H.J. (ed.), Munich.
Reifenberg P., van Hooff A. (eds.), 2001, *Gott für die Welt,* Mayence: Peter Grünewald.
Reisse J. *et al.*, 2006, "Qu'est-ce que la vie?", *Revue des questions scientifiques* 177/3–4.
Schellenbaum P., 1971, *Le Christ dans l'énergétique teilhardienne. Étude génétique*, Paris: Cerf.
Schönborn C., 2007, "Fides - Ratio - Scientia. Il dibattito sull'evoluzionismo", in Horne S. O., Wiedenhofer S. (eds.), *Creazione ed Evoluzione*, Bologna: Dehoniane, 75-96.
Solages B. de, 1967, *Teilhard de Chardin*, Toulouse: Privat.
Teilhard de Chardin P., 1913, *Der Mensch aller Zeiten*, Berlin, Munich, Vienne: Allgemeine Verlags-Geselleschaft,
Teilhard de Chardin P., 1913a, "Homme: IV: l'homme devant les enseignements de l'Église et devant la philosophie spiritualiste", in *Dictionnaire apologétique de la foi catholique*, vol. 2, Paris: Beauchesne, §§501–514.
Teilhard de Chardin P., 1956, *Lettres de voyages, 1923–1939*, Paris: Grasset [*Letters from a Traveller*, London: Collins, 1962].
Teilhard de Chardin P., 1957, *Nouvelles lettres de voyage (1939–1955)*, Paris: Grasset [*Letters from a Traveller*, London: Collins, 1962].
Teilhard de Chardin P., 1974, *Lettres intimes à Auguste Valensin, Bruno de Solages, Henri de Lubac, André Ravier, 1919–1955*, Introduction et notes par Henri de Lubac, Paris, Aubier Montaigne.
OT1 *Oeuvres de Pierre Teilhard de Chardin,* vol. 1, 1955, *Le Phénomène Humain*, Paris: Seuil [*The Phenomenon of Man*, London: Collins, 1959].
OT2 *Oeuvres de Pierre Teilhard de Chardin,* vol. 2, 1956 *L'Apparition de l'Homme*, Paris: Seuil [*The Appearance of Man*, London: Collins, 1965].
OT2a "La Paléontologie et l'Apparition de l'Homme", 51–81 ["Paleontology and the Appearance of Man", 33–57].
OT2b "La structure phylétique du Groupe Humain", 185–234 ["The Phyletic Structure of the Human Group", 132–71].
OT2c "Les singularités de l'espèce humaine", 293–369 ["The Singularities of the Human Species", 208–270].
OT2d "La singularité du Phénomeène chrétien", 371–74 ["Appendix: Complementary Remarks of the Nature of the point Omega", 271–76].
OT3 *Oeuvres de Pierre Teilhard de Chardin,* vol. 3, 1957, *La Vision du Passé*, Paris: Seuil [*The Vision of the Past*, London: Collins, 1966]
OT3a "Comment se pose aujourd'hui la Question du Transformisme" (1921), 15–40 ["How the Transformist Question Presents itself Today", 7–25].
OT3b "Sur la Loi d'irréversibilité en Évolution" (1923), 71–74 ["On the Law of Irreversibility in Evolution", 49–50].

OT3c "L'Hominisation" (1925), 75–112 ["Hominization", 51–79].
OT3d "Le Paradoxe Transformiste" (1925), 113–42 [The Transformist Paradox", 80–102].
OT3e "L'Histoire naturelle du Monde" (1925), 143–58 ["The Natural History of the World", 103–113].
OT3f"Sur l'Apparence nécessairement discosntinue de toute série évolutive" (1926), 159–62 ["On the Necessarily Discontinuous Appearance of Every Evolutionary Series", 114–15].
OT3g "Les fondements et le Fond de l'Idée d'Évolution" (1926), 163–97 ["The Basis and Foundations of the Idea of Evolution", 116–42].
OT3h "Que faut-il penser du Transformisme?" (1930), 211–24 ["What Should we Think of Transformism", 151–60].
OT3i "Le Phénomène Humain" (1930), 225–44 ["The Phenomenon of Man", 161–74].
OT3j "La Place de l'Homme dans la Nature" (1932), 245–56 ["Man's Place in Nature", 175–82].
OT3k "La Place de l'Homme dans l'Univers" (1942), 303–26 ["Man's Place in the Universe", 216–33].
OT3l "Évolution zoologique et invention" (1947), 327–31 ["Zoological Evolution and Invention", 234–6].
OT3m "Évolution de l'idée d'Évolution" (1950), 345–50 ["Evolution of the Idea of Evolution", 245–7].
OT3n "Note sur la Réalité actuelle et la Signification évolutive d'une Orthogénèse humaine" (1951), 351–62 ["Note on the Present Reality and Evolutionary Significance of a Human Orthogenesis", 248–55].
OT3o "Hominisation et Spéciation" (1952), 365–80 ["Hominization and Speciation", 256–67].
OT3p "Une défense de l'Orthogenèse à propos des figures de spéciation" (1955), 381–91 ["A Defence of Orthogenesis in the Matter of Patterns of Speciation", 268–74].
OT4 *Oeuvres de Pierre Teilhard de Chardin,* vol. 4, 1957, *Le Milieu Divin. Essai de Vie Intérieure*, Paris: Seuil [*The Divine Milieu. An Essay of Interior Life*, London: Collins, 1965].
OT5 *Oeuvres de Pierre Teilhard de Chardin,* vol. 5, 1959, *L'Avenir de l'Homme*, Paris: Seuil [*The Future of Man*, New York: Fontana-Collins 1969].
OT5a "Note sur le progrès" (1920), 21–37 ["A Note on Progress", 11–25].
OT5b "La Grande Option" (1939), 55–82 ["The Grand Option", 39–63].
OT5c "Sur les bases possibles d'un Credo humain" (1941), 101–106 ["On the Possible Bases of a Universal Human Creed", 79–84].
OT5d "Le cône du temps" (1942), 111–17 ["The Cone of Time", 86–92].
OT5e "La transposition conique de l'action" (1942), 118–26 ["The 'Conic' Transposition of Action", 92–100].
OT5f "Le rebondissement humain de l'Evolution et ses conséquences" (1947), 251–72 ["The Human Rebound of Evolution and its Consequences", 204–21].

OT5g "Les Directions et les Conditions de l'Avenir" (1948), 291–306 ["The Directions and Conditions of Future", 236–47].
OT5h "Agitation ou Genèse?" (1947), 273–89 ["Turmoil or Genesis?", 222–35].
OT5i "Comment concevoir que se réalise sur terre l'unanimisation humaine" (1950), 365–74 ["How may we Conceive and Hope that Human Unanimisation will be Realised on Earth?", 295–302].
OT6 *Oeuvres de Pierre Teilhard de Chardin,* vol. 6, 1962, *L'Énergie Humaine*, Paris: Seuil [*The Human Energy*, London: Collins, 1969].
OT6a "L'Esprit de la Terre" (9 mars 1931), 23–57 ["The Spirit of the Earth", 19–47].
OT6b "Esquisse d'un Univers Personnel" (4 mai 1936), 67–114 ["Sketch of a Personalistic Universe", 53–92].
OT6c "Le Phénomène Spirituel" (mars 1937), 115–39 ["The Phenomenon of Spirituality", 93–112].
OT6d "L'Énergie Humaine" (6 août 1937), 141–200 ["Human Energy", 113–62].
OT7 *Oeuvres de Pierre Teilhard de Chardin,* vol. 7, 1963, *L'Activation de l'Energie*, Paris: Seuil [*The Activaiton of Energy*, London: Collins, 1970].
OT7a "L'Atomisme de l'Esprit" (13 septembre 1941), 27–64 ["The Atomism of Spirit", 21–58].
OT7b "La Centrologie. Essai d'une dialectique de l'union" (13 décembre 1944), 103–134 ["Centrology", 97–128].
OT7c "Esquisse d'une Dialectique de l'Esprit" (1946), 147–58 ["Outline of a Dialectic of Spirit", 141–52].
OT7d "Un Phénomène de Contre-Évolution en Biologie Humaine ou la Peur de l'Existence" (26 janvier 1949), 187–202 ["A Phenomenon of Counter-Evolution in Human Biology", 181–96].
OT7e "Le Sens de l'Espèce chez l'Homme" (31 mai 1949), 203–210 ["The Sense of the Species in Man", 197–204].
OT7f "Pour y voir clair. Réflexions sur deux formes inverses d'Esprit" (25 juillet 1950), 223–236 ["A Clarification: Reflections on Two Converse Forms of Spirit", 215–227].
OT7g "L'Énergie spirituelle de la souffrance" (8 janvier 1950), 253–57 ["The Spiritual Energy of Suffering", 245–50].
OT7h "La Convergence de l'Univers" (23 juillet 1951), 293–309 ["The Convergence of the Universe", 281–96].
OT7i "Transformation et Prolongement en l'Homme du Mécanisme de l'Évolution" 19 novembre 1951, 311–23 ["The Transformation and Continuation in Man of the Mechanism of Evolution", 297–310].
OT7j "La Réflexion de l'Énergie", (27 avril 1952), 333–53 ["The Reflection of Energy", 319–38].
OT7k "L'Énergie d'Évolution", (24 mai 1953), 379–94 ["The Energy of Evolution", 359–72].

OT8 *Oeuvres de Pierre Teilhard de Chardin,* vol. 8, 1965, *La Place de l'Homme dans la Nature (le groupe zoologique humaine)*, Paris: Seuil.
OT9 *Oeuvres de Pierre Teilhard de Chardin,* vol. 9, 1965, *Science et Christ,* Paris: Seuil [*Science and Christ*, London: Collins, 1968].
OT9a "Science et Christ" (27 février 1921), 45–62 ["Science and Christ or Analysis and Synthesis", 21–36].
OT9b "Mon Univers" (25 mars 1924), 63–114 ["My Universe", 37–85].
OT9c "Le Phénomène Humain" (septembre 1928), 115–28 ["The Phenomenon of Man", 86–97].
OT9d "Le Christianisme dans le monde" (mai 1933), 129–46 ["Christianity in the World", 98–112].
OT9e "Action et Activation" (9 août 1945), 219–34 ["Action and Activation", 174–86].
OT9f "Catholicisme et Science" (août 19465), 235–42 ["Catholicism and Science", 187–91].
OT9g "Sur les degrés de certitude scientifique de l'idée d'Évolution" 15–20 novembre 1946), 243–49 ["Degrees of Scientific Certainty in the Idea of Evolution", 192–6].
OT10 *Oeuvres de Pierre Teilhard de Chardin,* vol. 10, 1969, *Comment je crois,* Paris: Seuil [*Christianity and Evolution*, London: Collins, 1969].
OT10a "Christologie et Évolution" (Noël 1933), 93–114 ["Christology and Evolution", 76–95].
OT10b "Comment je crois" (octobre 1934), 115–52 ["How I Believe", 96–132].
OT10c "Le Christ évoluteur" (8 octobre 1942), 161–176 ["Christ the Evolver", 138–150].
OT10d "Introduction au Christianisme" (29 juin 1944), 177–200 ["Introduction to the Christian Life", 151–72].
OT10e "Christianisme et Évolution" (11 novembre 1945), 201–216 ["Christianity and Evolution: Suggestions for a New Theology", 173–86].
OT10f "Le Dieu de l'Évolution" (25 octobre 1953), 283–91 ["The God of Evolution", 237–43].
OT11 *Oeuvres de Pierre Teilhard de Chardin,* vol. 11, 1973, *Les Directions de l'Avenir,* Paris: Seuil [*Toward the Future*, London: Collins 1975].
OT11a "Trois Choses que je vois" (février 1948), 161–75 ["Two Principles and a Corollary", 148–62].
OT11b "Comment je vois" (26 août 1948), 177–223 ["My Fundamental Vision", 163–208].
OT11c "Un Sommaire de ma Perspective 'Phénoménologique' du Monde" (14 janvier 1954), 231–36 ["A Summary of My 'Phenomenological' View of the World", 212–15].
OT12 *Oeuvres de Pierre Teilhard de Chardin,* vol. 12, 1975, *Écrits du Temps de la Guerre 1916–1919*, Paris: Seuil [*Writings in Time of War*, London: Collins, 1968].
OT12a "Le Millieu mystique" (1917), 153–92 ["The Mystical Milieu", 115–49].
OT12b "L'Union créatrice" (1917), 193–224 ["Creative Union", 151–76].

OT12c "L'Âme du monde" (1918), 243–59 ["The Soul of the World", 177–90].
OT12d "Mon Univers" (1918), 293–307 ["My Universe", OT13, 196–208].
OT12e "Note pour servir à l'évangélisation des temps nouveaux" (1917), 395–414 ["Note on the Presentation of the Gospel in a New Age", OT13, 209–24].
OT12f "La Puissance spirituelle de le Matière" (1919), 465–79 ["The Spiritual Power of Matter", in Pierre Teilhard de Chardin, *Hymn of the Universe,* London: Collins, 1965, 57–71].
OT13 *Oeuvres de Pierre Teilhard de Chardin,* vol. 13, 1976, *Le Coeur de la Matière,* Paris: Seuil [*The Heart of Matter*, London: Collins, 1978].
OT13a "Le Coeur de la Matière" (8 août 1919), 19–91 ["The Heart of Matter", 13–79].
OT13b "Le Christique" (mars 1955), 93–118 ["The Christic", 80–102].
Thomas J.-F., 1998, *Initiation à Teilhard de Chardin. Maître spirituel,* Paris: Cerf.
Virgoulay R., 1990, "La christologie philosophique de Maurice Blondel", in Folscheid D. (ed.), *Une dramatique de la Modernité,* Paris: Ed. Universitaries, 201-209.
Wagner J.-P., 2003, "Henri de Lubac et Pierre Teilhard de Chardin: le sens d'une hospitalité", *Henri de Lubac ou l'hospitalité de la théologie*, *Revue des sciences religieuses* 77/2: 199–214.
Worms F., 2004, *Bergson ou des deux sens de la vie*, Paris: PUF.

Theological debates around evolution

Robert John Russell

1. Introduction

How should Christian theology take into serious consideration the scientific theory of biological evolution? Given the input from the scientific and philosophical essays that precede this chapter, the theological question is now ripe for discussion.[1] To begin, I can think of no better place than the documents of Vatican II and key statements by Pope John Paul II.

Regarding the former, I am drawn in particular to the "Pastoral Constitution on the Church in the Modern World" *Gaudium et Spes* (Abbott 1966, 199–308) for its combination of pastoral and doctrinal concerns that so frequently touch on the issues raised by science for the Church. This document lifts up in general the fundamental questions which humanity asks about its place and role in the universe. It stresses the Church's task of "scrutinizing the signs of the times and interpreting them in the light of the gospel." It argues that "intellectual formation is ever increasingly based on the mathematical and natural sciences and on those dealing with man himself." It underscores the fact that humankind "has passed from a rather static concept of reality to a more dynamic, evolutionary one," a concept raising new problems that call for "new efforts of analysis and synthesis." While as a result "many persons are achieving a more vivid sense of God," it expresses deep concern over the rise of atheism in today's society, particularly when atheism draws on science and is presented as a "requirement of scientific progress." In response, picking up a theme from Vatican I, it affirms that science and faith are not mutually opposed and that they never truly conflict, "for earthly matters and the concerns of faith derive from the same God." (Abbott 1966, 200–205; 234)

These comments are reflected and extended profoundly in the Message of Pope John Paul II to the Director of the Vatican Observatory, Reverend George V. Coyne sj, delivered at the Vatican at the conclusion of a major international con-

---

[1] I am deeply grateful for extended conversations with three of my Roman Catholic graduate students as I prepared this chapter: Dr. John Braverman, Fr. Eraldo Cacchione, and Dr. Oliver Putz.

ference organized by the Observatory in 1987 and published in the volume *Physics, Philosophy and Theology: A Common Quest for Understanding* (John Paul II 1988). The following are some of the key points taken mostly verbatim (with slight editing) from the text:

The new and more nuanced interchange between religion and science must grow and deepen for its worldwide impact on the course of civilization cannot be overestimated.

- Theology … as *fides quaerens intellectum* … must be in vital interchange today with science … The hylomorphism of Aristotelian natural philosophy … was adopted by the medieval theologians … Theologians (today) might well ask, with respect to contemporary science … if they have accomplished this extraordinarily difficult process as well as did these medieval masters.
- If the cosmologies of the ancient Near Eastern world could be purified and assimilated into the first chapters of Genesis, might contemporary cosmology have something to offer to our reflections upon creation?
- Does an evolutionary perspective bring any light to bear upon theological anthropology, the meaning of the human person as the *imago Dei*, the problem of Christology …?
- What, if any, are the eschatological implications of contemporary cosmology, especially in light of the vast future of our universe?
- Can theological method fruitfully appropriate insights from scientific methodology and the philosophy of science?
- (S)ome theologians … should be sufficiently well-versed in the sciences to make authentic and creative use of … the best-established theories … (and not make) uncritical and overhasty … apologetics … of such recent theories as that of the "Big Bang" in cosmology.
- Science can purify religion from error and superstition; religion can purify science from idolatry and false absolutes. Each can draw the other into a wider world, a world in which both can flourish.
- … The Church and the scientific community will inevitably interact; their options do not include isolation … The only question is whether they will do this critically or unreflectively, with depth and nuance or with a shallowness that debases the Gospel and leaves us ashamed before history … Theology must not profess a pseudo-science and science must not become an unconscious theology. Our knowledge of each other can lead us to be more authentically ourselves.

The "good news" is that dozens of scholars around the world specializing in theology and science for the past three to five decades have painstakingly worked out many of the detailed arguments which undergird and implement the vision of Vatican II and Pope John Paul II. We have tested these arguments vigorously in international and ecumenical conferences, we have published the results widely in texts and journals, and made them available through online resources, and we have assessed their potential for creative research in doctoral courses and probed their pastoral fruitfulness in seminary classes.[2] Let us then turn to several key topics in this interdisciplinary field.

2. How should we appropriate evolutionary biology theologically?

2.1 Three philosophical guidelines.

I will first offer three philosophical guidelines among the many drawn from these many years of scholarship to guide our discussion of specific topics:

1) *Accept methodological naturalism* as the basis for the natural sciences, as Elliott Sober stressed in his chapter for this volume. The methodology underlying scientific research involves explanation strictly by natural causes. In this sense scientific rationality is a subset of rationality in general, appealing in most cases to only one of Aristotle's four causes: efficient cause.[3]

Historically, methodological naturalism is based on a combination of the doctrine of creation *ex nihilo* and the Greek *logos* tradition.[4] According to the former, God freely created the world out of no prior matter or laws, and thus the world is contingent. Hence knowledge about the world is empirical and empirical knowledge requires experimentation. According to the latter, the world is rational and intelligible, and its processes can be represented by mathematics. Contingency and rationality are cornerstones of the scientific method.

---

[2] The Center for Theology and the Natural Sciences offers many of these resources online at www.ctns.org. These include summaries of over ninety chapters of the five volume CTNS / Vatican Observatory series of reseach publications on "scientific perspectives on divine action." These resources also include syllabi of seventy-five grant-winning courses in our four year "Science and Religion Course Program" as well as hundreds of hours of videos of conferences recorded over the past decade by our colleagues at www.counterbalance.com.

[3] Actually, as is well known, classical physics can also be formulated in teleological, instead of efficient, terms via the principle of least action.

[4] For an early source of this insight see Foster (1969).

Note: ironically atheists who claim support from science (i.e., who endorse scientism) have forgotten that science as a method of inquiry presupposes the doctrine of creation and thus a partially theistic conception of nature.

2) *Affirm epistemological holism* against epistemological reductionism, the claim that only science offers genuine knowledge and that the rest of the academic disciplines, particularly theology, can be reduced to science without remainder. The scholarly literature is replete with extensive rebuttals of epistemological reductionism by such distinguished voices as Francisco Ayala and Arthur Peacocke dating back now over thirty years, and generally supporting, instead, a philosophy of "epistemic emergence," as William Stoeger reports in his chapter for this volume. They have urged that the academic disciplines form an epistemic hierarchy in which fields which study more complex phenomena, such as physiology or the neurosciences, describe processes and properties which cannot be reduced, without remainder, to fields studying simpler phenomena, such as biochemistry or anatomy, even while they are constrained by the results of the lower level disciplines. Theology, located at the top of the hierarchy, is maximally constrained by them all, and yet it is freed from reductionistic claims and given its own voice and partial autonomy.

3) *Reject scientism* in its various forms: metaphysical naturalism/materialism/scientism/"evolutionism." The claim that a scientific theory, such as evolutionary biology, is the sole basis for a worldview is to be rejected. It is of critical importance to stress that the acceptance of methodological naturalism does *not* require an acceptance of metaphysical naturalism, contrary to the attempts to so argue by many voices in the public arena. Serious scholars in theology and science not only reject "scientism"; they also seek to protect scientific theories from being attacked because metaphysical reductionists, such as Richard Dawkins, claim that science promotes and warrants an ideology of scientism. The slogan might well be: "Attack the victimizer (Dawkins), not the victim (evolutionary biology)"! As then Cardinal Joseph Ratzinger wrote in 1986:

> Today … "evolution" has been exalted above and beyond its scientific content and made into an intellectual model that claims to explain the whole of reality and thus has become a sort of "first philosophy" … Whereas faith today no longer has any difficulty in allowing the scientific hypothesis of evolution to develop in peace according to its own methods, the absolute claim of the philosophical explanatory model "evolution" is an all the more radical challenge to faith and theology (Horn and Wiedenhofer 2008, 8–11).

I would like to add that we should not rest content on rejecting the use of science to support a materialist ideology. We Christians must also deliver a convincing theological interpretation of evolutionary biology if we are to fulfill the doctrinal and pastoral imperatives of Vatican II. Now to a brief look at some of these constructive interpretations.

3. Overview of three broadly traveled ways

As an preliminary overview, I would like to suggest that there are three broadly-traveled ways that Christian theology can appropriate evolutionary biology. Each has its strengths and limitations which, if kept clearly in mind, allow all three to flourish and enrich the Church's engagement with the world. And at base, as Francis Schüssler Fiorenza writes, "For theology to overlook the growth in scientific knowledge and rationality is to withdraw into a ghetto." (Schüssler Fiorenza 1991, 69)

1) Inspiration from the beauty of nature. Beauty, joy and goodness radiate from the natural world around us, telling us we are beloved creatures of God. A spider web glistens with dew in a dark night. The Serengeti plains streams with numberless herds of wildebeests. Penguins nest by the thousands in the Antarctic. The melody of a parrot lilts through a tropical Hawaiian jungle. Chimpanzees emerge from their forest to stop and look upwards at a tremendous waterfall, seeming to experience a moment of sacred space. As the first chapter of Genesis proclaims, all that God creates is good, even very good.

But nature's beauty is ambiguous, revealing a "shadow side" to the world around us which leads us to cry out to God our Redeemer. The spider web hides the terror of the trapped insect watching the impending jaws of the spider. Across the Serengeti plain, wildebeests die wrenchingly in the jaws of lions just as Antarctic penguins are sported for and then devoured by leopard seals. Forrests in the Caribbean team with foraging ants and chimpanzees kill in groups. Indeed, in the food chain that marks the history of life on earth most animals die an agonizing death in the jaws of cunning predators. We seem to hear this "cry of the wild" in St. Paul epistle to the church at Rome, where he exclaims, "We know that the whole creation has been groaning in travail together until now" (Rom. 8: 22).

The upshot: There is no sure and unproblematic approach to natural theology as a technical discipline, although nature as a source of inspiration can certainly be efficacious to Christian religious experience.

2) Natural theology. The miriad wonders of nature often appear to be designed. Is there here the basis for a rational argument for the existence of God? St. Paul apparently assumed so in writing that:

For what can be known about God is plain to them, because God has shown it to them. Ever since the creation of the world his invisible nature, namely, his eternal power and deity, has been clearly perceived in the things that have been made. (Rom. 1: 19–20a)

Historically, profound philosophical design arguments have occurred throughout Western thought, from Aquinas' fifth proof of the existence of God from the design of objects in nature (*Sum. Theol.* I, q. 2, a. 3) to William Paley's watchmaker argument. But claims at refutation of them, from Hume against Aquinas to Darwin against Paley, undercut the force of these arguments. Evolutionary biology, in particular, does not dispute the fact that many biological phenomena seem to be designed for a functional purpose, what Francisco Ayala has termed "teleonomy." But evolutionary biology provides a natural explanation of the means of design: not by an agent or designer but by natural selection operating on random variation. Indeed while "design" is a frequent term in evolutionary biology, the fit between an organism and its environment is more often termed "adaptation." For this and other reasons, Ayala in his chapter here has referred to the neo-Darwinian synthesis as "completing the Copernican revolution," using the telling phrase "design without a designer."

The upshot: A natural theology drawing on evolutionary biology is no longer as promising a strategy as it was before Darwin, although it certainly deserves consideration. As William Stoeger writes in his chapter for this volume, science can be consonant with teleonomy and directionality, but it cannot do more in terms of teleology and finality. I would like to add an additional caveat: Science is a self-limited form of human reasoning, an agreement to limit the mode of explanation of effects to natural, secondary causes. Scientific rationality is not "reason" in its widest sense, reason which the traditional design arguments of natural theology assumed as open to and even leading to God as the ultimate explanation of events in the world. Because of this, if the science of evolutionary biology fails to provide a reliable basis for a natural theology, Christians should not accuse it of being atheistic or attempt to find scientific flaws in it, hoping hopelessly that a more Christian-friendly replacement will come along.

3) Theology of nature: theistic evolution. The method taken by most scholars in the dialogue between theology and science is frequently termed a "theology of nature." This approach follows the classical methodology of St. Anselm, *faith seeking understanding*, as recommended to us by John Paul II.

It starts with one of the many central theological doctrines, typically the doctrine of creation. In order to determine what light science might bring to our understanding of these doctrines, scholars recognize they must first of all "get the science right." To do so they turn to specialists in a given scientific field as has been done at the Gregorian conference in Rome this March, and learn from them both the broad sweep of discoveries to date and what issues are currently being debated.

Following this, a careful analysis of the philosophical implications of these discoveries and debates is required. The philosophical analysis can range from very focused issues raised by specific scientific theories to the way to appropriate an entire metaphysical system to embed these theories before bringing them into the theological conversation. An example of the former is the debate over ethics and evolution: can our moral codes be reduced to a mere product of evolution, as Michael Ruse argues, or is the capacity for morality, but not the content of our moral teaching, a product of evolution, leaving actual moral codes something which only revelation or philosophical ethics can give us, as Francisco Ayala has argued? An example of the latter is the use of neo-Thomistic philosophy or Whiteheadian metaphysics to interpret paleo-anthropology, the cognitive sciences, neuroscience, and so on.

Finally, scholars locate these philosophical analyses within their own theological traditions and seek what light they can shed on its key doctrines. My own writing then calls for a "creative mutual interaction" (CMI) between theology and science in which science not only affects theological reflection, but theology can play a creative role in suggesting new insights for scientific research (Russell, Murphy and Stoeger 2008).

Broadly speaking, most scholars in theology and science support what is widely termed "theistic evolution": biological evolution is how God creates the diversity of life on earth.[5] These scholars accept evolution as it is defined by the consensus of the scientific community: descent with modification as a result of selection pressures acting on the variations arising in heritable traits in a population of organisms. They accept the "deep time" of modern geology required by evolution:

---

[5] Biological evolution has little if anything to do with abiogenesis, the scientific study of how life on earth could have arisen from inanimate matter. Science so far does not have a definitive answer to the question of the origin of life. The emergence of catalytic RNA was probably a crucial early step. How that RNA came into being remains unknown. Debates continue between "genes first models" and "metabolism first models." One possibility is that life on Earth might have begun at deep-sea vents where molten lava boils through the ocean floor. Another is Lynn Margulis's endosymbiotic theory. Another possibility is "exogenesis," including the possibility that life originated elsewhere in the solar system, most likely Mars, and migrated to Earth via a meteorite (related to "panspermia").

given a natural history of some 3.8 Byrs, evolution has produced both change within species (microevolution) and the origin of new species (macroevolution). Some of them are also in conversations with scientific research in "evo-devo," the attempt to bring together the neo-Darwinian synthesis — evolutionary biology combined with population genetics — and developmental genetics.

Theistic evolution affirms that these natural processes are ultimately the creation of God whose purpose is that life, and in particular self-conscious creatures such as *Homo sapiens* on earth, may respond to God's gracious self-communication and enter into covenant with their Creator. A few[6] have considered the theological implications of the possibility of other self-conscious animals on earth, such as the great apes, the possibility of extra-intelligent life in the universe ("ETI") (Peters 2008), and in both cases God's gracious relation to them. Although united in their opposition to Intelligent Design and creation science, on the one hand, and more forcefully against scientistic, materialist and atheistic interpretations of evolution on the other hand, they represent an intriguing diversity of views on the theological and philosophical meaning of God's ongoing, creative activity in the evolution of life.

In a recent critical survey Lutheran theologian Ted Peters and Roman Catholic biologist Martinez Hewlett lay out a spectrum of positions within theistic evolution with examples of the scholars who take them up (Peters and Hewlett 2003). They range from the cautious but open approach of Princeton theologian B. B. Warfield at the turn of the century to the exuberance of Teilhard de Chardin's "cosmogenesis" of the mid-twentieth century. They include Brown University biologist Kenneth Miller, who assumes that scientific explanations of evolution are exhaustive and sufficient and yet affirms his faith in a personal Creator of the universe who reveals Godself to us. One of the most eloquent and widely read is the late biochemist and Anglican priest, Arthur Peacocke, who claimed that "God acts in, through, and under the processes of nature," and that biological complexity is partially the result of self-organization in non-linear thermodynamic systems. In direct response to atheists ranging from Jacques Monod to Richard Dawkins, Peacocke argued that God acts through both chance and law, resulting in a world of continuous creation. Roman Catholic theologian Denis Edwards picks up a theme from Peacocke and from the Lutheran theology of the Cross to affirm that God suffers with the creatures of this world. Deeply informed by the theology of

[6] Such the dissertation project of Dr. Oliver Putz, who is taking his second Ph. D. in theology and science at the GTU after his first Ph. D. in biology.

Richard of St. Victor and Karl Rahner, Edwards describes the Triune God as a God of mutual relations, a communion of love.

There are many other scholars who have contributed wonderfully to these conversation. These include, to name just a few, Ian Barbour, Charles Birch, Anne Clifford, John Cobb, Philip Clayton, Niels Henrik Gregersen, John Haught, Philip Hefner, Jürgen Moltmann, George Murphy, Nancey Murphy, Wolfhart Pannenberg, Holmes Rolston, Wentzel van Huyssteen. I would like to draw particular attention to Archbishop Józef Życiński's publication on evolution and Christian theology (Życiński 2006), both for its stunning scholarship and creative vision, and for the demanding example it poses for his fellow clergymen who wish to try their hand at interpreting evolution. According to Życiński, the universe is dynamic and evolutionary, and God provides both the universal laws of nature which give nature its stability and rationality (*logos*) and God participates in those events which seem to science as chance and which can magnify into major effects in nature (e.g., the "butterfly effect"). Thus the whole range of physical and biological processes should be taken as expressing God's immanence in nature. Życiński then boldly suggests that "a metaphysics of the future is necessary in order to explain the evolutionary present." He illustrates this idea of an eschatological aspect of evolution by drawing on analogies such as the "strange attractor" in non-linear dynamics and on the writings of Teilhard de Chardin, Karl Rahner, Jürgen Moltmann, Wolfhart Pannenberg, Ted Peters, and Alfred North Whitehead. I also want to single out for special attention Celia Deane-Drummond, a Catholic theologian holding a Ph. D. in biology. Deane-Drummond offers yet another excellent example of how Christian theology should be reconstructed in light of evolutionary biology. In her latest book (Deane-Drummond 2009), she claims that evolution is not just an epic of gradually unfolding potentials but a drama filled with tensions and tragedy, contingency and form. To make her case Deane-Drummond draws on such classical themes in Christology as Wisdom / Sophia and kenosis. She delights in "theo-drama" and draws deeply from Hans Urs von Balthasar.

Karl Rahner, of course, is probably the most influential example of how one can relate biological evolution to Catholic doctrine. In his widely read essay, "Christology within an evolutionary view of the world." Rahner describes his methodology as a correlation between Catholic Christology and evolutionary biology. Rahner understands humanity as "the being in whom the basic tendency of matter to find itself in the spirit by self-transcendence arrives at the point where it definitely breaks through … (This active self-transcendence) into God (is) made possible quite gratuitously by God, and by God's communication of himself (to us).” Such active self-transcendence finds its normative instantiation in the Incarnation of Christ which “appears as the necessary and permanent beginning of the divinization of the world as a whole.” “The Incarnation could be seen as the climax surpassing all the other world-realities arranged in ascending layers, so that it could be positively fitted into an evolutionary world-view...” (Rahner 1983, especially 160–61, 180; Rahner, 1978 178–203).

Rahner offers us a rich and complex philosophical system through which he achieves this correlation of evolutionary biology and Christology. It combines 1) crucial elements of neo-Thomism and the metaphysics of the “Great Chain of Being” now “temporalized” and historized to fit with an evolutionary perspective, with 2) a transcendental analysis of human experience and language rooted in the philosophy of Martin Heidegger and in a critical engagement with the philosophy of Immanuel Kant. It includes a theological anthropology in which “man is not an unnatural or merely temporary composite of spirit and matter but is a unity which is logically and objectively prior to the diversity of his distinguishable elements.” Rahner imbeds his understanding of the eschatological future of humanity in the “one (however ‘unimaginable’) consummation of his *whole* reality and that of the cosmos.”[7] Elsewhere Rahner argues for an “ontology in which created being is conceived in principle and in the very beginning as being which is in the process of becoming within an entire evolution of the cosmos, which is still in the process of evolution.” (Rahner 1988)

It might be instructive, here, to query how Rahner’s correlation between Christology and evolutionary theory, written some twenty-five years ago, fares given more recent science such as have been presented here at this conference. Let me offer one example among the many that should be pursued: How does the ongoing scientific debate over the degree to which radical contingencies shaped the course of evolution on earth relate to Rahner’s evolutionary worldview in which natural history has moved purposefully towards its completion in human life and the Incarnation?

Scientists like Simon Conway Morris have argued for convergence and predictability in the history of the evolution of species. Once life gets started on a

[7] Indeed, “we must also take into consideration the known history of the cosmos as it has been investigated and described by the modern natural sciences: this history is seen more and more as one homogeneous history of matter, life and man ... The history of Nature and spirit forms an inner, graded unity in which natural history develops towards man, continues in him as *his* history, is conserved and surpassed in him and hence reaches its proper goal with and in the history of the human spirit.” (Rahner 1983, 162–66)

earthlike planet, it is highly likely that creatures capable of self-consciousness would emerge. Such a position, while articulated scientifically, could support a teleological natural philosophy. This, in turn, could help Rahner make more persuasive, in light of science, his central theological claim that God created the universe for the purpose of God's self communication in grace to creatures like *Homo sapiens*. But what about the views of other scientists, such as Stephen Jay Gould or Francisco Ayala, who argue for a vastly greater degree of contingency in the evolution of life both at its origins and in the recent stages of the evolution of self-consciousness? If you "played back the tape" it would most likely never have resulted in any form of self-conscious life on earth. Does their position, still strictly scientific but clearly opposed to that of Conway Morris, necessarily lead to a dysteleological worldview such as Richard Dawkins promotes?

Interestingly, the answer is "no." One could still argue that we are, after all, here and that this is best explained — theologically — as the result of divine intent and purpose. Indeed, given the odds stacked against us being here, God must surely have been intimately involved in the processes of evolution, and we could develop a teleological worldview to explicate this belief in light of evolution. But this is clearly not a judgment which science alone can arrive at. It is instead "theology seeking understanding" in light of science. The lesson to be learned is that the choice between a philosophy of teleology or dysteleology is under-determined by science, even though it is in broad ways contoured by it.

Turning briefly to my own writings about theistic evolution, I start with quantum mechanics because this theory, unlike all others in physics, can be interpreted philosophically as pointing to real ontological indeterminism in nature's sub-atomic processes (Russell, Murphy and Stoeger, 2008 chs. 4–5; see also Russell 2001, 293–328). This was the view of Werner Heisenberg and is largely shared by physicists and philosophers today (Heisenberg 1958; see also Cushing 1994, 32).[8] By ontological indeterminism I mean the view that certain processes, like radioactive decay, occur without a sufficient efficient natural cause. The key point for theistic evolution is that quantum processes are directly involved in the occurrence of genetic mutations. These, in turn, are an important — though certainly not the only — contributing factor to the variation in heritable traits upon which natural selection acts. I have proposed that we can interpret God's continuous creation of life as divine action at the quantum level within genetic mutations. I have called

[8] Gennaro Auletta (2001, 509–11), who has written one of the best scholarly treatments of the multiple interpretations of quantum mechanics, is critical of such an interpretation.

this "non-interventionist objective divine action (NIODA)." Thus, in my version of theistic evolution we really can claim that God acts in, with and through the molecular biology in the production of new species in nature (Russell, Murphy and Stoeger 2008, ch. 6). This approach renders Intelligent Design unnecessary (even while it is already wrong-headed!) and it delivers on the "promissory note" of the theists I have cited above: God really is active in nature without in some heretical way being just another secondary cause. Instead God created the universe *ex nihilo* "fine-tuned" with just the right laws and constants to make non-interventionist divine action a delicate and universal part of the evolution of life in a way that would be compatible with science.

It is helpful to note the several challenges to my proposal: 1) Quantum mechanics is capable of multiple philosophical interpretations, and most of the other ones support ontological determinism. I agree: If new scientific evidence were to shift the balance of scholarly opinion in favor of one of these alternatives, my approach to divine action would be severely challenged. 2) If, and when, quantum mechanics is superseded by a new theory in physics, this too could mean a return to some philosophical form of ontological determinism, with similar problems for a non-interventionist account of divine action. Again, I agree. 3) If for philosophical reasons, as William Stoeger holds, one believes that the principle of sufficient reason is obeyed by nature, then one has philosophical reasons to reject the Heisenberg interpretation and favor a "hidden variables" view of nature.[9] Fair enough. Once again, the "correlation" that I seek, like that of Rahner, requires faithful, ongoing inspection and acknowledgment of the continuing problems as well as opportunities for deeper understanding. The task of theological engagement with culture, especially science, is never-ending.

4. Responses to the central challenge to theistic evolution: natural theodicy.

> I cannot persuade myself that a beneficent and omnipotent God would have designedly created the Ichneumonidae with the express intention of their feeding within the living bodies of Caterpillars. (Charles Darwin in a letter to Asa Gray)

While appreciating the many scholars who have sought to relate evolution and Christian theology, and despite the creativity of their approaches — particularly compared to the superficiality of those advanced by atheists like Richard Dawkins — I believe that the most challenging problem posed by evolutionary biology is

[9] I offer responses to these challenges in the previous references.

that of "natural evil." This term, though admittedly problematic, is commonly used to refer at the level of biology to suffering, disease, predation, death, and extinction, and at the level of physics, to tsunamis, earthquakes, meteor impacts, hurricanes, and so on.[10] Admittedly, these phenomena play an extraordinarily beneficial role as well in the evolution of life, but here I will focus on the way they contribute to the travail of creation.

Traditional theology has accounted for these facts as resulting from a combination of the Fall in Genesis and the rebellion of the angels depicted so movingly in Milton's *Paradise Lost.* But we now know that our human species has come very lately onto the biological stage, descending from pre-human primates some 1–2 Myrs ago, while fossils of cyanobacteria, the earliest known organisms for which "death" is a biological category, date back well over a thousand times earlier to the Archean era, 3.5Byrs ago, and simple animals capable of consciousness date as far back as 600 Myrs.[11] This means that biological death is not the *consequence* of the Fall. Instead death is *constitutive* of the evolution of life. So the challenge to any theology of nature in light of evolution is clear: If God creates life through the processes of nature, is God responsible for the panorama of natural evil intrinsic to the evolution of life?

[10] For an excellent, recent treatment see Christopher Southgate (2008).

[11] Biblical criticism over the past two centuries has suggested that the Genesis accounts ought not be seen as teaching science (the six days of the first account) or as describing human origins as an historical event (Adam and Eve and the Fall in the second account) but as proclaiming faith in Yahwah using the language of the ancient Babylonian cosmology. Partly in response to both biblical criticism and biological evolution, Pope Pius XII issued the encyclical, *Humani Generis* ("Of the Human Race") in 1950. It attacked the claim that Adam was not an individual but a population. It explicitly acknowledged the theory of evolution as a valid though unproven scientific hypothesis, and while Catholics were required to believe that the soul was created by God, they could consider the possibility that God created the human body through an evolutionary process.

The underlying problem was that, according to science, humanity was descended from a single population (humanity was "monophyletic" rather than "polyphyletic") but not from a single couple (humanity was "polygenetic" rather than "monogenetic"). The Church believed instead that humanity was both monophyletic and monogenetic as required by the doctrine of original sin. This in turn provided the foundation for the claim that all humans are sinful because they are descended from Adam (Paragraph 37). Note: Those later censured or restricted in expressing their views on this issue include Yves Congar, Marie-Dominique Chenu, Henri de Lubac, Karl Rahner, Pierre Teilhard de Chardin, Jean Daniélou, and John Courtney Murray. Still, Jean Daniélou was eventually made a cardinal by Pope Paul VI, as were Yves Congar and Henri de Lubac by Pope John Paul II.

4.1 God had no choice?

One might argue that even God had "no choice" but to opt for Darwinian evolution if God is to create life by natural means rather than by intervening in nature and creating each individual species. This means not only that God did not actively cause individual events of suffering and death in nature but even more significantly that natural evil itself is a necessary consequence of the evolutionary processes which even God could not eliminate. The "no choice" argument aims at lessening the intensity of natural theodicy by letting God "off the hook". Francisco Ayala makes this case cogently in his essay for this volume (see also Ayala 2006).[12] In an interesting turn of events, Michael Ruse also offers the "no choice" argument in defense of Christianity against Richard Dawkins by citing Dawkins' own reference to the fact that … "selection and only selection can do the job. No one — and presumably this includes God — could have got adaptive complexity without going the route of natural selection." (Ruse 2001, 136)[13]

I find this move rather superficial and I call it "Theodicy Lite." Basically it ignores a crucial underlying assumption: namely, that the laws of physics that undergird evolutionary biology are a "given." Admittedly, if God were to create life without intervening in the nature and if the laws of physics which govern these processes are taken for granted, then God may have had "no choice" other than Darwinian evolution. But if we push it one step further, the question returns at a more fundamental level in what I will call "cosmic theodicy." The point is that God created our universe *ex nihilo* with the specific laws of physics and constants of nature which make Darwinian evolution possible. Why then did God choose to create this universe with these laws and constants knowing that they would as a consequence make the full sweep of natural evil inevitable? In effect, the Ruse/Dawkins argument does not rescue God from blame, but merely places blame at a foundational level, leading to the Leibnizian challenge: is this the best of all possible universes? I address this challenge at length in the first volume of the new CTNS/VO series on scientific perspectives on the problem of natural evil (Russell 2007). My conclusion is that no past-looking theodicy such as these succeeds, and we are led instead toward the future: to Christian eschatology and the hope for the "new creation."

---

[12] See responses to Ayala and his responses in turn (Ayala 2008).

[13] Here Ruse cites Richard Dawkins (1983, 403–25). In this book Ruse gives a very thoughtful and promising account of what he believes is really at stake in the Christianity/evolution controversy.

4.2 The hiddenness of God

Another kind of response to natural evil is that God's action is hidden in the very processes where goodness and tragedy abound. This response takes on a variety of form.

1. Biblical response. One finds many places in the Biblical text where God is understood as hidden from us. Isaiah writes that "truly you are a God who hides himself, O God and Savior of Israel." (Isa. 45: 15) Deuteronomy speaks of God hiding his face when the people of Israel worship pagan gods, and the people will believe God has deserted them (Deut. 31: 16–18). The Psalmist refers to God as hidden when he writes "Why dost thou stand afar off, O LORD? Why dost thou hide thyself in times of trouble? (Ps. 10: 1) Most clearly, Job offers us a theodicy for whom God's will is, ultimately, inscrutable (Job 37-on).

2. Luther's theology of the cross (*deus absconditus*). Here Martin Luther argued that, while God is omnipotent, the form of divine power is the power of suffering love as witnessed normatively in the cross of Christ. Our God is the "hidden God" acting invisibly within the horror of the cross to save the world. In what is unthinkable to Jewish Messianism — a crucified Messiah — we see God's true power at work, not as the world works but in an act of dereliction and agonizing death infused with Jesus' trust in God's faithfulness.

3. Hick's "epistemic distance." Central to John Hick's theodicy, drawing on Irenaeus, is the argument that "pain and suffering are a necessary feature of a world that is to be the scene of a process of soul-making." (Hick 1966) Even their "haphazard and unjust distribution" and excess are ultimately beneficial, since "the right must be done for its own sake rather than a reward." God must be a hidden deity, veiled by His creation" so that, unlike our physical surroundings, God's presence is not coercively imposed on us. Thus the world must be "*etsi deus non daretur*" ("as if there were no God") because in such a world we have the necessary "cognitive freedom" by which faith and moral growth are possible.

4.3 Christology

1. Assumption Christology: According to one stream of traditional Christology, as represented by such Patristics as Gregory of Nazianzus and Origen, what Christ has not assumed he has not healed. The whole person would not have been saved unless Christ took upon himself the whole person. Contemporary scholars start with this idea, that Christ assumes all aspects of human nature, and extend it to the history of life on earth, citing our connection with all life through the common descent of species, their shared morphology, molecular biology, and the underlying, universal laws of physics which make evolution possible.

Karl Rahner develops this point beautifully in the essay I cited previously. Because in the Incarnation and the Hypostatic Union God takes matter into the divine being,

> The statement of God's *Incarnation* — of his becoming *material* — is the most basic statement of Christology ... In Jesus, the Logos bears the matter just as much as the soul, and this matter is a part of the reality and history of the cosmos ... We are perfectly entitled to think of the creation and of the Incarnation, not as two disparate, adjacent acts of God "*ad extra*", but as two moments and phases in the real world of the unique ... process of God's self-renunciation and self-expression into what is other than himself (Rahner 1983, 177–78).

Thus the Incarnation fits easily into an evolutionary view of the world: it is "the climax surpassing all the other world-realities arranged in ascending layers ... (and thus) as the goal and end of this world-reality..." (Rahner 1983, 180).

2. Kenotic Christology: God suffering with the world/with nature. According to a wide range of scholars in theology and science, including Ian Barbour, Celia Deane-Drummond, Arthur Peacocke, Holmes Rolston III, and Christopher Southgate, the suffering of Christ on the cross which takes up all human suffering should be extended to include the suffering of all living creatures. Thus all living creatures are part of God's redemption of the world, as we see previewed in Romans 8, and this has profound implications for Christian eschatology (Barbour 2001, 1–20; Deane-Drummond 2009; Peacocke 1998, 357–76; Rolston 2001).

4.4 Eschatology

A few scholars in theology and science have moved explicitly to eschatology as a final response to natural theodicy. These include Denis Edwards, Jürgen Moltmann, Ted Peters and John Polkinghorne, and I applaud their work in this regard (see for example Edwards 1999; Moltmann 1990; Peters 2006; Polkinghorne 2002). The Resurrection of Christ is seen as a transformation that includes elements of continuity (he could be recognized and touched) amidst radical discontinuity (he appeared in mysterious ways and he ascended to the Father). His Resurrection signals the beginning of the transformation of the universe into the New Creation, our eschatological hope.

I think that points the right direction. I have written about eschatology and cosmology in a way that learns its lessons from theodicy: it must at the very minimum respond to the challenges therein raised. But the biggest challenge to eschatology is cosmology with its dire predictions for the far future of the universe.

How can Christian theology continue the creative interaction here? I think it's possible to take some steps towards meeting this challenge, but that's for another essay (Russell, Murphy and Stoeger 2008, ch. 10).

Conclusions

I hope in this lecture to have indicated the many ways that scholars internationally and ecumenically in "theology and science" have creatively engaged the discoveries of evolutionary, molecular, and developmental biology as they focus on an explication of Christian theology, following the direction indicated repeatedly and persuasively by Pope John Paul II and with roots in the documents of Vatican II. The future clearly calls for our continued, robust engagement between Christian theology and the biological sciences. I am deeply inspired by, and grateful to, the contributors to and editors of this volume of essays, the STOQ project, the PCC, and the Pontifical Gregorian University, in pursuing this goal.

References

Abbott W. M. (ed.), 1966, *The Documents of Vatican* II, New York: America Press.

Auletta G., 2001, *Foundations and Interpretations of Quantum Mechanics: In the Light of a Critical-Historical Analysis of the Problems and of a Synthesis of the Results,* Singapore: World Scientific Publishing Co.

Ayala F. J., 2006, *Darwin and Intelligent Design,* Minneapolis: Fortress Press.

Ayala F. J., 2007, *Darwin's Gift to Science and Religion,* Washington DC: John Henry Press.

Ayala F. J., 2008, "Darwin's gift to science and religion: commentaries and responses," *Theology and Science* 6/2: 179–96.

Barbour I. G., 2001, "God's power: a process view," in Polkinghorne 2001, 1–20.

Cushing J. T., 1994, *Quantum Mechanics: Historical Contingency and the Copenhagen Hegemony,* Chicago: University of Chicago Press.

Dawkins R., 1983, "Universal Darwinism," in Bendall D. S. (ed.), *Molecules to Men,* Cambridge: Cambridge University Press, 403–25.

Deane-Drummond C., 2009, *Christ and Evolution: Wonder and Wisdom,* Minneapolis: Fortress Press.

Edwards D., 1999, *The God of Evolution: A Trinitarian Theology,* New York: Paulist Press.

Foster M., 1969, "The Christian doctrine of creation and the rise of modern science," in O'Connor D., Oakley F., 1969.

Heisenberg W., 1958, *Physics and Philosophy; the Revolution in Modern Science,* World Perspectives Series, No. 19, New York: Harper.

Hick J., 1966, *Evil and the God of Love*, Revised Edition, San Francisco: Harper & Row.

Horn S. O., Wiedenhofer S. (eds.), 2008, *Creation and Evolution: A Conference with Pope Benedict XVI at Castel Gandolfo*, trans. Michael J. Miller, San Francisco: Ignatius Press.

John Paul II, 1988, "Message to the Director of the Vatican Observatory (June 1, 1988), in Russell, Coyne and Stoeger 1988, M1-M14.

Moltmann J., 1990, *The Way of Jesus Christ: Christology in Messianic Dimensions*, trans. Margaret Kohl, London: SCM.

Murphy N., Russell R. J., Stoeger W. R. (eds.), 2007, *Scientific Perspectives on the Problem of Natural Evil,* Vatican City State; Berkeley: Vatican Observatory Publications; Center for Theology and the Natural Sciences.

O'Connor D., Oakley F., 1969, *Creation: The Impact of an Idea,* New York: Charles Scribner's Sons.

Peacocke A., 1998, "Biological evolution: a positive theological appraisal," in Russell, Stoeger and Ayala 1998, 357–76.

Peters T., Hewlett M., 2003, *Evolution From Creation to New Creation: The Controversy in Laboratory, Church, and Society,* Nashville: Abingdon Press.

Peters T., 2006, *Anticipating Omega: Science, Faith, and Our Ultimate Future,* Gottingen: Vandenhoeck & Ruprecht.

Peters T., 2008, *The Evolution of Terrestrial and Extraterrestrial Life,* Goshen: Pandora Press.

Polkinghorne J. (ed.), 2001, *The Work of Love: Creation as Kenosis*, Grand Rapids: William B. Eerdmans.

Polkinghorne J. 2002, *The God of Hope and the End of the World,* New Haven: Yale University Press.

Rahner K., 1978, *Foundations of Christian Faith: An Introduction to the Idea of Christianity*, trans. William V. Dych, New York: Crossroad.

Rahner K., 1983, "Why Does God Allow Us to Suffer?", *Theological Investigations* 19, New York: Crossroad.

Rahner K., 1988, "Natural Science and Reasonable Faith", *Theological Investigations* 21, New York: Crossroad.

Rolston H., 2001, "Kenosis and Nature", in Polkinghorne 2001,

Ruse M., 2001, *Can a Darwinian Be a Christian? The Relationship Between Science and Religion,* Cambridge: Cambridge University Press.

Russell R. J. *et al* (eds.), 2001, *Quantum Mechanics: Scientific Perspectives on Divine Action*, Vatican City State; Berkeley, California: Vatican Observatory Publications; Center for Theology and the Natural Sciences.

Russell R. J., Coyne G. C., Stoeger W. R., 1988, *Physics, Philosophy, and Theology: A Common Quest for Understanding*, Vatican City State: Vatican Observatory Publications.

Russell R. J., 2001, "Divine action and quantum mechanics: a fresh assessment," in Russell *et al.*, 2001, 293–328.

Russell R. J., 2007, "Physics, cosmology and the challenge to consequentialist natural theodicy," in Murphy, Russell and Stoeger 2007

Russell R. J., Stoeger, W. R., Ayala F. J. (eds.), 1998, *Evolutionary and Molecular Biology: Scientific Perspectives on Divine Action*, Vatican City State; Berkeley: Vatican Observatory Publications; Center for Theology and the Natural Sciences.

Russell R. J., Murphy N., Stoeger W. R. (eds.), 2008, *Scientific Perspectives on Divine Action: Twenty Years of Challenge and Progress,* Vatican City State; Berkeley, Calif.: Vatican Observatory Publications; Center for Theology and the Natural Sciences.

Schüssler Fiorenza F., Galvin J. P. (eds.), 1991, *Systematic Theology: Roman Catholic Perspectives,* Minneapolis: Fortress Press.

Schüssler Fiorenza F., 1991, "Systematic theology: task and methods," in Schüssler Fiorenza F., Galvin J.P., 1991.

Southgate C., 2008, *The Groaning of Creation: God, Evolution and the Problem of Evil,* Westminster: John Knox Press.

Życiński J., 2006, *God and Evolution: Fundamental Questions of Christian Evolutionism*, trans. Kenneth W. Kemp and Zuzanna Maslanka, Washington, D. C.: The Catholic University of America Press.

Appendix

From method to ideology:
The resistible ascent of reductionism in biology

Marcello Buiatti

1. Introduction

Scientific disciplines have a series of common methodological features but they may diverge in terms of the kind of "truth" searched for and, even more often, for their relationship with the "scientific imagery" or, in more general terms, with the ever changing "Spirit of times". Moreover, there is a very relevant difference between experimental sciences on one hand and mathematics or related disciplines on the other, particularly with reference to the "falsification" procedures. These are all internal in the case of mathematics, but involve the analysis of levels of coherence between expected and experimentally obtained results in the others. Another point of divergence influencing more and more the very structure and dynamics of sciences is the amount and quality of the relationship with derived technologies and through that with economics and politics itself. Within this frame of thought however, life sciences do not cluster with any other discipline as they deal with the "living state of matter", an object of study with very original and peculiar structure and dynamics. In the first place, at variance with other scientific disciplines, students of life sciences, being alive, are human observers and, at the same time, as living beings, objects of observation. In other the words, every biologist, studying the structure/functions of no matter which living being, may gain knowledge and develop concepts which may then be extrapolated to humans, that is to the observer himself and the human context surrounding him. This process inevitably leads to a very intense interaction between biology, the "spirit of times" and its prevalent concepts and even ideologies. This interaction is reciprocal in the sense that biologists are implicitly asked to confirm with scientific data prevalent opinions and ideologies, usually trying to respond in a positive way to the demands of the human context, but at the same time will observe and discover new features of life to be communicated, sometimes in a distorted manner influencing and changing the spirit of time. There is therefore a "circular" dynamic flux of ideas and knowledge between life sciences and the human contexts, each of them changing the other. To put it in another way, all sciences change with time but while in most of them the change essentially derives from the acquisition of new knowledge and the internal "falsification" process, changes in biology are also biased by the interaction with external concepts and ideologies.

The processes by which the "spirit of time" interferes with the development of life sciences are rather complex, derive from the very structure/dynamics of life, and influence different steps of the process of elaboration of new knowledge.

In the first place, living systems are "multiverse", that is they present apparently contradictory features at the same time (Buiatti and Buiatti 2008). For instance, life, as we shall discuss thoroughly later, has continuous and discrete features, is determined by genotypes and contexts, is always in an intermediate state between chance and necessity, its components may be independent and interacting in a non additive way etc. The spirit of time and the ideologies "suggest" which side of life should be privileged in the observation, hopefully leading to the development of "universal" concepts and theories. Such a choice implicitly entails the development of specific methodological concepts and tools which in turn influence the collection of data and their interpretation. From the methodological point of view for instance the choice between the holistic and reductionist approach has been very relevant for the kind of data obtained and the derived concepts and theories. As we shall see, the life sciences of the modern era have been based on the postulate of equivalence between living and non living objects or, often, machines. Now, machines are built following a single project through the assemblage of completely independent components. That means that every component has the same structure/function when present in the machine or isolated (a wheel of a car is the same in the car as it is outside it). If this were true, the knowledge of the whole object of study can be obtained through the assemblage of the data obtained through the study of components. If we apply the same concepts and the derived reductionist methodology to living systems, we shall certainly obtain interesting data more easily this way than we would through the study of whole objects. Moreover, the analysis of single components one by one, may lead to specialization into different disciplines looking at different sides (components) of the living systems, and therefore to different and sometimes opposite general conclusions and contrasting theories as it has happened throughout the twentieth century. In this paper I shall follow the interaction between the spirit of time and life sciences, from Darwin to the present years and what I call the third "millennium revolution", pointing out that, probably for the first time, the data from science are contradicting the spirit of time and apparently leading to more and more divergent paths in the future. The possible consequences of this possible pattern will be finally discussed in view of the dangers inherent to an increasing alienation of humans from the perception of the real nature of life and consequently of themselves.

2. Darwin, a naturalist of the nineteenth century

On the verge of the 150th anniversary of the first edition of the "Origin of Species" the debate on Darwin and his theory on evolution has been gaining pace and is often leading to a misrepresentation of the British naturalist. In many cases Darwin has been accused of proposing concepts not present in his writings but introduced by twentieth century population geneticists on the basis of what has been called by J. Huxley "the Modern Synthesis" in his famous book of 1942.

It should be stressed here that Charles Darwin was a nineteenth-century naturalist quite coherent with a "romantic" view of nature and a researcher well aware of his personal limits and therefore continuously open to falsification of his own data and theories which, indeed, he kept modifying from the beginning to the end of his scientific history and, particularly, throughout all the different editions of the "Origin". In his autobiography he wrote: "As far as I can judge....I have steadily endeavoured to keep my mind free so as to give up any hypothesis, however much beloved....I cannot remember a single first formed hypothesis which had not after a time to be given up or greatly modified". (Darwin 1958)

And indeed he has been changing his view throughout all the editions of the "Origin" particularly as far as the relative impacts on evolution of the environment was concerned.

Moreover, Darwin was an observer and not, although with some exceptions, an experimentalist fully adopting the Galilean method, nor, by any means, a reductionist. He also apparently disliked the usage of mathematics for the construction of models to the point of not being able sometimes to understand the value of the results obtained in some of his few experiments. One example of this behavior can be found in the treatise on the fertilization on plants where he showed that the progenies of crosses were more vigorous that the "pure lines" from which they had been derived but did not attribute to this discovery, anticipating the concept of heterosis very relevant in modern genetics, any particular interest and value.

Therefore, for him the "individuals" (the subjects of evolutionary change), were single (whole) organisms or species. His holistic vision led to the implicit concept of "multiversity" of life, in which antinomies like that between chance and necessity, discrete and continuous, genotype and environment etc. are present at the same time. It is for this reason that, particularly in the last edition but also before, he was upset when accused of considering selection the only factor of evolution.

In fact, this broad and pluralistic conception is confirmed by the Darwinian laws of evolution.

Those laws, taken in their wider meaning are:

- The law of growth through *reproduction*
- *heredity*, almost implicit in reproduction
- *variability due to the direct or indirect action of living conditions and use and disuse*
- a growth rhythm so high that it leads to a fight for existence and consequently to *natural selection* (Darwin 1872).

Again, he certainly did not attribute by any means evolution to the pure game of chance as many accuse him nowadays to have done, confusing as always Darwin with his self-defined followers, the neo-Darwinists, or better a current of them, the proponents of "neutralist" theories like Hugo De Vries at the beginning of the twentieth century or Motoo Kimura in 1961. On the contrary, Darwin did not believe in chance as total randomness but only as an intrinsic level of unpredictability by the observer of causes and effects within the frame of natural laws. This is very clearly stressed for instance in two sentences, one in the autobiography and the other in "The Origin of Variation of Domesticated Plants and Animals".

He wrote:

> I cannot anyhow be contented to view this wonderful universe, and especially the nature of man, and to conclude that everything is the result of brute force. I am inclined to look at everything as resulting from designed laws, with the details, whether good or bad, left to the working out of what we may call chance (Darwin 1958).
>
> Let an architect be compelled to build an edifice with uncut stones, fallen from a precipice. The shape of the fragments may be called accidental; yet the shape of each has been determined by the force of gravity, the nature of the rock, and the slope of the precipice, events and circumstances, all of which depend on natural laws; but there is no relation between these laws and the purpose for which each fragment is used by the builder. In the same manner the variations of each creature are determined by fixed and immutable laws but these bear no relation to the living structure which is slowly built up through the power of selection, whether this be natural or artificial selection (Darwin 1875, 236).

Finally, Darwin anticipated one of the most relevant differences between neo-Darwinism and the new conception of life, namely the fact that all living systems are networks of interacting elements rather than deriving from the assemblage of completely independent components:

> Variability is regulated by many unknown laws, the most important of which may be the law of correlated development. With this sentence I want to say that the different parts of an organism are so strictly connected that when small variations occur somewhere and are accumulated by natural selection, the other parts are subject to modification. (…) I mean by this expression that the whole organisation is so tied together during its growth and development, that when slight variations in any one part occur, and are accumulated through natural selection, other parts become modified.
>
> Although every variation is directly or indirectly caused by changes in external conditions, we should never forget that the nature of the organisation on which we act is by far the most important factor influencing the result. This is proven by the fact that different organisms in the same conditions will vary in a different way while similar organisms in different conditions will often undergo the same variation (Darwin 1875).

The concept of "correlated development" implied for Darwin that no new mutation could be "accepted" by an organism if not coherent with its "internal rules" previously fixed throughout evolution, thus anticipating the concept of "internal selection" proposed much later by Gregory Bateson in his famous "Mind and Nature" also allowing sudden accelerations of evolutionary change. These may occur when organisms survive to non fully coherent mutations (the "hopeful monsters" of Goldschmidt) and give rise to new species through a general re-arrangement of the whole genome: "In my own, strongly biased opinion, the problem of reconciling evident discontinuity in macroevolution with Darwinism is largely solved by the observation that small changes early in embryology accumulate through growth to yield profound differences among adults." (Gould 1977) It should be stressed here that the correlated variation is probably the concept which more clearly contradicts the unfortunately widespread opinion that Darwin envisaged evolution as a completely random process. The consequences drawn by Bateson and Gould moreover, fully justify the speed with which organization of living systems took place. This in fact has occurred because of the inevitable constraints, selecting only a part of the possible system variants not only by the environmental filter but by the necessary coherence of the systems.

3. The fragmentation of "modern" life sciences and the "central dogma"

The second half of the nineteenth century was marked by a strong acceleration in the industrial revolution, industries becoming the focus of human activities in many countries in Europe and in the United States. This led to the progressive concentration of a large part of the populations in industrial towns and an identification of the term progress with industrial development, life based agriculture activities being slowly dismissed and losing relevance in general economies. Also the spirit of time consequently changed and humanity started to firmly believe in its capacity to model the whole world according to human projects without the danger of unintended effects. Life sciences were heavily affected by this change and adopted for the first time the reductionist methodology thus opening the way to a mechanistic conception of life. The change was announced officially in 1847 by a group of scientists, mostly physicists and physiologists who published the *Manifesto of medical materialists* where they stated the substantial equivalence between living and non living systems. The Manifesto proposed the introduction of the reduction of living systems to components as the most powerful method in biology, thus "updating" life sciences to what had happened in physics and chemistry. This choice led to the usage of physical and chemical methods and tools in the study of systems through the analysis of their components. Consequently "Ockham's razor" was applied also to life proposing the simplest solutions of biological problems as always the most probable. The reductionist dissection of living systems into components led to an increase in specialization and fragmentation of Biological disciplines up to then united and to the breakdown of communication between them. Every discipline used specific tools and developed specific statutes built on a number of local truths all thought to be universal and therefore often conflicting with others. The Darwinian "multiversity" concept was abandoned and discussions were centered on the debate between false antinomies. Most of the writers of the Manifesto were chemists and physiologists but it has been a physicist who started using in his experiments the reductionist method. Gregor Mendel studied physics in Vienna and was an expert of probabilistic mathematics. He lived most of his life in the Moravian town of Brno, one of the first European centers of the industrial revolution. Being a physicist, he had the brilliant idea to apply his knowledge to the study of heredity, not looking at the structure/function of whole organisms but choosing on purpose single characters present in the chosen plant (*Pisum sativum*) with discrete variants. He then analyzed the results of a series of crosses between plants belonging to "pure lines" and showing discreet and different variants. He then simply calculated the frequencies of each variant at every generation after crossing and subsequent self-fertilization. The translation of frequencies into probabilities allowed then the formulation of "laws" which he hoped could apply to all characters in all organisms. To ascertain whether this was really the case, he asked a famous botanist, von Nägeli, to suggest him a plant different from *Pisum* for further experiments. Unfortunately, von Nägeli, in good faith, suggested a plant not endowed with sexual reproduction and therefore Mendel was not able to reproduce his earlier results. Disappointed, he completely abandoned his research on heredity dedicating the rest of his life to the study of bees and meteorological processes, as shown by the engraving on his burial stone: "Gregor Mendel, meteorologist and apiculturist". It is worth noting here that, although Mendel's work had been published, nobody seems to have understood its conceptual value, nor the possible consequences of his "laws" for evolution theory until the beginning of the twentieth century. An appealing explanation of this fact could be that the "mechanistic" spirit of time, present already in areas of biology near to chemistry like physiology, was not at all widespread in the community of life sciences where on the contrary the Darwinism of Charles Darwin was well rooted. This hypothesis finds some support in the fact that the concepts of discreteness, randomness, and independence of system components suddenly became widespread at the turn of the century not only in biology but also in many areas of human cultures such as music with the disruption of harmony by composers like Schonberg, paintings like in the case of divisionism and, later on, the Italian futurists, poetry, again with the futurists and their precursors etc. The twentieth-century revolution in such different areas of human thought certainly was a sign of the change in the "spirit of time" and of the triumph of the idea that progress was synonymous of the production of machines as stated openly in the Manifesto of Italian futurists. Humans began to equate the whole world to an immense machine liable to be "optimized" by our proud and clever species. More than that, world optimization was more and more considered to be the only real goal, just as the real goal of a car factory is the construction of an optimal machine liable to be produced and sold everywhere.

At the turn of the century the work of Mendel was re-discovered almost at the same time by three scientists, the Dutch Hugo de Vries, the Belgian Carl Correns, the Austrian Eric Tschermack von Seyssenegg. In 1903 de Vries published a book entitled *Die Mutationstheorie* where he challenged the Darwinian theory of evolution on the ground of Mendel's data. According to the Moravian founder of genetics the hereditary factors ("Anlage") were present in two copies in every organism carrying the same or different variants of the character considered (in the first case we call them now "homozygotes", in the second "heterozygotes"). The two copies were then transmitted independently at every generation by the parents to the progeny after the completely random assortment of their genotypes in the formation of gametes. Each member of the progeny, therefore, was endowed with a new, randomly assorted genetic complement, containing all the information needed for his development throughout the life cycle. Evolutionary processes therefore were stochastic/deterministic as the wholly stochastic assortment of gametes always followed a totally deterministic life cycle. In the view of de Vries, therefore there was no room left to selection nor to the action of environmental changes being all due to random completely "neutral" mutations. Evolution, moreover, was thought to proceed through discrete random "jumps" the Mendelian "Anlage" being discrete as previously discussed. Such a theory immediately originated a very harsh debate between the Mendelians and the followers of Francis Galton at that time convinced Darwinians believing in a continuous evolutionary dynamics directed by selection and the direct and indirect effects of environment.

Although his theory never became very popular and it was resumed only in the sixties by a Japanese scientist (M. Kimura), the main concepts of de Vries (chance, necessity and discreteness) have been the founding ground of the neo-Darwinian theories called by Julian Huxley in 1942 "the Modern Synthesis". Before going into the conceptual details of neo-Darwinism it is worth stressing here that the de Vries evolutionary theory was the first in the history of life sciences to have been built on results obtained with a reductionist approach and on the ground of a study on the hereditary behavior of only a few genes. These genes indeed apparently behaved as independent, deterministic "factors" and their alleles were randomly assorted at every generation. de Vries methodological innovation did not reside therefore in the acceptance of Mendel's data but in the extrapolation of his results to a universal theory. This is proven by the fact that even now the data obtained by Mendel have been never really challenged but it has been shown that the genes he chose were by no means a representative sample of the genome and only looked as if they were independent one from another and randomly assorted. In other words, life sciences of the third millennium have shown that not the experimental results obtained by Mendel but the extrapolation by de Vries was wrong due to the misleading usage of the reductionist methodology. Unfortunately however, this conclusion has not been clear for a long time and the whole population genetics has been based on stochastic determinism, starting with the mathematical generalization of Mendel's data by Hardy and Weinberg in 1908. The Modern Synthesis was the result of a sort of compromise between two opposed currents of thought on evolution, namely the followers of Mendel and de Vries on one hand, the heirs of C. R. Darwin (followers of F. Galton) on the other. Evolution for the Mendelians was thought to be dominated by chance and to proceed through random discrete "jumps" while the Galtonian group favored the "necessity" represented by selection and believed in a continuous evolutionary change. In other words the debate was based on two antinomies, namely on chance vs. necessity and continuity vs. discreteness. In both cases the chosen term of the antinomies completely excluded the other. The "compromise" offered by the Modern Synthesis was based on the co-presence of the two alternatives. The basic tenets of the new evolutionary theory can be therefore summarized with the following principles:

a) Total correspondence between gene and character as suggested by both Mendelians and Galtonians.

b) Additivity and independence of genes and alleles leading to phenotypes solely determined by the sum of allele effects, also present in both the older theories.

c) Although theoretically the unit of selection was the individual and therefore the phenotype, all laws of evolution were centered only on the genes while the influence of the environment on the phenotype, at least in the classical version of the Modern synthesis, was not taken into account.

d) Evolution was thought to involve *continuous* changes of the relative frequencies of *discrete* entities (the alleles).

e) Three main processes were considered to induce changes in allele frequencies, namely mutation (total chance), random drift (total chance), natural selection (total necessity).

As mentioned before, this version of neo-Darwinism was perfectly described by Sir R. A. Fisher, in his "Genetical theory of selection" published in 1930, in mathematical terms, the models proposed being based only on the results of Mendel without any further experimental evidence. As we shall see later the experimental work carried out on this basis did not fully confirm this theory but introduced always new exceptions leading to the fragmentation of the modern synthesis into a series of sub-theories all based on different sets of data coming from experiments planned with different criteria and within the frame of different disciplines. For instance the independence of genes was challenged in the early years of the twentieth century by the geneticists themselves as Bateson and Punnett who found that not in all cases the results obtained by Mendel could be confirmed, some combinations of alleles and genes being favored over others. Particularly, some experiments on the segregation in the second generation of a cross between plants containing two genes with two alleles each showed that some alleles tended to segregate together, the favored combinations being the parental ones. These results suggested the association of groups of genes which was confirmed by Thomas Hunt Morgan and shown to be due to the presence of more genes on single chromosomes. Physiological causes were also shown to determine non-Mendelian segregation because of interaction between genes in the control of development.

The achievements of genetics in the first decades of the twentieth century with their innovative content fully coherent with the spirit of times became soon very popular and consequently slowly led to the transformation of the reductionist method into an ideology. As the reader will recall, the reductionist method was transferred from physics and chemistry to biology as the only way to obtain fast and reliable results in experimental research through the analysis of single parts of systems, to be assembled at the end of the study allowing total knowledge of the system. In other words the reduction of the system to its components at the beginning did not imply that real systems were simply the addition of components but considered this postulate as a useful conceptual tool to analyze life complexity. Later on however the spirit of time tended to change the postulate from "living systems *can be considered with an experimentally useful approximation* as made of independent components" into " living systems *are* the sum of independent components". In other words a process of "reification" of ideas and concepts was initiated leading later on to the reification of the metaphor of "information" into a molecule, the DNA (Buiatti 1998; Gagliasso 1998). This result was obtained through the experimental progress of a new discipline, molecular biology, born from a further fragmentation of genetics with the input of biochemistry. As implicit in the name itself the objects of molecular biological research are molecules and particularly the nucleic acids and proteins. Also in this case the idea at the basis of this discipline was launched by a physicist, Erwin Schrödinger, who, in his famous book *What is Life*, a collection of lectures published in 1944, after a thorough discussion of the recent results obtained by geneticists, considered as a logical conclusion the suggestion, to be further proven, of a macro-molecule carrying the hereditary information to be translated into the molecular actors of life. His prophecy was confirmed and it was shown, with a series of successful experimental results, that the molecule is DNA. Furthermore, in 1953, the structure of crystallized DNA was discovered by Francis Crick and James Watson, and the first author proposed in 1958 the so-called "Central Dogma of Molecular Biology", the premise to the transformation of the reductionist method in a reductionist ideology. The principle of the "Dogma" is the wholly deterministic, co-linear and never ambiguous nature of the passage of information from DNA to protein through RNA, proposed by Schrodinger and translated into living matter by the discovery of the relationship between the three main macromolecules of life. The word "dogma" was consciously used by Francis Crick to stress the universal nature of the discovered "law", thought to hold true for all living systems from viruses to animals and humans. The immediate result has been, in the scientific imagery, the mechanistic "transformation" of living systems into computers endowed with only one program containing all the information needed for individual life cycles. This opened the way to the building of a sort of mechanistic informational ideology, offered to humanity in an exhaustive way by Jacques Monod in the famous book *Chance and Necessity* published in 1970. In that book, the French discoverer together with Francois Jacob wrote of the regulatory processes in bacteria:

- [The structure] realizes itself autonomously without external influences, without new information: information is already present although not expressed in its constituents. Its epigenetic construction is not a creation but rather a revelation.
- From them [protein sequences] (...) we can deduce today the general law: the law of chance. (...) Chance is taken up, conserved and reproduced through the invariance mechanism and transformed in order, rule, necessity.
- The fundamental biological invariant is DNA. (...) We cannot conceive any mechanism liable to transmit to DNA instruction or information. (...) As we can see it is fundamentally Cartesian and not Hegelian: a cell is really a machine.

With these principles the mechanistic/informational vision of life was completed and has been prevalent in the biological thought and in the world wide scientific imagery until the third millennium, although it has been challenged by several experimental findings and theories.

Just to quote some of the "heresies" (so they have been considered for a long time), the invariant nature of the genetic material, was shown not to be true as early as in 1951, just before the discovery of DNA basic structure, by Barbara McClintock with a brilliant series of cytological analyses of the dynamics of maize chromosomes. Her results have been considered heretic for thirty years and finally, in 1981 she received a Nobel prize for her work. The independent nature of the components of living systems was challenged by several authors among which Richard Goldschmidt, Sewall Wright, Michael Lerner, Richard Lewontin and Conrad Hal Waddington, may be considered the most representative of a far sighted current of thought which, in more complex form has been gaining attention and growing consensus in the third millennium. In the thirties Richard Goldschmidt built on the concept of the dynamic interactive network structure of living systems, a "saltationist" theory, ignored or attacked at that time, opposed to the continuous evolution of neo-Darwinism. According to Goldschmidt, evolutionary discontinuity would be caused by sudden changes of nodes with many relevant connections in living networks, causing the birth of what he called "hopeful monsters", liable to become possible ancestors of new living systems after the necessary establishment of a new equilibrium among components. Non additive interactions between components were proposed by Sewall Wright who developed interesting evolutionary mathematical models where the subjects of evolution were whole non additively interacting genotypes and not the single genes or alleles of the "classical version" *a la* Fisher of neo-Darwinism. I. M. Lerner stressed the positive value of the so-called hybrid vigour ("heterosis") showing that the advantage of heterozygosity (the contemporary presence of two different alleles of singe genes in one genotype) was due to a better buffering capacity of hybrid genotypes of changes induced by modifications in the environment. This concept allowed later on Richard Lewontin to build the "balance" theory of evolution stressing that heterotic reproductive advantage would be a positive factor for the necessary maintenance of useful genetic variability. Finally, Conrad Hal Waddington re-introduced the phenotype, deriving from the interaction between the genotype and the variable contexts encountered during life cycles and evolution, and called for a "phenotype paradigm" as opposed to the mechanistic "only DNA" one, largely prevalent in the second half of the twentieth century. Phenotype paradigm meant for Waddington a new attention to the effects of environment in building the phenotypes throughout lives and that way going back to the earlier intuitions of Charles Darwin. Waddington proposed in the fifties of the twentieth century the term "epigenetics", defining all the processes and acquirements of information from the context which had not "been written in DNA", that is the continuous changes in the individual "programs" induced by the non additive interactions between the living subjects and the context. In other words, for Waddington, the genotype contains the tools for a wide range of possible programs but not the program itself, being now thought as dynamic and changing with time and contexts.

4. The change of the second half of the twentieth century and the challenge of complex systems theories

In 1963 the deterministic, mechanistic, conception of nature was challenged by the meteorologist Edward Lorenz who re-discovered and further developed the findings of Henri Poincaré on dynamic instability of the solutions of non-linear equations. This prompted an explosion of studies on "deterministic chaos" and in general on complex dynamic systems in a number of different areas. In physics a whole new disciplinary area was opened in the field of non-linear physics and the discussion on the new vision was soon extended to other fields of knowledge. An incentive for this change of "eye-glasses" certainly was the general perception by many observers of nature that something not predicted and dangerous may be really occurring in the relationship between *Homo sapiens* and nature, as stressed for the first time by the Club of Rome whose predictions have been found to be true. The concept of a certain level of unavoidable unpredictability of natural processes and of the dangers for humanity in the lack of attention to it, were stressed by many observers and became a subject of general debate in the eighties. Biology was much slower in accepting the new view which has become prevalent only in the third millennium when biologists were confronted with the totally unpredicted results obtained through whole genomes sequencing and on the other hand with the effects on biodiversity, ecosystems, natural resources, economy, of the imposition of a human model to the Globe all based on its homogenization/optimization. Some doubts started to emerge on the mechanistic nature of the Planet and the idea of a progress only based on the further transformation of living and non living systems into man made machines was challenged if not in the society certainly in the scientific community, the first official position with a world wide impact being the conclusions by the Club of Rome coordinated by A. Peccei. In biology, the real abrupt change only started in the last decade of the twentieth century becoming a real conceptual revolution in the third millennium. The sudden acceleration of this process was certainly due on one hand to the increase of attention in the biological scientific community for the presence of unpredictable events in living systems dynamics certainly influenced by the modification of the spirit of times. However, as happens in science, the beginning of the revolution certainly was due to the usage of very powerful machines and methods of analysis and the unexpectedly fast increase in data fully contradicting the "central dogma" ideology. It all started with the analysis of the human genome showing that eukaryotic genes only cover a small part of the genome, are " ambiguous" (one gene – many proteins), and most development is controlled by regulatory, "non-gene" sequences, actively responding to external signals. We know now that the human genome only contains around 23000 thousand genes covering only 1.5% of total DNA. The remaining DNA, previously called "junk DNA", was on the other hand found to be active and transcribed into RNA essentially with regulatory properties but not leading to the synthesis of proteins. Moreover, our 23000 genes are now known to code for more than 4 millions proteins and are therefore, very highly ambiguous, thus contradicting again the central dogma. These early findings have led to a sudden burst of research in a series of new fields and even to the birth of entirely new biological disciplines such as the already mentioned epigenetics, all those included in the so called "Systems Biology". As discussed in two earlier papers (Buiatti 2007; Buiatti and Buiatti 2008), these and many other findings urgently call for a general revision of some of the basic concepts of the modern and for a post-modern synthesis of evolutionary theories. The basic tenets of the new vision can be summarized as follows:

a) No living system is completely random (equi-probability of all possible configurations of components) nor completely ordered (only one configuration possible) but, on the contrary, life is endowed with variability and constraints to randomness. To be sure, as we have shown, for Darwin himself chance is essentially unpredictability by human observers. The only real chance based evolutionary theories were those of H. de Vries and, later on, of M. Kimura, both being minority currents of biological thought.

b) Both variability and constraints are needed for the maintenance of the "living state of matter" through adaptive change within organization rules. Therefore all organisms developed during evolution tools and processes for variability generation, signal perception, choice of the part of variability to be used according to internal and external signals.

c) As proposed by Jablonka and Lamb, evolution "walks" within a four dimensional phase-space, and different groups of organisms use different strategies based on the differential usage of genetic, epigenetic, behavioral and symbolic variability (Jablonka and Lamb 2004). Therefore no unique evolutionary theory based on a single adaptation strategy can be built but, on the contrary, the "multiverse" nature of life has to be taken into account along with the fact that living systems often show the contemporary presence of apparently contradictory features (they are continuous and discontinuous, determined by genotype and environment, variable but with constraints etc.).

d) The different strategies employed by different groups of organisms to generate variability and choose context by context which part of it has to be chosen, are now known. Bacteria use mainly genetic variation (evolvability potential) and have developed sophisticated tools and processes to increase genetic variability in the presence of stress. This strategy derives from the fact that bacterial lives are very short and therefore individuals only have to face limited changes in the context. On the other hand plant survival is largely due to a very high capacity to change during single life cycles (epigenetic and plasticity potential) often extremely long as it happens with vegetatively propagated species. Animals use both, but also a higher capacity of choosing and changing the environment, through exploration and the usage of nervous systems. All the mentioned organisms but particularly plants and animals, play an active part in selection dynamics, modifying with their behavior the contexts sometimes in a very sophisticated manner although often stereotyped and transmitted as such throughout subsequent generations. In both cases there is convincing evidence for the existence of selection pressures for increased evolvability, plasticity and all the molecular tools needed for their improvement. Finally, humans, although using all the aforementioned tools and processes, have adopted a completely innovative adaptation strategy not relying on the selection by the context and consequent differentiation of the more favorable genotypes in different environments, but on the active modification of the contexts through the development of cultures.

e) Life is organized in a hierarchy of connected networks from molecules in cells, cells in colonies and tissues, tissues in organisms, organisms in populations, populations in species, species in ecosystems, ecosystems in the biosphere. In addition to that, as proposed for the first time by Vernadsky, humans, the most generalists of all species, have constructed their own, innovative networks of individuals, families, communities, ethnic groups and so on, covering the whole planet with the all-containing "noosphere" (Vernadsky 1926). All hierarchical levels are subjects/objects of evolutionary change. Network dynamic structures allow very fast and efficient communication between different components leading, through non-additive interactions to the formation of functional novelty (in life the interaction between a component A and a component B never gives AB but always leads to a novelty X).

f) A major part of constraints to randomness derives from the just discussed network structure where complementary structures/functions between interacting components are always developed within the frame of functional vital rules of all the different systems. This network structure in all cases and at all levels, necessarily leads, as also suggested by Darwin, to the amplification of single or few changes expanding their effects to other connected components. It is for this reason that not all mutations are "accepted" by the harmonic networks but may lead to the birth of "hopeful monsters" (Goldschmidt) where the whole genotype changes to adapt itself to the modifications occurred. In other words every mutation, before being fixed in a specific genome, has to be "accepted" by the host harmonic network ("internal selection" according to Gregory Bateson) before being challenged by environmental selection.

g) Selection is never a "choice" among *passive* variants but derives from "*active*" interactions potentially changing both the evolutionary unit and the selecting network.

5. The adaptive strategies of bacteria, plants and animals

The very fact that different organisms adopt different adaptive strategies implies that, due to the multiverse nature of life a single, detailed, universal theory predicting the histories and the fates of all living systems does not exist although the aforementioned basic rules are validated frames within which different organisms will display their lives and evolutions. It is worthwhile then to discuss with a little more detail the different types of activities which lead to adaptation in the different large groups of organisms.

Bacteria and viruses have indeed very short lives, individual life-cycle lasting generally less than ten minutes, their genomes, at variance with what happens in plants and animals, are very small and simple, all genes being present in only one chromosome and in only one copy instead of two in a single cell. This means that, while in eukaryotes a mutation in one of the two copies of a gene may be "masked" by the other copy and therefore may not be expressed, in prokaryotes every mutation is immediately active after the induction. Moreover bacteria can very easily change their genetic complement by simply "picking up" small or large fragments of DNA encountered accidentally. This process, known under the name of "transformation", is very frequent and may involve such large fractions of other genomes belonging to other species, to render in many cases impossible a correct taxonomy of these organisms. Finally, in the presence of stress, the regulatory gene RpoS is activated and induces the expression of about 10% of bacterial genes and inactivating others. The genes activated include several sequences capable of increasing the mutation frequencies thus increasing the probability that one of them will allow a sufficient response to the stress and the same effect is obtained through the repression of the "mismatch repair" genes MutS, MutL, MutH and the "protecting" gene Dps. Moreover, genes containing the information for possible positive response to the specific stress are also activated and activation in this case also increases mutation frequencies (Saint-Ruf and Matic 2006). Finally, a specific increase in the number of copies of potentially useful genes is also induced through RPOs action.

Plants and animals developed a series of tools and processes which allow them to change their "program" throughout life regulating from the quantitative and qualitative points of view the action of genes. The new branch of genetics called "epigenetics" following the already mentioned suggestion by Waddington, may be defined as the study of all the regulatory processes whose action is not directly regulated by DNA but by the incoming signals received by cells. The signals, coming from the external or the internal context of each cell are therefore directing the cell and organismal fate changing it throughout life and thus allowing permanent adaptation through plasticity. Within this conceptual frame therefore, genes are not anymore deterministic in the sense that they are carriers of one program, but rather they provide the information for a very high number of tools (the different proteins) to be produced in variable amounts according to the internal and external contexts. In other words genes are "ambiguous" as their information content allows the production of more than one protein per gene at variance with what was thought before. This high level of ambiguity is allowed by the development during evolution of a series of tools and processes through which the same DNA sequence can be "read" in many different ways interfering with all the steps of information flow, from DNA, to RNA, to proteins. At the level of transcription of DNA into RNA, RNA synthesis can start at different sites and be terminated at different termination sequences, giving rise to RNAs and therefore proteins with different lengths and to proteins with different numbers of amino acids. Moreover, in eukaryotes but not in prokaryotes, each RNA is not immediately translated as such but, on the contrary, between transcription and translation, is cut into parts, some of which (the "introns") are eliminated and the others (the "exons") are then assembled in more than one order giving rise again to qualitatively different proteins. The process leading to the elimination of introns and the "ligation" of exons is called "splicing" or, when RNA is alternatively processed, "alternative splicing". In all these cases alternative poly-peptides are produced following perception of different signals related to the internal situation of cells and to changes in the environment. Obviously not all genes are equally ambiguous but ambiguity levels are different according to the function. In animals for instance the highest ambiguity levels are found in genes used for signal perception and connections among neurons. This is for instance the case of neurexins, involved in neuron connection processes in humans (more than 2000 proteins from three genes) and DSCAM (38000 proteins from one gene) in *Drosophyla*. It has been recently shown that in brains alternative splicing is induced by spikes and leads to the production of different proteins. All the processes just described lead to ambiguity through differential "management" of DNA and RNA but also proteins are often ambiguous in the sense that many of them can perform different functions simply by changing their conformation. This is what happens for instance with the protein agent of the "mad cow" syndrome. In this case one particular protein may be present in two different conformations one of which has a relevant function in the brain, the other being not properly active. The infection process then is not due by any means to protein replication but to the fact that when one "good" protein molecule meets a "bad" one it will form a complex with it and thereby change its virtuous conformation into that of the new partner. In computational terms it can be said then that in this case there is deterministic information flow from one protein to the other. The change of function through conformational modification is only rarely associated with loss of function as it is happens for "prions" like the proteins involved in the case of the "mad cow" syndrome and in several other cases not only in higher eukaryotes but also in yeasts. On the contrary, complex "conformational landscapes" generally contribute to the potential plasticity of an organism allowing multiple functions performance by single proteins (calmodulin for instance, a key protein, can form complexes with a series of other molecules changing conformation and function at the same time).

Moreover, plants and animals are endowed with tools allowing a number of epigenetic, potentially heritable quali-quantitative processes, of both genetic and regulatory nature. The inheritance of somatic changes has been claimed in several instances in the past, but recently data obtained years ago have been confirmed and validated opening a wide field of innovative research. For instance, the heritable character of increases in the number of copies ("gene amplification") was discovered by Durrant and explained by Cullis and co-workers in flax many years ago (for an early review see Buiatti 1977). While somatic mutations have been known for a long time, the inheritance of wide range DNA "shuffling" was discovered very recently by Brunner *et al.* (2005) in plants and shown to be mediated by RNA molecules in *Paramecium*. Furthermore, since a long time the process of "horizontal transfer" of genes from one species to another has been shown to occur also in higher eukaryotes and particularly in plants, where one of the main causes of evolutionary change is inter-specific hybridization. For instance Intrieri and Buiatti (2001) showed that the integration of *Agrobacterium rhizogenes* genes into *Nicotiana* species has drastically influenced the evolution of the entire genus. Finally stable and even heritable changes of gene expression are now known to be induced through processes like DNA and histone methylation, histone acetylation or de-acetilation etc. It is worth noting that in animals, particularly methylation changes can be induced by exchanges of behavioral signals between members of the same species. In mice for instance, mothers cuddling their progeny will transmit this character to their daughters and to the following generations not through genetic changes but through de-methylation and therefore a higher expression level of the "gluco-corticoid receptor", a gene with wide and very specific affects on the hormonal system. Gene methylation, an at least partially reversible process, may differentiate into different phenotypes monozygotic twins with different life histories and bees castes after differential feeding. Some animals moreover, like social insects but also mammals not only respond to changes in the context through epigenetic plasticity but also modify it according to their needs, for instance constructing shelters of different kinds and even specific micro-ecosystems like in the case of "leaf cutter" ants.

6. The innovative human adaptive strategy

Our species although utilizing all the tools and processes present in bacteria, plants, and animals, in addition to those, has "invented" and uses as a priority a completely new strategy based on an extreme version of the aforementioned active changes in the environment induced by some animal species. With the notable difference that, while other animals keep stable throughout generations the models according to which they change the environment, humans have been changing them with increasing speed. The material basis of the extraordinary capacity of invention of our species is not genetic but has its material basis in the incredible organization of our brain. While our genome only contains 23000 genes allowing the synthesis of four millions proteins, our cortex contains hundred billions neurons potentially allowing the formation of one million billions synapses between neurons. This gives to our brain an incredible capacity of information storage and of connections between its components through very complex dynamic processes still very far from being completely understood. Recent data on the genetic differences between humans and the other primates show us the material basis for the evolutionary "jump" between us and our closer "cousins", namely chimpanzees, gorillas, orang-utans and bonobos. Very few genes (from 1% to 3%) separate us from the others having changed very rapidly in ours but not in their lineage. Almost all of these genes are expressed in the brain and their functions are related to the encephalization process leading to our specific brain organization and to the increased capacity to perceive and communicate signals throughout the neural network. Some of the specific genes involved are encephalin and ASPM, both related to encephalus growth, and neurexins, cadherins, reelin etc. all involved in signal perception and neuron connection. Our encephalus, as a consequence is much larger than in other animals and our neurons communicate with higher speed and efficiency allowing us to connect thoughts and thereby develop new ideas and inventions. In summary, our brain can be distinguished from that of other closely related primates through a series of parameters and particularly the encephalization index, the neuron number, the speed and efficiency of communication among them, the plasticity of neuronal organization, the level of perception and exchange of signals and the presence of wholly innovative areas like the area of Broca for languages. Of particular relevance is certainly our capacity to exchange information with other humans as shown by very recent experiments (Herrman *et al.* 2007) based on the comparisons of the performances in 2.5 years old children and chimpanzees of the equivalent age. It was shown this way that most basic faculties of manipulation and even the perception of numerosity (chimpanzees can count up to the number three) were more or less the same in both "children". However the human was found to be far better than the primates in exchanging information with other members of our species.

From what we know now, only the very basic organization of our brain, namely the material architecture of the few very large areas in which it is subdivided during fetal development, are controlled by specific genes whose malfunctioning leads to death or to very strong disturbances in mental functions. On the contrary, the brain is continuously changing and re-modelling itself from birth to death although its plasticity is continuously decreasing with age, and synapses are formed and maintained in chains only until signals are transmitted. If signals stop in the early days and years of life, synapses are destroyed ("pruned") and others may be formed. This has been shown by a large number of experiments. Just to quote one, carried out on mice, a little mouse kept without light for the first ten days of its life will remain blind for the rest of its life. The neural connections from the eyes to brain, present in normal animals, are destroyed but others are reinforced in a visible way from the brain to the moustaches, showing the high vicariance capacity of mammalian brain. In other words the neural connections leading to the eyes have been pruned as they do not receive light signals, but a vicariant set of synapses has been built helping the orientation of the animal. It is using the peculiar capacity of our brain to store and use for inventions an unbelievable amount of information, that we, the humans, developed an original strategy of adaptation. While animals use their nervous system to look and find convenient environments and are subject to genetic selection choosing genetic pools specific for every context, we, the humans, change the environments where we settle and develop different cultures. This is proven by the fact that our genetic variability is very low and the genetic variants are almost the same in all continents (the term "race" has therefore no biological meaning in our case), while we still speak some seven thousand languages, particularly in the subtropical areas of our Planet.

Our evolution therefore has gained an incredible pace not being anymore based on the slow genetic variability selection process, but on invention and changes of cultures (the symbolic dimension of evolution as Jablonka and Lamb call it in their *Evolution in Four Dimensions*). Probably the real starting point of this very fast process can be located at the time when the first utensils but more than that, the first paintings were developed by humans. Painting in fact means to elaborate and change in our brains the image of an external object changing it according to our will and feelings and to "impress" it on a stone or on other materials. What we are impressing therefore is not the original object but a transformed version of it, with a procedure in many ways very similar to what we do when we make a "product" of any kind, inventing and planning it and then using the external matter to construct it. This programming capacity allowed us therefore to rationally plan the environments where we settled after the invention of agricultures which occurred almost at the same time in different parts of the Globe. From that time on our extremely fast evolution gained pace and in a few thousand years we arrived at the industrial revolutions and the modern era where the mechanistic paradigm, as we showed, was largely prevalent both in societies and in biology. The mechanistic paradigm implicitly contains the concepts of "optimization" and homogenization as grounds of progress. In fact optimization is the aim of industries all working for the production of "optimal" machine to be all sold everywhere in the world. At the beginning the aim was to optimize the world to improve the quality of our lives and certainly for a long time it has had a considerable success in a part of the world although it failed to solve a number of problems, the first of which being famine which decreased for a number of years but is increasing again now rising in a few years from 800 million undernourished people to 950. Now, as we tried to show, the world or at least its living components are based on variability, plasticity, evolvability etc. and therefore an "optimal" and homogeneous living system is certainly a loser in evolution. Spreading homogenization therefore means moving towards death, as machines are indeed dead products based on one single project designed by human beings and incapable of adapting to the ever-changing contexts of the Planet. This conception, in a certain sense based on concepts proposed by Darwin but not by neo-Darwinists, is now widely accepted in life sciences but, probably for the first time it is not coherent with the prevalent spirit of time, where by this we mean the wide-spread feelings of the human populations. The refusal of the new conception of life, which by the way is in many aspects intuitive (we are all conscious that we are changing continuously our program of life), is in our opinion derived from a very dangerous tendency of humans to forget what being living really means thus altering the basic scale of values where life itself with its diversity, capacity of changing, plasticity and evolvability should be at the very top of our emotional and rational behavior. Unfortunately we are constructing a sort of virtual life (see Buiatti 2007) where even the concept of richness has been changed. Once upon a time to be rich meant to profit from abundant sources of food and have comfortable shelters, later on to lead industries producing goods, nowadays it only means to possess money as such. Not by chance the prevalent indicator of well-being is GNP, a parameter based on the amount of circulating money and not goods. This parameter will rise for instance when a catastrophic event happens (an earthquake, a typhoon, a war) which will increase money circulation for the necessary investments in re-construction. The same will happen if a national health service open to everybody collapses and privates have to enter in the area, making profit of a previously public service and in other similar cases where a decreased level of well-being leads to an increased GNP value. From a scientist point of view, the detachment from the very values of life substituting them with virtual money often bought and sold online is a suicidal behavior to be challenged particularly by scientists who should be fully conscious of the high level of responsibility in spreading and discussing structure and dynamics of real living systems. Our societies are been mislead by the identification of the progress in the continuous growth of money exchange, just as it was before, in the optimization period in the continuous increase in machine production. In the modern period a considerable damage to the conditions of living beings on this Planet was induced by the optimization paradigm. Now, the pace of production is further increasing but at the same time the value of life, be it human or of other living beings is being forgotten in the name of a "King Mida" virtual treasure. At the same time, the competition paradigm, increases the loneliness of human individuals and makes us forget that a peculiar character of our species is the capacity to exchange information between humans in blood and flesh and not only between computers as showed by the recent cases of alienation deriving from a distorted usage of web sites such as "Facebook" or "Second Life". Scientists aware of what is happening should feel the need to tell everybody that "the king is naked" not only in his body made of flesh and blood but also in his mind full of thoughts, feelings, anguishes and happiness, life and death.

References

Bateson G., 1980, *Mind and Nature*, McMillan: New York.

Brunner S., Fengler K., Morgante M., Tingey S., Rafalski A., 2005, "Evolution of DNA sequence non homologies among maize imbreds", *The Plant Cell* 17: 343–60.

Buiatti M., 1977, "DNA amplification and tissue cultures", in Reinert J., Bajaj Y. P. S. (eds.), *Applied and Fundamental Aspects of Plant Cell, Tissue and Organ Culture*, Berlin: Springer, 358–74

Buiatti M., 1998, "L'analogia informatica del dogma centrale e le conoscenze attuali della Biologia", in Continenza B., Gagliasso E. (eds.), *L'informazione e le scienze della vita*, Milano: F. Angeli.

Buiatti M., 2007, "Homo sapiens e la vita virtuale", *Iride* 1: 37–62.

Buiatti M., Buiatti M., 2008, "Chance vs. necessity in living systems: a false antinomy", *Biology Forum* 101: 29–66.

Crick F., 1970, "Central Dogma of molecular Biology", *Nature* 227: 561–63.

Darwin C. H., 1958, *The Autobiography of Ch. Darwin 1809-1882*, Barlow N. (ed.), London: Collins.

Darwin C. H., 1872, *The Origins of Species by Means of Natural Selection*, 6th edn., London: Murray

Darwin C. H., 1875, *Variations in Animals and Plants under Domestication*, London: Murray.

de Vries H., 1901–1903, *Die Mutationstheorie*, Leipzig: Veit.

Fisher R. A., 1930, *The Genetical Theory of Natural Selection*, Oxford: Clarendon.

Gagliasso E., 1998, "Metafore e realtà, il caso del concetto di informazione", in Continenza B., Gagliasso E. (eds.), *L'informazione e le scienze della vita*, Milano: F. Angeli, 173–89.

Goldschmidt R. B., 1940, *The Material Basis of Evolution*, New Haven: Yale University Press.

Herrman E., Call J., Hernandez-Lloreda M. V., Hare B., Tomasello M., 2007, "Humans have evolved specialized skills of social cognition: the cultural intelligence hypothesis", *Science* 317: 1360–66.

Huxley J., 1942, *Evolution: The Modern Synthesis*, New York: Harper.

Intrieri M., Buiatti M., 2001, "The horizontal transfer of Agrobacterium rhizogenes genes and the evolution of the genus Nicotiana", *Molecular Phylogenetics and Evolution* 20: 100–110.

Jablonka E., Lamb M., 2004, *Evolution in Four Dimensions*, Cambridge, Mass.: MIT Press.

Kimura M., 1961, "Some calculations on the mutational load", *Japanese Journal of Genetics* 36:179–89.

Lerner I. M., 1954, *Genetic Homeostasis*, London: Oliver and Boyd.

Lewontin R., 1974, *The Genetic Basis of Evolutionary Change*, New York: Columbia University Press.

McClintock B., 1951, "Chromosome organisation and gene expression", *Symposia on Quantitative Biology* 16: 13–47.

Monod J., 1970, *Le hasard et la necessité*, Seuil: Paris.

Saint-Ruf C., Matic I., 2006, "Environmental tuning of mutation rates", *Environmental Microbiology* 8: 193–99.

Schrödinger E., 1944, *What is life? The Physical Aspects of the Living Cell*, Cambridge: Cambridge University Press.
Vernadsky V. I.,1926, *The Biosphere*, Leningrad: Nauchtekhizdat.
Waddington C. H., 1975, *The Evolution of an Evolutionist*, Edinburgh: Edinburgh University Press.

Evolutionary developmental biology and intelligent design

Scott F. Gilbert

Biology is a young science, and it cannot discern Truth. However, in its attempts to explain the world's biodiversity, it has gathered evidence for a remarkable theory of change, called evolution. This theory of organic change has been tested numerous times, and it has survived each challenge. Currently, a challenge is being made by Intelligent Design, a set of ideas that is not science, but which frequently masquerades as such. The Church, and indeed any intelligent person, should be highly suspicious of the claims of Intelligent Design. First, the proponents of Intelligent Design (ID) do not present a counter-theory to evolution. Second, they celebrate ignorance rather than knowledge. Third, proponents of Intelligent Design often base their criticisms of evolutionary theory on assumptions or models that scientists have already disproved or abandoned (but which might sound reasonable to a person who has not been trained in science). And fourth, the theological argument from design is limited to proving the existence of a skillful mechanic, and not a God of mercy or judgment.

1. Some scientific reasons to be wary of intelligent design

Intelligent Design ignores the actual science of evolutionary biology and has not hypothesized anything to replace it. The new research into the mechanisms by which new anatomical forms can occur (evolutionary developmental biology) is being conspicuously evaded by the proponents of ID. Certain Intelligent Design proponents have realized that evolutionary developmental biology proposes a scientific explanation for the formation of large changes in animal form, and therefore that it constitutes a serious threat to design theorists. So this body of knowledge is either ignored or (as we will see below) reported in a remarkably creative and unscientific manner. Moreover, Intelligent Design proponents have not put forth any testable alternative to evolution; nor has ID any program to discover anything. Unlike earlier design theorists such as William Kirby (1836; see below), who used design to further science (the more detailed our knowledge of the contrivance, the more glory to its Creator), Intelligent Design is anti-science, an active denial that science can explain the natural world through non-teleological and non-supernatural processes.

One of the major reasons that scientists do not want to get involved in disputes about Intelligent Design is that there is nothing scientific to counter in such debates. In the trial in Dover, Pennsylvania, biochemist Michael Behe could not define what a scientific mechanism would be in Intelligent Design, and he ended up confessing that astrology would be considered a science under his framework. An evolutionary biologist who wants to understand how kestrels can catch voles from such a high altitude can point to a mutation in the kestrels' opsin gene, which enables the bird to see in the ultraviolet. This mutation would give those kestrels having it an advantage over those lacking it, since voles' urine trails fluoresce in the ultraviolet and their homes can thereby be identified from the air (see Carroll 2008). The Intelligent Design proponent can only say that the Designer wanted the kestrel to see vole urine and so changed its retinal pigments. So one major reason evolutionary biologists don't take ID seriously is that design advocates propose no scientific mechanisms to explain biodiversity or adaptation.

The second reason most scientists don't see ID as a scientific alternative to evolution is that it does not respect the rules of evidence. Scientists just can't make things up. If a hypothesis is shown to be false, it must be retracted or its proponents must demonstrate that the criticism is not conclusive. This is not the case for Intelligent Design. I'll again use the works of Michael Behe as an example, since he is one of the few scientists in this movement and was the chief witness for ID and the Dover Trial. Behe's book *Darwin's Black Box* (1996) makes certain declarative statements about nature and about scientists' ignorance concerning how bacterial flagella and the vertebrate eye could have possibly evolved. Each one of these assertions has been falsified, and people like Brown University biologist (and practicing Catholic) Kenneth Miller have repeatedly and publicly criticized Behe's work for being scientifically wrong (see The TalkOrigins Archive 2006; Miller and Levine 2002). Scientists actually know a lot about the things that Behe claims we don't know anything about. Behe states emphatically that the bacterial flagellum is so remarkably constructed that its component parts could not have evolved for separate functions. They would all have had to evolve concurrently, and this would indicate an "end", since they would have no function individually. Biologists actually have excellent evidence that each of the components of the bacterial flagellum actually had a separate function, and in some bacteria, they still perform these ancestral functions. Similarly, Behe claims that the component proteins of the complement cascade of the human immune system are so perfectly integrated that these component parts could not have existed separately and if you remove any one of them, the system falls apart. Scientists have known for years that the complement system is a jerry-rigged amalgamation of proteins that have had separate functions elsewhere, and that there are several organisms that have fully functional complement cascades despite lacking one or more of these parts. One of the basic principles of scientific evidence is that if you make a statement and it is subsequently proven wrong, you retract it (or at least do not include it in newer versions of one's hypotheses). Behe's refusal to consider the mountains of evidence that counter his statements makes it very difficult to take him seriously as a scientist.

Another scientist who plays unscientific games with data is Jonathan Wells. The Reverend Dr. Jonathan Wells (his ordination is from Sun Myung Moon's Unification Church) has a PhD in developmental biology, and he was told by Moon to enter biology in order to obtain evidence against evolution (Wells 2005; Pigliucci 2002). One of his pieces of evidence against evolution (and evolutionary developmental biology) is a remarkable experiment that he claims disproves the notion that genes have anything to do with development or evolution. According to his website (Wells 2005):

> My experiments focused on a reorganization of the egg cytoplasm after fertilization which causes the embryo to elongate into a tadpole; if I blocked the reorganization, the result was a ball of belly cells; if I induced a second reorganization after the first, I could produce a two-headed tadpole. Yet this reorganization had nothing to do with the egg's DNA, and proceeded quite well even in its absence (though the embryo eventually needed its DNA to supply it with additional proteins).
>
> So DNA does not program the development of the embryo.

Is this the undoing of evolutionary developmental biology? Does this demolish the Modern Synthesis that brought together genes and evolution? No. The actual explanation for this result is simple, but may not be obvious to a person who has not had biological training. Here's the trick: the early development of the frog doesn't have a thing to do with its nuclear genome. In the species he was using, the nuclear genome is not even active until the twelfth cell division. Rather, the early development of the frog is governed by its *mother's* genome! The mother's genes, active in the oocyte (while it is in the mother's ovary), encode the protein β-catenin. The localization of this transcription factor protein is the first step in forming the head. If you degrade β-catenin during the first division, no head will form; if you reposition it by rotation towards the end of first division, two heads will form, one in the original position and one in the new position (see Gilbert 2006 for details of these experiments). Thus, genes *are* critical in both evolution and development. Does Wells know about β-catenin and its role in forming heads? Yes — he is co-author on the paper that proved it (Rowning *et al.* 1997). However, he is trying to fool people into believing that genes play no role in development (and thus in evolution) by ignoring the data and pulling the wool over the eyes of citizens who have no idea what's going on inside the embryo. Certainly there is no absence of scientific knowledge in understanding how these frog embryos can get two heads; nor is there any hint of Intelligent Design; just good old-fashioned genes working in the oocyte as they have been known to work since the geneticist Alfred Sturtevant first proposed maternal effect genes in 1923.

Ironically, Wells, Behe, and others have claimed loudly and often that the diagrams of Ernst Haeckel (which show the similarities of all vertebrates during the early stages of their development) are fraudulent. Moreover, they claim that these diagrams get repeated so often in evolutionary biology textbooks because evolutionary biologists have no real evidence and must therefore be reduced to using these fraudulent pictures (see Wells 2000; Miller and Levine 2005). Of course, Haeckel's diagrams were teaching tools (from the late 1800!), and evolutionary biology was never based on them. Moreover, to say that evolution is false because someone's illustrations are wrong is like saying that Christianity is falsified by Titian's painting a much-revered Madonna whose red gown could never have been worn in first-century Judea.

The third reason why biologists do not particularly think that Intelligent Design is a scientific alternative to evolution is that it celebrates ignorance. ID proponents like to say that they are in a long tradition that goes back to natural theology. Natural Theology is a critically important philosophical and religious position, and it was one of the founding motivations for doing science. Reading a textbook such as Rev. William Kirby's *On the Power, Wisdom and Goodness of God as Manifested in the Creation of Animals and in Their History, Habits and Instincts* (Bridgewater Treatise 7; 1836) is to read good science based on careful observation. However, it differs from contemporary science in giving a final cause for these adaptations. Thus, one will find in Kirby a detailed analysis of the bird feather, ending with a paean to the Creator who made such feathers for the use of birds and for the enjoyment of mankind. In natural theology, the more detailed the science, the greater glory to the Creator. natural theology did not fear science: indeed it helped create science. Newton, Boyle, and other members of the Royal Society of London were prominent natural theologians.

In contrast, Duane Gish (1985), the leader of the Creation Science movement that was the direct predecessor to Intelligent Design (Forrest and Gross 2004), wrote, "*We do not know how the Creator created, what processes He used, or He used processes which are not now operating anywhere in the natural universe.* This is why we refer to creation as special creation. We cannot discover by scientific investigations anything about the creative processes used by the Creator." [Gish's italics]. While natural theology was a goad and motivation for discovery, Intelligent Design pronounces ignorance as our wisdom.

Having established ignorance as wisdom, Intelligent Design promotes superstition as science, generating ideas like teaching astrology in science classes. As Judge Jones summarized it in his decision that prevented ID from being taught as science:

> ID is predicated on supernatural causation, as we previously explained and as various expert testimony revealed... ID takes a natural phenomenon and, instead of accepting or seeking a natural explanation, argues that the explanation is supernatural... Further support for the conclusion that ID is predicated on supernatural causation is found in the ID reference book to which ninth grade biology students are directed, *Pandas*. *Pandas* states, in pertinent part, as follows: Darwinists object to the view of Intelligent Design because it does not give a natural cause explanation of how the various forms of life started in the first place. Intelligent Design means that various forms of life began abruptly, through an intelligent agency, with their distinctive features already intact – fish with fins and scales, birds with feathers, beaks, and wings, etc... Stated another way, ID posits that animals did not evolve naturally through evolutionary means but were created abruptly by a non-natural, or supernatural, designer... Professor Fuller agreed that ID aspires to "change the ground rules" of science and lead defense expert Professor Behe admitted that his broadened definition of science, which encompasses ID, would also embrace astrology.[1]

The fourth reason scientists do not feel that Intelligent Design is scientific is that it is based on fallacious assumptions. For instance, some of ID's most powerful arguments depend on a simple statistical fallacy: the assumption of an endpoint, or *telos*. William Dembski (1998), for instance, claims that it is impossible to evolve a particular protein because it has 100 amino acids and the chance of this occur-

[1] See www.pamd.uscourts.gov/kitzmiller/kitzmiller_342.pdf , 67–68.

ring randomly is 1 in 100^{20} since there can be any one of twenty amino acids in any position. Multiply that by all the proteins in the body and one can see that the body is indeed impossible. In fact, Dembski vastly underestimates the likelihood. Let's say that your mother ovulated 500 eggs during her life and that your father produced 10^{12} sperm. The chances of *you* being born, then, are 1 in 5×10^{14}. Now the chance that your father is who he is also is about 1 in 5×10^{14}, as is the chance of your specific mother being born. So the chances of your grandparents giving rise to you is 1 in 10^{28}. Taking this back even a few thousand years, and you have odds that are preposterous that you could ever exist. (So why should one argue with Intelligent Design proponents? They logically cannot exist!)

The fallacy behind their statistics is the assumed endpoint. Evolution does not have an endpoint. The chances of *someone* being born from the union of your mother and father are rather high. The probability that the person will be someone with your particular genotype is astronomically low. Similarly, evolution does not care what the sequence of a protein is as long as it does its function. In some species, a particular protein is a metabolic enzyme in the liver. In other species, the same protein can be used as a structural protein in the lens to focus light upon the retina. Some species of fly use a protein called Bicoid to specify that the part of the embryo closest to the nurse cells becomes the head. In other flies, a protein called Hunchback will do the same job equally well. If one deletes the β-galactosidase gene in bacteria, the bacteria will die if fed only lactose sugar. (The β-galactosidase protein encoded by this gene breaks down lactose into simple sugars that can be metabolized for energy.) However, if you first mutate these bacteria, some of them will gain the ability to metabolize lactose and will survive on the all-lactose diet. It turns out that some mutated genes will produce enzymes that have a small ability to metabolize lactose (see Miller and Levine 2005b). Within a few generations of selection, the bacteria can thrive on lactose. The enzymes have a different sequence than the original β-galactosidase that had been deleted. But they did the job.

Because of such supposed improbability, Dembski (2004) claims that "there are natural systems that cannot be adequately explained in terms of undirected natural forces and that exhibit features which in any other circumstance we would attribute to intelligence." Assuming that the purpose of science classes is to teach students how to do science, and admitting that the ability to do mathematics is useful for doing science, the only value of such cases from the ID literature is as examples of how to avoid common scientific errors.

2. Some theological reasons to be wary of intelligent design

Philosophers and Catholic theologians have a longstanding history of warning that design is a rickety pillar on which to ground one's beliefs. David Hume showed that the argument for God's existence from design was based on a selected set of analogies. A watch is to an intelligent human creator as a wing is to *x*. This *x* would have to be a remarkably higher intelligence. But Hume claimed that from such a finite analogy, one could at best extrapolate a powerful, but finite, god. Cardinal John Henry Newman (1870) extended this criticism, writing that "design teaches me power, skill, and goodness; not sanctity, not mercy, not a future judgment, which three are the essence of religion."

Second, if one looks at the particulars by which an organism comes into being, then God often turns out to be a rather incompetent mechanic. The human genome, for instance, is littered with the refuse of unused genes (see Carroll 2008). The noted paleontological anatomist (and practicing Catholic) Peter Dodson (2006) writes elegantly about both the brilliant and poor engineering of vertebrate bodies. The placement of certain nerves and blood vessels reveals poor and wasteful planning from an engineering perspective, but these are easily explained evolutionarily. The wanderings of the vagus nerve through the thorax and abdomen look bizarre, but these peregrinations can be explained by the nerves originally having a direct route in fish, which didn't have a neck. Similar evolutionary explanations explain numerous other "poorly designed" anatomical structures.

A related point concerns theodicy, and this gave Darwin particular trouble. Writing to his friend, Joseph Hooker, Darwin (1856) noted,

> What a book a devil's chaplain might write on the clumsy wasteful, blundering, low, and horribly cruel works of nature.

To another friend, Harvard biologist Asa Gray, Darwin (1860) confided:

> I cannot persuade myself that a beneficent and omnipotent God would have designedly created the Ichneumonidae with the express intention of their feeding within the living bodies of caterpillars.

To make God the designer of animals is to give the deity the attributes of incompetence and malice. Evolutionary explanations are not only better science, they are also, as Dodson (2006) notes, "Darwin's gift to theology."

Those theologians who would attempt to make God the director of evolution will often use conscious design as a substitute for chance and selection. This is a dangerous strategy, invoking a God of the gaps, who will become progressively smaller as science expands. This strategy has often been directed against some of the extreme reductionist and economically based models of evolution (see, for instance Rolston, 1999, writing against Dawkins). But such reductionist positions are becoming less tenable for scientific reasons (Gilbert and Epel 2009). Certainly, the idea that God the Designer duplicated the number of Hox genes so that vertebrates could evolve and then specifically duplicated them again for the emergence of teleost fishes (but not for amphibians, reptiles, or mammals) sounds ludicrous. The German/Brazilian naturalist Fritz Müller (1869), surveying some of the changes in crustacean body plans, wrote that such manipulations "would certainly never appear to anyone to be a pastime worthy of the Deity." Chance and selection play major roles in evolution. As Darwin steadfastly argued (against Lyell, Gray and others), Divine guidance was not compatible with an evolutionary biology based on natural selection. It may be a religious strategy for allowing science and religion to co-exist, but it is faith, not science.

3. Wonder: a modern alliance of science and religion

Natural theology had provided a way of bringing science and religion together. Are there ways now to unite these two areas in common cause, without sacrificing their differences? Let me suggest that there is a "common ancestor" of both science and religion, and that by acknowledging this common descent, science and religion can become allies. This common ancestor is "wonder," and according to the religious philosopher Abraham Joshua Heschel (1955; 1966), it is *wonder* that makes us human. It is our ability to be grateful and to appreciate. But wonder has a short half-life, and it decays rapidly into two critical principles, awe and curiosity. From awe, we derive our reverence and religion; from curiosity, we derive our philosophy and science. (Indeed, Aristotle and Plato agreed that wonder is the font of our curiosity about the world.) The English word "wonder" is used in both cases: (1) "I wonder…" (2) It is "wonderful." So both science and religion are related, having a common progenitor, wonder (Gilbert and Faber 1996). And science and religion, the products of wonder, are also specific human traits. This also means that science and religion must be allies in maintaining sources of wonder, for without such sources, we would cease being human. Indeed, in 1955 (p. 46), Heschel sounded a warning, "As civilization advances, the sense of wonder declines. Such decline is an alarming symptom of our state of mind. Humankind will not perish for want of information; but only for want of appreciation."

Few areas of experience are more imbued with wonder than development, especially human development. There appears to be something of the miraculous in every birth.

One person who analyzed this miraculous wonder was Moses Maimonides, the Jewish philosopher, rabbi, and physician. He wrote (1190), "A pious man of my time would say that an angel of God had to enter the womb of a pregnant woman to mold the organs of the fetus...This would constitute a miracle. But how much *more* of a miracle would it be if God had so empowered matter to be able to create the organs of a fetus without having to employ an angel for each pregnancy?"

If evolution is caused by changes in development, then we might wish to return to this analysis. If God had endowed matter with the ability to create embryos, then matter would also have the power to evolve. This seems in line with the model of self-assembly given by Stuart Kauffman (2008; this volume). Matter matters, and it has incredible formative abilities. Matter that can develop eggs into adults and that evolve into the world's biodiversity is truly something wonderful on scientific, philosophical, and theological levels. Intelligent Design, however, is poor science coupled with poor theology, and should not be trusted as a critique of evolutionary biology.

REFERENCES

Behe M. J., 1996, *Darwin's Black Box: The Biochemical Challenge to Evolution*, New York: The Free Press

Carroll S. B., 2008, *The Making of the Fittest: DNA and the Ultimate Forensic Record of Evolution*, Waltham: Quercus Press.

Darwin C., 1856, "Letter to Joseph D. Hooker, 13 July 1856", in *Darwin Correspondence Project*: http://www.darwinproject.ac.uk/entry-1924

Darwin C., 1860, "Letter to Asa Gray", in *The Correspondence of Charles Darwin* 8 (1993), Cambridge: Cambridge University Press.

Dembski W., 2004, *The Design Revolution*, Downer's Grove: InterVarsity Press.

Dodson P., 2010, "Is intelligent design really intelligent?", in Stewart M. Y. (ed.), *Science and Religion in Dialogue*, Oxford: Wiley-Blackwell, 237–42.

Forrest B., Gross P., 2004, *Creationism's Trojan Horse*, Oxford: Oxford University Press.

Gilbert S. F., Epel D., 2009, *Ecological Developmental Biology: Integrating Epigenetics, Medicine, and Evolution*, Sunderland: Sinauer Associates.

Gilbert S. F., Faber M., 1996, "Looking at embryos: the visual and conceptual aesthetics of emerging form", in Tabuer A. I. (ed.), *The Elusive Synthesis: Aesthetics and Science*, Dordrecht: Kluwer Academic Publishers, 125–51.

Gilbert S. F. and the Swarthmore College Evolution and Development Seminar, 2007, "The aerodynamics of flying carpets: why biologists are loath to 'teach the contro-

versy'", in Comfort N. (ed.), *The Panda's Black Box: Opening Up the Intelligent Design Controversy*, Baltimore: Johns Hopkins University Press, 40–62.
Gish D. T., 1985, *Evolution: The Challenge of the Fossil Record*, San Diego: Creation-Life Publishers.
Heschel J. A., 1955, *God in Search of Man*, New York: Harper and Row.
Heschel J. A., 1965, *Who Is Man?*, Stanford: Stanford University Press.
Jones J. E., 2005, *Kitzmiller T. et al. vs. The Dover Area School Board*, www.pamd.uscourts.gov/kitzmiller/kitzmiller_342.pdf
Kauffman S. A., 2008, *Reinventing the Sacred: A New View of Science, Reason, and Religion*, New York: Basic Books.
Kirby W., 1836, *On the Power, Wisdom and Goodness of God as Manifested in the Creation of Animals and in Their History, Habits and Instincts*, Philadelphia: Carey, Lea, & Blanchard.
Maimonides (Moshe ben Maimon), 1190, *A Guide for the Perplexed* (translation by Friedlander M., 1956), New York: Dover.
Miller K. R., 2000, *Finding Darwin's God: A Scientist's Search for Common Ground Between God and Evolution*, New York: Cliff Street Books.
Miller K. R., 2008, *Only a Theory: Evolution and the Battle for America's Soul*, New York: Penguin.
Miller K. R., 2002, "Answering the biochemical argument for Design", in http://www.millerandlevine.com/km/evol/design1/article.html
Miller K. R., 2005a, "Haeckel and his embryos", in http://www.millerandlevine.com/km/evol/embryos/Haeckel.html
Miller K. R., 2005b, "A true acid test", in http://www.millerandlevine.com/km/evol/DI/AcidTest.html
Müller F., 1869, *Facts and Arguments for Darwin*, London: John Murray.
Newman J., 1870, *An Essay in Aid of a Grammar of Assent*, reprinted 1955, Garden City: Doubleday Image.
Pigliucci M., 2002, *Denying Evolution: Creation, Scientism, and the Nature of Science*, Sunderland: Sinauer Associates.
Rolston H., 1999, *Genes, Genesis and God, Values and their Origins in Natural and Human History*, Cambridge: Cambridge University Press.
Rowning B. A., Wells J., Wu M., Gerhart J. C., Moon R. T., Larabell C. A., 1997, "Microtubule-mediated transport of organelles and localization of β-catenin to the future dorsal side of Xenopus eggs", *Proc. Natl. Acad. Sci. USA* 94: 1224–29.
The TalkOrigins Archive, 2006, http://www.talkorigins.org/faqs/behe.html
Wells J., 2000, *Icons of Evolution: Science or Myth?*, Washington DC: Regnery.
Wells J., 2005, *The Words of the Wells Family*, in http://www.tparents.org/library/unification/talks/wells/darwin.htm.

Evolution and the Emergence of the Human Person

William B. Hurlbut

Philosophy has succeeded, not without a struggle, in freeing itself from its obsession with the soul, only to find itself landed with something still more mysterious and captivating, the fact of man's bodiliness.

Fredrich Nietzsche (1901)

For Mercy has a human heart, Pity a human face, and Love the human form divine...

William Blake (1977)

1. Introduction

Three hundred and fifty years ago, at the dawn of the scientific revolution, the French philosopher and mathematician Blaise Pascal lamented, "When I commenced the study of man, I saw that these abstract sciences are not suited to man, and that I was wandering farther from my own state in examining them, than others in not knowing them." (Pascal 1958, 44) Today many feel the same disquiet, and nowhere is this more evident than in the controversies and social conflicts related to the theory of evolution.[1]

The eminent geneticist Theodosius Dobzhansky noted that "nothing in biology makes sense except in the light of evolution." (Dobzhansky 1973, 125) A century of discovery in molecular biology underscores this claim. With the sequencing of the human genome and our increasing knowledge of developmental biology we are gaining a greater appreciation of the unbroken lineage and intricate interrela-

[1] For a fuller discussion of these ideas see the author's chapter in The Oxford Handbook of Religion and Science (2006) on which portions of this essay are based.

tion of the whole of living nature. Yet as evolutionary theory has become the unifying principle of interpretation of the organic world, it has raised difficult questions concerning the source and significance of human life, questions that challenge our traditional concepts of the human person. Together with the recognition of our ancient heritage of molecular and cellular mechanisms has come an understanding that our emotions, conceptual categories, and broad capacities and inclinations of mind are likewise grounded in our evolutionary past. This insight has spawned the new multidisciplinary synthesis of "evolutionary psychology" that draws on genetics, neurobiology, anthropology and sociology in the search for a general description and explanation of universal aspects of human nature.

As a methodological program and theoretical framework that takes seriously the actual biological conditions of human life — its place within the seamless ecology of nature, its embodied being and evolved refinements of form and functions — evolutionary psychology holds promise for increasing our scientific understanding of human nature. Indeed, no theory more certainly affirms that there is a human nature. At the same time, this new theoretical perspective on human origins carries implications concerning the fundamental character of the natural world and calls into question our natural intuitions and traditional notions concerning the meaning and purpose of human life.

If we begin our discussion with a brief critical analysis of the presuppositions and implications of this theory, we can recognize both its legitimate contributions and its limitations. Most specifically, we can draw on evolutionary psychology to explore the inadequacy of a purely reductive mode of description and the error of inadvertently transforming our naturalistic methodologies into unjustified metaphysical assumptions.

We can then turn to reflect on the unique properties of living beings and their emerging complexity across the evolutionary panorama. Drawing on a truer scientific description (informed by insights from recent discoveries in biology), considered within a broader frame of human knowledge (including philosophical and theological reflections), we can seek a fuller, wider cosmology and anthropology that provides a richer and more integrated description of the human person and the world within which we have evolved.

2. Evolutionary psychology

In its fullest expression, evolutionary psychology is a theory about the origins and functional implications of the human mind. This theoretical perspective is grounded in the assumption that all human behaviors, like those of animals, are directed toward competitive advantage in the evolutionary struggle for life. Just as evolution has refined our anatomy and physiology for optimal performance, natural selection has shaped the general contours and practical impact of our behavior. The crucial filter for preservation is not mere survival, but "inclusive fitness": success in getting our genes into the next generation and beyond. Evolutionary psychology claims that a wide array of adaptive psychological mechanisms have been preserved, ranging from categories of heightened perceptual salience to specifics of social interaction and inclinations in mate choice. These adaptations extend beyond the realm of behaviors that relate in an obvious way to survival and reproduction, into the most subtle and complex manifestations of aesthetic preferences, religious ideation and practical moral judgment.

As a statement about human nature, evolutionary psychology challenges our most fundamental concepts and cultural traditions concerning freedom, morality, and spiritual purpose.[2] The individual is subsumed within a larger impersonal process of genetic proliferation; reproduction is the single goal for which human beings are designed, everything else is a means to that end. This concept reaches its most extreme expression in Richard Dawkins' idea of the "Selfish Gene", where we are just what he terms "survival machines — robot vehicles blindly programmed to preserve the selfish molecules known as genes." (Dawkins 1978) Genes are our true masters and human beings are at best unwitting accomplices or, indeed, victims in a physical process without transcendent or personal purpose. According to some theorists, even moral principles are delusions in the service of deeper evolutionary goals. Behavior that seems to our inward awareness as altruistic is only slightly veiled genetic self-interest — whether "kinship selection" (helping your genetic relatives) or "reciprocal altruism" (hoping to get something in return). As the author Robert Wright starkly puts it: "the question may be whether, after the new Darwinism takes root, the word moral can be anything but a joke." (Wright 1994, 325–6).[3]

---

[2] It is important to acknowledge the wide range of positions held by theorists of evolutionary psychology. The intention of this essay is not to deny the value of such an approach, but to seek a more thoughtful extension of its implications, one concordant with a broader reflection on the source and significance of human life.

[3] The media, picking up on the disturbing implications of this, has greatly disrupted the popular mind. One thinks, for example, of the cover of Time magazine a few years ago that proclaimed "Infidelity: it may be in our genes" — hardly a foundation for the self-control and self-sacrifice of genuine love.

There are some obvious and commonly noted criticisms of evolutionary psychology. Most fundamentally, it is far from clear that our entire psychological nature and functional behavior is oriented simply toward the propagation of genes. Of course the genes must be preserved and the generation of life must be sustained; there can be no continuity of species without these fundamental biological processes. But it is another thing to say that human life is ultimately about nothing more, in view of all its exuberant artistic expression, scientific curiosity, and passionate pursuit of goals beyond the span of individual existence. Likewise, to account reductionistically for all of human nature in terms of earlier and simpler evolutionary mechanisms ignores the evident emergence of biological complexity and its expression in genuine novelty and a widening range of flexible response and functional freedom. Throughout evolutionary history, new forms and functions arise that refine and revise the very operational principles of the evolutionary process and the creatures it produces. Human consciousness, symbolic language, and moral community are emergent phenomena not fully described nor understood by anything that has come before in the history of life.

Of much greater concern, however, is the fact that instead of simply serving as a reductive naturalistic methodology evolutionary psychology goes beyond the scope of science to include assumptions that represent philosophical and theological views. These assumptions extend the theoretical assertions of evolutionary psychology into an extreme form of philosophical naturalism. Without evidentiary justification it presupposes no design to nature, no direction or purpose and no intrinsic meaning to its process. But such assumptions preclude the very categories of values and virtues, at least as these terms are traditionally understood. The practical effect of this is to reduce all human behaviors to value-neutral adaptations having no genuine reference to transcendent truths, and to deny any ultimate significance to mind and moral culture. This amounts to a form of nihilism, corrosive to human concepts of motivation and meaningful action. Traditional categories of good and evil are reinterpreted as functional fictions generated for social cohesion, and human freedom is reduced to an illusion lacking actuality but useful for legislating and enforcing responsible behavior.

However much these conclusions may appear to conform to certain conceptions of modern science, such a perspective is blind to the evident testimony of the phenomenon of human life. Its methodological presuppositions obscure the most startling fact of evolution: the emergence of human persons and the manifestation within matter of freedom, consciousness, and moral awareness. What is needed is to release evolutionary psychology from its limiting assumptions and engage its explanatory power in exploring how freedom and the capacity for moral awareness are anchored in and arise from basic biology and yet beckon beyond to matters of transcendent meaning.

3. Emergence

If we pause to ponder anew the phenomenon of living nature, we are immediately struck by its mysterious character. How do we go from chemicals to consciousness, or, for that matter, even to the simplest living systems with their self-sustaining metabolic powers and their unique capacities for integrated self-development?

Clearly, there are essential bio-molecular mechanisms that are obedient to basic physio-chemical principles. Yet, somehow, "life elevates matter — it does new chemical things by means of doing the old chemical things" (Benignus 1947, 16) — it reconfigures matter in such a way that new properties not apparent in the parts are expressed. Moreover, these new properties appear to defy analysis in terms of the parts — new powers emerge that are not reducible to the intrinsic properties of the parts. Something within the structure and the relational interaction of the parts, is necessary for such powers to be expressed. Indeed, broadly speaking, not just the action of the parts, but their very existence is unintelligible except in relation to the whole — the whole appears as "end" and the parts as "means". Yet the whole, as the unified principle of life, precedes and produces the parts. "The whole acts as the cause of the very being of the parts before it actually exists as their sum" (Benignus 1947, 160). Notwithstanding Nietzsche's enigmatic assertion, some principle is in it (soul, or otherwise described) that cannot be reduced to its primary matter.

With the new perspective of Systems Biology (understood through insights from genomics, epigenetics, evolutionary developmental biology and ecology), we are gaining a greater appreciation of the natural basis for the emergent powers of living wholes. Indeed, the very word organism implies organization, an overarching principle of unity, a cooperative interaction of interdependent parts subordinated to the good of the whole. As a living being, an organism is an integrated, self-developing and self-maintaining unity under the governance of an immanent plan. Such a conception of the biological organism transcends the "nothing but its parts" of reductionism. It adds the understanding that a living being is not merely a mechanism but rather, a dynamic system, an interactive web of interdependent processes that expresses emergent properties not predictable from the biochemical parts.

Such a view liberates us from explanation framed exclusively in reductive and analytic terms — we can acknowledge and accept as phenomenologically real the new and previously unseen properties and causal powers of complex systems. Moreover, the concept of an autonomous system, with organizational and operational closure — a morphodynamic whole with a circular and self-referential network of relations — makes possible the concept of "identity" and "continuity within change" that is central to the actual existence of a "self" as a living being. There is a purposeful cohesion, a global integration of action toward an end, a "top-down"[4] causation that supercedes and subsumes the parts into an irreducible and unified whole. Not only are the parts unintelligible apart from the whole (what is an eye or a hand apart from the living body?) but the immanent powers and their actions are unintelligible without a purpose or goal. "Mechanism without implicit ends is a meaningless error of materialism — only purposeful unity makes sense of the mechanical powers of living beings" (Benignus 1947, 161). Neither mechanism nor materialism is the error of reductionism, it is, rather, ontological merism — reducing ultimate reality to the properties of parts without higher purpose.

Lifted from the error of reductionism, we can now consider as phenomenologically real the being and causal powers of the complex expressions, the multitudinous forms and functions of living nature. Liberated from the illusion of a simple and ultimate explanation of reality in terms of the properties of its elemental particulars, we are open to the world as we actually experience it, in all its proliferation of unexpected and unexplained manifestations of stratified and multi-leveled reality.

A crucial corollary of this is that we can now accept as valid the varied range of our ways of knowing. The flattened view at the empirical surface of reality is now complemented by other ways of knowing — tacit knowledge, pre-cognitive and intuitive comprehension of wholes, aesthetic impulse, moral and spiritual intimations and a sense of transcendent truth. We are liberated from the notion that somehow action at the microlevel is most ultimate as explanatory cause. As Huston Smith so eloquently states: "Inclusively, things are neither as science says they are nor as religion says they are. They are as science, and religion, and philosophy, and art, and common sense, and our deepest intuitions, and our practiced imaginations say they are." (Smith 2001, 43) Meaning and imagined ideals reemerge,

[4] While this concept of causation is often described as "hierarchical" or "top down", it is more rightly recognized as a "systems-causation", a seamless integration of action by the whole for the sake of the whole where causation is not exhaustively determined by the intrinsic properties of the particular parts.

along with a larger sense of personal significance within a narrative quest and its noble intentions.

From this wider and richer perspective of personal and cultural comprehension, we can now look back at the evolutionary panorama in a way that is at least open to consideration of possibilities previously closed off through the limitations of our naturalistic assumptions.

4. The ascent toward greater complexity and freedom

When we look back at the evolutionary process we are at once struck by both its continuity and creativity. At every stage and level the unfolding of the diversity of forms and functions reveals new and previously unseen dimensions of nature and so revises our understanding of the nature of nature. Even the earliest life forms exhibit a general characteristic of biological organisms, namely, that there is a constant dynamic balance between self-preservation and the production of variation. Life does not remain static and constrained; there is a continual thrusting outward, a probing for possibilities for ever-further proliferation into new realms of opportunity. Moreover, species do not only evolve for simple adaptation, but for flexibility of response; they evolve for evolvability, favoring adaptations which themselves provide the platform for further flexible evolution. We can see this tendency continuing all the way up to and including the emergence of human beings.

Adaptability is first expressed at the level of mutation, and such a strategy works very well in rapidly reproducing organisms. A single bacterium, for example, which has a limited ability to adjust to a changing environment, can produce tens of thousands of varied offspring within a few hours. This allows not only adaptation within a changing environment, but also extension into a wider range of environments with different conditions.

While early life forms adapted by mutation that created variation within a population, more complex systems of adaptation rapidly evolved that allowed adjustment to changing environmental conditions by an individual organism. Existing genes came under the control of coordinated epigenetic mechanisms, modulators of the quantity and chronology of their expression. Networks of interactive elements established self-maintaining regulatory systems that sustained or reestablished internal metabolic balance. Primary vital powers of awareness (primitive sentience) and action evolved allowing detection and response to environmental conditions of adversity or opportunity. The very capacity for life emerged as "bootstrapped being", immanent powers of self-production (autopoiesis) and self-governance — and the defining character of evolutionary process became an ever emerging complexity of innate activity in the service of life, distilled and bounded being with intrinsic executive control of its own existence.

Yet, even with the earliest living forms, the vital powers of life cannot be understood by reductive analysis. There is no simple direction of cause, but an interactive system, an entangled hierarchy of nested and multifaceted webs of causation. As Evelyn Fox Keller explains, "the notion of the gene as the atom of biology is very mistaken... we have to get away from the underlying assumption of the particulate units of inheritance that we seem so attached to." (Angier 2008) Genes do not provide the "blueprint of life", they are a component in a complex interweaving of circles within circles of causation. As Keller puts it, DNA "is a profoundly relational molecule...it has meaning only in the context of the cell." (Angier 2008) Cytoplasmic factors suppress or promote nuclear gene expression. Tiny microRNAs bind and degrade messenger RNA and thereby modulate protein production and cell development. And genes themselves are variably transcribed and their products respliced in complex adaptive patterns in the service of the living cell. Stephen Rose rightly states, there is no "nuclear board room" no molecular hegemony of privileged power (Rose 1997). Rather, there is molecular egalitarianism, every part being crucial for the well functioning whole. For the individual cell as a living organism this allows a multitude of modulation points with recursive adjustments for the fine-tuning of function. At the same time, this emerging multifaceted complexity opens additional avenues of evolutionary advance for the expression of new possibilities in the proliferation of living forms, lifting life to higher powers in breadth of awareness and flexibility of action.

Even with single celled organisms these higher powers are prefigured. Recent discoveries reveal a primitive presage of multicellularity. Adaptive awareness of the outside environment requires mechanisms of reception and response that in turn form the biological basis of cell-to-cell signaling. In an extraordinary expression of emergent powers, this communication becomes the mediating principle of cooperative communal action among a species of single celled amoeba. Journalist Liz Buchen explains:

> The social amoeba *Dictyostelium discoideum* usually lives alone, dining on bacteria in the forest soil. When food becomes scarce, tens of thousands of neighboring amoebae meld into a blob — about a third the length of an eyelash — that slithers much farther than any amoeba could on its own. When the slug reaches a warm, sunlit locale, the aggregate transforms into a fruiting body: About 20% of amoebae sacrifice themselves to form a rigid stalk, hoisting their comrades upward as a ball of spores. These lucky amoebae hitch rides

on the fur of passing mammals to reach greener pastures. The martyrs of the stalk wither and die (Buchen 2008).

She goes on to quote evolutionary biologist Richard Grosberg who says, "These findings contribute to our understanding of how some of the earliest organisms have balanced cooperation with self-interest…essential for social behavior." (Buchen 2008) It is equally evident that this constructive cooperation requires some kind of primitive mechanism of identity and the dissolution of self-protective boundaries that prefigure the highest reaches of altruistic action, possibilities that would remain latent in the history of life awaiting further refinements of form.

The path to this deeper sense of identity, reflexive self-awareness and the moral groundings of genuine altruism would be built along a bridge of new powers made manifest by organismal multicellularity. The coordinated cohesion of cells into a single unit of life was an extraordinary leap in the evolutionary process. The fossil record seems to indicate that for nearly three billion years single cells were the only life forms, though perhaps primitive commensualism and colonies were more common than we know.[5] During this long epoch of evolutionary process it appears that the rapid reproduction of single cell organisms within a wide range of environmental challenges provided the trillions upon trillions of biological experiments essential for the refinement and optimization of the fundamental intracellular mechanisms of life. Precise coordination of enzymes and substrates, complex recursive loops for regulatory balance and ever more intricate and responsive cell machinery set the stage for the specializations and integrated coordination of differentiated cell types that characterize multicellular organisms. Whatever one may think about implicit design or direction in evolution, it is truly remarkable that life's early "decisions" at the level of the single cell set the foundation for such an enormous proliferation of multicellular life forms with their extraordinary diversity of emergent properties. Once the momentum of selection for multicellular creatures developed during the Cambrian era, within a few tens of millions of years (a mere blink in the evolutionary journey) there were 850 new genera and a vast array of new multicellular plant and animal species of varied forms and powers.

Only at the level of multicellularity does the concept of organism as coherent organization and coordinated complexity find its fullest expression. Cell division

[5] There is some evidence that suggests very simple multicellular life forms as early as 1.2 billion years ago, but the major momentum of multicellular evolution is recognized only at the beginning of the Cambrian Period.

without spatial separation set the stage for the divergence of differentiated lineages of distinct cell types and the new biological functions of specialized tissues and organs within an integrated and coherent whole. Likewise, with multicellularity the significance of a systems biology perspective (as opposed to reductive analysis) is most evident. Properties and powers of the whole are sustained by the integrated interaction of the parts, while the overarching unity of form sculpts and entrains the developing parts in the service of the whole. Subtle shifts in intercellular relations based on proximity and position induce distinct patterns of gene expression that are magnified in non-linear fashion to produce divergent trajectories of differentiation. Highly refined niches of surrounding cells provide a unique microenvironment with spatio-temporal control of the rate and quality of stem cell divisions and the generation of progenitor cells of different tissue types. Even asymmetries of adhesion and physical pressure promote distinct cell fates. For example, primitive mesenchymal stem cells will form either bone or fat depending on their physical platform. These mechanisms of controlled and coordinated development, global actions of the whole expressed at local levels of organismal growth, assure the proper proportion and coordination of the parts. With the emergence of multicellular organisms, the flexible adaptive mechanisms of mutation and modulation at the level of individual cells is further extended by compartmentalization and modularity of specialized parts in the service of the whole.

The emergent powers made possible by multicellularity, and the significance of the whole in the entrainment and governance of the parts, are most evident in the evolution of the brain. With the appearance of coordinate neuronal networks more than 500 million years ago, the primary capacities of selective perception and locomotion in simple organisms were transcended by programs of integrated organismal response, innate reflex arcs of nerves and muscles triggered by external stimuli. This in turn allowed an enormous increase of adaptive behaviors that extended the reach and realm of life — and new challenges that provoked further complexity and capability. Whereas the oceans had provided a more or less stable chemical context and constant temperature, the ascent to dry land required more complex regulation of water and temperature but opened a vast new range of opportunities for the extension of life. This in turn led to the refinement of integrated motor and endocrine programs that coordinated body systems and formed the biological basis of what later in evolution became the emotions.

Within complex bodily forms, new powers of sentience distilled into an increasing adaptive awareness and sense of self. The emergence of affective life aided immediate survival but also pointed beyond it. Emotions had their evolutionary origins in the physiological processes of body regulation, such as postural and visceral changes that place the organism in a condition of readiness of response. For example, fear involves increased muscle tension and protective posture. As sensory perception and action became increasingly complex, organisms developed with a more integrated "inner" sense of subjective feelings associated with emotions.

Within a rising scale of feeling and self-awareness, sustained patterns of response came to be motivated and coordinated by an inwardly felt sense of aversion, appetite or desire. The philosopher of biology Hans Jonas considers this the essence of animal life: "[The animal] emancipates itself from its immersion in blind organic function and takes over an office of its own: its functions are the emotions. Animal being is thus essentially passionate being." (Jonas 1966, 106) The unconscious process of plant life becomes the immediate awareness and response of animal life.

In human beings these legacies of our animal ancestors are preserved, but they are largely subsumed under the control of voluntary actions guided by associative memory, analytical reason, and conscious aspiration, as a further extension of freedom within the phenomenon of life. Such refinement of internal mental capacities evolved in coordination with increasingly complex external morphology. Early in the evolutionary process the differentiation of the head region, with its organs of sensory perception and communication, was paralleled internally by cerebral structures capable of processing more complex impressions of the surrounding environment. In *Homo sapiens*, along with the transition to human upright posture, came coordinated revisions of body form, increased range of motion, and a radical cerebral reorganization that made possible new relations with the world. As sight replaced smell as the most prominent sense it allowed rapid perception of objects and actions at great distances. The cerebral processing and storage of visual images led to the emergence of imagination and its creative powers. Sight allowed insight.

Imagination in human beings conferred extraordinary adaptive benefit. The symbolic mind, which is capable of detaching image from object and thereby recombining images in new ways, can envision scenarios and sequences mentally, apart from immediate time and space, to anticipate their implications and outcomes. This is yet another powerful form of freedom in which the organism can imagine possibilities and try them out (in a kind of dress rehearsal) without the expense of time and risk of resources in the process.

The human capacity for imagination, however, goes far beyond adaptive anticipation. Imagination is not mere memory or imitation, but envisioned creation.

Forming mental images, maintaining them in the mind and achieving their realization, allows intention, planning, and implementation of ideals. Ian Tattersall explains, regarding the first production of the hand axe, that, "these symmetrical implements, shaped from large stone cores, were the first to conform to a "mental template" that existed in the toolmaker's mind." (Tattersall 2000, 61) This is perhaps the first intentional innovation: the bringing into being of an imagined ideal.

Whereas most creatures exist in an unbroken immediacy of life, humans have the freedom to draw the past into the present from learning stored as memory and the freedom to draw the future into the present though the creative imagination. Together with the ceaseless drive to organize the unexplained ("the cognitive imperative"), the capacities to calculate, extrapolate and recombine are used to reconfigure that which is into that which could be. While most creatures are pushed by biological and ecological circumstances, we are pulled into the future by our dreams and images of fullest flourishing and envisioned ideals. As Leon Kass writes, "desire, not DNA, is the deepest principle of life." (Kass 1994, 48)

The primitive strategies of life, refined by evolving life forms encountering ever more complex environmental challenges and opportunities, culminated in our human responsive ability to think and act. What was prefigured within nature, beginning at its most primary biological level, as a widening range of both responses to environmental stress and options for change, developed at the level of human life into what we internally experience as a sense of freedom. We are made not for a particular niche, but for unpredicted possibility, for comprehension and control, for flexibility and functional choice. We have adapted for adaptability itself.

Such freedom, and the creative novelty that humans have generated in the world through it, constitute genuinely emergent phenomena, neither predictable from nor reducible to the processes of less complex phenomenological levels. The brute fact of such emergence strongly argues against the sufficiency of reductionist views such as are represented by evolutionary psychology, and allows for a distinctive ontological and causal reality to higher phenomenological levels within nature.

5. The emergence of human personhood and development of moral awareness

A further implication of emergence is the open-ended possibility of genuine novelty, which argues against any claimed completeness or finality for a theory such as evolutionary psychology. Neither our goals nor our actions are rigidly and robotically fixed. For what we have seen of emergence at both the molecular and phylogenetic level can also be seen in the ontogeny (the process of development) of each individual person. Not only did the early "decisions" of single cell evolution set the platform for a vast proliferation of living forms, but the very process by which they evolved encoded in successive species the capacities for individualized adaptive response. The single cell that begins a human life carries the accumulated experience of over three billion years in the history of life. At every level of development, the human organism is shaped in its specific expression in adaptation to the environment in which it is formed. Even the stage at which we are born, early in neurological formation, allows a powerful role for environmental and social stimulation in shaping the developing mind.

To understand the emergence of the human person, it is essential to recognize the inseparable relationship between body and mind, and the way they are formed in interaction with the world. Even within the womb peripheral sensory perception of the body surface and proprioceptive awareness of the musculo-skeletal dynamics of body position and balance contribute to an inner web of self-awareness. This fundamental sense of self-awareness then serves as a standard against which any perturbation or alteration (experienced as sensory input) can be compared, thereby providing information and understanding of the world. From early infancy, these powers are further extended as the voluntary actions of the developing child reach out to probe the world in search of a richer understanding and capacity for control.

The mind, then, is not an abstract neurologic function, but is an activity of the whole body. We know the world not as a separate reality, but with reference to our embodied selves. As we grow and gain a more distilled sense of self, our personal identity is really a system, a unity of self-within-the-world. Even to the simplest sensory awareness we bring the entire weight of selective attention, contextual interpretation, and the images and ideals that guide our deepest goals. Likewise, the functional structure of rational process is tied to the fundamental evaluations and adaptive responses encoded within our emotions.

Our particular unique evolved form of human embodiment, and our similar experiences within the world, provides a common "language" of mental categories, emotional responses and shared needs that serve as the basis for intelligible communication. Our embodiment also provides a foundation for the desires and intentions that shape our shared system of values. With increasing organismal complexity, the central values of evolutionary success, survival and reproduction, are served by pleasurable intermediate activities that become emergent valued ends in themselves — biological "means" become personal "ends" in themselves. The most obvious of these are the pleasures associated with eating and with sexuality. These basic biological drives, in turn, point beyond survival to a shared realm of more transcendent values — the true, the good and the beautiful. Notwithstanding the great variety of traditions and diversity of personalities, there is a central core of common humanity that establishes a global community of shared values and aspirations.

All of these shared foundations of human existence, our particular form of embodied being and the common challenges of a similar environment in which our lives are embedded, provide the crucial underpinnings of human social life and culture. Human beings are, by their very nature, social beings. The intricate social interplay, first established between mother and infant, is built on a remarkable set of anatomic and physiologic adaptations that make possible the unique human expression of another great leap in the evolution of emergent powers, the capacity for empathy.

Simply defined, empathy is the ability to identify with and understand the situations, motives, and feelings of another. More than merely sympathy, it is a form of intersubjectivity in which the observer actually participates (as a kind of inward resonance) in the feelings of the other. The shared, bodily-based quality of emotions, manifested and experienced in a similar way by all people, makes possible the process of empathy. By perceiving and understanding the facial expressions and bodily movements of others, we come to understand something of their inner life.

This empathic intersubjectivity provides the patterning for personal identity and the platform for cultural awareness, and may provide a solution to that most difficult of questions: how do we leap beyond our subjective solipsistic self into genuine society with others? In a process of "social referencing" that builds a common set of values, the infant will point or gaze at an object to establish joint attention and then observe the mother's reaction. The mother's spoken responses, which at first convey to the baby only feelings — the shared affective language of posture and prosody — begin to carry specific semantic content. A web of meaning is formed within this linguistic system of empathetically grounded symbolic gestures, the coded concepts on which all human cultures are constructed. In a kind of "re-envoicement" the child begins to structure his understanding of the world, the very pattern of his thoughts, by the echo of the words of others.

The child psychologist Jerome Kagan describes how moral awareness develops together with empathetically grounded sociality (Kagan 1998, 158). In an orderly developmental progression, a child begins to crystallize a sense of self and other and discover the inner mental world of private beliefs and intentions. With conscious personal identity comes awareness of the distinct identity of others, and cognitive empathy, a willed and knowing stepping into the role of the other.

Between the second and third year of life children develop an appreciation of the symbolic categories of good and bad and learn to apply these to their own actions, thoughts, and feelings. With a growing understanding of the relationship between present actions and future outcomes a child begins to develop a conflict between acting on present desires and recognizing that to do so could have negative consequences on self and others. Before the age of five children have difficulty governing their actions, but by around six the sense of self-control, and therefore accountability, allows shame and guilt but also the happy sense of virtue, of consonant goodness of self.

Free choice becomes, increasingly, the central moral axis and, guided by the emotional pull of empathetic communion, leads to the poignant drama of the individual self in the quest for a sense of ethical worthiness that transcends any simple imperative of genetic propagation. As Richard Shweder notes, "the most powerful motive of human life is a desire to gain a feeling of virtue, the desire to be 'good' — a desire unique among sentient animals." (Schweder 1999, 798)

It is the nearly universal human testimony that this drama of moral choice is experienced subjectively as an exercise of genuine personal freedom. Such a view is central to how we understand both ourselves and our place within the world. Yet freedom of choice, experienced as open indeterminacy, is at once a powerful adaptive advantage and a potential liability; it must be governed or it dissolves into indecision or derails into destructiveness for individual and communal life. Moral principles provide the conceptual frame that structures and aligns our attitudes and actions in consonance with the most comprehensive vision of the fullness and flourishing of life. From the perspective of biology, the moral sense is an adaptation to guide and channel choice in conditions too complex or novel for the direct operation of our primary desires or fears or heritable patterns of control.

Moreover, there is also a sense in which moral action serves as an end as well as a means, and is itself an expression and an inwardly felt confirmation of the fullness of life. Moral actions unify and integrate our most primary biological needs, our call to honor and noble spirit, and our fullest sense of the meaning and purpose of human life. Moral ideals form the foundation of values, sustain social communion, and provide the coherent vision that constitutes and animates the cohesion of culture.

Indeed, the moral sense provides the very infrastructure of personal identity and unifies reason, emotion and intuition. It is to the mental life what the skeletal system is to the physical frame. It is the central core of the cognitive convergence we call consciousness, which is intrinsically interpersonal and moral. For as we develop our sense of self out of empathic interaction with others we must also develop our sense of how to relate to those others. As Pope John Paul II said, "When we look into the eyes of another person we know we have encountered a limit to our self-will."

The recognition of these foundational realities is encoded in the history of our words. In English consciousness and conscience share a common Latin root meaning "joint or mutual knowledge" — consciousness of moral sense. Yet even as empathy establishes individual identity and makes possible human community and cooperation by sensitizing individuals to the emotions and needs of others, the natural struggle for life may engage our empathic capacities for purposes of intimidation, dominance and exploitation. Indeed, as mentioned above, evolutionary psychology maintains that ultimately all adaptations, including the empathic agencies of social life, must redound to the benefit of the individual through selective advantage in the proliferation of his genes. From this perspective, consciousness itself is neither a capacity for moral clarity nor for comprehensive understanding, but an agency of purposes opaque to human awareness. Any concept of genuine freedom of thought or action is undercut and the imperative of moral truth is rendered irrelevant. Mercy, Pity and Love are seen as mere sentiments in the service of instrumental ends without transcendent referent or coherent human meaning. Human life is demoted to the status of just another "ordinary animal species", a hapless player in a ruthless and relentless process without purpose.

It is difficult to understand how such ideas could gain traction as a credible theory, much less provoke a widespread transformation in human self-understanding and social policy.

The reasoning at the root of evolutionary psychology would seem to undercut the very notion of reason itself. As Alfred North Whitehead observed, "scientists animated by the purpose of proving that they are purposeless constitute an interesting subject for study." (Whitehead 1929)

The source of the problem is a methodological error of modern thought, an epistemic fallacy built on unsupported assumptions concerning the very nature of ultimate reality and the categories of causation. The error is not naturalism per se, but an over-extension of selected evidence, and a systematic exclusion of the wider testimony of nature itself. The error is the prejudicing of material and efficient cause to the exclusion of a proper understanding of formal and final cause, or, more fundamentally, a failure to acknowledge the inseparable operation of all of these levels of cause.

If look anew at the actual manifest expression of living nature we can agree with Nietzsche at least enough to acknowledge the mysterious character of the reality we behold. No reductive ontology grounded in the fullest knowledge of physicochemical powers seems to even approximate a credible description or explanation of human consciousness. Rather, recognizing the holistic and irreducible relational character of living systems, we are compelled to acknowledge (joyfully!) the multilayered realities and emergent properties that constitute and continually extend the phenomenon of life.

The complement to this acknowledgement, is a renewed appreciation for the multiple perspectives that inform our understanding of life and the special powers of comprehension and control they provide. The philosopher Charles Taylor is adamant on this; we cannot back away from a trust in our phenomenological experience to some autonomous, experience-independent objectivity (Taylor 1989, 75). We must struggle though, critically and with perspicacity, continually revising and refining our understanding, but we cannot overrule reason itself, and we cannot escape ourselves to some supposedly neutral "view from nowhere".

Leon Kass, grounded in a scientific appreciation of human embodiment and the remarkable relation with the world made possible by its form, goes farther: "We have lost our way in the world partly because we no longer believe that our ordinary experience of life in the world may be the privileged road to the deepest truth." (Kass 1999, 8) We are the part of the whole that can think about the whole, and in contemplating its character search out the seamless unity of the material and the moral.

With these balancing corrections to our methodological approach, we can return to the study of the history of life and look again at the evolutionary process that has formed us. Drawing on the full range of human understanding and experience we can examine again questions concerning the source and significance of human life. And as we do so, we can reflect anew on the remarkable drama of life's ever emerging complexity where matter has assembled in such a way that it now ponders the very meaning of its being.

REFERENCES

Angier N., 2008, "Scientists and philosophers find that 'Gene' has a multitude of meanings", *New York Times*, Nov. 11.

Benignus Brother F.S.C., 1947, *Nature, Knowledge and God*, Milwaukee: The Bruce Publishing Company.

Blake W., 1977, "The Divine Image", in Ostriker A. (ed.), *William Blake: The Complete Poems*, London: Penguin Books.

Buchen L., 2008, "Science NOW", *Daily News*, Nov. 25.

Dawkins R., 1978, *The Selfish Gene*, London: Granada.

Dobzhansky T., 1973, "Nothing in biology makes sense except in the light of evolution", *The American Biology Teacher* 35: 125–29.
Jonas H., 1966, *The Phenomenon of Life: Toward a Philosophical Biology*, New York: Harper & Row.
Kagan J., 1998, *Three Seductive Ideas*, Cambridge, Mass.: Harvard University Press.
Kass L.,1994, *The Hungry Soul: Eating and the Perfecting of Our Nature*, New York: The Free Press.
Kass L., 1999, *The Hungry Soul: Eating and the Perfection of Our Nature* (reprinted with a new forward), Chicago: The University of Chicago Press.
Nietzsche F., 1901, *Der Wille zur Macht*, Leipzig: Naumann.
Pascal B., 1958, *Pensees*, New York: Dutton & Co.
Rose S., 1997, *Lifelines*, London: Penguin Books.
Shweder R., 1999, "Humans really are different", *Science* 283: 798–9.
Smith H., 2001, *Why Religion Matters: The Fate of the Human Spirit in an Age of Disbelief*, New York: HarperCollins.
Tattersall I., 2000, "Once we were not alone", *Scientific American* January: 56–62.
Taylor C., 1989, *Sources of the Self*, Cambridge, Mass.: Harvard University Press.
Whitehead A. N.,1929, *The Function of Reason*, Princeton: Princeton University Press.
Wright R.,1994, *The Moral Animal*, New York: Random House.

AUTHOR INDEX

Abbot F. E. 387, 390, 391
Abbott D. H. 276
Abbott W. M. 645
Abedin M. 174
Abrahamson W. G. 129
Abzhanov A. 160, 161
Agassiz L. 445, 518
Ahmanson H. F., Jr. 528
Ahlberg E. 42
Ahlquist J. 230, 231
Aiello L. C. 244, 257, 283
Alberch P. 169
Albert the Great, St 423
Alcorta C. 284
Alexander S. 395
Alexeev V. P. 295
Alighieri D. 602
Allgaier J. 533
Aloisi-Masella G. 599
Alszeghy Z. 591
Altner G. 412
Amerise M. 18,
Amundson R. 65, 157
Andrews P. 232
Andrieux P. 255
Angela da Foligno, Bl. 505
Angier N. 708
Anselm, St 650
Aquinas *see Thomas Aquinas, St*
Arambourg C. 219
Arber W. 75, 81, 85, 87, 88, 89, 90, 92
Aristotle 323, 342, 383, 404, 423, 425, 426, 441, 453, 466-475, 563, 647, 698
Arneodo A. 348
Arnold M. L. 121
Arnould J. 540, 573, 614, 632, 637
Arouet F.-M. 443, 472, 473
Arsuaga J. L. 206
Arthur W. 158
Artigas M. 592, 602
Asfaw B. 219
Ashburner M. 135
Astington J. W. 279
Atran S. 285
Augustine of Dacia 575
Augustine of Hippo, St 419, 420, 446, 576, 603
Auletta G. 488, 655
Avery O. T. 82
Ayala F. J. 25, 29, 40, 42, 43, 296, 302, 303, 469, 470, 487, 574, 648, 650, 651, 655, 658

Badyaev A. V. 171
Balcombe J. 437
Balmary M. 544
Balter M. 297
Baltes P. B. 411
Barbash D. A. 135
Barbour I. G. 585, 653, 660
Barbujani G. 236
Barlow N. 33
Barluenga M. 129
Barraclough T. G. 131
Barrow J. D. 299
Barth K. 581, 583
Barton N. H. 121, 127
Baruzi J. 626

Basil, St 419, 427
Basilides 42
Basolo A. L. 133
Bateson G. 329, 671, 675, 681
Bateson P. 162
Bateson W. 125, 137, 170
Baudry G.-H. 632
Beatty J. 462
Beauchamp P. 546
Becerra A. 107
Bégouën M. 625, 632
Bégouën S. 625, 632
Behe M. J. 31, 40, 41, 470, 528, 529, 692-695
Bell C. 31
Bell T. 45
Benci V. 513
Benedict XIV 595
Benedict XV 228, 595
Benedict XVI 228, 300, 570, 575, 585, 615, 617, 624, 625, 633, 636, 637, 648
Benignus Brother F. S. C. 705, 706
Benson J. 112
Bentley R. 581
Beraza B. 591
Bergson H. 292, 568, 633
Berlocher S. 130-133, 139
Bermudez de Castro J. M. 210
Berr H. 395
Berthoz A. 248
Bhat R. 173, 175, 178-181, 184
Biden J. 533
Biondi G. 196, 203, 225, 228-232, 234, 238
Birch C. 653
Birx H. J. 632
Blake W. 701
Blankenship J. R. 172
Blondel M. 620, 631
Bloom H. 42
Boesiger E. 303
Boethius 424
Bogart H. 360
Bogdanov A. 395
Bohm D. 324
Bohr N. 49, 350
Bois E. 269
Bole S. J. 520
Boltzmann L. 322
Bonaparte N. 374
Bonaventure, St 423
Bond C. 180
Bonis L. de 257
Bonomelli G. 593
Borges J. L. 42
Bostrom N. 415
Boule M. 497, 504, 505, 614
Bouligand Y. 350
Bourtzis K.
Boussac J. 504-507
Bouvier M. M. 124
Bowden R. M. 171
Bowen F. 461, 463, 473
Bowler P. J. 50, 53, 475, 590
Boyd R. 291, 308
Boyer C. 591
Boyer P. 285
Boyle R.
Branch G. 377
Brandi S. 592, 607, 608
Brandon R. N. 462, 469
Braverman J. 645
Brechtken J. 632
Bridgewater *see Egerton F. H.*
Broca P. 203
Bromage T. G. 210
Broom R. 204, 205, 219
Brown A. I. 520
Brown W. L., Jr. 99
Brucker J. 602
Bruner E. 206, 209, 211
Brunet M. 218, 219, 248, 258
Brunet-Lecomte P. 258
Brunner S. 684
Bryan W. J. 518

BUCHEN L. 708, 709
BUCKLAND W. 31, 472
BUIATTI MARCELLO 406, 667, 668, 676, 680, 684, 688
BUIATTI MARCO 668, 680
BUJANDA J. M. DE 593, 594
BURNS J. D. 517
BUONPENSIERE E. 604-606
BURKHARDT F. 50, 461
BURROW J. 464
BUSH G. L. 129, 130, 139
BUSK G. 45
BUTLER L. G. 524
BUTLER S. 110
BUTLIN R. K. 133
BYRNE R. W. 275, 276

CACCHIONE E. 645
CALLAHAN S. 335
CALLEBAUT W. 170
CALVIN J. 594
CAMPBELL B. G. 204
CAMPBELL D. T. 380, 387
CAMUS A. 495
CANN R. L. 232
CARAFA G. P. *SEE PAUL IV*
CARAMELLI D. 236
CARLIN L. 530
CARNAP R. 372
CARNOT S. 323
CARO P. 244
CARPENTER F. M. 99
CARROLL S. B. 155, 161, 164, 171, 173, 182, 692, 697
CASTLES D. 277
CAVALLI-SFORZA L. L. 291, 307, 309
CAVERNI R. 592, 600, 601
CEBRA-THOMAS J. 159
CECCONI E. 600
CELA-CONDE C. J. 43,
CHAISSON E. J. 329, 334
CHALINE J. 244, 247, 248, 258, 268, 269
CHANTRAINE G. 541, 613
CHAPMAN M. J. 107-109, 113
CHARLESWORTH B. 127
CHENU M.-D. 657
CHOW R. L. 157
CICOGNANI M. 598, 599, 603, 604, 606, 607
CLACK J. A. 42
CLARK G. 312
CLARK H. W. 517, 520
CLARK HOWELL F. 244
CLARKE R. J. 219
CLAYTON P. 332, 653
CLIFFORD A. 653
CLUTTON-BROCK T. H. 284
COBB J. 653
COHEN J. E. 327
COHN M. J. 159
COLEMAN S. 530
COLLARD M. 291
COLOSIMO P. F. 133
COMPER W. D. 171, 180
COMTE A. 446, 634
CONGAR Y. 657
CONWAY MORRIS S. 157, 174, 318, 337, 339, 465, 654, 655
COPERNICUS N. 27, 28, 626
COPPENS Y. 195, 206, 217, 219, 220, 222, 248, 255, 295, 298, 300
CORLISS J. 108
CORRENS C. 673
CORRIGAN M. 607
COTTIER G. M. 403, 419, 421
COTTON J. A. 57
COWLISHAW G. 285
COYNE G. 409, 645
COYNE J. A. 104, 121-125, 132, 133, 136-138, 149, 150
CRAIG T. P. 129,
CRICK F. H. C. 82, 178, 389, 676, 677
CUÉNOT C. 614, 626, 629
CULLIS C. A. 684

Currat M. 236
Cushing J. T. 655
Cuvier G. 473, 474

d'Armagnac C. 625, 626
D'Errico F. 296
Daeschler E. B. 100
Daleux A. 622
Dalla Porta N. 299
Dalton R. 43
Dambricourt Malassé A. 196, 243, 244, 247, 248, 254, 256, 258, 260, 262, 268, 269
Dambroski H. R. 130
Damen W. G. 179
Daniélou J. 657
Darlington C. D. 61
Darlington R. B. 275
Dart R. A. 204, 205, 219
Darwin C. R. 11, 26-29, 32-40, 42, 44-46, 49-52, 58-62, 64, 69, 77, 78, 81, 82, 90, 95, 96, 100, 102, 112, 114, 117-120, 124, 125, 129, 138, 139, 160, 169, 170, 184, 195, 199-200, 202, 227, 228, 263, 300, 301, 310, 318, 319, 321-323, 327, 332, 335, 337, 339, 365, 376, 379, 384, 385, 394, 403, 404, 409, 433-446, 448-452, 461-468, 470, 473, 474, 476, 495, 497, 500-502, 573, 580-582, 601, 650, 656, 669-671, 673, 675, 678, 680, 681, 687, 697, 698
Darwin E. 601
Darwin F. 444
Dautricourt J. 270
David B. 258
Davidson D. 384
Davidson E. H. 172
Davies P. 483, 487
Davis A. W. 135, 136
Davis P. 526, 527
Dawkins R. 8, 263, 310, 434, 456, 527, 565, 580, 648, 652, 655, 656, 658, 698, 703
Dawson C. 203
de Bary A. 109
de Buen O. 592
De Caro M. 384
De Filippi F. 202
de Gennes P. G. 175
de Lumley H. 248, 255, 294
De Rosa G. 228
de Saint Chéron M. 270
de Vries H. 670, 673, 674, 680
de Vries P. 527
Deacon T. W. 293, 294, 485, 488, 489
Dean M. C. 210
Deane-Drummond C. 541, 653, 660
Delattre A. 243
Dellian E. 322
Deloison Y. 258
Dembski W. 31, 528, 529, 695, 696
Democritus 467, 468
Dennett D. C. 279, 324, 395, 471, 474
Denton M. 580, 581, 583
Denzinger H. 425
Depew D. J. 322, 323, 341, 404, 461, 462, 469, 471-473
Dequéant M. L. 179
Derrida J. 637
Descartes R. 341, 409, 634
Deshayes M. J. 248, 254, 256
Devillers C. 258
Dewar D. 530
Dewey J. 382, 387, 394, 395
Diaconis P. 363, 364
Diamond J. 238
Dickens C. 442
Diderot D. 587
Diogenes 333
Dirac P. A. M. 347
Disse J. 622
Dobzhansky T. 25, 27, 61, 110, 120, 122, 125-128, 137, 139, 149, 200, 296, 297, 300, 303, 469, 701
Dod B. 136

DODSON P. 697
DOMENICHELLI T. 602, 603, 606
DOMOKOS G. 351,
DONNELLY M. J. 154
DOWNES J. A. 41
DRAPER J. W. 590, 592
DRAPER P. 369
DRÈS M. 132
DROSER M. L. 173
DRUMMOND A. J. 310
DUBOIS A. 258, 269
DUBOIS E. 204
DUHEM P. 371
DULLES A. 614
DUNBAR R. I. M. 196, 275, 276, 278, 280, 281, 283-286
DUNN C. W. 173
DUPRÉ J. 384
DUQUOC C. 577, 583
DURHAM W. H. 307
DURKHEIM E. 285
DURRANT A. 684

EDELMAN G. M. 393
EDEY M. 207
EDWARDS D. 541, 652, 653, 660
EGERTON F. H., EARL OF BRIDGEWATER 31, 581
EHEBAUER M. 176
EIGEN M. 352, 485, 489
EINSTEIN A. 49, 341, 347, 528, 584, 618, 632
ELDREDGE N. 32, 53-55, 65, 67, 207
ELGAR M. A. 41
ELIOT G. *SEE EVANS M.*
ELLIS G. F. R. 488, 489
ELSASSER W. M. 324-326
ELTON C. 298
EMPEDOCLES 30, 342, 441, 466-471
ENARD W. 99
ENDLER J. A. 132, 133
ENGELS E.-M. 589
ENQUIST M. 284
ENRIQUES F. 395
EPEL D. 154, 159, 160, 162, 163, 698
ERWIN D. H. 174
ETGES W. J. 137
EULER L. 443
EUVÉ F. 585
EVANS M. 435
EXCOFFIER L. 236

FACCHINI F. 196, 291, 293, 298, 302, 632
FALK D. 295
FAVARO A. 600
FEDER J. L. 77, 117, 129-135, 150
FEHR E. 284
FELDMAN M. W. 291, 307, 309
FELSENSTEIN J. 96
FENART R. 243
FEREMBACH D. 298
FERRIÈRE E. 592
FESSARD G. 636
FEYERABEND P. K. 495
FILCHAK K. E. 130
FILLER J. 377
FINLAY B. L. 275
FISHER E. 254
FISHER R. A. 61, 132, 169, 321, 322, 675, 678
FISKE J. 387, 390
FITZPATRICK B. M. 131
FLAGEOLLET J. 608
FLANNERY K. V. 311
FOKKELMAN J. 546
FONTANA E. 603
FORBES A. A. 130, 131
FORGACS G. 172, 174, 176
FORREST B. 649, 695
FORSTER P. 308, 310
FOSTER M. 647
FOX KELLER E. 347, 354, 355, 708
FRANÇAIS G. 607
FRANCESCHELLI O. 301

Frazer J. G. 394
Frege G. 319, 386
Freud S. A. 28, 394
Frisch H. L. 179
Fruchon P. 583
Frühwirth A. 603
Fuller S. 464, 465
Funk D. J. 133
Funkenstein A. 578, 579
Fusco G. 170
Futuyma D. J. 75, 76, 95, 99, 101, 154

Gagliasso E. 676
Galant R. 161
Galea A. M. 608
Galilei G. 28, 341, 379, 496, 498, 503, 540, 578, 582, 585, 589, 600, 602
Galimberti L. 599
Galleni L. 405, 493, 494, 496-498, 501, 503, 504, 506-508, 510, 512, 513, 626
Galton F. 53, 674, 675
Garcia-Ojalvo J. 178
Gardiner C. W. 349
Gatfield D. 178
Gaudefroy C. 508
Gaudry A. 263
Gauguin N. 569
Gavrilets S. 121, 127, 128
Gayon J. 49, 50, 61, 62, 64, 462
Gehlen A. 410
Gehling J. G. 173
Gehring W. J. 156
Gerhart J. C. 159
Geus A. 109
Ghiselin M. 474
Gibbons J. 607
Gibbs J. W. 322
Gibson G. 163
Gierer A. 176, 179
Gilbert S. F. 77, 78, 153-155, 159, 160, 162, 163, 173, 178, 338, 406, 472, 691, 694, 698
Gilbert W. 466
Gilson E. 421, 622
Girotto V. 301
Gish D. T. 530, 695
Giudicelli F. 179
Glasgow A. C. 88
Glässer A. 632
Glick T. F. 589, 592, 602
Gmelin J. 239
Goldbeter A. 178
Goldenfeld N. 169
Goldschmidt R. B. 66, 158, 169, 671, 677, 678, 681
Gomez C. 160
Goodman M. 97, 208, 238, 239
Goodman R. M. 171
Goosen C. 277
Gotthelf A. 467
Gotti G. M. 599
Gould S. J. 53-55, 65-69, 102, 184, 207, 209, 210, 244, 337, 339, 352, 391, 397, 463, 530, 584, 655, 671
Graham B. 522
Graniello G. M. 599
Grant B. R. 132, 160
Grant P. R. 132, 160
Graur D. 97
Gray A. 45, 50, 365, 404, 434, 436, 439, 441, 444-453, 461, 464, 465, 473, 656, 697, 698
Gray R. 471
Gray R. D. 310
Greene J. C. 239
Greenhill S. J. 310
Gregersen N. H. 653
Gregory of Nazianzus, St 659
Grib A. A. 503
Grice P. 279
Griffiths P. 469, 471
Grima R. 352
Groessens-Van Dyck M. C. 493, 503, 504
Grosberg R. 709

GROSS P. 695
GUDER C. 177
GULLAN P. J. 131

HADJOUIS D. 255
HAECKEL E. 203, 204, 394, 449, 601, 694
HAGADORN J. W. 177
HAGEN R. H. 136
HAILE SELASSIE Y. 218
HALDANE J. B. S. 61, 124, 135, 136
HALDER J. 157
HAM K. A. 530, 531
HAMILTON M. J. 278
HARRIS M. P. 159
HARRISON T. 203
HARTCOURT-SMITH W. E. H. 257
HARTL D. L. 136
HARVATI K. 203
HARVEY P. H. 117
HAUGHT J. F. 331, 335, 653
HAUSCHTECK-JUGEN E. 137
HAWKING S. W. 583, 584
HAWKS J. 260
HEDLEY J. C. 593
HEFNER P. 335, 653
HEGGARTY P. 310
HEIDEGGER M. 411, 654
HEIM K. 497
HEISENBERG W. 326, 655, 656
HELMHOLTZ H. 581
HELMS J. A. 160
HEMPEL C. 373
HENTSCHEL H. G. 179
HENZI S. P. 280
HERRMAN E. 686
HERSCHEL J. F. W. 32, 444, 446, 461, 463, 465
HEWLETT M. 541, 652
HICK J. 659
HIESTAND-NAUER R. 88
HIGLEY L. A. 520, 521
HILL R. A. 278
HODGE M. J. S. 60, 461, 462, 464, 472
HOFFDING H. 379, 384
HOGNESS D. S. 163
HOLLIMAN R. 533
HOLLOCHER H. 136
HOLSTEIN T. W. 178
HOOK S. 394
HOOKER J. 45, 697
HORN S. O. 615, 648
HÖSLE V. 403, 433-435, 457, 473
HOSTERT E. E. 121, 132
HOWARD D. J. 128
HÖXTERMANN E. 109
HUANG S. 344
HUGHES A. L. 104
HUGHES C. G. 157
HULL D. L. 42
HUME D. 383, 404, 444, 450, 453, 454, 650, 697
HUREL A. 504
HURLBUT W. D. 406, 701
HUSSERL E. 626
HUTTER P. 135
HUTTON J. 184, 444
HUXLEY J. 61, 153, 207, 500, 590, 618, 619, 669, 674
HUXLEY T. H. 53, 202, 203, 225, 226, 331, 375, 444, 466, 467, 476, 539, 601

IIDA S. 88
ILLIES C. 457
INTRIERI M. 684
IRELAND J. 606
IRVING S. 135
ISAYE G. 633
ISHIDA H. 218

JABLONKA E. 169, 171, 680, 686
JACKELEN A. 335, 504
JACOB F. 158, 677
JACOBI F. H. 440
JACOLLIOT L. 590, 592

Jaeger J. J. 244
Jaeger L. 488
Jaki S. 154
James W. 387
Jelinek K. 298
Jerison H. J. 275
Jernvall J. 179
Jiang T. X. 179
Jobling M. A. 235
Johanson D. C. 206, 207, 218
John Paul II 25, 228, 300, 301, 425, 541, 589, 590, 615, 645-647, 650, 657, 661, 716
Johnson P. E. 31, 371, 527, 528
Jonas H. 322, 412, 413, 587, 711
Jones C. D. 137
Jones J. E., III 32, 695
Jones J. M. 517
Jones S. 263
Joyce T. 533
Judson O. 41
Jungers W. L. 245
Juste R. 591

Kagan J. 714
Kageyama R. 178, 179
Kant I. 384, 390, 393, 404, 410, 411, 443, 448, 453-456, 458, 466, 469, 470, 473, 474, 634, 654
Karakash C. 585
Karner C. 176, 177
Kass L. R. 415, 712, 717
Kauffman S. 63, 318, 324, 337, 353, 513, 699
Kaufman M. 157, 354
Kavaloski V. C. 60
Kazmierczak J. 174
Ke Y. 234
Keane J. J. 607
Keller C. 335
Keller R. 177
Kempe S. 174
Kenyon D. H. 526, 527
Kepler J. 28, 414, 415
Keverne E. B. 277
Khakhina L. N. 109, 112
Kim J. 386, 394
Kimbel W. H. 206
Kimura M. 62, 98, 349, 353, 670, 674, 680
Kinderman P. 280
King M. C. 158
King N. 174
King W. 202
Kirby W. 691, 694
Kirkpatrick M. 121, 131, 133
Kirschner M. V. 159
Kitcher P. 382, 386, 387
Kizzia T. 533
Kleiber M. 349
Kobel H. R. 136
Kornegay J. R. 102
Kozo-Polyansky B. M. 109, 112
Krause J. 237
Krings M. 235
Krumlauf R. 252
Kuhn T. S. 52, 67
Kullmann W. 474
Kulp J. L. 521
Küng H. 584
Kunimatsu Y. 218
Küppers B. O. 485, 486, 489
Kutschera U. 532

L'Héritier P. 61
La Peyrère I. de 579
Lack D. L. 61
Lacordaire H. 602
Lacruz R. S. 210
Ladrière J. 632, 633
Lagrange M.-J. 575
Laguës M. 354
Lakatos I. 494, 495, 498
Laland K. N. 291
Lallouet F. 261

LALUEZA-FOX C. 236
LAMARCK J.-B. 405, 462, 498-500, 504, 512, 560, 614
LAMB M. J. 169, 171, 680, 686
LAMBERT D. 318, 347, 350, 616, 636, 637
LAMMERTS W. E. 521, 523, 524
LANDER A. D. 179
LANG B. F. 174
LANGE F. A. 387-390
LANGERHANS R. B. 133
LAPLACE P. S. 322, 323, 341, 376
LARGENT M. A. 433
LARROUX C. 173
LARSON E. J. 525, 526
LAUDAN L. 369, 370
LAUFER E. 159
LAURIN B. 258
LAWRENCE S. E. 41
LE GROS CLARK W. E. 203
LE ROY E. 506, 619
LEAKEY L. S. B. 205, 219, 294
LEAKEY M. G. 218
LECLERC M. 299, 619, 620, 633
LECOURT D. 573
LEDÓCHOWSKI M. 598
LEHMANN T. 278
LEIBNIZ W. G. 57, 404, 437, 443, 446, 451, 452, 457, 578, 658
LEIMAR O. 284
LENNOX J. G. 469, 473, 474
LEO XIII 592, 595, 599, 607
LEONARDI P. 503
LEPS G. 449
LERNER I. M. 677, 678
LEROY M.-D. 585, 592, 593, 602-605
LEROY P. 512
LESLIE A. M. 279, 280
LESNE A. 354, 355
LÉVÊQUE J. 550, 552
LEVI G. 250, 251
LEVINE J. 692, 694, 696
LEVINS R. 182
LEVY-LEBLOND J. M. 348
LEWENS T. 49, 470
LEWIS J. 178
LEWONTIN R. C. 62, 122, 126, 182, 318, 337-340, 343, 391, 677, 678
LI W.-H. 97
LICENT E. 506
LICHTENBERG G. C. 415
LIEBERMAN D. E. 211
LIJTMAER D. A. 124
LILEY N. R. 133
LINN C., JR. 130
LINNAEUS C. 225, 226, 239
LINNEN C. R. 154
LOCKE J. 390
LORENZ E. 380, 678
LORENZ K. 410
LOVELOCK J. 512
LOWRY D. B. 133
LUBAC H. DE 497, 541, 620, 622-634, 636, 637, 657
LUMSDEN C. J. 309
LUPORINI P. 178
LUTHER M. 581-583, 593, 659
LUYTEN P. H. 133
LYELL C. 45, 55, 184, 444, 446, 461, 462, 518
LYNCH J. D. 131

MACALLISTER J. 114
MACARTHUR D. 384
MACCHIARELLI R. 209, 210
MACE R. 309, 310
MACH E. 387
MACLEOD C. 82
MADDEN E. 386
MAIMONIDES M. 699
MAINI P. K. 352
MALAFOURIS L. 312
MALDAMÉ J.-M. 540, 557, 636
MALEBRANCHE N. 446
MALLET J. 132, 139

MALTHUS T. R. 500
MANZI G. 195, 199, 206, 209-211
MARCUS L. F. 209
MARÉCHAL J. 614
MARGULIS L. 58, 76, 97, 107-110, 112-114, 169, 651
MARITAIN J. 300, 301, 427, 432
MARSELLI N. 592
MARSH F. L. 524
MARSHACK A. 295
MARTELET G. 299, 619, 621, 634
MARTIN R. D. 263
MARTÍNEZ R. A. 540, 589, 592, 602
MASIDE X. R. 138
MAUPERTUIS P. L. M. DE 443
MAY R. M. 117
MAYER R. 379
MAYNARD SMITH J. 309
MAYR E. 61, 66, 82, 120, 123, 128, 132, 169, 170, 204, 382, 474, 475
MAZZELLA C. 591, 599, 603
MCCAIN J. 533
MCCARTY M. 82
MCCLAY D. R. 101
MCCLINTOCK B. 677
MCDOWELL J. 384
MCGINN C. 384
MCINERNEY J. O. 57
MCKEE J. K. 245
MCKENNA G. 377
MCLUHAN M. 322
MCMULLIN E. 452
MCPHAIL J. 133
MCPHERON B. A. 129
MCSHEA D. W. 102
MEIKLE W. E. 204
MEINHARDT H. 176, 179
MELLARS P. 211
MENDEL G. 82, 310, 504, 672-675
MENDELEIEV D. 503
MENDOZA M. 177
MERINO R. 159
MERLEAU-PONTY M. 586, 626
MESSENGER E. C. 589
MEYER A. 97, 139
MEYER-ABICH K. M. 412
MIDGLEY M. 415
MIKHAILOV A. S. 175
MILL J. S. 450
MILLER D. 377
MILLER K. 373, 652, 692, 694, 696
MILLER T. A. 110
MILLS S. 462
MILTON J. 434, 657
MINELLI A. 170
MITCHELL A. P. 172
MITHEN S. 308
MITTELSTRASS J. 403, 407, 409
MIVART G. J. 391, 405, 445, 498, 501-503, 508-510, 512, 513, 591, 593
MIXTER R. L. 521, 522
MOCZEK A. P. 170
MOGGI-CECCHI J. 210
MOLIÈRE *SEE POQUELIN J.-B.*
MOLLENHAUER D. 111
MOLLENHAUER R. 111
MOLTMANN J. 541, 653, 660
MONOD J. 359, 360, 376, 494, 495, 501, 508, 652, 677
MONTCHEUIL Y. DE 628
MONTELIUS O. 310
MONTGOMERY E. 387, 390
MONTON B. 377
MOODY J. W. T. 200
MOON S. M. 693
MOONEY C. F. 614, 616, 626, 631
MOORE A. 25
MOORE D. W. 517, 533
MORANGE M. 354
MORGAN T. H. 676
MORRIS H. M. 522-525, 527, 528, 530, 531
MOYLE L. C. 124
MÜLLER F. 698
MÜLLER G. B. 169-171, 182

MULLER H. J. 125, 136, 137
MÜLLER I. 109
MURPHY F. 533
MURPHY G. 653
MURPHY N. 651, 653, 655, 656, 661
MURRAY J. C. 657
MURRAY J. D. 352

NACHMAN M. W. 136
NAGEL E. 395
NAPIER J. R. 205
NAVEIRA H. F. 138
NEEDHAM J. 395
NEI M. 104
NETTLE D. 284
NEWELL N. D. 61
NEWMAN J. H. 503, 697
NEWMAN S. A. 78, 79, 169-182, 338, 343, 351, 472
NEWTON I. 28, 319, 321, 322, 330, 341, 342, 374, 379, 441, 461, 528, 695
NICHOLS S. A. 176-178
NIETZSCHE F. 411, 414, 701, 705, 716
NIJHOUT H. F. 163, 179
NOOR M. A. F. 121, 128, 134, 135, 150
NOSIL P. 133, 139
NOVAK M. 353
NUMBERS R. L. 376, 377, 405, 517-525, 527, 528, 530, 533, 573
NYHART L. 464

O'CONNELL D. J. 606
O'CONNELL S. 286
O'GRADY P. M. 137
O'TOOLE G. B. 528
OAKLEY K. P. 203
OBAMA B. 533
ODLING-SMEE F. J. 182
ODUM H. T. 324
OELSCHLÄGER H. H. 160
OGLE W. 467, 473
OKTAR A. 531

ORR H. A. 104, 124-126, 133, 135-138, 149, 150
ORSTROM E. 284, 285
OSBORN H. F. 55, 56
OXNARD C. E. 258
OYAMA S. 471
ÖZBUDAK E. M. 178

PÄÄBO S. 235, 236
PAGE S. L. 97
PAGNINI S. 601
PALEY W. 26, 29-32, 34, 36, 39, 40, 42, 436, 446, 454, 580, 581, 650
PALIN S. 533
PANNENBERG W. 583, 653
PARENTE P. 591
PAROCCHI L. M. 599
PASCAL B. 570, 618, 701
PASHLEY-PROWELL D. 136
PASTEUR L. 528
PAUL IV 594
PAUL VI 595, 657
PAYSEUR B. A. 136
PEACOCKE A. R. 25, 489, 648, 652, 660
PECCEI A. 679
PEIRCE B. 445
PEIRCE C. S. 387, 445
PENNOCK R. 362, 369-371, 373
PENROSE R. 350
PÉREZ-BARBERÍA J. 276
PERNER J. 279
PESCH C. 591
PETERS T. 541, 652, 653, 660
PETITOT J. 348
PHILIPPE H. 174
PIAGET J. 380, 395
PIATIGORSKY J. 101
PICKFORD M. 245, 260
PICO DELLA MIRANDOLA G. 410, 414, 415
PIEROTTI R. 599
PIETRONERO L. 355
PIEVANI T. 301

Pigliucci M. 171, 693
Pilbeam D. R. 263
Pinker S. 285
Pisani D. 57
Pitt Rivers A. H. L. 310
Pius IV 594
Pius V 594
Pius IX 425
Pius XII 228, 425, 589, 615, 657
Piveteau J. 248, 292, 297
Plantinga A. 367, 373, 397, 440, 456
Plato 379, 425, 563, 575, 698
Plessner H. 410, 414
Poincaré H. 678
Polkinghorne J. 660
Ponce de León M. S. 210
Poquelin J.-B. 376
Popper K. R. 371, 372, 380, 496, 560
Poulton E. B. 122, 127
Poupard P. 634
Powell B. 442
Presgraves D. C. 124, 135, 136, 138, 150
Price G. M. 405, 518-523, 529
Price T. D. 124
Procacci S. 395, 513
Prokopy R. 129, 139
Provine R. R. 284
Provine W. B. 61
Puddefoot J. C. 634
Pueyo J. I. 179
Putnam H. 384
Putz O. 645, 652

Quammen D. 153
Quine W. v. O. 319, 380, 382, 386, 394
Quinn P. 369

Rabut O. 622
Radman M. 85
Rahner K. 541, 577, 587, 591, 653-657, 660
Rak Y. 206, 244
Ramanuja 453
Ramirez Rozzi F. V. 210
Ramm B. 522
Ramsey J. 122
Randall J. H. 394
Ratzinger J. *see Benedict XVI*
Raup D. 64
Ravigne V. 121, 131
Ray J. 30
Raymond M. 154
Reati F. 503
Reich P. B. 349
Reichenbach H. 372
Reinke H. 178
Reinker S. 352
Reisse J. 633
Renfrew C. 197, 307-310, 312
Rentzsch F. 178
Rice S. 154, 155
Rice W. R. 121, 132
Richard of St Victor 653
Richards R. 464
Richardson M. D. 160
Richerson P. J. 291, 308
Richmond B. G. 245
Richmond M. L. 394
Richter M. 594
Rickards O. 196, 203, 225, 228-231, 234, 238
Ricoeur P. 293
Ridley M. 62
Riedel-Kruse I. H. 179
Ries J. 293
Rieseberg L. H. 121, 182
Rignano E. 395
Rimmer H. 520, 521
Robert J. S. 174
Robertson P. 586
Roche H. 294
Rohlf F. J. 209
Rokas A. 173
Rolston H., III 653, 660, 698
Ronshaugen M. 161

ROSE S. 708
ROSE S. M. 176
ROSEN R. 355
ROSENBERG A. 467
ROWNING B. A. 694
RUIZ DE LA PEÑA J. L. 592
RUNDLE H. D. 133
RUPKE N. A. 518
RUSE M. 60, 382, 474, 499, 526, 651, 658
RUSHDOONY R. J. 523, 528
RUSSELL B. 325, 331
RUSSELL R. J. 362, 377, 541, 645, 651, 655, 656, 658, 661
RUSSELL S. T. 124

SAETRE G. P. 136
SAGAN D. 58, 109, 110, 112, 169
SAINT-RUF C. 682
SALAZAR-CIUDAD I. 171, 173, 174, 179
SALT G. 132
SAPP J. 110
SARICH V. 229, 230
SASA M. M. 124
SAVOLAINEN V. 129
SAWAMURA K. 135
SCHÄFER J. 296
SCHÄFER-GUIGNIER O. 585
SCHEEBEN M. J. 591
SCHELER M. 410
SCHELLENBAUM P. 614, 618, 620, 626
SCHLEICHER A. 310
SCHLUTER D. 132-134
SCHMALHAUSEN I. I. 61, 162, 163, 182
SCHNEERSON M. M. 532
SCHNEIDER R. A. 160
SCHNELL S. 352
SCHÖNBORN C. 614, 633
SCHÖNMETZER A. 425
SCHOPENHAUER A. 437, 438, 453
SCHRAMM M. 433, 443, 451
SCHRÖDINGER E. 263, 341, 343, 676
SCHÜSSLER FIORENZA F. 649
SCHUSTER P. 352, 485, 489
SCHWALBE G. 394
SCHWARTZ D. 121
SCHWARTZ J. H. 207
SCHWENDENER S. 109
SCLAKE T. 352
SCOTT M. P. 178
SCRIBER J. M. 136
SEARLE J. 312
SECCO L. 299
SEEHAUSEN O. 121, 132
SEGNA F. 599
SELLARS R.W. 386, 394
SENGSTAG C. 87
SENUT B. 218, 245, 248, 260
SHAKESPEARE W. 280, 281, 283
SHALCHIAN-TABRIZI K. 174
SHAPIRO J. A. 87
SHELLEY M. 499
SHENNAN S. 307, 310
SHIPMAN P. 203
SHUBIN N. 99, 156, 157
SHULTZ S. 276, 283
SHWEDER R. 715
SIBLEY C. 230, 231
SICILIANI P. 592
SICK S. 179, 352
SIGMUND K. 353
SILK J. B. 276, 283
SILVER L. M. 415
SIMPSON G. G. 61, 170, 508
SIXTUS V 595
SLATKIN M. 132, 309
SLICE D. E. 209
SLOAN P. R. 444
SMITH H. 706
SMOCOVITIS V. B. 61
SMUTS B. B. 276
SMUTS J. C. 395
SOBER E. 63, 318, 359, 371-373, 376, 462, 466, 647
SOLAGES B. DE 637

Sonea S. 108
Sorressi M. 296
Sosis R. 284
Soulé C. 354
Southgate C. 657, 660
Spencer H. 601
Spinoza B. 362, 404, 435, 442-444, 458
Spoor F. 209, 245, 258, 261
Sprat T. 580
Srivastava M. 173, 177, 178
Stanley S. M. 57, 65, 67
Stanzione M. 319, 379, 386, 387, 395, 513
Starck D. 265
Staune J. 269
Stebbins G. L. 61
Stebbins R. E. 589
Steinberg M. S. 176
Steinhuber A. 599
Stenhouse J. 530
Sterelny K. 469
Stiling R. L. 519
Stiller J. 280
Stoeger W. R. 377, 404, 479, 481, 483, 484, 486, 490, 648, 650, 651, 655, 656, 661
Stohlgren T. J. 182
Stoneking M. 232
Stoppani A. 512
Strauss D. F. 435
Streicher J. 171
Stringer C. B. 203, 211, 232
Stroud B. 384
Sturtevant A. 122, 694
Suwa G. 218
Suzuki Y. 163
Szthamáry E. 309

Taieb M. 206
Takeichi M. 176
Takens F. 348
Tang S. 135, 138, 150
Tanquerey A. 591
Tanzella-Nitti G. 497
Tao Y. 136, 137
Tarsi T. 236
Tattersall I. 207, 295, 712
Taylor C. 717
Taylor E. B. 133
Taylor Parker S. 204
Teilhard de Chardin P. 248, 266, 268, 297, 405, 493, 494, 496-498, 503-513, 541, 568, 613-638, 652, 653, 657
Teillard-Chambon M. 631
Teissier G. 61
Tempier E. 581
Temple F. 442
Templeton A. R. 127, 234
Théry M. 133
Thom R. 269, 348
Thomas Aquinas, St 31, 42, 419-425, 427, 428, 430, 431, 446, 581, 582, 591, 635, 650
Thomas J.-F. 636
Thomas R. 354
Thompson D. W. 170, 348, 350, 351
Thomson R. 234
Thorne A. G. 232
Tickle C. 159
Tiezzi E. 333
Tilley S. G. 104
Tilliette X. 622
Timmer J. 352
Timofeeff-Ressovsky N. V. 61
Tindal M. 442
Tipler F. 299
Tobias P. V. 205, 219, 244, 248, 295
Toland J. 442
Tononi G. 393
Trinkaus E. 203
Tripepi L. 603
Trut L. 170
Tucker P. K. 136
Turelli M. 121, 131, 135, 136
Turing A. M. 179, 351

ULANOWICZ R. E. 317, 318, 321, 322, 324, 327, 329, 332-335, 338, 475
UNDERHILL P. A. 234
URIZ M. J. 180

VALENSIN A. 625
VAN HUYSSTEEN W. 653
VANHAEREN M. 296
VANNUTELLI S. 598, 607
VÁRKONYI P. L. 351
VAVILOV N. 503, 506, 508, 510
VEDEL V. 171
VERNADSKY V. I. 395, 506, 681
VIA S. 133
VIALET A. 504
VIALLETON L. 55, 56
VICKERS S. E. 113
VIITALA J. 154
VIRGOULAY R. 631
VLAMAKIS H. 172
VOGLER A. P. 131
VOGT M. 414
VOLTAIRE *SEE AROUET F.-M.*
VON BALTHASAR H. U. 653
VON NÄGELI K. W. 673
VON SEYSSENEGG E. T. 673
VRBA E. S. 391

WADDINGTON C. H. 61, 158, 162, 163, 182, 348, 350, 511, 512, 677, 678, 682
WAGGED P. N. 398
WAGNER H. 441
WAGNER J.-P. 622
WAGNER R. 85
WAINRIGHT P. O. 174
WALLACE A. R. 28, 45, 46, 125, 184, 200, 391, 445, 452, 498, 500-502, 504, 513
WALLIN I. E. 109
WANDSCHNEIDER D. 457
WANG H. 162
WARFIELD B. B. 652
WATANABE T. K. 135
WATSON E. E. 238
WATSON J. D. 82, 676
WATSON T., SR. 339
WATTERSON G. A. 349
WEATHERBEE S. D. 159
WEBER B. H. 322, 323, 462, 472
WEBER G. W. 210
WEBER M. 91
WEIL E. 583
WEINBERG S. 465
WEINER J. 160
WEINER J. S. 203
WEIS A. E. 129
WEISMANN A. 170
WELLS J. 693, 694
WÉNIN A. 539, 543, 544, 547, 549
WEST-EBERHARD M. J. 162, 164, 169, 171, 182
WHEELER P. T. 283
WHEWELL W. 376, 385, 444, 473
WHITCOMB J. C., JR. 522, 523
WHITE A. D. 590
WHITE E. G. 518
WHITE L. 307
WHITE T. D. 43, 207, 211, 218, 245, 248
WHITEHEAD A. N. 325, 342, 395, 653, 716
WHITEN A. 275, 276
WHITLOCK M. C. 133
WIEDENHOFER S. 615, 648
WIESEL E. 270
WIGNER E. 348
WILDMAN D. E. 238
WILKINS A. S. 171
WILLIAMSON D. I. 113
WILSON A. 229, 230, 232
WILSON A. C. 158
WILSON D. S. 63, 301
WILSON E. O. 99, 301, 309
WIMMER W. 180
WINKLER-OSWATITSCH R. 352
WOESE C. R. 57, 169
WOLDEGABRIEL G. 245

Wolpoff M. H. 232, 260
Wolters G. 414
Wood B. 206, 260, 291, 295
Wood T. K. 129
Woodward A. S. 203
Worms F. 633
Worthen F. M. 528
Wray G. A. 101
Wright C. 387, 390-394
Wright L. 469
Wright R. 703
Wright S. G. 61, 321, 322, 353, 677, 678
Wu C. I. 136, 138
Wu P. 160, 161

Yahya H. *see Oktar A.*
Yandel K. E. 453
Yin L. 177
Young C. C. 433

Zahm J. A. 591-593, 602, 605-608
Zardetti O. 605
Zardoya R. 97
Zeeman E. C. 348
Zeresenay A. 210
Zhou W.-X. 278
Zhu A. J. 178
Zigliara T. M. 600, 601
Zollikofer C. P. E. 209, 210, 258, 259
Zonneveld F. 245
Zuckerman S. 258
Życiński J. 541, 653

SUBJECT INDEX

A

accident 421, 463-473
acquired characters, inheritance of 78, 169-170, 183, 499-500, 504
act of being (*actus essendi*) 430-431
Acyrthosiphon 133
adaptation 33-36, 39-40, 60-66, 102, 121-137, 219-221, 243-244, 247, 262, 267, 294, 298, 338-339, 342-346, 385-390, 404, 437, 452, 456, 462-476, 650, 680-683, 686, 692-694, 703-704, 707, 713-716
  ecological adaptation 123-125, 129-132, 293
  adaptive radiation 132, 160
adjacent possible 340, 346
Aedes 136
Agelenopsis 133
age of the Earth 369-371, 518-521, 558
alleles 104, 163, 674-675, 678
  human alleles 104
altruism 110, 285, 302, 703, 709
American Scientific Affiliation (ASA) 405, 521-523
Americanism 605-607
amino acids 43, 100, 683, 695-696
  substitution 158, 162, 237
amoebae 109, 708
analytical philosophy 380, 382, 469
animals 76, 96, 108, 112, 157, 227, 302, 393, 412, 428, 432, 447, 544-547, 550-553, 566, 652, 657, 677, 680-686
Anolis 133
Anopheles 136, 154
anthropic principle 299, 481, 484, 490
anthropocentrism, anthropocentric 110, 413-414, 456, 475-476, 543, 623
apoptosis 159, 181
arboricolism 196, 244, 247
archaebacteria 107
Archaeopteryx 42
archaeprotista 108
Ardipithecus 201, 218, 246
  A. kadabba 239
  A. ramidus 219, 239, 248
Artemisia 133
art 203, 212, 270, 293, 311
arthropods 156, 161, 173, 179-180
artificial selection 33, 60, 112, 670
assortative mating 128, 131, 133
atheism 8, 366, 434, 446, 534, 580, 590, 645
atomism 323, 332, 334
attractor 268-269, 344, 513, 653
Australopithecus 201, 205, 207, 249, 292, 294, 301, 307-308
  A. afarensis 201, 206, 218-219, 239
  A. africanus 204-205, 219, 239, 260, 266
  A. anamensis 218-219, 239
  A. bahrelghazali 218-219, 239
  A. garhi 219, 239
autocatalysis 329-333
autopoiesis 489, 707

B

Bacillus cereus 108
Bacillus anthracis 108
bacteria 35, 38, 44, 57, 84-90, 97, 102-103, 107-108, 111, 114, 172, 677, 680, 682, 692, 696

barrier to gene flow 77, 89, 117, 119-127, 130-134, 139, 149
basicranial flexion 247, 253-254, 265
Bateson-Muller-Dobzhansky model 125
beak morphology 102, 159-161
Big Bang 424, 479-481, 483, 646
biface 294-295, 297
biocentrism 413
biodiversity 117, 120, 131, 164, 691-692, 699
bioethics 412-413, 634
biological function 86, 89, 391, 710
biologism 407-408, 410, 413, 415
bionts 109, 111, 113
biosphere 337, 339-340, 342, 346, 494, 504, 506-513, 560, 617-618, 632, 681
bipedality 196-197, 245-248, 256, 258, 261, 267
body plan 157-158, 170, 172-173, 182, 244, 246, 256-257, 262, 264, 268-269, 698
Bombina 133
bonobos (see also "*Pan paniscus*") 217, 226, 229, 231, 238, 685
Boolean networks 344
brain 40-41, 197, 211, 219-221, 282-286, 297, 345, 390, 457, 685-686, 710
  social brain hypothesis 275-277, 279
bush theory 55, 209, 227

C

cadherins 172, 175, 181, 685,
Cambrian 35, 78, 173, 709
  Cambrian explosion 172, 174, 177, 709
cancer cells 344
catastrophe theory 348, 350
Catholic Church 8, 300, 575, 590, 593, 605
  pragmatic policy 609
  public acceptance of evolution 589-590
  reaction to evolution 590-592, 608-610
causa dispositiva 427
cause 323, 342-343, 361-362, 381, 437, 444, 450, 453-456, 458, 500, 708
  efficient cause 342, 392, 443, 470, 472-473, 647, 655, 716
  final cause 342, 392, 443, 448, 450, 468-469, 472-474, 694, 716
  formal cause 342, 716
  material cause 323, 342, 470, 716
causal closure 317, 343, 381
causality 361, 387, 389-390, 422, 464, 468-470, 472, 563-564, 574
  bottom-up causality 486
  same-level causality 486
  top-down causality 323, 486
cell adhesion 156, 174-176, 179, 181
cell communication 155
cell death 159-160
cell polarization 176-177
cell state oscillations 177-178
cell types 172-176, 181, 344-345, 709-710
centripetality 330-331
chance 30, 39, 324, 326, 329, 332, 334, 389, 437, 449, 452, 461-467, 469, 471, 495, 501, 574, 652-653, 668-670, 674-675, 677, 680, 696-698
Chamaecrista 133
Chen caerulescens 104
Chen rossii 104
chimpanzees, chimps (see also "*Pan troglodytes*") 41, 44, 99-100, 104, 158, 196, 202, 217-218, 225-226, 229-231, 236-239, 245, 257, 259-260, 275, 283-284, 649, 685-686
choanozoa 171, 174
Christ 424, 441-442, 541, 578, 613-614, 620-621, 624, 626, 628-631, 634, 636-637, 654, 659-660
Christianity 577, 589-590, 621
Christology 653-654, 659-660
co-adaptation 151, 404, 463, 476
  organism-niche co-adaptation 182
coevolution 195, 197, 210, 301, 308-310
coherence, principle of 387
common ancestor 27, 43, 96-99, 103-104, 107,

113-114, 118-119, 155, 195, 207-208, 217-218, 230-231, 233, 235, 238, 698
community 112, 278, 282, 284-285, 298, 331, 704, 714
comparative anatomy 30, 42-43, 154, 231, 379, 391
complexity 12, 14, 30, 33, 37-38, 63, 102-103, 267, 269, 311, 318, 385, 480-481, 505, 507, 511-512, 528, 613, 618, 626, 652, 704, 708, 710, 713, 717,
 complexity theory 63, 337-338, 344, 513
concordism 403, 420, 427, 576, 628
conditio humana (human condition) 409-412, 415, 583
conscience 299, 302, 390, 392, 441, 716
consequence-etiology approach 470-471
consilience of induction 385
consciousness 196, 220-221, 383, 386, 391-394, 443, 480, 485, 490, 505, 512, 618, 623, 657, 704, 715-717
 reflexive consciousness 267-268, 270, 392, 635, 655
constraint 14, 262, 268, 318, 327, 343-344, 353-354, 671, 680-681
 environmental constraint 559, 565
 evolutionary constraint 263, 509
 formal constraint 63
 developmental constraint 65, 68
 morphological constraint 65, 68, 211, 255
constructive conceptualization 392
contingency 102, 332-333, 447, 465-466, 583, 647, 655
convergence: see "evolutionary convergence"
convergent extension 177, 181
Convoluta roscoffensis 111
cooperation 285, 302, 563, 709, 716
Copernican revolution 27-29, 650
correlated development 671
Costus 133
cranio-facial contraction 247-248, 253, 255, 267
Creation 14-15, 17, 36, 227-228, 419-429, 442-443, 446-467, 482, 490-491, 498, 539-540, 543-555, 561-566, 570-571, 575-576, 580, 591, 603, 609, 648, 652, 655, 659-660
 and origin 562-564
 and time 419, 422-424, 427-428, 562-564
 Creation out of nothing (*creatio ex nihilo*) 36, 421, 424, 647
 special Creation 396, 445, 518, 603, 695
creationism 8, 32, 227, 319, 359-362, 369-372, 405, 408-409, 420, 450, 475, 517-534, 539-540, 573-587
 evolutionary creationism 408
 scientific creationism 369, 525-526, 529
 special creationism 436-437, 442-443, 452
 young-earth creationism 322, 369-371, 408, 518, 523-527, 530-531, 534
Creator 8, 29-31, 227, 300, 381, 422-423, 426, 428, 439, 446, 453-454, 457, 482, 490-491, 493, 497-498, 501-502, 539, 544-546, 550, 552, 554, 564, 566, 570-571, 574-575, 580-582, 587, 634, 652, 694-695
creatureliness 540, 552
Cretaceous 64, 99
critical networks 344-345
Cro-Magnon 201, 202-204, 236
culture 13-15, 196-197, 220, 275, 281, 285, 291-303, 307-313, 686, 714-715
 cultural evolution 13, 197, 297, 302-303, 308-309 , 341
 culturalism 410

D

Darwin's finches 132, 140, 160-161, 338
Darwinian Centennial Celebration 319, 380, 385
Darwinism 25, 49-51, 61-69, 337-338, 379, 384-385, 389-395, 434, 445, 448-451, 453, 457, 462-470, 671-674

eclipse of 395, 590
molecular Darwinism 81-82, 91-92
deduction 342
deism, deist 361, 366, 441-442, 445, 499, 571
demarcation problem 371-372
descent with modification 27, 50-51, 53, 59, 75, 77, 95, 155, 157, 160, 651
design 26-27, 30-34, 38-41, 101, 448-452, 470-474, 566-569, 580, 650, 703-704, 709
argument from design 26, 29-32, 40-42, 396, 436, 444, 448-449, 453-454, 474, 691, 697
Desmognathus ochrophaeus 103
determinism 39-40, 268, 317, 323, 326-327, 332-334, 361-364, 392, 439, 467, 656
deterministic chaos 341, 513, 678
diploblasts 173
directionality 14, 404, 464, 475, 479-484, 487, 510, 513, 650, 708-709
discontinuity 14, 196, 221, 228, 232, 247, 261, 267-268, 296-301, 671, 678
disturbance of equilibrium 388
divine intervention 319, 335, 362, 367, 374, 561, 563-565, 567, 570-571
divine Providence 451
particular Providence 448
divine Revelation 435, 570, 576, 606, 610, 629
DNA 43-44, 62, 75, 81-92, 97-99, 155, 263, 348, 676-679, 693, 708, 712
acquisition 85, 88-90, 92
ancient DNA 234, 236-237
comparison of DNA sequences 43-44, 84, 97-99, 155
deletion 85, 87
hybridization 230
inversion 87-88
mitochondrial DNA 232-238
mispairing 85
nonfunctional DNA sequences 100
rearrangements 84-87, 90-92
sequence evolution 97-98
sequencing 155-157
Dover trial 692
Drosophila 124-126, 135-138, 352
D. mauritiana 137, 151
D. melanogaster 135-136, 138, 151
D. mojavensis 137
D. paulistorium 110
D. persimilis 128
D. pseudoobscura 126, 128, 137, 151
D. sechellia 137
D. simulans 135-138, 151
D. subobscura 137
dynamical patterning modules (DPMs) 171-172, 175, 177, 179, 181-183

E
ecology 449, 558
ecological ethics 412-413
ecological isolation 126, 129-133
ecological metaphysics 318, 332-334
ecological niche 12, 182, 218, 221, 298
ecological system, ecosystem 131, 221, 317-318, 321-322, 324, 326, 332, 472, 475, 681
Eden 518, 525, 546, 548,
Elysia chlorotica 111
Elysia viridis 111
emanationism 421
embryo, embryonic 100, 155-156, 159-164, 174, 176-179, 184, 196, 244-269, 352, 426-427, 512, 693-694, 696, 699
embryo-forming groups 108-109, 112
embryology 77, 338, 385, 558, 671
embryonic induction 178
emergence 14, 58, 67, 88, 113, 172-176, 220, 244, 260, 268, 310-313, 332-334, 346, 352, 390, 392, 395, 404-405, 463, 468, 471, 474, 479-491, 568, 648, 704-714, 717
of man 195-196, 208, 211, 217, 231-

232, 245, 257-258, 262, 308-309, 704, 707, 712-713
empathy 714, 716
encephalization 195, 206, 210-211, 685
Enchenopa binotata 129
endorphins 197, 277, 284-285
enhancer 159-161, 164
Enlightenment 225, 322-323, 346, 499, 504
entropy 262, 268, 270, 475-476, 558, 618, 632
environmentally induced modifications 162-164, 182, 499, 678, 683, 685
epigenetics, epigenetic 65, 171, 182, 253, 394, 404, 471-474, 677-678, 680, 682, 684-685, 705, 707
  epigenetic landscape 350, 511-512
eschatology 658, 660-661
Escherichia coli 84, 338
eternity 423-424, 570, 635
eubacteria 57, 107
eukaryosis 113-114
eukaryotes, eukarya 57-58, 76, 86, 91, 108-109, 486, 679, 682-684
Eurosta 133
  E. soligenesis 129
evangelicalism, evangelical Christians 519, 521-523, 528, 530, 532, 586-587, 589
evil 334-335, 437, 458, 577, 704
  natural evil 657-659
  no choice argument 658
evolution 8, 11-14, 25-69, 75, 83-86, 90-92, 95, 119, 200, 308-310, 321-322, 352-353, 384-385, 403-405, 426-427, 451, 463-466, 475, 479-480, 495, 557, 613-621, 650
  and ethics 301, 385, 411-414, 651
  and predictability 61, 99-100, 200-204, 234, 269, 318, 337-346, 347-349, 352, 355, 654-655, 670
  by integration 109, 113-114
  cultural evolution 13, 197, 297, 302-303, 309, 341, 685-686
  evidence for 26-27, 34, 42-44, 81-92, 95, 97, 103-104, 154, 157, 172, 519
  general fact of 51-58, 95
  gradual evolution 33, 36, 51-53, 58, 66, 79, 101, 103-104, 107, 124, 135, 207, 243-247, 260-262, 267, 352, 388, 447, 462, 469
  macroevolution 56-57, 62-63, 65-68, 79, 119, 171, 184, 224, 247, 251, 254, 257-258, 263-270, 652
  microevolution 62-63, 79, 119, 184, 247, 251, 256-257, 266-267, 298, 652
  mosaic evolution 76, 98, 101, 246
  of gene flow barriers 119, 122, 125
  rate of 97-99
  theistic evolution 528, 650-656
  evolution theory 12, 17, 25-69, 78, 155, 169, 337, 340-346, 359-363, 366-368, 403, 420, 466, 493-503, 557-561, 669-671
Evolution Protestant Movement (EPM) 530
evolutionary convergence 12, 75-76, 96, 102, 157, 337, 465, 505, 617-618, 628, 654
evolutionary developmental biology (Evo-Devo) 12, 65, 78, 157-159, 164, 170, 195, 210, 269, 351 475, 652, 691-694, 705
evolutionary epistemology 380, 387-388, 390-391, 395-396
evolutionary parallelism 337, 502-503, 505-506, 508-510, 512-513
evolutionary psychology 702-704, 712, 716
evolutionary trend 55, 61, 66-67, 76, 102-103, 297
evolutionism 12, 227-228, 424, 616-617, 620, 648
evolvability 75, 680-681, 687, 707
exaptation 211, 339-341, 346, 391-392
excluded middle 343
extinction 38-39, 64-66, 202, 207, 212, 464

mass extinction 64, 257
extracellular matrix (ECM) 179-181
eye 26, 36, 101, 156-157, 344, 422, 450, 467, 581, 686, 692
blind spot 101
development 156-157, 265
retina 101, 156-157, 696

F

faith 8-9, 17, 395-398, 420, 424-427, 458, 533, 569, 571, 575-577, 587, 589, 601, 604, 620-621, 626-629, 633, 645, 648, 650, 659
falsifiability 371-372
feedback 172, 177-178, 181, 329-330, 470, 486, 488, 490
finality 14, 390, 404, 443, 456, 479-482, 490, 565, 650
First Cause 421-423, 426-427, 437-438, 441, 452, 458, 609
First Vatican Council 8, 582, 645
fitness 63, 91, 98, 110, 122, 125, 128-129, 133, 135, 149-150, 157, 283, 342, 353, 368, 404, 462, 466, 470, 703
fitness landscape 352-353
flood geology 518, 520-525
fossils 42, 99-100, 204-206, 234, 243-244, 249, 251, 260, 263-264, 266, 268, 519, 558, 657
founder effect 127
free will 40, 282, 334, 383, 397, 439, 444, 451
freedom 196, 299-300, 302, 334, 393, 513, 617, 623-625, 630, 634, 659, 704, 707, 711-712, 715-716
fungi 108-110, 113, 177

G

Galerucella 133
game theory 353
Gelsemium 133
gene
Agouti 154
Bmp4 160-161
Distal-less 161
DKK 352
for ß-galactosidase 696
for C-type lectins 174
for Gremlin 159
for hemoglobin 154
FOXP2 99, 236-237
Hmr 133
Hox 157, 252, 261, 698
Lhr 133
MC1R 236
mhr 133
Nup96 150
Nup160 150
Ods 150
Ovd 150
Pax6 156-157
per 150
Prdm9 150
TGA1 162
Tinman 157
Ubx 161
Xmrk2 150
Zhr 133
167.7 133
gene amplification 684
gene expression 126, 156-159, 164, 178, 345, 684, 708, 710
gene flow 117, 119-132, 134, 136, 139, 149, 231
gene mapping 126-127, 134-135, 139
gene regulation 158, 171, 247
Genesis, Book of 408, 518-525, 527, 534, 539-540, 543-546, 550, 553, 562, 571, 575-576, 579, 602-603, 649, 657
genetic accommodation 182
genetic assimilation 162-163, 182
genetic drift 98, 121, 126-127, 137, 308, 311
genetic polymorphism 76, 85, 104, 237

genetic regulatory network 172, 177, 344-345
genetic variants 81, 84-85, 90-91, 153-154, 686
genetics 82, 153, 507, 676
  developmental genetics 155-157, 164, 652
  Mendelian genetics 12, 61, 119, 504, 673
  populations genetics 61-62, 66, 77, 98, 164, 298, 652, 674
  microbial genetics 84-85, 88, 91
  molecular genetics 76, 81-82, 92, 310
  of speciation 135-138, 164
genomics 82, 139, 155
geology 124, 461, 497, 505, 508, 519, 651
Geochelone elegans 351
geometric morphometrics 195, 209-210
Geosiphon pyriforme 111
Gilia 133
God-of-the-gaps 324, 473, 697
Gorilla 208, 229, 238, 243, 248, 251, 256-257, 260, 262
gorillas 100, 202-203, 225-226, 229-231, 238-239, 243-245, 249-250, 256, 258, 260, 685
Gospel 435, 442, 568, 621, 630, 645
gradation 428, 431, 447
great apes 210, 218, 246, 249, 256, 258, 260, 267, 275, 282, 286, 652
grooming 277-278, 284
group selection 62-63, 283, 385
Gunnera manicata 112
Gunnera monoeica 112

H

Haldane's rule 124, 135-136
Hardy-Weinberg law 349
Helianthus 133
Heliconius 133
Heliocidaris erythrogramma 101
Heliocidaris tubercolata 101
heterochrony 158, 160-161, 164
heterocyberny 162-164
heterometry 158, 160-161, 164
heterosis 669, 678
heterotopy 158-159, 161, 164
heterotypy 158, 161, 164
higher taxa 52, 54-55 101, 114, 170
history of life 57-58, 65, 68, 107, 298, 300, 366, 463-465, 476, 563-566, 569, 659, 704, 709, 713, 717
history of salvation 568-569
holism 395, 648
holobiontic species 111
Holy Ghost 570-571
Holy Scripture 419, 575
  figurative language of 602-603
  literal interpretation of 408, 569, 576, 591, 603
homeostasis 344, 486
Hominidae (see also "great apes") 195, 199, 204-212, 220, 244-245, 257, 260, 262, 293-298
  Homininae 208, 231-232, 239, 245-246, 259-260, 267, 309
hominization 228, 297-298, 301, 623-624
Homo 195-196, 200, 206-212, 219-221, 226, 239, 244-245, 251, 261-263, 291-292
  H. antecessor 231, 239
  H. erectus 201, 204, 209, 211, 221-222, 231, 239, 244, 262, 291-292, 295, 509
  H. ergaster 209, 231, 239, 244, 262, 291-292, 295
  H. floresiensis 222, 239
  H. georgicus 231, 239, 244, 262
  H. habilis 196, 201, 205-206, 221, 239, 244, 262, 291-292, 294-295, 297
  H. heidelbergensis 209, 231, 239, 284, 292
  H. neanderthalensis 202, 209, 211, 231, 234, 237, 239, 292
  H. rudolfensis 221, 239, 244, 249, 262, 291-292, 295, 297
  H. sapiens 38, 43, 99, 195-196, 199-201,

203, 209, 211-212, 225, 231-234, 236-237, 239, 244, 246-250, 254-256, 266, 268-270, 291-292, 296, 307-310, 409, 414-415, 568, 652, 655, 679, 711
homo faber 292, 414-415
homology 95-96
deep homology 156-157
hopeful monster 66, 169, 671, 678, 681, 684
horizontal gene transfer 57, 84-85, 88, 90, 92
horizontal transmission 309-310
human being 13, 221, 225, 270, 301, 411, 465, 475, 539-540, 546-547, 550-554, 559, 567-570, 589, 703, 711, 714
central position 543-544, 549-554, 569-570
dominion over the world 543-549, 554
male and female 545
human races 237-238
human soul 14, 227-228, 300, 403, 415, 424-426, 428-432, 518, 568, 591, 657
immortality 424, 435, 439, 451
hybrid 122, 128-129, 133-135, 137, 149-150, 182
inviability 104, 122-126, 134, 136-138, 149-150
lethality 138, 149-150
rescue genes 135
sterility 104, 122-126, 134-138, 149-150
hybridization 77, 121, 232, 684

I

imago Dei (image of God) 425, 540, 544-547, 554, 646
indeterminacy 325-328
indeterminism 268, 323, 655
Index of prohibited books 593-595, 600-601
Congregation of the Index 592-595, 601-602, 605-610
Congregation procedures 595-600
Index and evolution 592-593, 597, 600-601, 609
induction 386, 633
information 263, 485-487, 677-678, 683-684
genetic information 43-44, 81-83, 86-90, 268, 558, 682-683
intelligence 292, 300, 302, 480
Intelligent Design 13, 30-33, 40-41, 156, 263, 269, 317, 367, 405, 408, 463, 465, 470, 473-474, 526-534, 540, 574, 580-584, 586, 652, 656, 691-696, 699
intelligibility of nature 404, 45
intentionality 279-283, 286, 293, 490
Ipomopsis 133
Iris 133
irreversibility 57-58, 318, 323, 332, 475, 617
IS elements 87

K

Kenyanthropus 201, 218-219, 239
Kenyapithecus 257

L

Lamarckism, Lamarckian 108, 164, 170-171, 463, 499, 614
language 99, 237, 284, 299-300, 310, 393, 567-570, 633
last eukaryotic common ancestor (LECA) 113-114
lateral inhibition 175-176, 179, 181
laws of form 388
linguistic turn 386-387
Linnean Society of London 45, 445, 500
living being 29, 426, 499, 547, 560, 702, 705-706
Littorina 133
local falsification 496

M

Magisterium of the Catholic Church 424-426, 575, 577
mammals 55-56, 100, 250-251, 276-277, 685
Mastotermes darwiniensis
materia prima 419, 430-431
materialism 17, 322, 333-334, 366, 382, 397, 465, 467, 527, 531, 534, 562, 590, 604, 648, 706
mathematical Platonism 367-369
meiotic drive 121, 137-138, 150
meiotic sexuality 108-109
membrane 156, 173, 180, 343
meme 307, 310, 312,
mesenchyme 160-161, 173, 180, 710
metaphysics 14, 269-270, 322, 335, 383-385, 420-421, 431, 441, 458, 498, 504, 508, 564
Metazoa, metazoans 78, 157, 171-181, 183
Mimulus cardinalis 122, 150
Mimulus lewisii 122, 150
Miocene 217, 245, 247, 250, 256-257, 266-267
missing link 114, 203-204
Modern Synthesis 61-62, 66, 119, 153, 163-164, 171, 182-183, 207, 227, 504, 669, 674-676, 693
modularity 159, 486, 710
molecular biology 26-27, 42-44, 196, 321, 676, 701-702
  central dogma of 676, 679
molecular clock 75, 85, 97, 99, 104, 201, 229-230, 233, 235, 368
monism, monist 342-343
mono-monostatic body 351
Monosiga brevicollis 174
moral awareness 704-705, 712-715
moral sense 301, 715-716
morphogen 178-179, 181, 352, 510, 512
morphogenesis 171-175, 244-250, 253-261, 264-270, 343
  morphogenetic field 175, 178
multicellularity 169, 173-177, 708-710
Multiregional Continuity model 231-235, 237
multiscale systems 347, 354-355
multiverse model 458
Mus musculus 150
mutagenesis 81, 83, 84-86, 248, 261-262, 267
mutation 26, 33, 35-39, 75, 82-85, 87-89, 98-99, 125-126, 135, 154, 163, 171, 229-230, 233, 244, 298, 319, 349, 364-366, 464, 655, 671, 674-675, 681-682, 707
  macromutation 113, 169
  neutral mutation 62, 83, 98, 353, 674
  nonsynonymous mutation 98
  random DNA mutation 107, 109, 114
  spontaneous mutation 33, 83-84, 92, 464-465
  synonymous mutation 98-99
mutuality 318, 321, 331, 334

N

natural law 28-29, 339, 341, 346, 362, 387, 441-442, 446, 670
natural selection 26, 29-41, 45-46, 50, 59-68, 83-85, 98, 103-104, 113, 119, 150, 169, 199, 262, 268, 298, 343-344, 385-386, 436-439, 450, 452, 462-466, 471-472, 481, 485-487, 590, 650, 670-671
  as a process of elimination 26, 33-35, 39, 86, 98, 112-113
  as based on the environment 38, 82, 107, 121, 131, 137, 339, 462, 485, 500-501, 681
  as based on genomic conflicts 121
  canalizing selection 171
  disruptive selection 128, 131-132, 134
  positive selection 98, 137, 150
  stabilizing selection 171, 182
naturalism 13, 319, 380, 386, 392, 394-395, 407-408, 456, 527, 704, 716
  epistemological naturalism 380, 382, 387

metaphysical naturalism 359-360, 368, 380, 382, 527, 529, 648
methodological naturalism 13, 317-319, 334, 359-362, 367-370, 372-376, 380-381, 527-529, 647-648
naturalized epistemology 319, 380
nature 29, 322-323, 332-333, 359, 380, 414, 426-427, 436, 456-458, 464, 476, 479-480, 482, 494, 559, 580, 582, 637, 649, 658, 678, 707, 712, 716
Neanderthal man (see also "*H. neanderthalensis*") 201-204, 210, 212, 234-237, 292, 296
neo-Darwinian theory, neo-Darwinism 12, 81-82, 107, 246, 322, 337, 614, 671, 674-675, 678
neo-Thomism 654
neutral theory of molecular evolution 62-63, 98
network of life hypothesis 57-58
New Testament 435
Non Interventionist Objective Divine Action (NIODA) 362, 656
Non Overlapping Magisteria (NOMA) 397, 584
noosphere 395, 506, 541, 613, 619, 623, 681
Notch pathway 178-179, 181
nucleotide 43, 82, 85-87, 230, 235
additional insertion 85
deletion 85
sequence alteration 81-84
substitution 83, 85-86, 98, 158, 161

O

Ockham's razor 376, 672
Old Testament 435, 539, 540, 543, 547, 553
mythical-style narrative 543, 554
Omega point 626-628
On the Origin of Species 11-12, 27, 33-34, 37, 42-43, 50-51, 59, 76, 96, 114, 195, 199-200, 202, 227, 263, 385, 394, 433, 461, 539, 669
ontogenesis, ontogeny 183, 243-244, 246-247, 250, 256-261, 267, 269, 300-301, 324, 411, 463-464, 469, 471-473, 475, 712
ontogenetic revolution 195, 210-211
Orrorin 201, 218-219, 239, 245-248
orthogenesis 78, 169-172, 183, 510, 617, 619, 625
Out of Africa model 231-234

P

Palaeolithic 205, 208, 212, 295-296
Pan paniscus 217, 229, 238
Pan troglodytes 104, 217, 229, 238
panspermia 389, 651
pantheism 407-408, 446, 601, 614, 621, 631, 636
paracrine factor 155-156, 159-160
parallelism: see "evolutionary parallelism"
Paranthropus 201, 205, 207-208, 219, 249
P. aethipocus 239
P. robustus 205, 239, 266
P. boisei 205, 239
parasexual gene exchange 88-89
parasexual merger 111
pathocentrism 413
Pedicularis 133
Peking Man 497, 509-510
Penstemon 133
Perfilievia 113-114
person 270, 279, 383, 425, 447, 456, 566, 637, 646, 659, 702, 704, 712-716
phenotype 78, 82-83, 86, 122, 134-135, 154, 162-163, 182, 337-339, 343, 346, 675, 678
philosophical anthropology 409-410
philosophy of nature 405, 427, 557, 560, 633-634
Phlox 133
photosynthetic animals 111
phylogenesis, phylogeny 96, 97, 206, 210,

218, 234, 238, 244, 263-264, 300, 349, 352, 463-464, 476, 505
phylogenetic tree 58, 76, 96-97, 99, 195, 206-208, 233, 235, 237-238, 510
physicalism 317, 381
physiocentrism 413-414
Pisum sativum 672-673
Pithecanthropus 204
P. alalus 204
P. erectus 201, 204
planning capacity 196, 293, 297, 299
plants 100, 102, 108, 110-112, 123, 133, 162, 182, 329-330, 447, 669, 672-673, 675, 680, 684
plasticity 78-79, 164, 170, 180, 183-184, 353, 451, 471, 558, 680-681, 683-687
Plio-Pleistocene 249, 294
Pongo 230, 248, 250-251, 256-257, 262
population divergence 117, 121, 126, 131, 133, 234-235
posthumanism 415
preadaptation 318, 339-341
Precambrian 78, 173-174, 177
Ediacaran 173-174, 177
primates 38, 44, 97, 211, 217, 225-226, 229, 244, 247, 254, 261, 263, 275-277, 283-284, 685
classification of 238-239
probability revolution 462, 466
probability theory 326, 341
process, definition of 327
process ecology 318, 321, 333-335
Proconsul africanus 256
Proconsul major 257
Proconsul nyanze 257
prokaryotes 57-58, 75-76, 107, 113, 682-683
prosimians 264-265, 505
protein 43-44, 98, 100-101, 150, 155, 161, 230, 679, 683-685, 696
ß-catenin 693-694
ß-galactosidase 696
Bicoid 696
C-type lectins 174-175
calmodulin 684
DSL 176
Gremlin 159
opsin 154
TGA1 162
Ubx 161
Wnt 177-178
protist 172
protoctist 107-110, 113
punctuated equilibria 53, 55, 65-66, 195-197, 207-209, 211, 243-244, 246, 248, 257-258, 260, 262, 267, 352, 447

Q

quantitative trait loci (QTL) 135
quantum mechanics 49, 263, 317, 323, 341-343, 350, 367, 586, 655-656
qualitative leap 424, 427

R

Ramapithecus 229-230
randomness 39-40, 85, 92, 465, 670, 680-681
rationality 647, 649, 653
realism 384, 386-387, 390-392, 496
recapitulationism 475
recombination 86-90, 233
reductionism 13, 317, 323-324, 407-408, 467, 470, 474, 507, 648, 668, 672-676, 698, 704-706, 712
reductionist methodology 668, 672, 674
redundancy 486
reinforcement 125, 127-129
relationality 404, 480, 491
religion 275, 281-282, 284-285, 301-302, 440, 539, 587, 697
reproductive isolation 66, 84, 103-104, 110-111, 120-138, 149
reticulation 97
Rhagoletis 131, 133

R. mendax 130
R. pomonella 129-131
R. zephyria 130
Rhizobium 108
RNA 44, 151, 263, 268, 352-353, 485, 651, 676, 679, 683-684, 708
analysis 158
regulatory RNAs 172
RNAi interference 134

S

Saccharomyces bayanus 150
Saccharomyces cerevisiae 150
Sahelanthropus 201, 218, 239, 243-244, 248-249, 257, 259-260
saltationism, saltationist 53, 78, 169-171, 182-183, 244, 388, 447, 678
saltational process 183, 245, 261, 264, 268
Scala naturae 458, 472-473, 475
science and religion 8, 15, 317, 321, 397, 517, 541, 561, 574, 576, 584, 646, 698, 706
scientific evidence 226, 693
scientific explanation 13, 29, 382, 458, 559, 561
scientific fact 95, 118, 540, 557-558
scientific hypothesis 40, 99, 648, 657
scientific method 301, 395-396, 562, 573, 647
scientific principle 559
scientific research program 380, 498, 500
scientific theory 228, 368, 371-373, 375, 458, 495, 557
scientism 13, 375, 648
Second Vatican Council 8, 493, 497-498, 541, 568, 575, 645, 647, 649, 661
secondary cause 426, 442, 448, 458, 609, 650, 655
self 706, 714-715
self-awareness 300, 709-711, 713-714
self-consciousness 392-393, 457, 652, 655
self-determination 293, 296, 299, 301, 409-411, 415
self-organization 12, 63, 317-318, 338, 344-346, 472, 489, 502, 513, 558, 652
self-organizing process 92, 471
selfish gene 121, 283, 703
sexual selection 121, 127-128, 133, 136, 202, 257, 385
sign 392-394
Sivapithecus 230
sociality 283-284, 714
somitogenesis 179
speciation 53-54, 61, 66-67, 76-77, 107, 117-139, 164, 243, 524
allopatric speciation 127-128, 130-133, 137
by integration 113
ecological speciation 134-135
human speciation 211, 232, 294, 308-309
parapatric speciation 66, 127, 131-132
peripatric speciation 127
speciation genes 77, 126, 134-138
sympatric speciation 77, 127-129, 131-132
species, concept of 57-58, 76-77, 110, 119-121
splicing 683
struggle for life 59, 337, 385, 439, 449, 452, 463, 500-502, 565, 703, 716
substantial form 426
sufficient reason, principle of 443, 656
supernatural 13-14, 29, 41, 225, 281, 359-360, 365, 367, 369-371, 381, 391, 448, 518, 527
causation 529, 695
entity 362, 367-371, 373
supervenience 332, 457
survival of the fittest 153, 321, 501, 618
symbiogenesis 76, 108-114
symbiont 109, 113
symbiosis 57-58, 109-110, 476
endosymbiosis 92
symbol, symbolic 212, 291, 293, 300, 302, 307, 490, 680, 686, 704, 711, 714-715
symbolism, symbolization 196, 293-297, 299

sympatry 119-120, 125, 134, 209
synthetic theory of evolution 12, 66-67, 77-78, 395, 507-508, 616
systems biology 345, 680, 705, 710

T

Taraxacum officinale 100
taxonomy 60, 75, 77, 95, 101, 108, 120, 196, 208, 243, 291-292, 295, 682
  taxonomical levels 51-54, 120, 238, 275
technology 196, 291-293
teleology 102, 341, 387-390, 404-405, 427, 443, 449, 453, 455-458, 463, 466-476, 481-482, 487-488, 490, 565-566, 650, 655
teleonomy 456, 474, 486-487, 565, 650
theism, theist 360-361, 366, 375-376, 383, 396, 403-404, 434, 436-454, 465, 467, 571, 648
theodicy 334, 435, 437, 452, 458, 656, 658-660, 697
theologism 407-408
theology 15, 285, 408, 427, 498, 540, 575-578, 586-587, 646, 648-649, 655
  Anglican theology 434
  ethicotheology 455
  of Creation 554, 561, 565, 575
  of nature 650-651, 657
  natural theology 26, 29-30, 40, 383, 436-437, 442, 447-448, 451, 455, 472, 474, 564, 579-584, 649-650, 694-695, 698
  rational theology 442
theory of mind 277, 279-280
thermodynamics 317, 327, 558, 652
  second principle of 262, 268, 475
Tiktaalik 42, 100
Timema 133
transcendence 15, 564, 566-567, 627
  evolutionary transcendence 296-300, 303
  self-transcendence 653-654
transcription factor 156-157, 159, 161-162, 164, 172-173, 247, 693
transduction 89, 176
transgressive segregation 182
transhumanism 415
transposable elements 87
tree of life 44, 53, 57, 76, 118-120, 139, 567-568, 571
Trichoplax adhaerens 173, 177
Trigriopus californicus 150

U

unconditionality 392,
uniformitarianism 184, 444
uniparental reproduction 109-110

V

variation 33-35, 38-39, 50, 59-60, 103, 169, 308, 325, 439, 448-452, 462, 464-465, 469, 471, 501-502, 560, 650-651, 655, 671, 707
  genetic variation 75, 78, 82-86, 88-92, 129, 153-154, 162, 233, 236-237, 680
  hereditary variation 26, 32, 34-35, 39, 66, 112,
  selectable variation 78, 155, 159
vertebrates 97, 99-100, 133, 156, 159-160, 178-179, 694, 698
verticalization 196, 245, 248-249, 267
virus 84, 87-89, 107, 682

W

Wnt pathway 177, 181
Wolbachia 110-111

X

Xenopus 136
Xyphophorus helleri 150
Xyphophorus maculatus 150

Z

Zea mays 162
Zinjanthropus 219

ANALECTA GREGORIANA

311 COLAGÈ Ivan

Interazione e inferenza

Epistemologia scientifica ispirata al pensiero di Ch. S. Peirce

Premio Bellarmino 2010 - Filosofia

2010 • pp. 276
ISBN 978-88-7839-179-6 • € 29,00

310 CIAMPANELLI Filippo

«Hominem reducere ad Deum»

La funzione mediatrice del Verbo incarnato nella teologia di San Bonaventura

Premio Bellarmino 2010 - Teologia

2010 • pp. 468
ISBN 978-88-7839-178-9 • € 37,00

309 MIRANDA Leonel

«Ascensiones in corde»

Interpretación biblica y/o anábasis plotiniana. Estudio sobre el progreso espiritual en Agustín de Hipona

2010 • pp. 336
ISBN 978-88-7839-177-2 • € 32,00

308 GARCÍA GUILLÉN Domingo

"Padre es nombre de relación"

Dios Padre en la teologia de Gregorio Nacianceno

2010 • pp. 424
ISBN 978-88-7839-167-3 • € 32,00

Print in May 2011
Ugo Quintily Spa - Roma - Italy
[1.0]